Of The Elements

18
2 — 4.215 / 0.95 / 0.179¤ — **He** — Helium — 4.002 602 ±2

13	14	15	16	17	18
5 +3 / 4275 / 2300 / 2.34 — **B** — Boron — 10.811 ±7	**6** ±4,2 / 4470 / 4100 / 2.62 — **C** — Carbon — 12.010 7 ±8	**7** ±3,5,4,2 / 77.35 / 63.14 / 1.251¤ — **N** — Nitrogen — 14.006 74 ±7	**8** -2 / 90.18 / 50.35 / 1.429¤ — **O** — Oxygen — 15.999 4 ±3	**9** -1 / 84.95 / 53.48 / 1.696¤ — **F** — Fluorine — 18.998 403 2 ±5	**10** / 27.10 / 24.55 / 0.901¤ — **Ne** — Neon — 20.179 7 ±6
13 +3 / 2793 / 933.25 / 2.7 — **Al** — Aluminum — 26.981 538 ±2	**14** +4 / 3540 / 1685 / 2.33 — **Si** — Silicon — 28.085 5 ±3	**15** ±3,5,4 / 550 / 317.3 / 1.82 — **P** — Phosphorus — 30.973 761 ±2	**16** ±2,4,6 / 717.8 / 388.4 / 2.07 — **S** — Sulfur — 32.066 ±6	**17** ±1,3,5,7 / 239.1 / 172.2 / 3.17¤ — **Cl** — Chlorine — 35.452 7 ±9	**18** / 87.3 / 83.81 / 1.784¤ — **Ar** — Argon — 39.948

10	11	12					
28 +2,3 / 3187 / 1726 / 8.90 — **Ni** — Nickel — 58.693 4 ±2	**29** +2,1 / 2836 / 1358 / 8.96 — **Cu** — Copper — 63.546 ±3	**30** +2 / 1180 / 693 / 7.14 — **Zn** — Zinc — 65.39 ±2	**31** +3 / 2478 / 303 / 5.91 — **Ga** — Gallium — 69.723	**32** +4 / 3107 / 1210 / 5.32 — **Ge** — Germanium — 72.61 ±2	**33** ±3,5 / 876 / — / 5.72 — **As** — Arsenic — 74.921 60 ±2	**34** -2,4,6 / 958 / 494 / 4.80 — **Se** — Selenium — 78.96 ±3	**35** ±1,5 / 332.25 / 265.90 / 3.12 — **Br** — Bromine — 79.904	**36** / 119.80 / 115.78 / 3.74¤ — **Kr** — Krypton — 83.80
46 +2,4 / 3237 / 1825 / 12.0 — **Pd** — Palladium — 106.42	**47** +1 / 2436 / 1234 / 10.5 — **Ag** — Silver — 107.868 2 ±2	**48** +2 / 1040 / 594 / 8.65 — **Cd** — Cadmium — 112.411 ±8	**49** +3 / 2346 / 430 / 7.31 — **In** — Indium — 114.818 ±3	**50** +4,2 / 2876 / 505 / 7.30 — **Sn** — Tin — 118.710 ±7	**51** ±3,5 / 1860 / 904 / 6.68 — **Sb** — Antimony — 121.760	**52** -2,4,6 / 1261 / 723 / 6.24 — **Te** — Tellurium — 127.60 ±3	**53** ±1,5,7 / 458 / 387 / 4.92 — **I** — Iodine — 126.904 47 ±3	**54** / 165 / 161 / 5.89¤ — **Xe** — Xenon — 131.29 ±2
78 +2,4 / 4100 / 2045 / 21.4 — **Pt** — Platinum — 195.078 ±2	**79** +3,1 / 3130 / 1338 / 19.3 — **Au** — Gold — 196.966 55 ±2	**80** +2,1 / 630 / 234 / 13.5 — **Hg** — Mercury — 200.59 ±2	**81** +3,1 / 1746 / 577 / 11.85 — **Tl** — Thallium — 204.383 3 ±2	**82** +4,2 / 2023 / 601 / 11.4 — **Pb** — Lead — 207.2	**83** +3,5 / 1837 / 545 / 9.8 — **Bi** — Bismuth — 208.980 38 ±2	**84** +4,2 / 1235 / 527 / 9.4 — **Po** — Polonium — (~210)	**85** ±1,3,5,7 / 610 / 575 / — — **At** — Astatine — (~210)	**86** / 211 / 202 / 9.91¤ — **Rn** — Radon — (~222)

64 +3 / 3539 / 1585 / 7.89 — **Gd** — Gadolinium — 157.25 ±3	**65** +3 / 3496 / 1630 / 8.27 — **Tb** — Terbium — 158.925 34 ±2	**66** +3 / 2835 / 1682 / 8.54 — **Dy** — Dysprosium — 162.50 ±3	**67** +3 / 2968 / 1743 / 8.80 — **Ho** — Holmium — 164.930 32 ±2	**68** +3 / 3136 / 1795 / 9.05 — **Er** — Erbium — 167.26 ±3	**69** +3,2 / 2220 / 1818 / 9.33 — **Tm** — Thulium — 168.934 21 ±2	**70** +3,2 / 1467 / 1097 / 6.98 — **Yb** — Ytterbium — 173.04 ±3	**71** +3 / 3668 / 1936 / 9.84 — **Lu** — Lutetium — 174.967
96 +3 / 1340 / 13.51 — **Cm** — Curium — (247)	**97** +4,3 / 900 — **Bk** — Berkelium — (247)	**98** +3 — **Cf** — Californium — (251)	**99** — **Es** — Einsteinium — (254)	**100** — **Fm** — Fermium — (257)	**101** — **Md** — Mendelevium — (260)	**102** — **No** — Nobelium — (259)	**103** — **Lr** — Lawrencium — (262)

EXPLORING
CHEMICAL ANALYSIS

["The Experiment" by Sempé. Copyright C. Charillon, Paris.]

EXPLORING CHEMICAL ANALYSIS

Daniel C. Harris

Michelson Laboratory
China Lake, California

W. H. FREEMAN AND COMPANY
New York

Acquisitions Editor: Jessica Fiorillo
Marketing Manager: Claire Pearson
Project Editor: Mary Louise Byrd
Text and Cover Designer: Cambraia Fernandes
Illustration Coordinator: Bill Page
Illustrator: Network Graphics
Production Coordinator: Paul W. Rohloff
Composition: York Graphic Services
Manufacturing: RR Donnelley & Sons Company

Library of Congress Cataloging-in-Publication Data

Harris, Daniel C., 1948–
    Exploring chemical analysis / Daniel C. Harris.—2nd ed.
       p. cm.
    Includes index.
    ISBN 0-7167-3540-7
    1. Chemistry, Analytic. I. Title

    QD75.2 .H368 2000
    543–dc21   00-044292

Printed in the United States of America

First printing 2000

EXPLORING
CHEMICAL ANALYSIS

EXPLORING CHEMICAL ANALYSIS

My goal for this book is to provide an elementary, engaging introduction to analytical chemistry for students whose primary interests generally lie outside of chemistry. You might use this book in your analytical chemistry course or in a lower-level course such as freshman laboratory, when a brief introduction to quantitative analysis is required. You will find a distinct flavor of biological and environmental subject matter that I hope will capture your interest and give analytical chemistry a vital place in your studies.

Exploring Chemical Analysis covers modern analytical chemistry, not just wet chemical analysis. Major topics include statistics, volumetric analysis, gravimetric and combustion analysis, acid-base equilibrium, EDTA titrations, electrochemical methods, redox titrations, spectrophotometry, atomic spectroscopy, chromatography, and capillary electrophoresis. Laboratory experiments are provided for most major analytical methods, with emphasis on environmental and biological samples.

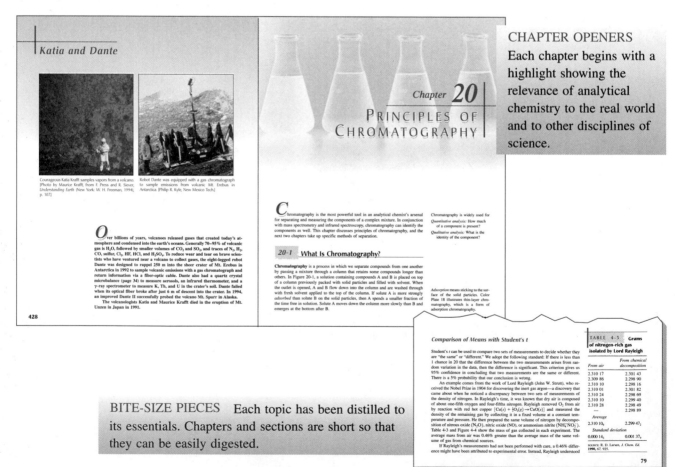

CHAPTER OPENERS Each chapter begins with a highlight showing the relevance of analytical chemistry to the real world and to other disciplines of science.

BITE-SIZE PIECES Each topic has been distilled to its essentials. Chapters and sections are short so that they can be easily digested.

DEMONSTRATIONS AND COLOR PLATES I can't come to your classroom to present chemical demonstrations, but I can tell you about some of my favorites and show you color photos of how they look. Color photos are located near the center of the book.

Demonstration 18-1

In Which Your Class Really Shines[1]

A fluorescent lamp is a glass tube filled with mercury vapor; the inner walls are coated with a *phosphor* (luminescent substance) consisting of a calcium halophosphate $(Ca_5(PO_4)_3F_{1-x}Cl_x)$ doped with Mn^{2+} and Sb^{3+}. (*Doping* means adding an intentional impurity, called a *dopant*.) The mercury atoms, promoted to an excited state by electric current passing through the lamp, emit mostly ultraviolet radiation at 254 and 185 nm. This radiation is absorbed by the Sb^{3+}, and some of the energy is passed on to Mn^{2+}. Sb^{3+} emits blue light and Mn^{2+} emits yellow light, with the combined emission appearing white. The emission spectrum is shown at the right. Fluorescent lamps are important energy-saving devices because they are more efficient than incandescent lamps in the conversion of electricity to light.

White fabrics are sometimes made "whiter" by treatment with a fluorescent dye. Turn on an ultraviolet lamp in a darkened classroom and illuminate some people standing at the front of the room. (*The victims should not look directly at the lamp*, because ultraviolet light is harmful to eyes.) You will discover a surprising amount of emission from white fabrics, including shirts, paints, shoelaces, and unmentionables. You may also be surprised to see fluorescence from teeth and from recently bruised areas of skin that show no surface damage.

A fluorescent whitener from laundry detergent

399

overloading can occur. When overloaded, the solute is so soluble in the concentrated part of the band that little solute trails behind the concentrated region.

Tailing is an asymmetric peak shape in which the trailing part of the band is elongated (Figure 20-11b). Such a peak occurs when there are strongly polar, highly adsorptive sites (such as exposed — OH groups) in the stationary phase that retain solute more strongly than other sites. To reduce tailing, we use a chemical treatment called *silanization* to convert polar — OH groups to nonpolar — $OSi(CH_3)_3$ groups.

Box 20-1 discusses *polarity*.

Ask Yourself

20-C. (a) Why is longitudinal diffusion a more serious problem in gas chromatography than in liquid chromatography? (Think about this one; the answer is not in the text.)
(b) (i) In Figure 20-8, what is the optimal flow rate for best separation of solutes with He mobile phase?
(ii) Why does plate height increase at high flow rate?
(iii) Why does plate height increase at low flow rate?
(c) Why does an open tubular gas chromatography column give better resolution than the same length of packed column?
(d) Why is it desirable to use a very long open tubular column? What difference between packed and open tubular columns allows much longer open tubular columns to be used?

439

ASK YOURSELF Sections end with a brief, straightforward problem entitled "Ask Yourself," which reinforces the main theme of the section. These are the first problems that you should do as you work your way through this course. Complex material is broken down into elementary steps in these questions. Complete solutions to "Ask Yourself" questions can be found at the back of the book.

SPREADSHEETS Spreadsheets are introduced as an optional tool. They are useful in this course and will serve you well in many ways outside this course.

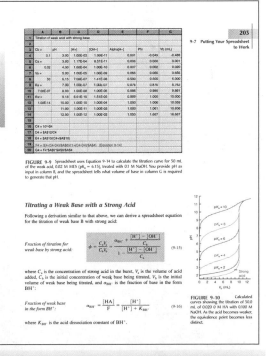

203

9-7 Putting Your Spreadsheet to Work

FIGURE 9-9 Spreadsheet uses Equation 9-14 to calculate the titration curve for 50 mL of the weak acid, 0.02 M MES ($pK_a = 6.15$), treated with 0.1 M NaOH. You provide pH as input in column B, and the spreadsheet tells what volume of base in column G is required to generate that pH.

Titrating a Weak Base with a Strong Acid

Following a derivation similar to that above, we can derive a spreadsheet equation for the titration of weak base B with strong acid:

Fraction of titration for weak base by strong acid:

$$\phi = \frac{C_a V_a}{C_b V_b} = \frac{\alpha_{BH^+} + \frac{[H^+] - [OH^-]}{C_b}}{1 - \frac{[H^+] - [OH^-]}{C_a}} \quad (9-15)$$

where C_a is the concentration of strong acid in the buret, V_a is the volume of acid added, C_b is the initial concentration of weak base being titrated, V_b is the initial volume of weak base being titrated, and α_{BH^+} is the fraction of base in the form BH^+:

Fraction of weak base in the form BH^+:

$$\alpha_{BH^+} = \frac{[HA]}{F} = \frac{[H^+]}{[H^+] + K_{BH^+}} \quad (9-16)$$

where K_{BH^+} is the acid dissociation constant of BH^+.

FIGURE 9-10 Calculated curves showing the titration of 50.0 mL of 0.020 M HA with 0.100 M NaOH. As the acid becomes weaker, the equivalence point becomes less distinct.

0-1   The Analytical Chemist's Job

Two students, Denby and Scott, began their quest at the library with a computer search for analytical methods. Searching through *Chemical Abstracts*, and using "caffeine" and "chocolate" as key words, they uncovered numerous articles in chemistry journals. One article, "High Pressure Liquid Chromatographic Determination of Theobromine and Caffeine in Cocoa and Chocolate Products,"[3] described a procedure suitable for the equipment available in their laboratory.

Sampling

The first step in any chemical analysis is procuring a representative sample to measure — a process called *sampling*. Is all chocolate the same? Of course not. Denby and Scott chose to buy chocolate in the neighborhood store and analyze pieces of it. If you wanted to make universal statements about "caffeine in chocolate," you would need to analyze a variety of chocolates from different manufacturers. You would also need to measure multiple samples of each type to determine the range of caffeine content in each kind of chocolate from the same manufacturer.

A pure chocolate bar is probably fairly **homogeneous**, which means that its composition is the same everywhere. It might be safe to assume that a piece from one end has the same caffeine content as a piece from the other end. A chocolate with a macadamia nut in the middle is an example of a **heterogeneous** material, which means that the composition differs from place to place: The nut is different from the chocolate. If you were sampling a heterogeneous material, you would need to use a strategy different from that used to sample a homogeneous material. You would need to know the average mass of chocolate and the average mass of nuts in many candies. You would need to know the average caffeine content of the chocolate and of the macadamia nut (if it has any caffeine). Only then could you make a statement about the average caffeine content of macadamia chocolate.

2

0 The Analytical Process

Chemical Abstracts is the most comprehensive source for locating articles published in chemistry journals.

Bold terms should be learned. They are listed at the end of the chapter and in the Glossary at the back of the book. *Italicized* words are less important, but many of their definitions are also found in the Glossary.

Homogeneous: same throughout
Heterogeneous: differs from region to region

MARGINAL NOTES Don't miss them! This learning tool provides emphasis and amplification of important points in the text and occasional questions for you to answer as you read this book.

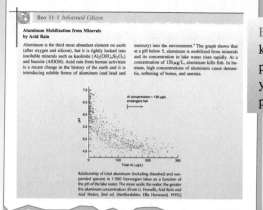
BOXES "Explanation" boxes expand or explain key points from the text. "Informed Citizen" boxes present interesting descriptive material that helps you see how analytical chemistry is applied to real problems of importance to real people.

Important Terms

adsorption chromatography	internal standard	partition chromatography
affinity chromatography	ion-exchange chromatography	plate height
chromatogram	liquid chromatography	polar compound
chromatography	mass spectrometer	resolution
co-chromatography	mobile phase	response factor
eluate	molecular exclusion chromatography	retention time
eluent	nonpolar compound	stationary phase
elution	open tubular column	theoretical plate
gas chromatography	packed column	van Deemter equation

VOCABULARY AND GLOSSARY The most important terms are in **bold** type and listed at the end of each chapter. Definitions are found in the Glossary at the back of the book. Less important new terms are in *italic*.

of 1.00 M HCl changed the pH from 8.61 to 8.41. Addition of 12.0 mL of 1.00 M HCl to 1.00 L of unbuffered solution would have lowered the pH to 1.93.

But *why does a buffer resist changes in pH?* **It does so because the strong acid or base is consumed by B or BH⁺.** If you add HCl to tris, B is converted to BH⁺. If you add NaOH, BH⁺ is converted to B. As long as you don't use up the B or BH⁺ by adding too much HCl or NaOH, the log term of the Henderson-Hasselbalch equation does not change very much and the pH does not change very much. Demonstration 8-1 illustrates what happens when the buffer does get used up. The buffer has its maximum capacity to resist changes of pH when pH = pK_a. We will return to this point later.

A buffer resists changes in pH . . .

. . . because the buffer "consumes" the added acid or base.

Ask Yourself

8-B. (a) What is the pH of a solution prepared by dissolving 10.0 g of tris plus 10.0 g of tris hydrochloride in 0.250 L water?
(b) What will the pH be if 10.5 mL of 0.500 M HClO₄ are added to (a)?
(c) What will the pH be if 10.5 mL of 0.500 M NaOH are added to (a)?

8-4   Preparing Buffers

Buffers are usually prepared by starting with a measured amount of either a weak acid (HA) or a weak base (B). Then OH⁻ is added to HA to make a mixture of HA and A⁻ (a buffer), or H⁺ is added to B to make a mixture of B and BH⁺ (a buffer).

EXAMPLE   **Calculating How to Prepare a Buffer Solution**

How many milliliters of 0.500 M NaOH should be added to 10.0 g of tris hydrochloride (BH⁺, Equation 8-3) to give a pH of 7.60 in a final volume of 250 mL?

SOLUTION The number of moles of tris hydrochloride in 10.0 g is (10.0 g)/(157.597 g/mol) = 0.063 5. We can make a table to help us solve the problem:

Reaction with OH⁻:	BH⁺	+	OH⁻	⟶	B
Initial moles	0.063 5		x		—
Final moles	0.063 5 − x				x

The Henderson-Hasselbalch equation allows us to find x, because we know pH and pK_a:

$$pH = pK_a + \log\left(\frac{\text{mol B}}{\text{mol BH}^+}\right)$$

169

WORKED EXAMPLES These appear frequently throughout the book to explain, amplify, and reinforce most types of calculations.

PROBLEMS AND ANSWERS Problems provide practice and build understanding of analytical chemistry. Short answers to numerical problems are at the back of the book. I think you are entitled to immediate feedback after struggling through a homework problem. "Did I get it right or have I missed something?"

Problems

11-1. What is an ionic atmosphere?

11-2. Explain why the solubility of an ionic compound increases as the ionic strength of the solution increases (at least up to ~0.5 M).

11-3. The figure shows the quotient of concentrations [CH₃CO₂⁻][H⁺]/[CH₃CO₂H] for the dissociation of acetic acid as a function of the concentration of KCl added to the solution. Explain the shape of the curve.

11-4. Which statements are true? In the ionic strength range 0–0.1 M, activity coefficients decrease with **(a)** increasing ionic strength; **(b)** increasing ionic charge; **(c)** decreasing hydrated radius.

11-5. Explain the following observations:
(a) Mg²⁺ has a greater hydrated radius than Ba²⁺.
(b) Hydrated radii decrease in the order Sn⁴⁺ > In³⁺ > Cd²⁺ > Rb⁺.
(c) H⁺ has a hydrated radius of 900 pm, whereas that of OH⁻ is just 350 pm.

11-6. State in words the meaning of the charge balance equation.

11-7. State the meaning of the mass balance equation.

11-8. Why does the solubility of a salt of a basic anion increase with decreasing pH? Write chemical reactions for the minerals galena (PbS) and cerussite (PbCO₃) to explain how acid rain mobilizes trace quantities of toxic metallic elements from relatively inert forms into the environment, where the metals can be taken up by plants and animals. Why are the minerals kaolinite and bauxite in Box 11-1 more soluble in acidic solution than in neutral solution?

11-9. Assuming complete dissociation of the salts, calculate the ionic strength of **(a)** 0.2 mM KNO₃; **(b)** 0.2 mM Cs₂CrO₄; **(c)** 0.2 mM MgCl₂ plus 0.3 mM AlCl₃.

11-10. Find the activity coefficient of each ion at the indicated ionic strength:
(a) SO₄²⁻   ($\mu = 0.01$ M)
(b) Sc³⁺   ($\mu = 0.005$ M)

HOW WOULD YOU DO IT? The last problem in each chapter is more open-ended and might involve interpretation of experimental data or applying your knowledge to a new situation. There may not be a unique answer to this question.

How Would You Do It?

4-19. Students at Eastern Illinois University[3] intended to prepare copper(II) carbonate by adding a solution of $CuSO_4 \cdot 5H_2O$ to a solution of Na_2CO_3.

$$CuSO_4 \cdot 5H_2O(aq) + Na_2CO_3(aq) \longrightarrow$$
$$\underset{\text{Copper(II) carbonate}}{CuCO_3(s)} + Na_2SO_4(aq) + 5H_2O(l)$$

After warming the mixture to 60°C, the gelatinous blue precipitate coagulated into an easily filterable pale green solid. The product was filtered, washed, and dried at 70°C. Copper in the product was measured by heating 0.4 g of solid in a stream of methane at high temperature to reduce the solid to pure Cu, which was weighed.

$$4CuCO_3(s) + CH_4(g) \xrightarrow{heat} 4Cu(s) + 5CO_2(g) + 2H_2O(g)$$

In 1995, 43 students found a mean value of 55.6 wt % Cu with a standard deviation of 2.7 wt %. In 1996, 39 students found 55.9 wt % with a standard deviation of 3.8 wt %. The instructor tried the experiment 9 times and measured 55.8 wt % with a standard deviation of 0.5 wt %. Was the product of the reaction probably $CuCO_3$? Could it have been a hydrate, $CuCO_3 \cdot xH_2O$?

KEY EQUATIONS Key equations are highlighted in the text and collected at the end of each chapter. If you understand how to use these equations, you have probably mastered much of the material in the chapter.

Key Equations

Diprotic acid equilibria	$H_2A \rightleftharpoons HA^- + H^+$	$K_{a1} = K_1$
	$HA^- \rightleftharpoons A^{2-} + H^+$	$K_{a2} = K_2$
Diprotic base equilibria	$A^{2-} + H_2O \rightleftharpoons HA^- + OH^-$	K_{b1}
	$HA^- + H_2O \rightleftharpoons H_2A + OH^-$	K_{b2}
Relation between K_a and K_b	Monoprotic system	$K_aK_b = K_w$
	Diprotic system	$K_{a1}K_{b2} = K_w$
		$K_{a2}K_{b1} = K_w$
	Triprotic system	$K_{a1}K_{b3} = K_w$
		$K_{a2}K_{b2} = K_w$
		$K_{a3}K_{b1} = K_w$

pH of H_2A (or BH_2^+) $H_2A \xrightarrow{K_{a1}} H^+ + HA^-$ $\dfrac{x^2}{F - x} = K_{a1}$

(This gives $[H^+]$, $[HA^-]$, and $[H_2A]$. You can solve for $[A^{2-}]$ from the K_{a2} equilibrium.)

pH of HA^- (or diprotic BH^+) $pH \approx \frac{1}{2}(pK_1 + pK_2)$

pH of A^{2-} (or diprotic B) $A^{2-} + H_2O \xrightarrow{K_{b1}} HA^- + OH^-$ $\dfrac{x^2}{F - x} = K_{b1} = \dfrac{K_w}{K_{a2}}$

(This gives $[OH^-]$, $[HA^-]$, and $[A^{2-}]$. You can find $[H^+]$ from the K_w equilibrium and $[H_2A]$ from the K_{a1} equilibrium.)

Diprotic buffer $pH = pK_1 + \log\left(\dfrac{[HA^-]}{[H_2A]}\right)$ $pH = pK_2 + \log\left(\dfrac{[A^{2-}]}{[HA^-]}\right)$

Both equations are always true and either can be used, depending on which set of concentrations you happen to know.

Titration of H_2A with OH^-		
	$V_b = 0$	Find pH of H_2A
	$V_b = \frac{1}{2}V_{e1}$	$pH = pK_1$
	$V_b = V_{e1}$	$pH \approx \frac{1}{2}(pK_1 + pK_2)$
	$V_b = \frac{3}{2}V_{e1}$	$pH = pK_2$
	$V_b = V_{e2}$	Find pH of A^{2-}
	$V_b > V_{e2}$	Find concentration of excess OH^-

Titration of diprotic B with H^+		
	$V_a = 0$	Find pH of B
	$V_a = \frac{1}{2}V_{e1}$	$pH = pK_{a2}$ (for BH^+)
	$V_a = V_{e1}$	$pH \approx \frac{1}{2}(pK_{a1} + pK_{a2})$
	$V_a = \frac{3}{2}V_{e1}$	$pH = pK_{a1}$ (for BH_2^+)
	$V_a = V_{e2}$	Find pH of BH_2^+
	$V_a > V_{e2}$	Find concentration of excess H^+

228

40   Chapter 8

8-20. Henderson-Hasselbalch spreadsheet

	A	B	C	D
1	pKa =	[A⁻]/[HA]	pH	log([A⁻]/[HA])
2	4	0.001	1.00	-3.0000
3		0.01	2.00	-2.0000
4		0.05	2.70	-1.3010
5		0.5	3.70	-0.3010
6		1	4.00	0.0000
7		2	4.30	0.3010
8		4	4.70	0.6990
9		10	5.00	1.0000
10		100	6.00	2.0000
11		1000	7.00	3.0000
12				
13	C2 = $A$2 + LOG(B2)			
14	D2 = LOG(B2)			

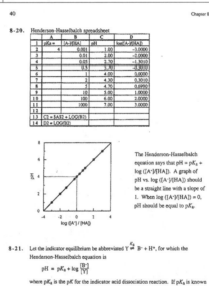

The Henderson-Hasselbalch equation says that $pH = pK_a + \log([A^-]/[HA])$. A graph of pH vs. log $([A^-]/[HA])$ should be a straight line with a slope of 1. When log $([A^-]/[HA]) = 0$, pH should be equal to pK_a.

8-21. Let the indicator equilibrium be abbreviated $Y \overset{K_a}{\rightleftharpoons} B^- + H^+$, for which the Henderson-Hasselbalch equation is

$$pH = pK_a + \log\frac{[B^-]}{[Y]}$$

where pK_a is the pK for the indicator acid dissociation reaction. If pK_a is known accurately, then we can find the pH of the solution by accurately measuring the quotient [B⁻]/[Y]. This quotient is available from measuring the visible absorbance of the solution at the two wavelengths of maximum absorbance. Regardless of what other equilibria are present in the solution, knowing the quotient [B⁻]/[Y] establishes the pH.

SOLUTIONS MANUAL Detailed solutions to all problems are in the *Solutions Manual for Exploring Chemical Analysis*. The manual is available to instructors upon request and can be made available to students through your bookstore.

EXPLORING CHEMICAL ANALYSIS
2nd Edition   Daniel C. Harris

This Web site is designed to help students review key concepts from the textbook through interactive exercises and learning tools. Resources are organized by chapter of the textbook and by content type. To access resources, please select a Chapter or Category below.

Select a Chapter:

Chapter 2: Tools of the Trade
Chapter 3: Math Toolkit
Chapter 4: Statistics
Chapter 5: Good Titrations
Chapter 6: Gravimetric and Combustion Analysis
Chapter 7: Introducing Acids and Bases
Chapter 8: Buffers

Select a Category:

Quizzes
Supplementary Topics
Spreadsheets to Support Supplementary Topics

WEB SITE (www.whfreeman.com/exploringchem) Here you will find quizzes for every chapter and coverage of supplementary topics with problems and solutions. Instructors will find electronic copies of illustrations and tables from the book. Feel free to make transparencies from the illustrations or tables for use in your classroom.

CONTENTS

What Am I Doing in This Course?

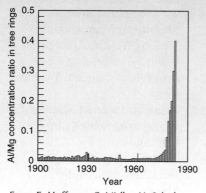

From E. Hoffman, C. Lüdke, H. Scholze, and H. Stephanowitz, *Fresenius J. Anal. Chem.* **1994**, *350*, 253.

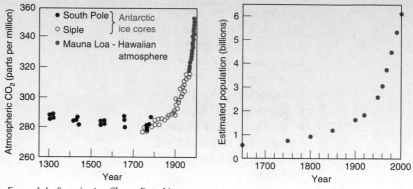

From J. L. Sarmiento, *Chem. Eng. News*, 31 May 1993, p. 30.

What is going on here? The figure at the left shows the recently increased availability of toxic Al^{3+} to a pine tree in Germany near a coal-burning power plant built in 1929. This increase is probably an effect of man-made acidity in rainfall, which mobilizes Al^{3+} from minerals. The center figure shows the growth of atmospheric CO_2, which threatens to warm the earth and induce climate changes. CO_2 comes from our burning of fossil fuel and destruction of forests. (The opening of Chapter 7 and Boxes 10-1 and 11-1 discuss these topics.) The graph at the right shows growth in world population. How long will our planet remain habitable if we do not control our population and our impact on the environment?

F or most of its users, analytical chemistry is a tool—not an end in itself. If you are interested in biology, medicine, environmental science, geology, oceanography, archeology, or paleontology, you will rely on chemical analyses to answer important questions. It is possible that you will make measurements yourself, but it is more likely that you will simply be a consumer of analytical results. One goal of this course is for you to understand how analysts derive the results that you will use. Another goal is for you to understand the limitations (the uncertainties) of the results. A third goal is for you to develop laboratory skills and careful work habits that you can apply in your field.

Scientists cannot prove theories. We can only disprove them. We cannot be sure that increasing atmospheric CO_2 is raising the temperature of the earth and will affect the climate. Nonetheless, politicians and engineers must make decisions every day to take actions in the absence of certainty. This course will help you, as an informed citizen, to appreciate how quantitative information is obtained and interpreted. And it should help give you a basis for positive actions in your own field of interest.

PREFACE

The first edition of this book was intended to provide a *short, interesting, elementary* introduction to analytical chemistry for students whose primary interests generally lie outside of chemistry. Surveys of users of the first edition led me to include topics in this second edition that had been ruthlessly rejected before. These include activity coefficients (Chapter 11), systematic treatment of equilibrium (Chapter 11), EDTA and redox titration calculations (Chapters 12 and 15), instrumental methods in electrochemistry (Chapter 16), and an expanded discussion of spectrophotometry (Chapter 18). Eight experiments were added to Chapter 23, including a block to illustrate a noteworthy laboratory approach developed at San Jose State University. Gravimetric analysis was moved to an early position, spectrophotometry was moved to a later position, and the discussion of acids and bases was consolidated.

This edition has 20 percent more end of chapter problems than the last edition. New "How Would You Do It?" problems might involve interpretation of experimental data or applying knowledge to a new situation. A new Web site accompanies the text. It includes more chapter review problems, additional topics, and all of the art from the text.

Jessica Fiorillo at W. H. Freeman and Company provided editorial guidance for this revision. Jodi Simpson's usual superb copy editing smoothed my prose and uncovered mistakes. Mary Louise Byrd is primarily responsible for converting my manuscript into the handsome book you are reading. I am grateful to the following people who provided comments on the first edition or the manuscript for the second edition: M. Dale Hawley (Kansas State University), Jean B. Smith (University of Nebraska, Lincoln), Stephen Goldberg (Adelphi University), T. Carter Gilmer (Bowling Green State University), Robert J. Racicot (U.S. Air Force Academy), L. Curtis Mehlhaff (University of Puget Sound), Simon J. Garrett (Michigan State University), Karen Creager (Clemson University), Philip D. Whitefield (University of Missouri, Rolla), Greg Szulczewski (University of Alabama, Tuscaloosa), Jack D. Cummins (Metropolitan State College of Denver), Larry T. Taylor (Virginia Tech), Vladimir Katovic (Wright State University), Nazir A. Khatri (Franklin College of Indiana), and John Murdzek. Sam Perone and Joe Pesek of San Jose State University and Jim Orenberg of San Francisco State University provided valuable input on their laboratory curricula. Corrections and comments from the first edition were received from Tom Eaton (St. Thomas University, Miami), Steve Creager (Clemson University), Gregorio Cruz Villalón (Spain), Chongmok Lee (Ewha Womans University, Korea), and Richard A. Foss (Daemen College). At Michelson Laboratory, Mike Seltzer and Eric Erickson provided many helpful discussions. Jed Mosenfelder and George Rossman at Caltech assisted me in locating the geological story that graces the back cover of this book.

My wife, Sally, worked on every aspect of this book and contributed greatly to its accuracy and readability. Solutions to problems were checked by Sara Avrantinis and Brett Helms of Harvey Mudd College.

I truly appreciate comments, criticism, and suggestions from students and teachers. You can reach me at the Chemistry and Materials Division, Michelson Laboratory, China Lake, CA 93555.

Dan Harris
December 2000

xvii

Capturing a Single Vesicle with "Optical Tweezers"

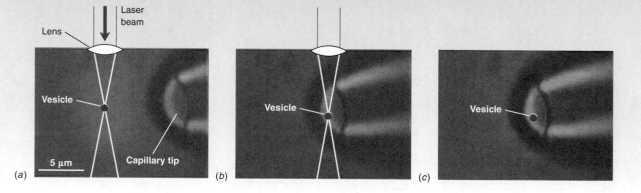

(a) A tiny vesicle is held in place by a focused infrared laser. The marker bar is 5×10^{-6} meters. (b) The open tip of a capillary electrophoresis tube is positioned next to the vesicle. (c) Vesicle enters the capillary under the influence of an electric field that draws solution into the tube. [From S. J. Lillard, D. T. Chiu, R. H. Scheller, R. N. Zare, S. E. Rodríquez-Cruz, E. R. Williams, O. Orwar, M. Sandberg, and J. A. Lundqvist, *Anal. Chem.* **1998**, *70*, 3517. The notation refers to the journal *Analytical Chemistry*, volume *70*, page 3517, published in the year **1998**.]

Many living cells that manufacture proteins and small molecules package them in tiny compartments called *vesicles*. The contents of a vesicle are secreted when the vesicle fuses with the outer cell membrane. At one frontier of analytical chemistry, chemists and cell biologists are now seeking to identify and measure the contents of a single vesicle.

The photos show the capture of a vesicle from the atrial gland of a gastropod, a marine animal related to the snail. This gland secretes peptides (small proteins) involved in reproductive behavior, and it also secretes small molecules. After breaking open the gland, the investigators isolated intact vesicles with a centrifuge. A single vesicle was then held stationary by a focused laser, called "optical tweezers."[1] With the aid of a microscope, the tip of a capillary electrophoresis tube was then positioned next to the vesicle, which was then coaxed into the capillary by an electric field. Inside the capillary, the vesicle was burst open by osmotic pressure and its contents separated by a strong electric field. Components were identified and measured by mass spectrometry. It was found that individual vesicles store different chemicals. This conclusion could not be reached from studies of whole populations of vesicles.

Later in this book we will discuss capillary electrophoresis and mass spectrometry, powerful instrumental techniques used in analytical chemistry. Prior to describing instrumental methods, we discuss basic "wet" chemical techniques and the chemical equilibria that underlie most analytical methods.

THE ANALYTICAL PROCESS

*C*hocolate has been the savior of many a student on the long night before a major assignment was due. My favorite chocolate bar, jammed with 33% fat and 47% sugar, propels me over mountains in California's Sierra Nevada. In addition to its high energy content, chocolate packs an extra punch with the stimulant caffeine and its biochemical precursor, theobromine.

Chocolate is great to eat but not so easy to analyze. [W. H. Freeman photo by K. Bendo.]

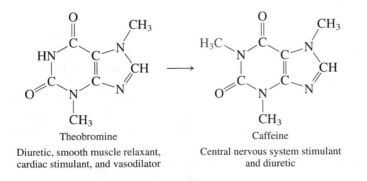

Theobromine
Diuretic, smooth muscle relaxant, cardiac stimulant, and vasodilator

Caffeine
Central nervous system stimulant and diuretic

A *diuretic* makes you urinate.

A *vasodilator* enlarges blood vessels.

Too much caffeine is harmful for many people, and even small amounts cannot be tolerated by some unlucky individuals. How much caffeine is in a chocolate bar? How does that amount compare with the quantity in coffee or soft drinks? At Bates College in Maine, Professor Tom Wenzel teaches students chemical problem solving through questions such as these.[2] How *do* you measure the caffeine content of a chocolate bar?

Footnotes and references are listed at the end of each chapter.

1

0-1 The Analytical Chemist's Job

Chemical Abstracts is the most comprehensive source for locating articles published in chemistry journals.

Two students, Denby and Scott, began their quest at the library with a computer search for analytical methods. Searching through *Chemical Abstracts,* and using "caffeine" and "chocolate" as key words, they uncovered numerous articles in chemistry journals. One article, "High Pressure Liquid Chromatographic Determination of Theobromine and Caffeine in Cocoa and Chocolate Products,"[3] described a procedure suitable for the equipment available in their laboratory.

Sampling

Bold terms should be learned. They are listed at the end of the chapter and in the Glossary at the back of the book. *Italicized* words are less important, but many of their definitions are also found in the Glossary.

Homogeneous: same throughout

Heterogeneous: differs from region
    to region

The first step in any chemical analysis is procuring a representative sample to measure—a process called **sampling.** Is all chocolate the same? Of course not. Denby and Scott chose to buy chocolate in the neighborhood store and analyze pieces of it. If you wanted to make universal statements about "caffeine in chocolate," you would need to analyze a variety of chocolates from different manufacturers. You would also need to measure multiple samples of each type to determine the range of caffeine content in each kind of chocolate from the same manufacturer.

A pure chocolate bar is probably fairly **homogeneous,** which means that its composition is the same everywhere. It might be safe to assume that a piece from one end has the same caffeine content as a piece from the other end. A chocolate with a macadamia nut in the middle is an example of a **heterogeneous** material, which means that the composition differs from place to place: The nut is different from the chocolate. If you were sampling a heterogeneous material, you would need to use a strategy different from that used to sample a homogeneous material. You would need to know the average mass of chocolate and the average mass of nuts in many candies. You would need to know the average caffeine content of the chocolate and of the macadamia nut (if it has any caffeine). Only then could you make a statement about the average caffeine content of macadamia chocolate.

Sample Preparation

So Denby and Scott simply analyzed a piece of chocolate from one chocolate bar. The first step in the procedure calls for weighing out a quantity of chocolate and extracting fat from it by dissolving the fat in a hydrocarbon solvent. Fat needs to be removed because it would interfere with chromatography later in the analysis. Unfortunately, if you just shake a chunk of chocolate with solvent, extraction is not very effective because the solvent has no access to the inside of the chocolate. So our resourceful students sliced the chocolate into fine pieces and put them into a mortar and pestle (Figure 0-1), thinking they would grind the solid into small particles.

Imagine trying to grind chocolate! The solid is too soft to be ground. So Denby and Scott froze the mortar and pestle with its load of sliced chocolate. Once the chocolate was cold, it was brittle enough to grind. Then the small pieces were placed in a preweighed 15-milliliter (mL) centrifuge tube, and their mass was noted.

Figure 0-2 outlines the next part of the procedure. A 10-mL portion of the organic solvent, petroleum ether, was added to the tube, and the top was capped with a stopper. The tube was shaken vigorously to dissolve fat from the solid chocolate into the solvent. Caffeine and theobromine are insoluble in this solvent. The mix-

Pestle

Mortar

FIGURE 0-1 Ceramic mortar and pestle, used to grind solids into fine powders.

ture of liquid and fine particles was then spun in a centrifuge to pack the chocolate at the bottom of the tube. The clear liquid, containing dissolved fat, could now be **decanted** (poured off) and discarded. Extraction with fresh portions of solvent was repeated twice more to ensure complete removal of fat from the chocolate. Residual solvent in the chocolate was finally removed by heating the centrifuge tube in a beaker of boiling water. By weighing the centrifuge tube plus its content of defatted chocolate residue and subtracting the known mass of the empty tube, the mass of chocolate residue could be calculated.

Substances being measured—caffeine and theobromine in this case—are called **analytes.** The next step in the sample preparation procedure was to make a **quantitative transfer** (a complete transfer) of the fat-free chocolate residue to an Erlenmeyer flask and to dissolve the analytes in water for the chemical analysis. If any residue were not transferred from the tube to the flask, then the final analysis would be in error because not all of the analyte would be present. To perform the quantitative transfer, Denby and Scott added a few milliliters of pure water to the centrifuge tube and used stirring and heating to dissolve or suspend as much of the chocolate as possible. The **slurry** (a suspension of solid in a liquid) was then poured from the tube into a 50-mL flask. They repeated the procedure several times with fresh portions of water to ensure that every last bit of chocolate was transferred from the centrifuge tube into the flask.

To complete the dissolution of analytes, Denby and Scott added water to bring the volume up to about 30 mL. They heated the flask in a boiling water bath to extract all the caffeine and theobromine from the chocolate into the water. To compute the quantity of analyte later, the total mass of solvent (water) must be accurately known. Denby and Scott knew the mass of chocolate residue in the centrifuge tube and they knew the mass of the empty Erlenmeyer flask. So they put the flask on a balance and added water drop by drop until exactly 33.3 g of water were in the flask. Later, they would compare known solutions of pure analyte in water with the unknown solution containing 33.3 g of water.

Before Denby and Scott could inject the unknown solution into a chromatograph for chemical analysis, they had to clean the unknown even more (Figure 0-3). The slurry of chocolate residue in water contained tiny solid particles that would surely clog their expensive chromatography column and ruin it. So they transferred a portion of the slurry to a centrifuge tube and centrifuged the mixture to pack as much of the solid as possible at the bottom of the tube. The cloudy, tan, **supernatant liquid** (liquid above the packed solid) was then filtered in a further attempt to remove tiny particles of solid from the liquid.

A solution of anything in water is called an **aqueous** solution.

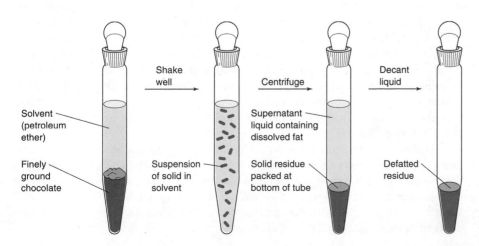

Solvent
(petroleum
ether)

Finely
ground
chocolate

Shake
well

Suspension
of solid in
solvent

Centrifuge

Supernatant
liquid containing
dissolved fat

Solid residue
packed at
bottom of tube

Decant
liquid

Defatted
residue

FIGURE 0-2 Extracting fat from chocolate to leave defatted solid residue for analysis.

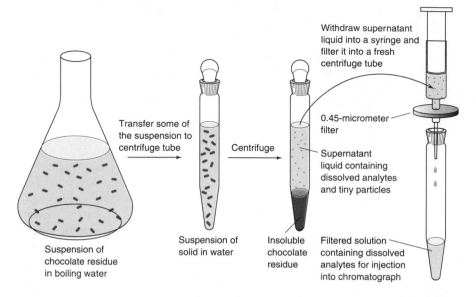

FIGURE 0-3 Centrifugation and filtration are used to separate unde-sired solid residue from the aqueous solution of analytes.

Transfer some of the suspension to centrifuge tube

Centrifuge

Withdraw supernatant liquid into a syringe and filter it into a fresh centrifuge tube

0.45-micrometer filter

Supernatant liquid containing dissolved analytes and tiny particles

Suspension of chocolate residue in boiling water

Suspension of solid in water

Insoluble chocolate residue

Filtered solution containing dissolved analytes for injection into chromatograph

Real-life samples rarely cooperate with you!

It is critical to avoid injecting solids into the chromatography column, but the tan liquid still looked cloudy. So Denby and Scott took turns between their classes to repeat the centrifugation and filtration steps five times. After each cycle in which the supernatant liquid was filtered and centrifuged, it became a little cleaner. But the liquid was never completely clear. Given enough time, more solid always seemed to precipitate from the filtered solution.

The tedious procedure described so far is called **sample preparation**—transforming sample into a state that is suitable for analysis. In this case, fat had to be removed from the chocolate, analytes had to be extracted into water, and residual solid had to be separated from the water.

The Chemical Analysis (At Last!)

Chromatography solvent is selected by a systematic trial-and-error process. The function of the acetic acid is to react with negatively charged oxygen atoms that lie on the silica surface and, when not neutralized, tightly bind a small fraction of caffeine and theobromine.

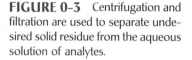

silica-O$^-$ $\xrightarrow{\text{acetic acid}}$ silica-OH

Binds analytes
very tightly

Does not bind
analytes as strongly

Only substances that absorb ultraviolet radiation at a wavelength of 254 nanometers are observed in Figure 0-5. By far, the major components in the aqueous extract are sugars, but they are not detected in this experiment.

Compromising with reality, Denby and Scott decided that the solution of analytes was as clean as they could make it in the time available. The next step was to inject solution into a *chromatography* column, which would separate the analytes and measure the quantity of each. The column in Figure 0-4a is packed with tiny particles of silica (SiO_2) on which are attached long hydrocarbon molecules. Twenty microliters (20.0×10^{-6} liters) of the solution of chocolate extract were injected into the column and washed through with a solvent made by mixing 79 mL of pure water, 20 mL of methanol, and 1 mL of acetic acid. Caffeine is more soluble than theobromine in the hydrocarbon on the silica surface. Therefore caffeine "sticks" to the coated silica particles in the column more strongly than theobromine does. When both analytes are flushed through the column by solvent, theobromine reaches the outlet before caffeine (Figure 0-4b).

Analytes are detected at the outlet by their ability to absorb ultraviolet radiation. As the compounds emerge from the column, they absorb radiation emitted from the lamp in Figure 0-4a and less radiation reaches the detector. The graph of detector response versus time in Figure 0-5 is called a *chromatogram*. Theobromine and caffeine are the major peaks in the chromatogram. The small peaks are other components of the aqueous extract from chocolate.

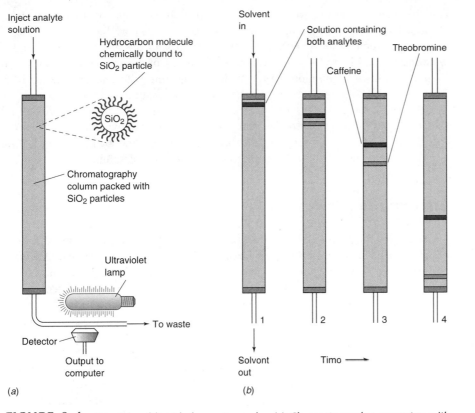

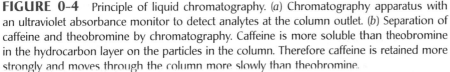

FIGURE 0-4 Principle of liquid chromatography. (*a*) Chromatography apparatus with an ultraviolet absorbance monitor to detect analytes at the column outlet. (*b*) Separation of caffeine and theobromine by chromatography. Caffeine is more soluble than theobromine in the hydrocarbon layer on the particles in the column. Therefore caffeine is retained more strongly and moves through the column more slowly than theobromine.

The chromatogram alone does not tell us what compounds are in the unknown. If you did not know beforehand what to expect, you would have to expend a great deal of effort to identify the peaks. One way to identify individual peaks is to measure spectral characteristics of each one as it emerges from the column. Another way is to add an authentic sample of either caffeine or theobromine to the unknown and see whether one of the peaks grows in magnitude.

Identifying *what* is in an unknown is called **qualitative analysis.** Identifying *how much* is present is called **quantitative analysis.** The vast majority of this book deals with quantitative analysis.

In Figure 0-5, the *area* under each peak is proportional to the quantity of that component passing through the detector. The best way to measure the area is with a computer that receives the output of the detector during the chromatography experiment. Denby and Scott did not have a computer linked to their chromatograph, so they measured the *height* of each peak instead.

Calibration Curves

In general, analytes with equal concentrations give different detector responses. Therefore, the response must be measured for known concentrations of each analyte. A graph showing detector response as a function of analyte concentration

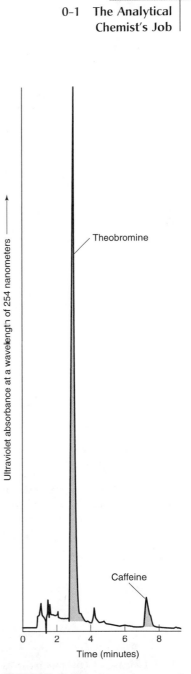

FIGURE 0-5 Chromatogram of 20.0 microliters of dark chocolate extract. A 4.6-mm-diameter × 150-mm-long column, packed with 5-micrometer particles of Hypersil ODS, was eluted (washed) with water:methanol:acetic acid (79:20:1 by volume) at a rate of 1.0 mL per minute.

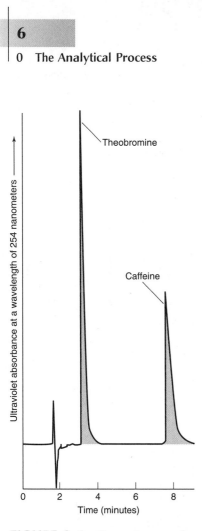

FIGURE 0-6 Chromatogram of 20.0 microliters of a standard solution containing 50.0 micrograms of theobromine and 50.0 micrograms of caffeine per gram of solution.

is called a **calibration curve** or a *standard curve*. To construct a calibration curve, **standard solutions** containing known concentrations of pure theobromine or caffeine were prepared and injected into the column, and the resulting peak heights were measured. Figure 0-6 is a chromatogram of one of the standard solutions, and Figure 0-7 shows calibration curves made by injecting solutions containing 10.0, 25.0, 50.0, or 100.0 micrograms of each analyte per gram of solution.

Straight lines drawn through the calibration points could then be used to find the concentrations of theobromine and caffeine in an unknown. Figure 0-7 shows that if the observed peak height of theobromine from an unknown solution is 15.0 cm, then the concentration is 76.9 micrograms per gram of solution.

Interpreting the Results

Knowing how much analyte is in the aqueous extract of the chocolate, Denby and Scott could calculate how much theobromine and caffeine were in the original chocolate. The results for dark and white chocolates are shown in Table 0-1. The quantities found in white chocolate are only about 2% as great as the quantities in dark chocolate.

The table also reports the *standard deviation* of three replicate measurements for each sample. Standard deviation, discussed in Chapter 4, is a measure of the reproducibility of the results. If three samples were to give identical results, the standard deviation would be 0. If the standard deviation is very large, then the results are not very reproducible. For theobromine in dark chocolate, the standard deviation (0.002) is less than 1% of the average (0.392), so we say the measurement is very reproducible. For theobromine in white chocolate, the standard deviation (0.007) is nearly as great as the average (0.010), so the measurement is not very reproducible.

The arduous path to reliable analytical results is not the end of the story. The purpose of the analysis is always to reach some interpretation or decision. Table 0-1 tells us that there is much more theobromine than caffeine in dark chocolate. It

FIGURE 0-7 Calibration curves, showing observed peak heights for known concentrations of pure compounds. One part per million is one microgram of analyte per gram of solution. Equations of the straight lines drawn through the experimental data points were determined by the *method of least squares* described in Chapter 4.

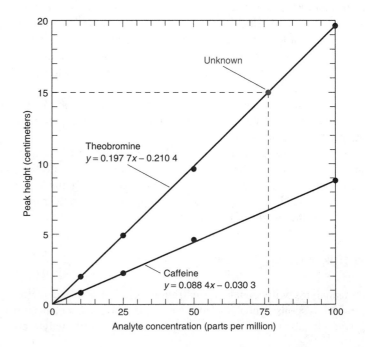

TABLE 0-1   Analyses of dark and white chocolate

Analyte	Grams of analyte per 100 grams of chocolate	
	Dark chocolate	White chocolate
Theobromine	0.392 ± 0.002	0.010 ± 0.007
Caffeine	0.050 ± 0.003	$0.000\,9 \pm 0.001\,4$

Uncertainties are the *standard deviation* of three replicate injections of each extract.

also tells us that there is very little of either substance in white chocolate. Dark chocolate is made from cocoa beans, known to contain theobromine and caffeine. White chocolate comes from carob, which Denby and Scott did not believe contains theobromine and caffeine.

Here are some possible explanations for the small peaks for theobromine and caffeine in white chocolate:

1. Both chocolates came from one shop. Perhaps dark chocolate left in the cookware contaminated the white chocolate.

2. Denby and Scott may have contaminated the white chocolate extract with dark chocolate extract.

3. Perhaps the very small peaks in the chromatogram of white chocolate are not theobromine and caffeine. When in doubt, we must identify peaks, usually by measuring their spectral characteristics.

The questions posed at the beginning of this chapter were "How much caffeine is in a chocolate bar?" and "How does it compare with the quantity in coffee or soft drinks?" After all this work, Denby and Scott discovered how much caffeine is in *one* particular chocolate bar that they analyzed. It would take a great deal more work to sample many chocolate bars of the same type and many different types of chocolate to gain a more universal view. Table 0-2 compares results from different kinds

TABLE 0-2   Caffeine content of beverages and foods

Source	Caffeine (milligrams per serving)	Serving size[a] (ounces)
Regular coffee	106–164	5
Decaffeinated coffee	2–5	5
Tea	21–50	5
Cocoa beverage	2–8	6
Baking chocolate	35	1
Sweet chocolate	20	1
Milk chocolate	6	1
Caffeinated soft drinks	36–57	12

a. 1 ounce = 28.35 grams.

SOURCE: Tea Association (http://www.chinamist.com/caffeine.htm).

of analyses of different sources of caffeine. A can of soft drink or a cup of tea contains about one-quarter to one-half of the caffeine in a small cup of coffee. Chocolate contains even less caffeine, but a hungry backpacker eating enough baking chocolate can get a pretty good jolt!

0-2   General Steps in a Chemical Analysis

The analytical process often begins with a question that is not phrased in terms of a chemical analysis. The question could be "Is this water safe to drink?" or "Does emission testing of automobiles reduce air pollution?" A scientist translates such questions into the need for particular measurements. An analytical chemist then must choose or invent a procedure to carry out those measurements.

When the analysis is complete, the analyst must translate the results into terms that can be understood by others—preferably by the general public. A most important feature of any result is its limitations. What is the statistical uncertainty in reported results? If you took samples in a different manner, would you obtain the same results? Is a tiny amount (a *trace*) of analyte found in a sample really there or is it contamination? Once all interested parties understand the results and their limitations, then they can draw conclusions and reach decisions.

We can now summarize general steps in the analytical process:

Formulating the question	Translate general questions into specific questions amenable to being answered through chemical measurements.
Selecting analytical procedures	Search the chemical literature to find appropriate procedures or, if necessary, devise new procedures to make the required measurements.
Sampling	*Sampling* is the process of selecting representative material to analyze. Box 0-1 provides some ideas on how to do this. If you begin with a poorly chosen sample, or if the sample changes between the time it is collected and the time it is analyzed, the results are meaningless. "Garbage in—garbage out!"
Sample preparation	*Sample preparation* is the process of converting a representative sample into a form suitable for chemical analysis, which usually means dissolving the sample. Samples with a low concentration of analyte may need to be concentrated prior to analysis. It may be necessary to remove, or *mask,* species that interfere with the chemical analysis. For a chocolate bar, sample preparation consisted of removing fat and dissolving the desired analytes. The reason for removing fat was that it would interfere with chromatography.

Chemists use the term **species** to refer to any chemical of interest. Species is both singular and plural. **Interference** occurs when a species other than analyte increases or decreases the response of the analytical method and makes it appear that there is more or less analyte than is actually present. **Masking** is the transformation of an interfering species into a form that is not detected. For example, Ca^{2+} in lake water can be measured with a reagent called EDTA. Al^{3+} interferes with this analysis, because it also reacts with EDTA. Therefore Al^{3+} is masked by treating the sample with excess F^- to form AlF_6^{3-}, which does not react with EDTA.

Analysis	Measure the concentration of analyte in several identical **aliquots** (portions). The purpose of *replicate measurements* (repeated measurements) is to assess the *variability* (uncertainty) in the analysis and to guard against a gross error in the analysis of a single aliquot. *The uncertainty of a measurement is as important as the measurement itself,* because it tells us how reliable the measurement is. If necessary, use different analytical methods on similar samples to make sure that all methods give the same result and that the choice of analytical method is not biasing the result. You may also wish to construct and analyze several different bulk samples to see what variations arise from your sampling procedure.
Reporting and interpretation	Deliver a clearly written, complete report of your results, highlighting any special limitations that you attach to them. Your report might be written to be read only by a specialist (such as your instructor), or it might be written for a general audience (perhaps your mother). Be sure the report is appropriate for its intended audience.
Drawing conclusions	Once a report is written, the analyst might or might not be further involved in what is done with the information, such as modifying the raw material supply for a factory or creating new laws to regulate food additives. The more clearly a report is written, the less likely it is to be misinterpreted by those who use it. The analyst should have the responsibility of ensuring that conclusions drawn from his or her data are consistent with the data.

Most of this book deals with measuring chemical concentrations in homogeneous aliquots of an unknown. The analysis is meaningless unless you have collected the sample properly, you have taken measures to ensure the reliability of the analytical method, and you communicate your results clearly and completely. The chemical analysis is only the middle portion of a process that begins with a question and ends with a conclusion.

Ask Yourself

0-A. After reading Box 0-1, answer the following questions:

(a) What is the difference between a *heterogeneous* and a *homogeneous* material?

(b) What is the difference between a *random* heterogeneous material and a *segregated* heterogeneous material?

(c) What is the difference between a *random* sample and a *composite* sample? When would each be used?

Box 0-1 *Explanation*

Constructing a Representative Sample

The diagram below summarizes the steps in transforming a complex substance into individual samples that can be analyzed. A *lot* is the total material (for example, a railroad car full of grain or a carton full of macadamia chocolates) from which samples are taken. A *bulk sample* (also called a *gross sample*) is taken from the lot for analysis or for *archiving* (storing for future reference). The bulk sample must be representative of the lot, and the choice of bulk sample is critical to producing a valid analysis. From the representative bulk sample, a smaller, homogeneous *laboratory sample* is formed that must have the same composition as the bulk sample. For example, we might obtain a laboratory sample by grinding an entire solid bulk sample to a fine powder, mixing thoroughly, and keeping one bottle of powder for testing. Small test portions (called *aliquots*) of the laboratory sample are used for individual analyses.

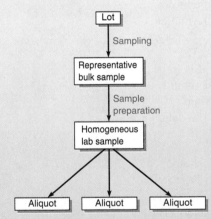

Sampling is the process of selecting a representative bulk sample from the lot. *Sample preparation* converts a bulk sample into a homogeneous laboratory sample. Sample preparation also refers to steps that eliminate interfering species or that concentrate the analyte.

To construct a representative sample from a heterogeneous material, you can first visually divide the material into segments. A **random sample** is collected by taking portions from the desired number of segments chosen at random. If you want to measure the magnesium content of the grass in the 10-meter × 20-meter field in panel *a*, you could divide the field into 20 000 small patches that are 10 centimeters (cm) on a side. After assigning a number to each small patch, you could use a computer program to pick 100 numbers from 1 to 20 000 at random. Then harvest and combine the grass from each of these 100 patches to construct a representative bulk sample for analysis.

For **segregated materials** (in which various regions have different compositions), a representative **composite sample** must be constructed. For example, the field in panel *b* has three different types of grass segregated into regions A, B, and C. You could draw a map of the field on graph paper and measure the area in each region. In this case, 66% of the area lies in region A, 14% lies in region B, and 20% lies in region C. To construct a representative bulk sample from this segregated material, take 66 of the small patches from region A, 14 from region B, and 20 from region C. You could do this by drawing random numbers from 1 to 20 000 to select patches until you have the desired number from each region.

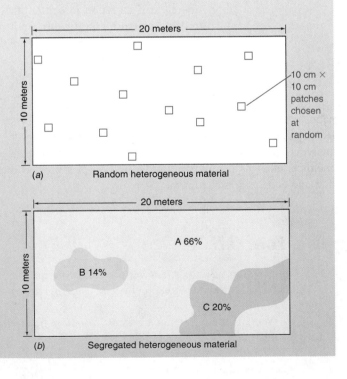

Important Terms[†]

aliquot	homogeneous	sample preparation
analyte	interference	sampling
aqueous	masking	segregated material
calibration curve	qualitative analysis	slurry
composite sample	quantitative analysis	species
decant	quantitative transfer	standard solution
heterogeneous	random sample	supernatant liquid

†Terms are introduced in **bold** type in the chapter and are also defined in the Glossary.

Problems

0-1. What is the difference between *qualitative* and *quantitative* analysis?

0-2. List the steps in a chemical analysis.

0-3. What does it mean to *mask* an interfering species?

0-4. What is the purpose of a calibration curve?

Notes and References

1. S. Chu, *Science* **1991,** *253,* 861. Optical tweezers work because the strong electric field of the focused laser induces a dipole (a separation of electric charge) in the vesicle. The dipole is attracted to the point where the electric field is strongest, which is the focal point of the laser. By this means, scientists can now manipulate micrometer-size particles suspended in liquid.

2. T. J. Wenzel, *Anal. Chem.* **1995,** *67,* 470A.

3. W. R. Kreiser and R. A. Martin, Jr., *J. Assoc. Off. Anal. Chem.* **1978,** *61,* 1424; W. R. Kreiser and R. A. Martin, Jr., *J. Assoc. Off. Anal. Chem.* **1980,** *63,* 591.

Further Reading

S. M. Gerber, ed., *Chemistry and Crime: From Sherlock Holmes to Today's Courtroom* (Washington, DC: American Chemical Society, 1983). Available in paperback from Oxford University Press, ISBN 0-8412-0785-2.

S. M. Gerber and R. Saferstein, eds., *More Chemistry and Crime: From Marsh Arsenic Test to DNA Profile* (Washington, DC: American Chemical Society, 1997). Available from Oxford University Press, ISBN 0-8412-3406-X.

Chemical Measurements

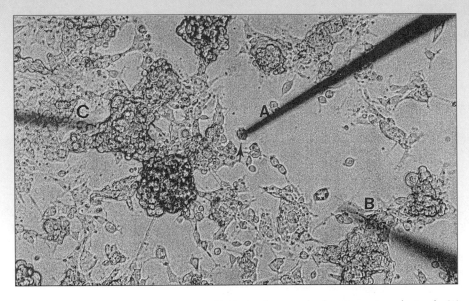

Optical micrograph showing microelectrode (A) and two micropipets (B and C) positioned near a single nerve cell. The electrode measures release of neurotransmitter chemicals from the cell when the cell is stimulated with nicotine from a pipet. [Courtesy A. G. Ewing, Pennsylvania State University. See T. K. Chen, G. Luo, and A. G. Ewing, *Anal. Chem.* **1994**, *66*, 3031.]

*E*very phenomenon in science is described in terms of a small set of units that enumerate quantities such as mass, length, time, temperature, electric current, and chemical concentration. To encompass the enormous range of measurements—from the size of a quark inside a proton to the distance between galaxies—we use prefixes such as *zepto* (10^{-21}) and *tera* (10^{+12}) to multiply the basic units.

When stimulated by nicotine (yes, the stuff in tobacco), the nerve cell in the photograph discharges approximately 20 bursts of neurotransmitter molecules per second. This is one means by which nerve cells communicate with each other. Each burst contains about 130 zeptomoles (130×10^{-21} moles or 80 000 molecules) of the neurotransmitter dopamine. Each discharge lasts about 10 milliseconds (10×10^{-3} seconds). To make such measurements, the electrode must respond to picoamperes (10^{-12} amperes, pronounced PEE-ko-amperes) of electric current. The first goal of this chapter is to acquaint you with units and prefixes used in chemical measurements.

Chapter 1

CHEMICAL MEASUREMENTS

Most people who practice analytical chemistry do not consider themselves to be analytical chemists. For example, chemical analysis is an essential tool used by biologists to understand how organisms function and by doctors to diagnose disease and monitor the response of a patient to treatment. Environmental scientists measure chemical changes in the atmosphere, water, and soil in response to the activities of both man and nature. Whether or not you aspire to be an analytical chemist, you are taking this course because you either will make chemical measurements yourself or you will need to understand analytical results reported by others. This chapter provides basic working knowledge of measurements and equilibrium for any scientist who needs to know about chemistry.

1-1   SI Units and Prefixes

SI units of measurement derive their name from the French *Système International d'Unités*. *Fundamental units* (base units) from which all others are derived are defined in Table 1-1. Standards of length, mass, and time are the *meter* (m), *kilogram* (kg), and *second* (s), respectively. Temperature is measured in *kelvins* (K), amount

TABLE 1-1   Fundamental SI units

Quantity	Unit (symbol)	Definition
Length	meter (m)	The meter is the distance light travels in a vacuum during $\frac{1}{299\ 792\ 458}$ of a second.
Mass	kilogram (kg)	One kilogram is the mass of the prototype kilogram kept at Sèvres, France.
Time	second (s)	The second is the duration of 9 192 631 770 periods of the radiation corresponding to a certain atomic transition of ^{133}Cs.
Electric current	ampere (A)	One ampere of current produces a force of 2×10^{-7} newtons per meter of length when maintained in two straight, parallel conductors of infinite length and negligible cross section, separated by one meter in a vacuum.
Temperature	kelvin (K)	Temperature is defined such that the triple point of water (at which solid, liquid, and gaseous water are in equilibrium) is 273.16 K, and the temperature of absolute zero is 0 K.
Luminous intensity	candela (cd)	Candela is a measure of luminous intensity visible to the human eye.
Amount of substance	mole (mol)	One mole is the number of particles equal to the number of atoms in exactly 0.012 kg of ^{12}C (approximately 6.022 141 99 $\times 10^{23}$).
Plane angle	radian (rad)	There are 2π radians in a circle.
Solid angle	steradian (sr)	There are 4π steradians in a sphere.

TABLE 1-2   SI-derived units with special names

Quantity	Unit	Symbol	Expression in terms of other units	Expression in terms of SI base units
Frequency	hertz	Hz		$1/s$
Force	newton	N		$m \cdot kg/s^2$
Pressure	pascal	Pa	N/m^2	$kg/(m \cdot s^2)$
Energy, work, quantity of heat	joule	J	$N \cdot m$	$m^2 \cdot kg/s^2$
Power, radiant flux	watt	W	J/s	$m^2 \cdot kg/s^3$
Quantity of electricity, electric charge	coulomb	C		$s \cdot A$
Electric potential, potential difference, electromotive force	volt	V	W/A	$m^2 \cdot kg/(s^3 \cdot A)$
Electric resistance	ohm	Ω	V/A	$m^2 \cdot kg/(s^3 \cdot A^2)$

Pressure is force per unit area.

1 pascal (Pa) = 1 N/m^2.

The pressure of the atmosphere is about 100 000 Pa.

of substance in *moles* (mol), and electric current in *amperes* (A). Table 1-2 lists other quantities that are defined in terms of the fundamental quantities. For example, force is measured in *newtons* (N), pressure is measured in *pascals* (Pa), and energy is measured in *joules* (J), each of which can be expressed in terms of the more fundamental units of length, time, and mass.

It is convenient to use the prefixes listed in Table 1-3 to express large or small quantities. For example, the pressure of dissolved oxygen in human arterial blood is approximately 1.3×10^4 Pa. Table 1-3 tells us that 10^3 is assigned the prefix k for "kilo." We can express the pressure in multiples of 10^3 as follows:

$$1.3 \times 10^4 \, \cancel{Pa} \times \frac{1 \text{ kPa}}{10^3 \, \cancel{(Pa)}} = 1.3 \times 10^1 \text{ kPa} = 13 \text{ kPa}$$

The unit kPa is read "kilopascals." It is a good idea to write units beside each number in a calculation and to cancel identical units in the numerator and denominator. This practice ensures that you know the units for your answer. If you intend to calculate pressure and your answer comes out with units other than pascals, something is wrong.

Question In the past, medical technologists referred to one microliter (μL) as one lambda (λ). How many microliters are in one milliliter?

EXAMPLE   Counting Neurotransmitter Molecules with an Electrode

Box 1-1 describes the process by which packets of neurotransmitters are released from a nerve cell in discrete bursts. Each neurotransmitter that diffuses to the surface of a nearby electrode releases two electrons. The average burst of charge measured at the electrode is 37 fC (femtocoulombs, 10^{-15} C). One coulomb of charge corresponds to 6.24×10^{18} electrons. How many molecules are released in an average burst?

SOLUTION Table 1-3 tells us that 1 fC equals 10^{-15} C. Therefore, 37 fC corresponds to

$$37 \, \cancel{fC} \times \left(\frac{10^{-15} \text{ C}}{\cancel{fC}} \right) = 37 \times 10^{-15} \text{ C}$$

The number of electrons in 37 fC is

$$(37 \times 10^{-15} \, \cancel{C}) \times \left(\frac{6.24 \times 10^{18} \text{ electrons}}{\cancel{C}} \right) = 2.3 \times 10^5 \text{ electrons}$$

Because each neurotransmitter molecule releases two electrons, the number of molecules in one burst is

$$(2.3 \times 10^5 \, \cancel{\text{electrons}}) \times \left(\frac{1 \text{ molecule}}{2 \, \cancel{\text{electrons}}} \right) = 1.2 \times 10^5 \text{ molecules}$$

TABLE 1-3		Prefixes
Prefix	*Symbol*	*Factor*
yotta	Y	10^{24}
zetta	Z	10^{21}
exa	E	10^{18}
peta	P	10^{15}
tera	T	10^{12}
giga	G	10^{9}
mega	M	10^{6}
kilo	k	10^{3}
hecto	h	10^{2}
deka	da	10^{1}
deci	d	10^{-1}
centi	c	10^{-2}
milli	m	10^{-3}
micro	μ	10^{-6}
nano	n	10^{-9}
pico	p	10^{-12}
femto	f	10^{-15}
atto	a	10^{-18}
zepto	z	10^{-21}
yocto	y	10^{-24}

Ask Yourself

1-A. What are the names and abbreviations for each of the prefixes from 10^{-24} to 10^{24}? Which abbreviations are capitalized?

Exocytosis of Neurotransmitters

Nerve signals are transmitted from the *axon* of one nerve cell (a *neuron*) to the *dendrite* of a neighboring neuron across a junction called a *synapse*. A change in electric potential at the axon causes tiny chemical-containing packages called *vesicles* to fuse with the cell membrane. This process, called *exocytosis,* releases neurotransmitter molecules stored in the vesicles. When neurotransmitters bind to receptors on the dendrite,

gates open up to allow cations to cross the dendrite membrane. Cations diffusing into the dendrite change its electric potential, thereby transmitting the nerve impulse into the second neuron. (Chemical analysis of a single vesicle from a cell was described at the opening of Chapter 0.)

The nerve cell in the photo at the opening of this chapter releases the neurotransmitter dopamine when

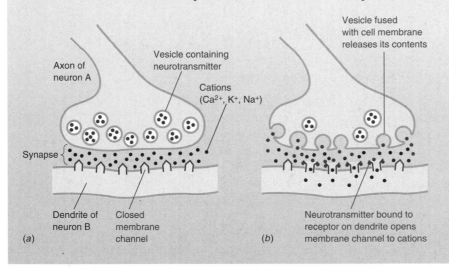

Action of neurotransmitters at a synapse. (*a*) Before release of neurotransmitter. (*b*) Neurotransmitters are released by *exocytosis* when vesicles merge with the neuron's outer cell membrane. (Cells can take molecules in by the reverse process, called *endocytosis.*)

1-2   Conversion Between Units

Although SI is the internationally accepted system of measurement in science, other units are encountered. Conversion factors are found in Table 1-4. For example, common non-SI units for energy are the *calorie* (cal) and the *Calorie* (with a capital C, which represents 1 000 calories, or 1 kcal). Table 1-4 states that 1 cal is exactly 4.184 J (joules).

You require approximately 46 Calories per hour (h) per 100 pounds (lb) of body mass to carry out basic functions required for life, such as breathing, pumping blood, and maintaining body temperature. This minimum energy requirement for a conscious person at rest is called the *basal metabolism.* A person walking at 2 miles per hour on a level path requires approximately 45 Calories per hour per 100 pounds of body mass beyond basal metabolism. The same person swimming at 2 miles per hour consumes 360 Calories per hour per 100 pounds.

1 *calorie* of energy will heat 1 gram of water from 14.5° to 15.5°C.

1 000 *joules* will raise the temperature of a cup of water by about 1°C.

1 cal = 4.184 J

1 pound (mass) ≈ 0.453 6 kg

1 mile ≈ 1.609 km

EXAMPLE   Unit Conversions

Express the rate of energy consumption for a walking person (46 + 45 = 91 Calories per hour per 100 pounds of body mass) in kilojoules per hour per kilogram of body mass.

stimulated by nicotine. The released dopamine can be measured by a microelectrode placed next to the cell. Each dopamine molecule that diffuses to the microelectrode gives up two electrons. The figure below shows pulses of electric current measured at the microelectrode. Each pulse arises from the dopamine released by one vesicle. The number of electrons in each pulse tells us how many dopamine molecules were released from one vesicle.

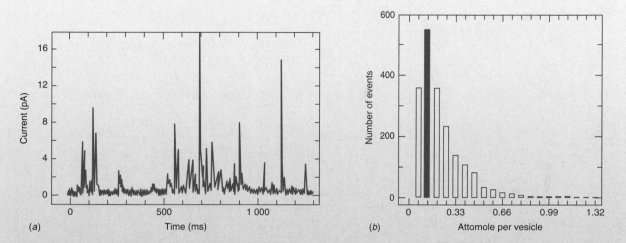

(a) Bursts of electric current from exocytosis of individual vesicles. The vertical axis (y-axis, called the **ordinate**) is measured in units of picoamperes (pA, 10^{-12} A). The horizontal axis (x-axis, called the **abscissa**, pronounced ab-sis-a) is measured in milliseconds (ms, 10^{-3} s). (b) *Histogram* showing relative frequencies of packets of neurotransmitter. Each bar covers a 0.066 amol (= 66 zmol) interval. The second bar (= $2 \times 66 = 132$ zmol) is the most common occurrence. The *average* packet size is greater than 132 zmol, however, because the distribution is not symmetric. [T. K. Chen, G. Luo, and A. G. Ewing, *Anal. Chem.* **1994**, *66*, 3031.]

TABLE 1-4   Conversion factors

Quantity	Unit	Symbol	SI equivalent[a]
Volume	liter	L	*10^{-3} m^3
	milliliter	mL	*10^{-6} m^3
Length	angstrom	Å	*10^{-10} m
	inch	in.	*0.025 4 m
Mass	pound	lb	*0.453 592 37 kg
Force	dyne	dyn	*10^{-5} N
Pressure	atmosphere	atm	*101 325 Pa
	bar	bar	*10^5 Pa
	torr	mm Hg	133.322 Pa
	pound/in.^{2}	psi	6 894.76 Pa
Energy	erg	erg	*10^{-7} J
	electron volt	eV	1.602 177 33 $\times$ 10^{-19} J
	calorie, thermochemical	cal	*4.184 J
	Calorie (with a capital C)	Cal	*1 000 cal = 4.184 kJ
	British thermal unit	Btu	1 055.06 J
Power	horsepower		745.700 W
Temperature	centigrade (= Celsius)	°C	*K − 273.15
	Fahrenheit	°F	*1.8(K − 273.15) + 32

a. An asterisk (*) indicates that the conversion is exact (by definition).

In 1999, the $125 million *Mars Climate Orbiter* spacecraft was lost when it entered the Martian atmosphere 100 km lower than planned. *The navigation error would have been avoided if people had labeled their units of measurement.* Engineers at the contractor that built the spacecraft calculated thrust in the English unit, pounds of force. Jet Propulsion Laboratory engineers thought they were receiving the information in the metric unit, newtons. Nobody caught the error.

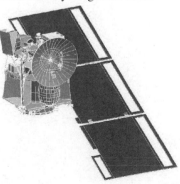

SOLUTION We will convert each non-SI unit separately. First, note that 91 Calories equals 91 kcal. Table 1-4 states that 1 cal = 4.184 J, or 1 kcal = 4.184 kJ, so

$$91 \ \text{kcal} \times \frac{4.184 \ \text{kJ}}{1 \ \text{kcal}} = 381 \ \text{kJ}$$

Correct use of significant figures is discussed in Chapter 3.

Table 1-4 also says that 1 lb is 0.453 6 kg; so 100 lb = 45.36 kg. The rate of energy consumption is, therefore,

$$\frac{91 \ \text{kcal/h}}{100 \ \text{lb}} = \frac{381 \ \text{kJ/h}}{45.36 \ \text{kg}} = 8.4 \ \frac{\text{kJ/h}}{\text{kg}}$$

You could have written this as one calculation with appropriate unit cancellations:

$$\text{rate} = \frac{91 \ \text{kcal/h}}{100 \ \text{lb}} \times \frac{4.184 \ \text{kJ}}{1 \ \text{kcal}} \times \frac{1 \ \text{lb}}{0.453 \ 6 \ \text{kg}} = 8.4 \ \frac{\text{kJ/h}}{\text{kg}}$$

EXAMPLE   Watts Measure Power (Energy per Second)

1 W = 1 J/s

One watt is 1 joule per second. The person in the previous example expends 8.4 kilojoules per hour per kilogram of body mass while walking. **(a)** How many watts per kilogram does the person use? **(b)** If the person's mass is 60 kg, how many watts does he expend?

SOLUTION (a) The person expends 8.4×10^3 J/h/kg. Because an hour contains 3 600 s (60 s/min $\times$ 60 min/h), the required power is

The complex unit joules per hour per kilogram (J/h/kg) is the same as the expression $\dfrac{\text{J}}{\text{h} \cdot \text{kg}}$.

$$8.4 \times 10^3 \frac{\text{J}}{\text{h} \cdot \text{kg}} \times \frac{3 \ 600 \ \text{s}}{1 \ \text{h}}$$

Oops! The units didn't cancel out. I guess we need to use the inverse conversion factor:

$$8.4 \times 10^3 \frac{\text{J}}{\text{h} \cdot \text{kg}} \times \frac{1 \ \text{h}}{3 \ 600 \ \text{s}} = 2.33 \ \frac{\text{J}}{\text{s} \cdot \text{kg}} = 2.33 \ \frac{\text{J/s}}{\text{kg}} = 2.33 \ \frac{\text{W}}{\text{kg}}$$

(b) Our intrepid walker has a mass of 60 kg. Therefore his power requirement is

$$2.33 \ \frac{\text{W}}{\text{kg}} \times 60 \ \text{kg} = 140 \ \text{W}$$

> ## Ask Yourself
>
> **1-B.** A 120-pound woman working in an office expends about 2.2×10^3 kcal/day, whereas the same woman climbing a mountain needs 3.4×10^3 kcal/day.
> > **(a)** How many joules per day does the woman expend in each activity?
> > **(b)** How many seconds are in one day?

(c) How many joules per second (= watts) does the woman expend in each activity?

(d) Which consumes more power (watts), the office worker or a 100-W light bulb?

1-3   Chemical Concentrations

The minor species in a solution is called the **solute** and the major species is the **solvent.** In this text, most discussions concern *aqueous* solutions, in which the solvent is water. **Concentration** refers to how much solute is contained in a given volume or mass.

Molarity and Molality

Molarity (M) is the number of moles of a substance per liter of solution. A mole is Avogadro's number of atoms or molecules or ions $(6.022 \times 10^{23}\ \text{mol}^{-1})$. A **liter** (L) is the volume of a cube that is 10 cm on each edge. Because 10 cm = 0.1 m, $1\ \text{L} = (0.1\ \text{m})^3 = (10^{-1})^3\ \text{m}^3 = 10^{-3}\ \text{m}^3$. In Figure 1-1, chemical concentrations in the ocean are expressed in micromoles per liter $(10^{-6}\ \text{mol/L} = \mu\text{M})$ and nanomoles per liter $(10^{-9}\ \text{mol/L} = \text{nM})$. The molarity of a species is usually designated by square brackets, as in $[\text{Cl}^-]$.

The **atomic mass** of an element is the number of grams containing Avogadro's number of that element's atoms. The **molecular mass** of a compound is the sum of atomic masses of the atoms in the molecule. It is the number of grams containing Avogadro's number of molecules of the compound.

$$\text{molarity (M)} = \frac{\text{moles of solute}}{\text{liters of solution}}$$

The liter is named after the Frenchman Claude Litre (1716–1778), who named his daughter Millicent. I suppose her friends called her Millie Litre.

"Mole Day" is celebrated at 6:02 A.M. on October 23 (10/23) at many schools.

EXAMPLE   Molarity of Salts in the Sea

(a) Typical seawater contains 2.7 g of salt (sodium chloride, NaCl) per deciliter (= dL = 0.1 L). What is the molarity of NaCl in the ocean? **(b)** $MgCl_2$ has a typical

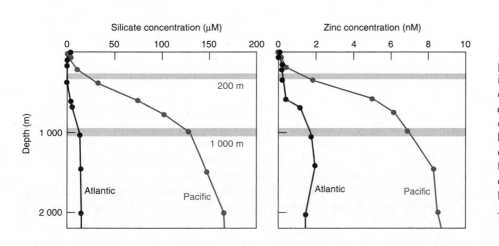

FIGURE 1-1 Concentration profiles of dissolved silicate and zinc in the northern Atlantic and northern Pacific oceans. Seawater is *heterogeneous:* Samples collected at depths of 200 m or 1 000 m do not have the same concentrations of each species. Living organisms near the ocean surface deplete seawater of both silicate and zinc. [Data from K. S. Johnson, K. H. Coale, and H. W. Jannasch, *Anal. Chem.* **1992**, *64,* 1065A.]

You can find atomic masses in the periodic table inside the front cover of this book.

concentration of 0.054 M in the ocean. How many grams of $MgCl_2$ are present in 25 mL of seawater?

SOLUTION   **(a)** The molecular mass of NaCl is 22.99 (Na) + 33.45 (Cl) = 58.44 g/mol. The moles of salt in 2.7 g are

$$\text{moles of NaCl} = \frac{(2.7 \text{ g})}{\left(58.44 \frac{\text{g}}{\text{mol}}\right)} = 0.046 \text{ mol}$$

so the molarity is

$$[\text{NaCl}] = \frac{\text{mol NaCl}}{\text{L of seawater}} = \frac{0.046 \text{ mol}}{0.1 \text{ L}} = 0.46 \text{ M}$$

(b) The molecular mass of $MgCl_2$ is 24.30 (Mg) + $[2 \times 35.45]$(Cl) = 95.20 g/mol, so the number of grams in 25 mL is

$$\text{grams of MgCl}_2 = 0.054 \frac{\text{mol}}{\text{L}} \times 95.20 \frac{\text{g}}{\text{mol}} \times (25 \times 10^{-3} \text{ L}) = 0.13 \text{ g}$$

Strong electrolyte: mostly dissociated into ions in solution

Weak electrolyte: partially dissociated into ions in solution

An *electrolyte* dissociates into ions in aqueous solution. Magnesium chloride is a **strong electrolyte,** which means that it is mostly dissociated into ions in most solutions. In seawater, about 89% of the magnesium is present as Mg^{2+} and 11% is found as the *complex ion,* $MgCl^{+}$.[1] The concentration of $MgCl_2$ molecules in seawater is close to 0. Sometimes the molarity of a strong electrolyte is referred to as **formal concentration** (F), to indicate that the substance is really converted to other species in solution. When we commonly, and inaccurately, say that the "concentration" of $MgCl_2$ is 0.054 M in seawater, we really mean that its formal concentration is 0.054 F. The "molecular mass" of a strong electrolyte is more properly called the **formula mass** (which we will abbreviate FM), because it is the sum of atomic masses in the formula, even though there may be few molecules with that formula.

A **weak electrolyte** such as acetic acid, CH_3CO_2H, is partially split into ions in solution:

		Formal concentration	Percent dissociated
		0.1 F	1.3%
		0.01	4.1
		0.001	12.4

A solution prepared by dissolving 0.010 00 mol of acetic acid in 1.000 L has a formal concentration of 0.010 00 F. The actual molarity of CH_3CO_2H is 0.009 59 M because 4.1% is dissociated into $CH_3CO_2^-$ and 95.9% remains as CH_3CO_2H. Nonetheless, we customarily say that the solution is 0.010 00 M acetic acid and understand that some of the acid is dissociated.

Molality (*m*) is a designation of concentration expressing the number of moles of substance per kilogram of solvent (not total solution). Unlike molarity, molality

Confusing abbreviations:

mol = moles

$$M = \text{molarity} = \frac{\text{mol solute}}{\text{L solution}}$$

$$m = \text{molality} = \frac{\text{mol solute}}{\text{kg solvent}}$$

is independent of temperature. Molarity changes with temperature because the volume of a solution usually increases when it is heated.

Percent Composition

The percentage of a component in a mixture or solution is usually expressed as a **weight percent** (wt %):

Definition of weight percent:
$$\text{weight percent} = \frac{\text{mass of solute}}{\text{mass of total solution or mixture}} \times 100 \quad (1\text{-}1)$$

A common form of ethanol (CH_3CH_2OH) is 95 wt %; it has 95 g of ethanol per 100 g of total solution. The remainder is water. Another common expression of composition is **volume percent** (vol %):

Definition of volume percent:
$$\text{volume percent} = \frac{\text{volume of solute}}{\text{volume of total solution}} \times 100 \quad (1\text{-}2)$$

Although "wt %" or "vol %" should always be written to avoid ambiguity, wt % is usually implied when you just see "%".

EXAMPLE   Converting Weight Percent to Molarity and Molality

Find the molarity of HCl in a reagent labeled "37.0 wt % HCl, density = 1.188 g/mL." The **density** of a substance is the mass per unit volume.

SOLUTION We need to find the moles of HCl per liter of solution. To find moles of HCl, we need to find the mass of HCl. The mass of HCl in 1 L is 37.0% of the mass of 1 L of solution. The mass of 1 L of solution is $(1.188 \text{ g/mL})(1\,000 \text{ mL/L}) = 1\,188$ g/L. The mass of HCl in 1 L is

$$\text{HCl}\left(\frac{g}{L}\right) = 1\,188\,\frac{\text{g solution}}{L} \times 0.370\,\frac{\text{g HCl}}{\text{g solution}} = 439.6\,\frac{\text{g HCl}}{L}$$

This is what 37.0 wt % HCl means.

The molecular mass of HCl is 36.46 g/mol, so the molarity is

$$\text{molarity} = \frac{\text{mol HCl}}{\text{L solution}} = \frac{439.6 \text{ g HCl/L}}{36.46 \text{ g HCl/mol}} = 12.1\,\frac{\text{mol}}{L} = 12.1 \text{ M}$$

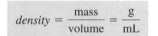

$$density = \frac{mass}{volume} = \frac{g}{mL}$$

A closely related dimensionless quantity is

Specific gravity =
$$\frac{\text{density of a substance}}{\text{density of water at } 4°C}$$

Because the density of water at 4°C is very close to 1 g/mL, specific gravity is nearly the same as density.

Parts per Million and Parts per Billion

Concentrations of trace components of a sample can be expressed as **parts per million** (ppm) or **parts per billion** (ppb), terms that mean grams of substance per million or billion grams of total solution or mixture.

Definition of parts per million:

$$ppm = \frac{\text{mass of substance}}{\text{mass of sample}} \times 10^6 \qquad (1\text{-}3)$$

Question What would be the definition of parts per trillion?

Definition of parts per billion:

$$ppb = \frac{\text{mass of substance}}{\text{mass of sample}} \times 10^9 \qquad (1\text{-}4)$$

Masses must be expressed in the same units in the numerator and denominator.

1 ppm $\approx$ 1 μg/mL

1 ppb $\approx$ 1 ng/mL

The symbol $\approx$ is read "is approximately equal to."

The density of a dilute aqueous solution is close to 1.00 g/mL, so *we frequently equate 1 g of water with 1 mL of water,* although this equivalence is only approximate. Therefore 1 ppm corresponds to 1 μg/mL (= 1 mg/L) and 1 ppb is 1 ng/mL (= 1 μg/L).

EXAMPLE   Converting Parts per Billion to Molarity

Hydrocarbons are compounds containing only hydrogen and carbon. Plants manufacture hydrocarbons (called fats or lipids) as components of the membranes of cells and vesicles. The biosynthetic pathway leads mainly to compounds with an odd number of carbon atoms. Figure 1-2 shows the concentrations of hydrocarbons washed from the air by rain in the winter and summer. The preponderance of odd-number hydrocarbons in the summer suggests that the source is mainly from plants. The more uniform distribution of odd- and even-number hydrocarbons in the winter suggests a man-made origin. The concentration of $C_{29}H_{60}$ in summer rainwater is 34 ppb. Find the molarity of this compound in nanomoles per liter (nM).

SOLUTION A concentration of 34 ppb means 34×10^{-9} g (= 34 ng) of $C_{29}H_{60}$ per gram of rainwater, which we equate to 34 ng/mL. To find moles per liter, we first find grams per liter:

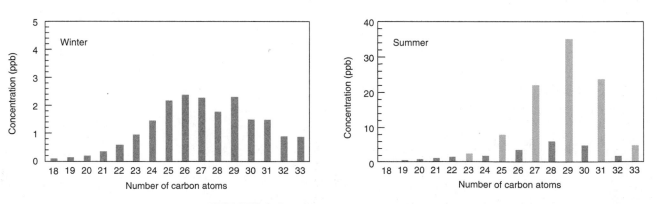

FIGURE 1-2 Concentrations of alkanes (hydrocarbons with the formula C_nH_{2n+2}) found in rainwater in Hannover, Germany, in the winter and summer in 1989 are measured in parts per billion (= μg hydrocarbon/L of rainwater). Summer concentrations are higher, and compounds with an odd number of carbon atoms (colored bars) predominate. Plants produce mainly hydrocarbons with an odd number of carbon atoms. [K. Levsen, S. Behnert, and H. D. Winkeler, *Fresenius J. Anal. Chem.* **1991**, *340*, 665.]

$$34 \times 10^{-9} \frac{g}{mL} \times \frac{1\ 000\ mL}{L} = 34 \times 10^{-6} \frac{g}{L}$$

Because the molecular mass of $C_{29}H_{60}$ is 408.8 g/mol, the molarity is

$$\text{molarity of } C_{29}H_{60} \text{ in rainwater} = \frac{34 \times 10^{-6}\ g/L}{408.8\ g/mol} = 8.3 \times 10^{-8}\ M$$

$$= 83 \times 10^{-9}\ M = 83\ nM$$

With reference to gases, ppm usually indicates volume rather than mass. For example, 8 ppm carbon monoxide in air means 8 μL of CO per liter of air. Always write the units to avoid confusion. Figure 1-3 shows gas concentration measured in *parts per trillion* by volume (picoliters per liter).

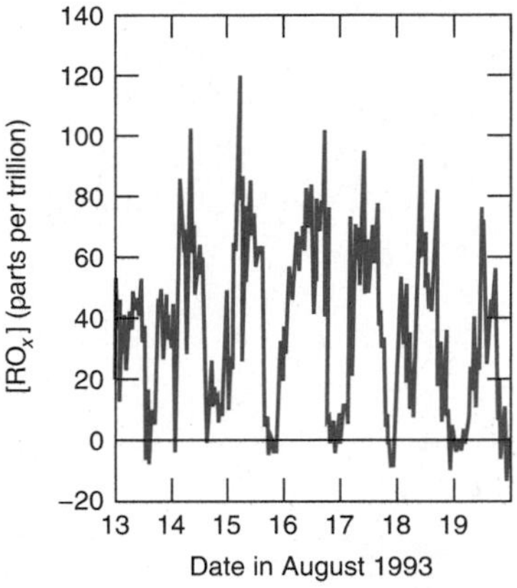

Ask Yourself

1-C. The density of 70.5 wt % aqueous perchloric acid is 1.67 g/mL. Note that grams refers to grams of *solution* (= g $HClO_4$ + g H_2O).
   (a) How many grams of solution are in 1.00 L?
   (b) How many grams of $HClO_4$ are in 1.00 L?
   (c) How many moles of $HClO_4$ are in 1.00 L? This is the molarity.

1-4   Preparing Solutions

To prepare a solution with a desired molarity, weigh out the correct mass of pure reagent, dissolve it in solvent in a *volumetric flask* (Figure 1-4), dilute with more solvent to the desired final volume, and mix well by inverting the flask 20 times.

EXAMPLE   Preparing a Solution with Desired Molarity

Cupric sulfate is commonly sold as the pentahydrate, $CuSO_4 \cdot 5H_2O$, which means that there are five moles of H_2O for each mole of $CuSO_4$ in the solid crystal. The formula mass of $CuSO_4 \cdot 5H_2O$ (= $CuSO_9H_{10}$) is 249.69 g/mol. How many grams of $CuSO_4 \cdot 5H_2O$ should be dissolved in a 250-mL volumetric flask to make a solution containing 8.00 mM Cu^{2+}?

SOLUTION An 8.00 mM solution contains 8.00×10^{-3} mol/L. Because 250 mL is 0.250 L, we need

$$8.00 \times 10^{-3} \frac{mol}{L} \times 0.250\ L = 2.00 \times 10^{-3}\ mol\ CuSO_4 \cdot 5H_2O$$

FIGURE 1-3 Concentration of RO_x (parts per trillion by volume = pL/L) measured outdoors at the University of Denver. RO_x refers to the combined concentrations of HO (hydroxyl radical), HO_2 (hydroperoxide radical), RO (alkoxy radical, where R is any organic group), and RO_2 (alkyl peroxide radical). These reactive species are created mainly by photochemical reactions driven by sunlight. Concentrations peak around 2:00 P.M. each day and fall close to 0 during the night. [J. Hu and D. H. Stedman, *Anal. Chem.* **1994**, *66*, 3384.]

Date in August 1993

[RO_x] (parts per trillion)

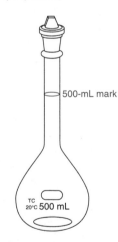

500-mL mark

TC 20°C 500 mL

FIGURE 1-4 A *volumetric flask* contains a specified volume when the liquid level is adjusted to the middle of the mark in the thin neck of the flask. Use of this flask is described in Chapter 2.

The required mass of reagent is

$$(2.00 \times 10^{-3} \ \text{mol})\left(249.69 \ \frac{\text{g}}{\text{mol}}\right) = 0.499 \ \text{g}$$

The procedure is to weigh 0.499 g of solid $CuSO_4 \cdot 5H_2O$ into a 250-mL volumetric flask, add about 200 mL of distilled water, and swirl to dissolve the reagent. Then dilute with distilled water up to the 250-mL mark and invert the stoppered flask 20 times to ensure complete mixing. The solution contains 8.00 mM Cu^{2+}.

Dilute solutions can be prepared from concentrated solutions. Typically, a desired volume or mass of the concentrated solution is transferred to a volumetric flask and diluted to the intended volume with solvent. The number of moles of reagent in V liters containing M moles per liter is the product $M \cdot V = (\text{mol/L})(L) = \text{mol}$. When a solution is diluted from a high concentration to a low concentration, the number of moles of solute is unchanged. Therefore we equate the number of moles in the concentrated (conc) and dilute (dil) solutions:

Dilution formula: $$\boxed{M_{conc} \cdot V_{conc} \ = \ M_{dil} \cdot V_{dil}}$$ (1-5)

Moles taken from   Moles placed in
concentrated solution   dilute solution

EXAMPLE   Preparing 0.1 M HCl

The symbol ~ is read "approximately."

The molarity of "concentrated" HCl purchased for laboratory use is ~12.1 M. How many milliliters of this reagent should be diluted to 1.00 L to make 0.100 M HCl?

SOLUTION The required volume of concentrated solution is found with Equation 1-5:

$$M_{conc} \cdot V_{conc} = M_{dil} \cdot V_{dil}$$

The symbol ⇒ is read "implies that."

$$(12.1 \ \text{M}) \cdot (x \ \text{mL}) = (0.100 \ \text{M}) \cdot (1 \ 000 \ \text{mL}) \Rightarrow x = 8.26 \ \text{mL}$$

It is all right to express both volumes in mL or in L. The important point is to use the same units for volume on both sides of the equation so that the units cancel. To make 0.100 M HCl, place 8.26 mL of concentrated HCl in a 1-L volumetric flask and add ~900 mL of water. After swirling to mix, dilute to the 1-L mark with water and invert the flask 20 times to ensure complete mixing.

EXAMPLE   A More Complicated Dilution Calculation

A solution of ammonia in water is called "ammonium hydroxide" because of the equilibrium

$$NH_3 \ + \ H_2O \ \rightleftharpoons \ NH_4^+ \ + \ OH^-$$

Ammonia   Ammonium   Hydroxide
                              ion   ion

The density of concentrated ammonium hydroxide, which contains 28.0 wt % NH_3 is 0.899 g/mL. What volume of this reagent should be diluted to 500 mL to make 0.250 M NH_3?

SOLUTION To use Equation 1-5, we need to know the molarity of the concentrated reagent. The density tells us that the reagent contains 0.899 grams of solution per milliliter of solution. The weight percent tells us that the reagent contains 0.280 grams of NH_3 per gram of solution. To find the molarity of NH_3 in the concentrated reagent, we need to know the number of moles of NH_3 in one liter:

$$\text{grams of } NH_3 \text{ per liter} = 899 \, \frac{\text{g solution}}{L} \times 0.280 \, \frac{g \, NH_3}{\text{g solution}} = 252 \, \frac{g \, NH_3}{L}$$

$$\text{molarity of } NH_3 = \frac{252 \, \dfrac{g \, NH_3}{L}}{17.03 \, \dfrac{g \, NH_3}{\text{mol } NH_3}} = 14.8 \, \frac{\text{mol } NH_3}{L} = 14.8 \, M$$

Now we use Equation 1-5 to find the volume of 14.8 M NH_3 required to prepare 500 mL of 0.250 M NH_3:

$$M_{conc} \cdot V_{conc} = M_{dil} \cdot V_{dil}$$

$$14.8 \, \frac{\text{mol}}{L} \times V_{conc} = 0.250 \, \frac{\text{mol}}{L} \times 0.500 \, L$$

$$\Rightarrow V_{conc} = 8.45 \times 10^{-3} \, L = 8.45 \, mL$$

The correct procedure is to place 8.45 mL of concentrated reagent in a 500-mL volumetric flask, add about 400 mL of water, and swirl to mix. Then dilute to exactly 500 mL with water and invert the stoppered flask 20 times to mix well.

EXAMPLE   Preparing a Parts per Million Concentration

Drinking water usually contains 1.6 ppm fluoride (F^-) for prevention of tooth decay. Consider a reservoir with a diameter of 450 m and a depth of 10 m. **(a)** How many liters of 0.10 M NaF should be added to produce 1.6 ppm F^-? **(b)** How many grams of solid NaF could be used instead?

SOLUTION **(a)** Assuming that the density of water in the reservoir is close to 1.00 g/mL, 1.6 ppm F^- corresponds to 1.6×10^{-6} g F^-/mL or

$$1.6 \times 10^{-6} \, \frac{g \, F^-}{mL} \times 1\,000 \, \frac{mL}{L} = 1.6 \times 10^{-3} \, \frac{g \, F^-}{L}$$

The atomic mass of fluorine is 19.00, so the desired molarity of fluoride in the reservoir is

$$\text{desired } [F^-] \text{ in reservoir} = \frac{1.6 \times 10^{-3} \, \dfrac{g \, F^-}{L}}{19.00 \, \dfrac{g \, F^-}{\text{mol}}} = 8.42 \times 10^{-5} \, M$$

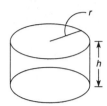

Volume of cylinder

= end area × height

= $\pi r^2 h$

The volume of the reservoir is $\pi r^2 h$, where r is the radius and h is the height.

$$\text{volume of reservoir} = \pi \times (225 \text{ m})^2 \times 10 \text{ m} = 1.59 \times 10^6 \text{ m}^3$$

To use the dilution formula, we need to express volume in liters. Table 1-4 told us that there are 1 000 L in one cubic meter. Therefore the volume of the reservoir in liters is

$$\text{volume of reservoir (L)} = 1.59 \times 10^6 \text{ m}^3 \times 1\,000 \frac{\text{L}}{\text{m}^3} = 1.59 \times 10^9 \text{ L}$$

Finally, we are in a position to use the dilution formula 1-5:

$$M_{\text{conc}} \cdot V_{\text{conc}} = M_{\text{dil}} \cdot V_{\text{dil}}$$

$$0.10 \frac{\text{mol}}{\text{L}} \times V_{\text{conc}} = \left(8.42 \times 10^{-5} \frac{\text{mol}}{\text{L}}\right) \times (1.59 \times 10^9 \text{ L})$$

$$\Rightarrow V_{\text{con}} = 1.3 \times 10^6 \text{ L}$$

We require 1.3 million liters of 0.10 M F^-. Note that our calculation assumed that the final volume of the reservoir is 1.59×10^9 L. Even though we are adding more than 10^6 L of reagent, this amount is small relative to 10^9 L. Therefore the approximation that the reservoir volume remains 1.59×10^9 L is pretty good.

(b) The number of moles of F^- in the reservoir is $(1.59 \times 10^9 \text{ L}) \times (8.42 \times 10^{-5} \text{ mol/L}) = 1.34 \times 10^5$ mol F^-. Because one mole of NaF provides one mole of F^-, we need $(1.34 \times 10^5 \text{ mol NaF}) \times (41.99 \text{ g NaF/mol NaF}) = 5.6 \times 10^6$ g NaF.

Ask Yourself

1-D. A 48 wt % solution of HBr in water has a density of 1.50 g/mL.
  (a) How many grams of solution are in 1.00 L?
  (b) How many grams of HBr are in 1.00 L?
  (c) What is the molarity of HBr?
  (d) How much solution is required to prepare 0.250 L of 0.160 M HBr?

1-5   The Equilibrium Constant

Equilibrium describes the state that a system will reach "if you wait long enough." Most reactions of interest in analytical chemistry reach equilibrium in times ranging from fractions of a second to many minutes.

If reactants A and B are converted to products C and D with the stoichiometry

$$a\text{A} + b\text{B} \rightleftharpoons c\text{C} + d\text{D} \tag{1-6}$$

we write the **equilibrium constant**, K, in the form

Equilibrium constant:

$$K = \frac{[C]^c[D]^d}{[A]^a[B]^b} \qquad (1\text{-}7)$$

where the small superscript letters denote stoichiometric coefficients and each capital letter stands for a chemical species. The symbol [A] stands for the concentration of A relative to its standard state (defined below). By definition, *a reaction is favored whenever K > 1.*

In deriving the equilibrium constant, each quantity in Equation 1-7 is expressed as the *ratio* of the concentration of a species to its concentration in its *standard state.* For solutes, the standard state is 1 M. For gases, the standard state is 1 bar (see Table 1-4); and for solids and liquids, the standard states are the pure solid or liquid.[2] It is understood (but rarely written) that the term [A] in Equation 1-7 really means [A]/(1 M) if A is a solute. If D is a gas, [D] really means (pressure of D in bars)/(1 bar). To emphasize that [D] means pressure of D, we usually write P_D in place of [D]. If C were a pure liquid or solid, the ratio [C]/(concentration of C in its standard state) would be unity (1) because the standard state *is* the pure liquid or solid. If [C] is a solvent, the concentration is so close to that of pure liquid C that the value of [C] is essentially 1. Each term of Equation 1-7 is dimensionless (because each is a ratio in which the units cancel); therefore all equilibrium constants are dimensionless.

The take-home lesson in this: In evaluating an equilibrium constant,

1. The concentrations of solutes should be expressed as moles per liter.

2. The concentrations of gases should be expressed in bars.

3. The concentrations of pure solids, pure liquids, and solvents are omitted because they are unity.

These conventions are arbitrary, but you must use them so that your results are consistent with tabulated values of equilibrium constants and standard reduction potentials.

The equilibrium constant is more correctly expressed as a ratio of *activities* rather than of concentrations. See Section 11-2.

1 bar = 10^5 Pa ≈ 0.987 atm

Equilibrium constants are dimensionless; but when specifying concentrations, you must use units of molarity (M) for solutes and bars for gases.

EXAMPLE   Writing an Equilibrium Constant

Write the equilibrium constant for the reaction

$$Zn(s) + 2NH_4^+(aq) \rightleftharpoons Zn^{2+}(aq) + H_2(g) + 2NH_3(aq)$$

(In chemical equations, *s* stands for solid, *aq* stands for *aqueous,* *g* stands for gas, and *l* stands for liquid.)

SOLUTION Omit the concentration of pure solid and express the concentration of gas as a pressure in bars:

$$K = \frac{[Zn^{2+}]P_{H_2}[NH_3]^2}{[NH_4^+]^2}$$

P_{H_2} stands for the pressure of $H_2(g)$ in bars.

Manipulating Equilibrium Constants

Consider the reaction of the acid HA that dissociates into H^+ and A^-:

Throughout this text, you should assume that all species in chemical equations are in aqueous solution, unless otherwise specified.

$$HA \xrightleftharpoons{K} H^+ + A^- \qquad K = \frac{[H^+][A^-]}{[HA]}$$

If the reverse reaction is written, the new K' is the reciprocal of the original K:

Equilibrium constant for reverse reaction:

$$H^+ + A^- \xrightleftharpoons{K'} HA \qquad K' = 1/K = \frac{[HA]}{[H^+][A^-]}$$

If reactions are added, the new K is the product of the original K's. The equilibrium of H^+ between the species HA and CH^+ can be derived by adding two equations:

$$\begin{aligned}
HA &\rightleftharpoons H^+ + A^- &\qquad K_1 \\
H^+ + C &\rightleftharpoons CH^+ &\qquad K_2 \\
\hline
HA + C &\rightleftharpoons A^- + CH^+ &\qquad K_3
\end{aligned}$$

If a reaction is reversed, then $K' = 1/K$. If two reactions are added, then $K_3 = K_1 K_2$.

Equilibrium constant for sum of reactions:

$$K_3 = K_1 K_2 = \frac{[H^+][A^-]}{[HA]} \cdot \frac{[CH^+]}{[H^+][C]} = \frac{[A^-][CH^+]}{[HA][C]}$$

If n reactions are added, the overall equilibrium constant is the product of all n individual equilibrium constants.

EXAMPLE Combining Equilibrium Constants

From the equilibria

H₂O is omitted from K because it is a pure liquid. Its concentration remains nearly constant.

$$H_2O \rightleftharpoons H^+ + OH^- \qquad K_w = [H^+][OH^-] = 1.0 \times 10^{-14}$$

$$NH_3(aq) + H_2O \rightleftharpoons NH_4^+ + OH^- \qquad K_{NH_3} = \frac{[NH_4^+][OH^-]}{[NH_3(aq)]} = 1.8 \times 10^{-5}$$

find the equilibrium constant for the reaction

$$NH_4^+ \rightleftharpoons NH_3(aq) + H^+$$

SOLUTION The third reaction is obtained by reversing the second reaction and adding it to the first reaction:

$$\begin{aligned}
H_2O &\rightleftharpoons H^+ + OH^- &\qquad K_1 &= K_w \\
NH_4^+ + OH^- &\rightleftharpoons NH_3(aq) + H_2O &\qquad K_2 &= 1/K_{NH_3} \\
\hline
NH_4^+ &\rightleftharpoons H^+ + NH_3(aq) &\qquad K_3 &= K_w \cdot \frac{1}{K_{NH_3}} = 5.6 \times 10^{-10}
\end{aligned}$$

Le Châtelier's Principle

Le Châtelier's principle states that if a system at equilibrium is disturbed, the direction in which the system proceeds back to equilibrium is such that the disturbance is partially offset.

Let's see what happens when we change the concentration of one species in the reaction:

$$\underset{\text{Bromate}}{BrO_3^-} + \underset{\text{Chromium(III)}}{2Cr^{3+}} + 4H_2O \rightleftharpoons \underset{\text{Bromide}}{Br^-} + \underset{\text{Dichromate}}{Cr_2O_7^{2-}} + 8H^+ \qquad (1\text{-}8)$$

for which the equilibrium constant is

$$K = \frac{[Br^-][Cr_2O_7^{2-}][H^+]^8}{[BrO_3^-][Cr^{3+}]^2} = 1 \times 10^{11} \quad \text{at } 25°C$$

H_2O is omitted from K because it is the solvent. Its concentration remains nearly constant.

In one particular equilibrium state of this system, the following concentrations exist:

$$[H^+] = 5.0 \text{ M} \qquad [Cr_2O_7^{2-}] = 0.10 \text{ M} \qquad [Cr^{3+}] = 0.003\ 0 \text{ M}$$

$$[Br^-] = 1.0 \text{ M} \qquad [BrO_3^-] = 0.043 \text{ M}$$

Suppose that equilibrium is disturbed by increasing the concentration of dichromate from 0.10 to 0.20 M. In what direction will the reaction proceed to reach equilibrium?

According to the principle of Le Châtelier, the reaction should go in the reverse direction to partially offset the increase in dichromate, which is a product in Reaction 1-8. We can verify this algebraically by setting up the *reaction quotient, Q,* which has the same form as the equilibrium constant. The only difference is that Q is evaluated with whatever concentrations happen to exist, even though the solution is not at equilibrium. When the system reaches equilibrium, $Q = K$. For Reaction 1-8,

Q has the same form as K, but the concentrations are generally not the equilibrium concentrations.

$$Q = \frac{(1.0)(0.20)(5.0)^8}{(0.043)(0.003\ 0)^2} = 2 \times 10^{11} > K$$

If $Q < K$, then the reaction must proceed to the right to reach equilibrium. If $Q > K$, then the reaction must proceed to the left to reach equilibrium.

Because $Q > K$, the reaction must go in reverse to decrease the numerator and increase the denominator, until $Q = K$.

In general,

1. If a reaction is at equilibrium and products are added (or reactants are removed), the reaction goes in the reverse direction (to the left).

2. If a reaction is at equilibrium and reactants are added (or products are removed), the reaction goes in the forward direction (to the right).

In equilibrium problems, we predict what must happen for a system to reach equilibrium, but not how long it will take. Some reactions are over in an instant; others do not reach equilibrium in a million years. A stick of dynamite will remain unchanged indefinitely, until a spark sets off the spontaneous, explosive decomposition. The size of an equilibrium constant tells us nothing about the rate of reaction. A large equilibrium constant does not imply that a reaction is fast.

Ask Yourself

1-E. **(a)** Show how the equations below can be rearranged and added to give the reaction $HOBr \rightleftharpoons H^+ + OBr^-$:

$$HOCl \rightleftharpoons H^+ + OCl^- \qquad\qquad K = 3.0 \times 10^{-8}$$
$$HOCl + OBr^- \rightleftharpoons HOBr + OCl^- \qquad K = 15$$

(b) Find the value of K for the reaction $HOBr \rightleftharpoons H^+ + OBr^-$.
(c) If the reaction $HOBr \rightleftharpoons H^+ + OBr^-$ is at equilibrium and a substance is added that consumes H^+, will the reaction proceed in the forward or reverse direction to reestablish equilibrium?

Key Equations

Molarity (M)
$$[A] = \frac{\text{moles of solute A}}{\text{liters of solution}}$$

Weight percent
$$\text{wt } \% = \frac{\text{mass of solute}}{\text{mass of solution or mixture}} \times 100$$

Volume percent
$$\text{vol } \% = \frac{\text{volume of solute}}{\text{volume of solution or mixture}} \times 100$$

Density
$$\text{density} = \frac{\text{grams of substance}}{\text{mL of substance}}$$

Parts per million
$$\text{ppm} = \frac{\text{mass of substance}}{\text{mass of sample}} \times 10^6$$

Parts per billion
$$\text{ppb} = \frac{\text{mass of substance}}{\text{mass of sample}} \times 10^9$$

Dilution formula
$$M_{conc} \cdot V_{conc} = M_{dil} \cdot V_{dil}$$

M_{conc} = concentration (molarity) of concentrated solution

M_{dil} = concentration of dilute solution

V_{conc} = volume of concentrated solution

V_{dil} = volume of dilute solution

Equilibrium constant
$$a A + b B \xrightleftharpoons{K} c C + d D \qquad K = \frac{[C]^c[D]^d}{[A]^a[B]^b}$$

Concentrations of solutes are in M and gases are in bar. Omit solvents and pure solids and liquids.

Reversed reaction
$$K' = 1/K$$

Add two reactions
$$K_3 = K_1 K_2$$

Le Châtelier's principle
1. Adding product (or removing reactant) drives reaction in reverse
2. Adding reactant (or removing product) drives reaction forward

Important Terms

abscissa

atomic mass

concentration

density

equilibrium constant

formal concentration

formula mass

Le Châtelier's principle

liter

molality

molarity

molecular mass

ordinate

parts per billion

parts per million

SI units

solute

solvent

strong electrolyte

volume percent

weak electrolyte

weight percent

Problems

1-1. (a) List the SI units of length, mass, time, electric current, temperature, and amount of substance. Write the abbreviation for each.

(b) Write the units and symbols for frequency, force, pressure, energy, and power.

1-2. Write the name and number represented by each symbol. For example, for kW you should write kW = kilowatt = 10^3 watts.

(a) mW   **(c)** kΩ   **(e)** TJ   **(g)** f g

(b) pm   **(d)** μC   **(f)** ns   **(h)** dPa

1-3. Express the following quantities with abbreviations for units and prefixes from Tables 1-1 through 1-3:

(a) 10^{-13} joules

(b) $4.317\ 28 \times 10^{-8}$ coulombs

(c) $2.997\ 9 \times 10^{14}$ hertz

(d) 10^{-10} meters

(e) 2.1×10^{13} watts

(f) 48.3×10^{-20} moles

1-4. Table 1-4 states that 1 horsepower = 745.700 watts. Consider a 100.0-horsepower engine. Express its power output in **(a)** watts; **(b)** joules per second; **(c)** calories per second; and **(d)** calories per hour.

1-5. (a) Refer to Table 1-4 and calculate how many meters are in 1 inch. How many inches are in 1 m?

(b) A mile is 5 280 feet and a foot is 12 inches. The speed of sound in the atmosphere at sea level is 345 m/s. Express the speed of sound in miles per second and miles per hour.

(c) There is a delay between lightning and thunder in a storm, because light reaches us almost instantaneously, but sound is slower. How many meters, kilometers, and miles away is lightning if the sound reaches you 3.00 s after the light?

1-6. Define the following measures of concentration:

(a) molarity

(b) molality

(c) density

(d) weight percent

(e) volume percent

(f) parts per million

(g) parts per billion

(h) formal concentration

1-7. What is the formal concentration (expressed as mol/L = M) of NaCl when 32.0 g are dissolved in water and diluted to 0.500 L?

1-8. If 0.250 L of aqueous solution with a density of 1.00 g/mL contains 13.7 μg of pesticide, express the concentration of pesticide in **(a)** ppm and **(b)** ppb.

1-9. The concentration of the sugar glucose $(C_6H_{12}O_6)$ in human blood ranges from about 80 mg/dL before meals up to 120 mg/dL after eating. Find the molarity of glucose in blood before and after eating.

1-10. Hot perchloric acid is a powerful (and potentially explosive) reagent used to decompose organic materials and solubilize them for chemical analysis.

(a) How many grams of perchloric acid, $HClO_4$, are contained in 100.0 g of 70.5 wt % aqueous perchloric acid?

(b) How many grams of water are in 100.0 g of solution?

(c) How many moles of $HClO_4$ are in 100.0 g of solution?

1-11. How many grams of boric acid $[B(OH)_3$, FM 61.83$]$ should be used to make 2.00 L of 0.050 0 M solution? (Remember that FM stands for "formula mass.")

1-12. Water is fluoridated to prevent tooth decay.

(a) How many liters of 1.5 M KF should be added to a reservoir with a diameter of 100 m and a depth of 20 m to give 1.2 ppm F^-?

(b) How many grams of solid KF should be added to the same reservoir to give 1.2 ppm F^-?

1-13. How many grams of 50 wt % NaOH (FM 40.00) should be diluted to 1.00 L to make 0.10 M NaOH?

1-14. A bottle of concentrated aqueous sulfuric acid, labeled 98.0 wt % H_2SO_4, has a concentration of 18.0 M.

(a) How many milliliters of reagent should be diluted to 1.00 L to give 1.00 M H_2SO_4?

(b) Calculate the density of 98.0 wt % H_2SO_4.

1-15. How many grams of methanol (CH_3OH, FM 32.04) are contained in 0.100 L of 1.71 M aqueous methanol?

1-16. A dilute aqueous solution containing 1 ppm of solute has a density of 1.00 g/mL. Express the concentration of solute in g/L, μg/L, μg/mL, and mg/L.

1-17. The concentration of $C_{20}H_{42}$ (FM 282.55) in winter rainwater in Figure 1-2 is 0.2 ppb. Assume that the density of rainwater is close to 1.00 g/mL; find the molar concentration of $C_{20}H_{42}$.

1-18. A 95.0 wt % solution of ethanol (CH_3CH_2OH, FM 46.07) in water has a density of 0.804 g/mL.

(a) Find the mass of 1.00 L of this solution and the grams of ethanol per liter.

(b) What is the molar concentration of ethanol in this solution?

1-19. (a) How many grams of nickel are contained in 10.0 g of a 10.2 wt % solution of nickel sulfate hexahydrate, $NiSO_4 \cdot 6H_2O$ (FM 262.85)?

(b) The concentration of this solution is 0.412 M. Find its density.

1-20. A 500.0-mL solution was prepared by dissolving 25.00 mL of methanol (CH_3OH, density = 0.791 4 g/mL) in chloroform. Find the molarity of the methanol.

1-21. Describe how to prepare exactly 100 mL of 1.00 M HCl from 12.1 M HCl reagent.

1-22. Cesium chloride is used to prepare dense solutions required for isolating cellular components with a centrifuge. A 40.0 wt % solution of CsCl (FM 168.36) has a density of 1.43 g/mL.

(a) Find the molarity of CsCl.

(b) How many milliliters of the concentrated solution should be diluted to 500 mL to make 0.100 M CsCl?

1-23. Protein and carbohydrates provide 4.0 Cal/g, whereas fat gives 9.0 Cal/g. (Remember that 1 Calorie, with a capital C, is really 1 kcal.) The weight percent of these components in some foods are

Food	Protein (wt %)	Carbohydrate (wt %)	Fat (wt %)
shredded wheat	9.9	79.9	—
doughnut	4.6	51.4	18.6
hamburger (cooked)	24.2	—	20.3
apple	—	12.0	—

Calculate the number of calories per gram and calories per ounce in each of these foods. (Use Table 1-4 to convert grams into ounces, remembering that there are 16 ounces in 1 pound.)

1-24. Even though you need to express concentrations of solutes in mol/L and the concentrations of gases in bar, why do we say that equilibrium constants are dimensionless?

1-25. Write the expression for the equilibrium constant for each of the following reactions. Write the pressure of a gaseous molecule, X, as P_X.

(a) $3Ag^+(aq) + PO_4^{3-}(aq) \rightleftharpoons Ag_3PO_4(s)$

(b) $C_6H_6(l) + \frac{15}{2}O_2(g) \rightleftharpoons 3H_2O(l) + 6CO_2(g)$

1-26. For the reaction $2A(g) + B(aq) + 3C(l) \rightleftharpoons D(s) + 3E(g)$, the concentrations at equilibrium are $P_A = 2.8 \times 10^3$ Pa, $[B] = 1.2 \times 10^{-2}$ M, $[C] = 12.8$ M, $[D] = 16.5$ M, and $P_E = 3.6 \times 10^4$ torr. (There are exactly 760 torr in 1 atm.)

(a) Gas pressure should be expressed in bars when writing the equilibrium constant. Express the pressures of A and E in bar.

(b) Find the numerical value of the equilibrium constant that would appear in a table of equilibrium constants.

1-27. Suppose that the reaction $Br_2(l) + I_2(s) + 4Cl^-(aq) \rightleftharpoons 2Br^-(aq) + 2ICl_2^-(aq)$ has come to equilibrium. If more $I_2(s)$ is added, will the concentration of ICl_2^- in the aqueous phase increase, decrease, or remain unchanged?

1-28. From the reactions

$$CuN_3(s) \rightleftharpoons Cu^+ + N_3^- \qquad K = 4.9 \times 10^{-9}$$
$$HN_3 \rightleftharpoons H^+ + N_3^- \qquad K = 2.2 \times 10^{-5}$$

find the value of K for the reaction $Cu^+ + HN_3 \rightleftharpoons CuN_3(s) + H^+$.

1-29. Consider the following equilibria in aqueous solution:

(1) $Ag^+ + Cl^- \rightleftharpoons AgCl(aq) \qquad K = 2.0 \times 10^3$

(2) $AgCl(aq) + Cl^- \rightleftharpoons AgCl_2^-$ $K = 93$

(3) $AgCl(s) \rightleftharpoons Ag^+ + Cl^-$ $K = 1.8 \times 10^{-10}$

(a) Find K for the reaction $AgCl(s) \rightleftharpoons AgCl(aq)$.

(b) Find $[AgCl(aq)]$ in equilibrium with excess $AgCl(s)$.

(c) Find K for $AgCl_2^- \rightleftharpoons AgCl(s) + Cl^-$.

How Would You Do It?

1-30. The ocean is a *heterogeneous* fluid with different concentrations of zinc at different depths, as shown in Figure 1-1. Suppose that you want to know how much zinc is contained in an imaginary cylinder of ocean water that is 1 m in diameter and 2 000 m deep. How would you construct a representative sample to measure the average concentration of zinc in the cylinder?

1-31. Chemical characteristics of the Naugatuck River in Connecticut were monitored by students from Sacred Heart University.[3] The mean flow of the river is 560 cubic feet (ft^3) per second. The concentration of nitrate anion (NO_3^-) in the river was reported to range from 2.05 to 2.50 milligrams of nitrate nitrogen per liter during dry weather and from 0.81 to 4.01 mg nitrate nitrogen per liter during wet weather at the outlet of the river. (The unit "mg nitrate nitrogen" refers to the mass of nitrogen in the nitrate anion.) Estimate how many metric tons of nitrate anion per year flow from the river (1 metric ton = 1 000 kg). State your assumptions.

Notes and References

1. S. J. Hawkes, *J. Chem. Ed.* **1996**, *73*, 421.

2. Prior to 1982, the standard state for a gas in an equilibrium constant was 1 atm. Tabulated equilibrium constants and electrode potentials are not changed very much by this small change in standard state. See R. S. Treptow, *J. Chem. Ed.* **1999**, *76*, 212; and R. D. Freeman, *J. Chem. Ed.* **1985**, *62*, 681.

3. J. Clark and E. Alkhatib, *Am. Environ. Lab.* February 1999, 421.

A Quartz-Crystal Tay-Sachs Disease Biosensor

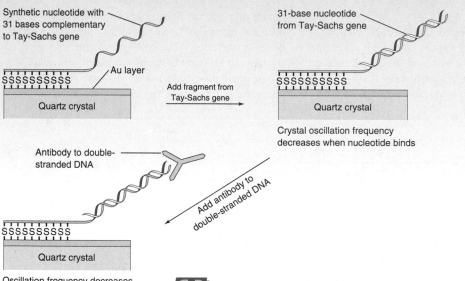

Synthetic nucleotide with 31 bases complementary to Tay-Sachs gene

Au layer

SSSSSSSSSS

Quartz crystal

Add fragment from Tay-Sachs gene →

31-base nucleotide from Tay-Sachs gene

SSSSSSSSSS

Quartz crystal

Crystal oscillation frequency decreases when nucleotide binds

Antibody to double-stranded DNA

Add antibody to double-stranded DNA

SSSSSSSSSS

Quartz crystal

Oscillation frequency decreases much more when heavy antibody binds

Tay-Sachs disease is a genetic disorder in which the body cannot metabolize a class of lipids found in the nervous system. The disease is most common in descendants of Jews from eastern Europe. Affected children become demented and blind by age two and do not live past age three. Genetic screening of a fetus is recommended if both parents carry the recessive Tay-Sachs gene.

A blood test that is being developed at Hebrew University in Jerusalem for genetic screening uses the same kind of vibrating quartz crystal that keeps time in your wristwatch and computer.[1] When a tiny amount of any substance is *adsorbed* (bound) on the crystal's surface, the vibrational frequency of the crystal decreases in proportion to the adsorbed mass.

The figure shows a short, synthetic strand of *deoxyribonucleic acid* (the genetic material, DNA) tightly bound through 10 sulfur atoms to a layer of gold on a quartz crystal. The synthetic strand is complementary to a section of the Tay-Sachs gene. In a genetic test, a digest of fetal DNA is added to the solution in which the crystal is immersed. If the Tay-Sachs gene is present, it forms a double helix with the strand on the crystal. The increased mass attached to the crystal lowers the crystal's oscillation frequency by a small amount. The signal can be amplified by adding an *antibody* that specifically binds to double-stranded DNA. It will only bind if the Tay-Sachs gene fragment was present to form a double strand. The frequency drops much more because the antibody is 15 times heavier than the fragment of Tay-Sachs gene.

The quartz crystal typically oscillates at a frequency near 10 MHz. If the Tay-Sachs gene fragment is present at a concentration of 0.6 μg/mL, the oscillation frequency drops by a barely detectable 2 ± 1 Hz. Addition of the antibody lowers the frequency by a clearly observable 14 ± 1 Hz.

2

TOOLS OF THE TRADE

A nalytical chemists use a range of equipment, from simple glassware to the most advanced instruments that measure spectroscopic or electrical properties of analytes. The object of the analysis might be as large as a vein of ore in a mountainside to a micrometer-size patch on the surface of a semiconductor viewed through an electron microscope. The intention of this course is to expose you to some of the instrumental techniques used in modern analytical chemistry. Along the way, you must gain some understanding and proficiency in "wet" laboratory operations with simple glassware. The most sophisticated instruments are useless if you cannot prepare accurate standards for calibration or accurate, representative samples of an unknown for analysis. This chapter discusses basic laboratory apparatus and manipulations associated with chemical measurements.

2-1   Safety with Chemicals and Waste

The primary safety rule is to do nothing that you (or your instructor) consider to be dangerous. If you believe that an operation is hazardous, discuss it with your instructor and do not proceed until sensible procedures and precautions are in place. If you still consider an activity to be too dangerous, don't do it.

   Before beginning work, you should be familiar with safety precautions appropriate to your laboratory. Goggles or safety glasses with side shields (Figure 2-1) are worn at all times to protect you from flying chemicals and glass, which visit labs from time to time. (Although you may be very careful, one of your neighbors will not be so careful.) A flame-resistant lab coat, long pants, and shoes that cover

FIGURE 2-1 Goggles or safety glasses with side shields are required in every laboratory.

your feet help protect you from spills and flames. Rubber gloves protect you when pouring concentrated acids, but organic solvents can penetrate rubber gloves. Food and chemicals should not mix: Don't bring food or drink into the lab.

Treat chemical spills on your skin *immediately* by flooding the affected area with water and then seeking medical attention. Clean up spills on the bench, floor, or reagent bottles immediately to prevent accidental contact by the next person who comes along.

Solvents and concentrated acids that produce harmful fumes should be handled in a fume hood that sweeps vapors away from you and out through a vent on the roof. The hood is not meant to transfer toxic vapors from the chemistry building to the cafeteria; it is designed to dilute low levels of pollutants to even lower levels. Never generate large quantities of toxic fumes in the hood.

Preservation of a habitable planet demands that we minimize waste production and responsibly dispose of chemical waste (Box 2-1). Recycling of chemicals is practiced in industry for economic as well as ethical reasons; it should be a component of pollution control in your lab.

All vessels should be labeled to show what they contain. If you don't do this, you may (no, you *will*) forget what is in some of your own containers and then you won't know what to do with them. Unlabeled waste is extremely expensive to discard, because you must analyze the contents before you can legally dispose of it.

Limitations of gloves: In 1997, popular Dartmouth College chemistry professor Karen Wetterhahn, age 48, died from a drop of dimethylmercury absorbed through the latex rubber gloves she was wearing. Many organic compounds readily penetrate rubber. Wetterhahn was an expert in the biochemistry of metals and the first female professor of chemistry at Dartmouth. She was a mother of two children and played a major role in bringing more women into science and engineering.

2-2   Your Lab Notebook

The lab notebook must

1. State what was done
2. State what was observed
3. Be understandable to someone else

The critical functions of your lab notebook are to state *what you did* and *what you observed,* and it should be *understandable by a stranger.* The greatest error, made even by experienced scientists, is writing ambiguous notes. After a few years, even the author of a notebook cannot interpret what was done or seen. Writing in *complete sentences* is an excellent way to reduce this problem.

Box 2-1 *Informed Citizen*

Disposal of Chemical Waste[2]

Many chemicals that we commonly use are harmful to plants, animals, and people if carelessly discarded. For each experiment, your instructor should establish safe procedures for waste disposal. Options include (1) pouring solutions down the drain and diluting with tap water, (2) saving the waste for disposal in an approved landfill, (3) treating waste to decrease the hazard and then pouring it down the drain or saving it for a landfill, and (4) recycling the chemical. *Chemically incompatible wastes should never be mixed with each other, and each waste container must be labeled to indicate the quantity and identity of its contents.* Labels should indicate whether the contents are flammable, toxic, corrosive, reactive, or have other dangerous properties.

A few examples illustrate different approaches to managing lab waste. Reduce dichromate ($Cr_2O_7^{2-}$) to Cr^{3+} with sodium hydrogen sulfite ($NaHSO_3$), treat the product with hydroxide to make insoluble $Cr(OH)_3$, and evaporate it to dryness for disposal in a landfill. Mix waste acid with waste base until nearly neutral (as determined with pH paper) and then pour the mixture down the drain. Reduce waste iodate (IO_3^-) to I^- with $NaHSO_3$, neutralize the solution with base, and pour it down the drain. Waste silver or gold should be chemically treated to recover the metals. If you use a toxic gas in a fume hood, bubble excess gas through a chemical trap or burn it in a flame to prevent its escape from the hood.

The measure of scientific "truth" is the ability to reproduce an experiment. A good lab notebook will allow you or anyone else to duplicate an experiment in the exact manner in which it was conducted the first time.

Beginning students find it useful (or required!) to write a complete description of an experiment, with sections describing the purpose, methods, results, and conclusions. Arranging your notebook to accept numerical data prior to coming to the lab is an excellent way to prepare for an experiment.

It is good practice to write a balanced chemical equation for every reaction you use. This helps you understand what you are doing and may point out what you do not understand.

Record in your notebook the names of computer disks and files where programs and data are stored. Printed copies of important data collected on a computer should be pasted into your notebook. The lifetime of a printed page is 10 to 100 times greater than that of a disk.

Without a doubt, somebody reading this book today is going to make an important discovery in the future and will seek a patent. The lab notebook is your legal record of your discovery. Therefore, each notebook page should be signed and dated. Anything of potential importance should also be signed and dated by a second person.

Do not rely on a computer for long-term storage of information. Even if a disk survives, software or hardware required to read the disk will become obsolete.

Ask Yourself

2-A. What are the three essential attributes of a lab notebook?

2-3   The Analytical Balance

Figure 2-2 shows a typical analytical **electronic balance** with a capacity of 100–200 g and a sensitivity of 0.01–0.1 mg. *Sensitivity* refers to the smallest increment of mass that can be measured. A *microbalance* weighs milligram quantities with a sensitivity of 0.1 μg. An electronic balance works by generating an electromagnetic force to balance the gravitational force acting on the object being weighed. The current required by the electromagnet is proportional to the mass being weighed.

Figure 2-3 shows the principle of operation of a single-pan **mechanical balance.** The balance beam is suspended on a sharp fulcrum called a *knife edge.* The mass of the pan hanging from the balance point (another knife edge) at the left is balanced by a counterweight at the right. After placing the object to be weighed on the pan, knobs are adjusted to remove standard weights from the bar above the pan. The balance beam is restored close to its original position when the weights removed from the bar are nearly equal to the mass on the pan. The slight difference from the original position is shown on an optical scale, whose reading is added to that of the knobs.

Using a Balance

A mechanical balance should be in its arrested position when you load or unload the pan and in the half-arrested position when you are dialing weights. This practice protects against the application of abrupt forces that wear down the knife edges and eventually decrease the sensitivity of the balance.

To weigh a chemical, place a clean receiving vessel on the balance pan. The mass of the empty vessel is called the **tare.** On most electronic balances, the tare

Balances and other lab equipment are delicate and expensive. Be gentle when you place objects on the pan and when you adjust the knobs.

FIGURE 2-2 Analytical electronic balance. Good-quality balances calibrate themselves with internal weights to correct for variations in the force of gravity, which can be as great as 0.3% from place to place. [Fisher Scientific, Pittsburgh, PA.]

can be set to 0 by pressing a button. Add chemical to the vessel and read the new mass. If there is no automatic tare operation, the mass of the empty vessel should be subtracted from that of the filled vessel. Chemicals should never be placed directly on the weighing pan. This precaution protects the balance from corrosion and allows you to recover all the chemical being weighed.

An alternate procedure, called "weighing by difference," is necessary for **hygroscopic** reagents, which rapidly absorb moisture from the air. First weigh a capped

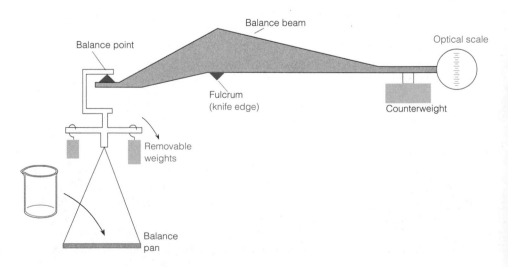

FIGURE 2-3 Single-pan mechanical balance. After placing an object on the pan, we detach removable weights until the balance beam is restored as near as possible to its original position. The remaining small difference is read on the optical scale.

bottle containing dry reagent. Then quickly pour some reagent from the weighing bottle into a receiver. Cap the weighing bottle and weigh it again. The difference is the mass of reagent.

Clean up spills on the balance and do not allow chemicals to get into the mechanism below the pan of an electronic balance. Use a paper towel or tissue to handle the vessel you are weighing, because fingerprints will change its mass. Samples should be at *ambient temperature* (the temperature of the surroundings) when weighed to avoid errors due to convective air currents. The doors of the balance in Figure 2-2 must be closed during weighing to prevent air currents from affecting the reading. A top-loading balance without sliding doors has a fence around the pan to deflect air currents. Sensitive balances should be placed on a heavy table, such as a marble slab, to minimize the effect of building vibrations on the reading. Use the bubble meter and adjustable feet of a balance to keep it level.

Buoyancy

The actual mass of an object (that measured in vacuum) is usually greater than the apparent mass measured in air. This effect arises from an object's **buoyancy** in air: A sample appears lighter than its actual mass by an amount equal to the mass of air that it displaces. The same effect applies to weights used to calibrate an electronic balance or removable weights in a mechanical balance. A buoyancy error occurs whenever the density of the object being weighed is not equal to the density of the standard weights.

If mass m' is read from a balance, the true mass m is

Buoyancy equation:
$$m = \frac{m'\left(1 - \dfrac{d_a}{d_w}\right)}{\left(1 - \dfrac{d_a}{d}\right)}$$
(2-1)

where d_a is the density of air (0.001 2 g/mL near 1 atm and 25°C); d_w is the density of balance weights (8.0 g/mL); and d is the density of the object being weighed.

| **EXAMPLE** | **Buoyancy Correction** |

Find the true mass of water (density = 1.00 g/mL) if the apparent mass is 100.00 g.

SOLUTION Equation 2-1 gives the true mass:

$$m = \frac{100.00 \text{ g}\left(1 - \dfrac{0.001\ 2 \text{ g/mL}}{8.0 \text{ g/mL}}\right)}{\left(1 - \dfrac{0.001\ 2 \text{ g/mL}}{1.00 \text{ g/mL}}\right)} = 100.11 \text{ g}$$

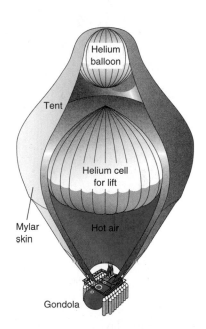

Helium balloon

Tent

Helium cell for lift

Mylar skin

Hot air

Gondola

Buoyancy The *Breitling Orbiter 3* in 1999 was the first balloon to fly around the world. Previous balloons could not carry enough propane fuel for such a trip. The design of the *Breitling Orbiter 3* keeps the temperature of the large, inner helium cell as constant as possible, so the buoyancy variation between the warm day and the cold night is minimal. During the day, the sun heats the helium cell, which expands, thereby increasing its buoyancy and its ability to keep the vessel aloft. If the temperature rises too much, a solar-powered fan brings in cool air to keep the vessel from rising to an undesired altitude. At night, the helium cell cools and shrinks, which reduces its buoyancy. To keep the balloon aloft at night, heat from burning propane is required. However, the double-wall design reduces radiational cooling of the helium cell and decreases the requirement for propane.

The buoyancy error for water is 0.11%, which is significant for many purposes. For NaCl with a density of 2.16 g/mL, the error is 0.04%.

Ask Yourself

2-B. (a) Buoyancy corrections are most critical when you calibrate glassware such as a volumetric flask to see how much volume it actually holds. Suppose that you fill a 25-mL volumetric flask with distilled water and find that the mass of water in the flask measured in air is 24.913 g. What is the true mass of the water?

   (b) You made the measurement when the lab temperature was 21°C, at which temperature the density of water is 0.998 00 g/mL. What is the true volume of water contained in the volumetric flask?

2-4   Burets[3]

Operating a buret:

- Read bottom of concave meniscus
- Estimate reading to 1/10 of a division
- Avoid parallax
- Account for graduation thickness in readings
- Drain liquid slowly
- Wash buret with new solution
- Deliver fraction of a drop near end point
- Eliminate air bubble before use

A **buret** is a precisely manufactured glass tube with graduations enabling you to measure the volume of liquid delivered through the stopcock (the valve) at the bottom (Figure 2-4a). The numbers on the buret increase from top to bottom (with 0 mL near the top). A volume measurement is made by reading the level before and after draining liquid from the buret and subtracting the first reading from the second reading. The graduations of Class A burets (the most accurate grade) are certified to meet the tolerances in Table 2-1. For example, if the reading of a 50-mL buret

FIGURE 2-4 (*a*) Glass buret with Teflon stopcock. It is a good idea to cover your buret with a loose-fitting cap to keep dust out and reduce evaporation. [From Fisher Scientific, Pittsburgh, PA.] (*b*) Digital titrator with its plastic cartridge containing reagent carries out the same function as a buret for analyses in the field. [Hach Co., Loveland, CO.]

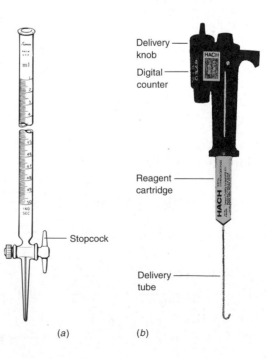

(*a*)   (*b*)

TABLE 2-1	Tolerances of Class A burets	
Buret volume (mL)	*Smallest graduation (mL)*	*Tolerance (mL)*
5	0.01	±0.01
10	0.05 *or* 0.02	±0.02
25	0.1	±0.03
50	0.1	±0.05
100	0.2	±0.10

is 32.50 mL, the true volume can be anywhere in the range 32.45 to 32.55 mL and still be within the manufacturer's stated tolerance of ±0.05 mL.

When reading the liquid level in a buret, your eye should be at the same height as the top of the liquid. If your eye is too high, the liquid seems to be higher than it actually is. If your eye is too low, the liquid appears too low. The error that occurs when your eye is not at the same height as the liquid is called **parallax error.**

The surface of most liquids forms a concave **meniscus,** shown in Figure 2-5. It is helpful to use black tape on a white card as a background for locating the precise position of the meniscus. Align the top of the tape with the bottom of the meniscus and read the position on the buret. Highly colored solutions may appear to have a double meniscus. In such case, either one may be used. Because volumes are determined by subtracting one reading from another, the most important point is to read the position of the meniscus reproducibly. Always estimate the reading to the nearest tenth of a division between marks.

The thickness of a graduation line on a 50-mL buret corresponds to about 0.02 mL. To use the buret most accurately, consider the *top* of a graduation line to be 0. When the meniscus is at the bottom of the same graduation line, the reading is 0.02 mL greater.

Near the end point of a titration, try to deliver less than one drop at a time from the buret. This practice permits a finer location of the end point. (A drop from a 50-mL buret is about 0.05 mL.) To deliver a fraction of a drop, carefully open the stopcock until part of a drop is hanging from the buret tip. Then touch the inside glass wall of the receiving flask to the buret tip to transfer the droplet to the wall of the flask. Carefully tip the flask so that the main body of liquid washes over the newly added droplet. Then swirl the flask to mix the contents. Near the end of a titration, the flask should be tipped and rotated often to ensure that droplets on the wall containing unreacted analyte contact the bulk solution.

Liquid should drain evenly down the wall of a buret. The tendency of liquid to stick to the glass is reduced by draining the buret slowly (< 20 mL/min). If many droplets stick to the wall, the buret should be cleaned with detergent and a buret brush. If this cleaning is insufficient, the buret should be soaked in peroxydisulfate-sulfuric acid cleaning solution prepared by your instructor.[4] Cleaning solution eats clothing and people, as well as grease in the buret. Volumetric glassware should not be soaked in alkaline solutions, which attack glass. (A 5 wt % NaOH solution at 95°C dissolves Pyrex glass at a rate of 9 μm/h.)

A common error in using a buret is caused by failure to expel the bubble of air often found directly beneath the stopcock (Figure 2-6). A bubble present at the start of the titration may be filled with liquid during the titration. Therefore some

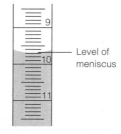

FIGURE 2-5 Buret with the meniscus at 9.68 mL. Estimate the reading of any scale to the nearest tenth of a division. Because this buret has 0.1-mL divisions, we estimate the reading to 0.01 mL.

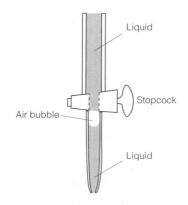

FIGURE 2-6 An air bubble trapped beneath the stopcock of a buret should be expelled before you use the apparatus.

FIGURE 2-7 Battery-operated electronic buret with digital readout delivers 0.01-mL increments from a reagent bottle. This device can be used for accurate titrations in the field. [Cole-Parmer Co., Niles, IL.]

Adsorption: to bind a substance on the surface

Absorption: to bind a substance internally

volume that drained out of the graduated portion of the buret did not reach the titration vessel. Usually the bubble can be dislodged by draining the buret for a second or two with the stopcock wide open. A tenacious bubble can be expelled by carefully shaking the buret while draining it into a sink.

When you fill a buret with fresh solution, it is a wonderful idea to rinse the buret several times with small portions of the new solution, discarding each wash. It is not necessary to fill the entire buret with wash solution. Simply tilt the buret so that its whole surface contacts the wash liquid. This same technique should be used with any vessel (such as a spectrophotometer cuvet or a pipet) that is reused without drying.

The *digital titrator* in Figure 2-4b is more convenient and portable, but less accurate, than the glass buret. The digital titrator is useful for conducting titrations in the field where samples are collected. The counter tells how much reagent from the cartridge has been dispensed by rotation of the delivery knob. Its accuracy of 1% is 10 times poorer than the accuracy of a glass buret, but many measurements do not require higher accuracy. The *electronic buret* in Figure 2-7 takes the challenge out of titrations. This battery-operated dispenser fits on a reagent bottle and delivers up to 99.99 mL in 0.01-mL increments. The volume dispensed is displayed on a digital readout.

2-5 Volumetric Flasks

A **volumetric flask** (Figure 2-8, Table 2-2) is calibrated to contain a particular volume of solution at 20°C when the bottom of the meniscus is adjusted to the center of the mark on the neck of the flask. Most flasks bear the label "TC 20°C," which means *to contain* at 20°C. (Other types of glassware may be calibrated *to deliver*, "TD," their indicated volume.) The temperature of the container is relevant because both liquid and glass expand when heated.

We use a volumetric flask to prepare a solution of known volume. Typically, reagent is weighed into the flask, dissolved, and diluted to the mark. The mass of reagent and final volume are therefore known. The reagent is first dissolved in *less* than the final volume of liquid. More liquid is added and the solution is mixed again. The final volume adjustment is done with as much well-mixed liquid in the flask as possible. (When two different liquids are mixed, there is generally a small volume change. The total volume is *not* the sum of the two volumes that were mixed. By swirling the liquid in a nearly full volumetric flask before the liquid reaches the thin neck, you minimize the change in volume when the last liquid is added.) For best control, add the final drops of liquid with a pipet, *not a squirt bottle.* After adjusting the liquid to the correct level, hold the cap firmly in place and invert the flask 20 times to ensure complete mixing.

Figure 2-9 shows how liquid appears when it is at the *center* of the mark of a volumetric flask or a pipet. Adjust the liquid level while viewing the flask from above or below the level of the mark. The front and back of the mark will describe an ellipse with the meniscus at the center when the level is right.

Glass is notorious for *adsorbing* traces of chemicals—especially cations. **Adsorption** means to stick to the surface. (In contrast, **absorption** means to take inside, as a sponge takes up water.) For critical work, **acid wash** the glassware to replace low concentrations of cations on the glass surface with H^+. To do this, soak already thoroughly cleaned glassware in 3–6 M HCl (in a fume hood) for >1 h,

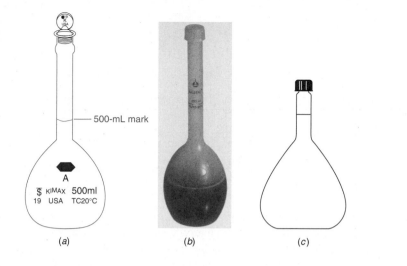

(a) (b) (c)

FIGURE 2-8 (a) Class A glass volumetric flask meets tolerances listed in Table 2-2. [A. H. Thomas Co., Philadelphia, PA.] (b) Class B polypropylene plastic flask for trace analysis (ppb concentrations) in which analyte might be lost by adsorption (sticking) on glass or contamined with previously adsorbed species. Class B flasks are less accurate than Class A flasks, with twice the tolerances of Table 2-2. [Fisher Scientific, Pittsburgh, PA.] (c) Short-form volumetric flask with Teflon-lined screw cap fits on analytical balance. Teflon prevents solutions from attacking the inside of the cap.

TABLE 2-2	Tolerances of Class A volumetric flasks		
Flask capacity (mL)	*Tolerance (mL)*	*Flask capacity (mL)*	*Tolerance (mL)*
1	±0.02	100	±0.08
2	±0.02	200	±0.10
5	±0.02	250	±0.12
10	±0.02	500	±0.20
25	±0.03	1 000	±0.30
50	±0.05	2 000	±0.50

followed by several rinses with distilled water and a final soak in distilled water. The HCl can be reused many times, as long as it is only used for soaking clean glassware.

FIGURE 2-9 Proper position of the meniscus: at the center of the ellipse formed by the front and back of the calibration mark when viewed from above or below. Volumetric flasks and transfer pipets are calibrated to this position.

Ask Yourself

2-C. How would you use a volumetric flask to prepare 250.0 mL of 0.150 0 M K_2SO_4?

2-6 Pipets and Syringes

Pipets deliver known volumes of liquid. Four common types are shown in Figure 2-10. The *transfer pipet* is calibrated to deliver one fixed volume. The last drop of liquid does not drain out of the pipet; *it should not be blown out*. The *measuring pipet* is calibrated to deliver a variable volume, which is the difference between the

Do not blow the last drop out of a transfer pipet.

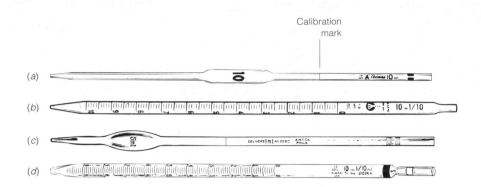

FIGURE 2-10 Common pipets: (a) transfer; (b) measuring (Mohr); (c) Ostwald-Folin (blow out last drop); (d) serological (blow out last drop). [A. H. Thomas Co., Philadelphia, PA.]

TABLE 2-3

Tolerances of Class A transfer pipets

Volume (mL)	Tolerance (mL)
0.5	±0.006
1	±0.006
2	±0.006
3	±0.01
4	±0.01
5	±0.01
10	±0.02
15	±0.03
20	±0.03
25	±0.03
50	±0.05
100	±0.08

Accuracy refers to the difference between the delivered volume and the desired volume. *Precision* refers to the reproducibility of replicate deliveries.

volumes indicated before and after delivery. The measuring pipet in Figure 2-10 could be used to deliver 5.6 mL by starting delivery at the 1-mL mark and terminating at the 6.6-mL mark. The *Ostwald-Folin pipet* is similar to the transfer pipet, except that the last drop *should* be blown out. *Serological pipets* are measuring pipets calibrated all the way to the tip; the last drop *should* be blown out.

A transfer pipet is more accurate than a measuring pipet. The manufacturer's tolerance for a pipet is the allowed uncertainty in the volume that is actually delivered. Tolerances for Class A (the most accurate grade) transfer pipets are given in Table 2-3.

Using a Transfer Pipet

Using a rubber bulb, *not your mouth,* suck liquid up past the calibration mark. It is a good idea to discard one or two pipet volumes of liquid to remove traces of previous reagents from the pipet. After taking up a third volume past the calibration mark, quickly replace the bulb with your index finger at the end of the pipet. The liquid should still be above the mark after this maneuver. Pressing the pipet against the bottom of the vessel while removing the rubber bulb helps prevent liquid from draining while you put your finger in place. Wipe the excess liquid off the outside of the pipet with a clean tissue. *Touch the tip of the pipet to the side of a beaker* and drain the liquid until the bottom of the meniscus just reaches the center of the mark, as shown in Figure 2-9. The purpose of touching the beaker wall while draining liquid is to draw liquid out of the pipet without leaving part of a drop hanging from the pipet when the level reaches the calibration mark.

Transfer the pipet to the desired receiving vessel and drain it *while holding the tip against the wall of the vessel.* After the pipet stops draining, hold it against the wall for a few more seconds to complete draining. *Do not blow out the last drop.* The pipet should be nearly vertical at the end of delivery. When you finish with a pipet, it should be rinsed with distilled water or soaked in a pipet container until it is cleaned. Solutions should never be allowed to dry inside a pipet because removing internal deposits is very difficult.

Micropipets

Plastic micropipets, such as that in Figure 2-11a, are used to deliver volumes in the 1 to 1 000 μL range ($1 \mu L = 10^{-6}$ L). The liquid is contained in the disposable plastic tip. The accuracy is 1–2%, and precision may be as good as 0.5%. Micropipets may have a metal barrel on the inside that can be corroded by pipetting volatile

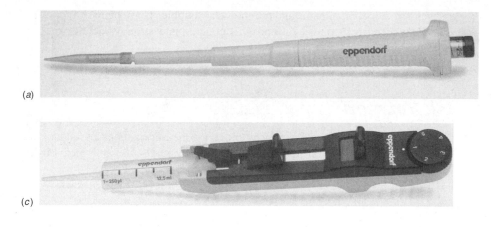

(a)

(c)

(b)

FIGURE 2-11 (a) Microliter pipet with disposable plastic tip. (b) Volume selection dial of microliter pipet. (c) Repeater pipet delivers preset volumes between 10 μL and 5 mL up to 48 times at 1-second intervals without refilling. [Brinkman Instruments, Westbury, NY.]

acids such as concentrated HCl. Corrosion will eventually diminish the accuracy of the pipet.

To use a micropipet, place a fresh tip tightly on the barrel. Tips are contained in a package or dispenser so that you do not handle (and contaminate) the points with your fingers. Set the desired volume with the knob at the top of the pipet. Depress the plunger to the first stop, which corresponds to the selected volume. Hold the pipet *vertically,* dip it 3–5 mm into the reagent solution, and *slowly* release the plunger to suck up liquid. Withdraw the tip from the liquid by sliding it along the wall of the vessel to remove liquid from the outside of the tip. To dispense liquid, touch the micropipet tip to the wall of the receiver and gently depress the plunger to the first stop. After a few seconds to allow liquid to drain down the wall of the pipet tip, depress the plunger further to squirt out the last liquid. It is a good idea to clean and wet a fresh tip by taking up and discarding two or three squirts of reagent first. The tip can be discarded or rinsed well with a squirt bottle and reused.

The volume of liquid taken into the tip depends on the angle at which the pipet is held and how far beneath the surface of reagent the tip is held during uptake. Each person will attain slightly different precision and accuracy with a micropipet.

A microliter *syringe* dispenses tiny volumes. Syringes such as that in Figure 2-12a come in sizes from 1 to 500 μL and have an accuracy and precision near 1%. The digital dispenser in Figure 2-12b improves the accuracy and precision to 0.5%. When using a syringe, take up and discard several volumes of liquid to wash the

Needle   Barrel   Plunger

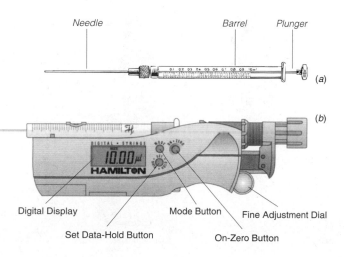

(a)

(b)

Digital Display   Mode Button   Fine Adjustment Dial

Set Data-Hold Button   On-Zero Button

FIGURE 2-12 (a) Hamilton syringe with a volume of 1 μL and graduations of 0.01 μL on the glass barrel. (b) Digital dispenser for syringes with volumes of 0.5 to 500 μL provides accuracy and precision of 0.5%. [Hamilton Co., Reno, NV.]

glass walls free of contaminants and to remove air bubbles from the barrel. The steel needle is attacked by strong acid and will contaminate strongly acidic solutions with iron.

Ask Yourself

2-D. Which is more accurate, a transfer pipet or a measuring pipet? What would you do differently in delivering 1.00 mL of liquid from a 1-mL serological pipet instead of a 1-mL measuring pipet?

2-7   Filtration

In **gravimetric analysis,** the mass of product from a reaction is measured to determine how much unknown was present. Precipitates from gravimetric analyses are collected by filtration, washed, and then dried. Most precipitates are collected in a *fritted-glass funnel;* suction is used to speed filtration (Figure 2-13). The porous glass plate in the funnel allows liquid to pass but retains solids. The empty crucible is first dried at 110°C and weighed. After collecting solid and drying again, the crucible and its contents are weighed a second time to determine the mass of solid.

Liquid from which a substance precipitates or crystallizes is called the **mother liquor.** Liquid that passes through the filter is called **filtrate.**

In some gravimetric procedures, **ignition** (heating at high temperature over a burner or in a furnace) is used to convert a precipitate to a known, constant composition. For example, Fe^{3+} precipitates as an ill-defined hydrated form of $Fe(OH)_3$ with variable composition. Ignition converts it to Fe_2O_3 prior to weighing. When a gravimetric precipitate is to be ignited, it is collected in **ashless filter paper,** which leaves little residue when burned.

To use filter paper with a conical glass funnel, fold the paper into quarters, tear off one corner (to allow a firm fit into the funnel), and place the paper in the funnel

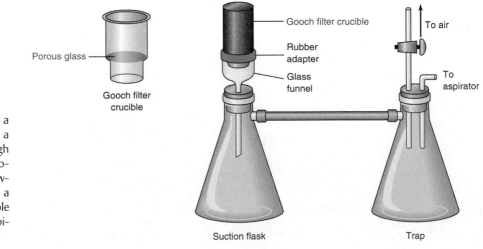

FIGURE 2-13 Filtration with a Gooch filter crucible that has a porous (*fritted*) glass disk through which liquid can pass. Suction is provided by an *aspirator* that uses flowing water from a tap to create a vacuum. The trap prevents possible backup of tap water from the aspirator into the suction flask.

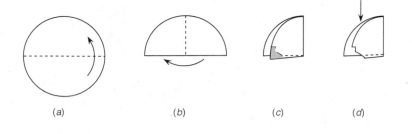

(a)   (b)   (c)   (d)

(Figure 2-14). The filter paper should fit snugly and be seated with some distilled water. When liquid is poured in, an unbroken stream of liquid should fill the stem of the funnel. The weight of liquid in the stem helps speed filtration.

For filtration, pour the slurry of precipitate in the mother liquor down a glass rod to prevent splattering (Figure 2-15). (A *slurry* is a suspension of solid in liquid.) Dislodge any particles adhering to the beaker or rod with a **rubber policeman,** which is a flattened piece of rubber at the end of a glass rod. Use a jet of appropriate wash liquid from a squirt bottle to transfer particles from the rubber and glassware to the filter. If the precipitate is going to be ignited, particles remaining in the beaker should be wiped onto a small piece of moist filter paper, which is then added to the filter to be ignited.

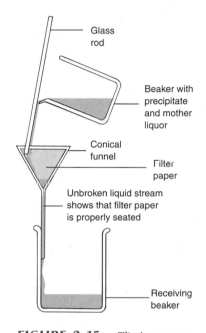

Glass rod

Beaker with precipitate and mother liquor

Conical funnel

Filter paper

Unbroken liquid stream shows that filter paper is properly seated

Receiving beaker

FIGURE 2-15 Filtering a precipitate.

2-8 Drying

Reagents, precipitates, and glassware are conveniently dried in an oven at 110°C. (Some chemicals require other temperatures.) Anything that you put in the oven should be labeled. A beaker and watchglass (Figure 2-16) minimize contamination by dust during drying. It is good practice to cover all vessels on the benchtop to prevent dust contamination.

We measure the mass of a gravimetric precipitate by weighing a dry, empty filter crucible before the procedure and reweighing the same crucible filled with dry product after the procedure. To weigh the empty crucible, first bring it to "constant mass" by drying it in the oven for 1 h or longer and then cooling it for 30 min in a desiccator. Weigh the crucible and then heat it again for about 30 min. Cool it and reweigh it. When successive weighings agree to ± 0.3 mg, the filter has reached "constant mass." A kitchen microwave oven can be used instead of an electric oven for drying reagents and crucibles. Try an initial heating time of 4 min, with subsequent 2-min heatings.

A **desiccator** (Figure 2-17) is a closed chamber containing a drying agent called a **desiccant.** The lid is greased to make an airtight seal. Desiccant is placed in the bottom beneath the perforated disk. Common desiccants in approximate order of decreasing efficiency are magnesium perchlorate ($Mg(ClO_4)_2$) > barium oxide (BaO) ≈ alumina (Al_2O_3) ≈ phosphorus pentoxide (P_4O_{10}) ≫ calcium chloride ($CaCl_2$) ≈ calcium sulfate ($CaSO_4$, called Drierite) ≈ silica gel (SiO_2). After placing a hot object in the desiccator, leave the lid cracked open for a minute until the object has cooled slightly. This practice prevents the lid from popping open when the air inside warms up. To open a desiccator, slide the lid sideways rather than trying to pull it straight up.

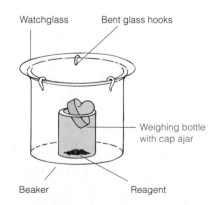

Watchglass   Bent glass hooks

Weighing bottle with cap ajar

Beaker   Reagent

FIGURE 2-16 Use of watchglass as a dustcover while drying reagent or crucible in the oven.

FIGURE 2-17 (a) Ordinary desiccator. (b) Vacuum desiccator, which can be evacuated through the sidearm and then sealed by rotating the joint containing the sidearm. Drying is more efficient at low pressure. Drying agents (*desiccants*) are placed at the bottom of each desiccator below the porous porcelain plate. [A. H. Thomas Co., Philadelphia, PA.]

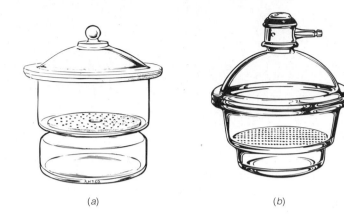

(a)   (b)

2-9   Calibration of Volumetric Glassware

Volumetric glassware can be calibrated to measure the volume that is actually contained in or delivered by a particular piece of equipment. Calibration is done by measuring the mass of water contained or delivered and using Table 2-4 to convert mass to volume:

$$\text{true volume} = (\text{mass of water}) \times (\text{correction factor in Table 2-4}) \qquad (2\text{-}2)$$

To calibrate a 25-mL transfer pipet, first weigh an empty weighing bottle like the one in Figure 2-16. Then fill the pipet to the mark with distilled water, drain it into the weighing bottle, and put on the lid to prevent evaporation. Weigh the bottle again to find the mass of water delivered from the pipet. Use Equation 2-2 to convert mass to volume.

TABLE 2-4

Correction factors for volumetric calibration

Temperature (°C)	Correction factor (mL/g)[a]
15	1.002 0
16	1.002 1
17	1.002 3
18	1.002 5
19	1.002 7
20	1.002 9
21	1.003 1
22	1.003 3
23	1.003 5
24	1.003 8
25	1.004 0
26	1.004 3
27	1.004 6
28	1.004 8
29	1.005 1
30	1.005 4

a. Factors are based on the density of water and are corrected for buoyancy with Equation 2-1.

EXAMPLE   Calibration of a Pipet

An empty weighing bottle had a mass of 10.283 g. After adding water from a 25-mL pipet, the mass was 35.225 g. If the lab temperature was 23°C, find the volume of water delivered by the pipet.

SOLUTION The mass of water is 35.225 − 10.283 = 24.942 g. From Equation 2-2 and Table 2-4, the volume of water is (24.942 g)(1.003 5 mL/g) = 25.029 mL.

Ask Yourself

2-E. A 10-mL pipet delivered 10.000 0 g of water at 15°C to a weighing bottle. What is the true volume of the pipet?

According to the flow chart for the analytical process in Box 0-1, after a representative bulk sample is selected, a homogeneous laboratory sample must be prepared. We usually homogenize materials by grinding them to a fine powder or by dissolving an entire sample. Solids can be ground with a **mortar and pestle** like the set in Figure 2-18.

Dissolving Inorganic Materials with Strong Acids

The acids HCl, HBr, HF, H_3PO_4, and dilute H_2SO_4 dissolve most metals (M) with heating by the reaction

$$M(s) + nH^+(aq) \longrightarrow M^{n+}(aq) + \tfrac{n}{2}H_2(g)$$

Many other inorganic substances can also be dissolved. Some anions react with H^+ to form **volatile** products (species that evaporate easily), which are lost from hot solutions in open vessels. Examples include carbonate ($CO_3^{2-} + 2H^+ \rightarrow H_2CO_3 \rightarrow CO_2 + H_2O$) and sulfide ($S^{2-} + 2H^+ \rightarrow H_2S$). Hot hydrofluoric acid dissolves silicates found in most rocks. HF also attacks glass, so it is used in Teflon, polyethylene, silver, or platinum vessels. Teflon is inert to attack by most chemicals and can be used up to 260°C.

Substances that do not dissolve in the acids above may dissolve as a result of oxidation by HNO_3 or concentrated H_2SO_4. Nitric acid attacks most metals, but not Au and Pt, which dissolve in the 3:1 (vol/vol) mixture of $HCl:HNO_3$ called *aqua regia*.

Acid dissolution is conveniently carried out with a Teflon-lined **bomb** (a sealed vessel) (Figure 2-19) in a microwave oven, which heats the contents to 200°C in a minute. The bomb cannot be made of metal, which absorbs microwaves. The bomb should be cooled prior to opening to prevent loss of volatile products.

FIGURE 2-18 Agate mortar and pestle. The mortar is the base and the pestle is the grinding tool. Agate is very hard and expensive. Less expensive porcelain mortars are widely used, but they are somewhat porous and easily scratched. These properties can lead to contamination of the sample by porcelain particles or by traces of previous samples embedded in the porcelain. [Thomas Scientific, Swedesboro, N.J.]

HCl	hydrochloric acid
HBr	hydrobromic acid
HF	hydrofluoric acid
H_3PO_4	phosphoric acid
H_2SO_4	sulfuric acid
HNO_3	nitric acid

HF is extremely harmful to touch or breathe. Flood the affected area with water, coat the skin with calcium gluconate (or another calcium salt), and seek medical help.

Teflon is a *polymer* (a chain of repeating units) with the structure

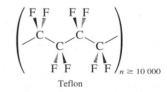

Carbon atoms are in the plane of the page. A solid wedge is a bond coming out of the page toward you, and a dashed wedge is a bond going behind the page.

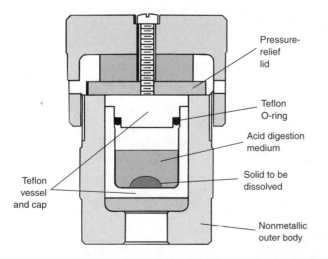

Pressure-relief lid

Teflon O-ring

Acid digestion medium

Solid to be dissolved

Teflon vessel and cap

Nonmetallic outer body

FIGURE 2-19 Microwave digestion bomb lined with Teflon. A typical 23–mL vessel can be used to digest up to 1 g of inorganic material (or 0.1 g of organic material, which releases a great deal of gaseous CO_2) in up to 15 mL of concentrated acid. The outer container maintains its strength up to 150°C, but rarely rises above 50°C. If the internal pressure exceeds 80 atm, the cap deforms and releases the excess pressure. [Parr Instrument Co., Moline, IL.]

Fusion

Inorganic substances that do not dissolve in acid can usually be dissolved by a hot, molten inorganic **flux,** examples of which are lithium tetraborate ($Li_2B_4O_7$) and sodium hydroxide (NaOH). Mix the finely powdered unknown with 2 to 20 times its mass of solid flux, and **fuse** (melt) the mixture in a platinum-gold alloy crucible at 300° to 1 200°C in a furnace or over a burner. When the sample is homogeneous, carefully pour the molten flux into a beaker containing 10 wt % aqueous HNO_3 to dissolve the product.

Digestion of Organic Substances

To analyze elements such as N, P, halogens (F, Cl, Br, I), and metals in an organic compound, first decompose the compound by combustion (described in Section 6-4) or by *digestion.* **Digestion** is the process in which a substance is decomposed by a reactive liquid and dissolved. For this purpose, add sulfuric acid or a mixture of H_2SO_4 and HNO_3 to an organic substance and gently boil the mixture (or heat it in a microwave bomb) for 10 to 20 min until all particles have dissolved and the solution has a uniform black appearance. After cooling, discharge the dark color by adding hydrogen peroxide (H_2O_2) or HNO_3, and apply heat again. Analyze the constituents of the decomposed sample after it is digested.

Extraction

In **extraction,** analyte is dissolved in a solvent that does not dissolve the entire sample and does not decompose the analyte. In a typical extraction of pesticides from soil, a mixture of soil and the solvents acetone and hexane is placed in a Teflon-lined bomb and heated by microwaves to 150°C. This temperature is 50° to 100° higher than the boiling points of the individual solvents in an open vessel at atmospheric pressure. Soluble pesticides dissolve, but most of the soil remains behind. To complete the analysis, analyze the solution by chromatography, which is described in Chapters 20–22.

Ask Yourself

2-F. Lead sulfide (PbS) is a black solid that is very sparingly soluble in water but dissolves in concentrated HCl. If such a solution is boiled to dryness, white, crystalline lead chloride ($PbCl_2$) remains. What happened to the sulfide?

Key Equation

Buoyancy

$$m = m'\left(1 - \frac{d_a}{d_w}\right) \bigg/ \left(1 - \frac{d_a}{d}\right)$$

m = true mass; m' = mass measured in air

d_a = density of air (0.001 2 g/mL near 1 atm and 25°C)

d_w = density of balance weights (8.0 g/mL)

d = density of object being weighed

Important Terms

absorption	electronic balance	mortar and pestle
acid wash	extraction	mother liquor
adsorption	filtrate	parallax error
ashless filter paper	flux	pipet
bomb	fusion	rubber policeman
buoyancy	gravimetric analysis	tare
buret	hygroscopic	volatile
desiccant	ignition	volumetric flask
desiccator	mechanical balance	
digestion	meniscus	

Problems

2-1. What do the symbols TD and TC mean on volumetric glassware?

2-2. When would it be preferable to use a plastic volumetric flask instead of a glass flask?

2-3. What is the purpose of the trap in Figure 2-13? What does the watchglass do in Figure 2-16?

2-4. Distinguish absorption from adsorption. When you heat glassware in a drying oven, are you removing absorbed or adsorbed water?

2-5. What is the difference between digestion and extraction?

2-6. What is the true mass of water if the mass measured in air is 5.397 4 g?

2-7. Pentane (C_5H_{12}) is a liquid with a density of 0.626 g/mL. Find the true mass of pentane when the mass weighed in air is 14.82 g.

2-8. Ferric oxide $(Fe_2O_3$, density = 5.24 g/mL) obtained from ignition of a gravimetric precipitate weighed 0.296 1 g in the atmosphere. What is the true mass in vacuum?

2-9. Your professor has recruited you to work in her lab to help her win the Nobel Prize. It is therefore critical that your work be as accurate as possible. Rather than using the stated volumes of glassware in the lab, you decide to calibrate each piece. An empty 10-mL volumetric flask weighed 10.263 4 g. When filled to the mark with distilled water at 20°C, it weighed 20.214 4 g. What is the true volume of the flask?

2-10. Water from a 5-mL pipet was drained into a weighing bottle whose empty mass was 9.974 g to give a new mass of 14.974 g at 26°C. Find the volume of the pipet.

2-11. Water was drained from a buret between the 0.12- and 15.78-mL marks. The apparent volume was 15.78 − 0.12 = 15.66 mL. Measured in air at 25°C, the mass of water delivered was 15.569 g. What was the true volume?

Notes and References

1. A. Bardea, A. Dagan, I. Ben-Dov, B. Amit, and I. Willner, *Chem. Commun.* **1998,** 839.

2. Handling and disposing of chemicals is described in *Prudent Practices for Handling Hazardous Chemicals in Laboratories* (1983) and *Prudent Practices for Disposal of Chemicals in Laboratories* (1983), both available from National Academy Press, 2101 Constitution Avenue N.W., Washington, DC 20418. See also G. Lunn and E. B. Sansone, *Destruction of Hazardous Chemicals in the Laboratory* (New York: Wiley, 1994), and M. A. Armour, *Hazardous Laboratory Chemical Disposal Guide* (Boca Raton, FL: CRC Press, 1991).

3. A 40-min videotape entitled "Quantitative Techniques in Volumetric Analysis" demonstrates basic laboratory techniques.

[J. Zimmerman and J. J. Jacobsen, *J. Chem. Ed.* **1996,** *73,* 1117.]

4. Prepare cleaning solution by dissolving 36 g of ammonium peroxydisulfate, $(NH_4)_2S_2O_8$, in a *loosely stoppered* 2.2-L ("one gallon") bottle of 98 wt % sulfuric acid. Add ammonium peroxydisulfate every few weeks to maintain the oxidizing strength. The commercial cleaning solution EOSULF is an effective alternative to acid-based cleaning solutions for removing proteins and other residues from glassware in a biochemistry lab. EOSULF contains the metal binder EDTA and a sulfonate detergent. It can be safely poured down the drain. [P. L. Manske, T. M. Stimpfel, and E. L. Gershey, *J. Chem. Ed.* **1990,** *67,* A280.]

Experimental Error

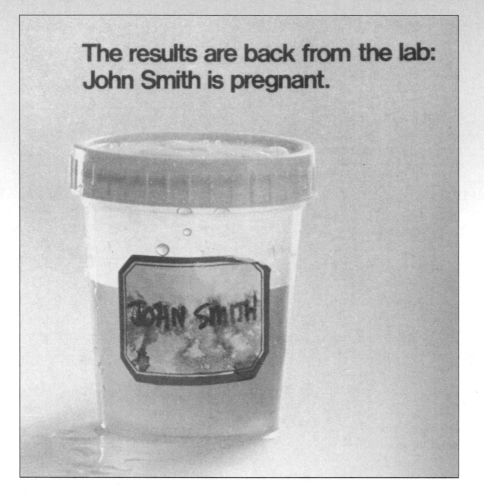

The results are back from the lab:
John Smith is pregnant.

Courtesy 3M Company, St. Paul, MN.

Some laboratory errors are more obvious than others, but there is error associated with every measurement. There is no way to measure the "true value" of anything. The best we can do in a chemical analysis is to carefully apply a technique that experience tells us is reliable. Repetition of one type of measurement several times tells us the reproducibility (*precision*) of the measurement. Measuring the same quantity by different methods gives us confidence of nearness to the "truth" (*accuracy*), if the results agree with one another.

Chapter 3

MATH TOOLKIT

Suppose that you measure the density of a mineral by finding its mass $(4.635 \pm 0.002$ g) and its volume $(1.13 \pm 0.05$ mL). Density is mass per unit volume: $4.635/1.13 = 4.101\ 8$ g/mL. The uncertainties in measured mass and volume are ±0.002 g and ±0.05 mL, but what is the uncertainty in the computed density? And how many significant figures should be used for the density? This chapter answers these questions and introduces you to spreadsheets—a powerful numerical tool that will be invaluable to you in and out of this course.

3-1 Significant Figures

The number of **significant figures** is the minimum number of digits needed to write a given value in scientific notation without loss of accuracy. The number 142.7 has four significant figures, because it can be written 1.427×10^2. If you write $1.427\ 0 \times 10^2$, you imply that you know the value of the digit after 7, which is not the case for the number 142.7. The number $1.427\ 0 \times 10^2$ has five significant figures.

Significant figures: Minimum number of digits required to express a value in scientific notation without loss of accuracy.

The number 6.302×10^{-6} has four significant figures, because all four digits are necessary. You could write the same number as 0.000 006 302, which also has just *four* significant figures. The zeros to the left of the 6 are merely holding decimal places. The number 92 500 is ambiguous. It could mean any of the following:

9.25×10^4	3 significant figures
9.250×10^4	4 significant figures
$9.250\,0 \times 10^4$	5 significant figures

You should write one of the three numbers above, instead of 92 500, to indicate how many figures are actually known.

Zeros are significant when they occur (1) in the middle of a number or (2) at the end of a number on the right-hand side of a decimal point.

The last (farthest to the right) significant figure in a measured quantity always has some associated uncertainty. The minimum uncertainty is ± 1 in the last digit. The scale of a Spectronic 20 spectrophotometer is drawn in Figure 3-1. The needle in the figure appears to be at an absorbance value of 0.234. We say that this number has three significant figures because the numbers 2 and 3 are completely certain and the number 4 is an estimate. The value might be read 0.233 or 0.235 by other people. The percent transmittance is near 58.3. Because the transmittance scale is smaller than the absorbance scale at this point, there is more uncertainty in the last digit of transmittance. A reasonable estimate of the uncertainty might be 58.3 ± 0.2. There are three significant figures in the number 58.3.

Significant zeros below are **bold**:

10**6**   0.0**10** 6   0.**10**6   0.**106 0**

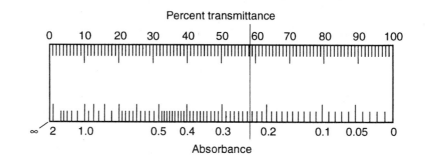

FIGURE 3-1   Scale of a Bausch and Lomb Spectronic 20 spectrophotometer. Percent transmittance is a linear scale and absorbance is a logarithmic scale.

Interpolation: Estimate all readings to the nearest tenth of the distance between scale divisions.

When reading the scale of any apparatus, interpolate between the markings. Try to estimate to the nearest tenth of the distance between two marks. Thus, on a 50-mL buret, which is graduated to 0.1 mL, read the level to the nearest 0.01 mL. When using a ruler calibrated in millimeters, estimate distances to the nearest tenth of a millimeter.

Ask Yourself

3-A. How many significant figures are there in each number below?
  (a) 1.903 0   **(b)** 0.039 10   **(c)** 1.40×10^4

We now address the question of how many digits to retain in the answer after you have performed arithmetic operations with your data. Rounding should only be done on the *final answer* (not intermediate results), to avoid accumulating round-off errors.

Addition and Subtraction

If the numbers to be added or subtracted have equal numbers of digits, the answer goes to the *same decimal place* as in any of the individual numbers:

$$
\begin{array}{r}
1.362 \times 10^{-4} \\
+\ 3.111 \times 10^{-4} \\
\hline
4.473 \times 10^{-4}
\end{array}
$$

The number of significant figures in the answer may exceed or be less than that in the original data.

$$
\begin{array}{r}
5.345 \\
+\ 6.728 \\
\hline
12.073
\end{array}
\qquad
\begin{array}{r}
7.26 \times 10^{14} \\
-\ 6.69 \times 10^{14} \\
\hline
0.57 \times 10^{14}
\end{array}
$$

If the numbers being added do not have the same number of significant figures, we are limited by the least certain one. For example, in a calculation of the molecular mass of KrF_2, the answer is known only to the second decimal place, because we are limited by our knowledge of the atomic mass of Kr.

$$
\begin{array}{rl}
18.998\ 403\ 2 & \text{(F)} \\
+\ 18.998\ 403\ 2 & \text{(F)} \\
+\ 83.80 & \text{(Kr)} \\
\hline
121.796\ 806\ 4 &
\end{array}
$$

Not significant

The number 121.796 806 4 should be rounded to 121.80 as the final answer.

When rounding off, look at *all* the digits *beyond* the last place desired. In the example above, the digits 6 806 4 lie beyond the last significant decimal place. Because this number is more than halfway to the next higher digit, we round the 9 up to 10 (i.e., we round up to 121.80 instead of down to 121.79). If the insignificant figures were less than halfway, we would round down. For example, 121.794 8 is correctly rounded to 121.79.

In the special case where the number is exactly halfway, round to the nearest *even* digit. Thus, 43.550 00 is rounded to 43.6, if we can only have three significant figures. If we are retaining only three figures, 1.425×10^{-9} becomes 1.42×10^{-9}. The number $1.425\ 01 \times 10^{-9}$ would become 1.43×10^{-9}, because 501 is more than halfway to the next digit. The rationale for rounding to an even digit is to avoid systematically increasing or decreasing results through successive round-off errors. Half the round-offs will be up and half down.

Rules for rounding off numbers.

Addition and subtraction: Express all numbers with the same exponent and align all numbers with respect to the decimal point. Round off the answer according to the number of decimal places in the number with the fewest decimal places.

In adding or subtracting numbers expressed in scientific notation, all numbers should first be expressed with the same exponent:

$$
\begin{array}{rl}
1.632 \times 10^5 & \\
+\ 4.107 \times 10^3 & \Rightarrow \\
+\ 0.984 \times 10^6 &
\end{array}
\qquad
\begin{array}{rl}
1.632 & \times 10^5 \\
+\ 0.041\ 07 & \times 10^5 \\
+\ 9.84 & \times 10^5 \\
\hline
11.51 & \times 10^5
\end{array}
$$

Challenge Show that the answer has four significant figures even if all numbers are expressed as multiples of 10^4 instead of 10^5.

The sum $11.513\ 07 \times 10^5$ is rounded to 11.51×10^5 because the number 9.84×10^5 limits us to two decimal places when all numbers are expressed as multiples of 10^5.

Multiplication and Division

In multiplication and division, we are normally limited to the number of digits contained in the number with the fewest significant figures:

$$
\begin{array}{c}
3.26 \times 10^{-5} \\
\times\ 1.78 \\
\hline
5.80 \times 10^{-5}
\end{array}
\qquad
\begin{array}{c}
4.317\ 9 \times 10^{12} \\
\times\ 3.6 \quad \times 10^{-19} \\
\hline
1.6 \quad \times 10^{-6}
\end{array}
\qquad
\begin{array}{c}
34.60 \\
\div\ 2.462\ 87 \\
\hline
14.05
\end{array}
$$

The power of 10 has no influence on the number of figures that should be retained.

Logarithms and Antilogarithms

The base 10 **logarithm** of n is the number a, whose value is such that $n = 10^a$:

Logarithm of n: $\qquad \boxed{n = 10^a \text{ means that } \log n = a} \qquad$ (3-1)

$$10^{-3} = \frac{1}{10^3} = \frac{1}{1\ 000} = 0.001$$

For example, 2 is the logarithm of 100 because $100 = 10^2$. The logarithm of 0.001 is -3 because $0.001 = 10^{-3}$. To find the logarithm of a number with your calculator, enter the number and press the *log* function.

In Equation 3-1, the number n is said to be the **antilogarithm** of a. That is, the antilogarithm of 2 is 100 because $10^2 = 100$ and the antilogarithm of -3 is 0.001 because $10^{-3} = 0.001$. Your calculator may have a *10^x* key or an *antilog* key or an *INV log* key. To find the antilogarithm of a number, enter it in your calculator and press *10^x* (or *antilog* or *INV log*).

A logarithm is composed of a **characteristic** and a **mantissa.** The characteristic is the integer part and the mantissa is the decimal part:

$$
\underbrace{\log 339}_{} = \underbrace{2}_{} . \underbrace{530}_{}
$$
Characteristic = 2 Mantissa = 0.530

$$
\log 3.39 \times 10^{-5} = -4.470
$$
Characteristic = −4 Mantissa = 0.470

Number of digits in **mantissa** of log x = number of significant figures in x:

$$\underbrace{\log (5.403 \times 10^{-8})}_{4 \text{ digits}} = \underbrace{-7.267\ 4}_{4 \text{ digits}}$$

The number 339 can be written 3.39×10^2. *The number of digits in the mantissa of log 339 should equal the number of significant figures in 339.* The logarithm of 339 is properly expressed as 2.530. The *characteristic,* 2, corresponds to the exponent in 3.39×10^2.

To see that the third decimal place is the last significant place, consider the following results:

$$10^{2.531} = 340 \quad (339.6)$$
$$10^{2.530} = 339 \quad (338.8)$$
$$10^{2.529} = 338 \quad (338.1)$$

The numbers in parentheses are the results prior to rounding to three figures. Changing the exponent by one digit in the third decimal place changes the answer by one digit in the last (third) place of 339.

In converting a logarithm to its antilogarithm, *the number of significant figures in the antilogarithm should equal the number of digits in the mantissa.* Thus

$$\text{antilog} (-3.42) = 10^{-3.42} = 3.8 \times 10^{-4}$$

2 digits   2 digits   2 digits

Number of digits in antilog $x = (10^x) =$ number of significant figures in **mantissa** of x:

$$10^{6.142} = 1.39 \times 10^6$$

3 digits   3 digits

Here are some examples showing the proper use of significant figures:

$$\log 0.001\,237 = -2.907\,6$$
$$\log 1\,237 = 3.092\,4$$
$$\log 3.2 - 0.51$$

$$\text{antilog } 4.37 = 2.3 \times 10^4$$
$$10^{4.37} = 2.3 \times 10^4$$
$$10^{-2.600} = 2.51 \times 10^{-3}$$

Ask Yourself

3-B. How would you express each answer with the correct number of digits?
   (a) $1.021 + 2.69 = 3.711$
   (b) $12.3 - 1.63 = 10.67$
   (c) $4.34 \times 9.2 = 39.928$
   (d) $0.060\,2 \div (2.113 \times 10^4) = 2.849\,03 \times 10^{-6}$
   (e) $\log (4.218 \times 10^{12}) = ?$
   (f) $\text{antilog} (-3.22) = ?$
   (g) $10^{2.384} = ?$

3-3   Types of Errors

Every measurement has some uncertainty, which is called *experimental error.* Scientific conclusions can be expressed with a high or low degree of confidence, but never with complete certainty. Experimental error is classified as either *systematic* or *random.*

Systematic Error

A **systematic error,** also called a **determinate error,** is repeatable if you make the measurement over again in the same way. In principle, a systematic error can be

Systematic error is a consistent error that can be detected and corrected. Box 3-1 describes Standard Reference Materials designed to reduce systematic errors.

Ways to detect systematic error:

1. Analyze samples of known composition, such as a Standard Reference Material. Your method should reproduce the known answer. (See Box 14-1 for an example.)

2. Analyze "blank" samples containing none of the analyte being sought. If you observe a nonzero result, your method responds to more than you intend.

3. Use different analytical methods to measure the same quantity. If the results do not agree, there is error in one (or more) of the methods.

4. *Round robin* experiment: Identical samples are analyzed in several laboratories by other persons using the same or different methods. Disagreement beyond the expected random error is systematic error.

Random error cannot be eliminated, but it might be reduced by a better experiment.

discovered and corrected. For example, using a pH meter that has been standardized incorrectly produces a systematic error in your results. Suppose you think that the pH of the buffer used to standardize the meter is 7.00, but it is really 7.08. If the meter is otherwise working properly, all your pH readings will be 0.08 pH unit too low. When you read a pH of 5.60, the actual pH of the sample is 5.68. This systematic error could be discovered by using another buffer of known pH to test the meter.

Another systematic error arises when you use an uncalibrated buret. The manufacturer's tolerance for a Class A 50-mL buret is ±0.05 mL. When you think you have delivered 29.43 mL, the real volume could be 29.40 mL and still be within tolerance. One way to correct for an error of this type is by constructing an experimental calibration curve (Figure 3-2). To do this, distilled water is delivered from the buret into a flask and weighed. You can determine the volume of water from its mass by using Table 2-4. The graph tells us to apply a correction factor of −0.03 mL to the measured value of 29.43 mL to reach the correct value of 29.40 mL.

Systematic error may be positive in some regions and negative in others. The key feature of systematic error is that, with care and cleverness, you can detect and correct it.

Random Error

Random error, also called **indeterminate error,** arises from limitations on our ability to make physical measurements. Random error has an equal chance of being positive or negative. It is always present and cannot be corrected. One random error is that associated with reading a scale. Different people reading the scale in Figure 3-1 report a range of values representing their subjective interpolation between the markings. One person reading the same instrument several times might report several different readings. Another indeterminate error results from random electrical noise in an instrument. Positive and negative fluctuations occur with approximately equal frequency and cannot be completely eliminated.

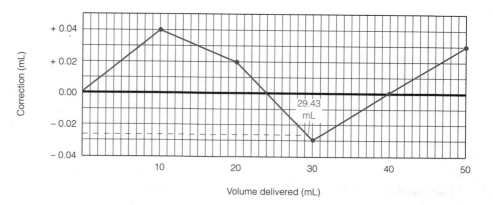

FIGURE 3-2   Calibration curve for a 50-mL buret.

What Are Standard Reference Materials?

Inaccurate laboratory measurements can mean wrong medical diagnosis and treatment, lost production time, wasted energy and materials, manufacturing rejects, and product liability problems. To minimize errors in laboratory measurements, the U.S. National Institute of Standards and Technology distributes more than 1 000 standard reference materials, such as metals, chemicals, rubber, plastics, engineering materials, radioactive substances, and environmental and clinical standards that can be used to test the accuracy of analytical procedures used in different laboratories.

For example, in treating patients with epilepsy, physicians depend on laboratory tests to measure blood serum concentrations of anticonvulsant drugs. Drug lev-els that are too low lead to seizures; high levels are toxic. Tests of identical serum specimens at different laboratories gave an unacceptably wide range of results. Therefore, the National Institute of Standards and Technology developed a standard reference material containing known levels of antiepilepsy drugs in serum. The reference material allows different laboratories to detect and correct errors in their assay procedures.

Before introduction of this reference material, five laboratories analyzing identical samples reported a range of results with relative errors of 40 to 110% of the expected value. After distribution of the reference material, the error was reduced to 20 to 40%.

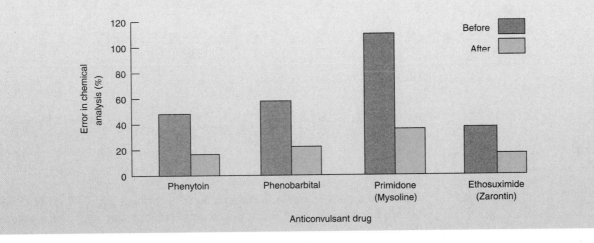

Precision and Accuracy

Precison is a measure of the reproducibility of a result. **Accuracy** refers to how close a measured value is to the "true" value.

A measurement might be reproducible, but wrong. For example, if you made a mistake preparing a solution for a titration, the solution would not have the desired concentration. You might then do a series of reproducible titrations but report an incorrect result because the concentration of the titrating solution was not what you intended. In this case, the precision is good but the accuracy is poor. Conversely, it is possible to make poorly reproducible measurements clustered around the correct value. In this case, the precision is poor but the accuracy is good. An ideal procedure is both precise and accurate.

Accuracy is defined as nearness to the "true" value. The word *true* is in quotes because somebody must *measure* the "true" value, and there is error associated with *every* measurement. The "true" value is best obtained by an experienced person

Precision: reproducibility

Accuracy: nearness to the "truth"

using a well-tested procedure. It is desirable to test the result by using different procedures, because, even though each method might be precise, systematic error could lead to poor agreement between methods. Good agreement among several methods affords us confidence, but never proof, that the results are "true."

Absolute and Relative Uncertainty

Absolute uncertainty expresses the margin of uncertainty associated with a measurement. If the estimated uncertainty in reading a calibrated buret is ± 0.02 mL, we say that ± 0.02 mL is the absolute uncertainty associated with the reading.

   Relative uncertainty compares the size of the absolute uncertainty to the size of its associated measurement. The relative uncertainty of a buret reading of 12.35 ± 0.02 mL is a dimensionless quotient:

Relative uncertainty:

$$\text{Relative uncertainty} = \frac{\text{absolute uncertainty}}{\text{magnitude of measurement}} \tag{3-2}$$

$$= \frac{0.02 \text{ mL}}{12.35 \text{ mL}} = 0.002$$

The percent relative uncertainty is simply

Percent relative uncertainty:

$$\text{Percent relative uncertainty} = 100 \times \text{relative uncertainty} \tag{3-3}$$

$$= 100 \times 0.002 = 0.2\%$$

If the absolute uncertainty in reading a buret is constant at ± 0.02 mL, the percent relative uncertainty is 0.2% for a volume of 10 mL and 0.1% for a volume of 20 mL.

Ask Yourself

3-C. Cheryl, Cynthia, Carmen, and Chastity shot these targets at Girl Scout camp. Match each target with the proper description.

   (a) accurate and precise   **(c)** precise but not accurate
   (b) accurate but not precise   **(d)** neither precise nor accurate

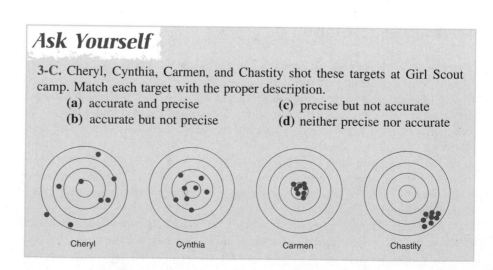

Cheryl   Cynthia   Carmen   Chastity

We can usually estimate or measure the random error associated with a measurement, such as the length of an object or the temperature of a solution. The uncertainty might be based on how well we can read an instrument or on our experience with a particular method. If possible, uncertainty is expressed as the *standard deviation* or as a *confidence interval;* these parameters are based on a series of replicate measurements. The following discussion applies only to random error. We assume that systematic error has been detected and corrected.

In most experiments, it is necessary to perform arithmetic operations on several numbers, each of which has an associated random error. The most likely uncertainty in the result is not simply the sum of the individual errors, because some of these are likely to be positive and some negative. We expect some cancellation of errors.

Standard deviation and confidence interval are discussed in Chapter 4.

Addition and Subtraction

Suppose you wish to perform the following arithmetic, in which the experimental uncertainties, designated e_1, e_2, and e_3, are given in parentheses.

$$
\begin{array}{ll}
1.76\,(\pm0.03) & \leftarrow e_1 \\
+\ 1.89\,(\pm0.02) & \leftarrow e_2 \\
-\ 0.59\,(\pm0.02) & \leftarrow e_3 \\
\hline
3.06\,(\pm e_4) &
\end{array}
\tag{3-4}
$$

The arithmetic answer is 3.06; but what is the uncertainty associated with this result?

For addition and subtraction, the uncertainty in the answer is obtained from the *absolute uncertainties* of the individual terms as follows:

Uncertainty in addition and subtraction:
$$
e_4 = \sqrt{e_1^2 + e_2^2 + e_3^2}
\tag{3-5}
$$

For addition and subtraction, use *absolute* uncertainty.

For the sum in expression 3-4, we can write

$$
e_4 = \sqrt{(0.03)^2 + (0.02)^2 + (0.02)^2} = 0.04_1
$$

The absolute uncertainty e_4 is ±0.04, and we can write the answer as 3.06 ± 0.04. Although there is only one significant figure in the uncertainty, we wrote it initially as 0.04_1, with the first insignificant figure subscripted. We retain one or more insignificant figures to avoid introducing round-off errors into later calculations through the number 0.04_1. The insignificant digit was subscripted to remind us where the last significant figures should be at the conclusion of the calculations.

To find the percent relative uncertainty in the sum of expression 3-4, we write

$$
\text{percent relative uncertainty} = \frac{0.04_1}{3.06} \times 100 = 1._3\%
$$

For addition and subtraction, use absolute uncertainty. Relative uncertainty can be found at the end of the calculation.

The uncertainty, 0.04_1, is $1._3\%$ of the result, 3.06. The subscript 3 in $1._3\%$ is not significant. It is sensible to drop the insignificant figures now and express the final result as

$$3.06 \, (\pm 0.04) \qquad \text{(absolute uncertainty)}$$
$$3.06 \, (\pm 1\%) \qquad \text{(relative uncertainty)}$$

EXAMPLE   Uncertainty in a Buret Reading

The volume delivered by a buret is the difference between the final reading and the initial reading. If the uncertainty in each reading is ± 0.02 mL, what is the uncertainty in the volume delivered?

SOLUTION Suppose that the initial reading is $0.05 \, (\pm 0.02)$ mL and the final reading is $17.88 \, (\pm 0.02)$ mL. The volume delivered is the difference:

$$
\begin{array}{r}
17.88 \, (\pm 0.02) \\
- \; 0.05 \, (\pm 0.02) \\
\hline
17.83 \, (\pm e)
\end{array}
\qquad e = \sqrt{0.02^2 + 0.02^2} = 0.03
$$

Regardless of the initial and final readings, if the uncertainty in each one is ± 0.02 mL, the uncertainty in volume delivered is ± 0.03 mL.

Multiplication and Division

For multiplication and division, first convert all uncertainties to percent relative uncertainties. Then calculate the error of the product or quotient as follows:

For multiplication and division, use percent relative uncertainty.

Uncertainty in multiplication and division:

$$e_4 = \sqrt{(\% e_1)^2 + (\% e_2)^2 + (\% e_3)^2} \qquad (3\text{-}6)$$

For example, consider the following operations:

$$\frac{1.76 \, (\pm 0.03) \times 1.89 \, (\pm 0.02)}{0.59 \, (\pm 0.02)} = 5.64 \pm e_4$$

First convert absolute uncertainties to percent relative uncertainties:

$$\frac{1.76 \, (\pm 1._7\%) \times 1.89 \, (\pm 1._1\%)}{0.59 \, (\pm 3._4\%)} = 5.64 \pm e_4$$

Then find the percent relative uncertainty of the answer by using Equation 3-6.

$$\% e_4 = \sqrt{(1._7)^2 + (1._1)^2 + (3._4)^2} = 4._0\%$$

Advice: Retain one or more extra insignificant figures until you have finished your entire calculation. Then round off to the correct number of digits. When storing intermediate results in a calculator, keep all digits without rounding.

The answer is $5.6_4 \, (\pm 4._0\%)$.

To convert relative uncertainty to absolute uncertainty, find $4._0\%$ of the answer:

$$4._0\% \times 5.6_4 = 0.04_0 \times 5.6_4 = 0.2_3$$

The answer is $5.6_4\,(\pm0.2_3)$. Finally, drop the insignificant digits:

$$5.6\,(\pm0.2) \qquad \text{(absolute uncertainty)}$$
$$5.6\,(\pm4\%) \qquad \text{(relative uncertainty)}$$

The denominator of the original problem, 0.59, limits the answer to two digits.

For multiplication and division, use percent relative uncertainty. Absolute uncertainty can be found at the end of the calculation.

Mixed Operations

Now consider an operation containing subtraction and division:

$$\frac{[1.76\,(\pm0.03) - 0.59\,(\pm0.02)]}{1.89\,(\pm0.02)} = 0.619_0 \pm ?$$

First work out the difference in the numerator, using absolute uncertainties:

$$1.76\,(\pm0.03) - 0.59\,(\pm0.02) = 1.17\,(\pm0.03_6)$$

because $\sqrt{(0.03)^2 + (0.02)^2} = 0.03_6$.

Then convert to percent relative uncertainties:

$$\frac{1.17\,(\pm0.03_6)}{1.89\,(\pm0.02)} = \frac{1.17\,(\pm3._1\%)}{1.89\,(\pm1._1\%)} = 0.619_0\,(\pm3._3\%)$$

because $\sqrt{(3._1\%)^2 + (1._1\%)^2} = 3._3\%$.

The percent relative uncertainty is $3._3\%$, so the absolute uncertainty is $0.03_3 \times 0.619_0 = 0.02_0$. The final answer can be written as follows:

$$0.619\,(\pm0.02_0) \qquad \text{(absolute uncertainty)}$$
$$0.619\,(\pm3._3\%) \qquad \text{(relative uncertainty)}$$

Because the uncertainty begins in the 0.01 decimal place, it is reasonable to round the result to the 0.01 decimal place:

$$0.62\,(\pm0.02) \qquad \text{(absolute uncertainty)}$$
$$0.62\,(\pm3\%) \qquad \text{(relative uncertainty)}$$

The result of a calculation ought to be written in a manner consistent with the uncertainty in the result.

The Real Rule for Significant Figures

The first uncertain figure of the answer is the last significant figure. For example, in the quotient

$$\frac{0.002\ 364\,(\pm0.000\ 003)}{0.025\ 00\,(\pm0.000\ 05)} = 0.094\ 6\,(\pm0.000\ 2)$$

The real rule: The first uncertain figure is the last significant figure.

the uncertainty ($\pm 0.000\,2$) occurs in the fourth decimal place. Therefore the answer is properly expressed with *three* significant figures, even though the original data have four figures. The first uncertain figure of the answer is the last significant figure. The quotient

$$\frac{0.002\,664\,(\pm 0.000\,003)}{0.025\,00\,(\pm 0.000\,05)} = 0.106\,6\,(\pm 0.000\,2)$$

is expressed with *four* significant figures because the uncertainty occurs in the fourth place. The quotient

$$\frac{0.821\,(\pm 0.002)}{0.803\,(\pm 0.002)} = 1.022\,(\pm 0.004)$$

is expressed with *four* figures even though the dividend and divisor each have *three* figures.

EXAMPLE Significant Figures in Laboratory Work

You prepared a 0.250 M NH_3 solution by diluting 8.45 (± 0.04) mL of 28.0 (± 0.5) wt % NH_3 [density = 0.899 (± 0.003) g/mL] up to 500.0 (± 0.2) mL. Find the uncertainty in 0.250 M. Consider the molecular mass of NH_3, 17.031 g/mol, to have negligible uncertainty.

SOLUTION To find the uncertainty in molarity, you need to find the uncertainty in moles delivered to the 500-mL flask. The concentrated reagent contains 0.899 (± 0.003) g of solution per milliliter. The weight percent tells you that the reagent contains 0.280 (± 0.005) g of NH_3 per gram of solution. In the following calculations, you should retain extra insignificant digits and round off only at the end.

For multiplication and division, convert absolute uncertainty to percent relative uncertainty.

$$\begin{aligned}\text{grams of } NH_3 \text{ per mL} \atop \text{in concentrated reagent} &= 0.899\,(\pm 0.003)\,\frac{\text{g solution}}{\text{mL}} \times 0.280\,(\pm 0.005)\,\frac{\text{g } NH_3}{\text{g solution}} \\ &= 0.899\,(\pm 0.334\%)\,\frac{\text{g solution}}{\text{mL}} \times 0.280\,(\pm 1.79\%)\,\frac{\text{g } NH_3}{\text{g solution}} \\ &= 0.251\,7\,(\pm 1.82\%)\,\frac{\text{g } NH_3}{\text{mL}}\end{aligned}$$

because $\sqrt{(0.334\%)^2 + (1.79\%)^2} = 1.82\%$.

Next, find the moles of ammonia contained in 8.45 (± 0.04) mL of concentrated reagent. The relative uncertainty in volume is $\pm 0.04/8.45 = \pm 0.473\%$.

$$\text{mol } NH_3 = \frac{0.251\,7\,(\pm 1.82\%)\,\dfrac{\text{g } NH_3}{\text{mL}} \times 8.45\,(\pm 0.473\%)\,\text{mL}}{17.031\,\dfrac{\text{g } NH_3}{\text{mol}}}$$

$$= 0.124\,9\,(\pm 1.88\%)\,\text{mol}$$

because $\sqrt{(1.82\%)^2 + (0.473\%)^2 + (0\%)^2} = 1.88\%$.

This much ammonia was diluted to 0.500 0 ($\pm$0.000 2) L. The relative uncertainty in final volume is $\pm$0.000 2/0.500 0 = $\pm$0.04%. The diluted molarity is

$$\frac{\text{mol NH}_3}{\text{L}} = \frac{0.124\,9\,(\pm 1.88\%)\,\text{mol}}{0.500\,0\,(\pm 0.04\%)\,\text{L}}$$

$$= 0.249\,8\,(\pm 1.88\%)\,\text{M}$$

because $\sqrt{(1.88\%)^2 + (0.04\%)^2} = 1.88\%$. The absolute uncertainty is 1.88% of 0.249 8 M = 0.018 8 $\times$ 0.249 8 M = 0.004 7 M. The uncertainty in molarity is in the third decimal place, so your final, rounded answer is

$$[\text{NH}_3] = 0.250\,(\pm 0.005)\,\text{M}$$

Ask Yourself

3-D. To help identify an unknown mineral in your geology class, you measured its mass and volume and found them to be 4.635 $\pm$ 0.002 g and 1.13 $\pm$ 0.05 mL.
    (a) Find the percent relative uncertainty in the mass and in the volume.
    (b) Write the density (= mass/volume) and its uncertainty with the correct number of digits.

3-5   Introducing Spreadsheets

Spreadsheets are powerful tools for manipulating quantitative information with a computer. They allow us to conduct "what if" experiments in which we investigate effects such as changing acid strength or concentration on the shape of a titration curve. Any spreadsheet is suitable for the exercises in this book. Our specific instructions apply to Microsoft Excel, which is very commonly available. You will need directions for your particular software. Although this book can be used with no loss of continuity if you skip the spreadsheet exercises, they will enrich your understanding and give you a tool that is valuable outside of this course.

A Spreadsheet for Temperature Conversions

Let's prepare a spreadsheet to convert temperature from degrees Celsius to kelvins and degrees Fahrenheit by using formulas derived from Table 1-4:

$$K = {}^{\circ}C + C_0 \tag{3-7a}$$

$${}^{\circ}F = \left(\tfrac{9}{5}\right){}^{*\circ}C + 32 \tag{3-7b}$$

where C_0 is the constant 273.15.
    Figure 3-3a shows a blank spreadsheet as it would appear on your computer screen. Rows are numbered 1, 2, 3, ... and columns are lettered A, B, C, ... Each

rectangular box is called a *cell*. The fourth cell down in the second column, for example, is designated cell B4.

We adopt a standard format in this book in which constants are collected in column A. Select cell A1 and type "Constant:" as a column heading. Select cell A2 and type "C0 =" to indicate that the constant C_0 will be written in the next cell down. Now select cell A3 and type the number 273.15. Your spreadsheet should now look like the one in Figure 3-3b.

In cell B1, type the label "°C" (or "Celsius" or whatever you like). For illustration, we will enter the numbers −200, −100, 0, 100, and 200 in cells B2 through B6. This is our *input* to the spreadsheet. The *output* will be computed values of kelvins and °F in columns C and D. (If you want to enter very large or very small numbers, you can write, for example, 6.02E23 for 6.02×10^{23} and 2E−8 for 2×10^{-8}.)

Label column C "kelvin" in cell C1. In cell C2, we enter our first *formula*—an entry beginning with an equals sign. Select cell C2 and type "= B2+$A$3". This expression tells the computer to calculate the contents of cell C2 by taking the contents of cell B2 and adding the contents of cell A3 (which contains the constant, 273.15). We will explain the dollar signs shortly. When this formula is entered, the computer responds by calculating the number 73.15 in cell C2. This is the kelvin equivalent of −200°C.

Now comes the beauty of a spreadsheet. Instead of typing many similar formulas, select cells C2, C3, C4, C5, and C6 all together and select the FILL DOWN

The formula "= B2+$A$3" in cell C2 is equivalent to writing K = °C + C_0.

(a)

	A	B	C	D
1				
2				
3				
4		cell B4		
5				
6				
7				
8				
9				
10				

(b)

	A	B	C	D
1	Constant:			
2	C0 =			
3	273.15			
4				
5				
6				
7				
8				
9				
10				

(c)

	A	B	C	D
1	Constant:	°C	kelvin	°F
2	C0 =	−200	73.15	−328
3	273.15	−100	173.15	−148
4		0	273.15	32
5		100	373.15	212
6		200	473.15	392
7				
8				
9				
10				

(d)

	A	B	C	D
1	Constant:	°C	kelvin	°F
2	C0 =	−200	73.15	−328
3	273.15	−100	173.15	−148
4		0	273.15	32
5		100	373.15	212
6		200	473.15	392
7				
8	Formulas:			
9	C2 = B2+$A$3			
10	D2 = (9/5)*B2+32			

FIGURE 3-3   Constructing a spreadsheet for temperature conversions.

command from the menu. (Other software uses the command COPY, instead of FILL DOWN.) This command tells the computer to do the same thing in cells C3 through C6 that was done in cell C2. The numbers 173.15, 273.15, 373.15, and 473.15 will appear in cells C3 through C6.

When computing the output in cell C3, the computer automatically uses input from cell B3 instead of cell B2. The reason for the dollar signs in $A$3 is that we do not want the computer to go down to cell A4 to find input for cell C3. $A$3 is called an *absolute reference* to cell A3. No matter what cell uses the constant C_0, we want it to come from cell A3. The reference to cell B2 is a *relative reference*. Cell C2 will use the contents of cell B2. Cell C6 will use the contents of cell B6. In general, references to constants in column A will be absolute (with dollar signs). References to numbers in the remainder of a spreadsheet will usually be relative (without dollar signs).

In cell D1, enter the label "°F". In cell D2, type the formula "= (9/5)*B2+32". This is equivalent to writing °F = (9/5)*°C + 32. The slash (/) is a division sign and the asterisk (*) is a multiplication sign. Parentheses are used to make the computer do what we intend. Operations inside parentheses are carried out before operations outside the parentheses. The computer responds to this formula by writing −328 in cell D2. This is the Fahrenheit equivalent to −200°C. Select cells D2 through D6 all together and use FILL DOWN to complete the table shown in Figure 3-3c.

Absolute reference: $A$3

Relative reference: B2

Order of Operations

The arithmetic operations in a spreadsheet are addition, subtraction, multiplication, division, and exponentiation (which uses the symbol ^). The order of operations in formulas is ^ first, followed by * and / (evaluated in order from left to right as they appear), finally followed by + and − (also evaluated from left to right). Make liberal use of parentheses to be sure that the computer does what you intend. The contents of parentheses are evaluated first, before carrying out operations outside the parentheses. Here are some examples:

$$9/5*100+32 = (9/5)*100+32 = (1.8)*100+32 = (1.8*100)+32 = (180) + 32 = 212$$

$$9/5*(100+32) = 9/5*(132) = (1.8)*(132) = 237.6$$

$$9+5*100/32 = 9+(5*100)/32 = 9+(500)/32 = 9+(500/32) = 9 + (15.625) = 24.625$$

$$9/5^2+32 = 9/(5^2)+32 = (9/25)+32 = (0.36)+32 = 32.36$$

When in doubt about how an expression will be evaluated by the computer, use parentheses to force it to do what you intend.

Documentation and Readability

If you look at your spreadsheet next month, you will probably not know what formulas were used. Therefore, we *document* the spreadsheet to show how it works by adding the text in cells A8, A9, and A10 in Figure 3-3d. In cell A8, write "Formulas:". In cell A9, write "C2 = B2+$A$3", and in cell A10, write "D2 = (9/5)*B2+32". Documentation is an excellent practice for every spreadsheet. As you learn to use your spreadsheet, you should use CUT and PASTE commands to copy the formulas used in cells C2 and D2 into the text in cells A9 and A10. This saves time and reduces transcription errors. Another basic form of documentation

If your spreadsheet cannot be read by another person without your help, it needs better documentation. (The same is true of your lab notebook!)

that we will add to future spreadsheets is a title in cell A1. A title such as "Temperature Conversions" tells us immediately what spreadsheet we are looking at.

For additional readability, you should learn to control how many decimal places are displayed in a given cell or column, even though the computer retains more digits for calculations. It does not throw away the digits that are not displayed. You can also control whether numbers are displayed in decimal or exponential notation.

Ask Yourself

3-E. Can you reproduce the spreadsheet in Figure 3-3 on your computer? The boiling point of nitrogen (N_2) at 1 atm pressure is $-196°C$. Use your spreadsheet to find the kelvin and degrees Fahrenheit equivalents of $-196°C$. Check your answers with your calculator.

3-6   Spreadsheet + Graph = Power

Although a spreadsheet has great power for crunching numbers, humans require a visual display to appreciate the relationship between two columns of numbers. Therefore, results of a spreadsheet calculation are plotted directly by the spreadsheet program or can be pasted into a graphing program for display.

Figure 3-4 shows a spreadsheet for computing the density of water as a function of temperature (°C) with the equation

$$\text{density (g/mL)} = a_0 + a_1*T + a_2*T^2 + a_3*T^3 \tag{3-8}$$

where $a_0 = 0.999\ 89$, $a_1 = 5.332\ 2 \times 10^{-5}$, $a_2 = -7.589\ 9 \times 10^{-6}$, and $a_3 = 3.671\ 9 \times 10^{-8}$. After writing a title in cell A1, the constants a_0 to a_3 are entered in column A. Column B is labeled "Temp (°C)" and column C is labeled "Density (g/mL)". Enter values of temperature in column B. In cell C4, type the formula "= $A$5+($A$7*B4)+($A$9*B4^2)+($A$11*B4^3)", which uses the exponent

Equation 3-8 is accurate to five decimal places over the range 4° to 40°C.

	A	B	C
1	Density of Water		
2			
3	Constants:	Temp (°C)	Density (g/mL)
4	a0 =	5	0.99997
5	0.99989	10	0.99970
6	a1 =	15	0.99911
7	5.3322E-05	20	0.99821
8	a2 =	25	0.99705
9	−7.5899E-06	30	0.99565
10	a3 =	35	0.99403
11	3.6719E-08	40	0.99223
12			
13	Formula:		
14	C4 = $A$5 + ($A$7*B4) + ($A$9*B4^2) + ($A$11*B4^3)		

FIGURE 3-4 Spreadsheet for computing the density of water as a function of temperature.

symbol ^ to compute T^2 and T^3. When you enter the formula, the number 0.999 97 is computed in cell C4. The remainder of column C is completed with a FILL DOWN (or COPY) command. The spreadsheet is not finished until it is documented by entering text in cells A13 and A14 to show what formula was used in column C.

To see the relationship between the input in column B and the output in column C, we need a graph. At this point you will need instruction for using your spreadsheet software or a graphing program to plot column C on the y-axis (the ordinate) and column B on the x-axis (the abscissa). Results displayed in Figure 3-5 show that the change in density from 5° to 15°C is small relative to the change in density between 30° and 40°C.

A spreadsheet is not complete until it is documented.

Ask Yourself

3-F. Can you reproduce the spreadsheet in Figure 3-4 on your computer? Can you use a graphing program to reproduce the graph in Figure 3-5?

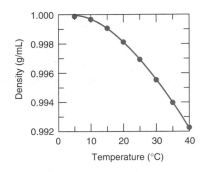

FIGURE 3-5 Density of water computed with the spreadsheet in Figure 3-4.

Key Equations

Definition of logarithm	If $n = 10^a$, then a is the logarithm of n.
Definition of antilogarithm	If $n = 10^a$, then n is the antilogarithm of a.
Relative uncertainty	$$\text{relative uncertainty} = \frac{\text{absolute uncertainty}}{\text{magnitude of measurement}}$$
Percent relative uncertainty	percent relative uncertainty $= 100 \times$ relative uncertainty
Uncertainty in addition and subtraction	$e_4 = \sqrt{e_1^2 + e_2^2 + e_3^2}$ (use absolute uncertainties)
	$e_4 =$ uncertainty in final answer
	$e_1, e_2, e_3 =$ uncertainty in individual terms
Uncertainty in multiplication and division	$\%e_4 = \sqrt{\%e_1^2 + \%e_2^2 + \%e_3^2}$ (use percent relative uncertainties)

Important Terms

absolute uncertainty	indeterminate error	relative uncertainty
accuracy	logarithm	significant figure
antilogarithm	mantissa	systematic error
characteristic	precision	
determinate error	random error	

Problems

3-1. Round each number as indicated:

(a) 1.236 7 to 4 significant figures

(b) 1.238 4 to 4 significant figures

(c) 0.135 2 to 3 significant figures

(d) 2.051 to 2 significant figures

(e) 2.005 0 to 3 significant figures

3-2. Round each number to three significant figures: **(a)** 0.216 74 **(b)** 0.216 5 **(c)** 0.216 500 3

3-3. Indicate how many significant figures there are in **(a)** 0.305 0 **(b)** 0.003 050 **(c)** 1.003×10^4

3-4. Write each answer with the correct number of digits:

(a) $1.0 + 2.1 + 3.4 + 5.8 = 12.300\ 0$

(b) $106.9 - 31.4 = 75.500\ 0$

(c) $107.868 - (2.113 \times 10^2) + (5.623 \times 10^3) =$ 5 519.568

(d) $(26.14/37.62) \times 4.38 = 3.043\ 413$

(e) $(26.14/37.62 \times 10^8) \times (4.38 \times 10^{-2}) =$ $3.043\ 413 \times 10^{-10}$

(f) $(26.14/3.38) + 4.2 = 11.933\ 7$

(g) $\log (3.98 \times 10^4) = 4.599\ 9$

(h) $10^{-6.31} = 4.897\ 79 \times 10^{-7}$

3-5. Write each answer with the correct number of digits:

(a) $3.021 + 8.99 = 12.011$

(b) $12.7 - 1.83 = 10.87$

(c) $6.345 \times 2.2 = 13.959\ 0$

(d) $0.030\ 2 \div (2.114\ 3 \times 10^{-3}) = 14.283\ 69$

(e) $\log (2.2 \times 10^{-18}) = ?$

(f) antilog $(-2.224) = ?$

(g) $10^{-4.555} = ?$

3-6. Find the formula mass of **(a)** $BaCl_2$ and **(b)** $C_{31}H_{32}O_8N_2$ with the correct number of significant figures.

3-7. Why do we use quotation marks around the word *true* in the statement that accuracy refers to how close a measured value is to the "true" value?

3-8. **(a)** Explain the difference between systematic and random errors. State whether the errors in **(b)**–**(e)** are random or systematic:

(b) A 25-mL transfer pipet holds 25.031 ± 0.009 mL when it is filled to the mark.

(c) A 10-mL buret consistently delivers 1.98 ± 0.01 mL when drained from exactly 0 to exactly 2 mL and consistently delivers 2.03 mL $\pm$ 0.02 mL when drained from 2 to 4 mL.

(d) A 10-mL buret delivered 1.983 9 g of water when drained from exactly 0.00 to 2.00 mL. The next time I delivered water from the 0.00 to the 2.00 mL mark, the delivered mass was 1.990 0 g.

(e) Four consecutive 20.0-μL injections of a solution into a chromatograph were made (as in Figure 0-6) and the area of a particular peak was 4 383, 4 410, 4 401, and 4 390 units.

(f) A clean funnel that had been in a drawer in the lab since last semester had a mass of 15.432 9 g. When filled with a solid precipitate and dried thoroughly in the oven at 110°C, the mass was 15.845 6 g. The calculated mass of precipitate was therefore $15.845\ 6 - 15.432\ 9 = 0.412\ 7$ g. What systematic and/or random error is there in the mass of precipitate?

3-9. Rewrite the number 3.123 56 ($\pm 0.167\ 89\%$) in the forms **(a)** number ($\pm$absolute uncertainty) and **(b)** number ($\pm$percent relative uncertainty) with an appropriate number of digits.

3-10. Write each answer with the correct number of digits. Find the absolute uncertainty and percent relative uncertainty for each answer.

(a) $6.2\ (\pm 0.2) - 4.1\ (\pm 0.1) = ?$

(b) $9.43\ (\pm 0.05) \times 0.016\ (\pm 0.001) = ?$

(c) $[6.2\ (\pm 0.2) - 4.1\ (\pm 0.1)] \div 9.43\ (\pm 0.05) = ?$

(d) $9.43\ (\pm 0.05) \times \{[6.2\ (\pm 0.2) \times 10^{-3}] + [4.1\ (\pm 0.1) \times 10^{-3}]\} = ?$

3-11. Write each answer with a reasonable number of figures. Find the absolute uncertainty and percent relative uncertainty for each answer.

(a) $[12.41\ (\pm 0.09) \div 4.16\ (\pm 0.01)] \times 7.068\ 2\ (\pm 0.000\ 4) = ?$

(b) $[3.26\ (\pm 0.10) \times 8.47\ (\pm 0.05)] - 0.18\ (\pm 0.06) = ?$

(c) $6.843\ (\pm 0.008) \times 10^4 \div [2.09\ (\pm 0.04) - 1.63\ (\pm 0.01)] = ?$

3-12. Write each answer with the correct number of digits. Find the absolute uncertainty and percent relative uncertainty for each answer.

(a) $9.23\ (\pm 0.03) + 4.21\ (\pm 0.02) - 3.26\ (\pm 0.06) = ?$

(b) $91.3\ (\pm 1.0) \times 40.3\ (\pm 0.2)/21.2\ (\pm 0.2) = ?$

(c) $[4.97\ (\pm 0.05) - 1.86\ (\pm 0.01)]/21.2\ (\pm 0.2) = ?$

(d) $2.016\ 4\ (\pm 0.000\ 8) + 1.233\ (\pm 0.002) + 4.61\ (\pm 0.01) = ?$

(e) $2.016\ 4\ (\pm 0.000\ 8) \times 10^3 + 1.233\ (\pm 0.002) \times 10^2 + 4.61\ (\pm 0.01) \times 10^1 = ?$

3-13. Find the absolute and percent relative uncertainty and express each answer with a reasonable number of significant figures.

(a) $3.4 \, (\pm 0.2) + 2.6 \, (\pm 0.1) = ?$

(b) $3.4 \, (\pm 0.2) \div 2.6 \, (\pm 0.1) = ?$

(c) $[3.4 \, (\pm 0.2) \times 10^{-8}] \div [2.6 \, (\pm 0.1) \times 10^3] = ?$

(d) $[3.4 \, (\pm 0.2) - 2.6 \, (\pm 0.1)] \times 3.4 \, (\pm 0.2) = ?$

3-14. Express the molecular mass ($\pm$ uncertainty) of benzene, C_6H_6, with the correct number of significant figures. The periodic table inside the cover of this book has a note in the legend about uncertainties in atomic mass.

3-15. Express the molecular mass of $C_6H_{13}B$ with the correct number of significant figures and find its uncertainty.

3-16. (a) Show that the formula mass of NaCl is 58.442 5 ($\pm 0.000\,9$) g/mol.

(b) To prepare a solution of NaCl, you weigh out 2.634 (± 0.002) g and dissolve it in a volumetric flask with a volume of 100.00 (± 0.08) mL. Express the molarity of the resulting solution, along with its uncertainty, with an appropriate number of digits.

3-17. (a) For use in an iodine titration, you prepare a solution from 0.222 2 ($\pm 0.000\,2$) g of KIO_3 [FM 214.001 0 ($\pm 0.000\,9$)] in 50.00 (± 0.05) mL. Find the molarity and its uncertainty with an appropriate number of significant figures.

(b) Would your answer be affected significantly if the reagent were only 99.9% pure?

3-18. Your instructor has asked you to prepare 2.00 L of 0.169 M NaOH from a stock solution of 53.4 (± 0.4) wt % NaOH with a density of 1.52 (± 0.01) g/mL.

(a) How many milliliters of stock solution will you need?

(b) If the uncertainty in delivering the NaOH is ± 0.10 mL, calculate the absolute uncertainty in the molarity (0.169 M). Assume negligible uncertainty in the molecular mass of NaOH and in the final volume, 2.00 L.

3-19. Create a spreadsheet to convert energy from joules (in column B) into calories (in column C), British thermal units (in column D), and electron volts (in column E). In column A, write constants from Table 1-4 to convert one set of units to another. For input in column B, use 2×10^{-20}, 1, 100, 1 000, and 2×10^{20} J. Document your spreadsheet with readable formulas at the bottom.

How Would You Do It?

3-20. Here are two methods you might use to prepare a dilute silver nitrate solution:

Method 1: Weigh out 0.046 3 g $AgNO_3$ and dissolve it in a 100-mL volumetric flask.

Method 2: Weigh out 0.463 0 g $AgNO_3$ and dissolve it in a 100-mL volumetric flask. Then pipet 10 mL of this solution into a fresh 100-mL flask and dilute to the mark.

The uncertainty in the balance is ± 3 in the last decimal place. Which method is more accurate?

Further Reading

E. J. Billo, *Microsoft Excel for Chemists* (New York: Wiley, 1997).

D. Diamond and V. Hanratty, *Spreadsheet Applications in Chemistry Using Microsoft Excel* (New York: Wiley, 1997).

B. V. Liengme, *A Guide to Microsoft Excel for Scientists and Engineers* (New York: Wiley, 1997).

Is My Red Blood Cell Count High Today?

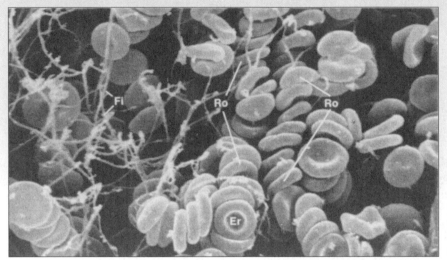

Red blood cells (erythrocytes, Er) tangled in fibrin threads (Fi) in a blood clot. Stacks of erythrocytes in a clot are called a rouleaux formation (Ro). [From R. H. Kardon, *Tissues and Organs* (San Francisco: W. H. Freeman, 1978), p. 39.]

All measurements contain experimental error, so it is impossible to be completely certain of a result. Nevertheless, we seek to answer questions such as "Is my red blood cell count today higher than usual?" If today's count is twice as high as usual, it is probably truly higher than normal. But what if the "high" count is not excessively above "normal" counts?

Count on "normal" days	Today's count
5.1	
5.3	
4.8 ⟩ × 10^6 cells/μL	5.6 × 10^6 cells/μL
5.4	
5.2	

The number 5.6 is higher than the five normal values, but the random variation in normal values might lead us to expect that 5.6 will be observed on some "normal" days.

The study of statistics allows us to say that over a long period, today's value will be observed on 1 out of 20 normal days. It is still up to you to decide what to do with this information.

*E*xperimental measurements always have some random error, so no conclusion can be drawn with complete certainty. However, statistics gives us tools to accept conclusions that have a high probability of being correct and to reject conclusions that do not. This chapter describes basic statistical tests and introduces the method of least squares for creating calibration curves.

Statistics deals only with random error—not determinate error. We must be ever vigilant and try to detect and prevent systematic errors. Good ways to detect systematic errors are to use different methods of analysis and see whether the results agree and to analyze certified standards to see whether our method gives the expected results.

4-1   The Gaussian Distribution

Nerve cells communicate with muscle cells by releasing neurotransmitter molecules adjacent to the muscle. As shown in Figure 4-1, neurotransmitters bind to membrane proteins of the muscle cell and open up channels that permit cations to diffuse into the muscle cell. Ions entering the cell trigger contraction of the muscle.

FIGURE 4-1 (*a*) In the absence of neurotransmitter, the ion channel is closed and cations cannot enter the muscle cell. (*b*) In the presence of neurotransmitter, the channel opens, cations enter the cell, and muscle action is initiated.

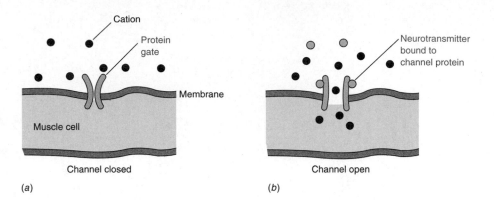

(*a*)   (*b*)

The study of signal transmission at the neuromuscular junction led to the Nobel Prize in Medicine or Physiology for Bert Sakmann and Erwin Neher in 1991.

Channels are all the same size, so each should allow a similar rate of ion passage across the membrane. Because ions are charged particles, the flow of ions is equivalent to a flow of electricity across the membrane. Of the 922 ion channel responses recorded for Figure 4-2, 190 are in the narrow range 2.64 to 2.68 pA (picoamperes, 10^{-12} amperes), represented by the tallest bar at the center of the chart. The next most probable responses fall in the range just to the right of the tallest bar, and the third most probable responses fall in the range just to the left of the tallest bar.

The bar graph in Figure 4-2 is typical of many laboratory measurements: The most probable response is at the center, and the probability of observing other responses decreases as the distance from the center increases. The smooth, bell-shaped curve superimposed on the data in Figure 4-2 is called a **Gaussian distribution.**

FIGURE 4-2 Bar chart showing observed cation current passing through individual channels of a frog muscle cell. The smooth line is the Gaussian curve that has the same mean and standard deviation as the measured data. [Data from Nobel Lecture of B. Sakmann, *Angew. Chem. Int. Ed. Engl.* **1992,** *31,* 830.]

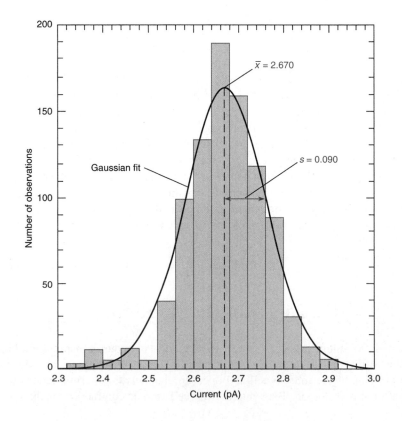

The more measurements made on any physical system, the closer the bar chart comes to the smooth curve.

Mean and Standard Deviation

A Gaussian distribution is characterized by a *mean* and a *standard deviation*. The mean is the *center* of the distribution, and the standard deviation is a measure of the *width* of the distribution.

The arithmetic **mean,** $\bar{x}$, also called the **average,** is the sum of the measured values divided by the number of measurements.

Mean locates center of distribution.

Standard deviation measures width of distribution.

Mean:

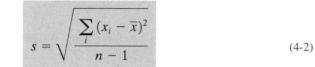

$$\bar{x} = \frac{\sum_i x_i}{n} = \frac{1}{n}(x_1 + x_2 + x_3 + \ldots + x_n) \qquad (4\text{-}1)$$

where each x_i is a measured value. A capital Greek sigma, Σ, is the symbol for a sum. In Figure 4-2, the mean value is indicated by the dashed line at 2.670 pA.

The **standard deviation,** s, is a measure of the width of the distribution. *The smaller the standard deviation, the narrower the distribution.*

The smaller the standard deviation, the more *precise* (reproducible) the results. Greater precision does not necessarily imply greater accuracy, which means nearness to the "truth."

Standard deviation:

$$s = \sqrt{\frac{\sum_i (x_i - \bar{x})^2}{n - 1}} \qquad (4\text{-}2)$$

In Figure 4-2, $s = 0.090$ pA. Figure 4-3 shows that if the standard deviation were doubled, the Gaussian curve for the same number of observations would be shorter and broader.

The *relative standard deviation* is the standard deviation divided by the average. It is usually expressed as a percentage. For $s = 0.090$ pA and $\bar{x} = 2.670$ pA, the relative standard deviation is $(0.090/2.670) \times 100 = 3.4\%$.

The quantity $n - 1$ in the denominator of Equation 4-2 is called the *degrees of freedom.* Initially, we have n independent data points, which represent n pieces of information. After computing the average, there are only $n - 1$ independent pieces of information left, because it is possible to calculate the nth data point if you know $n - 1$ data points and the average.

The symbols $\bar{x}$ and s apply to a finite set of measurements. For the Gaussian curve describing an infinite set of data, the true mean (called the *population mean*) is designated μ (Greek mu) and the true standard deviation is denoted by σ (lowercase Greek sigma).

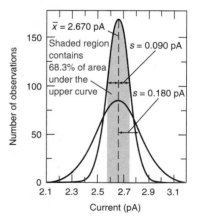

FIGURE 4-3 Gaussian curves showing the effect of doubling the standard deviation. The number of observations described by each curve is the same.

EXAMPLE   Mean and Standard Deviation

Find the mean, standard deviation, and the relative standard deviation for the set of measurements (7, 18, 10, 15).

SOLUTION The mean is

$$\bar{x} = \frac{7 + 18 + 10 + 15}{4} = 12._5$$

To avoid accumulating round-off errors, retain one more digit for the mean and the standard deviation than was present in the original data. The standard deviation is

$$s = \sqrt{\frac{(7 - 12._5)^2 + (18 - 12._5)^2 + (10 - 12._5)^2 + (15 - 12._5)^2}{4 - 1}} = 4._9$$

The mean and the standard deviation should both end at the *same decimal place.* For $\bar{x} = 12._5$, we write $s = 4._9$. The relative standard deviation is $(4._9/12._5) \times 100 = 39\%$.

Your calculator may have mean and standard deviation functions. Now is the time to learn to use them. Check to see whether your calculator reproduces the results in this example.

Other terms you should know are the median and the range. The *median* is the middle number in a series of measurements. When the measurements (8, 17, 11, 14, 12) are ordered from lowest to highest to give (8, 11, 12, 14, 17), the middle number (12) is the median. For an even number of measurements, the median is the average of the two middle numbers. For (8, 11, 12, 14), the median is 11.5. Some people prefer to report the median instead of the average, because the median is less influenced by outlying data. The *range* is the difference between the highest and lowest values. The range of (8, 17, 11, 14, 12) is $17 - 8 = 9$.

Standard Deviation and Probability

For an ideal Gaussian distribution, about two-thirds of the measurements (68.3%) lie within one standard deviation on either side of the mean (i.e., in the interval $\mu \pm \sigma$). That is, 68.3% of the area beneath a Gaussian curve lies in the interval $\mu \pm \sigma$, as shown in Figure 4-3. The percentage of measurements lying in the interval $\mu \pm 2\sigma$ is 95.5% and the percentage in the interval $\mu \pm 3\sigma$ is 99.7%. For real data with a standard deviation s, about 1 in 20 measurements (4.5%) will lie outside the range $\bar{x} \pm 2s$, and only 3 in 1 000 measurements (0.3%) will lie outside the range $\bar{x} \pm 3s$. Table 4-1 shows the good correspondence between ideal Gaussian behavior and the observations in Figure 4-2.

The Gaussian distribution is symmetric. If 4.5% of measurements lie outside the range $\mu \pm 2\sigma$, 2.25% of measurements are above $\mu \pm 2\sigma$ and 2.25% are below $\mu - 2\sigma$.

TABLE 4-1

Percentage of observations in Gaussian distribution

Range	Gaussian distri- bution	Observed in Figure 4-2
$\mu \pm 1\sigma$	68.3%	71.0%
$\mu \pm 2\sigma$	95.5	95.6
$\mu \pm 3\sigma$	99.7	98.5

Question What fraction of observations in a Gaussian distribution is expected to be below $\mu - 3\sigma$?

Ask Yourself

4-A. What are the mean, standard deviation, relative standard deviation, median, and range for the numbers 821, 783, 834, and 855? Express each to the nearest tenth.

Student's *t* is the statistical tool used to express confidence intervals and to compare results from different experiments. You can use it to evaluate the probability that your red blood cell count will be found in a certain range on "normal" days.

Confidence Intervals

From a limited number of measurements, it is impossible to find the true mean, μ, or the true standard deviation, σ. What we can determine are $\bar{x}$ and s, the sample mean and the sample standard deviation. The **confidence interval** is a range of values within which there is a specified probability of finding the true mean. We say that the true mean, μ, is likely to lie within a certain distance from the measured mean, $\bar{x}$. The confidence interval ranges from $-ts/\sqrt{n}$ below $\bar{x}$ to $+ts/\sqrt{n}$ above $\bar{x}$:

Confidence interval:
$$\mu = \bar{x} \pm \frac{ts}{\sqrt{n}}$$
(4-3)

where s is the measured standard deviation, n is the number of observations, and t is Student's *t*, taken from Table 4-2. Remember that in this table the *degrees of freedom* are equal to $n - 1$. If there are five data points, there are four degrees of freedom.

EXAMPLE   Calculating Confidence Intervals

In replicate analyses, the carbohydrate content of a glycoprotein (a protein with sugars attached to it) is found to be 12.6, 11.9, 13.0, 12.7, and 12.5 g of carbohydrate per 100 g of protein. Find the 50% and 90% confidence intervals for the carbohydrate content.

SOLUTION First we calculate $\bar{x} = 12.5_4$ and $s = 0.4_0$ for the five measurements. To find the 50% confidence interval, look up *t* in Table 4-2 under 50 and across from *four* degrees of freedom (degrees of freedom = $n - 1$). The value of *t* is 0.741, so the confidence interval is

$$\mu(50\%) = \bar{x} \pm \frac{ts}{\sqrt{n}} = 12.5_4 \pm \frac{(0.741)(0.4_0)}{\sqrt{5}} = 12.5_4 \pm 0.1_3$$

The 90% confidence interval is

$$\mu(90\%) = \bar{x} \pm \frac{ts}{\sqrt{n}} = 12.5_4 \pm \frac{(2.132)(0.4_0)}{\sqrt{5}} = 12.5_4 \pm 0.3_8$$

These calculations mean that there is a 50% chance that the true mean, μ, lies in the range $12.5_4 \pm 0.1_3$ (12.4_1 to 12.6_7). There is a 90% chance that μ lies in the range $12.5_4 \pm 0.3_8$ (12.1_6 to 12.9_2).

"Student" was the pseudonym of W. S. Gossett, whose employer, the Guinness Breweries of Ireland, restricted publications for proprietary reasons. Because of the importance of his work, Gossett published it under an assumed name in 1908.

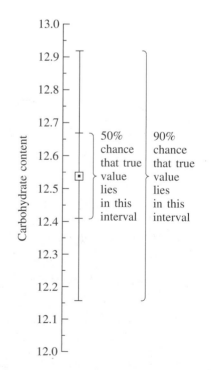

You might appreciate Box 4-1 at this time.

TABLE 4-2   Values of Student's *t*

Degrees of freedom	Confidence level (%)						
	50	90	95	98	99	99.5	99.9
1	1.000	6.314	12.706	31.821	63.657	127.32	636.619
2	0.816	2.920	4.303	6.965	9.925	14.089	31.598
3	0.765	2.353	3.182	4.541	5.841	7.453	12.924
4	0.741	2.132	2.776	3.747	4.604	5.598	8.610
5	0.727	2.015	2.571	3.365	4.032	4.773	6.869
6	0.718	1.943	2.447	3.143	3.707	4.317	5.959
7	0.711	1.895	2.365	2.998	3.500	4.029	5.408
8	0.706	1.860	2.306	2.896	3.355	3.832	5.041
9	0.703	1.833	2.262	2.821	3.250	3.690	4.781
10	0.700	1.812	2.228	2.764	3.169	3.581	4.587
15	0.691	1.753	2.131	2.602	2.947	3.252	4.073
20	0.687	1.725	2.086	2.528	2.845	3.153	3.850
25	0.684	1.708	2.068	2.485	2.787	3.078	3.725
30	0.683	1.697	2.042	2.457	2.750	3.030	3.646
40	0.681	1.684	2.021	2.423	2.704	2.971	3.551
60	0.679	1.671	2.000	2.390	2.660	2.915	3.460
120	0.677	1.658	1.980	2.358	2.617	2.860	3.373
∞	0.674	1.645	1.960	2.326	2.576	2.807	3.291

Note: In calculating confidence intervals, σ may be substituted for s in Equation 4-3 if you have a great deal of experience with a particular method and have therefore determined its "true" population standard deviation. If σ is used instead of s, the value of t to use in Equation 4-3 comes from the bottom row of Table 4-2.

Improving the Reliability of Your Measurements

We wish to be as accurate and precise as possible. Systematic errors reduce the accuracy of a measurement. If a pH meter is not calibrated correctly, it will give inaccurate readings, no matter how precise (reproducible) they are. Making a measurement by two different methods is a good way to detect systematic errors. If the measurements do not agree within the expected experimental uncertainty, systematic error is probably present.

Better measurements give smaller confidence intervals. That is, being 90% sure that a quantity lies in the range 62.3 ± 0.5 is better than being 90% sure that the quantity is in the range 62.3 ± 1.3. Equation 4-3, $\mu = \bar{x} \pm ts/\sqrt{n}$, tells us that to reduce the size of the confidence interval we either make more measurements (increase n) or decrease the standard deviation (s). The only way to reduce s is to improve your experimental procedure. In the absence of a procedural change, the way to improve the accuracy of a result is to increase the number of replicate measurements. Doubling the number of measurements decreases the factor $1/\sqrt{n}$ by $1/\sqrt{2} = 0.71$.

Replicate measurements improve reliability:

If $s = 2.0\%$, 3 measurements give a 95% confidence interval of

$$\frac{\pm ts}{\sqrt{n}} = \frac{(4.303)(2.0\%)}{\sqrt{3}} = \pm 5.0\%$$

Making 9 measurements reduces the 95% confidence interval to

$$\frac{\pm ts}{\sqrt{n}} = \frac{(2.306)(2.0\%)}{\sqrt{9}} = \pm 1.5\%$$

(Values of t came from Table 4-2.)

Box 4-1 *Informed Citizen*

Analytical Chemistry and the Law

As a person who will either produce or use analytical results, you should be aware of this warning published in 1983 in a report entitled "Principles of Environmental Analysis":[1]

> Analytical chemists must always emphasize to the public that *the single most important characteristic of any result obtained from one or more analytical measurements is an adequate statement of its uncertainty interval.* Lawyers usually attempt to dispense with uncertainty and try to obtain unequivocal statements; therefore, an uncertainty interval must be clearly defined in cases involving litigation and/or enforcement proceedings. Otherwise, a value of 1.001 without a specified uncertainty, for example, may be viewed as legally exceeding a permissible level of 1.

Some legal limits make no scientific sense. The Delaney Amendment to the U.S. Food, Drug, and Cosmetic Act of 1958 stated that "no additive [in processed food] shall be deemed to be safe if it is found to induce cancer when ingested by man or animal. ... " This statement meant that no detectable level of any carcinogenic (cancer-causing) pesticide may remain in processed foods, even if the level is far below that which can be shown to cause cancer. The law was passed at a time when the sensitivity of analytical procedures was relatively poor. As procedures became more sensitive, detectable levels of chemical residues decreased by 10^3 to 10^6. A concentration that may have been acceptable in 1965 was 10^6 times above the legal limit in 1995, regardless of whether there was any evidence that such a low level is harmful. In 1996, Congress finally changed the law to take a health-based approach to pesticide regulation. Pesticides are supposed to be allowed at a level where there is a "reasonable certainty of no harm." For carcinogens, the allowed level will probably be set at a concentration that produces less than one excess cancer per million persons exposed. Since 1996, the Environmental Protection Agency has had an exceedingly difficult time trying to set safe limits because of the dearth of experimental data on health effects from low-level exposure.

Comparison of Means with Student's t

Student's *t* can be used to compare two sets of measurements to decide whether they are "the same" or "different." We adopt the following standard: If there is less than 1 chance in 20 that the difference between the two measurements arises from random variation in the data, then the difference is significant. This criterion gives us 95% confidence in concluding that two measurements are the same or different. There is a 5% probability that our conclusion is wrong.

An example comes from the work of Lord Rayleigh (John W. Strutt), who received the Nobel Prize in 1904 for discovering the inert gas argon—a discovery that came about when he noticed a discrepancy between two sets of measurements of the density of nitrogen. In Rayleigh's time, it was known that dry air is composed of about one-fifth oxygen and four-fifths nitrogen. Rayleigh removed O_2 from air by reaction with red hot copper $[Cu(s) + \frac{1}{2}O_2(g) \rightarrow CuO(s)]$ and measured the density of the remaining gas by collecting it in a fixed volume at a constant temperature and pressure. He then prepared the same volume of nitrogen by decomposition of nitrous oxide (N_2O), nitric oxide (NO), or ammonium nitrite ($NH_4^+NO_2^-$). Table 4-3 and Figure 4-4 show the mass of gas collected in each experiment. The average mass from air was 0.46% greater than the average mass of the same volume of gas from chemical sources.

If Rayleigh's measurements had not been performed with care, a 0.46% difference might have been attributed to experimental error. Instead, Rayleigh understood

TABLE 4-3 Grams of nitrogen-rich gas isolated by Lord Rayleigh

From air	From chemical decomposition
2.310 17	2.301 43
2.309 86	2.298 90
2.310 10	2.298 16
2.310 01	2.301 82
2.310 24	2.298 69
2.310 10	2.299 40
2.310 28	2.298 49
—	2.298 89
Average	
$2.310\ 10_9$	$2.299\ 47_2$
Standard deviation	
$0.000\ 14_3$	$0.001\ 37_9$

SOURCE: R. D. Larsen, *J. Chem. Ed.* **1990,** *67,* 925.

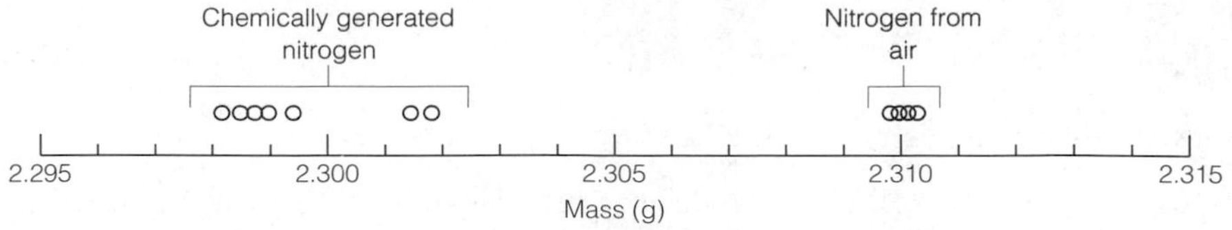

FIGURE 4-4 Lord Rayleigh's measurements of the mass of nitrogen isolated from air or generated by decomposition of nitrogen compounds. Rayleigh recognized that the difference between the two data sets was too great to be due to experimental error. He deduced that a heavier component, which turned out to be argon, was present in nitrogen isolated from air.

Challenge Rayleigh found a systematic error by comparing two different methods of measurement. Which method had the systematic error? Did it overestimate or underestimate the mass of nitrogen in air?

that the discrepancy was outside his margin of error, and he postulated that nitrogen from the air was mixed with a heavier gas, which turned out to be argon.

Let's see how to use the **t test** to decide whether nitrogen isolated from air is "significantly" heavier than nitrogen isolated from chemical sources. For two sets of data consisting of n_1 and n_2 measurements (with averages $\bar{x}_1$ and $\bar{x}_2$), we calculate a value of t from the formula

t Test for comparison of means:

$$t = \frac{|\bar{x}_1 - \bar{x}_2|}{s_{\text{pooled}}} \sqrt{\frac{n_1 n_2}{n_1 + n_2}} \qquad (4\text{-}4)$$

where

$$s_{\text{pooled}} = \sqrt{\frac{s_1^2(n_1 - 1) + s_2^2(n_2 - 1)}{n_1 + n_2 - 2}} \qquad (4\text{-}5)$$

Here s_{pooled} is a *pooled* standard deviation making use of both sets of data. The absolute value of $\bar{x}_1 - \bar{x}_2$ is used in Equation 4-4 so that t is always positive. The value of t from Equation 4-4 is to be compared with the value of t in Table 4-2 for $(n_1 + n_2 - 2)$ degrees of freedom. *If the calculated t is greater than the tabulated t at the 95% confidence level, the two results are considered to be significantly different.*

If $t_{\text{calculated}} > t_{\text{table}}$ (95%), then the difference is significant.

EXAMPLE Is Rayleigh's N_2 from Air Denser than N_2 from Chemicals?

The average mass of nitrogen from air in Table 4-3 is $\bar{x}_1 = 2.310\ 10_9$ g, with a standard deviation of $s_1 = 0.000\ 14_3$ (for $n_1 = 7$ measurements). The average mass from chemical sources is $\bar{x}_2 = 2.299\ 47_2$ g, with a standard deviation of $s_2 = 0.001\ 37_9$ (for $n_2 = 8$ measurements). Are the two masses significantly different?

SOLUTION To answer this question, we calculate s_{pooled} with Equation 4-5,

$$s_{\text{pooled}} = \sqrt{\frac{0.000\ 14_3{}^2(7 - 1) + 0.001\ 37_9{}^2(8 - 1)}{7 + 8 - 2}} = 0.001\ 01_7$$

and t with Equation 4-4:

$$t = \frac{2.310\ 10_9 - 2.299\ 47_2}{0.001\ 01_7}\sqrt{\frac{7 \cdot 8}{7 + 8}} = 20.2$$

For $7 + 8 - 2 = 13$ degrees of freedom in Table 4-2, t lies between 2.228 and 2.131 at the 95% confidence level. The observed value ($t = 20.2$) is greater than the tabulated t, so the difference is significant.[2] In fact, the tabulated value of t for 99.9% confidence is about 4.3. The difference is significant beyond the 99.9% confidence level. Our eyes do not lie to us in Figure 4-4: N_2 from the air is undoubtedly denser than N_2 from chemical sources. This observation led Rayleigh to discover argon as a heavy constituent of air.

Statistical tests give us only probabilities. They do not relieve us of the ultimate subjective decision to accept or reject a conclusion.

Ask Yourself

4-B. A reliable assay of ATP (adenosine triphosphate) in a certain type of cell gives a value of $111._0$ μmol/100 mL, with a standard deviation of $2._8$ in four replicate measurements. You have developed a new assay, which gave the following values in replicate analyses: 117, 119, 111, 115, 120 μmol/100 mL.

(a) Find the mean and standard deviation of your new analysis.

(b) Can you be 95% confident that your method produces a result different from the "reliable" value?

4-3 Q Test for Bad Data

There was always somebody in my lab section with four thumbs. Freshmen at Phillips University perform an experiment in which they dissolve the zinc from a galvanized nail and measure the mass lost by the nail to tell how much of the nail was zinc. Several students performed the experiment in triplicate and pooled their results:

Mass loss (%): 10.2, 10.8, 11.6   9.9, 9.4, 7.8   10.0, 9.2, 11.3   9.5, 10.6, 11.6

Sidney   Cheryl   Tien   Dick

It appears that Cheryl might be the person with four thumbs, because her value 7.8 looks out of line with the other data. Should the group reject the value 7.8 before averaging the rest of the data, or should 7.8 be retained?

We answer this question with the **Q test.** To apply the Q test, arrange the data in order of increasing value and calculate Q, defined as

Q test for discarding data:
$$Q = \frac{\text{gap}}{\text{range}}$$
(4-6)

If $Q_{calculated} > Q_{table}$, then reject the questionable point.

Gap = 1.4

(7.8)   9.2 9.4 9.5 9.9 10.0 10.2 10.6 10.8 11.3 11.6 11.6

Range = 3.8

Questionable value
(too low?)

The *range* is the total spread of the data. The *gap* is the difference between the questionable point and the nearest value. *If Q calculated from Equation 4-6 is greater than Q in Table 4-4, the questionable point should be discarded.*

For the numbers above, $Q_{\text{calculated}} = 1.4/3.8 = 0.37$. The value Q_{table} at the 90% confidence limit is 0.38 for 12 observations in Table 4-4. Because $Q_{\text{calculated}}$ is smaller than Q_{table}, the questionable point should be retained. There is more than a 10% chance that the value 7.8 is a member of the same population as the other measurements.

Common sense must always prevail. If Cheryl knows that her measurement was low because she spilled some of her unknown, then the probability that the result is wrong is 100% and the datum should be discarded. Any datum based on a faulty procedure should be discarded, no matter how well it fits the rest of the data.

TABLE 4-4 Values of Q for rejection of data

Q (90% confidence)[a]	Number of observations
0.76	4
0.64	5
0.56	6
0.51	7
0.47	8
0.44	9
0.41	10
0.39	11
0.38	12
0.34	15
0.30	20

a. Q = gap/range. If Q (observed) > Q (tabulated), the value in question can be rejected with 90% confidence.

SOURCE: R. B. Dean and W. J. Dixon, *Anal. Chem.* **1951,** *23,* 636; and D. R. Rorabacher, *Anal. Chem.* **1991,** *63,* 139.

Ask Yourself

4-C. Would you reject the value 216 from the set of results 192, 216, 202, 195, and 204?

4-4   Finding the "Best" Straight Line

The *method of least squares* finds the "best" straight line through experimental data points. We will apply this procedure to analytical chemistry calibration curves in the next section.

The equation of a straight line is

Equation of straight line: $$y = mx + b \qquad (4\text{-}7)$$

in which m is the **slope** and b is the **y-intercept** (Figure 4-5). If we measure between two points that lie on the line, the slope is $\Delta y/\Delta x$, which is constant for any pair of points on the line. The y-intercept is the point at which the line crosses the y-axis.

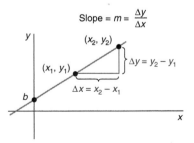

FIGURE 4-5 Parameters of a straight line:

equation: $y = mx + b$

slope: $m = \dfrac{\Delta y}{\Delta x} = \dfrac{y_2 - y_1}{x_2 - x_1}$

y-intercept: b = crossing point on y-axis

Method of Least Squares

The **method of least squares** finds the "best" line by adjusting the line to minimize the vertical deviations between the points and the line (Figure 4-6). The reasons for minimizing only the vertical deviations are that (1) experimental uncertainties in

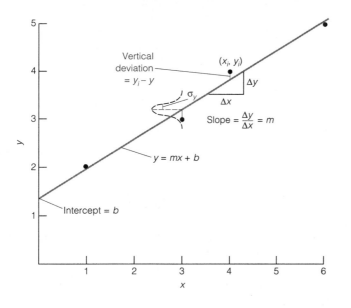

FIGURE 4-6 Least-squares curve fitting minimizes the sum of the squares of the vertical deviations of the measured points from the line. The Gaussian curve drawn over the point (3, 3) is a schematic indication of the distribution of measured y values about the straight line. The most probable value of y falls on the line, but there is a finite probability of measuring y some distance from the line.

y values are often greater than uncertainties in x values and (2) the calculation for minimizing the vertical deviations is relatively simple.

In Figure 4-6, the vertical deviation for the point (x_i, y_i) is $y_i - y$, where y is the ordinate of the straight line when $x = x_i$.

$$\text{vertical deviation} = d_i = y_i - y = y_i - (mx_i + b) \qquad (4\text{-}8)$$

The *ordinate* is the value of y that really lies on the line; y_i is the measured value that does not lie exactly on the line.

Some of the deviations are positive and some are negative. To minimize the magnitude of the deviations irrespective of their signs, we square the deviations to create positive numbers:

$$d_i^2 = (y_i - y)^2 = (y_i - mx_i - b)^2$$

Because we minimize the squares of the deviations, this is called the *method of least squares*.

When we use such a procedure to minimize the sum of squares of the vertical deviations, the slope and the intercept of the "best" straight line fitted to n points turn out to be

Least-squares slope: $\qquad m = \dfrac{n\Sigma(x_i y_i) - \Sigma x_i \Sigma y_i}{D} \qquad (4\text{-}9)$

Least-squares intercept: $\qquad b = \dfrac{\Sigma(x_i^2)\Sigma y_i - \Sigma(x_i y_i)\Sigma x_i}{D} \qquad (4\text{-}10)$

Remember that Σ means summation: $\Sigma x_i = x_1 + x_2 + x_3 + \ldots$

where the denominator, D, is given by

$$D = n\Sigma(x_i^2) - (\Sigma x_i)^2 \qquad (4\text{-}11)$$

These equations are not as terrible as they appear. Table 4-5 sets out an example in which the four points ($n = 4$) in Figure 4-6 are treated. The first two columns list x_i and y_i for each point. The third column gives the product $x_i y_i$, and the fourth column lists the square x_i^2. At the bottom of each column is the sum for

TABLE 4-5 Calculations for least-squares analysis

x_i	y_i	$x_i y_i$	x_i^2	$d_i (= y_i - mx_i - b)$	d_i^2
1	2	2	1	0.038 462	0.001 479
3	3	9	9	−0.192 308	0.036 982
4	4	16	16	0.192 308	0.036 982
6	5	30	36	−0.038 462	0.001 479
$\Sigma x_i = 14$	$\Sigma y_i = 14$	$\Sigma(x_i y_i) = 57$	$\Sigma(x_i^2) = 62$		$\Sigma(d_i^2) = 0.076\ 923$

that column. That is, beneath column 1 is Σx_i and beneath column 3 is $\Sigma(x_i y_i)$. The last two columns at the right will be used later.

With the sums from Table 4-5, we compute the slope and intercept by substituting into Equations 4-11, 4-9, and 4-10:

$$D = n\Sigma(x_i^2) - (\Sigma x_i)^2 = 4 \cdot 62 - 14^2 = 52$$

$$m = \frac{n\Sigma(x_i y_i) - \Sigma x_i \Sigma y_i}{D} = \frac{4 \cdot 57 - 14 \cdot 14}{52} = 0.615\ 38$$

$$b = \frac{\Sigma(x_i^2)\Sigma y_i - \Sigma(x_i y_i)\Sigma x_i}{D} = \frac{62 \cdot 14 - 57 \cdot 14}{52} = 1.346\ 15$$

The equation of the best straight line through the points in Figure 4-6 is therefore

$$y = 0.615\ 38x + 1.346\ 15$$

We address significant figures for m and b in the next section.

How Reliable Are Least-Squares Parameters?

The uncertainties in m and b are related to the uncertainty in measuring each value of y. Therefore, we first estimate the standard deviation describing the population of y values. This standard deviation, s_y, characterizes the little Gaussian curve inscribed in Figure 4-6. The deviation of each y_i from the center of its Gaussian curve is $d_i = y_i - y = y_i - (mx_i + b)$ (Equation 4-8). The standard deviation of these vertical deviations is

$$s_y \approx \sqrt{\frac{\Sigma(d_i^2)}{n - 2}} \tag{4-12}$$

Analysis of uncertainty for Equations 4-10 and 4-11 leads to the following results:

Standard deviation of slope:
$$s_m = s_y\sqrt{\frac{n}{D}} \tag{4-13}$$

Standard deviation of intercept:
$$s_b = s_y\sqrt{\frac{\Sigma(x_i^2)}{D}} \tag{4-14}$$

where s_y is given by Equation 4-12, and D is given by Equation 4-11.

At last, we can address significant figures for the slope and the intercept of the line in Figure 4-6. In Table 4-5, we see that $\Sigma(d_i^2) = 0.076\ 923$. Inserting this value into Equation 4-12 gives

$$s_y = \sqrt{\frac{0.076\ 923}{4 - 2}} = 0.196\ 12$$

Now, we can plug numbers into Equations 4-13 and 4-14 to find

$$s_m = s_y\sqrt{\frac{n}{D}} = (0.196\ 12)\sqrt{\frac{4}{52}} = 0.054\ 394$$

$$s_b = s_y\sqrt{\frac{\Sigma(x_i^2)}{D}} = (0.196\ 12)\sqrt{\frac{62}{52}} = 0.214\ 15$$

Combining the results for m, s_m, b, and s_b, we write

Slope: $\quad \dfrac{0.615\ 38}{\pm 0.054\ 39} = 0.62 \pm 0.05 \quad$ or $\quad 0.61_5 \pm 0.05_4$

The first digit of the uncertainty is the last significant figure.

Intercept: $\quad \dfrac{1.346\ 15}{\pm 0.214\ 15} - 1.3 \pm 0.2 \quad$ or $\quad 1.3_5 \pm 0.2_1$

where the uncertainties represent one standard deviation. *The first decimal place of the standard deviation is the last significant figure of the slope or intercept.*

Ask Yourself

4-D. Construct a table analogous to Table 4-5 to calculate the equation of the best straight line going through the points (1, 3), (3, 2), and (5, 0). Express your answer in the form $y(\pm s_y) = [m(\pm s_m)]x + [b(\pm s_b)]$, with a reasonable number of significant figures.

4-5 Constructing a Calibration Curve

Real data from a spectrophotometric analysis are given in Table 4-6. In this procedure, a color develops in proportion to the amount of protein in the sample. Color is measured by the absorbance of light recorded on a spectrophotometer. The first row in Table 4-6 shows readings obtained when no protein was present. The nonzero values arise from color in the reagents themselves. A result obtained with zero analyte is called a **blank** (or *reagent blank*), because it measures effects due to the analytical reagents. The second row shows three readings obtained with 5 μg of protein. Successive rows give results for 10, 15, 20, and 25 μg of protein. A solution containing a known quantity of analyte (or other reagent) is called a **standard solution.**

A **calibration curve** is a graph showing how the experimentally measured property (absorbance) depends on the known concentrations of the standards. To

TABLE 4-6   Spectrophotometer readings for protein analysis by the Lowry method

Sample (μg)	Absorbance of three independent samples			Range	Corrected absorbance (after subtracting average blank)			
0	0.099	0.099	0.100	0.001	-0.000_3	-0.000_3	0.000_7	
5	0.185	0.187	0.188	0.003	0.085_7	0.087_7	0.088_7	Data used
10	0.282	0.272	0.272	0.010	0.182_7	0.172_7	0.172_7	for calibration
15	0.392	0.345	0.347	0.047	—	0.245_7	0.247_7	curve
20	0.425	0.425	0.430	0.005	0.325_7	0.325_7	0.330_7	
25	0.483	0.488	0.496	0.013	0.383_7	0.388_7	0.396_7	

construct the calibration curve in Figure 4-7, we first subtract the average absorbance of the blanks (0.099_3) from those of the standards to obtain *corrected absorbance*. When all points are plotted in Figure 4-7 and a rough straight line is drawn through them, two features stand out:

Inspect your data and use judgment before mindlessly using a computer to draw a calibration curve!

1. One data point for 15 μg of protein (shown by the square in Figure 4-7) is ridiculously high. When we inspect the range of values for each set of 3 measurements in Table 4-6, we discover that the range for 15 μg is almost four times greater than the next greatest range. The absorbance value 0.392 is clearly out of line, so we discard it as "bad data." Perhaps the glassware was contaminated with protein from a previous experiment?

2. All three points at 25 μg lie slightly below the straight line through the remaining data. The only way to know whether or not this effect is real is to repeat the experiment. Many repetitions show that these points are consistently below the straight line. Therefore the *linear range* for this determination extends from 0 to 20 μg, but not to 25 μg.

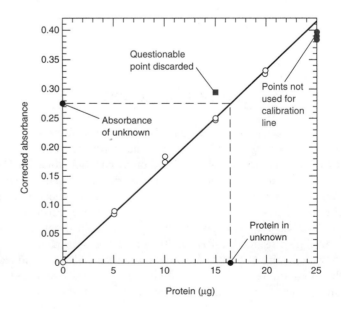

FIGURE 4-7 *Calibration curve* showing the average absorbance values in Table 4-6 versus micrograms of protein analyzed. The average *blank value* for 0 μg of protein has been subtracted from each point.

In view of these observations, we discard the 0.392 absorbance value and we do not use the three points at 25 μg for the least-squares straight line.

To construct the calibration curve (the straight line) in Figure 4-7, we use the method of least squares with $n = 14$ data points from Table 4-6 (including three blank values) covering the range 0 to 20 μg of protein. The results of applying Equations 4-9 through 4-14 are

$$m = 0.016\ 3_0 \qquad\qquad s_m = 0.000\ 2_2$$
$$b = 0.004_7 \qquad\qquad s_b = 0.002_6$$
$$s_y = 0.005_9$$

87

4-5 Constructing a
Calibration Curve

Equation of calibration line:
$$y\,(\pm s_y) = [m\,(\pm s_m)]x + [b\,(\pm s_b)]$$
$$y = [0.016\ 3_0\,(\pm 0.000\ 2_2)]x + [0.004_7\,(\pm 0.002_6)]$$

Finding the Protein in an Unknown

Suppose that the measured absorbance of an unknown sample is 0.373. How many micrograms of protein does it contain, and what uncertainty is associated with the answer?

The first question is easy. The equation of the calibration line is

$$y = mx + b = (0.016\ 3_0)x + 0.004_7$$

where y is corrected absorbance ($=$ observed absorbance $-$ blank absorbance) and x is micrograms of protein. If the absorbance of unknown is 0.373, its corrected absorbance is $0.373 - 0.099_3 = 0.273_7$. Plugging this value in for y in the equation above permits us to solve for x:

$$0.273_7 = (0.016\ 3_0)x + (0.004_7) \qquad\qquad (4\text{-}15a)$$

$$x = \frac{0.273_7 - 0.004_7}{0.016\ 3_0} = 16.50\ \mu\text{g of protein} \qquad\qquad (4\text{-}15b)$$

But what is the uncertainty in 16.50 μg?

The uncertainty in x in Equation 4-15 turns out to be:

$$\text{uncertainty in } x = \frac{s_y}{|m|}\sqrt{1 + \frac{x^2 n}{D} + \frac{\Sigma(x_i^2)}{D} - \frac{2x\Sigma x_i}{D}} \qquad (4\text{-}16)$$

where $|m|$ is the absolute value of the slope and D is given by Equation 4-11. If you measure k values of y, the uncertainty in x will be reduced. In this case, use the average value of y and change the first term from 1 to $1/k$ in the square root.

Applying Equation 4-16 to the problem at hand gives an uncertainty of ± 0.38 for x. The final answer is therefore expressed with a reasonable number of significant figures as

$$x = 16.5\,(\pm 0.4)\ \mu\text{g of protein}$$

Example: If 4 replicate samples of an unknown have an average absorbance of 0.376, use the value $0.376 - 0.099_3 = 0.276_7$ on the left side of Equation 4-15a. Use the value 1/4 in place of 1 for the first term of the square root in Equation 4-16 because you measured $k = 4$ replicate unknowns. The value of n in Equation 4-16 is 14 because there are 14 points on the calibration curve (Table 4-6).

Ask Yourself

4-E. Using your results from Problem 4-D, find the value of x (and its uncertainty) corresponding to $y = 1.00$.

4-6   A Spreadsheet for Least Squares

Figure 4-8 translates the least-squares computations of Table 4-5 into a spreadsheet. Values of x and y are entered in columns B and C, and the total number of points ($n = 4$) is entered in cell A10. The products xy and x^2 are computed in columns D and E. The sums of columns B through G appear on line 9. For example, the sum of xy values in cells D4 through D7 is computed with the formula " = Sum(D4:D7)" in cell D9. The least-squares parameters D, m, and b are computed in cells A12, A14, and A16 from Equations 4-9 through 4-11. The vertical deviations, d, in column F make use of the slope and intercept. Column G contains the squares of the deviations. Standard deviations s_y, s_m, and s_b are then computed in cells B12, B14, and B16, using Equations 4-12 through 4-14. As should always be your practice, formulas used in the spreadsheet are documented in Figure 4-8.

At the bottom of the spreadsheet, we use Equation 4-16 to evaluate uncertainty in a derived value of x. A measured value of y is entered in cell A22 and derived values of x and its uncertainty are computed in cells C22 and F22. Thus, for a measured value of $y = 2.72$ in Figure 4-6, the value of x is $2.2_3 \pm 0.3_7$.

Ask Yourself

4-F. Reproduce the spreadsheet in Figure 4-8 to solve linear least-squares problems. Use a graphics program to plot the observed data points and the least-squares straight line. (*Note:* Most graphics packages can draw the least-squares line automatically. Use this feature routinely to examine the quality of your data.)

	A	B	C	D	E	F	G
1	Least-Squares Spreadsheet						
2							
3	n	x	y	xy	x^2	d	d^2
4	1	1	2	2	1	0.038462	0.001479
5	2	3	3	9	9	-0.192308	0.036982
6	3	4	4	16	16	0.192308	0.036982
7	4	6	5	30	36	-0.038462	0.001479
8		----------------------Column Sums [Example: D9 = Sum(D4:D7)]----------------------					
9	n=	14	14	57	62	0.0000	0.076923
10	4						
11	D=	StdDev(y) =		A12: D = $A$10*E9-B9*B9			
12	52.0000	0.1961		A14: m = (D9*$A$10-B9*C9)/$A$12			
13	m=	StdDev(m) =		A16: b = (E9*C9-D9*B9)/$A$12			
14	0.6154	0.0544		F4 = C4-$A$14*B4-$A$16			
15	b=	StdDev(b) =		B12: StdDev(y) = SQRT(G9/($A$10-2))			
16	1.3462	0.2141		B14: StdDev(m) = B12*SQRT($A$10/$A$12)			
17				B16: StdDev(b) = B12*SQRT(E9/$A$12)			
18							
19	Propagation of uncertainty calculation with Equation 4-16:						
20							
21	Measured y =		Derived x =		Uncertainty in x =		
22	2.72		2.2325			0.3735	
23							
24	Derived x in cell C22 = ($A$22-$A$16)/$A$14						
25	Uncertainty in x in cell F22 =						
26	($B$12/$A$14)*SQRT(1 + (C22*C22*$A$10/$A/$12) + (E9/$A$12) - (2*C22*$B$9/$A$12))						

FIGURE 4-8 Spreadsheet for least-squares calculations.

Mean or average	$\bar{x} = \dfrac{1}{n}\sum_i x_i = \dfrac{1}{n}(x_1 + x_2 + x_3 + \cdots + x_n)$

$(x_i = $ individual observation, $n = $ number of observations$)$

Standard deviation	$s = \sqrt{\sum_i (x_i - \bar{x})^2/(n-1)}$ $(\bar{x} = $ average$)$

Confidence interval	$\mu = \bar{x} \pm \dfrac{ts}{\sqrt{n}}$ $(\mu = $ true mean$)$

Student's t comes from Table 4-2 for $n - 1$ degrees of freedom at the selected confidence level.

t test	$t = \dfrac{\bar{x}_1 - \bar{x}_2}{s_{\text{pooled}}}\sqrt{\dfrac{n_1 n_2}{n_1 + n_2}}$ $s_{\text{pooled}} = \sqrt{\dfrac{s_1^2(n_1 - 1) + s_2^2(n_2 - 1)}{n_1 + n_2 - 2}}$

If $t_{\text{calculated}} > t_{\text{table}}$ (for 95% confidence and $n_1 + n_2 - 2$ degrees of freedom), the difference is significant.

Q test	$Q = $ gap/range. You should know how to apply this equation for rejection of questionable data.
Straight line	$y = mx + b$
	$(m = $ slope $= \Delta y/\Delta x, b = y$-intercept$)$
Least-squares equations	You should know how to use Equations 4-9 through 4-14 to derive least-squares slope and intercept and uncertainties.
Calibration curve	You should be able to use Equation 4-16 to find the uncertainty in a result derived from a calibration curve.

Important Terms

average	mean	standard solution
blank	method of least squares	Student's t
calibration curve	Q test	t test
confidence interval	slope	y-intercept
Gaussian distribution	standard deviation	

Problems

4-1. What is the relation between the standard deviation and the precision of a procedure? What is the relation between standard deviation and accuracy?

4-2. What fraction of observations in an ideal Gaussian distribution lie within $\mu \pm \sigma$? Within $\mu \pm 2\sigma$? Within $\mu \pm 3\sigma$?

4-3. The ratio of the number of atoms of the isotopes ^{69}Ga

and ^{71}Ga in samples from different sources is listed below:

Sample	^{69}Ga/^{71}Ga	Sample	^{69}Ga/^{71}Ga
1	1.526 60	5	1.528 94
2	1.529 74	6	1.528 04
3	1.525 92	7	1.526 85
4	1.527 31	8	1.527 93

(a) Find the mean value of $^{69}Ga/^{71}Ga$.

(b) Find the standard deviation and relative standard deviation.

(c) Sample 8 was analyzed seven times, with $\bar{x} = 1.527\ 93$ and $s = 0.000\ 07$. Find the 99% confidence interval for sample 8.

4-4. What is the meaning of a confidence interval?

4-5. For the numbers 116.0, 97.9, 114.2, 106.8, and 108.3, find the mean, standard deviation, and 90% confidence interval for the mean.

4-6. The calcium content of a mineral was analyzed five times by each of two methods. Are the mean values significantly different at the 95% confidence level?

Method	Ca (wt %)				
1	0.027 1	0.028 2	0.027 9	0.027 1	0.027 5
2	0.027 1	0.026 8	0.026 3	0.027 4	0.026 9

4-7. Find the 95% and 99% confidence intervals for the mean mass of nitrogen from chemical sources given in Table 4-3.

4-8. Two methods were used to measure the specific activity (units of enzyme activity per milligram of protein) of an enzyme. One unit of enzyme activity is defined as the amount of enzyme that catalyzes the formation of one micromole of product per minute under specified conditions.

Method	Enzyme activity (five replications)				
1	139	147	160	158	135
2	148	159	156	164	159

Is the mean value of method 1 significantly different from the mean value of method 2 at the 95% confidence level?

4-9. Students measured the concentration of HCl in a solution by various titrations in which different indicators were used to find the end point.

Indicator	Mean HCl concentration (M) ($\pm$ standard deviation)	Number of measurements
Bromothymol blue	0.095 65 $\pm$ 0.002 25	28
Methyl red	0.086 86 $\pm$ 0.000 98	18
Bromocresol green	0.086 41 $\pm$ 0.001 13	29

Is the difference between indicators 1 and 2 significant at the 95% confidence level? Answer the same question for indicators 2 and 3.

4-10. The calcium content of a person's urine was determined on two different days.

Day	[Ca] (mg/L) Average $\pm$ standard deviation	Number of measurements
1	238 $\pm$ 8	4
2	255 $\pm$ 10	5

Are the average values significantly different at the 95% confidence level?

4-11. Using the Q test, decide whether the value 0.195 should be rejected from the set of results 0.217, 0.224, 0.195, 0.221, 0.221, 0.223.

4-12. Students at the University of North Dakota measured visible light absorbance of various food colorings. Replicate measurements of the absorbance of a solution of the drink Kool-Aid at a wavelength of 502 nm gave the following values: 0.189, 0.169, 0.187, 0.183, 0.186, 0.182, 0.181, 0.184, 0.181, and 0.177. Identify the outlier and decide whether or not to exclude it from the data set.

4-13. Find the values of m and b in the equation $y = mx + b$ for the straight line going through the points $(x_1, y_1) = (6, 3)$ and $(x_2, y_2) = (8, -1)$. You can do this by writing

$$m = \frac{\Delta y}{\Delta x} = \frac{(y_2 - y_1)}{(x_2 - x_1)} = \frac{(y - y_1)}{(x - x_1)}$$

and rearranging to the form $y = mx + b$. Sketch the curve and satisfy yourself that the value of b is a sensible y-intercept.

4-14. A straight line is drawn through the points $(3.0, -3.87 \times 10^4)$, $(10.0, -12.99 \times 10^4)$, $(20.0, -25.93 \times 10^4)$, $(30.0, -38.89 \times 10^4)$, and $(40.0, -51.96 \times 10^4)$ by using the method of least squares. The results are $m = -1.298\ 72 \times 10^4$, $b = 256.695$, $s_m = 13.190$, $s_b = 323.57$, and $s_y = 392.9$. Express the slope and intercept and their uncertainties with the correct significant figures.

4-15. Consider the least-squares problem illustrated in Figure 4-6. Suppose that a single new measurement produces a y value of 2.58.

(a) Calculate the corresponding x value and its uncertainty.

(b) Suppose you measure y four times and the average is 2.58. Calculate the uncertainty in x based on four measurements, not one.

4-16. In a common protein analysis, a dye binds to the protein and the color of the dye changes from brown to blue. The intensity of blue color is proportional to the amount of protein present.

Protein (μg):	0.00	9.36	18.72	28.08	37.44
Absorbance:	0.466	0.676	0.883	1.086	1.280

(a) After subtracting the blank absorbance (0.466) from the remaining absorbances, use the method of least squares to determine the equation of the best straight line through these 5 points ($n = 5$). Use the standard deviation of the slope and intercept to express the equation in the form $y (\pm s_y) = [m (\pm s_m)]x + [b (\pm s_b)]$ with a reasonable number of significant figures.

(b) Make a graph showing the experimental data and the calculated straight line.

(c) An unknown gave an observed absorbance of 0.973. Calculate the number of micrograms of protein in the unknown, and estimate its uncertainty.

4-17. *Using statistical functions in a spreadsheet.* Almost every spreadsheet has functions such as the average and standard deviation built into it. In Excel, these functions are called "Average" and "Stdev." To try using them, enter the numbers 1, 2, 3, and 4 in cells A1 through A4 of a blank spreadsheet. In cell A5, enter the formula "= Average(A1:A4)"; and in cell A6, enter the formula "Stdev(A1:A4)". Use the average and standard deviation functions on your calculator to check the results in cells A5 and A6. Search the menus of your spreadsheet to find a list of the many many functions that are available for you to use.

4-18. The equation for a Gaussian curve is

$$y = \frac{1}{\sigma \sqrt{2\pi}} e^{-(x-\mu)^2/2\sigma^2}$$

The square root function on most spreadsheets is Sqrt and the exponential function is Exp. To find $e^{-3.4}$, write Exp(-3.4). The exponential function above is written

$$e^{-(x-\mu)^2/2\sigma^2} = \text{Exp}(-(x - \mu)\text{^}2/(2*\sigma\text{^}2))$$

(a) Suppose that $\mu = 10.0$ and $\sigma = 1.0$. Use a spreadsheet to compute values of y for the range $4 \leq x \leq 16$. For x, use the values 4.0, 4.4, 4.8, . . . , 16.0.

(b) Repeat the calculation for $\sigma = 2.0$.

(c) Use graphics software to plot the results of (a) and (b) on one graph.

(d) For the curve in (c) with $\sigma = 2.0$, shade in the region that should contain 68.3% of all observations. Use different shading to show the region that should contain 95.5% of observations.

How Would You Do It?

4-19. Students at Eastern Illinois University[3] intended to prepare copper(II) carbonate by adding a solution of $CuSO_4 \cdot 5H_2O$ to a solution of Na_2CO_3.

$$CuSO_4 \cdot 5H_2O(aq) + Na_2CO_3(aq) \longrightarrow$$
$$\underset{\text{Copper(II) carbonate}}{CuCO_3(s)} + Na_2SO_4(aq) + 5H_2O(l)$$

After warming the mixture to 60°C, the gelatinous blue precipitate coagulated into an easily filterable pale green solid. The product was filtered, washed, and dried at 70°C. Copper in the product was measured by heating 0.4 g of solid in a stream of methane at high temperature to reduce the solid to pure Cu, which was weighed.

$$4CuCO_3(s) + CH_4(g) \xrightarrow{\text{heat}}$$
$$4Cu(s) + 5CO_2(g) + 2H_2O(g)$$

In 1995, 43 students found a mean value of 55.6 wt % Cu with a standard deviation of 2.7 wt %. In 1996, 39 students found 55.9 wt % with a standard deviation of 3.8 wt %. The instructor tried the experiment 9 times and measured 55.8 wt % with a standard deviation of 0.5 wt %. Was the product of the reaction probably $CuCO_3$? Could it have been a hydrate, $CuCO_3 \cdot xH_2O$?

Notes and References

1. L. H. Keith, W. Crummett, J. Deegan, Jr., R. A. Libby, J. K. Taylor, and G. Wentler, *Anal. Chem.* **1983**, *55*, 2210.

2. Strictly speaking, the test we described for comparing means is only valid if the standard deviations of the two methods are similar to each other. This is clearly not the case in Figure 4-4, where the data for nitrogen from air has a very small spread.

The more correct way to compare the means is beyond the scope of this text, but it is described in that marvelous text by D. C. Harris, *Quantitative Chemical Analysis,* 5th ed. (New York: W. H. Freeman, 1999), Equations 4-8a and 4-9a.

3. D. Sheeran, *J. Chem. Ed.* **1998**, *75*, 453. See also H. Gamsjäger and W. Preis, *J. Chem. Ed.* **1999**, *76*, 1339.

Further Reading

J. C. Miller and J. N. Miller, *Statistics for Analytical Chemistry,* 2nd ed. (Chichester, England: Ellis Horwood, 1992).

The Earliest Known Buret

Good Titrations—Ode to a Lab Partner
Kurt Wood and Jeff Lederman
(*University of California, Davis, 1977*)
(Sung to the tune of *Good Vibrations* by the Beach Boys)

Ah! I love the color of pink you get,
And the way the acid drips from your buret.
All the painful things in life seem alien
As I mix in several drops of phenolphthalein.
   I'm pickin' up good titrations
   She's given' me neutralizations
      Dew drop drop, good titrations . . .

I look at you and drift away;
The ruby red turns slowly to rosé.
You gaze at me and light my fire
As drop on drop falls to the Erlenmeyer . . .
   I'm pickin' up good titrations
   She's given' me neutralizations
      Dew drop drop, good titrations . . .

I look longingly to your eyes,
But you stare down at the lab bench in surprise.
Gone our love before it grew much sweeter,
'Cause we passed the end point by a milliliter.
   I'm pickin' up back titrations
   She's prayin' for neutralizations
      Dew drop drop, good titrations . . .

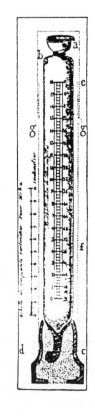

The buret, invented by F. Descroizilles in the early 1800s, was used in the same manner as a graduated cylinder is used today. The stopcock was introduced in 1846. This buret and its progeny have terrorized generations of analytical chemistry students. [From E. R. Madsen, *The Development of Titrimetric Analysis 'till 1806* (Copenhagen: E. E. C. Gad Publishers, 1958).]

Chapter **5**

GOOD TITRATIONS

*I*n **volumetric analysis,** the volume of a known reagent needed to react with analyte is measured. From this volume, we calculate how much analyte is in the unknown. In this chapter we discuss general principles that apply to any volumetric procedure, and then we illustrate some analyses based on precipitation reactions. Along the way, we introduce the solubility product as a means of understanding precipitation reactions.

5-1 Principles of Volumetric Analysis

A **titration** is a procedure in which increments of the known reagent solution—the **titrant**—are added to *analyte* until the reaction is complete. Titrant is usually delivered from a buret, as shown in Figure 5-1. Each increment of titrant should be completely and quickly consumed by reaction with analyte until the analyte is used up. Common titrations are based on acid-base, oxidation-reduction, complex formation, or precipitation reactions.

Methods of determining when analyte has been consumed include (1) detecting a sudden change in voltage or current between a pair of electrodes, (2) observing an

We will study end point detection methods later:

electrodes: Section 9-4 and Chapter 14

indicators: Sections 5-6, 8-6, 9-4, 12-3, and 15-2

absorbance: Section 18-3

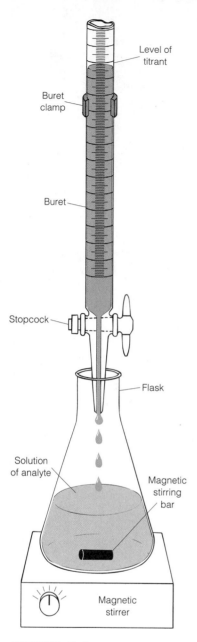

FIGURE 5-1 Typical setup for a titration. The analyte is contained in the flask, and the titrant in the buret. The stirring bar is a magnet coated with Teflon, which is inert to almost all solutions. The bar is spun by a rotating magnet inside the stirrer.

Question When we use the purple color of permanganate to locate an end point, is the titration error systematic or random?

indicator color change (Color Plate 1), and (3) monitoring the absorbance of light by species in the reaction. An **indicator** is a compound with a physical property (usually color) that changes abruptly when the titration is complete. The change is caused by the disappearance of analyte or the appearance of excess titrant.

The **equivalence point** occurs when the quantity of titrant added is the exact amount necessary for stoichiometric reaction with the analyte. For example, five moles of oxalic acid react with two moles of permanganate in hot acidic solution:

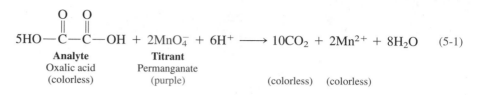

$$5HO\!-\!\overset{\overset{O}{\|}}{C}\!-\!\overset{\overset{O}{\|}}{C}\!-\!OH \;+\; 2MnO_4^- \;+\; 6H^+ \longrightarrow 10CO_2 \;+\; 2Mn^{2+} \;+\; 8H_2O \qquad (5\text{-}1)$$

Analyte   **Titrant**
Oxalic acid Permanganate
(colorless)   (purple)   (colorless)   (colorless)

If the unknown contains 5.00 mmol of oxalic acid, the equivalence point is reached when 2.00 mmol of MnO_4^- have been added.

The equivalence point is the ideal result we seek in a titration. What we actually measure is the **end point,** which is marked by a sudden change in a physical property of the solution. For Reaction 5-1, a convenient end point is the abrupt appearance of the purple color of permanganate in the flask. Up to the equivalence point, all the added permanganate is consumed by oxalic acid, and the titration solution remains colorless. After the equivalence point, unreacted MnO_4^- ion builds up until there is enough to see. The *first trace* of purple color marks the end point. The better your eyes, the closer the measured end point will be to the true equivalence point. The end point cannot exactly equal the equivalence point because extra MnO_4^-, more than that needed to react with oxalic acid, is required to create perceptible purple color.

The difference between the end point and the equivalence point is an inescapable **titration error.** By choosing an appropriate physical property, in which a change is easily observed (such as indicator color, optical absorbance of a reactant or product, or pH), it is possible to have an end point that is very close to the equivalence point. It is also possible to estimate the titration error with a **blank titration,** in which the same procedure is carried out without analyte. For example, a solution containing no oxalic acid could be titrated with MnO_4^- to see how much titrant is needed to create observable purple color. This volume of MnO_4^- is then subtracted from the volume observed in the titration of unknown.

The validity of an analytical result depends on knowing the amount of one of the reactants used. A **primary standard** is a reagent that is pure enough to be weighed out and used directly to provide a known number of moles. For example, if you want to titrate unknown hydrochloric acid with base, you could weigh out primary standard grade sodium carbonate and dissolve it in water to make titrant:

$$2HCl \;+\; Na_2CO_3 \longrightarrow H_2CO_3 \;+\; 2NaCl \qquad (5\text{-}2)$$

        Sodium carbonate
        FM 105.99
(unknown)   (primary standard)

Two moles of HCl react with one mole of Na_2CO_3, which has a mass of 105.99 g. You could not carry out the same procedure beginning with solid NaOH

$$HCl \;+\; NaOH \longrightarrow H_2O \;+\; NaCl$$

Sodium hydroxide
FM 40.00

because the solid is not pure. NaOH is normally contaminated with some Na_2CO_3 (from reaction with CO_2 in the air) and H_2O (also from the air). If you weighed out 40.00 g of sodium hydroxide, it would not contain exactly one mole.

A primary standard should be 99.9% pure, or better. It should not decompose under ordinary storage, and it should be stable when dried by heating or vacuum, because drying is required to remove traces of adsorbed water from the atmosphere.

In most cases, titrant is not available as a primary standard. Instead, a solution having approximately the desired concentration is used to titrate a weighed, primary standard. From the volume of titrant required to react with the primary standard, we calculate the concentration of titrant. The process of titrating a standard to determine the concentration of titrant is called **standardization.** We say that a solution whose concentration is known is a *standard solution*. The validity of the analytical result ultimately depends on knowing the composition of some primary standard.

In a **direct titration,** titrant is added to analyte until the end point is observed.

Direct titration: analyte   +   titrant $\longrightarrow$ product
                      Unknown   Known

In a **back titration,** a known *excess* of a standard reagent is added to the analyte. Then a second standard reagent is used to titrate the excess of the first reagent.

analyte   +   reagent 1 $\longrightarrow$ product   +   excess reagent 1   (5-3a)
Unknown   Known   Unknown quantity

Back titration: excess reagent 1   +   reagent 2 $\longrightarrow$ product   (5-3b)
                    Unknown   Known

Back titrations are useful when the end point of the back titration is clearer than the end point of the direct titration or when an excess of the first reagent is required for complete reaction with the analyte.

The addition of permanganate titrant to oxalic acid analyte in Reaction 5-1 is a direct titration. As an example of a back titration, an *excess* (but known amount) of permanganate could be added to consume the oxalic acid. Then the excess permanganate might be back-titrated with standard Fe^{2+} to measure how much permanganate was left unreacted. The difference between the initial quantity of permanganate and the amount left unreacted tells us how much was consumed by the oxalic acid.

Box 3-1 describes Standard Reference Materials that allow different laboratories to test the accuracy of their procedures.

Ask Yourself

5-A. (a) Why does the validity of an analytical result ultimately depend on knowing the composition of a primary standard?

(b) How does a blank titration reduce titration error?

(c) What is the difference between a direct titration and a back titration?

(d) Suppose that the uncertainty in locating the equivalence point in a titration is ±0.04 mL. Why is it more accurate to use enough of an unknown to require ~40 mL in a titration with a 50-mL buret instead of ~20 mL?

For interpreting the results of a direct titration, the key steps are

1. From the volume of titrant, calculate the number of moles of titrant consumed.

2. From the stoichiometry of the titration reaction, relate the unknown moles of analyte to the known moles of titrant.

1:1 Stoichiometry

Consider the titration of an unknown chloride solution with standard Ag^+:

When K is large, we sometimes write $\rightarrow$ instead of $\rightleftharpoons$:

$$Ag^+ + Cl^- \longrightarrow AgCl(s)$$

$$Ag^+ + Cl^- \overset{K}{\rightleftharpoons} AgCl(s) \qquad (5\text{-}4)$$

The reaction is rapid and the equilibrium constant is large ($K = 5.6 \times 10^9$), so the reaction essentially goes to completion with each addition of titrant. White AgCl precipitate forms as soon as the two reagents are mixed.

Suppose that 10.00 mL of unknown chloride solution (measured with a transfer pipet) requires 22.97 mL of 0.052 74 M $AgNO_3$ solution (delivered from a buret) for a complete reaction. What is the concentration of Cl^- in the unknown?

Following the two-step recipe, we first find the moles of Ag^+:

Retain an extra, insignificant digit until the end of the calculations to avoid round-off error.

$$\text{mol } Ag^+ = \text{volume} \times \text{molarity} = (0.022\ 97\ L)\left(0.052\ 74\ \frac{\text{mol}}{L}\right) = 0.001\ 211_4\ \text{mol}$$

Next, relate the unknown moles of Cl^- to the known moles of Ag^+. We know that 1 mol of Cl^- reacts with 1 mol of Ag^+. If $0.001\ 211_4$ mol of Ag^+ is required, then $0.001\ 211_4$ mol of Cl^- must have been in 10.00 mL of unknown. Therefore

$$[Cl^-] \text{ in unknown} = \frac{\text{mol } Cl^-}{L \text{ of unknown}} = \frac{0.001\ 211_4\ \text{mol}}{0.010\ 00\ L} = 0.121\ 1_4\ M$$

The solution we just titrated was made by dissolving 1.004 g of an unknown solid in a total volume of 100.0 mL. What is the weight percent of chloride in the solid? We know that 10.00 mL of unknown solution contain $0.001\ 211_4$ mol of Cl^-. Therefore 100.0 mL must contain 10 times as much, or $0.012\ 11_4$ mol of Cl^-. This much Cl^- weighs $(0.012\ 11_4\ \text{mol } Cl^-)(35.452\ 7\ \text{g/mol } Cl^-) = 0.429\ 4_7\ \text{g } Cl^-$. The weight percent of Cl^- in the unknown is

$$\text{wt \% } [Cl^-] = \frac{\text{g } Cl^-}{\text{g unknown}} \times 100 = \frac{0.429\ 4_7\ \text{g } Cl^-}{1.004\ \text{g unknown}} \times 100 = 42.78\ \text{wt \%}$$

EXAMPLE   A Case Involving a Dilution

(a) Standard Ag^+ solution was prepared by dissolving 1.224 3 g of dry $AgNO_3$ (FM 169.87) in water in a 500.0-mL volumetric flask. A dilution was made by delivering 25.00 mL of solution with a pipet to a second 500.0-mL volumetric flask and diluting to the mark. Find the concentration of Ag^+ in the dilute standard solution. **(b)** A 25.00-mL aliquot of unknown solution containing Cl^- was titrated with the dilute Ag^+ solution, and the equivalence point was reached when 37.38 mL of Ag^+ solution had been delivered. Find the concentration of Cl^- in the unknown.

SOLUTION **(a)** The concentration of the initial $AgNO_3$ solution is

$$[Ag^+] = \frac{(1.224\ 3\ g)/(169.868\ 2\ g/mol)}{0.500\ 0\ L} = 0.014\ 41_5\ M$$

To find the concentration of the dilute solution, use the dilution formula 1-5:

$$[Ag^+]_{conc} \cdot V_{conc} = [Ag^+]_{dil} \cdot V_{dil} \qquad (1\text{-}5)$$

$$(0.014\ 41_5\ M) \cdot (25.00\ mL) = [Ag^+]_{dil} \cdot (500.0\ mL)$$

$$[Ag^+]_{dil} = \left(\frac{25.00\ mL}{500.0\ mL}\right)(0.014\ 41_5\ M) = 7.207_3 \times 10^{-4}\ M$$

Notice that the general form of all dilution problems is

$$[X]_{final} = \frac{V_{initial}}{V_{final}} \cdot [X]_{initial}$$

(b) One mole of Cl^- requires one mole of Ag^+. The number of moles of Ag^+ required to reach the equivalence point is

$$mol\ Ag^+ = (7.207_3 \times 10^{-4}\ M)(0.037\ 38\ L) = 2.694_1 \times 10^{-5}\ mol$$

The concentration of Cl^- in 25.00 mL of unknown is therefore

$$[Cl^-] = \frac{2.694_1 \times 10^{-5}\ mol}{0.025\ 00\ L} = 1.078 \times 10^{-3}\ M$$

Silver ion titrations are especially nice because $AgNO_3$ is a primary standard. After drying at 110°C for 1 h to remove residual moisture, the solid has the exact composition $AgNO_3$. Methods for finding the end point in silver titrations are described in Section 5-6. Silver compounds and solutions should be stored in the dark to protect against photodecomposition and should never be exposed to direct sunlight.

$AgNO_3$ solution is an antiseptic. If you spill $AgNO_3$ on yourself, your skin will turn black for a few days until the affected skin is shed.

Photodecomposition of white $AgCl(s)$:

$$AgCl(s) \xrightarrow{light} Ag(s) + \tfrac{1}{2}Cl_2(g)$$

Finely divided $Ag(s)$ imparts a faint violet color to the solid.

x:y Stoichiometry

In Reaction 5-1, five moles of oxalic acid ($H_2C_2O_4$) react with two moles of permanganate (MnO_4^-). If an unknown quantity of $H_2C_2O_4$ consumed 2.345×10^{-4} mol of MnO_4^-, there must have been

$$mol\ H_2C_2O_4 = (mol\ MnO_4^-)\left(\frac{5\ mol\ H_2C_2O_4}{2\ mol\ MnO_4^-}\right)$$

$$= (2.345 \times 10^{-4}\ mol\ MnO_4^-)\left(\frac{5\ mol\ H_2C_2O_4}{2\ mol\ MnO_4^-}\right) = 5.862 \times 10^{-4}\ mol$$

Ask Yourself

5-B. Vitamin C (ascorbic acid) from foods can be measured by titration with I_3^-:

$$\underset{\substack{\text{Ascorbic acid} \\ \text{FM 176.126}}}{C_6H_8O_6} + \underset{\text{Triiodide}}{I_3^-} + H_2O \longrightarrow C_6H_8O_7 + 3I^- + 2H^+$$

Starch is used as an indicator in the reaction. The end point is marked by the appearance of a deep blue starch-iodine complex when unreacted I_3^- is present.

   (a) If 29.41 mL of I_3^- solution are required to react with 0.197 0 g of pure ascorbic acid, what is the molarity of the I_3^- solution?

   (b) A vitamin C tablet containing ascorbic acid plus inert binder was ground to a powder, and 0.424 2 g was titrated by 31.63 mL of I_3^-. How many moles of vitamin C are present in the 0.424 2-g sample?

   (c) Find the weight percent of ascorbic acid in the tablet.

Linda A. Hughes

5-3 Chemistry in a Fishtank

Students of Professor Kenneth Hughes at Georgia Institute of Technology study analytical chemistry by measuring chemical changes in a saltwater aquarium in their laboratory. One of the chemicals measured is nitrite, NO_2^-, which is a key species in the natural cycle of nitrogen (Figure 5-2). Box 5-1 shows concentrations of ammonia (NH_3), nitrite, and nitrate (NO_3^-) measured in the aquarium. Concentrations are expressed in parts per million of nitrogen, which means micrograms (μg) of

NH_3 (FM 17.031) contains 82.24 wt % N. Therefore, a solution containing 1.216 g NH_3/mL contains

$$\left(1.216\ \frac{g\ NH_3}{mL}\right)\left(0.822\ 4\ \frac{g\ N}{g\ NH_3}\right) =$$
$$1.000\ \mu g\ N/mL = 1\ ppm\ N$$

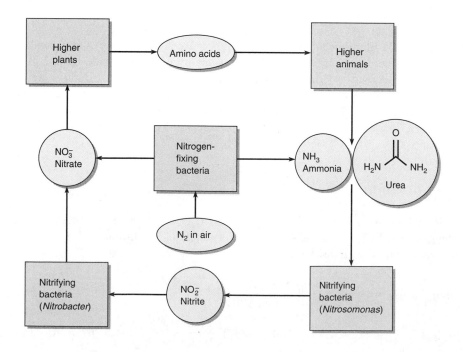

FIGURE 5-2 Nitrogen is exchanged among different forms of life through the *nitrogen cycle*. Only a few organisms, such as blue–green algae, are able to use N_2 directly from the air. Our existence depends on the health of all organisms in the nitrogen cycle.

nitrogen per gram of seawater. Because 1 g of water $\approx$ 1 mL, we will consider 1 ppm to be 1 μg/mL.

The nitrite measurement was done by a spectrophotometric procedure described in Section 17-4. Because there is no convenient primary standard for nitrite, a titration is used to standardize a $NaNO_2$ solution that serves as the standard for the spectrophotometric procedure. As always, the validity of any analytical procedure ultimately depends on knowing the composition of a primary standard, which is sodium oxalate in this case.

Three solutions are required for the measurement of nitrite:

a. Prepare ~0.018 M $NaNO_2$ (FM 68.995) by dissolving 1.25 g of $NaNO_2$ in 1.00 L of distilled water. Standardization of this solution is described later.

b. Prepare ~0.025 00 M $Na_2C_2O_4$ (FM 134.00) by dissolving ~3.350 g of primary standard grade $Na_2C_2O_4$ in water and diluting to 1.000 L. It is not necessary to weigh out exactly 3.350 g. What is important is that you know exactly how much you weigh out so that you can calculate the molarity of the reagent.

c. Prepare ~0.010 M $KMnO_4$ (FM 158.03) by dissolving 1.6 g of $KMnO_4$ in 1.00 L of distilled water. $KMnO_4$ is not pure enough to be a primary standard. Also, traces of organic impurities in distilled water consume some of the freshly dissolved MnO_4^- to produce $MnO_2(s)$. Therefore, dissolve $KMnO_4$ in distilled water to give the approximately desired concentration and boil the solution for 1 h to complete the reaction between MnO_4^- and organic impurities. Filter the mixture through a sintered-glass filter (not a paper filter, which is organic) to remove $MnO_2(s)$, cool the solution, and standardize it against primary standard sodium oxalate ($Na_2C_2O_4$):

The concentrations of $NaNO_2$ and $KMnO_4$ are only approximate. The solutions will be standardized in subsequent titrations.

$$5C_2O_4^{2-} + 2MnO_4^- + 16H^+ \longrightarrow 10CO_2 + 2Mn^{2+} + 8H_2O \qquad (5\text{-}5)$$
Oxalate   Permanganate

For best results, treat the oxalic acid solution at 25°C with 90–95% of the expected volume of $KMnO_4$. Then heat the solution to 60°C and complete the titration.

EXAMPLE   **Standardizing $KMnO_4$ by a Direct Titration of Oxalate**

Standard oxalate solution was prepared by dissolving 0.356 2 g of $Na_2C_2O_4$ (FM 134.00) in 1.0 M H_2SO_4 in a 250.0-mL volumetric flask. A 10.00-mL aliquot of this solution required 48.39 mL of $KMnO_4$ solution for titration, and a blank titration required 0.03 mL of $KMnO_4$. Find the molarity of $KMnO_4$.

SOLUTION The number of moles of $Na_2C_2O_4$ dissolved in 250.0 mL is 0.356 2 g/134.00 g/mol = 0.002 658$_2$ mol. The concentration of oxalate solution is

$$[Na_2C_2O_4] = \frac{0.002\ 658_2\ \text{mol}}{0.250\ 0\ \text{L}} = 0.010\ 63_3\ \text{M}$$

Box 5-1 *Informed Citizen*

Studying a Marine Ecosystem

A saltwater aquarium was set up at Georgia Tech to study the chemistry of a marine ecosystem. When fish and food were introduced into the aquarium, bacteria began to grow and to metabolize organic compounds into ammonia (NH_3). Ammonia is toxic to marine animals when the level exceeds 1 ppm; but, fortunately, it is removed by *Nitrosomonas* bacteria, which oxidize ammonia to nitrite (NO_2^-). Alas, nitrite is also toxic at levels above 1 ppm, but it is further oxidized to nitrate (NO_3^-) by *Nitrobacter* bacteria:

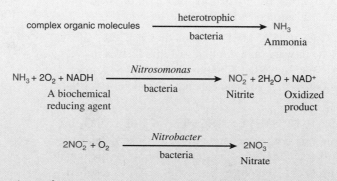

$$\text{complex organic molecules} \xrightarrow[\text{bacteria}]{\text{heterotrophic}} \underset{\text{Ammonia}}{NH_3}$$

$$\underset{\substack{\text{A biochemical}\\\text{reducing agent}}}{NH_3 + 2O_2 + NADH} \xrightarrow[\text{bacteria}]{\textit{Nitrosomonas}} \underset{\text{Nitrite}}{NO_2^-} + 2H_2O + \underset{\substack{\text{Oxidized}\\\text{product}}}{NAD^+}$$

$$2NO_2^- + O_2 \xrightarrow[\text{bacteria}]{\textit{Nitrobacter}} \underset{\text{Nitrate}}{2NO_3^-}$$

Some actions of nitrogen-metabolizing bacteria. Heterotrophic bacteria require complex organic molecules from the breakdown of other organisms for nourishment. By contrast, autotrophic bacteria can utilize CO_2 as their carbon source for biosynthesis.

The graph shows ammonia and nitrite concentrations observed by students at Georgia Tech. About 18 days after introducing fish, significant levels of ammonia were observed. The first dip in NH_3 concentration occurred during a 48-h period when no food was added to the tank. The third peak in NH_3 concentration arose when changing flow patterns through the aquarium filter exposed fresh surfaces devoid of the bacteria that remove ammonia. When the population of *Nitrosomonas* bacteria was sufficient, ammonia levels decreased but nitrite became perilously high. After 60 days, the population of *Nitrobacter* bacteria was great enough to convert most nitrite to nitrate.

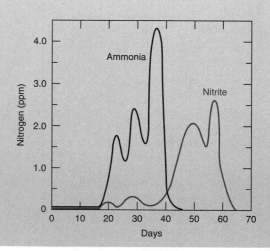

Ammonia and nitrite concentrations in a saltwater aquarium at Georgia Tech after fish and food were introduced into the "sterile" tank, beginning on day 0. Concentrations are expressed in parts per million of nitrogen (i.e., μg of N per mL of solution). [From K. D. Hughes, *Anal. Chem.* **1993**, *65*, 883A.]

The $C_2O_4^{2-}$ in 10.00 mL is $(0.010\ 63_3\ M)(0.010\ 00\ L) = 1.063_3 \times 10^{-4}$ mol. Reaction 5-5 requires 2 mol of permanganate for 5 mol of oxalate. Therefore

$$\text{mol MnO}_4^- = (\text{mol C}_2\text{O}_4^{2-})\left(\frac{2\ \text{mol MnO}_4^-}{5\ \text{mol C}_2\text{O}_4^{2-}}\right) = 4.253_1 \times 10^{-5}\ \text{mol}$$

The equivalence volume of $KMnO_4$ is $48.39 - 0.03 = 48.36$ mL. The concentration of MnO_4^- titrant is

$$[\text{MnO}_4^-] = \frac{4.253_1 \times 10^{-5}\ \text{mol}}{0.048\ 36\ \text{L}} = 8.794_7 \times 10^{-4}\ \text{M}$$

You can use either ratio,

$$\frac{5\ \text{mol C}_2\text{O}_4^{2-}}{2\ \text{mol MnO}_4^-} \quad \text{or} \quad \frac{2\ \text{mol MnO}_4^-}{5\ \text{mol C}_2\text{O}_4^{2-}}$$

whenever you please, as long as the units work out.

Here is the procedure for standardizing the solution of $NaNO_2$ prepared in step (a):

1. Pipet 50.00 mL of standard $KMnO_4$, ~5 mL of concentrated (18 M) H_2SO_4, and 50.00 mL of $NaNO_2$ solution into a glass-stoppered flask. Mix gently and warm to 70°–80°C on a hot plate. At this point, the nitrite is consumed and there is excess permanganate in the solution:

$$5\text{NO}_2^- + 2\text{MnO}_4^- + 6\text{H}^+ \longrightarrow 5\text{NO}_3^- + 2\text{Mn}^{2+} + 3\text{H}_2\text{O} \qquad (5\text{-}6)$$

2. Add standard $Na_2C_2O_4$ solution in 10.00-mL aliquots until the purple color of MnO_4^- disappears. Now there is excess $H_2C_2O_4$.

3. *Back-titrate* the excess $H_2C_2O_4$ in the warm solution with standard $KMnO_4$.

4. Perform a *blank titration* with water in place of the 50 mL of $KMnO_4$, the 50 mL of $NaNO_2$, and the $Na_2C_2O_4$. The purpose is to see how much $KMnO_4$ must be added to observe purple color.

Knowing how much $Na_2C_2O_4$ and how much $KMnO_4$ were used allows us to find the concentration of $NaNO_2$.

EXAMPLE   Standardizing NaNO₂ by a Back Titration

In step 1, 50.00 mL of 0.010 54 M $KMnO_4$, 5 mL of H_2SO_4, and 50.00 mL of $NaNO_2$ solution were mixed and warmed to take Reaction 5-6 to completion. Addition of 10.00 mL of 0.025 00 M $Na_2C_2O_4$ in step 2 did not decolorize the permanganate after a few minutes of heating. Addition of a second 10.00-mL aliquot of $Na_2C_2O_4$ left the solution colorless, meaning that permanganate was used up and there was excess oxalate. In step 3, the excess oxalate was back-titrated, which required 2.11 mL of 0.010 54 M $KMnO_4$ to give perceptible purple color. The blank titration in step 4 required 0.06 mL of $KMnO_4$. This volume should be subtracted from the volume required in step 3, which is therefore $2.11 - 0.06 = 2.05$ mL. Find the concentration of $NaNO_2$ solution.

SOLUTION This problem is not nearly as complicated as it seems. Our strategy is to compute the total moles of $KMnO_4$ used and subtract the moles that reacted with $Na_2C_2O_4$. The remainder must have reacted with $NaNO_2$. So here goes:

total volume of $KMnO_4$ = 50.00 (step 1) + 2.05 (steps 3 and 4) = 52.05 mL

total moles of $KMnO_4$ = (0.010 54 M)(0.052 05 L) = $5.486_0 \times 10^{-4}$ mol

volume of $Na_2C_2O_4$ = 20.00 mL

moles of $Na_2C_2O_4$ = (0.025 00 M)(0.020 00 L) = $5.000_0 \times 10^{-4}$ mol

From the stoichiometry of Reaction 5-5, we know that $5.000_0 \times 10^{-4}$ mol $Na_2C_2O_4$ reacts with

mol $KMnO_4$

$$= (5.000_0 \times 10^{-4}\ \text{mol}\ Na_2C_2O_4)\left(\frac{2\ \text{mol}\ KMnO_4}{5\ \text{mol}\ C_2O_4^{2-}}\right) = 2.000_0 \times 10^{-4}\ \text{mol}$$

Therefore the amount (in moles) of $KMnO_4$ that reacted with $NaNO_2$ was

$KMnO_4$ reacting with $NaNO_2$ = total $KMnO_4$ − $KMnO_4$ reacting with $Na_2C_2O_4$
$$= 5.486_0 \times 10^{-4}\ \text{mol} - 2.000_0 \times 10^{-4}\ \text{mol}$$
$$= 3.486_0 \times 10^{-4}\ \text{mol}$$

From Reaction 5-6, we can say that $3.486_0 \times 10^{-4}$ mol of $KMnO_4$ reacts with

mol $NaNO_2$ reacting with $KMnO_4$

$$= (3.486_0 \times 10^{-4}\ \text{mol}\ KMnO_4)\left(\frac{5\ \text{mol}\ NaNO_2}{2\ \text{mol}\ KMnO_4}\right) = 8.715_0 \times 10^{-4}\ \text{mol}$$

The volume of $NaNO_2$ containing this many moles is 50.00 mL, so the concentration is

$$[NaNO_2] = \frac{8.715_0 \times 10^{-4}\ \text{mol}}{0.050\ 00\ \text{L}} = 0.017\ 43\ \text{M}$$

Ask Yourself

5-C. The procedure above used 0.009 22 M $KMnO_4$. Step 2 required 30.00 mL of $Na_2C_2O_4$, step 3 required 9.68 mL of $KMnO_4$, and step 4 required 0.07 mL of $KMnO_4$.

(a) Find the total moles of $KMnO_4$ in the reaction. Remember to subtract the blank volume of step 4 before computing the total moles of $KMnO_4$.

(b) Find the total moles of $Na_2C_2O_4$ in the reaction.

(c) How many moles of $KMnO_4$ reacted with $Na_2C_2O_4$?

(d) How many moles of $KMnO_4$ reacted with $NaNO_2$?

(e) How many moles of $NaNO_2$ were delivered to the reaction?

(f) What was the molarity of the $NaNO_2$ stock solution?

The **solubility product,** K_{sp}, is the equilibrium constant for the reaction in which a solid *salt* (an ionic compound) dissolves to give its constituent ions in solution. The concentration of the solid is omitted from the equilibrium constant because the solid is in its standard state. We will need the solubility product to discuss precipitation titrations later in Section 5-5.

Calculating the Solubility of an Ionic Compound

Consider the dissolution of lead(II) iodide in water:

$$PbI_2(s) \rightleftharpoons Pb^{2+} + 2I^- \qquad K_{sp} = [Pb^{2+}][I^-]^2 = 7.9 \times 10^{-9} \qquad (5-7)$$

Lead(II) iodide   Lead(II)   Iodide

Pure solid is omitted from the equilibrium constant because the concentration of PbI_2 in the solid does not change.

for which the solubility product is listed in Appendix A. A solution that contains all the solid capable of being dissolved is said to be **saturated.** What is the concentration of Pb^{2+} in a solution saturated with PbI_2?

Reaction 5-7 produces two I^- ions for each Pb^{2+} ion. If the concentration of dissolved Pb^{2+} is x M, the concentration of dissolved I^- must be $2x$ M. We can show this relationship neatly in a little table:

	$PbI_2(s)$	$\rightleftharpoons$	Pb^{2+}	$+$	$2I^-$
Initial concentration	solid		0		0
Final concentration	solid		x		$2x$

Putting these concentrations into the solubility product gives

$$[Pb^{2+}][I^-]^2 = (x)(2x)^2 = 7.9 \times 10^{-9}$$
$$4x^3 = 7.9 \times 10^{-9}$$
$$x = \left(\frac{7.9 \times 10^{-9}}{4}\right)^{1/3} = 0.001\ 3 \text{ M}$$

To find the cube root of a number with a calculator, raise the number to the 0.333 333 33 . . . power with the y^x key.

The concentration of Pb^{2+} is 0.001 3 M and the concentration of I^- is $2x = (2)(0.001\ 3) = 0.002\ 6$ M.

The physical meaning of the solubility product is this: If an aqueous solution is left in contact with excess solid PbI_2, solid dissolves until the condition $[Pb^{2+}][I^-]^2 = K_{sp}$ is satisfied. Thereafter, no more solid dissolves. Unless excess solid remains, there is no guarantee that $[Pb^{2+}][I^-]^2 = K_{sp}$. If Pb^{2+} and I^- are mixed together (with appropriate counterions) such that the product $[Pb^{2+}][I^-]^2$ exceeds K_{sp}, then $PbI_2(s)$ will precipitate (Figure 5-3).

The solubility product does not tell the entire story of the solubility of ionic compounds. The concentration of *undissociated* species may be significant. In a solution of calcium sulfate, for example, about half of the dissolved material dissociates to Ca^{2+} and SO_4^{2-} and half dissolves as undissociated $CaSO_4(aq)$ (a tightly bound *ion pair*). In the case of lead(II) iodide, species such as PbI^+ and PbI_3^- also contribute to the total solubility.

The Common Ion Effect

Consider what happens when we add a second source of I^- to a solution saturated with $PbI_2(s)$. Let's add 0.030 M NaI, which dissociates completely to Na^+ and I^-. What is the concentration of Pb^{2+} in this solution?

			$PbI_2(s)$	$\rightleftharpoons$	Pb^{2+}	+	$2I^-$	(5-8)
Initial concentration			solid		0		0.030	
Final concentration			solid		x		$2x + 0.030$	

The initial concentration of I^- is from dissolved NaI. The final concentration of I^- has contributions from NaI and PbI_2.

The solubility product is

$$[Pb^{2+}][I^-]^2 = (x)(2x + 0.030)^2 = K_{sp} = 7.9 \times 10^{-9} \quad (5\text{-}9)$$

But think about the size of x. With no added I^-, we found $x = 0.001\,3$ M. In the present case, we anticipate that x will be smaller than $0.001\,3$ M, because of Le Châtelier's principle. Addition of the product I^- to Reaction 5-8 displaces the reaction in the reverse direction. In the presence of extra I^-, there will be less dissolved Pb^{2+}. This application of Le Châtelier's principle is called the **common ion effect**. *A salt will be less soluble if one of its constituent ions is already present in the solution.*

Getting back to Equation 5-9, we suspect that $2x$ may be much smaller than 0.030. As an approximation, we will ignore $2x$ in comparison to 0.030. The equation simplifies to

$$(x)(0.030)^2 = K_{sp} = 7.9 \times 10^{-9}$$
$$x = 7.9 \times 10^{-9}/(0.030)^2 = 8.8 \times 10^{-6}$$

Because $2x = 1.8 \times 10^{-5} \ll 0.030$, it was justifiable to neglect $2x$ to solve the problem. The answer also illustrates the common ion effect. In the absence of added I^-, the solubility of Pb^{2+} was $0.001\,3$ M. In the presence of 0.030 M I^-, $[Pb^{2+}]$ is reduced to 8.8×10^{-6} M.

What is the maximum I^- concentration at equilibrium in a solution in which the concentration of Pb^{2+} is somehow *fixed* at 1.0×10^{-4} M? Our concentration table looks like this now:

		$PbI_2(s)$	$\rightleftharpoons$	Pb^{2+}	+	$2I^-$
Initial concentration		solid		1.0×10^{-4}		0
Final concentration		solid		1.0×10^{-4}		x

$[Pb^{2+}]$ is not x in this example, so there is no reason to set $[I^-] = 2x$. The problem is solved by plugging each concentration into the solubility product:

$$[Pb^{2+}][I^-]^2 = K_{sp}$$
$$(1.0 \times 10^{-4})(x)^2 = 7.9 \times 10^{-9}$$
$$x = [I^-] = 8.9 \times 10^{-3}\ \text{M}$$

If I^- is added above a concentration of 8.9×10^{-3} M, then $PbI_2(s)$ will precipitate.

FIGURE 5-3 The yellow solid, lead(II) iodide (PbI_2), precipitates when a colorless solution of lead nitrate ($Pb(NO_3)_2$) is added to a colorless solution of potassium iodide (KI). [Photo by Chip Clark.]

It is important to confirm at the end of the calculation that the approximation $2x \ll 0.030$ is valid. Box 5-2 discusses approximations.

Common ion effect: A salt is less soluble if one of its ions is already present in the solution.

The compound silver ferrocyanide dissociates into Ag^+ and $Fe(CN)_6^{4-}$ in solution. Find the solubility of $Ag_4Fe(CN)_6$ in water expressed as **(a)** moles of $Ag_4Fe(CN)_6$ per liter and **(b)** ppb Ag^+ ($\approx$ ng Ag^+/mL).

SOLUTION **(a)** We start off with a concentration table:

$Ag_4Fe(CN)_6(s)$ Silver ferrocyanide FM 643.43	$\rightleftharpoons$	$4Ag^+$	$+$	$Fe(CN)_6^{4-}$ Ferrocyanide
Initial concentration solid		0		0
Final concentration solid		$4x$		x

The solubility product of $Ag_4Fe(CN)_6$ in Appendix A is 8.5×10^{-45}. Therefore,

$$[Ag^+]^4[Fe(CN)_6^{4-}] = (4x)^4(x) = 8.5 \times 10^{-45}$$
$$256x^5 = 8.5 \times 10^{-45}$$
$$x^5 = \frac{8.5 \times 10^{-45}}{256}$$
$$x = \left(\frac{8.5 \times 10^{-45}}{256}\right)^{1/5} = 5.0_6 \times 10^{-10} \text{ M}$$

There are $5.0_6 \times 10^{-10}$ moles of $Ag_4Fe(CN)_6$ dissolved per liter. The concentration of Ag^+ is $4x$ and the concentration of $Fe(CN)_6^{4-}$ is x.

(b) The concentration of Ag^+ is $4x = 4(5.0_6 \times 10^{-10} \text{ M}) = 2.0_2 \times 10^{-9} \text{ M}$. Use the atomic mass of Ag to convert mol/L to g/L:

$$\left(2.0_2 \times 10^{-9} \frac{mol}{L}\right)\left(107.868\ 2 \frac{g}{mol}\right) = 2.1_8 \times 10^{-7} \text{ g/L}$$

Atomic mass of Ag

To find ppb, we need units of ng/mL:

$$\left(2.1_8 \times 10^{-7} \frac{g}{L}\right)\left(\frac{1}{1\ 000}\frac{L}{mL}\right)\left(10^9 \frac{ng}{g}\right) = 0.21_8 \frac{ng}{mL} = 0.22 \text{ ppb}$$

Ask Yourself

5-D. What is the concentration of Pb^{2+} in **(a)** a saturated solution of $PbBr_2$ in water or **(b)** in a saturated solution of $PbBr_2$ in which $[Br^-]$ is somehow fixed at 0.10 M?

💡 **Box 5-2** *Explanation*

The Logic of Approximations

Many problems are difficult to solve without judicious approximations. For example, rather than solving the equation

$$(x)(2x + 0.030)^2 = 7.9 \times 10^{-9}$$

we hoped and prayed that $2x \ll 0.030$, and therefore solved the much simpler equation

$$(x)(0.030)^2 = 7.9 \times 10^{-9}$$

But how can we be sure that our solution fits the original problem?

When we use an approximation, we assume it is true. *If the assumption is true, it does not create a contradiction. If the assumption is false, it leads to a contradiction.* You can test an assumption by using it and seeing whether you are right or wrong afterward.

Gail's lake

You may object to this reasoning, feeling "How can the truth of an assumption be tested by using the assumption?" Suppose you wish to test the statement "Gail can swim 100 meters." To see whether or not the statement is true, you can assume it is. If Gail can swim 100 meters, then you could dump her in the middle of a lake with a radius of 100 meters and expect her to swim to shore. If she comes ashore alive, then your assumption was correct and no contradiction is created. If she does not make it to shore, then there is a contradiction. Your assumption must have been wrong. There are only two possibilities: Either the assumption is correct and using it is correct, or the assumption is wrong and using it is wrong. (A third possibility in this case is that there are freshwater sharks in the lake.)

Example 1. $(x)(2x + 0.030)^2 = 7.9 \times 10^{-9}$

$$(x)(0.030)^2 = 7.9 \times 10^{-9}$$
$$(\text{assuming } 2x \ll 0.030)$$

$$x = (7.9 \times 10^{-9})/(0.030)^2 = 8.8 \times 10^{-6}$$

No contradiction:
$$2x = 1.76 \times 10^{-5} \ll 0.030$$

The assumption is true.

Example 2. $(x)(2x + 0.030)^2 = 7.9 \times 10^{-5}$

$$(x)(0.030)^2 = 7.9 \times 10^{-5}$$
$$(\text{assuming } 2x \ll 0.30)$$

$$x = (7.9 \times 10^{-5})/(0.030)^2 = 0.088$$

A contradiction:
$$2x = 0.176 > 0.030$$

The assumption is false.

In Example 2, the assumption leads to a contradiction, so the assumption cannot be correct. When this happens, you must solve the cubic equation $(x)(2x + 0.030)^2 = 7.9 \times 10^{-5}$.

A reasonable way to solve this equation is by trial-and-error, as shown in the table. You can create this table by hand or, even more easily, with a spreadsheet. In cell A1, enter a guess for x. In cell A2, enter the formula "$= A1*(2*A1+0.030)^2$". When you guess x correctly in cell A1, cell A2 will have the value 7.9×10^{-5}.

Guess	$x(2x + 0.030)^2$	Result is
$x = 0.01$	2.5×10^{-5}	too low
$x = 0.02$	9.8×10^{-5}	too high
$x = 0.015$	5.4×10^{-5}	too low
$x = 0.018$	7.84×10^{-5}	too low
$x = 0.019$	8.79×10^{-5}	too high
$x = 0.018\ 1$	7.93×10^{-5}	too high
$x = 0.018\ 05$	7.89×10^{-5}	too low
$x = 0.018\ 06$	7.90×10^{-5}	not bad!

5-5 Titration of a Mixture

When Ag^+ is added to a solution containing Cl^- and I^-, the less soluble $AgI(s)$ precipitates first:

$$Ag^+ + I^- \longrightarrow AgI(s) \quad\left.\right\} \quad AgI\ (K_{sp} = 8.3 \times 10^{-17})\ precipitates \qquad (5\text{-}10a)$$

$$Ag^+ + Cl^- \longrightarrow AgCl(s) \quad\left.\right\} \quad before\ AgCl\ (K_{sp} = 1.8 \times 10^{-10}) \qquad (5\text{-}10b)$$

Because the two solubility products are sufficiently different, the first precipitation is nearly complete before the second commences.

Figure 5-4 shows how the reaction is monitored with a silver electrode to find both end points. We will learn how this electrode responds to silver ion concentration in Section 14-1. Figure 5-5 shows experimental curves for the titration of I^- or a mixture of I^- plus Cl^- by Ag^+.

In the titration of I^- (curve b in Figure 5-5), the voltage remains almost constant near 650 mV until the equivalence point (23.76 mL) when I^- is used up. At this point, there is an abrupt decrease in the electrode potential. The reason for the abrupt change is that the electrode is responding to the concentration of Ag^+ in the solution. Prior to the equivalence point, virtually all the added Ag^+ reacts with I^- to precipitate $AgI(s)$. The concentration of Ag^+ in solution is very low and nearly

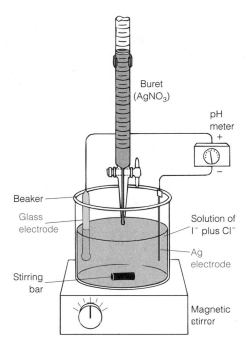

FIGURE 5-4 Apparatus for measuring the titration curves in Figure 5-5. The silver electrode responds to changes in Ag^+ concentration, and the glass electrode provides a constant reference potential in this experiment. The voltage changes by approximately 59 mV for each factor-of-10 change in [Ag]. All solutions, including $AgNO_3$, were maintained at pH 2.0 by using 0.010 M sulfate buffer prepared from H_2SO_4 and KOH.

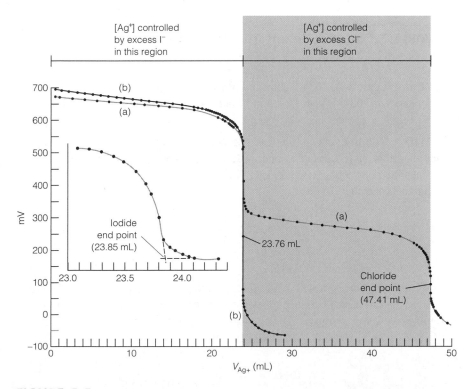

FIGURE 5-5 Experimental titration curves. The ordinate (y-axis) is the electric potential difference (in millivolts) between the two electrodes in Figure 5-4. (*a*) Titration of KI plus KCl with 0.084 5 M $AgNO_3$. The inset is an expanded view of the region near the first equivalence point. (*b*) Titration of I^- with 0.084 5 M Ag^+.

constant. When I^- has been consumed, the concentration of Ag^+ suddenly increases because Ag^+ is being added from the buret and is no longer consumed by I^-. This change gives rise to the abrupt decrease in electrode potential.

In the titration of $I^- + Cl^-$ (curve a in Figure 5-5), there are two abrupt changes in electrode potential. The first occurs when I^- is used up, and the second comes when Cl^- is used up. Prior to the first equivalence point, the very low concentration of Ag^+ is governed by the solubility of AgI. Between the first and second equivalence points, essentially all I^- has precipitated and Cl^- is in the process of being consumed. The concentration of Ag^+ is still small but governed by the solubility of AgCl, which is greater than that of AgI. After the second equivalence point, the concentration of Ag^+ shoots upward as Ag^+ is added from the buret. Therefore we observe two abrupt changes of electric potential in this experiment.

The I^- end point is taken as the intersection of the steep and nearly horizontal curves shown at 23.85 mL in the inset of Figure 5-5. The reason for using the intersection is that the precipitation of I^- is not quite complete when Cl^- begins to precipitate. Therefore the end of the steep portion (the intersection) is a better approximation of the equivalence point than is the middle of the steep section. The Cl^- end point is taken as the midpoint of the second steep section, at 47.41 mL. The moles of Cl^- in the sample correspond to the moles of Ag^+ delivered between the first and second end points. That is, it requires 23.85 mL of Ag^+ to precipitate I^-, and $(47.41 - 23.85) = 23.56$ mL of Ag^+ to precipitate Cl^-.

EXAMPLE   Extracting Results from Figure 5-5

The elements F, Cl, Br, I, and At are called *halogens*. Their anions are called *halides*.

In curve a of Figure 5-5, 40.00 mL of unknown solution containing both I^- and Cl^- were titrated with 0.084 5 M Ag^+. Find the concentrations of each halide ion.

SOLUTION The inset shows the first end point at 23.85 mL. Reaction 5-10a tells us that one mole of I^- consumes one mole of Ag^+. The moles of Ag^+ delivered at this point are $(0.084\ 5\ M)(0.023\ 85\ L) = 2.015 \times 10^{-3}$ mol. The molarity of iodide in the unknown is therefore

$$[I^-] = \frac{2.015 \times 10^{-3}\ \text{mol}}{0.040\ 00\ L} = 0.050\ 3_8\ M$$

The second end point is at 47.41 mL. The quantity of Ag^+ titrant required to react with Cl^- is the difference between the two end points: $(47.41 - 23.85) = 23.56$ mL. The moles of Ag^+ required to react with Cl^- are $(0.084\ 5\ M)(0.023\ 56\ L) = 1.991 \times 10^{-3}$ mol. The molarity of chloride in the unknown is

$$[Cl^-] = \frac{1.991 \times 10^{-3}\ \text{mol}}{0.040\ 00\ L} = 0.049\ 7_7\ M$$

Ask Yourself

5-E. A 25.00 mL solution containing Br^- and Cl^- was titrated with 0.033 33 M $AgNO_3$.

   (a) Write the two titration reactions and use solubility products to find which occurs first.

   (b) In an experiment analogous to that in Figures 5-4 and 5-5, the first end point was observed at 15.55 mL. Find the concentration of the first halide that precipitated. Is it Br^- or Cl^-?

   (c) The second end point was observed at 42.23 mL. Find the concentration of the other halide.

5-6   Titrations Involving Silver Ion

We now introduce two widely used indicator methods for titrations involving Ag^+, which are called *argentometric titrations*. The methods are

1. **Volhard titration:** formation of a soluble, colored complex at the end point

2. **Fajans titration:** adsorption of a colored indicator on the precipitate at the end point

The Latin word for silver is *argentum*, from which the symbol Ag is derived.

Volhard Titration

The Volhard titration is actually a titration of Ag^+. To determine Cl^-, a back titration is necessary. First, the Cl^- is precipitated by a known, excess quantity of standard $AgNO_3$:

$$Ag^+ + Cl^- \longrightarrow AgCl(s)$$

The AgCl is isolated, and excess Ag^+ is titrated with standard KSCN in the presence of Fe^{3+}:

$$Ag^+ + SCN^- \longrightarrow AgSCN(s)$$

When all the Ag^+ has been consumed, the SCN^- reacts with Fe^{3+} to form a red complex:

$$Fe^{3+} + SCN^- \longrightarrow \underset{\text{Red}}{FeSCN^{2+}}$$

Because the Volhard method is a titration of Ag^+, it can be adapted for the determination of any anion that forms an insoluble silver salt.

The appearance of red color signals the end point. Knowing how much SCN^- was required for the back titration tells us how much Ag^+ was left over from the reaction with Cl^-. Because the total amount of Ag^+ is known, the amount consumed by Cl^- can then be calculated.

In the analysis of Cl^- by the Volhard method, the end point slowly fades because AgCl is more soluble than AgSCN. The AgCl slowly dissolves and is replaced by AgSCN. To prevent this secondary reaction from happening, two techniques are commonly used. One is to filter off the AgCl and titrate only the Ag^+ in the filtrate. An easier procedure is to shake a few milliliters of nitrobenzene, $C_6H_5NO_2$, with the precipitated AgCl prior to the back titration. Nitrobenzene coats the AgCl and effectively isolates it from attack by SCN^-. Br^- and I^-, whose silver salts are *less* soluble than AgSCN, can be titrated by the Volhard method without isolating the silver halide precipitate.

Fajans Titration

The Fajans titration uses an **adsorption indicator.** To see how this works, consider the electric charge of a precipitate. When Ag^+ is added to Cl^-, there is excess Cl^- in solution prior to the equivalence point. Some Cl^- is selectively adsorbed on the AgCl surface, imparting a negative charge to the crystal surface (Figure 5-6a). After the equivalence point, there is excess Ag^+ in solution. Adsorption of Ag^+

TABLE 5-1   **Applications of precipitation titrations**

Species analyzed	Notes
	VOLHARD METHOD
Br^-, I^-, SCN^-, CNO^-, AsO_4^{3-}	Precipitate removal is unnecessary.
Cl^-, PO_4^{3-}, CN^-, $C_2O_4^{2-}$, CO_3^{2-}, S^{2-}, CrO_4^{2-}	Precipitate removal required.
	FAJANS METHOD
Cl^-, Br^-, I^-, SCN^-, $Fe(CN)_6^{4-}$	Titrate with Ag^+. Detection with dyes such as fluorescein, dichlorofluorescein, eosin, bromophenol blue.
F^-	Titrate with $Th(NO_3)_4$ to produce ThF_4. End point detected with alizarin red S.
Zn^{2+}	Titrate with $K_4Fe(CN)_6$ to produce $K_2Zn_3[Fe(CN)_6]_2$. End point detection with diphenylamine.
SO_4^{2-}	Titrate with $Ba(OH)_2$ in 50 vol % aqueous methanol, using alizarin red S as indicator.
Hg_2^{2+}	Titrate with NaCl to produce Hg_2Cl_2. End point detected with bromophenol blue.
PO_4^{3-}, $C_2O_4^{2-}$	Titrate with $Pb(CH_3CO_2)_2$ to give $Pb_3(PO_4)_2$ or PbC_2O_4. End point detected with dibromofluorescein (PO_4^{3-}) or fluorescein ($C_2O_4^{2-}$).

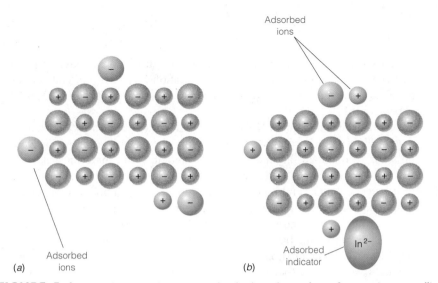

(a) Adsorbed
ions

(b) Adsorbed
indicator

Adsorbed
ions

In^{2-}

FIGURE 5-6 Ions from a solution are adsorbed on the surface of a growing crystallite. (*a*) A crystal growing in the presence of excess lattice anions (anions that belong in the crystal) has a negative charge because anions are predominantly adsorbed. (*b*) A crystal growing in the presence of excess lattice cations has a positive charge and can therefore adsorb a negative indicator ion. Anions and cations in the solution that do not belong in the crystal lattice are less likely to be adsorbed than are ions belonging to the lattice.

cations on the crystal surface creates a positive charge on the particles of precipitate (Figure 5-6b). The abrupt change from negative charge to positive charge occurs at the equivalence point.

Common adsorption indicators are anionic (negatively charged) dyes, which are attracted to the positively charged precipitate produced immediately after the equivalence point. Adsorption of the dye on the surface of the solid precipitate changes the color of the dye by interactions that are not well understood. The color change signals the end point in the titration. Because the indicator reacts with the precipitate surface, it is desirable to have as much surface area as possible. In other words, the titration is performed under conditions that tend to keep the particles as

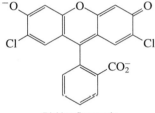

Dichlorofluorescein

Notation for drawing structures of organic compounds is discussed in Box 6-1.

Demonstration 5-1

Fajans Titration

The Fajans titration of Cl^- with Ag^+ demonstrates indicator end points in precipitation titrations. Dissolve 0.5 g of NaCl plus 0.15 g of dextrin in 400 mL of water. The purpose of the dextrin is to retard coagulation of the AgCl precipitate. Add 1 mL of dichlorofluorescein indicator solution containing 1 mg/mL of dichlorofluorescein in 95 wt % aqueous ethanol or 1 mg/mL of the sodium salt in water. Titrate the NaCl solution with a solution containing 2 g of $AgNO_3$ in 30 mL of water. About 20 mL are required to reach the end point.

Color Plate 1a shows the yellow color of the indicator in the NaCl solution prior to the titration. Color Plate 1b shows the milky-white appearance of the AgCl suspension during titration, before the end point is reached. The pink suspension in Color Plate 1c appears at the end point, when the anionic indicator becomes adsorbed to the cationic particles of precipitate.

small as possible, because small particles have more surface area than an equal volume of large particles. Low electrolyte concentration helps prevent coagulation of the precipitate and maintain small particle size. As in all silver titrations, strong light should be avoided.

The indicator most commonly used for AgCl is dichlorofluorescein, which has a greenish yellow color in solution but turns pink when adsorbed on AgCl (Demonstration 5-1). To maintain the indicator in its anionic form, there must not be too much H^+ in the solution. The dye eosin is useful in the titration of Br^-, I^-, and SCN^-. It gives a sharper end point than dichlorofluorescein and is more sensitive (i.e., less halide can be titrated). It cannot be used for AgCl because the eosin anion is more strongly bound to AgCl than is Cl^- ion. Eosin will bind to the AgCl crystallites even before the particles become positively charged.

Applications of precipitation titrations are listed in Table 5-1. Whereas the Volhard method is specifically for argentometric titrations, the Fajans method has wider application. Because the Volhard titration is carried out in acidic solution (typically 0.2 M HNO_3), it avoids certain interferences that would affect other titrations. Silver salts of anions such as CO_3^{2-}, $C_2O_4^{2-}$, and AsO_4^{3-} are soluble in acidic solution, so these anions do not interfere with the analysis.

Challenge Consider the equilibrium

$$\underset{\text{Carbonate}}{CO_3^{2-}} + H^+ \rightleftharpoons \underset{\substack{\text{Hydrogen carbonate} \\ \text{(also called bicarbonate)}}}{HCO_3^-}$$

Use Le Châtelier's principle to explain why carbonate salts are soluble in acidic solution (which contains a high concentration of H^+).

Ask Yourself

5-F. **(a)** Why is nitrobenzene used in the Volhard titration of chloride?

(b) Why does the surface charge of a precipitate change sign at the equivalence point?

(c) In the Fajans titration of Zn^{2+} in Table 5-1, do you expect the charge on the precipitate to be positive or negative after the equivalence point?

Key Equations

Stoichiometry   For the reaction $a\text{A} + b\text{B} \longrightarrow$ products, use the ratio $\left(\dfrac{a \text{ mol A}}{b \text{ mol B}}\right)$ for stoichiometry calculations.

Solubility product $PbI_2(s) \overset{K_{sp}}{\rightleftharpoons} Pb^{2+} + 2I^-$ $K_{sp} = [Pb^{2+}][I^-]^2$

Common ion effect: A salt is less soluble in the presence of one of its constituent ions.

Important Terms

adsorption indicator	blank titration	direct titration
back titration	common ion effect	end point

equivalence point	solubility product	titration
Fajans titration	standardization	titration error
indicator	standard solution	Volhard titration
primary standard	titrant	volumetric analysis
saturated solution		

Problems

5-1. Distinguish the terms *end point* and *equivalence point*.

5-2. For Reaction 5-1, how many milliliters of 0.165 0 M $KMnO_4$ are needed to react with 108.0 mL of 0.165 0 M oxalic acid? How many milliliters of 0.165 0 M oxalic acid are required to react with 108.0 mL of 0.165 0 M $KMnO_4$?

5-3. A 10.00-mL aliquot of unknown oxalic acid solution required 15.44 mL of 0.011 17 M $KMnO_4$ solution to reach the purple end point. A blank titration of 10 mL of similar solution containing no oxalic acid required 0.04 mL to exhibit detectable color. Find the concentration of oxalic acid in the unknown.

5-4. Ammonia reacts with hypobromite, OBr^-, according to

$$2NH_3 + 3OBr^- \longrightarrow N_2 + 3Br^- + 3H_2O$$

Find the molarity of OBr^- if 1.00 mL of OBr^- solution reacts with 1.69 mg of NH_3 (FM 17.03)?

5-5. How many milliliters of 0.100 M KI are needed to react with 40.0 mL of 0.040 0 M $Hg_2(NO_3)_2$ if the reaction is $Hg_2^{2+} + 2I^- \rightarrow Hg_2I_2(s)$?

5-6. Cl^- in blood serum, cerebrospinal fluid, or urine can be measured by titration with mercuric ion: $Hg^{2+} + 2Cl^- \rightarrow HgCl_2(aq)$. When the reaction is complete, excess Hg^{2+} reacts with the indicator, diphenylcarbazone, which forms a violet-blue color.

(a) Mercuric nitrate was standardized by titrating a solution containing 147.6 mg of NaCl (FM 58.442), which required 28.06 mL of $Hg(NO_3)_2$ solution. Find the molarity of the $Hg(NO_3)_2$.

(b) When this same $Hg(NO_3)_2$ solution was used to titrate 2.000 mL of urine, 22.83 mL were required. Find the concentration of Cl^- (mg/mL) in the urine.

5-7. *Volhard titration.* A 30.00-mL solution of unknown I^- was treated with 50.00 mL of 0.365 0 M $AgNO_3$. The precipitated AgI was filtered off, and the filtrate (plus Fe^{3+}) was titrated with 0.287 0 M KSCN. When 37.60 mL had been added, the solution turned red. How many milligrams of I^- were present in the original solution?

5-8. How many milligrams of oxalic acid dihydrate, $H_2C_2O_4 \cdot 2H_2O$ (FM 126.07), will react with 1.00 mL of 0.027 3 M ceric sulfate $(Ce(SO_4)_2)$ if the reaction is $H_2C_2O_4 + 2Ce^{4+} \rightarrow 2CO_2 + 2Ce^{3+} + 2H^+$?

5-9. Arsenic (III) oxide (As_2O_3) is available in pure form and is a useful (and poisonous) primary standard for standardizing many oxidizing agents, such as MnO_4^-. The As_2O_3 is first dissolved in base and then titrated with MnO_4^- in acidic solution. A small amount of iodide (I^-) or iodate (IO_3^-) is used to catalyze the reaction between H_3AsO_3 and MnO_4^-. The reactions are

$$As_2O_3 + 4OH^- \rightleftharpoons 2HAsO_3^{2-} + H_2O$$
$$HAsO_3^{2-} + 2H^+ \rightleftharpoons H_3AsO_3$$
$$5H_3AsO_3 + 2MnO_4^- + 6H^+ \longrightarrow$$
$$5H_3AsO_4 + 2Mn^{2+} + 3H_2O$$

(a) A 3.214-g aliquot of $KMnO_4$ (FM 158.034) was dissolved in 1.000 L of water, heated to cause any reactions with impurities to occur, cooled, and filtered. What is the theoretical molarity of this solution if no MnO_4^- was consumed by impurities?

(b) What mass of As_2O_3 (FM 197.84) would be sufficient to react with 25.00 mL of the $KMnO_4$ solution in part **(a)**?

(c) It was found that 0.146 8 g of As_2O_3 required 29.98 mL of $KMnO_4$ solution for the faint color of unreacted MnO_4^- to appear. In a blank titration, 0.03 mL of MnO_4^- was required to produce enough color to be seen. Calculate the molarity of the permanganate solution.

5-10. This problem describes a gravimetric titration in which the *mass* of titrant is measured instead of the *volume*. A solution of NaOH was standardized by titration of a known quantity of the primary standard, potassium hydrogen phthalate:

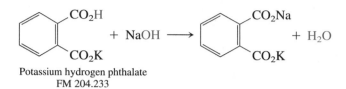

Potassium hydrogen phthalate
FM 204.233

The NaOH was then used to find the concentration of an unknown solution of H_2SO_4:

$$H_2SO_4 + 2NaOH \longrightarrow Na_2SO_4 + H_2O$$

(a) Titration of 0.824 g of potassium hydrogen phthalate required 38.314 g of NaOH solution to reach the end point detected by phenolphthalein indicator. Find the concentration of NaOH expressed as mol NaOH/kg solution.

(b) A 10.00-mL aliquot of H_2SO_4 solution required 57.911 g of NaOH solution to reach the phenolphthalein end point. Find the molarity of H_2SO_4.

5-11. An unknown molybdate (MoO_4^{2-}) solution (50.00 mL) was passed through a column containing $Zn(s)$ to convert molybdate to Mo^{3+}. One mole of MoO_4^{2-} gives one mole of Mo^{3+}. The resulting sample required 22.11 mL of 0.012 34 M $KMnO_4$ to reach a purple end point from the reaction

$$3MnO_4^- + 5Mo^{3+} + 4H^+ \longrightarrow 3Mn^{2+} + 5MoO_2^{2+} + 2H_2O$$

A blank required 0.07 mL. Find the molarity of molybdate in the unknown.

5-12. A 25.00-mL sample of La^{3+} was treated with excess $Na_2C_2O_4$ to precipitate $La_2(C_2O_4)_3$, which was washed, dissolved in acid, and titrated with 12.34 mL of 0.004 321 M $KMnO_4$ to reach a purple end point. Find the molarity of La^{3+} in the unknown.

5-13. A glycerol solution weighing 153.2 mg was treated with 50.0 mL of 0.089 9 M Ce^{4+} in 4 M $HClO_4$ at 60°C for 15 min to convert glycerol to formic acid:

$$C_3H_8O_3 + 8Ce^{4+} + 3H_2O \longrightarrow$$
Glycerol
FM 92.095
$$3HCO_2H + 8Ce^{3+} + 8H^+$$
Formic acid

The excess Ce^{4+} required 10.05 mL of 0.043 7 M Fe^{2+} for a back titration by the reaction $Ce^{4+} + Fe^{2+} \rightarrow Ce^{3+} + Fe^{3+}$. What was the weight percent of glycerol in the unknown?

5-14. *Propagation of uncertainty.* Consider the titration of 50.00 ($\pm$0.05) mL of a mixture of I^- and SCN^- with 0.068 3 ($\pm$0.000 1) M Ag^+. Look up the solubility products of AgI and AgSCN to decide which precipitate is formed first. The first equivalence point is observed at 12.6 ($\pm$0.4) mL, and the second occurs at 27.7 ($\pm$0.3) mL. Find the molarity and the uncertainty in molarity of thiocyanate in the original mixture.

5-15. Calculate the solubility of CuBr (FM 143.45) in water expressed as **(a)** moles per liter and **(b)** grams per 100 mL. (This question presumes that Cu^+ and Br^- are the only significant soluble species. $CuBr_2^-$ is negligible and we suppose that the ion pair CuBr(aq) is also negligible.)

5-16. Find the solubility of silver chromate (FM 331.73) in water. Express your answer as **(a)** moles of chromate per liter and **(b)** ppm Ag^+ ($\approx \mu g\ Ag^+/mL$).

5-17. Ag^+ at 10–100 ppb (ng/mL) disinfects swimming pools. One way to maintain an appropriate concentration of Ag^+ is to add a slightly soluble silver salt to the pool. Calculate the ppb of Ag^+ at equilibrium in saturated solutions of AgCl, AgBr, and AgI.

5-18. The mercury(I) ion (Hg_2^{2+}, also called mercurous ion) is a diatomic ion with a charge of $+2$. Mercury(I) iodate dissociates into three ions:

$$Hg_2(IO_3)_2(s) \rightleftharpoons Hg_2^{2+} + 2IO_3^- \quad K_{sp} = [Hg_2^{2+}][IO_3^-]^2$$
FM 750.99

Find the concentrations of Hg_2^{2+} and IO_3^- in **(a)** a saturated solution of $Hg_2(IO_3)_2(s)$ and **(b)** a 0.010 M solution of KIO_3 saturated with $Hg_2(IO_3)_2(s)$.

5-19. If a solution containing 0.10 M Cl^-, Br^-, I^-, and CrO_4^{2-} is treated with Ag^+, in what order will the anions precipitate?

How Would You Do It?

5-20. *Thermometric titration.* Almost every chemical reaction liberates or absorbs heat. One physical property that can be measured to determine the equivalence point in a titration is temperature rise. The figure shows temperature changes measured

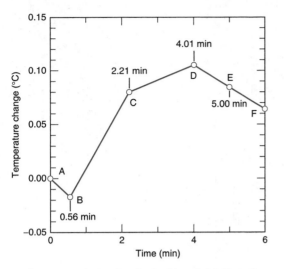

Thermometric titration for Problem 5–20. [Data from L. D. Hansen, D. Kenney, W. M. Litchman, and E. A. Lewis, *J. Chem. Ed.* **1971**, *48*, 851.]

by an electrical resistor during the titration of a mixture of the bases "tris" and pyridine by HCl:

$$tris + H^+ \longrightarrow trisH^+ \qquad (1)$$
$$pyridine + H^+ \longrightarrow pyridineH^+ \qquad (2)$$

Prior to adding H^+, the temperature drifted down from points A to B. Then 1.53 M HCl was added at a rate of 0.198 mL/min to the 40-mL solution between points B and E. The temperature rose rapidly during Reaction 1 between points B and C and less rapidly during Reaction 2 between points C and D. At point D, the pyridine was used up and the temperature of the solution drifted down steadily as more HCl continued to be added between points D and E. After point E, the flow of HCl was stopped, but the temperature continued to drift down.

(a) How many mmol of tris and how many mmol of pyridine were present?

(b) Reaction 1 liberates 47.5 kJ/mol. Does Reaction 2 absorb or liberate heat?

Combustion Analysis Reveals Pollutant Buildup in Trees

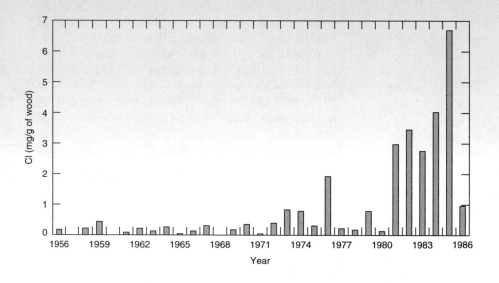

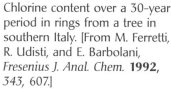

Chlorine content over a 30-year period in rings from a tree in southern Italy. [From M. Ferretti, R. Udisti, and E. Barbolani, *Fresenius J. Anal. Chem.* **1992**, *343*, 607.]

$\mathbf{A}$ combustion procedure was employed to measure Cl, P, and S in pine tree rings from southern Italy. Wood was taken from the tree with a core drill, and rings were sliced apart and dried at 75°C for 3 days. Samples were wrapped in filter paper, leaving an unfolded tail as shown in Figure 6-1. The wrapped sample was then placed in the platinum basket of the disassembled *Schöniger flask*.

The 250-mL flask was loaded with 20 mL of 4.4 mM hydrogen peroxide (H_2O_2) and flushed with O_2 gas. The paper tail was ignited, and the stopper-sample assembly was immediately inserted into the flask. The flask was inverted to prevent gaseous products from leaking out. Following combustion, 30 min were allowed for absorption of the products by the solution. The elements Cl, P, and S in the wood were converted to chloride (Cl^-), phosphate (PO_4^{3-}), and sulfate (SO_4^{2-}) in the oxidizing solution.

Analysis of the solution by ion chromatography (Chapter 22) showed that the average chlorine uptake by the tree in the period 1971–1986 was 10 times greater than the average uptake in the period 1956–1970. Uptake of phosphorus and sulfur increased by a factor of 2 in the recent 15-year period. The chemical content of the rings presumably mirrors chemical changes in the environment.

Chapter 6

GRAVIMETRIC AND COMBUSTION ANALYSIS

*I*n **gravimetric analysis,** the mass of a product is used to calculate the quantity of original analyte. In the early 1900s, Nobel Prize–winning gravimetric analysis by T. W. Richards and his colleagues measured the atomic mass of Ag, Cl, and N to six-figure accuracy and formed the basis for accurate atomic mass determinations. In **combustion analysis,** a sample is burned in excess oxygen and the

Gravimetric procedures were the mainstay of chemical analyses of ores and industrial materials in the eighteenth and nineteenth centuries, long before the chemical basis for the procedures was understood.

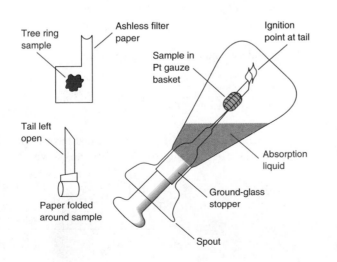

FIGURE 6-1 Schöniger combustion procedure, using 10- to 100-mg tree ring samples. The Pt gauze basket helps catalyze complete combustion of the sample, which is carried out behind a blast shield. *Ashless* filter paper is a special grade that leaves little residue when it is burned. Prior to analysis of tree rings, the filter paper had to be washed several times with distilled water to remove traces of Cl^- and SO_4^{2-}.

117

products are measured. Combustion is typically used to determine the amounts of C, H, N, S, and halogens in organic compounds.

Gravimetric analysis has been largely replaced by instrumental methods of analysis, which are faster and less labor intensive. However, gravimetric determinations carried out by a skilled analyst remain among the most accurate methods available for producing standards used in instrumental analysis. Students are still exposed to some gravimetric analysis early in their laboratory career because gravimetric procedures demand good laboratory technique to produce accurate and precise results.

6-1 Examples of Gravimetric Analysis

An important and widely used industrial gravimetric analysis is the Rose Gottlieb method for measuring fat in food products.[1] First, a weighed food sample is dissolved in an appropriate manner to solubilize the protein. Ammonia and ethanol are then added to break up microscopic droplets of fat, which are then extracted into an organic solvent. Proteins and carbohydrates remain in the aqueous phase. After separating the organic phase from the aqueous phase, the organic phase is evaporated to dryness at 102°C and the residue from the organic solvent is weighed. The residue consists of the fats from the food.

A simple gravimetric analysis that you are likely to encounter is the determination of Cl^- by precipitation with Ag^+:

$$Ag^+ + Cl^- \longrightarrow AgCl(s)$$

The mass of AgCl product tells us how many moles of AgCl were produced. For every mole of AgCl, there must have been one mole of Cl^- in the unknown solution.

EXAMPLE   A Simple Gravimetric Calculation

A 10.0-mL solution containing Cl^- was treated with excess $AgNO_3$ to precipitate 0.436 8 g of AgCl (FM 143.321). What was the molarity of Cl^- in the unknown?

SOLUTION A precipitate weighing 0.463 8 g contains

$$\frac{0.436\ 8\ \text{g AgCl}}{143.321\ \text{g AgCl/mol AgCl}} = 3.048 \times 10^{-3}\ \text{mol AgCl}$$

Because 1 mol of AgCl contains 1 mol of Cl^-, there must have been 3.048×10^{-3} mol of Cl^- in the unknown. The molarity of Cl^- in the unknown is therefore

$$[Cl^-] = \frac{3.048 \times 10^{-3}\ \text{mol}}{0.010\ 00\ \text{L}} = 0.304\ 8\ \text{M}$$

TABLE 6-1   Representative gravimetric analyses

Species analyzed	Precipitated form	Form weighed	Some interfering species
K^+	$KB(C_6H_5)_4$	$KB(C_6H_5)_4$	NH_4^+, Ag^+, Hg^{2+}, Tl^+, Rb^+, Cs^+
Mg^{2+}	$Mg(NH_4)PO_4 \cdot 6H_2O$	$Mg_2P_2O_7$	Many metals except Na^+ and K^+
Ca^{2+}	$CaC_2O_4 \cdot H_2O$	$CaCO_3$ or CaO	Many metals except Mg^{2+}, Na^+, K^+
Ba^{2+}	$BaSO_4$	$BaSO_4$	Na^+, K^+, Li^+, Ca^{2+}, Al^{3+}, Cr^{3+}, Fe^{3+}, Sr^{2+}, Pb^{2+}, NO_3^-
Cr^{3+}	$PbCrO_4$	$PbCrO_4$	Ag^+, NH_4^+
Mn^{2+}	$Mn(NH_4)PO_4 \cdot H_2O$	$Mn_2P_2O_7$	Many metals
Fe^{3+}	$Fe(HCO_2)_3$	Fe_2O_3	Many metals
Co^{2+}	Co(1-nitroso-2-naphtholate)$_3$	$CoSO_4$ (by reaction with H_2SO_4)	Fe^{3+}, Pd^{2+}, Zr^{4+}
Ni^{2+}	Ni(dimethylglyoximate)$_2$	Same	Pd^{2+}, Pt^{2+}, Bi^{3+}, Au^{3+}
Cu^{2+}	$CuSCN$	$CuSCN$	NH_4^+, Pb^{2+}, Hg^{2+}, Ag^+
Zn^{2+}	$Zn(NH_4)PO_4 \cdot H_2O$	$Zn_2P_2O_7$	Many metals
Al^{3+}	Al(8-hydroxyquinolate)$_3$	Same	Many metals
Sn^{4+}	Sn(cupferron)$_4$	SnO_2	Cu^{2+}, Pb^{2+}, As(III)
Pb^{2+}	$PbSO_4$	$PbSO_4$	Ca^{2+}, Sr^{2+}, Ba^{2+}, Hg^{2+}, Ag^+, HCl, HNO_3
NH_4^+	$NH_4B(C_6H_5)_4$	$NH_4B(C_6H_5)_4$	K^+, Rb^+, Cs^+
Cl^-	$AgCl$	$AgCl$	Br^-, I^-, SCN^-, S^{2-}, $S_2O_3^{2-}$, CN^-
Br^-	$AgBr$	$AgBr$	Cl^-, I^-, SCN^-, S^{2-}, $S_2O_3^{2-}$, CN^-
I^-	AgI	AgI	Cl^-, Br^-, SCN^-, S^{2-}, $S_2O_3^{2-}$, CN^-
SCN^-	$CuSCN$	$CuSCN$	NH_4^+, Pb^{2+}, Hg^{2+}, Ag^+
CN^-	$AgCN$	$AgCN$	Cl^-, Br^-, I^-, SCN^-, S^{2-}, $S_2O_3^{2-}$
F^-	$(C_6H_5)_3SnF$	$(C_6H_5)_3SnF$	Many metals (except alkali metals), SiO_4^{4-}, CO_3^{2-}
ClO_4^-	$KClO_4$	$KClO_4$	
SO_4^{2-}	$BaSO_4$	$BaSO_4$	Na^+, K^+, Li^+, Ca^{2+}, Al^{3+}, Cr^{3+}, Fe^{3+}, Sr^{2+}, Pb^{2+}, NO_3^-
PO_4^{3-}	$Mg(NH_4)PO_4 \cdot 6H_2O$	$Mg_2P_2O_7$	Many metals except Na^+, K^+
NO_3^-	Nitron nitrate	Nitron nitrate	ClO_4^-, I^-, SCN^-, CrO_4^{2-}, ClO_3^-, NO_2^-, Br^-, $C_2O_4^{2-}$

Representative analytical precipitations are listed in Table 6-1. Potentially interfering substances listed in the table may need to be removed prior to analysis. A few common organic **precipitants** (agents that cause precipitation) are listed in Table 6-2.

For those who are not familiar with drawing structures of organic compounds, Box 6-1 provides a primer.

Ask Yourself

6-A. A 50.00-mL solution containing NaBr was treated with excess $AgNO_3$ to precipitate 0.214 6 g of AgBr (FM 187.772).

(a) How many moles of AgBr product were isolated?

(b) What was the molarity of NaBr in the solution?

TABLE 6-2 Common organic precipitating agents

Name	Structure	Some ions precipitated
Dimethylglyoxime		Ni^{2+}, Pd^{2+}, Pt^{2+}
Cupferron		Fe^{3+}, VO_2^+, Ti^{4+}, Zr^{4+}, Ce^{4+}, Ga^{3+}, Sn^{4+}
8-Hydroxyquinoline (oxine)		Mg^{2+}, Zn^{2+}, Cu^{2+}, Cd^{2+}, Pb^{2+}, Al^{3+}, Fe^{3+}, Bi^{3+}, Ga^{3+}, Th^{4+}, Zr^{4+}, UO_2^{2+}, TiO^{2+}
1-Nitroso-2-naphthol		Co^{2+}, Fe^{3+}, Pd^{2+}, Zr^{4+}
Nitron		NO_3^-, ClO_4^-, BF_4^-, WO_4^{2-}
Sodium tetraphenylborate	$Na^+B(C_6H_5)_4^-$	K^+, Rb^+, Cs^+, NH_4^+, Ag^+, organic ammonium ions
Tetraphenylarsonium chloride	$(C_6H_5)_4As^+Cl^-$	$Cr_2O_7^{2-}$, MnO_4^-, ReO_4^-, MoO_4^{2-}, WO_4^{2-}, ClO_4^-, I_3^-

6-2 Precipitation

The ideal gravimetric precipitate should be insoluble, be easily filtered, and possess a known, constant composition. The precipitate should be stable when you heat it to remove the last traces of solvent. Although few substances meet these requirements, techniques described in this section help optimize the properties of precipitates.

Precipitate particles should be large enough to be collected by filtration; they should not be so small that they clog or pass through the filter. Large crystals also have less surface area to which foreign species may become attached. At the other extreme is a *colloid,* whose particles are so small (1–100 nm) that they pass through most filters (Demonstration 6-1).

Colloids and Dialysis

Colloids are particles with diameters in the range 1–100 nm. They are larger than molecules, but too small to precipitate. Colloids remain in solution indefinitely, suspended by the Brownian motion (random movement) of solvent molecules.

To make a colloid, heat a beaker containing 200 mL of distilled water to 70°–90°C and leave an identical beaker of water at room temperature. Add 1 mL of 1 M $FeCl_3$ to each beaker and stir. The warm solution turns brown-red in a few seconds, whereas the cold solution remains yellow (Color Plate 2a). The yellow color is characteristic of low molecular mass Fe^{3+} compounds. The red color results from colloidal aggregates of Fe^{3+} ions held together by hydroxide, oxide, and some chloride ions. These particles have a molecular mass of $\sim 10^5$ and a diameter of ~ 10 nm and contain $\sim 10^3$ atoms of Fe.

To demonstrate the size of colloidal particles, we perform a **dialysis** experiment in which two solutions are separated by a *semipermeable membrane*. The membrane has pores through which small molecules, but not large molecules and colloids, can diffuse. Cellulose dialysis tubing (such as catalog number 3787 from A.H. Thomas Co.) has 1–5 nm pores.

Pour some of the brown-red colloidal Fe solution into a dialysis tube knotted at one end; then tie off the other end. Drop the tube into a flask of distilled water to show that the color remains entirely within the bag even after several days (Color Plate 2b and 2c). For comparison, leave an identical bag containing a dark blue solution of 1 M $CuSO_4 \cdot 5H_2O$ in another flask. Cu^{2+} diffuses out of the bag and the solution in the flask becomes light blue in 24 h. Alternatively, the yellow food coloring, tartrazine, can be used in place of Cu^{2+}. If dialysis is conducted in hot water, it is completed during one class period.

Dialysis is used to treat patients suffering from kidney failure. Their blood is run over a dialysis membrane having a very large surface area. Small metabolic waste products in the blood diffuse across the membrane and are diluted into a large volume of liquid going out as waste. Large proteins, which are a necessary part of the blood plasma, cannot cross the membrane and are retained in the blood.

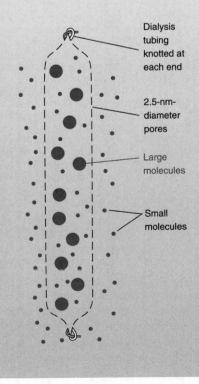

Dialysis tubing knotted at each end

2.5-nm-diameter pores

Large molecules

Small molecules

Large molecules remain trapped inside a dialysis bag, whereas small molecules diffuse through the membrane in both directions. The opening of Chapter 21 describes how a *microdialysis probe* is used for sampling fluids in living organisms.

Box 6-1 *Explanation*

Shorthand for Organic Structures

Chemists and biochemists use simple conventions for drawing structures of carbon-containing compounds to avoid drawing every atom. Each vertex of a structure is understood to be a carbon atom, unless otherwise labeled. In the shorthand, we usually omit bonds from carbon to hydrogen. Carbon forms four chemical bonds. If you see carbon forming fewer than four bonds, the remaining bonds are assumed to go to hydrogen atoms that are not written. Here are some examples:

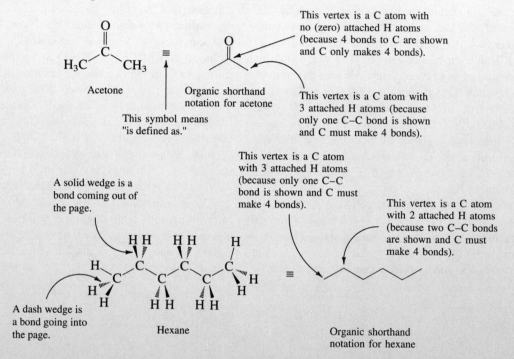

This vertex is a C atom with no (zero) attached H atoms (because 4 bonds to C are shown and C only makes 4 bonds).

This vertex is a C atom with 3 attached H atoms (because only one C–C bond is shown and C must make 4 bonds).

This symbol means "is defined as."

Organic shorthand notation for acetone

Acetone

This vertex is a C atom with 3 attached H atoms (because only one C–C bond is shown and C must make 4 bonds).

This vertex is a C atom with 2 attached H atoms (because two C–C bonds are shown and C must make 4 bonds).

A solid wedge is a bond coming out of the page.

A dash wedge is a bond going into the page.

Hexane

Organic shorthand notation for hexane

Atoms other than carbon and hydrogen are always shown. Hydrogen atoms attached to atoms other than carbon are always shown. Oxygen and sulfur normally make two bonds. Nitrogen makes three bonds if it is neutral and four bonds if it is a cation. Here are some examples:

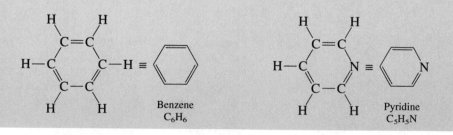

Benzene
C_6H_6

Pyridine
C_5H_5N

Crystal Growth

Crystallization occurs in two phases: nucleation and particle growth. During **nucleation,** dissolved molecules come together randomly and form small aggregates. In *particle growth,* more molecules add to a nucleus to form a crystal.

A solution containing more dissolved solute than should be present at equilibrium is said to be **supersaturated.** Nucleation proceeds faster than particle growth in a highly supersaturated solution, creating tiny particles or, worse, a colloid. In a more dilute solution, nucleation is slower, so the nuclei have a chance to grow into

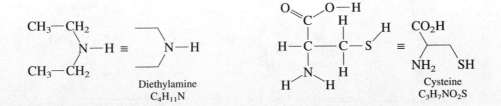

Diethylamine
$C_4H_{11}N$

Cysteine
$C_3H_7NO_2S$

Because of the two equivalent resonance structures of a benzene ring, the alternating single and double bonds are often replaced by a circle:

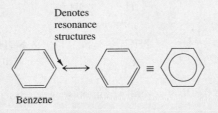

Denotes resonance structures

Benzene

Exercise: Write the chemical formula (e.g., C_4H_8O) for each structure below.

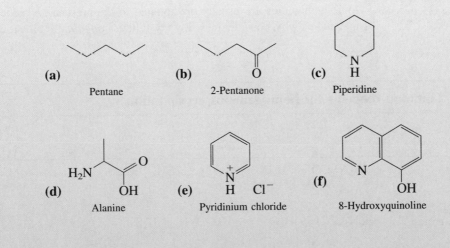

(a) Pentane

(b) 2-Pentanone

(c) Piperidine

(d) Alanine

(e) Pyridinium chloride

(f) 8-Hydroxyquinoline

Answers: (a) C_5H_{12}; (b) $C_5H_{10}O$; (c) $C_5H_{11}N$; (d) $C_3H_7NO_2$; (e) C_5H_6NCl; (f) C_9H_7NO

larger, more tractable particles. Techniques that promote particle growth include

1. raising the temperature to increase solubility and thereby decrease supersaturation

2. adding precipitant slowly with vigorous mixing, to avoid high local supersaturation where the stream of precipitant first enters the analyte

3. keeping the volume of solution large so that the concentrations of analyte and precipitant are low

123

Homogeneous Precipitation

In **homogeneous precipitation,** precipitant is generated slowly by a chemical reaction. For example, urea decomposes in boiling water to produce OH^-:

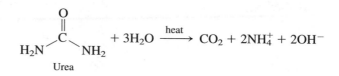

By using urea, we can raise the pH of a solution very gradually, and the slow OH^- formation enhances the particle size of iron(III) formate precipitate:

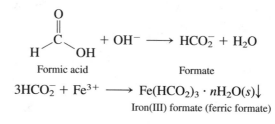

$$3HCO_2^- + Fe^{3+} \longrightarrow Fe(HCO_2)_3 \cdot nH_2O(s)\downarrow$$

Iron(III) formate (ferric formate)

Table 6-3 lists several common reagents for homogeneous precipitation.

Precipitation in the Presence of Electrolyte

An *electrolyte* is a compound that dissociates into ions when it dissolves. We say that the electrolyte ionizes when it dissolves.

Ionic compounds are usually precipitated in the presence of added electrolyte. To understand why, consider how tiny crystallites *coagulate* (come together) into larger

TABLE 6-3   Common reagents for homogeneous precipitation

Precipitant	Reagent	Reaction	Some elements precipitated
OH^-	Urea	$(H_2N)_2CO + 3H_2O \longrightarrow CO_2 + 2NH_4^+ + 2OH^-$	Al, Ga, Th, Bi, Fe, Sn
S^{2-}	Thioacetamide[a]	$CH_3\overset{S}{\overset{\|}{C}}NH_2 + H_2O \longrightarrow CH_3\overset{O}{\overset{\|}{C}}NH_2 + H_2S$	Sb, Mo, Cu, Cd
SO_4^{2-}	Sulfamic acid	$H_3\overset{+}{N}SO_3^- + H_2O \longrightarrow NH_4^+ + SO_4^{2-} + H^+$	Ba, Ca, Sr, Pb
$C_2O_4^{2-}$	Dimethyl oxalate	$CH_3O\overset{O\,O}{\overset{\|\;\|}{CC}}OCH_3 + 2H_2O \longrightarrow 2CH_3OH + C_2O_4^{2-} + 2H^+$	Ca, Mg, Zn
PO_4^{3-}	Trimethyl phosphate	$(CH_3O)_3P{=}O + 3H_2O \longrightarrow 3CH_3OH + PO_4^{3-} + 3H^+$	Zr, Hf

a. Hydrogen sulfide is volatile and toxic; it should be handled only in a well-vented hood. Thioacetamide is a carcinogen that should be handled with gloves. If thioacetamide contacts your skin, wash yourself thoroughly immediately. Leftover reagent is destroyed by heating at 50°C with 5 mol of NaOCl per mole of thioacetamide, and then washing the products down the drain.

crystals. We will illustrate the case of AgCl, which is commonly formed in the presence of 0.1 M HNO$_3$.

Figure 6-2 shows a colloidal particle of AgCl growing in a solution containing excess Ag$^+$, H$^+$, and NO$_3^-$. The particle has an excess positive charge due to **adsorption** of extra silver ions on exposed chloride ions. (To be adsorbed means to be attached to the surface. In contrast, **absorption** involves penetration beyond the surface, to the inside.) The positively charged surface of the solid attracts anions and repels cations from the *ionic atmosphere* in the liquid surrounding the particle.

Colloidal particles must collide with each other to coalesce. However, the negatively charged ionic atmospheres of the particles repel one another. The particles must have enough kinetic energy to overcome electrostatic repulsion before they can coalesce. Heating promotes coalescence by increasing the particles' kinetic energy.

Increasing electrolyte concentration (HNO$_3$ for AgCl) decreases the volume of the ionic atmosphere and allows particles to approach closer together before repulsion becomes significant. Therefore, most gravimetric precipitations are done in the presence of electrolyte.

Although it is common to find the excess common ion adsorbed on the crystal surface, it is also possible to find other ions selectively adsorbed. In the presence of citrate and sulfate, there is more citrate than sulfate adsorbed on a particle of BaSO$_4$.

Digestion

Digestion is the process of allowing a precipitate to stand in contact with the *mother liquor* for some period of time, usually with heating. Digestion promotes slow recrystallization of the precipitate. Particle size increases and impurities tend to be expelled from the crystal.

Mother liquor is the solution from which a substance crystallized.

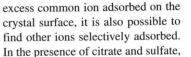

Boundary of
ionic atmosphere

FIGURE 6-2 Schematic diagram showing a colloidal particle of AgCl in a solution containing excess Ag$^+$, H$^+$, and NO$_3^-$. The particle has a net positive charge because of adsorbed Ag$^+$ ions. The region of solution surrounding the particle is called the *ionic atmosphere*. It has a net negative charge, because the particle attracts anions and repels cations.

Adsorbed impurities are bound to the surface of a crystal. *Absorbed* impurities (within the crystal) are classified as *inclusions* or *occlusions*. Inclusions are impurity ions that randomly occupy sites in the crystal lattice normally occupied by ions that belong in the crystal. Inclusions are more likely when the impurity ion has a size and charge similar to those of one of the ions that belongs to the product. Occlusions are pockets of impurity that are literally trapped inside the growing crystal.

Adsorbed, occluded, and included impurities are said to be **coprecipitated.** That is, the impurity is precipitated along with the desired product, even though the solubility of the impurity has not been exceeded. Coprecipitation tends to be worst in colloidal precipitates (which have a large surface area), such as $BaSO_4$, $Al(OH)_3$, and $Fe(OH)_3$. Some procedures call for washing away the mother liquor, redissolv-

---

Box 6-2 *Explanation*

A Case Study in Coprecipitation and Digestion[2]

This box requires an explanation from you instead of from Dan. Students at the University of Tulsa experimented with procedural variations to learn the effects of coprecipitation and digestion in the gravimetric analysis of barium by precipitation of $BaSO_4$. The basic procedure, described below, was carried out in duplicate:

1. Pipet 50.0 mL of standard (~0.05 M) $BaCl_2$ into a 400-mL beaker and add 100 mL of H_2O.

2. Add 1 mL of 16 M HCl, which increases the solubility of $BaSO_4$ and thereby decreases supersaturation during precipitation. Solubility is increased because the reaction $SO_4^{2-} + H^+ \rightleftharpoons HSO_4^-$ consumes some dissolved sulfate.

3. Heat to 90°C to increase the solubility of $BaSO_4$ and promote crystallization during precipitation.

4. Add 50 mL of 0.15 M Na_2SO_4 all at once.

5. Digest the precipitate at 90°C for 1 h in the mother liquor.

6. Decant the mother liquor (pour it off), wash the precipitate with hot water, and decant the wash water.

7. Collect the precipitate by filtration through ashless filter paper and wash it with hot water until no more Cl^- is detected in the filtrate (by precipitation with Ag^+).

8. Transfer the paper and precipitate to a porcelain crucible and ignite it over a Bunsen burner to burn away the paper, leaving pure $BaSO_4$.

9. Cool the crucible in a desiccator and weigh it. Reheat and reweigh until successive weighings agree within ±0.3 mg.

Randy and Khang each carried out the basic procedure twice and obtained an average of 0.497 g of $BaSO_4$ product (range = 0.495 to 0.499).

Coprecipitation of Ca^{2+}: Antonio followed the same procedure, but the initial solution contained 0.5 mmol of $CaCl_2$ in addition to $BaCl_2$. The average mass of his product was 0.505 g (0.503 and 0.507 g). How would you explain this?

Prolonged digestion: Celine carried out the same procedure as Antonio (with $CaCl_2$ present) but digested her product for 20 h, instead of 1 h, at 90°C. She obtained an average mass of 0.497 g (0.495 and 0.499 g) of product. Why did her results agree with those of Randy and Khang even though Ca^{2+} was present in her initial sample?

ing the precipitate, and *reprecipitating* the product. During the second precipitation, the concentration of impurities in the solution is lower than during the first precipitation, and the degree of coprecipitation therefore tends to be lower. Box 6-2 illustrates effects of coprecipitation and digestion.

Occasionally, a trace component that is too dilute to be measured is intentionally concentrated by coprecipitation with a major component of the solution. The procedure is called **gathering,** and the precipitate used to collect the trace component is said to be a *gathering agent.* When the precipitate is dissolved in a small volume of solvent, the concentration of the trace component is high enough for accurate analysis.

Some impurities can be treated with a **masking agent,** which prevents them from reacting with the precipitant. In the gravimetric analysis of Be^{2+}, Mg^{2+}, Ca^{2+}, or Ba^{2+} with the reagent *N-p*-chlorophenylcinnamohydroxamic acid (designated RH), impurities such as Ag^+, Mn^{2+}, Zn^{2+}, Cd^{2+}, Hg^{2+}, Fe^{2+}, and Ga^{3+} are kept in solution by excess KCN.

$$Ca^{2+} + 2RH \longrightarrow CaR_2(s)\!\downarrow + 2H^+$$
Analyte   Precipitate

$$Mn^{2+} + 6CN^- \longrightarrow Mn(CN)_6^{4-}$$
Impurity   Masking agent   Stays in solution

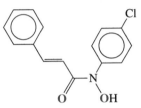

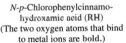

N-p-Chlorophenylcinnamo-
hydroxamic acid (RH)
(The two oxygen atoms that bind
to metal ions are bold.)

Impurities might collect on the product while it is standing in the mother liquor. This process is called *postprecipitation* and usually involves a supersaturated impurity that does not readily crystallize. An example is the crystallization of magnesium oxalate (MgC_2O_4) on calcium oxalate (CaC_2O_4).

Washing precipitate on a filter helps remove droplets of liquid containing excess solute. Some precipitates can be washed with water, but many require electrolyte to maintain coherence. For these precipitates, the ionic atmosphere is required to neutralize the surface charge of the tiny particles. If electrolyte is washed away with water, the charged solid particles repel one another and the product breaks up. This breaking up, called **peptization,** results in loss of product through the filter. Silver chloride peptizes if washed with water, so it is washed with dilute HNO_3 instead. Volatile electrolytes including HNO_3, HCl, NH_4NO_3, NH_4Cl, and $(NH_4)_2CO_3$ are used for washing because they evaporate during drying.

Product Composition

The final product must have a known, stable composition. A **hygroscopic substance** is one that picks up water from the air and is therefore difficult to weigh accurately. Many precipitates contain a variable quantity of water and must be dried under conditions that give a known (possibly zero) stoichiometry of H_2O.

Ignition (strong heating) is used to change the chemical form of some precipitates that do not have a constant composition after drying at moderate temperatures. For example, $Fe(HCO_2)_3 \cdot nH_2O$ is ignited at 850°C for 1 hr to give Fe_2O_3, and $Mg(NH_4)PO_4 \cdot 6H_2O$ is ignited at 1 100°C to give $Mg_2P_2O_7$.

In **thermogravimetric analysis,** a sample is heated, and its mass is measured as a function of temperature. Figure 6-3 shows how the composition of calcium salicylate changes in four stages:

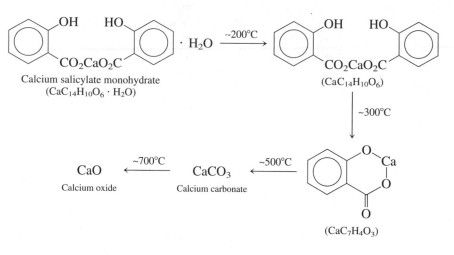

The composition of the product depends on the temperature and duration of heating.

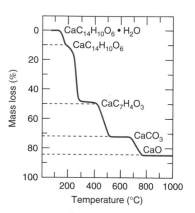

FIGURE 6-3 Thermogravimetric curve for calcium salicylate. [From G. Liptay, ed., *Atlas of Thermoanalytical Curves* (London: Heyden and Son, 1976).]

Ask Yourself

6-B. See if you have digested this section by answering these questions:

  (a) What is the difference between absorption and adsorption?

  (b) How is an inclusion different from an occlusion?

  (c) What are desirable properties of a gravimetric precipitate?

  (d) Why is high supersaturation undesirable in a gravimetric precipitation?

  (e) How can you decrease supersaturation during a precipitation?

  (f) Why are many ionic precipitates washed with electrolyte solution instead of pure water?

  (g) Why is it less desirable to wash AgCl precipitate with aqueous $NaNO_3$ than with HNO_3 solution?

  (h) Why would a reprecipitation be employed in a gravimetric analysis?

  (i) What is done in thermogravimetric analysis?

6-3 Examples of Gravimetric Calculations

We now illustrate how to relate the mass of a gravimetric precipitate to the quantity of original analyte. *The general approach is to relate the moles of product to the moles of reactant.*

EXAMPLE   Relating Mass of Product to Mass of Reactant

The piperazine content of an impure commercial material can be determined by precipitating and weighing piperazine diacetate:

$$:NH \quad HN: + 2CH_3CO_2H \longrightarrow H_2\overset{+}{N} \quad \overset{+}{N}H_2(CH_3CO_2^-)_2$$

Piperazine	Acetic acid	Piperazine diacetate
FM 86.136	FM 60.052	FM 206.240

(6-1)

In one experiment, 0.312 6 g of the sample was dissolved in 25 mL of acetone, and 1 mL of acetic acid was added. After 5 min, the precipitate was filtered, washed with acetone, dried at 110°C, and found to weigh 0.712 1 g. What is the weight percent of piperazine in the commercial material?

If you were performing this analysis, it would be important to determine that impurities in the piperazine are not also precipitated.

SOLUTION For each mole of piperazine in the impure material, 1 mol of product is formed.

$$\text{moles of piperazine} = \text{moles of product} = \frac{0.712\ 1\ \text{g product}}{206.240\ \dfrac{\text{g product}}{\text{mol product}}}$$

$$= 3.453 \times 10^{-3}\ \text{mol}$$

This many moles of piperazine corresponds to

grams of piperazine =

$$(3.453 \times 10^{-3}\ \text{mol piperazine})\left(86\ 136\ \frac{\text{g piperazine}}{\text{mol piperazine}}\right) = 0.297\ 4\ \text{g}$$

which gives

$$\text{wt \% piperazine in analyte} = \frac{0.297\ 4\ \text{g piperazine}}{0.312\ 6\ \text{g unknown}} \times 100 = 95.14\%$$

Reminder:

$$\text{wt \%} = \frac{\text{mass of analyte}}{\text{mass of unknown}} \times 100$$

EXAMPLE   When the Stoichiometry Is Not 1:1

Solid residue weighing 8.444 8 g from an aluminum refining process was dissolved in acid, treated with 8-hydroxyquinoline, and ignited to give Al_2O_3 weighing 0.855 4 g. Find the weight percent of Al in the original mixture.

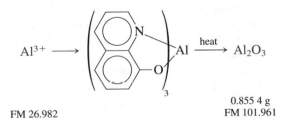

Al³⁺ ⟶ ... ⟶ Al₂O₃

FM 26.982

0.855 4 g
FM 101.961

SOLUTION Each mole of product (Al_2O_3) contains two moles of Al. The mass of product tells us the moles of product, and from this we can find the moles of Al. The moles of product are $(0.855\ 4\ \text{g})/(101.961\ \text{g/mol}) = 0.008\ 389_5\ \text{mol Al}_2O_3$.

A note from Dan: Whether or not I show it, I keep at least one extra, insignificant figure in my calculations and do not round off until the final answer. Usually, I keep all the digits in my calculator.

Because each mole of product contains two moles of Al, there must have been

$$\text{moles of Al in unknown} = \frac{2 \text{ mol Al}}{\text{mol Al}_2O_3} \times 0.008\ 389_5 \text{ mol Al}_2O_3 = 0.016\ 77_9 \text{ mol Al}$$

The mass of Al is $(0.016\ 77_9 \text{ mol})(26.982 \text{ g/mol}) = 0.452\ 7_3 \text{ g Al}$. The weight percent of Al in the unknown is

$$\text{wt \% Al} = \frac{0.452\ 7_3 \text{ g Al}}{8.444\ 8 \text{ g unknown}} \times 100 = 5.361\%$$

EXAMPLE   Calculating How Much Precipitant to Use

(a) To measure the nickel content in steel, the steel is dissolved in 12 M HCl and neutralized in the presence of citrate ion, which binds iron and keeps it in solution. The slightly basic solution is warmed, and dimethylglyoxime (DMG) is added to precipitate the red DMG-nickel complex. The product is filtered, washed with cold water, and dried at 110°C.

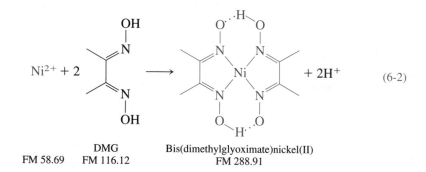

	DMG	Bis(dimethylglyoximate)nickel(II)
FM 58.69	FM 116.12	FM 288.91

$$\text{Ni}^{2+} + 2 \longrightarrow + 2\text{H}^+ \qquad (6\text{-}2)$$

1.0 wt % DMG means
$$\frac{1.0 \text{ g DMG}}{100 \text{ g solution}}$$

Density means
$$\frac{\text{grams of solution}}{\text{mL of solution}}$$

If the nickel content is known to be near 3 wt % and you wish to analyze 1.0 g of the steel, what volume of 1.0 wt % DMG in alcohol solution should be used to give a 50% excess of DMG for the analysis? Assume that the density of the alcohol solution is 0.79 g/mL.

SOLUTION Our strategy is to estimate the moles of Ni in 1.0 g of steel. Equation 6-2 tells us that two moles of DMG are required for each mole of Ni. After finding the required number of moles of DMG, we will multiply it by 1.5 to get a 50% excess, to be sure we have enough.

The Ni content of the steel is around 3%, so 1.0 g of steel contains about $(0.03)(1.0 \text{ g}) = 0.03 \text{ g}$ of Ni, which corresponds to $(0.03 \text{ g Ni})/(58.69 \text{ g/mol Ni}) = 5.1 \times 10^{-4} \text{ mol Ni}$. This amount of Ni requires

$$2\left(\frac{\text{mol DMG}}{\text{mol Ni}}\right)(5.1 \times 10^{-4} \text{ mol Ni})\left(116.12 \frac{\text{g DMG}}{\text{mol DMG}}\right) = 0.12 \text{ g DMG}$$

A 50% excess of DMG would be $(1.5)(0.12 \text{ g}) = 0.18 \text{ g}$.

The DMG solution is 1.0 wt %, which means that there are 0.010 g of DMG per gram of solution. The required mass of solution is

$$\left(\frac{0.18 \text{ g DMG}}{0.010 \text{ g DMG/g solution}}\right) = 18 \text{ g solution}$$

The volume of solution is found from the mass of solution and the density:

$$\text{volume} = \frac{\text{mass}}{\text{density}} = \frac{18 \text{ g solution}}{0.79 \text{ g solution/mL}} = 23 \text{ mL}$$

$$\text{density} = \frac{\text{mass}}{\text{volume}}$$

(b) If 1.163 4 g of steel gave 0.179 5 g of $Ni(DMG)_2$ precipitate, what is the weight percent of Ni in the steel?

SOLUTION Here is the strategy: From the mass of precipitate, we can find the moles of precipitate. We know that one mole of precipitate comes from one mole of Ni in Equation 6-2. From the moles of Ni, we can compute the mass of Ni and its weight percent in the steel:

First, find the moles of precipitate in 0.179 5 g of precipitate:

$$\frac{0.179 \text{ 5 g Ni(DMG)}_2}{288.91 \text{ g Ni(DMG)}_2/\text{mol Ni(DMG)}_2} = 6.213 \times 10^{-4} \text{ mol Ni(DMG)}_2$$

There must have been 6.213×10^{-4} mol of Ni in the steel. The mass of Ni in the steel is

$$(6.213 \times 10^{-4} \text{ mol Ni})\left(58.69 \frac{\text{g}}{\text{mol Ni}}\right) = 0.036 \text{ 46 g}$$

and the weight percent of Ni in steel is

$$\text{wt \% Ni} = \frac{0.036 \text{ 46 g Ni}}{1.163 \text{ 4 g steel}} \times 100 = 3.134\%$$

Ask Yourself

6-C. The element cerium, discovered in 1839 and named for the asteroid Ceres, is a major component of flint lighters. To find the Ce^{4+} content of a solid, an analyst dissolved 4.37 g of the solid and treated it with excess iodate to precipitate $Ce(IO_3)_4$. The precipitate was collected, washed, dried, and ignited to produce 0.104 g of CeO_2.

$$Ce^{4+} + 4IO_3^- \longrightarrow Ce(IO_3)_4(s) \xrightarrow{\text{heat}} CeO_2(s)$$
$$\text{FM 172.115}$$

(a) How much cerium is contained in 0.104 g of CeO_2?
(b) What was the weight percent of Ce in the original solid?

6-4   Combustion Analysis

A historically important form of gravimetric analysis is *combustion analysis,* used to determine the carbon and hydrogen content of organic compounds burned in excess O_2. Modern combustion analyzers use thermal conductivity, infrared absorption, or electrochemical methods to measure the products.

Gravimetric Combustion Analysis

In gravimetric combustion analysis (Figure 6-4), partially combusted product is passed through catalysts such as Pt gauze, CuO, PbO_2, or MnO_2 at elevated temperature to complete the oxidation to CO_2 and H_2O. The products are flushed through a chamber containing P_4O_{10} ("phosphorus pentoxide"), which absorbs water, and then through a chamber of Ascarite (NaOH on asbestos), which absorbs CO_2. The increase in mass of each chamber tells how much hydrogen and carbon, respectively, were initially present. A guard tube prevents atmospheric H_2O or CO_2 from entering the chambers from the exit.

EXAMPLE   Combustion Analysis Calculations

A compound weighing 5.714 mg produced 14.414 mg of CO_2 and 2.529 mg of H_2O upon combustion. Find the weight percent of C and H in the sample.

SOLUTION One mole of CO_2 contains one mole of carbon. Therefore

$$\text{moles of C in sample} = \text{moles of } CO_2 \text{ produced}$$

$$= \frac{14.414 \times 10^{-3} \text{ g } CO_2}{44.010 \text{ g } CO_2/\text{mol}} = 3.275 \times 10^{-4} \text{ mol}$$

$$\text{mass of C in sample} = (3.275 \times 10^{-4} \text{ mol C})\left(12.010\,7\,\frac{g}{\text{mol C}}\right) = 3.934 \text{ mg}$$

$$\text{wt \% C} = \frac{3.934 \text{ mg C}}{5.714 \text{ mg sample}} \times 100 = 68.84\%$$

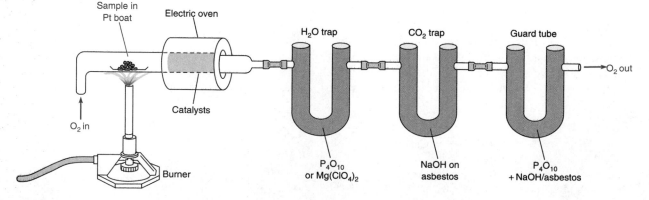

FIGURE 6-4   Gravimetric combustion analysis for carbon and hydrogen.

One mole of H_2O contains two moles of H. Therefore

$$\text{moles of H in sample} = 2(\text{moles of } H_2O \text{ produced})$$

$$= 2\left(\frac{2.529 \times 10^{-3} \text{ g } H_2O}{18.015 \text{ g } H_2O/\text{mol}}\right) = 2.808 \times 10^{-4} \text{ mol}$$

$$\text{mass of H in sample} = (2.808 \times 10^{-4} \text{ mol H})\left(1.007\ 94 \frac{\text{g}}{\text{mol H}}\right) = 2.830 \times 10^{-4} \text{ g}$$

$$\text{wt \% H} = \frac{0.283\ 0 \text{ mg H}}{5.714 \text{ mg sample}} \times 100 = 4.952\%$$

Combustion Analysis Today

Figure 6-5 shows how C, H, N, and S are measured in a single operation. An accurately weighed 2-mg sample is sealed in a tin or silver capsule. The analyzer is swept with He gas that has been treated to remove traces of O_2, H_2O, and CO_2. At the start of a run, a measured excess volume of O_2 is added to the He stream. Then the sample capsule is dropped into a preheated ceramic crucible, where the capsule melts and the sample is rapidly oxidized.

$$C, H, N, S \xrightarrow[O_2]{1\ 050°C} CO_2(g) + H_2O(g) + N_2(g) + \underbrace{SO_2(g) + SO_3(g)}_{95\% \ SO_2}$$

The products pass through a hot WO_3 catalyst to complete the combustion of carbon to CO_2. In the next zone, metallic Cu at 850°C reduces SO_3 to SO_2 and removes excess O_2:

$$Cu + SO_3 \xrightarrow{850°C} SO_2 + CuO(s)$$

$$Cu + \tfrac{1}{2}O_2 \xrightarrow{850°C} CuO(s)$$

Elemental analyzers use an *oxidation catalyst* to complete the oxidation of sample and a *reduction catalyst* to carry out any required reduction and to remove excess O_2.

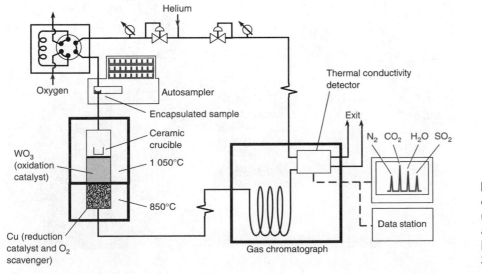

FIGURE 6-5 Schematic diagram of C,H,N,S elemental analyzer that uses gas chromatographic separation and thermal conductivity detection. [From E. Pella, *Am. Lab.* August 1990, 28.]

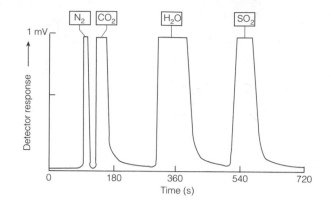

FIGURE 6-6 Gas chromatographic trace from elemental analyzer, showing substantially complete separation of combustion products. The area of each peak (when they are not off-scale) is proportional to the mass of each product. [From E. Pella, *Am. Lab.* August 1990, 28.]

The mixture of CO_2, H_2O, N_2, and SO_2 is separated by gas chromatography, and each component is measured with a thermal conductivity detector described in Section 21-1 (Figure 6-6). Figure 6-7 shows a different C,H,N,S analyzer, which uses infrared absorbance to measure CO_2, H_2O, and SO_2 and thermal conductivity to measure N_2.

A key to elemental analysis is *dynamic flash combustion,* which creates a short burst of gaseous products, instead of slowly bleeding products out over several minutes. This feature is important because chromatographic analysis requires that the whole sample be injected at once. Otherwise, the injection zone is so broad that the products cannot be separated.

In dynamic flash combustion, the sample is encapsulated in tin and dropped into the preheated furnace shortly after the flow of a 50 vol % O_2/50 vol % He mix-

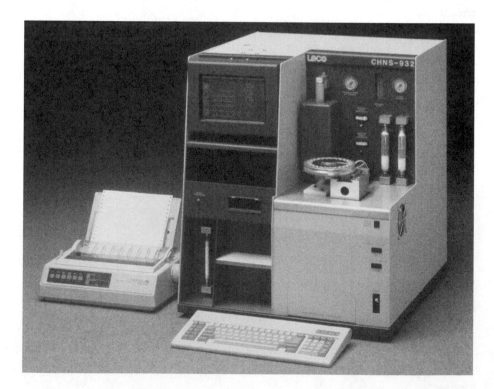

FIGURE 6-7 Combustion analyzer that uses infrared absorbance to measure CO_2, H_2O, and SO_2 and thermal conductivity to measure N_2. Three separate infrared cells in series are equipped with filters to isolate wavelengths absorbed by one of the products. Absorbance is integrated over time as the combustion product mixture is swept through each cell. [Leco Corp., St. Joseph, MI.]

ture is started (Figure 6-8). The Sn capsule melts at 235°C and is instantly oxidized to SnO_2, a process that liberates 594 kJ/mol and heats the sample to 1 700°–1 800°C. Because the sample is dropped in before very much O_2 is present (Figure 6-8), decomposition of the sample occurs prior to oxidation, which minimizes the formation of nitrogen oxides.

Oxygen analysis requires a different strategy. The sample is thermally decomposed (a process called **pyrolysis**) without adding oxygen. The gaseous products are passed through nickel-coated carbon at 1 075°C to convert oxygen from the analyte into CO (not CO_2). Other products include N_2, H_2, CH_4, and hydrogen halides. Acidic products are absorbed by NaOH/asbestos, and the remaining gases are separated and measured by gas chromatography with a thermal conductivity detector.

For halogen analysis, the combustion product contains HX (X = Cl, Br, I). HX is trapped in aqueous solution and titrated with Ag^+ ions by an automated electrochemical process.

Table 6-4 shows analytical results for pure acetanilide obtained with two different commercial instruments. Chemists consider a result within ±0.3 of the theoretical percentage of an element to be good evidence that the compound has the expected formula. For N in acetanilide, ±0.3 corresponds to a relative error of 0.3/10.36 = 3%, which is not hard to achieve. For C, ±0.3 corresponds to a relative error of 0.3/71.09 = 0.4%, which is not so easy. The standard deviation for C in Instrument 1 is 0.41/71.17 = 0.6%; and for Instrument 2, it is 1.1/71.22 = 1.5%.

The Sn capsule is oxidized to SnO_2, which

1. liberates heat to vaporize and crack (decompose) sample
2. uses available oxygen immediately
3. ensures that sample oxidation occurs in gas phase
4. acts as an oxidation catalyst

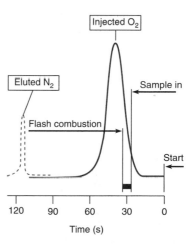

FIGURE 6-8 Sequence of events in dynamic flash combustion. [From E. Pella, *Am. Lab.* February 1990, 116.]

Ask Yourself

6-D. (a) What is the difference between combustion and pyrolysis?
(b) What is the purpose of the WO_3 and Cu in Figure 6-5?
(c) Why is tin used to encapsulate a sample for combustion analysis?
(d) Why is sample dropped into the preheated furnace before the oxygen concentration reaches its peak in Figure 6-8?
(e) What is the balanced equation for the combustion of $C_8H_7NO_2SBrCl$ in a C,H,N,S elemental analyzer?

TABLE 6-4   C, H, and N in acetanilide: $C_6H_5NHCCH_3$ with O double-bonded to the carbonyl carbon

Element	Theoretical value (wt %)	Instrument 1	Instrument 2
C	71.09	71.17 ± 0.41	71.22 ± 1.1
H	6.71	6.76 ± 0.12	6.84 ± 0.10
N	10.36	10.34 ± 0.08	10.33 ± 0.13

Uncertainties are standard deviations from five replicate determinations.
SOURCE: E. M. Hodge, H. P. Patterson, M. C. Williams, and E. S. Gladney, *Anal. Chem.* **1994**, *66*, 1119.

Important Terms

absorption

adsorption

colloid

combustion analysis

coprecipitation

dialysis

digestion

gathering

gravimetric analysis

homogeneous precipitation

hygroscopic substance

ignition

masking agent

nucleation

peptization

precipitant

pyrolysis

supersaturated solution

thermogravimetric analysis

Problems

6-1. An organic compound with a molecular mass of 417 was analyzed for ethoxyl (CH_3CH_2O—) groups by the reactions

$$ROCH_2CH_3 + HI \longrightarrow ROH + CH_3CH_2I$$
$$(R = \text{remainder of molecule})$$
$$CH_3CH_2I + Ag^+ + OH^- \longrightarrow AgI(s) + CH_3CH_2OH$$

A 25.42-mg sample of compound produced 29.03 mg of AgI (FM 234.77). How many ethoxyl groups are there in each molecule?

6-2. A 0.050 02-g sample of impure piperazine contained 71.29 wt % piperazine. How many grams of product will be formed if this sample is analyzed by Reaction 6-1?

6-3. A 1.000-g sample of unknown analyzed by Reaction 6-2 gave 2.500 g of bis(dimethylglyoximate)nickel(II). Find the wt % of Ni in the unknown.

6-4. How many milliliters of 2.15 wt % dimethylglyoxime solution should be used to provide a 50.0% excess for Reaction 6-2 with 0.998 4 g of steel containing 2.07 wt % Ni? The density of the dimethylglyoxime solution is 0.790 g/mL.

6-5. A solution containing 1.263 g of unknown potassium compound was dissolved in water and treated with excess sodium tetraphenylborate, $Na^+B(C_6H_5)_4^-$ solution to precipitate 1.003 g of insoluble $K^+B(C_6H_5)_4^-$ (FM 358.33). Find the wt % of K in the unknown.

6-6. Twenty dietary iron tablets with a total mass of 22.131 g were ground and mixed thoroughly. Then 2.998 g of the powder were dissolved in HNO_3 and heated to convert all the iron to Fe^{3+}. Addition of NH_3 caused quantitative precipitation of $Fe_2O_3 \cdot xH_2O$, which was ignited to give 0.264 g of Fe_2O_3 (FM 159.69). What is the average mass of $FeSO_4 \cdot 7H_2O$ (FM 278.01) in each tablet?

6-7. *The man in the vat problem.* Long ago, a workman at a dye factory fell into a vat containing a hot concentrated mixture of sulfuric and nitric acids. He dissolved completely! Because nobody witnessed the accident, it was necessary to prove that he fell in so that the man's wife could collect his insurance money. The man weighed 70 kg, and a human body contains about 6.3 parts per thousand phosphorus. The acid in the vat was analyzed for phosphorus to see if it contained a dissolved human.

(a) The vat had 8.00×10^3 L of liquid, and 100.0 mL were analyzed. If the man did fall into the vat, what is the expected quantity of phosphorus in 100.0 mL?

(b) The 100.0-mL sample was treated with a molybdate reagent that precipitates ammonium phosphomolybdate, $(NH_4)_3[P(Mo_{12}O_{40})] \cdot 12H_2O$. This substance was dried at 110°C to remove waters of hydration and heated to 400°C until it reached a constant composition corresponding to the formula $P_2O_5 \cdot 24MoO_3$, which weighed 0.371 8 g. When a fresh mixture of the same acids (not from the vat) was treated in the same manner, 0.033 1 g of $P_2O_5 \cdot 24MoO_3$ (FM 3 596.46) was produced. This *blank determination* gives the amount of phosphorus in the starting reagents. The $P_2O_5 \cdot 24MoO_3$ that could have come from the dissolved man is therefore 0.371 8 − 0.033 1 = 0.338 7 g. How much phosphorus was present in the 100.0-mL sample? Is this quantity consistent with a dissolved man?

6-8. Consider a mixture of the two solids, $BaCl_2 \cdot 2H_2O$ (FM 244.26) and KCl (FM 74.551), in an unknown ratio. When the unknown is heated to 160°C for 1 h, the water of crystallization is driven off:

$$BaCl_2 \cdot 2H_2O(s) \xrightarrow{160°C} BaCl_2(s) + 2H_2O(g)$$

A sample originally weighing 1.783 9 g weighed 1.562 3 g after heating. Calculate the weight percent of Ba, K, and Cl in the original sample. (*Hint:* The mass loss tells how much water was lost, which tells how much $BaCl_2 \cdot 2H_2O$ was present. The remainder of the sample is KCl.)

6-9. Write a balanced equation for the combustion of benzoic acid, $C_6H_5CO_2H$, to give CO_2 and H_2O. How many milligrams of CO_2 and of H_2O will be produced by the combustion of 4.635 mg of benzoic acid?

6-10. Combustion of 8.732 mg of an unknown organic compound gave 16.432 mg of CO_2 and 2.840 mg of H_2O.

(a) Find the wt % of C and H in the substance.

(b) Find the smallest reasonable integer mole ratio of C:H in the compound.

6-11. Combustion analysis of a compound known to contain just C, H, N, and O demonstrated that it contains 46.21 wt % C, 9.02 wt % H, 13.74 wt % N, and, by difference, $100 - 46.21 - 9.02 - 13.74 = 31.03$ wt % O. This means that 100 g of unknown would contain 46.21 g of C, 9.02 g of H, etc. Find the atomic ratio C:H:N:O. Then divide each stoichiometry coefficient by the smallest one and express the atomic composition in the lowest reasonable integer ratio ($C_xH_yN_zO_w$, where x, y, z, and w are integers and one of them is 1).

6-12. A method for measuring organic carbon in seawater involves oxidation of the organic materials to CO_2 with $K_2S_2O_8$, followed by gravimetric determination of the CO_2 trapped by a column of NaOH-coated asbestos. A water sample weighing 6.234 g produced 2.378 mg of CO_2 (FM 44.010). Calculate the ppm carbon in the seawater.

6-13. The reagent nitron forms a fairly insoluble salt with nitrate, the product having a solubility of 0.99 g/L near 20°C. Sulfate and acetate do not precipitate with nitron, but many anions including ClO_4^-, ClO_3^-, I^-, SCN^-, and $C_2O_4^{2-}$ do precipitate and interfere with the analysis. A 50.00-mL unknown solution containing KNO_3 and $NaNO_3$ was treated with 1 mL of acetic acid and heated to near boiling; 10 mL of solution containing excess nitron was added with stirring. After cooling to 0°C for 2 h, the crystalline product was filtered, washed with three 5-mL portions of ice-cold, saturated nitron nitrate solution, and finally washed with two 3-mL portions of ice-cold water. The product weighed 0.513 6 g after drying at 105°C for 1 h.

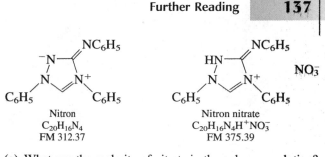

Nitron
$C_{20}H_{16}N_4$
FM 312.37

Nitron nitrate
$C_{20}H_{16}N_4H^+NO_3^-$
FM 375.39

(a) What was the molarity of nitrate in the unknown solution?

(b) Some of the nitron nitrate dissolves in the final ice-cold wash water. Does this dissolution lead to a random or a systematic error in the analysis?

6-14. A mixture of $Al_2O_3(s)$ and $CuO(s)$ weighing 18.371 mg was heated under $H_2(g)$ in a thermogravimetric experiment. Upon reaching a temperature of 1 000°C, the mass was 17.462 mg and the final products were $Al_2O_3(s)$, $Cu(s)$, and $H_2O(g)$. Find the weight percent of Al_2O_3 in the original solid mixture.

6-15. Use the uncertainties from Instrument 1 in Table 6-4 to estimate the uncertainties in the stoichiometry coefficients in the formula $C_8H_{h\pm x}N_{n\pm y}$.

How Would You Do It?

6-16. The fat content of homogenized whole milk was measured by the Rose Gottlieb method described at the beginning of Section 6-1. Replicate measurements were made by a manual method and by an automated method. Do the two methods give the same or statistically different results?

Weight percent fat in milk				
Manual method		Automated method		
2.934	2.925	2.967	2.958	3.022
2.981	2.948	3.034	3.052	2.974
2.906	2.981	3.022	2.983	2.946
2.976	2.913	2.982	2.966	2.997
2.958	2.881	2.992	3.006	3.027
2.945	2.847	2.950	2.982	2.979
2.893	2.880	2.965	2.951	3.047

Notes and References

1. A. R. Matheson and P. Otten, *Am. Lab.* March 1999, 13.

2. T. M. Harris, *J. Chem. Ed.* **1995,** *72,* 355.

Further Reading

C. M. Beck, "Classical Analysis: A Look at the Past, Present, and Future," *Anal. Chem.* **1994,** *66,* 224A.

I. M. Kolthoff, "Analytical Chemistry in the USA in the First Quarter of This Century," *Anal. Chem.* **1994,** *66,* 241A.

Acid Rain

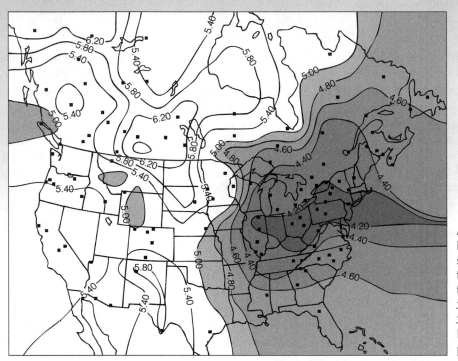

Average pH of precipitation over North America (from a 1981 study). Any pH below 7 is acidic; the lower the pH, the more acidic the water. [From R. Semonin, *Study of Atmospheric Pollution Scavenging* (Washington, DC: U.S. Department of Energy, 1981), pp. 56–119.]

$$\underbrace{NO + NO_2}_{\substack{\text{Nitrogen oxides}\\\text{designated } NO_x}} \xrightarrow[\text{H}_2\text{O}]{\text{oxidation}} \underset{\text{Nitric acid}}{HNO_3} \qquad \underset{\text{Sulfur dioxide}}{SO_2} \xrightarrow[\text{H}_2\text{O}]{\text{oxidation}} \underset{\text{Sulfuric acid}}{H_2SO_4}$$

*C*ombustion products from automobiles and factories include nitrogen oxides and sulfur dioxide, which react with water and oxidizing agents in the atmosphere to produce acids.

Acid rain in North America is most severe in the east, downwind of many coal-burning power plants and factories. The pH of rain over northern Europe is in the range 4.2 to 4.5. Acid rain threatens forests and lakes around the world. For example, half of the essential nutrients Ca^{2+} and Mg^{2+} have been leached from the soil of Sweden's forests since 1950. Acid rain increases the solubility of toxic Al^{3+} and other metals in groundwater. The area over which fish populations have declined or disappeared in acidic lakes in Norway has doubled since 1970.

Chapter 7

INTRODUCING ACIDS AND BASES

*T*he chemistry of acids and bases is probably the most important topic you will study in chemical equilibrium. It is difficult to have a meaningful discussion of subjects ranging from protein folding to the weathering of rocks without understanding acids and bases.

7-1   What Are Acids and Bases?

In aqueous chemistry, an **acid** is a substance that increases the concentration of H_3O^+ (**hydronium ion**). Conversely, a **base** decreases the concentration of H_3O^+ in aqueous solution. As we shall see shortly, a decrease in H_3O^+ concentration necessarily requires an increase in OH^- concentration. Therefore, a base is also a substance that increases the concentration of OH^- in aqueous solution.

The species H^+ is called a *proton* because a proton is all that remains when a hydrogen atom loses its electron. Hydronium ion, H_3O^+, is a combination of H^+ with H_2O (Figure 7-1). Although H_3O^+ is a more accurate representation than H^+ for the hydrogen ion in aqueous solution, we will use H_3O^+ and H^+ interchangeably in this book.

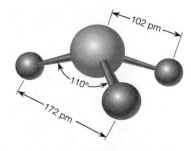

FIGURE 7-1   Structure of hydronium ion, H_3O^+.

139

A more general definition of acids and bases given by Brønsted and Lowry is that an *acid* is a *proton donor* and a *base* is a *proton acceptor*. This definition includes the one stated above. For example, HCl is an acid because it donates a proton to H_2O to form H_3O^+:

$$HCl + H_2O \rightleftharpoons H_3O^+ + Cl^-$$

Brønsted-Lowry acid: proton donor
Brønsted-Lowry base: proton
    acceptor

The Brønsted-Lowry definition can be extended to nonaqueous solvents and to the gas phase:

$$\underset{\substack{\text{Hydrochloric acid} \\ \text{(acid)}}}{HCl(g)} + \underset{\substack{\text{Ammonia} \\ \text{(base)}}}{NH_3(g)} \rightleftharpoons \underset{\substack{\text{Ammonium chloride} \\ \text{(salt)}}}{NH_4^+Cl^-(s)}$$

Salts

Any ionic solid, such as ammonium chloride, is called a **salt.** In a formal sense, a salt can be thought of as the product of an acid-base reaction. When an acid and a base react stoichiometrically, they are said to **neutralize** each other. Most salts are *strong electrolytes,* meaning that they dissociate almost completely into their component ions when dissolved in water. Thus, ammonium chloride gives NH_4^+ and Cl^- in aqueous solution:

$$NH_4^+Cl^-(s) \longrightarrow NH_4^+(aq) + Cl^-(aq)$$

Conjugate Acids and Bases

The products of a reaction between an acid and a base are also acids and bases:

A solid wedge is a bond coming out of the page toward you. A dashed wedge is a bond going behind the page.

Conjugate acids and bases are related by the gain or loss of one proton.

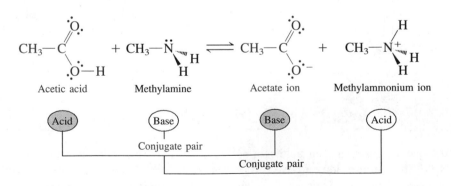

Acetate is a base because it can accept a proton to make acetic acid. The methylammonium ion is an acid because it can donate a proton and become methylamine. Acetic acid and the acetate ion are said to be a **conjugate acid-base pair.** Methylamine and methylammonium ion are likewise conjugate. *Conjugate acids and bases are related to each other by the gain or loss of one H^+.*

Ask Yourself

7-A. When an acid and base react, they are said to _____ each other. Acids and bases related by the gain or loss of one proton are said to be _____.

7-2   Relation Between [H$^+$], [OH$^-$], and pH

In **autoprotolysis,** one substance acts as both an acid and a base:

Autoprotolysis of water:

$$H_2O + H_2O \xrightleftharpoons{K_w} H_3O^+ + OH^- \qquad (7\text{-}1a)$$
$$\text{Hydronium ion} \quad \text{Hydroxide ion}$$

We abbreviate Reaction 7-1a in the following manner:

$$H_2O \xrightleftharpoons{K_w} H^+ + OH^- \qquad (7\text{-}1b)$$

and we designate its equilibrium constant (the autoprotolysis constant) as K_w.

Autoprotolysis constant for water:

$$K_w = [H^+][OH^-] = 1.0 \times 10^{-14} \text{ at } 25°C \qquad (7\text{-}2)$$

Equation 7-2 provides a tool with which we can find the concentration of H$^+$ and OH$^-$ in pure water. Also, given that the product $[H^+][OH^-]$ is constant, we can always find the concentration of either species if the concentration of the other is known. Because the product is a constant, *as the concentration of H$^+$ increases, the concentration of OH$^-$ necessarily decreases, and vice versa.*

EXAMPLE   Concentration of H$^+$ and OH$^-$ in Pure Water at 25°C

Calculate the concentrations of H$^+$ and OH$^-$ in pure water at 25°C.

SOLUTION H$^+$ and OH$^-$ are produced in a 1:1 mole ratio in Reaction 7-1b. Calling each concentration x, we write

$$K_w = 1.0 \times 10^{-14} = [H^+][OH^-] = [x][x] \implies x = 1.0 \times 10^{-7} \text{ M}$$

The concentrations of H$^+$ and OH$^-$ are both 1.0×10^{-7} M.

EXAMPLE   Finding [OH$^-$] When H$^+$ Is Known

What is the concentration of OH$^-$ if $[H^+] = 1.0 \times 10^{-3}$ M at 25°C?

SOLUTION Setting $[H^+] = 1.0 \times 10^{-3}$ M in Equation 7-2 gives

$$K_w = [H^+][OH^-] \implies [OH^-] = \frac{K_w}{[H^+]} = \frac{1.0 \times 10^{-14}}{1.0 \times 10^{-3}} = 1.0 \times 10^{-11} \text{ M}$$

When $[H^+] = 1.0 \times 10^{-3}$ M, $[OH^-] = 1.0 \times 10^{-11}$ M. A concentration of $[OH^-] = 1.0 \times 10^{-3}$ M would give $[H^+] = 1.0 \times 10^{-11}$ M. As one concentration increases, the other decreases.

To simplify the writing of H^+ concentration, we define **pH** as

> pH is really defined in terms of the *activity* of H^+, not the concentration. Section 11-2 discusses activity.

Approximate definition of pH: $\boxed{pH = -\log[H^+]}$ (7-3)

Here are some examples:

$$[H^+] = 10^{-3} \text{ M} \implies pH = -\log(10^{-3}) = 3$$
$$[H^+] = 10^{-10} \text{ M} \implies pH = -\log(10^{-10}) = 10$$
$$[H^+] = 3.8 \times 10^{-8} \text{ M} \implies pH = -\log(3.8 \times 10^{-8}) = 7.42$$

A solution is **acidic** if $[H^+] > [OH^-]$. A solution is **basic** if $[H^+] < [OH^-]$. An earlier example demonstrated that in pure water (which is neither acidic nor basic and is said to be *neutral*), $[H^+] = [OH^-] = 10^{-7}$ M, so the pH is $-\log(10^{-7}) = 7$. At 25°C, *an acidic solution has a pH below 7, and a basic solution has a pH above 7* (Figure 7-2).

Example: An acidic solution has pH = 4. This means $[H^+] = 10^{-4}$ M and $[OH^-] = K_w/[H^+] = 10^{-10}$ M. Therefore, $[H^+] > [OH^-]$.

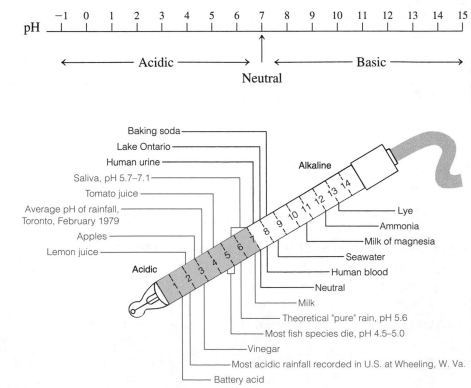

FIGURE 7-2 pH values of various substances. The most acidic rainfall in the United States is more acidic than lemon juice.

Although pH generally falls in the range 0 to 14, these are not limits. A pH of -1, for example, means $-\log[H^+] = -1$, or $[H^+] = 10^{+1} = 10$ M. This pH is attained in a concentrated solution of a strong acid such as HCl.

Ask Yourself

7-B. A solution of 0.050 M Mg^{2+} is treated with NaOH until $Mg(OH)_2$ precipitates.

    (a) At what concentration of OH^- does this occur? (Remember the solubility product in Section 5-4.)

    (b) At what pH does this occur?

7-3   Strengths of Acids and Bases

Acids and bases are classified as strong or weak, depending on whether they react "completely" or only "partly" to produce H^+ or OH^-. Because there is a continuous range for a "partial" reaction, there is no sharp distinction between weak and strong. However, some compounds react so completely that they are unquestionably strong acids or bases—and everything else is defined as weak.

Strong Acids and Bases

Common strong acids and bases are listed in Table 7-1, which you must memorize. Note that even though HCl, HBr, and HI are strong acids, HF is *not*. A **strong acid** or **strong base** is completely dissociated in aqueous solution. That is, the equilibrium constants for the following reactions are very large:

$$HCl(aq) \longrightarrow H^+ + Cl^-$$

$$KOH(aq) \longrightarrow K^+ + OH^-$$

Virtually no undissociated HCl or KOH exists in aqueous solution. Demonstration 7-1 shows one consequence of the strong-acid behavior of HCl.

Weak Acids and Bases

All **weak acids,** HA, react with water by donating a proton to H_2O:

$$HA + H_2O \xrightarrow{K_a} H_3O^+ + A^-$$

which means exactly the same as

Dissociation of weak acid: $HA \xrightleftharpoons{K_a} H^+ + A^- \qquad K_a = \dfrac{[H^+][A^-]}{[HA]}$ (7-4)

TABLE 7-1

Common strong acids and bases

Formula	Name
ACIDS	
HCl	Hydrochloric acid (hydrogen chloride)
HBr	Hydrogen bromide
HI	Hydrogen iodide
H_2SO_4[a]	Sulfuric acid
HNO_3	Nitric acid
$HClO_4$	Perchloric acid
BASES	
LiOH	Lithium hydroxide
NaOH	Sodium hydroxide
KOH	Potassium hydroxide
RbOH	Rubidium hydroxide
CsOH	Cesium hydroxide
R_4NOH[b]	Quaternary ammonium hydroxide

a. For H_2SO_4, only the first proton ionization is complete. Dissociation of the second proton has an equilibrium constant of 1.0×10^{-2}.
b. This is a general formula for any hydroxide salt of an ammonium cation containing four organic groups. An example is tetrabutylammonium hydroxide: $(CH_3CH_2CH_2CH_2)_4N^+OH^-$.

HCl Fountain

The complete dissociation of HCl into H^+ and Cl^- makes HCl(g) extremely soluble in water.

$$HCl(g) \rightleftharpoons HCl(aq) \qquad \text{(A)}$$

$$HCl(aq) \longrightarrow H^+(aq) + Cl^-(aq) \qquad \text{(B)}$$

Reaction B consumes the product of Reaction A, thereby pulling Reaction A to the right.

An HCl fountain is assembled as shown below.[1] In panel a, an inverted 250-mL round-bottom flask containing air is set up with its inlet tube leading to a source of HCl(g) and its outlet tube directed into an inverted bottle of water. As HCl is admitted to the flask, air is displaced. When the bottle is filled with air, the flask is filled mostly with HCl(g).

The hoses are disconnected and replaced with a beaker of indicator and a rubber bulb (panel b). For an indicator, we use slightly alkaline methyl purple, which is green above pH 5.4 and purple below pH 4.8. When 1 mL of water is squirted from the rubber bulb into the flask, a vacuum is created and indicator solution is drawn up into the flask, creating a colorful fountain (Color Plate 3).

Questions Why is vacuum created when water is squirted into the flask? Why does the indicator change color when it enters the flask?

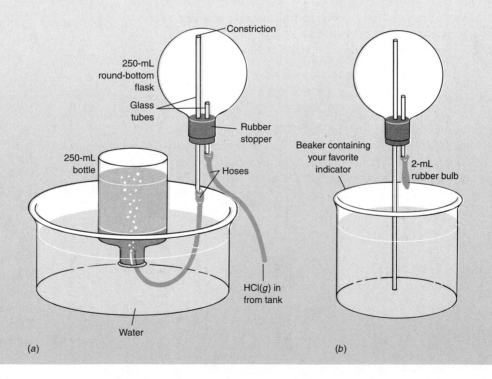

(a)

(b)

The equilibrium constant, K_a, is called the **acid dissociation constant.** A weak acid is only partially dissociated in water. This definition means that, for a weak acid, K_a is "small."

Weak bases, B, react with water by abstracting (grabbing) a proton from H_2O:

Base hydrolysis:

$$B + H_2O \xrightleftharpoons{K_b} BH^+ + OH^- \qquad K_b = \frac{[BH^+][OH^-]}{[B]} \qquad (7\text{-}5)$$

The equilibrium constant K_b is called the **base hydrolysis constant.** A weak base is one for which K_b is "small."

Carboxylic Acids Are Weak Acids and Amines Are Weak Bases

Acetic acid is a typical weak acid:

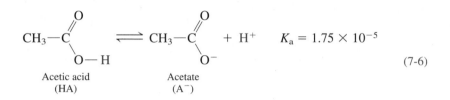

Acetic acid   Acetate
(HA)   (A⁻)

$K_a = 1.75 \times 10^{-5}$

(7-6)

Acetic acid is representative of carboxylic acids, which have the general structure shown below, where R is an organic substituent. *Most **carboxylic acids** are weak acids, and most **carboxylate anions** are weak bases.*

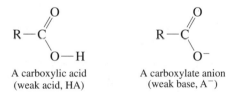

A carboxylic acid   A carboxylate anion
(weak acid, HA)   (weak base, A⁻)

Methylamine is a typical weak base. It forms a bond to H^+ by sharing the lone pair of electrons from the nitrogen atom of the amine:

Carboxylic acids (RCO_2H) and ammonium ions (R_3NH^+) are weak acids. Carboxylate anions (RCO_2^-) and amines (R_3N) are weak bases.

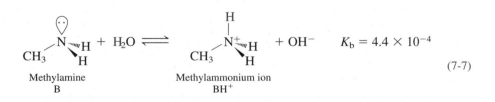

Methylamine   Methylammonium ion
B   BH⁺

$K_b = 4.4 \times 10^{-4}$

(7-7)

Methylamine is a representative **amine,** a nitrogen-containing compound:

$R\ddot{N}H_2$ a primary amine RNH_3^+

$R_2\ddot{N}H$ a secondary amine $R_2NH_2^+$ } **ammonium ions**

$R_3\ddot{N}$ a tertiary amine R_3NH^+

Weak acids: **HA** and **BH$^+$**

Weak bases: **A$^-$** and **B**

Methylammonium chloride is a weak acid because

1. It dissociates into $CH_3NH_3^+$ and Cl^-.
2. $CH_3NH_3^+$ is a weak acid, being conjugate to CH_3NH_2, a weak base.
3. Cl^- has no basic properties. It is conjugate to HCl, a strong acid. That is, HCl dissociates completely.

Challenge Phenol (C_6H_5OH) is a weak acid. Explain why a solution of the ionic compound potassium phenolate ($C_6H_5O^-K^+$) will be basic.

Amines are weak bases, and ammonium ions are weak acids. The "parent" of all amines is ammonia, NH_3. When methylamine reacts with water, the product is the conjugate acid. That is, the methylammonium ion produced in Reaction 7-7 is a weak acid:

$$\overset{+}{CH_3}\overset{}{N}H_3 \xrightarrow{K_a} CH_3\ddot{N}H_2 + H^+ \qquad K_a = 2.3 \times 10^{-11} \qquad (7\text{-}8)$$
$$\underset{BH^+}{} \qquad\qquad \underset{B}{}$$

The methylammonium ion is the conjugate acid of methylamine.

You should learn to recognize whether a compound is acidic or basic. For example, the salt methylammonium chloride dissociates completely in water to give methylammonium cation and chloride anion:

$$\underset{\substack{\text{Methylammonium}\\\text{chloride}}}{CH_3\overset{+}{N}H_3Cl^-(s)} \longrightarrow CH_3\overset{+}{N}H_3(aq) + Cl^-(aq)$$

The methylammonium ion, being the conjugate acid of methylamine, is a weak acid (Reaction 7-8). The chloride ion is neither an acid nor a base. It is the conjugate base of HCl, a strong acid. In other words, Cl^- *has virtually no tendency to associate with* H^+; otherwise, HCl would not be classified as a strong acid. We predict that methylammonium chloride is acidic, because the methylammonium ion is an acid and Cl^- is not a base.

Relation Between K_a and K_b

An important relationship exists between K_a and K_b of a conjugate acid-base pair in aqueous solution. We can derive this result with the acid HA and its conjugate base A$^-$.

$$HA \rightleftharpoons H^+ + A^- \qquad K_a = \frac{[H^+][A^-]}{[HA]}$$

$$\underline{A^- + H_2O \rightleftharpoons HA + OH^-} \qquad K_b = \frac{[HA][OH^-]}{[A^-]}$$

$$H_2O \rightleftharpoons H^+ + OH^- \qquad K_w = K_a \cdot K_b$$
$$= \frac{[H^+][A^-]}{[HA]}\frac{[HA][OH^-]}{[A^-]}$$

When the reactions above are added, their equilibrium constants must be multiplied, thereby obtaining a most useful result:

$K_a \cdot K_b = K_w$ for a conjugate acid-base pair in aqueous solution.

Relation between K_a and K_b for a conjugate pair: $\boxed{K_a \cdot K_b = K_w}$ (7-9)

Equation 7-9 applies to any acid and its conjugate base in aqueous solution.

EXAMPLE   Finding K_b for the Conjugate Base

The value of K_a for acetic acid is 1.75×10^{-5} (Reaction 7-6). Find K_b for the acetate ion.

SOLUTION
$$K_b = \frac{K_w}{K_a} = \frac{1.0 \times 10^{-14}}{1.75 \times 10^{-5}} = 5.7 \times 10^{-10}$$

EXAMPLE   Finding K_a for the Conjugate Acid

K_b for methylamine is 4.4×10^{-4} (Reaction 7-7). Find K_a for methylammonium ion.

SOLUTION
$$K_a = \frac{K_w}{K_b} = 2.3 \times 10^{-11}$$

Ask Yourself

7-C. Which is the stronger acid, **A** or **B**? Write the K_a reaction for each.

$$
\begin{array}{cc}
& O \\
& \parallel \\
\textbf{A}\quad Cl_2HCCOH & \textbf{B}\quad ClH_2CCOH
\end{array}
$$

A Cl_2HCCOH	**B** ClH_2CCOH
Dichloroacetic acid	Chloroacetic acid
$K_a = 5.0 \times 10^{-2}$	$K_a = 1.36 \times 10^{-3}$

Which is the stronger base, **C** or **D**? Write the K_b reaction for each.

C H_2NNH_2	**D** H_2NCNH_2
Hydrazine	Urea
$K_b = 3.0 \times 10^{-6}$	$K_b = 1.5 \times 10^{-14}$

7-4   pH of Strong Acids and Bases

The principal components that make rainfall acidic are nitric and sulfuric acids, which are strong acids. Each mole of *strong acid* or *strong base* in aqueous solution dissociates completely to provide one mole of H^+ or OH^-. Nitric acid is a strong acid, so the reaction

$$\underset{\text{Nitric acid}}{HNO_3} \longrightarrow H^+ + \underset{\text{Nitrate}}{NO_3^-}$$

goes to completion. In the case of sulfuric acid, one proton is completely dissociated, but the second is only partially dissociated (depending on conditions):

$$\underset{\text{Sulfuric acid}}{H_2SO_4} \longrightarrow H^+ + \underset{\substack{\text{Hydrogen sulfate} \\ \text{(also called bisulfate)}}}{HSO_4^-} \overset{K_a = 0.010}{\rightleftharpoons} H^+ + \underset{\text{Sulfate}}{SO_4^{2-}}$$

pH of a Strong Acid

Because HBr is completely dissociated, the pH of 0.010 M HBr is

$$pH = -\log[H^+] = -\log(0.010) = 2.00$$

Acid: pH < 7
Base: pH > 7

Is pH 2 sensible? (Always ask yourself that question at the end of a calculation.) Yes—because pH values below 7 are acidic and pH values above 7 are basic.

EXAMPLE   pH of a Strong Acid

Find the pH of 4.2×10^{-3} M $HClO_4$.

SOLUTION An easy one! The pH is simply

$$pH = -\log[H^+] = -\log(\underbrace{4.2}_{\substack{\text{2 significant} \\ \text{figures}}} \times 10^{-3}) = 2.\underbrace{38}_{\substack{\text{2 digits} \\ \text{in mantissa}}}$$

How about significant figures? The two significant figures in the mantissa of the logarithm correspond to the two significant figures in the number 4.2×10^{-3}.

For consistency in working problems in this book, *we are generally going to express pH values to the 0.01 decimal place regardless of what is justified by significant figures.* Real pH measurements are rarely more accurate than ±0.02, although differences in pH between two solutions can be accurate to ±0.002 pH units.

pH of a Strong Base

Now we ask, "What is the pH of 4.2×10^{-3} M KOH?" The concentration of OH^- is 4.2×10^{-3} M, and we can calculate $[H^+]$ from the K_w equation, 7-2:

Keep at least one extra insignificant figure (or all the digits in your calculation) in the middle of a calculation to avoid round-off errors in the final answer.

$$[H^+] = \frac{K_w}{[OH^-]} = \frac{1.0 \times 10^{-14}}{4.2 \times 10^{-3}} = 2.3_8 \times 10^{-12} \text{ M}$$

$$pH = -\log[H^+] = -\log(2.3_8 \times 10^{-12}) = 11.62$$

Here is a trick question: What is the pH of 4.2×10^{-9} M KOH? By our previous reasoning, we might first say

$$[H^+] = \frac{K_w}{[OH^-]} = \frac{1.0 \times 10^{-14}}{4.2 \times 10^{-9}} = 2.3_8 \times 10^{-6} \, M \quad \Rightarrow \quad pH = 5.62$$

Is this reasonable? Can we dissolve base in water and obtain an acidic pH (< 7)? No way!

The fallacy is that we neglected the contribution of the reaction $H_2O \rightleftharpoons H^+ + OH^-$ to the concentration of OH^-. Pure water creates 10^{-7} M OH^-, which is more than the KOH that we added. The pH of water plus added KOH cannot fall below 7. The pH of 4.2×10^{-9} M KOH is very close to 7. Similarly, the pH of 10^{-10} M HNO_3 is very close to 7, not 10. Figure 7-3 shows how pH depends on concentration for a strong acid and a strong base. In a very dilute solution of acid or base, the chemistry of dissolved carbon dioxide ($CO_2 + H_2O \rightleftharpoons HCO_3^- + H^+$) overwhelms the effect of the added acid or base.

Water Almost Never Produces 10^{-7} M H^+ and 10^{-7} M OH^-

The ion concentrations 10^{-7} M H^+ and 10^{-7} M OH^- occur *only* in extremely pure water with no added acid or base. In a 10^{-4} M solution of HBr, for example, the pH is 4. The concentration of OH^- is $K_w/[H^+] = 10^{-10}$ M. But the source of $[OH^-]$ is dissociation of water. If water produces only 10^{-10} M OH^-, it must also produce only 10^{-10} M H^+, because it makes one H^+ for every OH^-. In a 10^{-4} M HBr solution, water dissociation produces only 10^{-10} M OH^- and 10^{-10} M H^+.

Acids and bases suppress water ionization, as predicted by Le Châtelier's principle.

Question What concentrations of H^+ and OH^- are produced by H_2O dissociation in 10^{-2} M NaOH?

Ask Yourself

7-D. (a) What is the pH of (i) 1.0×10^{-3} M HBr and (ii) 1.0×10^{-2} M KOH?
(b) Calculate the pH of (i) 3.2×10^{-5} M HI and (ii) 7.7 mM LiOH.
(c) Find the concentration of H^+ in a solution whose pH is 4.44.
(d) Find $[H^+]$ in 7.7 mM LiOH solution. What is the source of this H^+?
(e) Find the pH of 3.2×10^{-9} M tetramethylammonium hydroxide, $(CH_3)_4N^+OH^-$.

7-5   Tools for Dealing with Weak Acids and Bases

By analogy to the definition of pH, we define **pK** as the negative logarithm of an equilibrium constant. For the acid dissociation constant in Equation 7-4 and the base hydrolysis constant in Equation 7-5, we can write

$$pK_a = -\log K_a \qquad pK_b = -\log K_b \qquad (7\text{-}10)$$

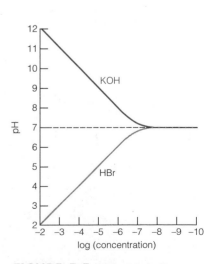

FIGURE 7-3 Calculated pH as a function of the concentration of a strong acid or strong base dissolved in water.

The *stronger* an acid, the *smaller* its pK_a.

Stronger acid	Weaker acid
$K_a = 10^{-4}$	$K_a = 10^{-8}$
$pK_a = 4$	$pK_a = 8$

Remember from Equation 7-9 the very important relation between K_a and K_b for a conjugate acid-base pair, which are related by gain or loss of a single proton: $K_a \cdot K_b = K_w$.

Weak Is Conjugate to Weak

> The conjugate base of a weak acid is a weak base. The conjugate acid of a weak base is a weak acid.

> *Weak is conjugate to weak.*

The conjugate base of a weak acid is a weak base. The conjugate acid of a weak base is a weak acid. Consider a weak acid, HA, with $K_a = 10^{-4}$. The conjugate base, A^-, has $K_b = K_w/K_a = 10^{-10}$. That is, if HA is a weak acid, A^- is a weak base. If the K_a value were 10^{-5}, then the K_b value would be 10^{-9}. As HA becomes a weaker acid, A^- becomes a stronger base (but never a strong base). Conversely, the greater the acid strength of HA, the less the base strength of A^-. However, if either A^- or HA is weak, so is its conjugate. If HA is strong (such as HCl), its conjugate base (Cl^-) is *so* weak that it is not a base at all in water.

Using Appendix B

CH_3NH_2	$CH_3NH_3^+$
Methylamine	Methylammonium ion

Acid dissociation constants appear in Appendix B. Each compound is shown in its *fully protonated form*. Methylamine, for example, is shown as $CH_3NH_3^+$, which is really the methylammonium ion. The value of K_a (2.3×10^{-11}) given for methylamine is actually K_a for the methylammonium ion. To find K_b for methylamine, we write $K_b = K_w/K_a = (1.0 \times 10^{-14})/(2.3 \times 10^{-11}) = 4.3 \times 10^{-4}$.

For polyprotic acids and bases, several K_a values are given, beginning with the most acidic group. Pyridoxal phosphate is given in its fully protonated form as follows:

Pyridoxal phosphate is derived from vitamin B_6, which is essential in amino acid metabolism in your body. Box 6-1 discussed the drawing of organic structures.

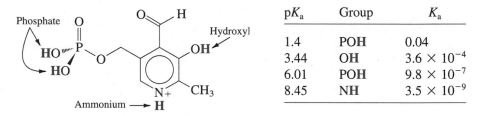

pK_a	Group	K_a
1.4	POH	0.04
3.44	OH	3.6×10^{-4}
6.01	POH	9.8×10^{-7}
8.45	NH	3.5×10^{-9}

pK_1 (1.4) is for dissociation of one of the phosphate protons, and pK_2 (3.44) is for the hydroxyl proton. The third most acidic proton is the other phosphate proton, for which $pK_3 = 6.01$, and the NH^+ group is the least acidic ($pK_4 = 8.45$).

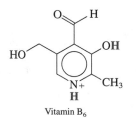

Vitamin B_6

Ask Yourself

7-E. **(a)** Write the acid dissociation reaction for formic acid, HCO_2H.
(b) What is the conjugate base of formic acid?
(c) Write the K_a equilibrium expression for formic acid and look up its value.
(d) Write the K_b equilibrium expression for formate ion, HCO_2^-.
(e) Find the base hydrolysis constant for formate.

Let's find the pH and composition of a solution containing 0.020 0 mol benzoic acid in 1.00 L of water.

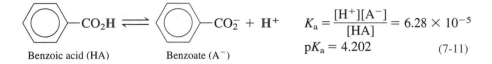

Benzoic acid (HA) Benzoate (A^-)

$$K_a = \frac{[H^+][A^-]}{[HA]} = 6.28 \times 10^{-5}$$

$$pK_a = 4.202 \qquad (7\text{-}11)$$

For every mole of A^- formed by dissociation of HA, one mole of H^+ is created. That is, $[A^-] = [H^+]$. (For reasonable concentrations of weak acids of reasonable strength, the contribution of H^+ from the acid is much greater than the contribution from $H_2O \rightleftharpoons H^+ + OH^-$.) Abbreviating the formal concentration of HA as F, and using x for the concentration of H^+, we can make a table showing concentrations before and after dissociation of the weak acid:

	HA	$\rightleftharpoons$	A^-	+	H^+
Initial concentration	F		0		0
Final concentration	F $- x$		x		x

Putting these values into the K_a expression in Equation 7-11 gives

$$K_a = \frac{[H^+][A^-]}{[HA]} = \frac{(x)(x)}{F - x} \qquad (7\text{-}12)$$

Setting F $= 0.020\ 0$ M and $K_a = 6.28 \times 10^{-5}$ gives

$$\frac{x^2}{0.020\ 0 - x} = 6.28 \times 10^{-5} \qquad (7\text{-}13)$$

The first step in solving Equation 7-13 for x is to cross multiply the numerator on one side by the denominator on the other side:

$$\frac{x^2}{0.020\ 0 - x} = \frac{6.28 \times 10^{-5}}{1}$$

$$(x^2)(1) = (6.28 \times 10^{-5})(0.020\ 0 - x) = (1.25_6 \times 10^{-6}) - (6.28 \times 10^{-5})x$$

Collecting terms gives a quadratic equation (in which the highest power is x^2):

$$x^2 + (6.28 \times 10^{-5})x - (1.25_6 \times 10^{-6}) = 0 \qquad (7\text{-}14)$$

Equation 7-14 has two solutions (called *roots*) described in Box 7-1. One root is positive and one is negative. Because a concentration cannot be negative, we reject the negative solution:

$$x = 1.09 \times 10^{-3} \quad \text{(negative root rejected)}$$

You should **check your answer** by plugging it back into Equation 7-12 and seeing whether the equation is satisfied.

Quadratic Equations and Successive Approximations

A quadratic equation of the general form $ax^2 + bx + c = 0$ has two solutions:

$$x = \frac{-b + \sqrt{b^2 - 4ac}}{2a} \qquad\qquad x = \frac{-b - \sqrt{b^2 - 4ac}}{2a}$$

The solutions to Equation 7-14,

$$\underbrace{(1)[H^+]^2}_{a = 1} + \underbrace{(6.28 \times 10^{-5})[H^+]}_{b = 6.28 \times 10^{-5}} - \underbrace{(1.256 \times 10^{-6})}_{c = -1.25_6 \times 10^{-6}} = 0$$

are

$$[H^+] = \frac{-(6.28 \times 10^{-5}) + \sqrt{(6.28 \times 10^{-5})^2 - 4(1)(-1.25_6 \times 10^{-6})}}{2(1)} = 1.09 \times 10^{-3}\ M$$

and

$$[H^+] = \frac{-(6.28 \times 10^{-5}) - \sqrt{(6.28 \times 10^{-5})^2 - 4(1)(-1.25_6 \times 10^{-6})}}{2(1)} = -1.09 \times 10^{-3}\ M$$

Because the concentration of $[H^+]$ cannot be negative, we reject the negative solution and choose 1.09×10^{-3} M as the correct answer.

When you solve a quadratic equation, retain all the digits in your calculator at intermediate stages of the computation, or serious round-off errors can occur in some cases. If you have a programmable calculator, we strongly recommend writing a program to solve quadratic equations and using it whenever one arises.

Another way to solve Equation 7-13 is by successive approximations. This approach is often faster than using the quadratic equation. As a first approximation, we suppose that x in the denominator is much smaller than 0.020 0 M and neglect x:

$$\frac{x^2}{0.020\ 0 - x} = 6.28 \times 10^{-5} \implies \frac{x^2}{0.020\ 0} \approx 6.28 \times 10^{-5}$$

$$x^2 \approx (6.28 \times 10^{-5})(0.020\ 0) \implies x_1 \approx \sqrt{(6.28 \times 10^{-5})(0.020\ 0)} = 1.12 \times 10^{-3}$$

From the value of x, we can find concentrations and pH:

For uniformity, we are going to express pH to the 0.01 decimal place, even though significant figures in this problem justify another digit.

$$[H^+] = [A^-] = x = 1.09 \times 10^{-3}\ M$$

$$[HA] = F - x = 0.020\ 0 - (1.09 \times 10^{-3}) = 0.018\ 9\ M$$

$$pH = -\log x = 2.96$$

Was the approximation $[H^+] \approx [A^-]$ justified? The concentration of H^+ is 1.09×10^{-3} M, which means $[OH^-] = K_w/[H^+] = 9.20 \times 10^{-12}$ M.

In a weak-acid solution, H^+ is derived almost entirely from HA, not from H_2O.

$$[H^+]\ \text{from HA dissociation} = [A^-]\ \text{from HA dissociation} = 1.09 \times 10^{-3}\ M$$

$$[H^+]\ \text{from } H_2O\ \text{dissociation} = [OH^-]\ \text{from } H_2O\ \text{dissociation} = 9.20 \times 10^{-12}\ M$$

We call this first approximate solution x_1. To find a second approximation, x_2, plug the value of x_1 back into the denominator of the original equation:

$$\frac{x_2^2}{0.020\ 0 - (1.12 \times 10^{-3})} = 6.28 \times 10^{-5}$$

$$\Rightarrow \quad x_2 = \sqrt{(6.28 \times 10^{-5})[0.020\ 0 - (1.12 \times 10^{-3})]} = 1.09 \times 10^{-3}$$

Now use x_2 to obtain a third approximation:

$$\frac{x_3^2}{0.020\ 0 - (1.09 \times 10^{-3})} = 6.28 \times 10^{-5}$$

$$\Rightarrow \quad x_3 = \sqrt{(6.28 \times 10^{-5})[0.020\ 0 - (1.09 \times 10^{-3})]} = 1.09 \times 10^{-3}$$

The third approximation is the same as the second approximation, so we must be there.

If x_1 comes out to $\leq 0.5\%$ of F in the denominator of Equation 7-12, then x_1 is good enough and there is no need to continue. When x_1 is not small relative to F, then many iterations (cycles of calculation) are required and successive approximations oscillate between a high and a low value. Choosing an approximation midway between the high and low values will get you close to the correct solution. Here are two examples to try:

$x_1 \leq 0.5\%$ of F: $\qquad \dfrac{x^2}{0.100 - x} = 1.00 \times 10^{-6}$ $\qquad$ (answer $= 0.000\ 316$)

x_1 is not much smaller than F: $\qquad \dfrac{x^2}{0.100 - x} = 0.050\ 0$ $\qquad$ (answer $= 0.050\ 0$)

The assumption that H^+ is derived mainly from HA is excellent because 1.09×10^{-3} M $\gg 9.20 \times 10^{-12}$ M.

Fraction of Dissociation

What fraction of HA is dissociated? If the total concentration of acid $(= [HA] + [A^-])$ is 0.020 0 M and the concentration of A^- is 1.09×10^{-3} M, then the *fraction of dissociation* is

Fraction of dissociation of an acid: $\qquad \dfrac{[A^-]}{[A^-] + [HA]} = \dfrac{1.09 \times 10^{-3}}{0.020\ 0} = 0.054$ $\qquad$ (7-15)

The acid is indeed weak. It is only 5.4% dissociated.

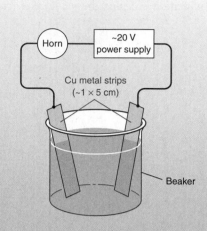

Demonstration 7-2

Conductivity of Weak Electrolytes

The relative conductivity of strong and weak acids is directly related to their different degrees of dissociation in aqueous solution. To demonstrate conductivity, we use a Radio Shack piezo alerting buzzer, but any kind of buzzer or light bulb could be substituted for the electric horn. The voltage required will depend on the buzzer or light chosen.

When a conducting solution is placed in the beaker, the horn sounds. First show that distilled water and sucrose solution are nonconductive. Solutions of the strong electrolytes NaCl or HCl are conductive. Compare strong and weak electrolytes by demonstrating that 1 mM HCl gives a loud sound, whereas 1 mM acetic acid gives little or no sound. With 10 mM acetic acid, the strength of the sound varies noticeably as the electrodes are moved away from each other in the beaker.

Apparatus for demonstrating the conductivity of electrolyte solutions.

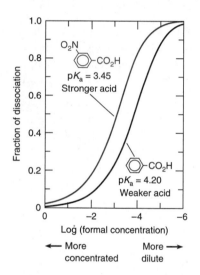

FIGURE 7-4 The fraction of dissociation of a weak acid increases as it is diluted. The stronger of the two acids (pK_a = 3.45 is stronger than pK_a = 4.20) is more dissociated at any given concentration.

Figure 7-4 compares the fraction of dissociation of two weak acids as a function of formal concentration. As the solution becomes more dilute, the fraction of dissociation increases. The stronger acid has a greater fraction of dissociation at any formal concentration. Demonstration 7-2 compares properties of strong and weak acids.

The Essence of a Weak-Acid Problem

When faced with finding the pH of a weak acid, you should immediately realize that $[H^+] = [A^-] = x$ and proceed to set up and solve the equation

Equation for weak acids:

$$\frac{[H^+][A^-]}{[HA]} = \frac{x^2}{F - x} = K_a \qquad (7\text{-}16)$$

where F is the formal concentration of HA. The approximation $[H^+] = [A^-]$ would be poor only if the acid were very dilute or very weak, neither of which constitutes a practical problem.

EXAMPLE Finding the pH of a Weak Acid

Find the pH of 0.100 M trimethylammonium chloride.

$$\left[\begin{array}{c} H \\ | \\ H_3C\cdots N - CH_3 \\ H_3C \end{array} \right]^+ \quad Cl^- \quad \text{Trimethylammonium chloride}$$

SOLUTION We must first realize that salts of this type are *completely dissociated* to give $(CH_3)_3NH^+$ and Cl^-. We then recognize that trimethylammonium ion is a weak acid, being conjugate to trimethylamine, $(CH_3)_3N$, a weak base. The ion Cl^- has no basic or acidic properties and should be ignored. In Appendix B, we find trimethylammonium ion listed under the name trimethylamine but drawn as the trimethylammonium ion. The value of pK_a is 9.800, so

$$K_a = 10^{-pK_a} = 10^{-9.800} = 1.58 \times 10^{-10}$$

It's all downhill from here:

$$(CH_3)_3NH^+ \underset{F-x}{\overset{K_a}{\rightleftharpoons}} \underset{x}{(CH_3)_3N} + \underset{x}{H^+}$$

$$\frac{x^2}{0.100 - x} = 1.58 \times 10^{-10}$$

$$x = 3.97 \times 10^{-6}\,M \quad \Rightarrow \quad pH = -\log(3.97 \times 10^{-6}) = 5.40$$

| **EXAMPLE** Finding pK_a of a Weak Acid

A 0.100 M solution of hydrazoic acid has pH = 2.83. Find pK_a for this acid.

$$\underset{\substack{\text{Hydrazoic acid} \\ 0.100 - x}}{\overset{-}{:}\!N{=}\overset{+}{N}{=}\overset{..}{N}{-}H} \overset{K_a}{\rightleftharpoons} \underset{\substack{\text{Azide } (N_3^-) \\ x}}{\overset{-}{:}\!N{=}\overset{+}{N}{=}\overset{..}{N}\!:^-} + \underset{x}{H^+}$$

SOLUTION We know that $[H^+] = 10^{-pH} = 10^{-2.83} = 1.4_8 \times 10^{-3}$ M. Because $[N_3^-] = [H^+]$ in this solution, $[N_3^-] = 1.48 \times 10^{-3}$ M and $[HN_3] = 0.100 - 1.4_8 \times 10^{-3} = 0.098\,5$ M. From these concentrations, we calculate K_a and pK_a:

$$K_a = \frac{[N_3^-][H^+]}{[HN_3]} = \frac{(1.48 \times 10^{-3})^2}{0.098\,5} = 2.2_2 \times 10^{-5}$$

$$\Rightarrow \quad pK_a = -\log(2.2_2 \times 10^{-5}) = 4.65$$

Ask Yourself

7-F. (a) What is the pH and fraction of dissociation of a 0.100 M solution of the weak acid HA with $K_a = 1.00 \times 10^{-5}$?

 (b) A 0.045 0 M solution of HA has a pH of 2.78. Find pK_a for HA.

 (c) A 0.045 0 M solution of HA is 0.60% dissociated. Find pK_a for HA.

The treatment of weak bases is almost the same as that of weak acids.

$$B + H_2O \xrightleftharpoons{K_b} BH^+ + OH^- \qquad K_b = \frac{[BH^+][OH^-]}{[B]}$$

Nearly all OH^- comes from the reaction of $B + H_2O$, and little comes from dissociation of H_2O. Setting $[OH^-] = x$, we must also set $[BH^+] = x$, because one BH^+ is produced for each OH^-. Setting $F = [B] + [BH^+]$, we can write

$$[B] = F - [BH^+] = F - x$$

Plugging these values into the K_b equilibrium expression, we get

A weak-base problem has the same algebra as a weak-acid problem, except $K = K_b$ and $x = [OH^-]$.

Equation for weak base: $\qquad \dfrac{[BH^+][OH^-]}{[B]} = \dfrac{x^2}{F - x} = K_b \qquad$ (7-17)

which looks a lot like a weak-acid problem, except that now $x = [OH^-]$.

EXAMPLE   Finding the pH of a Weak Base

Find the pH of a 0.037 2 M solution of the commonly encountered weak base, cocaine:

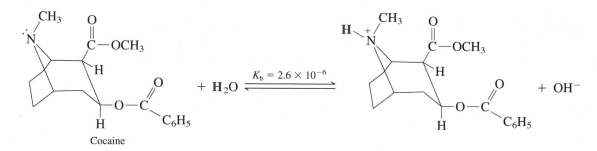

Cocaine

SOLUTION Designating cocaine as B, we formulate the problem as follows:

$$\underset{0.037\ 2 - x}{B} + H_2O \rightleftharpoons \underset{x}{BH^+} + \underset{x}{OH^-}$$

$$\frac{x^2}{0.037\ 2 - x} = 2.6 \times 10^{-6} \implies x = 3.1_0 \times 10^{-4}$$

Question   Were we justified in neglecting water dissociation as a source of OH^-? What concentration of OH^- is produced by H_2O dissociation in this solution?

Because $x = [OH^-]$, we can write

$$[H^+] = K_w/[OH^-] = (1.0 \times 10^{-14})/(3.1_0 \times 10^{-4}) = 3.2_2 \times 10^{-11}$$

$$pH = -\log[H^+] = 10.49$$

This is a reasonable pH for a weak base.

What fraction of cocaine reacted with water?

Fraction of association
of a base:
$$\frac{[BH^+]}{[BH^+] + [B]} = \frac{3.1_0 \times 10^{-4}}{0.037\,2} = 0.008\,3 \qquad (7\text{-}18)$$

because $[BH^+] = [OH^-] = 3.1 \times 10^{-4}$ M. Only 0.83% of the base has reacted.

Physical properties of a neutral amine, such as cocaine (B), and its protonated product, cocaine hydrochloride (BH^+Cl^-), are different. Neutral cocaine is the extremely dangerous substance called "crack." It is soluble in organic solvents and cell membranes and has a moderate vapor pressure. The ionic hydrochloride is soluble in water, not in cell membranes, and has much lower vapor pressure. Its physiological action is slower than that of "crack."

Conjugate Acids and Bases—Revisited

The conjugate base of a weak acid is a weak base, and the conjugate acid of a weak base is a weak acid. The exceedingly important relation between the equilibrium constants for a conjugate acid-base pair is

HA and A⁻ are a conjugate acid-base pair. So are BH⁺ and B.

$$K_a \cdot K_b = K_w$$

Reaction 7-11 featured benzoic acid, designated HA. Now consider 0.050 M sodium benzoate, Na^+A^-, which contains the conjugate base of benzoic acid. When this salt dissolves in water, it dissociates to Na^+ and A^-. Na^+ does not react with water, but A^- is a weak base:

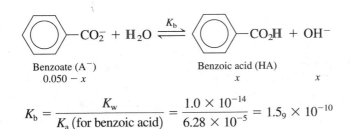

Benzoate (A⁻)   Benzoic acid (HA)
0.050 − x   x   x

$$K_b = \frac{K_w}{K_a \text{ (for benzoic acid)}} = \frac{1.0 \times 10^{-14}}{6.28 \times 10^{-5}} = 1.5_9 \times 10^{-10}$$

To find the pH of this solution, we write

$$\frac{[HA][OH^-]}{[A^-]} = \frac{x^2}{0.050 - x} = 1.5_9 \times 10^{-10} \Rightarrow x = [OH^-] = 2.8 \times 10^{-6} \text{ M}$$

$$[H^+] = K_w/[OH^-] = 3.5 \times 10^{-9} \text{ M} \Rightarrow \text{pH} = 8.45$$

This is a reasonable pH for a solution of a weak base.

EXAMPLE A Weak-Base Problem

Find the pH of 0.10 M ammonia.

SOLUTION When ammonia is dissolved in water, its reaction is

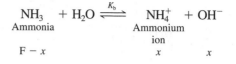

$$NH_3 + H_2O \overset{K_b}{\rightleftharpoons} NH_4^+ + OH^-$$

Ammonia   Ammonium
ion

$$F - x \qquad\qquad\qquad x \qquad x$$

In Appendix B, we find the ammonium ion, NH_4^+, listed next to ammonia. K_a for ammonium ion is 5.70×10^{-10}. Therefore K_b for NH_3 is

$$K_b = \frac{K_w}{K_a} = \frac{1.0 \times 10^{-14}}{5.70 \times 10^{-10}} = 1.7_5 \times 10^{-5}$$

To find the pH of 0.10 M NH_3, we set up and solve the equation

$$\frac{[NH_4^+][OH^-]}{[NH_3]} = \frac{x^2}{0.10 - x} = K_b = 1.7_5 \times 10^{-5}$$

$$x = [OH^-] = 1.3_1 \times 10^{-3} \text{ M}$$

$$[H^+] = \frac{K_w}{[OH^-]} = 7.6_1 \times 10^{-12} \text{ M} \quad \Rightarrow \quad pH = -\log[H^+] = 11.12$$

Ask Yourself

7-G. (a) What is the pH and fraction of association of a 0.100 M solution of a weak base with $K_b = 1.00 \times 10^{-5}$?

$$\text{fraction of association} = \frac{[BH^+]}{[BH^+] + [B]}$$

(b) A 0.10 M solution of a base has pH = 9.28. Find K_b.
(c) A 0.10 M solution of a base is 2.0% associated. Find K_b.

Key Equations

Conjugate acids and bases

$$HA + B \rightleftharpoons A^- + BH^+$$
Acid Base   Base   Acid

Autoprotolysis of water

$$H_2O \overset{K_w}{\rightleftharpoons} H^+ + OH^- \quad K_w = [H^+][OH^-] = 1.0 \times 10^{-14} \text{ (25°C)}$$

Finding $[OH^-]$ from $[H^+]$

$$[OH^-] = K_w/[H^+]$$

Definition of pH

$$pH = -\log[H^+]$$ (This relationship is only approximate—but it is the one we use in this book.)

Acid dissociation constant	$HA \xrightleftharpoons{K_a} H^+ + A^-$	$K_a = \dfrac{[H^+][A^-]}{[HA]}$

Base hydrolysis constant	$B + H_2O \xrightleftharpoons{K_b} BH^+ + OH^-$	$K_b = \dfrac{[BH^+][OH^-]}{[B]}$

Relation between K_a and K_b for conjugate acid-base pair

$$K_a \cdot K_b = K_w$$

Common weak acids

$$\underset{\text{Carboxylic acid}}{RCO_2H} \qquad \underset{\text{Ammonium ion}}{R_3NH^+}$$

Common weak bases

$$\underset{\text{Carboxylate anion}}{RCO_2^-} \qquad \underset{\text{Amine}}{R_3N}$$

Definitions of pK

$$pK_a = -\log K_a \qquad pK_b = -\log K_b$$

Weak-acid equilibrium

$$\underset{F-x}{HA} \xrightleftharpoons{K_a} \underset{x}{H^+} + \underset{x}{A^-} \qquad K_a = \dfrac{[H^+][A^-]}{[HA]} = \dfrac{x^2}{F-x}$$

(F = formal concentration of weak acid or weak base)

Weak-base equilibrium

$$\underset{F-x}{B} + H_2O \xrightleftharpoons{K_b} \underset{x}{BH^+} + \underset{x}{OH^-} \qquad K_b = \dfrac{[BH^+][OH^-]}{[B]} = \dfrac{x^2}{F-x}$$

Fraction of dissociation of HA

$$\text{fraction of dissociation} = \dfrac{[A^-]}{[A^-] + [IIA]}$$

Fraction of association of B

$$\text{fraction of association} = \dfrac{[BH^+]}{[BH^+] + [B]}$$

Important Terms

acid	base hydrolysis constant	pH
acid dissociation constant	basic solution	pK
acidic solution	carboxylate anion	salt
amine	carboxylic acid	strong acid
ammonium ion	conjugate acid-base pair	strong base
autoprotolysis	hydronium ion	weak acid
base	neutralization	weak base

Problems

7-1. Identify the conjugate acid-base pairs in the following reactions:

(a) $CN^- + HCO_2H \rightleftharpoons HCN + HCO_2^-$

(b) $PO_4^{3-} + H_2O \rightleftharpoons HPO_4^{2-} + OH^-$

(c) $HSO_3^- + OH^- \rightleftharpoons SO_3^{2-} + H_2O$

7-2. A solution is *acidic* if _____.
A solution is *basic* if _____.

7-3. Find the pH of a solution containing

(a) 10^{-4} M H^+ (c) 5.8×10^{-4} M H^+

(b) 10^{-5} M OH^- (d) 5.8×10^{-5} M OH^-

7-4. The concentration of H^+ in your blood is 3.5×10^{-8} M.

(a) What is the pH of blood?

(b) Find the concentration of OH^- in blood.

7-5. Sulfuric acid is the principal acidic component of acid rain. The mean pH of rainfall in southern Norway in the 1970s was 4.3. What concentration of H_2SO_4 will produce this pH by the reaction $H_2SO_4 \rightleftharpoons 2H^+ + SO_4^{2-}$?

7-6. An acidic solution containing 0.010 M La^{3+} is treated with NaOH until $La(OH)_3$ precipitates. Use the solubility product for $La(OH)_3$ to find the concentration of OH^- when La^{3+} first precipitates. At what pH does this occur?

7-7. Make a list of the common strong acids and strong bases. Memorize this list.

7-8. Write the structures and names for two classes of weak acids and two classes of weak bases.

7-9. Calculate the concentration of H^+ and the pH of the following solutions:

(a) 0.010 M HNO_3 **(d)** 3.0 M $HClO_4$

(b) 0.035 M KOH   **(e)** 0.010 M $[(CH_3)_4N^+]OH^-$

(c) 0.030 M HCl   Tetramethylammonium hydroxide

7-10. (a) Write the K_a reaction for trichloroacetic acid, Cl_3CCO_2H ($K_a = 0.22$), and for the anilinium ion,

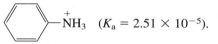
$(K_a = 2.51 \times 10^{-5})$.

(b) Which is the stronger acid?

7-11. (a) Write the K_b reactions for pyridine, sodium 2-mercaptoethanol, and cyanide.

Pyridine   Sodium 2-mercaptoethanol   Cyanide

$HOCH_2CH_2\ddot{S}:^-Na^+$ CN^-

H^+ binds to S in sodium 2-mercaptoethanol and to C in cyanide.

(b) Values of K_a for the conjugate acids are shown below. Which base in part **(a)** is strongest?

Pyridinium ion   2-Mercaptoethanol   Hydrogen cyanide
$K_a = 5.90 \times 10^{-6}$ $K_a = 1.91 \times 10^{-10}$ $K_a = 6.2 \times 10^{-10}$

7-12. Write the autoprotolysis reaction of H_2SO_4, whose

structure is

7-13. Write the K_b reactions for piperidine and benzoate.

Piperidine   Benzoate

7-14. Hypochlorous acid has the structure $H-O-Cl$. Write the base hydrolysis reaction of hypochlorite, OCl^-. Given that K_a for HOCl is 3.0×10^{-8}, find K_b for hypochlorite.

7-15. Calculate the pH of 3.0×10^{-5} M $Mg(OH)_2$, which is completely dissociated to Mg^{2+} and OH^-.

7-16. Find the concentration of H^+ in a solution whose pH is 11.65.

7-17. Find the pH and fraction of dissociation of a 0.010 0 M solution of the weak acid HA with $K_a = 1.00 \times 10^{-4}$.

7-18. Find the pH and fraction of dissociation in a 0.150 M solution of hydroxybenzene (also called phenol).

7-19. Calculate the pH of 0.085 0 M pyridinium bromide, $C_5H_5NH^+Br^-$. Find the concentrations of pyridine (C_5H_5N), pyridinium ion ($C_5H_5NH^+$), and Br^- in the solution.

7-20. A 0.100 M solution of the weak acid HA has a pH of 2.36. Calculate pK_a for HA.

7-21. A 0.022 2 M solution of HA is 0.15% dissociated. Calculate pK_a for this acid.

7-22. Find the pH and concentrations of $(CH_3)_3N$ and $(CH_3)_3NH^+$ in a 0.060 M solution of trimethylammonium chloride.

7-23. Calculate the pH and fraction of dissociation of **(a)** $10^{-2.00}$ M and **(b)** $10^{-10.00}$ M barbituric acid.

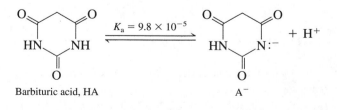

Barbituric acid, HA A^-

7-24. $BH^+ClO_4^-$ is a salt formed from the base B ($K_b = 1.00 \times 10^{-4}$) and perchloric acid. It dissociates into

BH^+, a weak acid, and ClO_4^-, which is neither an acid nor a base. Find the pH of 0.100 M $BH^+ClO_4^-$.

7-25. Find K_a for cyclohexylammonium ion and K_b for cyclohexylamine.

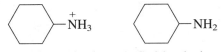

Cyclohexylammonium ion   Cyclohexylamine

7-26. Write the chemical reaction whose equilibrium constant is **(a)** K_b for aniline and **(b)** K_a for anilinium ion.

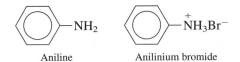

Aniline   Anilinium bromide

7-27. Find the pH and concentrations of $(CH_3)_3N$ and $(CH_3)_3NH^+$ in a 0.060 M solution of trimethylamine.

7-28. Calculate the pH and fraction of association of 1.00×10^{-1}, 1.00×10^{-2}, and 1.00×10^{-12} M sodium acetate.

7-29. Find the pH of 0.050 M NaCN.

7-30. Find the pH and fraction of association of 0.026 M NaOCl.

7-31. If a 0.030 M solution of a base has pH = 10.50, find K_b for the base.

7-32. If a 0.030 M solution of a base is 0.27% hydrolyzed ($\alpha_{BH^+} = 0.002\ 7$), find K_b for the base.

7-33. The smell (and taste) of fish arise from amine compounds. Suggest a reason why adding lemon juice (which is acidic) reduces the "fishy" smell.

7-34. Cr^{3+} (and most metal ions with charge ≥ 2) is acidic by virtue of the hydrolysis reaction

$$Cr^{3+} + H_2O \rightleftharpoons Cr(OH)^{2+} + H^+ \qquad K = 10^{-3.80}$$

[Further reactions that we will neglect produce $Cr(OH)_2^+$, $Cr(OH)_3$, and $Cr(OH)_4^-$.] Find the pH of 0.010 M $Cr(ClO_4)_3$. What fraction of chromium is in the form $Cr(OH)^{2+}$?

7-35. Create a spreadsheet to solve the equation $x^2/(F - x) = K$. The input will be F and K. The output is the positive value of x. Use your spreadsheet to check your answer to Ask Yourself Problem 7-F, part **(a)**.

How Would You Do It?

7-36. *Weighted average pH in precipitation.* The following data were reported for rainfall at the Philadelphia airport in 1990:[2]

Season	Precipitation (cm)	Weighted average pH
Winter	17.3	4.40
Spring	30.5	4.68
Summer	17.8	4.68
Fall	14.7	5.10

Suggest a procedure to calculate the average pH for the entire year. This would be the pH observed if all the rain for the whole year were pooled in one container.

Notes and References

1. For a related demonstration in which an oscillating ammonium chloride cloud is created, see M. D. Alexander, *J. Chem. Ed.* **1999**, *76*, 210.

2. E. Cerceo, *Am. Environmental Lab.* February 1999, 26.

Measuring pH
Inside Single Cells

R. Kopelman, University of Michigan

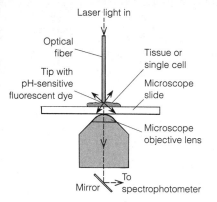

Laser light in

Optical fiber

Tip with pH-sensitive fluorescent dye

Tissue or single cell

Microscope slide

Microscope objective lens

Mirror

To spectrophotometer

Micrograph shows light emitted from the tip of an optical fiber inserted into a rat embryo. The emission spectrum (not shown) indicates that the pH of fluid surrounding the fiber is 7.50 ± 0.05 in 10 embryos. Prior to the development of this microscopic method, 1 000 embryos had to be homogenized for a single measurement.

*P*H controls the rate and thermodynamics of every biological process. To measure pH inside embryos and even single cells, an optical fiber with a pH-sensitive, fluorescent dye bound to its tip can be used.[1] The fiber is inserted into the specimen and laser light is directed down the fiber. Dye molecules at the fiber tip absorb the laser light and then emit light whose wavelength depends on the pH of the surrounding medium. The pH of your blood is closely regulated at 7.40 ± 0.05 by the buffering action of carbonate, phosphate, and proteins.

Chapter 8

BUFFERS

A buffered solution resists changes in pH when small amounts of acids or bases are added or when dilution occurs. The **buffer** consists of a mixture of an acid and its conjugate base. Biochemists are particularly interested in buffers because the functioning of biological systems is critically dependent on pH. For example, Figure 8-1 shows how the rate of the following reaction varies with pH:

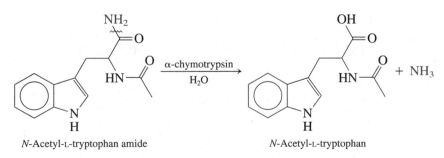

N-Acetyl-L-tryptophan amide   N-Acetyl-L-tryptophan

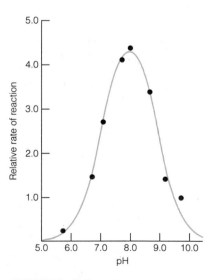

FIGURE 8-1 α-Chymotrypsin-catalyzed cleavage of the C—N bond has a maximum rate near pH 8 that is twice as great as the rate near pH 7 or pH 9.

Enzymes are proteins that *catalyze* (increase the rate of) selected chemical reactions. In the absence of the enzyme α-chymotrypsin, the rate of this reaction is negligible. For any organism to survive, it must control the pH of each subcellular compartment so that each of its enzyme-catalyzed reactions can proceed at the proper rate.

163

When you mix a weak acid with its conjugate base, you get what you mix!

F is the formal concentration of HA, which is 0.100 M in this example.

The approximation that what you mix is what you get breaks down if the solution is too dilute or the acid or base is too weak. We will not consider these cases.

8-1   What You Mix Is What You Get

If you mix A moles of a weak acid with B moles of its conjugate base, the moles of acid remain close to A and the moles of base remain close to B. Little reaction occurs to change either concentration.

To understand why this should be so, look at the K_a and K_b reactions in terms of Le Châtelier's principle. Consider an acid with $pK_a = 4.00$ and its conjugate base with $pK_b = 10.00$. We will calculate the fraction of acid that dissociates in a 0.100 M solution of HA.

$$\underset{0.100-x}{HA} \rightleftharpoons \underset{x}{H^+} + \underset{x}{A^-} \qquad pK_a = 4.00$$

$$\frac{x^2}{F-x} = K_a = 1.0 \times 10^{-4} \implies x = 3.1 \times 10^{-3} \text{ M}$$

$$\text{Fraction of dissociation} = \frac{[A^-]}{[A^-] + [HA]} = \frac{x}{F} = 0.031$$

The acid is only 3.1% dissociated under these conditions.

In a solution containing 0.100 mol of A^- dissolved in 1.00 L, the extent of reaction of A^- with water is even smaller:

$$\underset{0.100-x}{A^-} + H_2O \rightleftharpoons \underset{x}{HA} + \underset{x}{OH^-} \qquad pK_b = 10.00$$

$$\frac{x^2}{F-x} = K_b = 1.0 \times 10^{-10} \implies x = 3.2 \times 10^{-6} \text{ M}$$

$$\text{Fraction of association} = \frac{[HA]}{[A^-] + [HA]} = \frac{x}{F} = 3.2 \times 10^{-5}$$

HA dissociates very little, and Le Châtelier's principle tells us that adding extra A^- to the solution will make the HA dissociate even less. Similarly, A^- does not react very much with water, and adding extra HA makes A^- react even less. If 0.050 mol of A^- plus 0.036 mol of HA are added to water, there will be close to 0.050 mol of A^- and close to 0.036 mol of HA in the solution at equilibrium.

8-2   The Henderson-Hasselbalch Equation

The central equation for buffers is the **Henderson-Hasselbalch equation,** which is merely a rearranged form of the K_a equilibrium expression:

$$K_a = \frac{[H^+][A^-]}{[HA]}$$

Reminder:

$\log xy = \log x + \log y$

$$\log K_a = \log\left(\frac{[H^+][A^-]}{[HA]}\right) = \log[H^+] + \log\left(\frac{[A^-]}{[HA]}\right)$$

$$\underbrace{-\log[H^+]}_{pH} = \underbrace{-\log K_a}_{pK_a} + \log\left(\frac{[A^-]}{[HA]}\right)$$

Henderson-Hasselbalch
equation for an acid:

$$pH = pK_a + \log\left(\frac{[A^-]}{[HA]}\right) \qquad (8\text{-}1)$$

Henderson and Hasselbalch recognized that the concentrations $[A^-]$ and $[HA]$ can be set equal to their formal concentrations, and they were among the first people to apply Equation 8-1 to practical problems.

The Henderson-Hasselbalch equation tells us the pH of a solution, provided we know the ratio of concentrations of conjugate acid and base, as well as pK_a for the acid.

If a solution is prepared from the weak base B and its conjugate acid, the analogous equation is

Henderson-Hasselbalch
equation for a base:

$$pH = pK_a + \log\left(\frac{[B]}{[BH^+]}\right) \qquad (8\text{-}2)$$

pK_a applies to *this* acid

where pK_a is the acid dissociation constant of the weak acid BH^+. The important features of Equations 8-1 and 8-2 are (1) that the base (A^- or B) appears in the numerator and (2) pK_a applies to the acid in the denominator.

When $[A^-] = [HA]$, pH = pK_a

When the concentrations of A^- and HA are equal in Equation 8-1, the log term is 0 because $\log(1) - 0$. Therefore, when $[A^-] - [HA]$, $pH = pK_a$.

$$pH = pK_a + \log\left(\frac{[A^-]}{[HA]}\right) = pK_a + \log(1) = pK_a$$

$\log(1) = 0$ because $10^0 - 1$

Regardless of how complex a solution may be, whenever $pH = pK_a$, $[A^-]$ must equal $[HA]$. This relation is true because *all equilibria must be satisfied simultaneously in any solution at equilibrium.* If there are 10 different acids and bases in the solution, the 10 forms of Equation 8-1 must all give the same pH, because **there can be only one concentration of H^+ in a solution.**

When $[A^-]/[HA]$ Changes by a Factor of 10, the pH Changes by One Unit

Another feature of the Henderson-Hasselbalch equation is that for every power-of-10 change in the ratio $[A^-]/[HA]$, the pH changes by one unit (Table 8-1). As the concentration of base (A^-) increases, the pH goes up. As the concentration of acid (HA) increases, the pH goes down. For any conjugate acid-base pair, you can say, for example, that if $pH = pK_a - 1$, there must be 10 times as much HA as A^-. Therefore ten-elevenths is in the form HA and one-eleventh is in the form A^-.

TABLE 8-1

Change of pH with change of $[A^-]/[HA]$

$[A^-]/[HA]$	pH
100:1	$pK_a + 2$
10:1	$pK_a + 1$
1:1	pK_a
1:10	$pK_a - 1$
1:100	$pK_a - 2$

If $pH = pK_a$, $[HA] = [A^-]$.
If $pH < pK_a$, $[HA] > [A^-]$.
If $pH > pK_a$, $[HA] < [A^-]$.

When we say that the solution is buffered to pH 6.20, we mean that unspecified acids and bases were used to set the pH to 6.20. If buffering is sufficient, then adding a little more acid or base does not change the pH appreciably.

EXAMPLE Using the Henderson-Hasselbalch Equation

Sodium hypochlorite (NaOCl, the active ingredient of bleach) was dissolved in a solution buffered to pH 6.20. Find the ratio $[OCl^-]/[HOCl]$ in this solution.

SOLUTION OCl^- is the conjugate base of hypochlorous acid, HOCl. In Appendix B, we find that $pK_a = 7.53$ for HOCl. Knowing the pH, we calculate the ratio $[OCl^-]/[HOCl]$ from the Henderson-Hasselbalch equation:

$$HOCl \rightleftharpoons H^+ + OCl^- \qquad pH = pK_a + \log\left(\frac{[OCl^-]}{[HOCl]}\right)$$

$$6.20 = 7.53 + \log\left(\frac{[OCl^-]}{[HOCl]}\right)$$

$$-1.33 = \log\left(\frac{[OCl^-]}{[HOCl]}\right)$$

It doesn't matter how the pH got to 6.20. If we know that pH = 6.20, the Henderson-Hasselbalch equation tells us the ratio $[OCl^-]/[HOCl]$.

If $a = b$, then $10^a = 10^b$.

Remember that $10^{\log a} = a$.

To solve this equation, raise 10 to the power shown on each side:

$$10^{-1.33} = 10^{\log([OCl^-]/[HOCl])} = \frac{[OCl^-]}{[HOCl]}$$

$$0.047 = \frac{[OCl^-]}{[HOCl]}$$

To find $10^{-1.33}$ with my calculator, I use the *10^x* function, with $x = -1.33$. If you have the *antilog* function instead of *10^x* on your calculator, you should compute antilog (-1.33).

Finding the ratio $[OCl^-]/[HOCl]$ requires only pH and pK_a. We do not care what else is in the solution, how much NaOCl was added, or what the volume of the solution is.

Ask Yourself

8-A. (a) What is the pH of a buffer prepared by dissolving 0.100 mol of the weak acid HA ($K_a = 1.0 \times 10^{-5}$) plus 0.050 mol of its conjugate base Na^+A^- in 1.00 L?

(b) Write the Henderson-Hasselbalch equation for a solution of formic acid, HCO_2H. What is the quotient $[HCO_2^-]/[HCO_2H]$ at pH 3.00, 3.745, and 4.00?

8-3   A Buffer in Action

For illustration, we choose a widely used buffer called "tris."

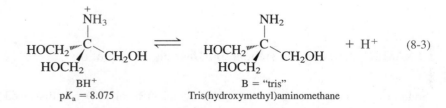

BH$^+$
$pK_a = 8.075$

B = "tris"
Tris(hydroxymethyl)aminomethane

$+ H^+$ (8-3)

In Appendix B, we find pK_a for BH^+, the conjugate acid of tris, to be 8.075. An example of a salt containing BH^+ is tris hydrochloride, which is BH^+Cl^-. When BH^+Cl^- is dissolved in water, it dissociates completely to BH^+ and Cl^-. To find the pH of a known mixture of B and BH^+, simply plug their concentrations into the Henderson-Hasselbalch equation.

EXAMPLE   A Buffer Solution

Find the pH of a solution prepared by dissolving 12.43 g of tris (FM 121.136) plus 4.67 g of tris hydrochloride (FM 157.597) in 1.00 L of water.

SOLUTION The concentrations of B and BH^+ added to the solution are

$$[B] = \frac{12.43 \text{ g/L}}{121.136 \text{ g/mol}} = 0.102\ 6 \text{ M} \qquad [BH^+] = \frac{4.67 \text{ g/L}}{157.597 \text{ g/mol}} = 0.029\ 6 \text{ M}$$

Assuming that what we mixed stays in the same form, we insert the concentrations into the Henderson-Hasselbalch equation to find the pH:

$$pH = pK_a + \log\left(\frac{[B]}{[BH^+]}\right) = 8.075 + \log\left(\frac{0.102\ 6}{0.029\ 6}\right) = 8.61$$

Don't panic over significant figures in pH. For consistency, we are almost always going to express pH to the 0.01 place.

Notice that *the volume of solution is irrelevant to finding the pH,* because volume cancels in the numerator and denominator of the log term:

$$pH = pK_a + \log\left(\frac{\text{moles of B/L of solution}}{\text{moles of } BH^+/L \text{ of solution}}\right)$$

$$= pK_a + \log\left(\frac{\text{moles of B}}{\text{moles of } BH^+}\right)$$

The pH of a buffer is nearly independent of volume.

If strong acid is added to a buffered solution, some of the buffer base will be converted into the conjugate acid and the quotient $[B]/[BH^+]$ changes. If a strong base is added to a buffered solution, some BH^+ is converted into B. Knowing how much strong acid or base is added, we can compute the new quotient $[B]/[BH^+]$ and the new pH.

EXAMPLE   Effect of Adding Acid to a Buffer

If we add 12.0 mL of 1.00 M HCl to the solution in the previous example, what will the new pH be?

SOLUTION The key to this problem is to realize that **when a strong acid is added to a weak base, they react completely to give BH^+** (see Box 8-1). In the present example, we are adding 12.0 mL of 1.00 M HCl, which contain

Strong Plus Weak Reacts Completely

A strong acid reacts with a weak base "completely" because the equilibrium constant is large:

$$B + H^+ \rightleftharpoons BH^+ \qquad K = \frac{1}{K_a \text{(for BH}^+\text{)}}$$

Weak base   Strong acid

If B is tris, then the equilibrium constant for reaction with HCl is

$$K = \frac{1}{K_a} = \frac{1}{10^{-8.075}} = 1.2 \times 10^8 \leftarrow \text{A big number}$$

A strong base reacts "completely" with a weak acid because the equilibrium constant is, again, very large:

$$OH^- + HA \rightleftharpoons A^- + H_2O \qquad K = \frac{1}{K_b \text{(for A}^-\text{)}}$$

Strong base   Weak acid

If HA is acetic acid, then the equilibrium constant for reaction with NaOH is

$$K = \frac{1}{K_b} = \frac{K_a \text{(for HA)}}{K_w} = 1.7 \times 10^9 \leftarrow \text{Another big number}$$

The reaction of a strong acid with a strong base is even more complete than a strong plus weak reaction:

$$H^+ + OH^- \rightleftharpoons H_2O \qquad K = \frac{1}{K_w} = 10^{14}$$

Strong acid   Strong base   ↑ Ridiculously big!

If you mix a strong acid, a strong base, a weak acid, and a weak base, the strong acid and base will neutralize each other until one is used up. The remaining strong acid or base will then react with the weak base or weak acid.

$(0.012\ 0\ \text{L})(1.00\ \text{mol/L}) = 0.012\ 0$ mol of H^+. The H^+ consumes $0.012\ 0$ mol of B to create $0.012\ 0$ mol of BH^+:

	B tris	+	H⁺ from HCl	⟶	BH⁺
Initial moles	0.102 6		0.012 0		0.029 6
Final moles	0.090 6		—		0.041 6
	(0.102 6 − 0.012 0)				(0.029 6 + 0.012 0)

The table contains enough information for us to calculate the pH.

$$pH = pK_a + \log\left(\frac{\text{moles of B}}{\text{moles of BH}^+}\right)$$

$$= 8.075 + \log\left(\frac{0.090\ 6}{0.041\ 6}\right) = 8.41$$

Question Does the pH change in the right direction when HCl is added?

The volume of solution is irrelevant.

The preceding example illustrates that *the pH of a buffer does not change very much when a limited amount of a strong acid or base is added.* Addition of 12.0 mL of 1.00 M HCl changed the pH from 8.61 to 8.41. Addition of 12.0 mL of 1.00 M HCl to 1.00 L of unbuffered solution would have lowered the pH to 1.93.

But *why* does a buffer resist changes in pH? **It does so because the strong acid or base is consumed by B or BH$^+$.** If you add HCl to tris, B is converted to BH$^+$. If you add NaOH, BH$^+$ is converted to B. As long as you don't use up the B or BH$^+$ by adding too much HCl or NaOH, the log term of the Henderson-Hasselbalch equation does not change very much and the pH does not change very much. Demonstration 8-1 illustrates what happens when the buffer does get used up. The buffer has its maximum capacity to resist changes of pH when pH = pK_a. We will return to this point later.

A buffer resists changes in pH . . .

. . . because the buffer "consumes" the added acid or base.

Ask Yourself

8-B. (a) What is the pH of a solution prepared by dissolving 10.0 g of tris plus 10.0 g of tris hydrochloride in 0.250 L water?
   (b) What will the pH be if 10.5 mL of 0.500 M HClO$_4$ are added to **(a)**?
   (c) What will the pH be if 10.5 mL of 0.500 M NaOH are added to **(a)**?

8-4   Preparing Buffers

Buffers are usually prepared by starting with a measured amount of either a weak acid (HA) or a weak base (B). Then OH$^-$ is added to HA to make a mixture of HA and A$^-$ (a buffer), or H$^+$ is added to B to make a mixture of B and BH$^+$ (a buffer).

EXAMPLE   Calculating How to Prepare a Buffer Solution

How many milliliters of 0.500 M NaOH should be added to 10.0 g of tris hydrochloride (BH$^+$, Equation 8-3) to give a pH of 7.60 in a final volume of 250 mL?

SOLUTION The number of moles of tris hydrochloride in 10.0 g is (10.0 g)/(157.597 g/mol) = 0.063 5. We can make a table to help solve the problem:

Reaction with OH$^-$:	BH$^+$	+	OH$^-$	$\longrightarrow$	B
Initial moles	0.063 5		x		—
Final moles	0.063 5 − x		—		x

The Henderson-Hasselbalch equation allows us to find x, because we know pH and pK_a:

$$\text{pH} = pK_a + \log\left(\frac{\text{mol B}}{\text{mol BH}^+}\right)$$

How Buffers Work

A buffer resists changes in pH because the added acid or base is consumed by the buffer. As the buffer is used up, it becomes less resistant to changes in pH.

In this demonstration,[2] a mixture containing a 10:1 mole ratio of $HSO_3^-:SO_3^{2-}$ is prepared. Because pK_a for HSO_3^- is 7.2, the pH should be approximately

$$pH = pK_a + \log\left(\frac{[SO_3^{2-}]}{[HSO_3^-]}\right) = 7.2 + \log\left(\frac{1}{10}\right) = 6.2$$

When formaldehyde is added, the net reaction is the consumption of HSO_3^- but not of SO_3^{2-}.

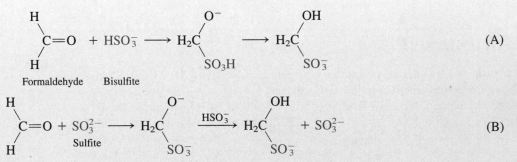

(A)

Formaldehyde   Bisulfite

Sulfite

(B)

(Sequence A consumes bisulfite. In sequence B, the net reaction is destruction of HSO_3^-, with no change in the SO_3^{2-} concentration.)

We can prepare a table showing how the pH should change as the HSO_3^- reacts:

Percentage of reaction completed	$[SO_3^{2-}]:[HSO_3^-]$	Calculated pH
0	1 : 10	6.2
90	1 : 1	7.2
99	1 : 0.1	8.2
99.9	1 : 0.01	9.2
99.99	1 : 0.001	10.2

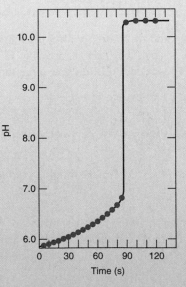

Graph of pH versus time in the formaldehyde clock reaction.

You can see that through 90% completion, the pH rises by just one unit. In the next 9% of the reaction, the pH rises by another unit. At the end of the reaction, the change in pH is very abrupt.

In the formaldehyde clock reaction, formaldehyde is added to a solution containing HSO_3^-, SO_3^{2-}, and phenolphthalein indicator. Phenolphthalein is colorless below pH 8.5 and red above this pH. We observe that the solution remains colorless for more than a minute after the addition of formaldehyde. Suddenly the pH shoots up and the liquid turns pink. Monitoring the pH with a glass electrode gave the results shown in the figure.

Procedure: Phenolphthalein solution is prepared by dissolving 50 mg of solid indicator in 50 mL of ethanol and diluting with 50 mL of water. The following solutions should be fresh: Dilute 9 mL of 37 wt % formaldehyde to 100 mL. Dissolve 1.5 g $NaHSO_3$ and 0.18 g Na_2SO_3 in 400 mL of water and add 1 mL of phenolphthalein indicator solution. To initiate the clock reaction, add 23 mL of formaldehyde solution to the well-stirred buffer solution. Reaction time is affected by the temperature, concentrations, and volume.

$$7.60 = 8.075 + \log\left(\frac{x}{0.063\ 5 - x}\right)$$

$$-0.475 = \log\left(\frac{x}{0.063\ 5 - x}\right)$$

To solve for x, raise 10 to the power of the terms on both sides, remembering that $10^{\log z} = z$:

$$10^{-0.475} = 10^{\log[x/(0.063\ 5 - x)]}$$

$$0.335 = \frac{x}{0.063\ 5 - x} \quad \Rightarrow \quad x = 0.015\ 9 \text{ mol}$$

If your calculator has the *antilog* function, solve the equation $-0.475 = \log z$ by calculating $z = \text{antilog}(-0.475)$.

This many moles of NaOH is contained in

$$\frac{0.015\ 9 \text{ mol}}{0.500 \text{ mol/L}} = 0.031\ 8 \text{ L} = 31.8 \text{ mL}$$

Our calculation tells us to mix 31.8 mL of 0.500 M NaOH with 10.0 g of tris hydrochloride to give a pH of 7.60.

Preparing a Buffer in Real Life

When you mix calculated quantities of acid and base to make a buffer, the pH *does not* come out exactly as expected. The main reason for the discrepancy is that pH is governed by the *activities* of the conjugate acid-base pair, not by the concentrations. (Activity is discussed in Section 11-2.) If you really want to prepare tris buffer at pH 7.60, you should use a pH electrode to get exactly what you need.

Suppose you wish to prepare 1.00 L of buffer containing 0.100 M tris at pH 7.60. You have available solid tris hydrochloride and ~1 M NaOH. Here's how to do it.

1. Weigh out 0.100 mol tris hydrochloride and dissolve it in a beaker containing about 800 mL water.

2. Place a pH electrode in the solution and monitor the pH.

3. Add NaOH solution until the pH is exactly 7.60. The electrode does not respond instantly. Be sure to allow the pH reading to stabilize after each addition of reagent.

4. Transfer the solution to a volumetric flask and wash the beaker a few times. Add the washings to the volumetric flask.

5. Dilute to the mark and mix.

The reason for using 800 mL of water in the first step is so that the volume will be close to the final volume during pH adjustment. Otherwise, the pH will change slightly when the sample is diluted to its final volume and the *ionic strength* changes.

Ionic strength is a measure of the total concentration of ions in a solution. Changing ionic strength changes activities of ionic species such as H^+ and A^-, even though their concentrations are constant. Diluting a buffer with water changes the ionic strength and therefore changes the pH slightly.

Ask Yourself

8-C. (a) How many milliliters of 1.20 M HCl should be added to 10.0 g of tris (B, Equation 8-3) to give a pH of 7.60 in a final volume of 250 mL?

(b) What operations would you perform to prepare exactly 100 mL of 0.200 M acetate buffer, pH 5.00, starting with pure liquid acetic acid and solutions containing ~3 M HCl and ~3 M NaOH?

8-5   Buffer Capacity

The greater the buffer capacity, the less the pH changes when H^+ or OH^- is added. Buffer capacity is maximum when $pH = pK_a$.

Buffer capacity measures how well a solution resists changes in pH when acid or base is added. The greater the buffer capacity, the less the pH changes. We will find that *the maximum buffer capacity occurs when* $pH = pK_a$ *for the buffer.*

Figure 8-2 shows an experiment conducted with a spreadsheet to see how a buffer responds to small additions of H^+ or OH^-. A buffer was made by mixing HA (with acid dissociation constant $K_a = 10^{-5}$) with A^-. The total moles of HA + A^- were set equal to 1. The relative quantities of HA and A^- were varied to give initial pH values from 3.4 to 6.6. Then 0.01 mol of either H^+ or OH^- was added to the solution and the new pH was calculated. Figure 8-2 shows the change in pH, that is, $\Delta(pH)$, as a function of the initial pH of the buffer.

For example, mixing 0.038 3 mol of A^- and 0.961 7 mol of HA gives an initial pH of 3.600:

$$pH = pK_a + \log\left(\frac{\text{mol A}^-}{\text{mol HA}}\right) = 5.000 + \log\left(\frac{0.038\ 3}{0.961\ 7}\right) = 3.600$$

When 0.010 0 mol of OH^- is added to this mixture, the concentrations and pH change in a manner that you are now smart enough to calculate:

Reaction with OH^-:	HA	+	OH^-	$\longrightarrow$	A^-
Initial mole	0.961 7		0.010 0		0.038 3
Final moles	0.961 7 − 0.010 0		—		0.038 3 + 0.010 0
	0.951 7				0.048 3

$$pH = pK_a + \log\left(\frac{\text{mol A}^-}{\text{mol HA}}\right) = 5.000 + \log\left(\frac{0.048\ 3}{0.951\ 7}\right) = 3.705$$

The change in pH is $\Delta(pH) = 3.705 - 3.600 = +0.105$. This is the value plotted in Figure 8-2 on the upper curve for an initial pH of 3.600. If the same starting mixture had been treated with 0.010 0 mol of H^+, the pH would change by -0.136, which is shown on the lower curve.

We see in Figure 8-2 that the magnitude of the pH change is smallest when the initial pH is equal to pK_a for the buffer. That is, *the buffer capacity is greatest when* $pH = pK_a$.

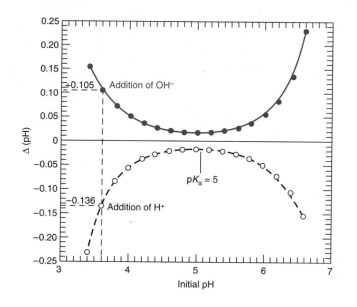

FIGURE 8-2 *Buffer capacity:* Effect of adding 0.01 mol of H^+ or OH^- to a buffer containing HA and A^- (total quantity of HA + A^- = 1 mol). The minimum change in pH occurs when the initial pH of the buffer equals pK_a for HA. That is, the maximum buffer capacity occurs when pH = pK_a.

In choosing a buffer for an experiment, you should *seek one whose pK$_a$ is as close as possible to the desired pH. The useful pH range of a buffer is usually considered to be pK$_a$ ± 1 pH unit.* Outside this range, there is not enough of either the weak acid or the weak base to react with added base or acid. Clearly, buffer capacity increases with increasing concentration of the buffer. To maintain a stable pH, you must use a concentration of buffer high enough to react with the anticipated quantity of acid or base to be encountered.

Table 8-2 lists pK_a values for common buffers. Some of the buffers have more than one acidic proton, so more than one pK is listed. We will consider polyprotic acids and bases in Chapter 10.

Choose a buffer whose pK$_a$ is close to the desired pH.

Buffer pH Depends on Temperature and Ionic Strength

Most buffers exhibit a noticeable dependence of pK_a on temperature. Tris has an exceptionally large dependence, approximately -0.031 pK_a units per degree, near 25°C. A solution of tris made up to pH 8.08 at 25°C will have pH ≈ 8.7 at 4°C and pH ≈ 7.7 at 37°C. When a 0.5 M stock solution of phosphate buffer at pH 6.6 is diluted to 0.05 M, the pH rises to 6.9, because the ionic strength and activities of the buffering species, $H_2PO_4^-$, and HPO_4^{2-}, change.

Summary

A buffer consists of a mixture of a weak acid and its conjugate base. The buffer is most useful when pH ≈ pK_a. Over a reasonable range of concentrations, the pH of a buffer is nearly independent of concentration. A buffer resists changes in pH because it reacts with added acids or bases. If too much acid or base is added, the buffer will be consumed and no longer resists changes in pH.

TABLE 8-2 Structures and pK_a values for common buffers

Name	Structure[a]	pK_a (~25°C)	Formula mass
Phosphoric acid	$\mathbf{H_3PO_4}$	2.15 (pK_1)	97.995
Citric acid	$HO_2CCH_2\overset{\displaystyle OH}{\underset{\displaystyle CO_2H}{C}}CH_2CO_2\mathbf{H}$	3.13 (pK_1)	192.125
Citric acid	$\mathbf{H_2}$(citrate)$^-$	4.76 (pK_2)	192.125
Acetic acid	$CH_3CO_2\mathbf{H}$	4.76	60.053
2-(N-Morpholino)ethanesulfonic acid (MES)	$O\overbrace{}^{}\overset{+}{N}\mathbf{H}CH_2CH_2SO_3^-$	6.15	195.240
Citric acid	$\mathbf{H}$(citrate)$^{2-}$	6.40 (pK_3)	192.125
Piperazine-N,N'-bis(2-ethanesulfonic acid) (PIPES)	$^-O_3SCH_2CH_2\overset{+}{N}\mathbf{H}\quad\overset{+}{\mathbf{H}N}CH_2CH_2SO_3^-$	6.80	302.373
3-(N-Morpholino)-2-hydroxypropanesulfonic acid (MOPSO)	$O\overbrace{}^{}\overset{+}{N}\mathbf{H}CH_2\overset{\displaystyle OH}{C}HCH_2SO_3^-$	6.95	225.265
Imidazole hydrochloride	$\overset{+}{\mathbf{H}N}\diagdown\text{...}\underset{\underset{\mathbf{H}}{N}}{} \quad Cl^-$	6.99	104.539
Phosphoric acid	$\mathbf{H_2}PO_4^-$	7.20 (pK_2)	97.995
N-2-Hydroxyethylpiperazine-N'-2-ethanesulfonic acid (HEPES)	$HOCH_2CH_2N\overbrace{}^{}\overset{+}{N}\mathbf{H}CH_2CH_2SO_3^-$	7.56	238.308
Tris(hydroxymethyl)aminomethane hydrochloride (tris hydrochloride)	$(HOCH_2)_3C\overset{+}{N}\mathbf{H_3}\quad Cl^-$	8.08	157.597
Glycylglycine	$\mathbf{H_3}\overset{+}{N}CH_2\overset{\displaystyle O}{\overset{\|}{C}}NHCH_2CO_2^-$	8.40	132.11[9]
Boric acid	$\mathbf{B(OH)_3}$	9.24 (pK_1)	61.833
Cyclohexylaminoethanesulfonic acid (CHES)	$\bigcirc\!\!-\overset{+}{N}\mathbf{H_2}CH_2CH_2SO_3^-$	9.50	207.294
3-(Cyclohexylamino)propanesulfonic acid (CAPS)	$\bigcirc\!\!-\overset{+}{N}\mathbf{H_2}CH_2CH_2CH_2SO_3^-$	10.40	221.321
Phosphoric acid	$\mathbf{H}PO_4^{2-}$	12.15 (pK_3)	97.995
Boric acid	$OB(O\mathbf{H})_2^-$	12.74 (pK_2)	61.833

a. The protonated form of each molecule is shown. Acidic hydrogen atoms are shown in **bold** type.

Ask Yourself

8-D. (a) Look up pK_a for each of the following acids and decide which would be best for preparing a buffer of pH 3.10:

    (i) hydroxybenzene   **(iii)** cyanoacetic acid
    (ii) propanoic acid   **(iv)** sulfuric acid

    (b) Why does buffer capacity increase as the concentration of buffer increases?

    (c) Based on the K_b values listed below, which of the following bases would be best for preparing a buffer of pH 9.00?

    (i) NH_3 (ammonia, $K_b = 1.75 \times 10^{-5}$)
    (ii) $C_6H_5NH_2$ (aniline, $K_b = 3.99 \times 10^{-10}$)
    (iii) H_2NNH_2 (hydrazine, $K_b = 3.0 \times 10^{-6}$)
    (iv) C_5H_5N (pyridine, $K_b = 1.69 \times 10^{-9}$)

8-6   How Indicators Work

An acid-base **indicator** is itself an acid or base whose various protonated species have different colors. In the next chapter, we will learn how to choose an indicator to find the end point for a titration. For now, let's try to use the Henderson-Hasselbalch equation to understand the pH range over which color changes are observed.

    Consider bromocresol green as an example. We will call pK_a of the indicator pK_{HIn}, to distinguish it from pK_a of an acid being titrated.

> An *indicator* is an acid or a base whose various protonated forms have different colors.

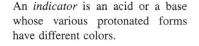

Yellow (Y)
Bromocresol green

Blue (B⁻)

pK_{HIn} for bromocresol green is 4.7. Below pH 4.7, the predominant species is yellow (Y); above pH 4.7, the predominant species is blue (B⁻).

    The equilibrium between Y and B⁻ can be written

$$Y \rightleftharpoons B^- + H^+ \qquad K_{HIn} = \frac{[B^-][H^+]}{[Y]}$$

for which the Henderson-Hasselbalch equation is

$$pH = pK_{HIn} + \log\left(\frac{[B^-]}{[Y]}\right) \qquad (8\text{-}4)$$

pH	[B⁻]:[Y]	Color
3.7	1:10	yellow
4.7	1:1	green
5.7	10:1	blue

At pH = pK_a = 4.7, there will be a 1:1 mixture of the yellow and blue species, which will appear green. As a crude rule of thumb, we can say that the solution will appear yellow when $[Y]/[B^-] \gtrsim 10/1$ and blue when $[B^-]/[Y] \gtrsim 10/1$. (The symbol $\gtrsim$ means "is approximately greater than.") From Equation 8-4, we can see that the solution will be yellow when $pH \lesssim pK_{HIn} - 1$ (=3.7) and blue when $pH \gtrsim pK_{HIn} + 1$ (=5.7). By comparison, Table 8-3 lists bromocresol green as yellow below pH 3.8 and blue above pH 5.4. Between pH 3.8 and 5.4, various shades of green are seen. Demonstration 8-2 illustrates indicator color changes, and Box 8-2 describes an everyday application of indicators.

Several indicators are listed twice in Table 8-3, with two different sets of colors. An example is thymol blue, which loses one proton with a pK of 1.7 and a second proton with a pK of 8.9:

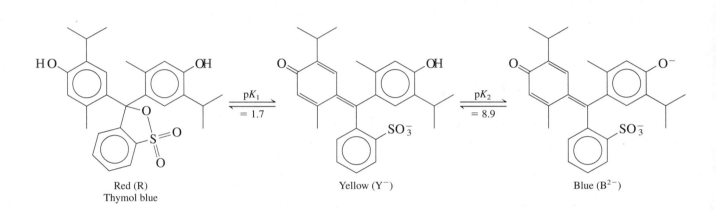

Red (R)
Thymol blue

Yellow (Y⁻)

Blue (B²⁻)

Below pH 1.7, the predominant species is red (R); between pH 1.7 and pH 8.9, the predominant species is yellow (Y⁻); and above pH 8.9, the predominant species is

---

TABLE 8-3 Common indicators

Indicator	Transition range (pH)	Acid color	Base color	Indicator	Transition range (pH)	Acid color	Base color
Methyl violet	0.0–1.6	Yellow	Violet	Litmus	5.0–8.0	Red	Blue
Cresol red	0.2–1.8	Red	Yellow	Bromothymol blue	6.0–7.6	Yellow	Blue
Thymol blue	1.2–2.8	Red	Yellow	Phenol red	6.4–8.0	Yellow	Red
Cresol purple	1.2–2.8	Red	Yellow	Neutral red	6.8–8.0	Red	Yellow
Erythrosine, disodium	2.2–3.6	Orange	Red	Cresol red	7.2–8.8	Yellow	Red
				α-Naphtholphthalein	7.3–8.7	Pink	Green
Methyl orange	3.1–4.4	Red	Yellow	Cresol purple	7.6–9.2	Yellow	Purple
Congo red	3.0–5.0	Violet	Red	Thymol blue	8.0–9.6	Yellow	Blue
Ethyl orange	3.4–4.8	Red	Yellow	Phenolphthalein	8.0–9.6	Colorless	Red
Bromocresol green	3.8–5.4	Yellow	Blue	Thymolphthalein	8.3–10.5	Colorless	Blue
Methyl red	4.8–6.0	Red	Yellow	Alizarin yellow	10.1–12.0	Yellow	Orange-red
Chlorophenol red	4.8–6.4	Yellow	Red	Nitramine	10.8–13.0	Colorless	Orange-brown
Bromocresol purple	5.2–6.8	Yellow	Purple	Tropaeolin O	11.1–12.7	Yellow	Orange
p-Nitrophenol	5.6–7.6	Colorless	Yellow				

Indicators and Carbonic Acid[3]

This one is just plain fun. Fill two 1-L graduated cylinders with 900 mL of water and place a magnetic stirring bar in each. Add 10 mL of 1 M NH_3 to each. Then put 2 mL of phenolphthalein indicator solution in one and 2 mL of bromothymol blue indicator solution in the other. Both indicators will have the color of their basic species.

Drop a few chunks of Dry Ice (solid CO_2) into each cylinder. As the CO_2 bubbles through each cylinder, the solutions become more acidic. First the pink phenolphthalein color disappears. After some time, the pH

drops just low enough for bromothymol blue to change from blue to its pale green intermediate color. The pH does not go low enough to turn bromothymol blue to its yellow color.

Add about 20 mL of 6 M HCl to *the bottom* of each cylinder, using a length of Tygon tubing attached to a funnel. Then stir each solution for a few seconds on a magnetic stirrer. Explain what happens. The sequence of events is shown in Color Plate 4.

When CO_2 dissolves in water, it makes carbonic acid, which has two acidic protons:

$$CO_2(g) \rightleftharpoons CO_2(aq) \qquad K = \frac{[CO_2(aq)]}{P_{CO_2}} = 0.034\,4$$

$$CO_2(aq) + H_2O \rightleftharpoons \underset{\text{Carbonic acid}}{\text{HO}-\overset{\overset{\text{O}}{\|}}{\text{C}}-\text{OH}} \qquad K = \frac{[H_2CO_3]}{[CO_2(aq)]} \approx 0.002$$

$$H_2CO_3 \rightleftharpoons \underset{\text{Bicarbonate}}{HCO_3^-} + H^+ \quad K_{a1} = 4.45 \times 10^{-7}$$

$$HCO_3^- \rightleftharpoons \underset{\text{Carbonate}}{CO_3^{2-}} + H^+ \qquad K_{a2} = 4.69 \times 10^{-11}$$

The value $K_{a1} = 4.45 \times 10^{-7}$ applies to the equation

$$\underset{(= CO_2(aq) + H_2CO_3)}{\text{all dissolved } CO_2} \rightleftharpoons HCO_3^- + H^+$$

$$K_{a1} = \frac{[HCO_3^-][H^+]}{[CO_2(aq) + H_2CO_3]} = 4.45 \times 10^{-7}$$

Only about 0.2% of dissolved CO_2 is in the form H_2CO_3. If the true value of $[H_2CO_3]$ were used instead of $[H_2CO_3 + CO_2(aq)]$, the equilibrium constant would be $\sim 2 \times 10^{-4}$.

blue (B^{2-}). The sequence of color changes for thymol blue is shown in Color Plate 5.

The equilibrium between R and Y^- can be written

$$R \rightleftharpoons Y^- + H^+ \qquad K_1 = \frac{[Y^-][H^+]}{[R]}$$

$$pH = pK_1 + \log\left(\frac{[Y^-]}{[R]}\right) \qquad (8\text{-}5)$$

At pH = 1.7 ($=pK_1$), there will be a 1:1 mixture of the yellow and red species, which appears orange. We expect that the solution will appear red when $[R]/[Y^-] \gtrsim 10/1$ and yellow when $[Y^-]/[R] \gtrsim 10/1$. From the Henderson-Hasselbalch equation 8-5, we can see that the solution will be red when

pH	$[Y^-]:[R]$	Color
0.7	1:10	red
1.7	1:1	orange
2.7	10:1	yellow

177

Box 8-2 *Informed Citizen*

The Secret of Carbonless Copy Paper[4]

Back in the days when dinosaurs roamed the earth and I was a child, people would insert a messy piece of carbon paper between two pages so that a copy of what was written on the upper page would appear on the lower page. Then carbonless copy paper was invented to perform the same task without the extra sheet of carbon paper.

The secret of carbonless copy paper is an acid-base indicator contained inside polymeric microcapsules stuck to the underside of the upper sheet of paper. When you write on the upper sheet, the pressure of your pen breaks open the microcapsules on the bottom side, which then release their indicator.

The indicator becomes adsorbed on the upper surface of the lower page, which is coated with microscopic particles of an acidic material such as the clay known as bentonite. This clay contains negatively charged alu-

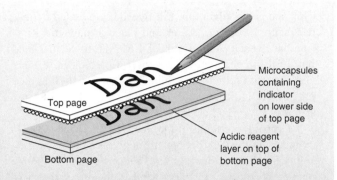

minosilicate layers with hydronium ion (H_3O^+) between the layers to balance the negative charge. The adsorbed indicator reacts with the hydronium ion to give a colored product that appears as a copy of your writing on the lower page.

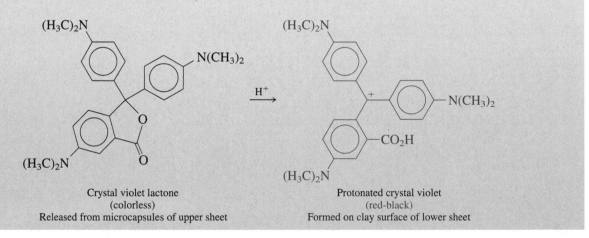

Crystal violet lactone	Protonated crystal violet
(colorless)	(red-black)
Released from microcapsules of upper sheet	Formed on clay surface of lower sheet

pH $\lesssim$ pK_1 − 1 (=0.7) and yellow when pH $\gtrsim$ pK_1 + 1 (=2.7). In Table 8-3, thymol blue is listed as red below pH 1.2 and yellow above pH 2.8. Thymol blue has another transition, from yellow to blue, between pH 8.0 and pH 9.6. In this range, various shades of green are seen.

Ask Yourself

8-E. (a) What is the origin of the rule of thumb that indicator color changes occur at pK_{HIn} ± 1?

(b) What color do you expect to observe for cresol purple indicator (Table 8-3) at the following pH values: 1.0, 2.0, 3.0?

Key Equations

Henderson-Hasselbalch

$$pH = pK_a + \log\left(\frac{[A^-]}{[HA]}\right)$$

$$pH = pK_a + \log\left(\frac{[B]}{[BH^+]}\right) \quad \leftarrow pK_a \text{ applies to } this \text{ acid}$$

Important Terms

buffer   Henderson-Hasselbalch equation   indicator

Problems

8-1. Explain what happens when acid is added to a buffer and the pH does not change very much.

8-2. A solution contains 63 different conjugate acid-base pairs. Among them is acrylic acid and acrylate ion, with the ratio [acrylate]/[acrylic acid] = 0.75. What is the pH of the solution?

$$H_2C = CHCO_2H \qquad pK_a = 4.25$$
Acrylic acid

8-3. Table 8-1 shows the relation of pH to the quotient [A⁻]/[HA].

(a) Use the Henderson-Hasselbalch equation to show that pH = pK_a + 2 when [A⁻]/[HA] = 100.

(b) Find the quotient [A⁻]/[HA] when pH = pK_a − 3.

(c) Find the pH when [A⁻]/[HA] = 10^{-4}.

8-4. Given that pK_b for iodate ion (IO_3^-) is 13.83, find the quotient [HIO₃]/[IO₃⁻] in a solution of sodium iodate at

(a) pH 7.00; **(b)** pH 1.00.

8-5. Given that pK_b for nitrite ion (NO_2^-) is 10.85, find the quotient [HNO₂]/[NO₂⁻] in a solution of sodium nitrite at **(a)** pH 2.00; **(b)** pH 10.00.

8-6. Write the Henderson-Hasselbalch equation for a solution of methylamine. Calculate the quotient [CH₃NH₂]/[CH₃NH₃⁺] at **(a)** pH 4.00; **(b)** pH 10.64; **(c)** pH 12.00.

8-7. Why is the pH of a buffer nearly independent of concentration?

8-8. Find the pH of a solution prepared from 2.53 g of oxoacetic acid, 5.13 g of potassium oxoacetate, and 103 g of water. Use Appendix B to find pK_a for oxoacetic acid.

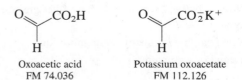

Oxoacetic acid
FM 74.036

Potassium oxoacetate
FM 112.126

8-9. (a) Find the pH of a solution prepared by dissolving 1.00 g of glycine amide hydrochloride plus 1.00 g of glycine amide in 0.100 L.

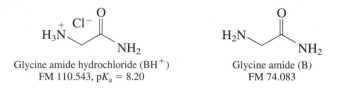

Glycine amide hydrochloride (BH^+)   Glycine amide (B)
FM 110.543, $pK_a = 8.20$ FM 74.083

(b) How many grams of glycine amide should be added to 1.00 g of glycine amide hydrochloride to give 100 mL of solution with pH 8.00?

(c) What would be the pH if the solution in **(a)** is mixed with 5.00 mL of 0.100 M HCl?

(d) What would be the pH if the solution in **(c)** is mixed with 10.00 mL of 0.100 M NaOH?

8-10. (a) Write the chemical reactions whose equilibrium constants are K_b and K_a for imidazole and imidazole hydrochloride, respectively.

(b) Calculate the pH of a 100-mL solution containing 1.00 g imidazole (FM 68.078) and 1.00 g of imidazole hydrochloride (FM 104.539).

(c) Calculate the pH of the solution if 2.30 mL of 1.07 M $HClO_4$ are added to the solution.

(d) How many milliliters of 1.07 M $HClO_4$ should be added to 1.00 g of imidazole to give a pH of 6.993?

8-11. (a) Calculate the pH of a solution prepared by mixing 0.080 0 mol of chloroacetic acid plus 0.040 0 mol of sodium chloroacetate in 1.00 L of water.

(b) Using first your head, and then the Henderson-Hasselbalch equation, find the pH of a solution prepared by dissolving all of the following in a total volume of 1.00 L: 0.180 mol $ClCH_2CO_2H$, 0.020 mol $ClCH_2CO_2Na$, 0.080 mol HNO_3, and 0.080 mol $Ca(OH)_2$. Assume that $Ca(OH)_2$ dissociates completely.

8-12. How many milliliters of 0.246 M HNO_3 should be added to 213 mL of 0.006 66 M ethylamine to give a pH of 10.52?

8-13. Calculate how many milliliters of 0.626 M KOH should be added to 5.00 g of HEPES (Table 8-2) to give a pH of 7.40.

8-14. How many milliliters of 0.113 M HBr should be added to 52.2 mL of 0.013 4 M morpholine to give a pH of 8.00?

8-15. Which buffer system will have the greatest buffer capacity at pH 9.0?

  (i) dimethylamine / dimethylammonium ion

  (ii) ammonia / ammonium ion

  (iii) hydroxylamine / hydroxylammonium ion

  (iv) 4-nitrophenol / 4-nitrophenolate ion

8-16. (a) Describe the correct procedure for preparing 0.250 L of 0.050 0 M HEPES (Table 8-2), pH 7.45.

(b) Would you need NaOH or HCl to bring the pH to 7.45?

8-17. (a) Calculate how many milliliters of 0.100 M HCl should be added to how many grams of sodium acetate dihydrate ($NaOAc \cdot 2H_2O$, FM 118.06) to prepare 250.0 mL of 0.100 M buffer, pH 5.00.

(b) If you mixed what you calculated, the pH would not be 5.00. Describe how you would actually prepare this buffer in the lab.

8-18. Consider the buffer in Figure 8-2, which has an initial pH of 6.200 and pK_a of 5.000. It is made from p mol of A^- plus q mol of HA such that $p + q = 1$ mol.

(a) Setting mol HA $= q$ and mol $A^- = 1 - q$, use the Henderson-Hasselbalch equation to find the values of p and q.

(b) Calculate the pH change when 0.010 0 mol of H^+ is added to the buffer. Did you get the same value shown in Figure 8-2? For the sake of this problem, use three decimal places for pH.

8-19. Cresol red has *two* transition ranges listed in Table 8-3. What color would you expect it to be at the following pH values?

(a) 0   **(b)** 1   **(c)** 6   **(d)** 9

8-20. *Henderson-Hasselbalch spreadsheet.* Prepare a spreadsheet with the constant $pK_a = 4$ in column A. Type the heading $[A^-]/[HA]$ for column B and enter values ranging from 0.001 to 1 000, with a selection of values in between. In column C, compute pH from the Henderson-Hasselbalch equation, using pK_a from column A and $[A^-]/[HA]$ from column B. In column D, compute $\log([A^-]/[HA])$. Use the values from columns C and D to prepare a graph of pH versus $\log([A^-]/[HA])$. Explain the shape of the resulting graph.

How Would You Do It?

8-21. We can measure the concentrations of the two forms of an indicator such as bromocresol green in a solution by measuring the absorption of visible light at two appropriate wavelengths. One wavelength would be where the yellow species has maximum absorption and the other wavelength would be where the blue species has maximum absorption. If we know how much each species absorbs at each wavelength, we can deduce the concentrations of both species from the two measurements. Suggest a procedure for using an optical absorption measurement to find the pH of a solution.

Notes and References

1. W. Tan, S.-Y. Shi, S. Smith, D. Birnbaum, and R. Kopelman, *Science* **1992,** *258,* 778; W. Tan, S.-Y. Shi, and R. Kopelman, *Anal. Chem.* **1992,** *64,* 2985.

2. R. L. Barrett, *J. Chem. Ed.* **1955,** *32,* 78.

3. You can find more indicator demonstrations in J. T. Riley, *J. Chem. Ed.* **1977,** *54,* 29. The chemistry of carbonic acid is discussed by M. Kern, *J. Chem. Ed.* **1960,** *37,* 14.

4. M. A. White, *J. Chem. Ed.* **1998,** *75,* 1119.

World Record Small Titration

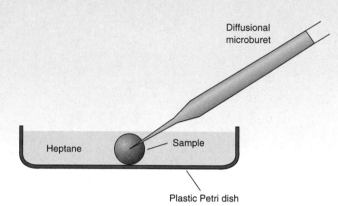

Diffusional microburet

Heptane

Sample

Plastic Petri dish

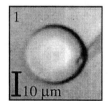

1

10 μm

Before end point

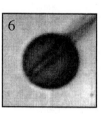

6

After end point

Titration of femtoliter samples. The frames at the bottom show an 8.7-pL droplet before and after the end point; the microburet is making contact at the right side. [Courtesy M. Gratzl, Case Western Reserve University.]

*T*he titration of 29 fmol (f = femto = 10^{-15}) of HNO_3 in a 1.9-pL (p = pico = 10^{-12}) water drop under a layer of hexane in a Petri dish was carried out with 2% precision by using KOH delivered by diffusion from the 1-μm-diameter tip of a glass capillary tube.[1] A mixture of the indicators bromothymol blue and bromocresol purple, with a distinct yellow-to-purple transition at pH 6.7, allowed the end point to be observed through a microscope by a video camera.

A more practical automated system was developed to titrate larger—but still very tiny—volumes.[2] The procedure uses an electronically controlled microburet and a fluorescent indicator monitored by a fiber optic to achieve 1% precision in titrations of 9-nL (n = nano = 10^{-9}) volumes.

Chapter 9

ACID–BASE TITRATIONS

*I*n a *titration*, we measure the quantity of a known reagent required to react with an unknown sample. From this quantity, we deduce the concentration of analyte in the unknown. Titrations of acids and bases are among the most widespread procedures in chemical analysis. For example, at the end of this chapter, we will see how acid-base titrations are used to measure the nitrogen content of foods.

9-1   Titration of Strong Acid with Strong Base

For each type of titration in this chapter, *our goal is to construct a graph showing how the pH changes as titrant is added.* If you can do this, then you understand what is happening during the titration, and you will be able to interpret an experimental titration curve.

The first step is to write the chemical reaction between titrant and analyte. Then use that reaction to calculate the composition and pH after each addition of titrant.

First write the reaction between *titrant* and *analyte*.

183

The titration reaction.

Let's consider the titration of 50.00 mL of 0.020 00 M KOH with 0.100 0 M HBr. The chemical reaction between titrant and analyte is merely

$$H^+ + OH^- \longrightarrow H_2O \qquad K = \frac{1}{K_w} = \frac{1}{10^{-14}} = 10^{14}$$

Because the equilibrium constant for this reaction is 10^{14}, it is fair to say that it "goes to completion." Prior to the equivalence point, *any amount of H^+ added will consume a stoichiometric amount of OH^-.*

Equivalence point: when moles of added titrant are exactly sufficient for stoichiometric reaction with analyte.

A useful starting point is to calculate the volume of HBr (V_e) needed to reach the equivalence point:

$$\underbrace{(V_e(\text{mL}))(0.100\ 0\ \text{M})}_{\substack{\text{mmol of HBr} \\ \text{at equivalence point}}} = \underbrace{(50.00\ \text{mL})(0.020\ 00\ \text{M})}_{\substack{\text{mmol of } OH^- \\ \text{being titrated}}} \Rightarrow V_e = 10.00\ \text{mL}$$

$$\boxed{\text{mL} \times \frac{\text{mol}}{\text{L}} = \text{mmol}}$$

When 10.00 mL of HBr have been added, the titration is complete. Prior to V_e, excess, unreacted OH^- is present. After V_e, there is excess H^+ in the solution.

In the titration of a strong base with a strong acid, three regions of the titration curve require different kinds of calculations:

1. Before the equivalence point, the pH is determined by excess OH^- in the solution.

2. At the equivalence point, added H^+ is just sufficient to react with all the OH^- to make H_2O. The pH is determined by the dissociation of water.

3. After the equivalence point, pH is determined by excess H^+ in the solution.

We will do one sample calculation for each region. Complete results are shown in Table 9-1 and Figure 9-1 (page 186).

Region 1: Before the Equivalence Point

Before the equivalence point, there is excess OH^-.

Before adding any HBr titrant from the buret, the flask of analyte contains 50.00 mL of 0.020 00 M KOH, which amounts to $(50.00\ \text{mL})(0.020\ 00\ \text{M}) = 1.000$ mmol of OH^-. Remember that $\text{mL} \times (\text{mol/L}) = \text{mmol}$.

If we add 3.00 mL of HBr, we add $(3.00\ \text{mL})(0.100\ 0\ \text{M}) = 0.300$ mmol of H^+, which consumes 0.300 mmol of OH^-.

$$OH^- \text{ remaining} = \underbrace{1.000\ \text{mmol}}_{\text{Initial } OH^-} - \underbrace{0.300\ \text{mmol}}_{\substack{OH^- \text{ consumed} \\ \text{by HBr}}} = 0.700\ \text{mmol}$$

The total volume in the flask is now $50.00\ \text{mL} + 3.00\ \text{mL} = 53.00\ \text{mL}$. Therefore the concentration of OH^- in the flask is

$$\frac{\text{mmol}}{\text{mL}} = \frac{\text{mol}}{\text{L}} = \text{M}$$

$$[OH^-] = \frac{0.700\ \text{mmol}}{53.00\ \text{mL}} = 0.013\ 2\ \text{M}$$

TABLE 9-1 **Calculation of the titration curve for 50.00 mL of 0.020 00 M KOH treated with 0.100 0 M HBr**

	mL HBr added (V_a)	Concentration of unreacted OH^- (M)	Concentration of excess H^+ (M)	pH
Region 1 (excess OH^-)	0.00	0.020 0		12.30
	1.00	0.017 6		12.24
	2.00	0.015 4		12.18
	3.00	0.013 2		12.12
	4.00	0.011 1		12.04
	5.00	0.009 09		11.95
	6.00	0.007 14		11.85
	7.00	0.005 26		11.72
	8.00	0.003 45		11.53
	9.00	0.001 69		11.22
	9.50	0.000 840		10.92
	9.90	0.000 167		10.22
	9.99	0.000 016 6		9.22
Region 2	10.00	—	—	7.00
Region 3 (excess H^+)	10.01		0.000 016 7	4.78
	10.10		0.000 166	3.78
	10.50		0.000 826	3.08
	11.00		0.001 64	2.79
	12.00		0.003 23	2.49
	13.00		0.004 76	2.32
	14.00		0.006 25	2.20
	15.00		0.007 69	2.11
	16.00		0.009 09	2.04

From the concentration of OH^-, it is easy to find the pH:

$$[H^+] = \frac{K_w}{[OH^-]} = \frac{1.0 \times 10^{-14}}{0.013\ 2} = 7.5_8 \times 10^{-13}\ M \implies pH = 12.12$$

(If you were crazy) you could reproduce all the calculations prior to the equivalence point in Table 9-1 in the same way we calculated the 3.00-mL point. The volume of acid added is designated V_a in Table 9-1, and pH is expressed to the 0.01 decimal place, regardless of what is justified by significant figures. We do this for the sake of consistency and also because 0.01 is near the limit of accuracy in pH measurements.

Challenge Calculate $[OH^-]$ and pH when 6.00 mL of HBr have been added. Check your answers against Table 9-1.

Region 2: At the Equivalence Point

At the equivalence point, enough H^+ has been added to react with all the OH^-. We could prepare the same solution by dissolving KBr in water. The pH is determined by the dissociation of water:

$$H_2O \rightleftharpoons \underset{x}{H^+} + \underset{x}{OH^-}$$

$$K_w = x^2 = 1.0 \times 10^{-14} \implies x = 1.0 \times 10^{-7}\,M \implies pH = 7.00$$

At the equivalence point, pH = 7.00, but *only* in a strong acid-strong base reaction.

The pH at the equivalence point in the titration of any strong base (or acid) with strong acid (or base) will be 7.00 at 25°C.

As we will soon discover, *the pH is **not** 7.00 at the equivalence point in the titration of weak acids or bases.* The pH is 7.00 only if the titrant and analyte are both strong.

Region 3: After the Equivalence Point

After the equivalence point, there is excess H^+.

Beyond the equivalence point, excess HBr is present. For example, at the point where 10.50 mL of HBr have been added, there is an excess of $10.50 - 10.00 = 0.50$ mL. The excess H^+ amounts to

$$\text{excess } H^+ = (0.50 \text{ mL})(0.100\,0 \text{ M}) = 0.050 \text{ mmol}$$

Using the total volume of solution ($50.00 + 10.50 = 60.50$ mL), we can find the pH:

$$[H^+] = \frac{0.050 \text{ mmol}}{60.50 \text{ mL}} = 8.2_6 \times 10^{-4}\,M \implies pH = -\log(8.2_6 \times 10^{-4}) = 3.08$$

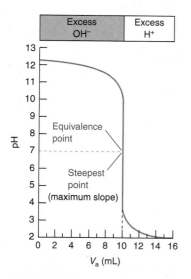

The Titration Curve

The titration curve in Figure 9-1 is a graph of pH versus V_a, the volume of acid added. The sudden change in pH near the equivalence point is characteristic of all analytically useful titrations. The curve is steepest at the equivalence point, which means that the slope is greatest. The pH at the equivalence point is 7.00 *only* in a strong acid-strong base titration. If one or both of the reactants is weak, the equivalence point pH is *not* 7.00.

FIGURE 9-1 Calculated titration curve showing how pH changes as 0.100 0 M HBr is added to 50.00 mL of 0.020 00 M KOH. At the equivalence point, the curve is steepest. The first derivative reaches a maximum and the second derivative is 0. We discuss derivatives later in this chapter.

EXAMPLE   Titration of Strong Acid with Strong Base

Find the pH when 12.74 mL of 0.087 42 M NaOH have been added to 25.00 mL of 0.066 66 M $HClO_4$.

SOLUTION The titration reaction is $H^+ + OH^- \rightarrow H_2O$. The equivalence point is

$$(V_e(\text{mL}))(0.087\ 42\ \text{M}) = (25.00\ \text{mL})(0.066\ 66\ \text{M}) \quad \Rightarrow \quad V_e = 19.06\ \text{mL}$$

mmol of NaOH   mmol of HClO$_4$
at equivalence point   being titrated

At V_b (volume of base) = 12.74 mL, there is excess acid in the solution:

$$\text{H}^+ \text{ remaining} = \underbrace{(25.00\ \text{mL})(0.066\ 66\ \text{M})}_{\substack{\text{Initial mmol} \\ \text{of HClO}_4}} - \underbrace{(12.74\ \text{mL})(0.087\ 42\ \text{M})}_{\substack{\text{Added mmol} \\ \text{of NaOH}}} = 0.553\ \text{mmol}$$

$$[\text{H}^+] = \frac{0.553\ \text{mmol}}{(25.00 + 12.74)\ \text{mL}} = 0.014\ 7\ \text{M}$$

$$\text{pH} = -\log(0.014\ 7) = 1.83$$

Ask Yourself

9-A. What is the equivalence volume in the titration of 50.00 mL of 0.010 0 M NaOH with 0.100 M HCl? Calculate the pH at the following points: V_a = 0.00, 1.00, 2.00, 3.00, 4.00, 4.50, 4.90, 4.99, 5.00, 5.01, 5.10, 5.50, 6.00, 8.00, and 10.00 mL. Make a graph of pH versus V_a.

9-2 Titration of Weak Acid with Strong Base

The titration of a weak acid with a strong base puts all our knowledge of acid-base chemistry to work. The example we treat is the titration of 50.00 mL of 0.020 00 M MES with 0.100 0 M NaOH. MES is an abbreviation for 2-(N-morpholino)ethanesulfonic acid, which is a weak acid with $pK_a = 6.15$. It is widely used in biochemistry as a buffer for the pH 6 region.

The *titration reaction* is

$$\text{O} \bigcirc \overset{+}{\text{N}}\text{HCH}_2\text{CH}_2\text{SO}_3^- + \text{OH}^- \longrightarrow \text{O} \bigcirc \text{NCH}_2\text{CH}_2\text{SO}_3^- + \text{H}_2\text{O} \qquad (9\text{-}1)$$

First write the titration reaction.

HA   A$^-$
MES, $pK_a = 6.15$

Reaction 9-1 is the reverse of the K_b reaction for the base A^-. The equilibrium constant is $1/K_b = 1/(K_w/K_{\text{HA}}) = 7.1 \times 10^7$. The equilibrium constant is so large that we can say that the reaction goes "to completion" after each addition of OH^-. As we saw in Box 8-1, *strong + weak react completely*.

strong + weak $\longrightarrow$ complete reaction

It is helpful first to calculate the volume of base needed to reach the equivalence point. Because one mole of OH^- reacts with one mole of MES, we can say

$$(V_e(\text{mL}))(0.100\ 0\ \text{M}) = (50.00\ \text{mL})(0.020\ 00\ \text{M}) \quad \Rightarrow \quad V_e = 10.00\ \text{mL}$$

mmol of base   mmol of HA

The four regions of the titration curve are important enough to be shown inside the back cover of this book.

The titration calculations for this problem are of four types:

1. Before any base is added, the solution contains just HA in water. This is a weak-acid problem in which the pH is determined by the equilibrium

$$HA \xrightleftharpoons{K_a} H^+ + A^-$$

2. From the first addition of NaOH until immediately before the equivalence point, there is a mixture of unreacted HA plus the A^- produced by Reaction 9-1. *Aha! A buffer!* We can use the Henderson-Hasselbalch equation to find the pH.

3. At the equivalence point, "all" the HA has been converted to A^-. The problem is the same as if the solution had been made by merely dissolving A^- in water. We have a weak-base problem in which the pH is determined by the reaction

$$A^- + H_2O \xrightleftharpoons{K_b} HA + OH^-$$

4. Beyond the equivalence point, excess NaOH is being added to a solution of A^-. We calculate the pH as if we had simply added excess NaOH to water. We will neglect the very small effect from having A^- present as well.

Region 1: Before Base Is Added

The initial solution contains just the *weak acid* HA.

Before adding any base, we have a solution of 0.020 00 M HA with $pK_a = 6.15$. This is simply a weak-acid problem.

F is the formal concentration of HA, which is 0.020 00 M.

$$\begin{array}{ccc} HA & \rightleftharpoons & H^+ + A^- \\ F - x & & x \quad x \end{array} \qquad K_a = 10^{-6.15}$$

$$\frac{x^2}{0.020\ 00 - x} = K_a \Rightarrow x = 1.19 \times 10^{-4} \Rightarrow pH = 3.93$$

Region 2: Before the Equivalence Point

Before the equivalence point, there is a mixture of HA and A^-, which is a *buffer. Aha! A buffer!*

Once we begin to add OH^-, a mixture of HA and A^- is created by the titration reaction (9-1). This mixture is a buffer whose pH can be calculated with the Henderson-Hasselbalch equation (8-1) once we know the quotient $[A^-]/[HA]$.

Henderson-Hasselbalch equation:

$$pH = pK_a + \log\left(\frac{[A^-]}{[HA]}\right) \tag{8-1}$$

Consider the point where 3.00 mL of OH^- have been added:

Titration reaction:	HA	+	OH⁻	⟶	A⁻	+	H₂O
Initial mmol	1.000		0.300		—		
Final mmol	0.700		—		0.300		

Once we know the *quotient* $[A^-]/[HA]$ in any solution, we know its pH:

$$\text{pH} = \text{p}K_a + \log\left(\frac{[A^-]}{[HA]}\right) = 6.15 + \log\left(\frac{0.300}{0.700}\right) = 5.78$$

The Henderson-Hasselbalch equation needs only mmol because volumes cancel in the quotient $[A^-]/[HA]$.

The point at which the volume of titrant is $\frac{1}{2}V_e$ is a special one in any titration.

Titration reaction:	HA	+	OH$^-$	$\longrightarrow$	A$^-$	+	H$_2$O
Initial mmol	1.000		0.500		—		
Final mmol	0.500		—		0.500		

$$\text{pH} = \text{p}K_a + \log\left(\frac{0.500}{0.500}\right) = \text{p}K_a$$

Landmark point:

When $V_b = \frac{1}{2}V_e$, pH = pK_a.

[When activities are taken into account (Section 11-2), this statement is not exactly true, but it is a good approximation.]

When the volume of titrant is $\frac{1}{2}V_e$, pH = pK_a for the acid HA. From the experimental titration curve, you can find pK_a by reading the pH when $V_b = \frac{1}{2}V_e$, where V_b is the volume of added base.

Advice. As soon as you recognize a mixture of HA and A$^-$ in any solution, *you have a buffer!* You can calculate the pH from the quotient $[A^-]/[HA]$.

Region 3: At the Equivalence Point

At the equivalence point ($V_b = 10.00$ mL), the quantity of NaOH is exactly enough to consume the HA.

At the equivalence point, HA has been converted to A$^-$, a *weak base*.

Titration reaction:	HA	+	OH$^-$	$\longrightarrow$	A$^-$	+	H$_2$O
Initial mmol	1.000		1.000		—		
Final mmol	—		—		1.000		

The resulting solution contains "just" A$^-$. We could have prepared the same solution by dissolving the salt Na$^+$A$^-$ in distilled water. *Na$^+$A$^-$ is a weak base.*

To compute the pH of a weak base, we write the reaction of the weak base with water:

$$\underset{F'-x}{A^-} + H_2O \rightleftharpoons \underset{x}{HA} + \underset{x}{OH^-} \qquad K_b = \frac{K_w}{K_a} \qquad (9\text{-}2)$$

The only tricky point is that the formal concentration of A$^-$ is no longer 0.020 00 M, which was the initial concentration of HA. The initial 1.000 mmol of HA in 50.00 mL has been diluted with 10.00 mL of titrant:

$$[A^-] = \frac{1.000 \text{ mmol}}{(50.00 + 10.00) \text{ mL}} = 0.016\ 67\ M \equiv F'$$

189

Designating the formal concentration of A^- as F', we can find the pH from Reaction 9-2:

$$\frac{x^2}{F' - x} = K_b = \frac{K_w}{K_a} = 1.43 \times 10^{-8} \Rightarrow x = 1.54 \times 10^{-5} \text{ M}$$

$$pH = -\log[H^+] = -\log\left(\frac{K_w}{x}\right) = 9.18$$

The pH is higher than 7 at the equivalence point in the titration of a weak acid with a strong base.

The pH at the equivalence point in this titration is 9.18. **It is not 7.00.** The equivalence-point pH will *always* be above 7 for the titration of a weak acid with a strong base, because the acid is converted to its conjugate base at the equivalence point.

Region 4: After the Equivalence Point

Here we assume that the pH is governed by the excess OH^-.

Now we are adding NaOH to a solution of A^-. The base NaOH is so much stronger than the base A^- that it is a fair approximation to say that the pH is determined by the concentration of excess OH^- in the solution.

Let's calculate the pH when $V_b = 10.10$ mL, which is just 0.10 mL past V_e. The quantity of excess OH^- is $(0.10 \text{ mL})(0.100\,0 \text{ M}) = 0.010$ mmol, and the total volume of solution is $50.00 + 10.10$ mL $= 60.10$ mL.

Challenge Compare the concentration of OH^- from excess titrant at $V_b = 10.10$ mL to the concentration of OH^- from hydrolysis of A^-. Satisfy yourself that it is fair to neglect the contribution of A^- to the pH after the equivalence point.

$$[OH^-] = \frac{0.010 \text{ mmol}}{50.00 + 10.10 \text{ mL}} = 1.66 \times 10^{-4} \text{ M}$$

$$pH = -\log\left(\frac{K_w}{[OH^-]}\right) = 10.22$$

The Titration Curve

Landmarks in a titration:

At $V_b = V_e$, curve is steepest.

At $V_b = \frac{1}{2}V_e$, pH = pK_a and the slope is minimal.

The *buffer capacity* measures the ability of the solution to resist changes in pH.

A summary of the calculations for the titration of MES with NaOH is shown in Table 9-2. The titration curve in Figure 9-2 has two easily identified points. One is the equivalence point, which is the steepest part of the curve. The other landmark is the point where $V_b = \frac{1}{2}V_e$ and pH = pK_a. This latter point has the minimum slope, which means that the pH changes least for a given addition of NaOH. This is another way of saying that the maximum *buffer capacity* occurs when pH = pK_a and $[HA] = [A^-]$.

Ask Yourself

9-B. Write the reaction between formic acid (Appendix B) and KOH. What is the equivalence volume (V_e) in the titration of 50.0 mL of 0.050 0 M formic acid with 0.050 0 M KOH? Calculate the pH at the points $V_b = 0.0, 10.0, 20.0, 25.0, 30.0, 40.0, 45.0, 48.0, 49.0, 49.5, 50.0, 50.5, 51.0, 52.0, 55.0,$ and 60.0 mL. Draw a graph of pH versus V_b. Without doing any calculations, what should the pH be at $V_b = \frac{1}{2}V_e$? Does your calculated result agree with the prediction?

TABLE 9-2   **Calculation of the titration curve for 50.00 mL of 0.020 00 M MES treated with 0.100 0 M NaOH**

	mL base added (V_b)	pH
Region 1 (weak acid)	0.00	3.93
Region 2 (buffer)	0.50	4.87
	1.00	5.20
	2.00	5.55
	3.00	5.78
	4.00	5.97
	5.00	6.15
	6.00	6.33
	7.00	6.52
	8.00	6.75
	9.00	7.10
	9.50	7.43
	9.90	8.15
Region 3 (weak base)	10.00	9.18
Region 4 (excess OH⁻)	10.10	10.22
	10.50	10.91
	11.00	11.21
	12.00	11.50
	13.00	11.67
	14.00	11.79
	15.00	11.88
	16.00	11.95

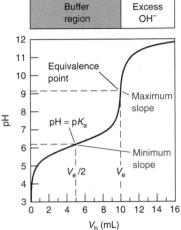

FIGURE 9-2 Calculated titration curve for the reaction of 50.00 mL of 0.020 00 M MES with 0.100 0 M NaOH. Landmarks occur at half of the equivalence volume (pH = pK_a) and at the equivalence point, which is the steepest part of the curve.

9-3   Titration of Weak Base with Strong Acid

The titration of a weak base with a strong acid is just the reverse of the titration of a weak acid with a strong base. The *titration reaction* is

$$B + H^+ \longrightarrow BH^+$$

Because the reactants are weak + strong, the reaction goes essentially to completion after each addition of acid. There are four distinct regions of the titration curve:

1. Before acid is added, the solution contains just the weak base, B, in water. The pH is determined by the K_b reaction:

$$\underset{F-x}{B} + H_2O \xrightleftharpoons{K_b} \underset{x}{BH^+} + \underset{x}{OH^-}$$

When $V_a = 0$, we have a *weak-base* problem.

When $0 < V_a < V_e$, we have a *buffer.*

2. Between the initial point and the equivalence point, there is a mixture of B and BH^+—*Aha! A buffer!* The pH is computed by using

$$pH = pK_a \text{ (for } BH^+) + \log\left(\frac{[B]}{[BH^+]}\right)$$

At the special point where $V_a = \frac{1}{2}V_e$, $pH = pK_a$ (for BH^+).

3. At the equivalence point, B has been converted into BH^+, a weak acid. The pH is calculated by considering the acid dissociation reaction of BH^+:

When $V_a = V_e$, the solution contains the *weak acid* BH^+.

$$\underset{F' - x}{BH^+} \rightleftharpoons \underset{x}{B} + \underset{x}{H^+} \qquad K_a = \frac{K_w}{K_b}$$

The formal concentration of BH^+, F', is not the same as the original formal concentration of B, because some dilution has occurred. Because the solution contains BH^+ at the equivalence point, it is acidic. *The pH at the equivalence point must be below 7.*

For $V_a > V_e$, there is excess *strong acid.*

4. After the equivalence point, there is excess strong acid in the solution. We treat this problem by considering only the concentration of excess H^+ and neglecting the contribution of weak acid, BH^+.

EXAMPLE Titration of Pyridine with HCl

Consider the titration of 25.00 mL of 0.083 64 M pyridine with 0.106 7 M HCl.

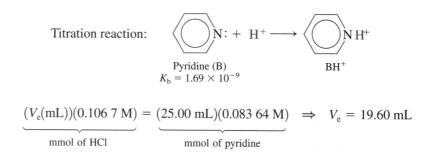

Titration reaction:

Pyridine (B)
$K_b = 1.69 \times 10^{-9}$

BH^+

$$\underbrace{(V_e(mL))(0.106\ 7\ M)}_{\text{mmol of HCl}} = \underbrace{(25.00\ mL)(0.083\ 64\ M)}_{\text{mmol of pyridine}} \quad \Rightarrow \quad V_e = 19.60\ mL$$

Find the pH when $V_a = 4.63$ mL, which is before the equivalence point.

SOLUTION Part of the pyridine has been neutralized, so there is a mixture of pyridine and pyridinium ion—*Aha! A buffer!* The initial mmol of pyridine are $(25.00\ mL)(0.083\ 64\ M) = 2.091$ mmol. The added H^+ is $(4.63\ mL) \times (0.106\ 7\ M) = 0.494$ mmol. Therefore we can write

Titration reaction:	B	+	H^+	$\longrightarrow$	BH^+
Initial mmol	2.091		0.494		—
Final mmol	1.597		—		0.494

$$\text{pH} = \text{p}K_{\text{BH}^+} + \log\left(\frac{[\text{B}]}{[\text{BH}^+]}\right) = 5.23 + \log\left(\frac{1.597}{0.494}\right) = 5.74$$

$$\underset{-\log(K_\text{w}/K_\text{b})}{\uparrow}$$

$\text{p}K_{\text{BH}^+}$ is the *acid* $\text{p}K_\text{a}$ for the *acid* BH^+.

Ask Yourself

9-C. (a) Why is the equivalence point pH necessarily below 7 when a weak base is titrated with strong acid?

(b) What is the equivalence volume in the titration of 100.0 mL of 0.100 M cocaine (Section 7-7, $K_\text{b} = 2.6 \times 10^{-6}$) with 0.200 M HNO_3? Calculate the pH at $V_\text{a} = 0.0$, 10.0, 20.0, 25.0, 30.0, 40.0, 49.0, 49.9, 50.0, 50.1, 51.0, and 60.0 mL. Draw a graph of pH versus V_a.

9-4   Finding the End Point

The *equivalence point* in a titration is defined by the stoichiometry of the reaction. The *end point* is the abrupt change of a physical property (such as pH) that we measure to locate the equivalence point. Indicators and pH measurements are commonly used to find the end point in an acid-base titration.

Using Indicators to Find the End Point

In Section 8-6, we learned that an indicator is an acid or base whose various protonated species have different colors. For the weak-acid indicator HIn, the solution takes on the color of HIn when pH $\lesssim K_{\text{HIn}} - 1$ and has the color of In$^-$ when pH $\gtrsim \text{p}K_{\text{HIn}} + 1$. In the interval $\text{p}K_{\text{HIn}} - 1 \lesssim \text{pH} \lesssim \text{p}K_{\text{HIn}} + 1$, a mixture of both colors is observed.

A titration curve for which pH = 5.54 at the equivalence point is shown in Figure 9-3. The pH drops steeply (from 7 to 4) over a small volume interval. An indicator with a color change in this pH interval would provide a fair approximation to the equivalence point. The closer the point of color change is to pH 5.54, the more accurate will be the end point. The difference between the observed end point (color change) and the true equivalence point is called the **indicator error.**

If you dump half a bottle of indicator into your reaction, you will introduce a different indicator error. Because indicators are acids or bases, they consume analyte or titrant. We use indicators under the assumption that the moles of indicator are negligible relative to the moles of analyte. Never use more than a few drops of dilute indicator solution.

Many indicators in Table 8-3 would be useful for the titration in Figure 9-3. For example, if bromocresol purple were used, we would use the purple-to-yellow color change as the end point. The last trace of purple should disappear near pH 5.2, which is quite close to the true equivalence point in Figure 9-3. If bromocresol

Choose an indicator whose color change comes as close as possible to the theoretical pH of the equivalence point.

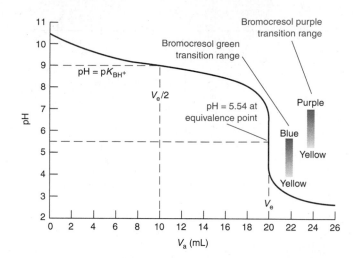

FIGURE 9-3 Calculated titration curve for the reaction of 100 mL of 0.010 0 M base ($pK_b = 5.00$) with 0.050 0 M HCl. As in the titration of HA with OH^-, $pH = pK_{BH^+}$ when $V_a = \frac{1}{2}V_e$.

One of the most common indicators is phenolphthalein, which changes from colorless in acid to pink in base:

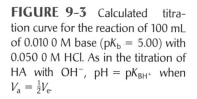

Colorless in acid

$2H^+ \upharpoonleft\downharpoonright 2OH^-$

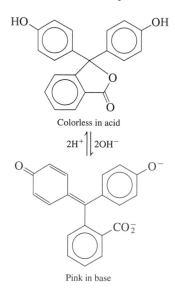

Pink in base

The end point has maximum slope.

green were used as the indicator, a color change from blue to green (= yellow + blue) would mark the end point.

In general, *we seek an indicator whose transition range overlaps the steepest part of the titration curve as closely as possible.* The steepness of the titration curve near the equivalence point in Figure 9-3 ensures that the indicator error caused by the noncoincidence of the color change and equivalence point will not be large. For example, if the indicator color change were at pH 6.4 (instead of 5.54), the error in V_e would be only 0.25% in this particular case.

Using a pH Electrode to Find the End Point

Figure 9-4 shows experimental results for the titration of the weak acid, H_6A, with NaOH. Because the compound is difficult to purify, only a tiny amount was available for titration. Just 1.430 mg was dissolved in 1.00 mL of water and titrated with microliter (μL) quantities of 0.065 92 M NaOH delivered with a Hamilton syringe.

When H_6A is titrated, we might expect to see an abrupt change in pH at all six equivalence points. The curve in Figure 9-4 shows two clear breaks, near 90 and 120 μL, which correspond to titration of the *third* and *fourth* protons of H_6A.

$$H_4A^{2-} + OH^- \longrightarrow H_3A^{3-} + H_2O \qquad (\sim 90 \ \mu L \text{ equivalence point})$$
$$H_3A^{3-} + OH^- \longrightarrow H_2A^{4-} + H_2O \qquad (\sim 120 \ \mu L \text{ equivalence point})$$

The first two and last two equivalence points give unrecognizable end points, because they occur at pH values that are too low or too high.

The end point is where the slope of the titration curve is greatest. The slope is the change in pH (ΔpH) between two points divided by the change in volume (ΔV) between the points:

Slope of titration curve: $$\text{slope} = \frac{\Delta pH}{\Delta V} \qquad (9\text{-}3)$$

The slope (which is also called the *first derivative*) displayed in the middle of Figure 9-4 is calculated in Table 9-3. The first two columns of this table give experimental

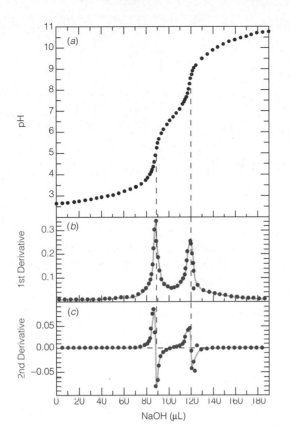

FIGURE 9-4 (*a*) Experimental points in the titration of 1.430 mg of xylenol orange, a hexaprotic acid, dissolved in 1.000 mL of 0.10 M $NaNO_3$. The titrant was 0.065 92 M NaOH. (*b*) The first derivative, $\Delta pH/\Delta V$, of the titration curve. (*c*) The second derivative, $\Delta(\Delta pH/\Delta V)/\Delta V$, which is the derivative of the curve in panel (*b*). Derivatives for the first end point are calculated in Table 9-3. End points are taken as maxima in the derivative curve and zero crossings of the second derivative.

TABLE 9-3   **Computation of first and second derivatives for a titration curve**

		First derivative		Second derivative	
$\mu L\ NaOH$	pH	μL	$\dfrac{\Delta pH}{\Delta \mu L}$	μL	$\dfrac{\Delta(\Delta pH/\Delta \mu L)}{\Delta \mu L}$
85.0	4.245				
		85.5	0.155		
86.0	4.400			86.0	0.071 0
		86.5	0.226		
87.0	4.626			87.0	0.081 0
		87.5	0.307		
88.0	4.933			88.0	0.033 0
		88.5	0.340		
89.0	5.273			89.0	−0.083 0
		89.5	0.257		
90.0	5.530			90.0	−0.068 0
		90.5	0.189		
91.0	5.719			91.25	−0.039 0
		92.0	0.130		
93.0	5.980				

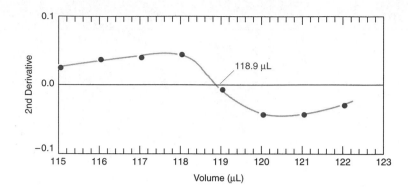

FIGURE 9-5 Enlargement of the second end point in the second derivative curve shown in Figure 9-4c.

volumes and pH measurements. (The pH meter was precise to three digits, even though accuracy ends in the second decimal place.) To compute the first derivative, each pair of volumes is averaged and the quantity $\Delta pH / \Delta V$ is calculated.

The last two columns of Table 9-3 and Figure 9-4c give the slope of the slope (called the second derivative), computed as follows:

> The slope of the slope (the second derivative) is 0 at the end point.

Second derivative:
$$\frac{\Delta(\text{slope})}{\Delta V} = \frac{\Delta(\Delta pH / \Delta V)}{\Delta V} \tag{9-4}$$

The end point is the volume at which the second derivative is 0. A graph on the scale of Figure 9-5 allows us to make a good estimate of the end point volume.

EXAMPLE   Computing Derivatives of a Titration Curve

Let's see how the first and second derivatives in Table 9-3 are calculated.

SOLUTION The first number in the third column, 85.5, is the average of the first two volumes (85.0 and 86.0) in the first column. The slope (first derivative) $\Delta pH / \Delta V$ is calculated from the first two pH values and the first two volumes:

> Retain extra insignificant digits for these calculations.

$$\frac{\Delta pH}{\Delta V} = \frac{4.400 - 4.245}{86.0 - 85.0} = 0.15_5$$

The coordinates ($x = 85.5$, $y = 0.15_5$) are one point in the graph of the first derivative in Figure 9-4b.

The second derivative is computed from the first derivative. The first volume in the fifth column of Table 9-3 is 86.0, which is the average of 85.5 and 86.5. The second derivative is found as follows:

$$\frac{\Delta(\Delta pH / \Delta V)}{\Delta V} = \frac{0.226 - 0.15_5}{86.5 - 85.5} = 0.071_0$$

The coordinates ($x = 86.0$, $y = 0.071_0$) are plotted in the second derivative graph in Figure 9-4c. These calculations are tedious by hand, but they are not bad with a spreadsheet.

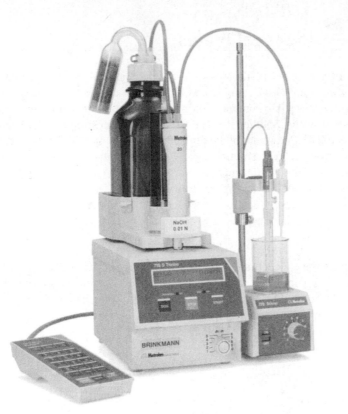

FIGURE 9-6 Autotitrator delivers titrant from the bottle at the back to the beaker of analyte on the stirring motor at the right. The electrode immersed in the beaker monitors pH or the concentrations of specific ions. Volume and pH readings can go directly to the spreadsheet program in a computer. [Brinkman Instruments, Westbury, NY.]

Figure 9-6 shows an *autotitrator* in which the titration is performed automatically and the results sent directly to a computer spreadsheet.[3] Titrant from the plastic bottle at the rear is dispensed in small increments by a syringe pump, and the pH is measured by the electrode in the beaker of analyte on the stirrer. The instrument waits for the pH to stabilize after each addition, before adding the next increment.

Ask Yourself

9-D. (a) Select indicators from Table 8-3 that would be useful for the titrations in Figures 9-1, 9-2, and the $pK_a = 8$ curve in Figure 9-10 (see page 203). Select a different indicator for each titration and state what color change you would use as the end point.

(b) Shown here are data points around the second apparent end point in Figure 9-4. Prepare a table analogous to Table 9-3 showing the first and second derivatives. Plot both derivatives versus V_b and locate the end point from each plot.

V_b (μL)	pH	V_b (μL)	pH	V_b (μL)	pH	V_b (μL)	pH
107	6.921	114	7.457	117	7.878	120	8.591
110	7.117	115	7.569	118	8.090	121	8.794
113	7.359	116	7.705	119	8.343	122	8.952

NaOH and KOH must be standardized with primary standards.

Acids and bases listed in Table 9-4 can be purchased in forms pure enough to be used as *primary standards*. NaOH and KOH are not primary standards, because the reagent-grade materials contain carbonate (from reaction with atmospheric CO_2) and adsorbed water. Solutions of NaOH and KOH must be standardized against a primary standard. Potassium hydrogen phthalate is among the most convenient compounds for this purpose. Solutions of NaOH for titrations are prepared by diluting a stock solution of 50 wt % aqueous NaOH. Sodium carbonate is relatively insoluble in this stock solution and settles to the bottom.

The guard tube on top of the bottle in Figure 9-6 contains a CO_2 absorbent.

Alkaline solutions (such as the NaOH solution in the bottle in Figure 9-6) must be protected from the atmosphere because they absorb CO_2:

$$OH^- + CO_2 \longrightarrow HCO_3^-$$

CO_2 changes the concentration of base over a period of time and reduces the sharpness of the end point in the titration of weak acids. If base is kept in a tightly capped polyethylene bottle, it can be used for weeks with little change. Strong base attacks glass and should not be kept in a buret longer than necessary.

TABLE 9-4	Primary standards	
Compound	*Formula mass*	*Notes*
ACIDS		
Potassium hydrogen phthalate	204.233	The pure solid is dried at 105°C and used to standardize base. A phenolphthalein end point is satisfactory.
$KH(IO_3)_2$ Potassium hydrogen iodate	389.912	This is a strong acid, so any indicator with an end point between ~5 and ~9 is adequate.
BASES		
$H_2NC(CH_2OH)_3$ Tris(hydroxymethyl)aminomethane (also called tris or tham)	121.136	The pure solid is dried at 100°–103°C and titrated with strong acid. The end point is in the range pH 4.5–5. $$H_2NC(CH_2OH)_3 + H^+ \longrightarrow H_3\overset{+}{N}C(CH_2OH)_3$$
Na_2CO_3 Sodium carbonate	105.989	Primary standard grade Na_2CO_3 is titrated with acid to an end point of pH 4–5. Just before the end point, the solution is boiled to expel CO_2.
$Na_2B_4O_7 \cdot 10H_2O$ Borax	381.367	The recrystallized material is dried in a chamber containing an aqueous solution saturated with NaCl and sucrose. This procedure gives the decahydrate in pure form. The standard is titrated with acid to a methyl red end point. $$\text{``}B_4O_7^{2-} \cdot 10H_2O\text{''} + 2H^+ \longrightarrow 4B(OH)_3 + 5H_2O$$

Ask Yourself

9-E. (a) Give the name and formula of a primary standard used to standardize **(i)** HCl and **(ii)** NaOH.

   (b) Referring to Table 9-4, determine how many grams of potassium hydrogen phthalate should be used to standardize ~ 0.05 M NaOH if you wish to use ~ 30 mL of base for the titration.

9-6 Kjeldahl Nitrogen Analysis

Developed in 1883, the **Kjedahl nitrogen analysis** remains one of the most widely used methods for determining nitrogen in organic substances such as protein, cereal, and flour. The solid is *digested* (decomposed and dissolved) in boiling sulfuric acid to convert nitrogen to ammonium ion, NH_4^+:

Kjeldahl digestion: organic C, H, N $\xrightarrow[H_2SO_4]{boiling}$ $NH_4^+ + CO_2 + H_2O$

Each atom of nitrogen in the unknown is converted into one NH_4^+ ion.

Mercury, copper, and selenium compounds catalyze the digestion process. To speed the rate of reaction, the boiling point of concentrated (98 wt %) sulfuric acid (338°C) is raised by adding K_2SO_4. Digestion is carried out in a long-neck *Kjeldahl flask* (Figure 9-7) that prevents loss of sample by spattering. (An alternative to the Kjeldahl flask is a microwave bomb containing H_2SO_4 and H_2O_2, like that in Figure 2-19.)

   After digestion is complete, the solution containing NH_4^+ is made basic, and the liberated NH_3 is distilled (with a large excess of steam) into a receiver containing a known amount of HCl (Figure 9-8). Excess, unreacted HCl is titrated with standard NaOH to determine how much HCl was consumed by NH_3.

Protein source	Wt % nitrogen
meat	16.0
blood plasma	15.3
milk	15.6
flour	17.5
egg	14.9

D. J. Holme and H. Peck, *Analytical Biochemistry,* 3rd ed. (New York: Addison Wesley Longman, 1998), p. 388.

Neutralization of NH_4^+: $NH_4^+ + OH^- \longrightarrow NH_3(g) + H_2O$ (9-5)

Distillation of NH_3 into standard HCl: $NH_3 + H^+ \longrightarrow NH_4^+$ (9-6)

Titration of unreacted HCl with NaOH: $H^+ + OH^- \longrightarrow H_2O$ (9-7)

EXAMPLE   Kjeldahl Analysis

A typical protein contains 16.2 wt % nitrogen. A 0.500-mL aliquot of protein solution was digested, and the liberated NH_3 was distilled into 10.00 mL of 0.021 40 M HCl. The unreacted HCl required 3.26 mL of 0.019 8 NaOH for complete titration. Find the concentration of protein (mg protein/mL) in the original sample.

SOLUTION The original amount of HCl in the receiver was $(10.00\text{ mL}) \times (0.021\,40\text{ M}) = 0.214\,0$ mmol. The NaOH required for titration of unreacted HCl in Reaction 9-7 was $(3.26\text{ mL})(0.019\,8\text{ M}) = 0.064\,5$ mmol. The difference, $0.214\,0 - 0.064\,5 = 0.149\,5$ mmol, must equal the quantity of NH_3 produced in Reaction 9-5 and distilled into the HCl.

   Because 1 mmol of nitrogen in the protein gives rise to 1 mmol of NH_3, there must have been 0.149 5 mmol of nitrogen in the protein, corresponding to

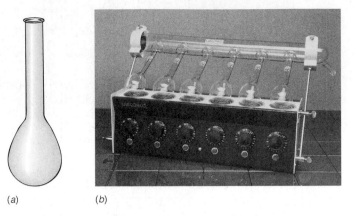

(a)   (b)

FIGURE 9-7 (a) The Kjeldahl digestion flask has a long neck to minimize loss by spattering. (b) Six-port manifold for multiple samples provides for exhaust of fumes. [Fisher Scientific, Pittsburgh, PA.]

$$(0.149\,5\ \text{mmol})\left(14.006\,74\ \frac{\text{mg N}}{\text{mmol}}\right) = 2.093\ \text{mg N}$$

If the protein contains 16.2 wt % N, there must be

$$\frac{2.093\ \text{mg N}}{0.162\ \text{mg N/mg protein}} = 12.9\ \text{mg protein}$$

$$\frac{12.9\ \text{mg protein}}{0.500\ \text{mL}} = 25.8\ \frac{\text{mg protein}}{\text{mL}}$$

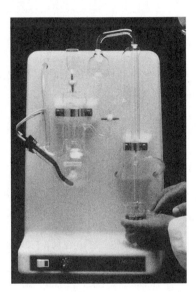

FIGURE 9-8 Kjeldahl distillation unit employs electric immersion heater in flask at left to carry out distillation in 5 min. Beaker at right collects liberated NH_3 in standard HCl. [Fisher Scientific, Pittsburgh, PA.]

Ask Yourself

9-F. The Kjeldahl procedure was used to analyze 256 μL of a solution containing 37.9 mg protein/mL. The liberated NH_3 was collected in 5.00 mL of 0.033 6 M HCl, and the remaining acid required 6.34 mL of 0.010 M NaOH for complete titration.
  (a) How many moles of NH_3 were liberated?
  (b) How many grams of nitrogen are contained in the NH_3 in **(a)**?
  (c) How many grams of protein were analyzed?
  (d) What is the weight percent of nitrogen in the protein?

9-7   Putting Your Spreadsheet to Work

In Sections 9-1 to 9-3, we calculated titration curves because they helped us understand the chemistry behind a titration curve. Now we will see how a spreadsheet and graphics program decrease the agony and mistakes of titration calculations. First we must derive equations relating pH to volume of titrant for use in the spreadsheet.

Charge Balance

The **charge balance** states that in any solution, the sum of positive charges must equal the sum of negative charges, because the solution must have zero net charge.

For a solution of the weak acid HA plus NaOH, the charge balance is

Charge balance: $$[H^+] + [Na^+] = [A^-] + [OH^-] \qquad (9\text{-}8)$$

The sum of the positive charges of H^+ and Na^+ equals the sum of the negative charges of A^- and OH^-.

If the solution contained HA and $Ca(OH)_2$, the charge balance would be:

$$[H^+] + 2[Ca^{2+}] = [A^-] + [OH^-]$$

because one mole of Ca^{2+} provides two moles of charge. If $[Ca^{2+}] = 0.1$ M, the positive charge it contributes is 0.2 M.

Titrating a Weak Acid with a Strong Base

Consider the titration of a volume V_a of acid HA (initial concentration C_a) with a volume V_b of NaOH of concentration C_b. The concentration of Na^+ is just the moles of NaOH ($C_b V_b$) divided by the total volume of solution ($V_a + V_b$):

$$[Na^+] = \frac{C_b V_b}{V_a + V_b} \qquad (9\text{-}9)$$

Similarly, the formal concentration of the weak acid is

$$F = [HA] + [A^-] = \frac{C_a V_a}{V_a + V_b} \qquad (9\text{-}10)$$

because we have diluted $C_a V_a$ moles of HA to a total volume of $V_a + V_b$.

Now we introduce two equations that are derived in Section 11-5:

Fraction of weak acid in the form HA:
$$\alpha_{HA} = \frac{[HA]}{F} = \frac{[H^+]}{[H^+] + K_a} \qquad (9\text{-}11)$$

Fraction of weak acid in the form A^-:
$$\alpha_{A^-} = \frac{[A^-]}{F} = \frac{K_a}{[H^+] + K_a} \qquad (9\text{-}12)$$

Equations 9-11 and 9-12 say that if a weak acid has a formal concentration F, the concentration of HA is $\alpha_{HA} \cdot F$ and the concentration of A^- is $\alpha_{A^-} \cdot F$. These fractions must add up to 1.

$$\alpha_{HA} + \alpha_{A^-} = 1$$

Getting back to our titration, we can write an expression for the concentration of A^- by combining Equation 9-12 with Equation 9-10:

$$[A^-] = \alpha_{A^-} \cdot F = \frac{\alpha_{A^-} \cdot C_a V_a}{V_a + V_b} \qquad (9\text{-}13)$$

Substituting for $[Na^+]$ (Equation 9-9) and $[A^-]$ (Equation 9-13) in the charge balance (Equation 9-8) gives

$$[H^+] + \frac{C_b V_b}{V_a + V_b} = \frac{\alpha_{A^-} \cdot C_a V_a}{V_a + V_b} + [OH^-]$$

which can be rearranged to the form

$\phi = C_b V_b / C_a V_a$ is the fraction of the way to the equivalence point:

ϕ	Volume of base
0.5	$V_b = \frac{1}{2}V_e$
1	$V_b = V_e$
2	$V_b = 2V_e$

2-(*N*-Morpholino)ethanesulfonic acid
MES, p$K_a = 6.15$

The spreadsheet in Figure 9-9 can be used to find the pH of a weak acid. Just search for the pH at which $V_b = 0$.

To home in on an exact volume (such as V_e), set the spreadsheet to show extra digits in the cell of interest. Tables in this book have been formatted to reduce the number of digits.

Fraction of titration for
weak acid by strong base:

$$\phi = \frac{C_b V_b}{C_a V_a} = \frac{\alpha_{A^-} - \dfrac{[H^+] - [OH^-]}{C_a}}{1 + \dfrac{[H^+] - [OH^-]}{C_b}} \qquad (9\text{-}14)$$

At last! Equation 9-14 is really useful. It relates the volume of titrant (V_b) to the pH. The quantity ϕ ($= C_b V_b / C_a V_a$) is the fraction of the way to the equivalence point, V_e. When $\phi = 1$, the volume of base added, V_b, is equal to V_e. Equation 9-14 works backward from the way you are accustomed to thinking, because you need to put in pH (on the right) to get out volume (on the left).

Let's set up a spreadsheet for Equation 9-14 to calculate the titration curve for 50.00 mL of the weak acid 0.020 00 M MES with 0.100 0 M NaOH, which was shown in Figure 9-2 and Table 9-2. The equivalence volume is $V_e = 10.00$ mL. The quantities in Equation 9-14 are

$$C_b = 0.1 \qquad\qquad [H^+] = 10^{-pH}$$
$$C_a = 0.02 \qquad\qquad [OH^-] = K_w/[H^+]$$
$$V_a = 50$$
$$K_a = 7.0_8 \times 10^{-7} \qquad \alpha_{A^-} = \frac{K_a}{[H^+] + K_a}$$
$$K_w = 10^{-14}$$

pH is the input

$$V_b = \frac{\phi C_a V_a}{C_b} \text{ is the output}$$

The input to the spreadsheet in Figure 9-9 is pH in column B and the output is V_b in column G. From the pH, the values of $[H^+]$, $[OH^-]$, and α_{A^-} are computed in columns C, D, and E. Equation 9-14 is used in column F to find the fraction of titration, ϕ. From this value, we calculate the volume of titrant, V_b, in column G.

How do we know what pH values to put in? Trial and error allows us to find the starting pH, by putting in a pH and seeing whether V_b is positive or negative. In a few tries, it is easy to home in on the pH at which $V_b = 0$. In Figure 9-9, we see that a pH of 3.00 is too low, because ϕ and V are both negative. Input values of pH are spaced as closely as you like, so that you can generate a smooth titration curve. To save space, we only show a few points in Figure 9-9, including the midpoint (pH 6.15 $\Rightarrow$ $V_b = 5.00$ mL) and the end point (pH 9.18 $\Rightarrow$ $V_b = 10.00$ mL). The spreadsheet reproduces Table 9-2 without dividing the titration into different regions that use different approximations.

The Power of a Spreadsheet

By changing K_a in cell A10 of Figure 9-9, we can calculate a family of curves for different acids. Figure 9-10 shows how the titration curve depends on the acid dissociation constant of HA. The strong-acid curve at the bottom of Figure 9-10 was computed with a large value of K_a ($K_a = 10^3$) in cell A10. Figure 9-10 shows that, as K_a decreases (pK_a increases), the pH change near the equivalence point decreases, until it becomes too shallow to detect. Similar behavior occurs as the concentrations of analyte and titrant decrease. *It is not practical to titrate an acid or a base when its strength is too weak or its concentration too dilute.*

	A	B	C	D	E	F	G
1	Titration of weak acid with strong base						
2							
3	Cb =	pH	[H+]	[OH–]	Alpha[A–]	Phi	Vb (mL)
4	0.1	3.00	1.00E-03	1.00E-11	0.001	-0.049	-0.488
5	Ca =	3.93	1.17E-04	8.51E-11	0.006	0.000	0.001
6	0.02	4.00	1.00E-04	1.00E-10	0.007	0.002	0.020
7	Va =	5.00	1.00E-05	1.00E-09	0.066	0.066	0.656
8	50	6.15	7.08E-07	1.41E-08	0.500	0.500	5.000
9	Ka =	7.00	1.00E-07	1.00E-07	0.876	0.876	8.762
10	7.08E-07	8.00	1.00E-08	1.00E-06	0.986	0.986	9.861
11	Kw =	9.18	6.61E-10	1.51E-05	0.999	1.000	10.000
12	1.00E-14	10.00	1.00E-10	1.00E-04	1.000	1.006	10.059
13		11.00	1.00E-11	1.00E-03	1.000	1.061	10.606
14		12.00	1.00E-12	1.00E-02	1.000	1.667	16.667
15							
16	C4 = 10^-B4						
17	D4 = $A$12/C4						
18	E4 = $A$10/(C4+$A$10)						
19	F4 = (E4-(C4-D4)/$A$6)/(1+(C4-D4)/$A$4) [Equation 9-14]						
20	G4 – F4*$A$6*$A$8/$A$4						

FIGURE 9-9 Spreadsheet uses Equation 9-14 to calculate the titration curve for 50 mL of the weak acid, 0.02 M MES (pK$_a$ = 6.15), treated with 0.1 M NaOH. You provide pH as input in column B, and the spreadsheet tells what volume of base in column G is required to generate that pH.

Titrating a Weak Base with a Strong Acid

Following a derivation similar to that above, we can derive a spreadsheet equation for the titration of weak base B with strong acid:

Fraction of titration for weak base by strong acid:

$$\phi = \frac{C_aV_a}{C_bV_b} = \frac{\alpha_{BH^+} + \dfrac{[H^+] - [OH^-]}{C_b}}{1 - \dfrac{[H^+] - [OH^-]}{C_a}} \qquad (9\text{-}15)$$

where C_a is the concentration of strong acid in the buret, V_a is the volume of acid added, C_b is the initial concentration of weak base being titrated, V_b is the initial volume of weak base being titrated, and α_{BH^+} is the fraction of base in the form BH^+:

Fraction of weak base in the form BH$^+$:

$$\alpha_{BH^+} = \frac{[HA]}{F} = \frac{[H^+]}{[H^+] + K_{BH^+}} \qquad (9\text{-}16)$$

where K_{BH^+} is the acid dissociation constant of BH^+.

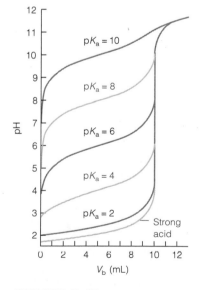

FIGURE 9-10 Calculated curves showing the titration of 50.0 mL of 0.020 0 M HA with 0.100 M NaOH. As the acid becomes weaker, the equivalence point becomes less distinct.

Ask Yourself

9-G. (a) *Effect of pK$_a$ in the titration of weak acid with strong base.* Use the spreadsheet in Figure 9-9 to compute and plot the family of curves in Figure 9-10. For strong acid, use $K_a = 10^3$.

 (b) *Effect of concentration in the titration of weak acid with strong base.* Use your spreadsheet to prepare a family of titration curves for pK$_a$ = 6, with the following combinations of concentrations: **(i)** C_a = 20 mM, C_b = 100 mM; **(ii)** C_a = 2 mM, C_b = 10 mM; and **(iii)** C_a = 0.2 mM, C_b = 1 mM.

Key Equations

Useful shortcut

$$\text{mL} \times \frac{\text{mol}}{\text{L}} = \text{mmol}$$

Equivalence volume (V_e)

$$\underbrace{C_a V_a = C_b V_e}_{\substack{\text{Titrating acid} \\ \text{with base}}} \quad \text{or} \quad \underbrace{C_a V_e = C_b V_b}_{\substack{\text{Titrating base} \\ \text{with acid}}}$$

C_a = acid concentration C_b = base concentration

V_a = acid volume V_b = base volume V_e = equivalence volume

Titration of weak acid
(see inside back cover)

1. Initial solution—weak acid

$$\underset{\substack{F - x}}{\text{HA}} \overset{K_a}{\rightleftharpoons} \underset{x}{\text{H}^+} + \underset{x}{\text{A}^-} \qquad \frac{x^2}{F - x} = K_a$$

2. Before equivalence point—buffer

Titration reaction tells how much HA and A$^-$ are present

$$\text{pH} = \text{p}K_a + \log\left(\frac{[\text{A}^-]}{[\text{HA}]}\right)$$

3. Equivalence point—weak base—pH > 7

$$\underset{\substack{F' - x}}{\text{A}^-} + \text{H}_2\text{O} \overset{K_b}{\rightleftharpoons} \underset{x}{\text{HA}} + \underset{x}{\text{OH}^-} \qquad K_b = \frac{K_w}{K_a}$$

F′ is diluted concentration

4. After equivalence point—excess strong base

$$\text{pH} = -\log(K_w/[\text{OH}^-]_{\text{excess}})$$

Titration of weak base
(see inside back cover)

1. Initial solution—weak base

$$\underset{\substack{F - x}}{\text{B}} + \text{H}_2\text{O} \overset{K_b}{\rightleftharpoons} \underset{x}{\text{BH}^+} + \underset{x}{\text{OH}^-} \qquad \frac{x^2}{F - x} = K_b$$

2. Before equivalence point—buffer

Titration reaction tells how much B and BH^+ are present

$$pH = pK_{BH^+} + \log\left(\frac{[B]}{[BH^+]}\right)$$

3. Equivalence point—weak acid—pH < 7

$$\underset{F'-x}{BH^+} \overset{K_{BH^+}}{\rightleftharpoons} \underset{x}{B} + \underset{x}{H^+}$$

4. After equivalence point—excess strong acid

$$pH = -\log\left([H^+]_{excess}\right)$$

Spreadsheet titration equations	Use Equations 9-14 and 9-15
	Input is pH and output is volume
Choosing indicator	Use indicator with color change close to theoretical pH at equivalence point of titration
Using electrodes for end point	End point has greatest slope: $\Delta pH/\Delta V$ is maximum
	End point has zero second derivative: $\dfrac{\Delta(\Delta pH/\Delta V)}{\Delta V} = 0$

Important Terms

charge balance   indicator error   Kjeldahl nitrogen analysis

Problems

9-1. Explain what chemistry occurs in each region of the titration of OH^- with H^+. State how you would calculate the pH in each region.

9-2. Explain what chemistry occurs in each region of the titration of the weak base, A^-, with strong acid, H^+. State how you would calculate the pH in each region.

9-3. Why is the titration curve in Figure 9-3 steepest at the equivalence point?

9-4. Why do we use the maximum of the first derivative curve or the zero crossing of the second derivative curve to locate the end points in Figure 9-4?

9-5. Consider the titration of 100.0 mL of 0.100 M NaOH with 1.00 M HBr. What is the equivalence volume? Find the pH at the following volumes of HBr and make a graph of pH versus V_a: $V_a = 0, 1.00, 5.00, 9.00, 9.90, 10.00, 10.10,$ and 12.00 mL.

9-6. Consider the titration of 25.0 mL of 0.050 0 M $HClO_4$ with 0.100 M KOH. Find the equivalence volume. Find the pH at the following volumes of KOH and plot pH versus V_b:

$V_b = 0, 1.00, 5.00, 10.00, 12.40, 12.50, 12.60,$ and 13.00 mL.

9-7. A 50.0-mL volume of 0.050 0 M weak acid HA ($pK_a = 4.00$) was titrated with 0.500 M NaOH. Write the titration reaction and find V_e. Find the pH at $V_b = 0, 1.00, 2.50, 4.00, 4.90, 5.00, 5.10,$ and 6.00 mL and plot pH versus V_b.

9-8. When methylammonium chloride is titrated with tetramethylammonium hydroxide, the titration reaction is

$$\underset{\substack{BH^+ \\ \text{Weak acid}}}{CH_3NH_3^+} + \underset{\substack{\text{From} \\ (CH_3)_4N^+OH^-}}{OH^-} \longrightarrow \underset{\substack{B \\ \text{Weak base}}}{CH_3NH_2} + H_2O$$

Find the equivalence volume in the titration of 25.0 mL of 0.010 0 M methylammonium chloride with 0.050 0 M tetramethylammonium hydroxide. Calculate the pH at $V_b = 0, 2.50, 5.00,$ and 10.00 mL. Sketch the titration curve.

9-9. Write the reaction for the titration of 100 mL of 0.100 M anilinium bromide ("aminobenzene · HBr") with 0.100 M NaOH. Sketch the titration curve for the points $V_b = 0, 0.100V_e, 0.500V_e, 0.900V_e, V_e,$ and $1.200V_e$.

9-10. What is the pH at the equivalence point when 0.100 M hydroxyacetic acid is titrated with 0.050 0 M KOH?

9-11. When 16.24 mL of 0.064 3 M KOH were added to 25.00 mL of 0.093 8 M weak acid, HA, the observed pH was 3.62. Find pK_a for the acid.

9-12. When 22.63 mL of aqueous NaOH were added to 1.214 g of CHES (FM 207.29, structure in Table 8-2) dissolved in 41.37 mL of water, the pH was 9.24. Calculate the molarity of the NaOH.

9-13. **(a)** When 100.0 mL of weak acid HA were titrated with 0.093 81 M NaOH, 27.63 mL were required to reach the equivalence point. Find the molarity of HA.

(b) What is the formal concentration of A$^-$ at the equivalence point?

(c) The pH at the equivalence point was 10.99. Find pK_a for HA.

(d) What was the pH when only 19.47 mL of NaOH had been added?

9-14. A 100.0-mL aliquot of 0.100 M weak base B (pK_b = 5.00) was titrated with 1.00 M HClO$_4$. Find V_e and calculate the pH at V_a = 0, 1.00, 5.00, 9.00, 9.90, 10.00, 10.10, and 12.00 mL and make a graph of pH versus V_a.

9-15. A solution of 100.0 mL of 0.040 0 M sodium propanoate (the sodium salt of propanoic acid) was titrated with 0.083 7 M HCl. Find V_e and calculate the pH at V_a = 0, $\frac{1}{4}V_e$, $\frac{1}{2}V_e$, $\frac{3}{4}V_e$, V_e, and 1.1V_e.

9-16. A 50.0-mL solution of 0.031 9 M benzylamine was titrated with 0.050 0 M HCl.

(a) What is the equilibrium constant for the titration reaction?

(b) Find V_e and calculate the pH at V_a = 0, 12.0, $\frac{1}{2}V_e$, 30.0, V_e, and 35.0 mL.

9-17. Don't ever mix acid with cyanide (CN$^-$) because it liberates poisonous HCN(g). But, just for fun, calculate the pH of a solution made by mixing 50.00 mL of 0.100 M NaCN with

(a) 4.20 mL of 0.438 M HClO$_4$

(b) 11.82 mL of 0.438 M HClO$_4$

(c) What is the pH at the equivalence point with 0.438 M HClO$_4$?

9-18. Would the indicator bromocresol green, with a transition range of pH 3.8–5.4, ever be useful in the titration of a weak acid with a strong base? Why?

9-19. Consider the titration in Figure 9-2, for which the pH at the equivalence point is calculated to be 9.18. If thymol blue is used as an indicator, what color will be observed through most of the titration prior to the equivalence point? At the equivalence point? After the equivalence point?

9-20. Why would an indicator end point not be very useful in the titration curve for pK_a = 10.00 in Figure 9-10?

9-21. Phenolphthalein is used as an indicator for the titration of HCl with NaOH.

(a) What color change is observed at the end point?

(b) The basic solution just after the end point slowly absorbs CO$_2$ from the air and becomes more acidic by virtue of the reaction CO$_2$ + OH$^-$ $\rightleftharpoons$ HCO$_3^-$. This causes the color to fade from pink to colorless. If you carry out the titration too slowly, does this reaction lead to a systematic or a random error in finding the end point?

9-22. In the titration of 0.10 M pyridinium bromide (the salt of pyridine plus HBr) by 0.10 M NaOH, the pH at 0.99V_e is 7.22. At V_e, pH = 8.96, and at 1.01V_e, pH = 10.70. Select an indicator from Table 8-3 that would be suitable for this titration and state what color change will be used.

9-23. Prepare a graph of the second derivative to find the end point from the titration data below:

mL NaOH	pH	mL NaOH	pH
10.679	7.643	10.729	5.402
10.696	7.447	10.733	4.993
10.713	7.091	10.738	4.761
10.721	6.700	10.750	4.444
10.725	6.222	10.765	4.227

9-24. Borax (Table 9-4) was used to standardize a solution of HNO$_3$. Titration of 0.261 9 g borax required 21.61 mL. What is the molarity of the HNO$_3$?

9-25. A 10.231-g sample of window cleaner containing ammonia was diluted with 39.466 g of water. Then 4.373 g of solution was titrated with 14.22 mL of 0.106 3 M HCl to reach a bromocresol green end point.

(a) What fraction of the 10.231-g sample of window cleaner is contained in the 4.373 g that was analyzed?

(b) How many grams of NH$_3$ (FM 17.031) were in the 4.373-g sample?

(c) Find the weight percent of NH$_3$ in the cleaner.

9-26. In the Kjeldahl nitrogen analysis, the final product is NH$_4^+$ in HCl solution. It is necessary to titrate the HCl without titrating the NH$_4^+$ ion.

(a) Calculate the pH of pure 0.010 M NH$_4$Cl.

(b) The steep part of the titration curve when HCl is titrated with NaOH runs from pH $\approx$ 4 up to pH $\approx$ 10. Select an indicator that would allow you to titrate HCl but not NH$_4^+$.

9-27. Prepare a spreadsheet like the one in Figure 9-9, but using Equation 9-15, to reproduce the titration curve in Figure 9-3.

9-28. *Effect of pK_b in the titration of weak base with strong acid.* Use the spreadsheet from the previous problem to compute and plot a family of curves analogous to Figure 9-10 for the titration of 50.0 mL of 0.020 0 M B (pK_b = −2.00, 2.00, 5.00, 8.00, and 10.00) with 0.100 M HCl. (pK_b = −2.00 corresponds to $K_b = 10^{+2.00}$, which represents a strong base.)

How Would You Do It?

9-29. The table below gives data for 100.0 mL of solution of a single unknown base titrated with 0.111 4 M HCl. Provide an argument for how many protons the base is able to accept (Is it monoprotic, diprotic, etc.?) and find the molarity of the base. A high-quality pH meter provides pH to the 0.001 place, even though accuracy is limited to the 0.01 place. How could you deliver volumes up to 50 mL with a precision of 0.001 mL?

Coarse titration				Fine data near 1st end point				Fine data near 2nd end point			
mL	pH	mL	pH	mL	pH	mL	pH	mL	pH	mL	pH
0.595	12.148	29.157	5.785	26.939	8.217	27.481	7.228	40.168	3.877	41.542	2.999
1.711	12.006	31.512	5.316	27.013	8.149	27.501	7.158	40.403	3.767	41.620	2.949
3.540	11.793	33.609	5.032	27.067	8.096	27.517	7.103	40.498	3.728	41.717	2.887
5.250	11.600	36.496	4.652	27.114	8.050	27.537	7.049	40.604	3.669	41.791	2.845
7.258	11.390	38.222	4.381	27.165	7.987	27.558	6.982	40.680	3.618	41.905	2.795
9.107	11.179	39.898	3.977	27.213	7.916	27.579	6.920	40.774	3.559	42.033	2.735
11.557	10.859	40.774	3.559	27.248	7.856	27.600	6.871	40.854	3.510	42.351	2.617
13.967	10.486	41.791	2.845	27.280	7.791	27.622	6.825	40.925	3.457	42.709	2.506
16.042	10.174	42.709	2.506	27.309	7.734	27.649	6.769	40.994	3.407	43.192	2.401
18.474	9.850	45.049	2.130	27.338	7.666	27.675	6.717	41.057	3.363	43.630	2.312
20.338	9.627	47.431	1.937	27.362	7.603	27.714	6.646	41.114	3.317		
22.136	9.402	49.292	1.835	27.386	7.538	27.747	6.594	41.184	3.263		
24.836	8.980			27.406	7.485	27.793	6.535	41.254	3.210		
26.216	8.608			27.427	7.418	27.846	6.470	41.329	3.150		
27.013	8.149			27.444	7.358	27.902	6.411	41.406	3.093		
27.969	6.347			27.463	7.287	27.969	6.347	41.466	3.047		

Notes and References

1. M. Gratzl and C. Yi, *Anal. Chem.* **1993,** *65,* 2085; C. Yi and M. Gratzl, *Anal. Chem.* **1994,** *66,* 1976; C. Yi, D. Huang, and M. Gratzl, *Anal. Chem.* **1996,** *68,* 1580; H. Xie and M. Gratzl, *Anal. Chem.* **1996,** *68,* 3665.

2. E. Litborn, M. Stjernström, and J. Roeraade, *Anal. Chem.* **1998,** *70,* 4847.

3. For student experiments with autotitrators, see K. R. Williams, *J. Chem. Ed.* **1998,** *75,* 1133.

Disappearing Statues and Teeth

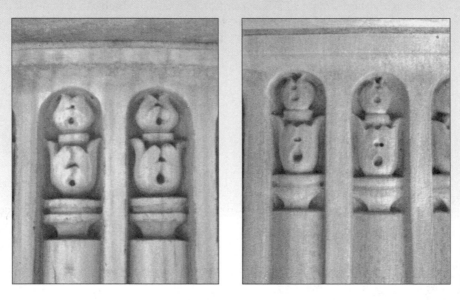

Erosion of carbonate stone. The columns at the left are on the inside of a monument, protected from rain. The columns at the right are on the outside and beginning to wear out from exposure to acid rain. The opening of Chapter 7 showed rainfall pH in North America.

P. A. Baedecker, U.S. Geological Survey

*T*he main constituent of limestone and marble is calcite, a crystalline form of calcium carbonate. This mineral is insoluble in neutral or basic solution but dissolves in acid by virtue of two *coupled equilibria,* in which the product of one reaction is consumed in the next reaction:

$$CaCO_3(s) \rightleftharpoons Ca^{2+} + CO_3^{2-}$$

Calcite   Carbonate

$$CO_3^{2-} + H^+ \rightleftharpoons HCO_3^-$$

Bicarbonate

Le Châtelier's principle tells us that if we remove a product of the first reaction, we will draw the reaction to the right, making calcite more soluble.

Tooth enamel contains the mineral hydroxyapatite, a calcium hydroxyphosphate. This mineral also dissolves in acid, because both PO_4^{3-} and OH^- react with H^+:

$$Ca_{10}(PO_4)_6(OH)_2 + 14H^+ \rightleftharpoons 10Ca^{2+} + 6H_2PO_4^- + 2H_2O$$

Hydroxyapatite

Bacteria residing on your teeth metabolize sugar into lactic acid, thereby creating a pH below 5 at the surface of a tooth. Acid dissolves the hydroxyapatite, creating tooth decay.

$$\underset{CH_3CHCO_2H}{\overset{OH}{|}}$$

Lactic acid

Chapter 10

POLYPROTIC ACIDS AND BASES

*C*arbonic acid from marble, phosphoric acid from teeth, and amino acids from proteins are all **polyprotic acids**—those having more than one acidic proton. This chapter extends our discussion of acids, bases, and buffers to polyprotic systems encountered throughout nature.

10-1 Amino Acids Are Polyprotic

The **amino acids** from which proteins are built have an acidic carboxylic acid group, a basic amino group, and a variable substituent designated R in the structure below:

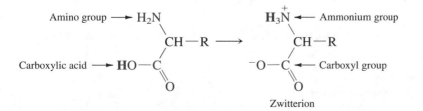

Zwitterion

A *zwitterion* is a molecule with positive and negative charges.

Because the amino group is more basic than the carboxyl group, the acidic proton resides on nitrogen of the amino group instead of on oxygen from the carboxyl group. The resulting structure, with positive and negative sites, is called a **zwitterion.**

209

At low pH, both the ammonium group and the carboxyl group are protonated. At high pH, neither is protonated. The substituent may also have acidic or basic properties. Acid dissociation constants of the 20 common amino acids are given in Table 10-1, in which each substituent (R) is shown in its fully protonated form. For example, Table 10-1 shows the amino acid cysteine, which has three acidic protons:

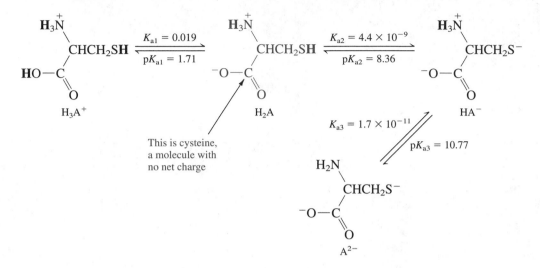

This is cysteine, a molecule with no net charge

K_{a1} applies to the most acidic proton. The subscript a in K_{a1} and K_{a2} is customarily omitted, and we will write K_1 and K_2 throughout most of this chapter.

In general, a *diprotic* acid has two acid dissociation constants, designated K_{a1} and K_{a2} (where $K_{a1} > K_{a2}$):

$$H_2A \xrightleftharpoons{K_{a1}} HA^- + H^+ \qquad\qquad HA^- \xrightleftharpoons{K_{a2}} A^{2-} + H^+$$

The two base association constants are designated K_{b1} and K_{b2} ($K_{b1} > K_{b2}$):

$$A^{2-} + H_2O \xrightleftharpoons{K_{b1}} HA^- + OH^- \qquad\qquad HA^- + H_2O \xrightleftharpoons{K_{b2}} H_2A + OH^-$$

Relation Between K_a and K_b

If you add the K_{a1} reaction to the K_{b2} reaction, the sum is $H_2O \rightleftharpoons H^+ + OH^-$—the K_w reaction. By this means, you can derive a most important set of relationships between the acid and base equilibrium constants:

Challenge Add the K_{a1} and K_{b2} reactions to prove that $K_{a1} \cdot K_{b2} = K_w$

Relation between K_a and K_b for diprotic system:

$$K_{a1} \cdot K_{b2} = K_w$$
$$K_{a2} \cdot K_{b1} = K_w$$

(10-1)

For a *triprotic* system, with three acidic protons, the corresponding relationships are

Relation between K_a and K_b for triprotic system:

$$K_{a1} \cdot K_{b3} = K_w$$
$$K_{a2} \cdot K_{b2} = K_w$$
$$K_{a3} \cdot K_{b1} = K_w$$

(10-2)

The standard notation for successive acid dissociation constants of a polyprotic acid is K_1, K_2, K_3, and so on, with the subscript a usually omitted. We retain or

TABLE 10-1 Acid dissociation constants of amino acids

Amino acid[a]	Substituent	Carboxylic acid pK_a	Ammonium pK_a	Substituent pK_a	Molecular mass
Alanine (A)	$-CH_3$	2.348	9.867		89.09
Arginine (R)	$-CH_2CH_2CH_2NHC\begin{smallmatrix}+NH_2\\\\NH_2\end{smallmatrix}$	1.823	8.991	12.48	174.20
Asparagine (N)	$-CH_2\overset{O}{\overset{\|}{C}}NH_2$	2.14	8.72		132.12
Aspartic acid (D)	$-CH_2CO_2\mathbf{H}$	1.990	10.002	3.900	133.10
Cysteine (C)	$-CH_2\mathbf{SH}$	1.71	10.77	8.36	121.16
Glutamic acid (E)	$-CH_2CH_2CO_2\mathbf{H}$	2.23	9.95	4.42	147.13
Glutamine (Q)	$-CH_2CH_2\overset{O}{\overset{\|}{C}}NH_2$	2.17	9.01		146.15
Glycine (G)	$-H$	2.350	9.778		75.07
Histidine (H)	$-CH_2$ (imidazole)	1.7	9.08	6.02	155.16
Isoleucine (I)	$-CH(CH_3)(CH_2CH_3)$	2.319	9.754		131.17
Leucine (L)	$-CH_2CH(CH_3)_2$	2.329	9.747		131.17
Lysine (K)	$-CH_2CH_2CH_2CH_2NH_3^+$	2.04	9.08	10.69	146.19
Methionine (M)	$-CH_2CH_2SCH_3$	2.20	9.05		149.21
Phenylalanine (F)	$-CH_2-$ (phenyl)	2.20	9.31		165.19
Proline (P)	(Structure of entire amino acid)	1.952	10.640		115.13
Serine (S)	$-CH_2OH$	2.187	9.209		105.09
Threonine (T)	$-CH(CH_3)(OH)$	2.088	9.100		119.12
Tryptophan (W)	$-CH_2-$ (indole)	2.35	9.33		204.23
Tyrosine (Y)	$-CH_2-$ (phenol) $-OH$	2.17	9.19	10.47	181.19
Valine (V)	$-CH(CH_3)_2$	2.286	9.718		117.15

a. Standard abbreviations are shown in parentheses. Acidic protons are **bold.** Each substituent is written in its fully protonated form.

omit the subscript a as dictated by clarity. For successive base hydrolysis constants, we retain the subscript b. *K_{a1} (or K_1) refers to the acidic species with the most protons, and K_{b1} refers to the basic species with no acidic protons.*

Ask Yourself

10-A. (a) Each ion below has two consecutive acid-base reactions (called the stepwise acid-base reactions) when placed in water. Write the reactions and the correct symbol (e.g., K_2 or K_{b1}) for the equilibrium constant for each. Use Appendix B to find the numerical value of each acid or base equilibrium constant.

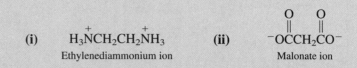

(i) $\overset{+}{H_3}NCH_2CH_2\overset{+}{N}H_3$
Ethylenediammonium ion

(ii) $^-OCCH_2CO^-$ (with two C=O groups)
Malonate ion

(b) Starting with the fully protonated species shown below, write the stepwise acid dissociation reactions of the amino acids aspartic acid and arginine. Be sure to remove the protons in the correct order based on the pK_a values in Table 10-1. Remember that the proton with the lowest pK_a (i.e., the greatest K_a) comes off first. Label the neutral molecules that we call aspartic acid and arginine.

(iii)
Aspartic acid cation

(iv)
Arginine cation

10-2 Finding the pH in Diprotic Systems

Consider the amino acid leucine, designated HL:

The side chain in leucine is an isobutyl group: R = $-CH_2CH(CH_3)_2$

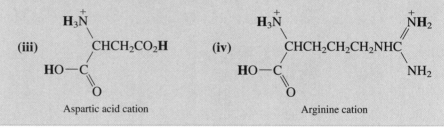

$$\underset{H_2L^+}{\overset{R}{H_3\overset{+}{N}CHCO_2H}} \underset{pK_{a1}=2.329}{\rightleftharpoons} \underset{\underset{Leucine}{HL}}{\overset{R}{H_3\overset{+}{N}CHCO_2^-}} \underset{pK_{a2}=9.747}{\rightleftharpoons} \underset{L^-}{\overset{R}{H_2NCHCO_2^-}}$$

The equilibrium constants apply to the following reactions:

Diprotic acid: $H_2L^+ \rightleftharpoons HL + H^+$ $K_{a1} \equiv K_1$ (10-3)

$HL \rightleftharpoons L^- + H^+$ $K_{a2} \equiv K_2$ (10-4)

Diprotic base: $L^- + H_2O \rightleftharpoons HL + OH^-$ K_{b1} (10-5)

$HL + H_2O \rightleftharpoons H_2L^+ + OH^-$ K_{b2} (10-6)

We now set out to calculate the pH and composition of individual solutions of 0.050 0 M H_2L^+, 0.050 0 M HL, and 0.050 0 M L^-. Our methods do not depend on the charge of the acids and bases. We would use the same procedure to find the pH of the diprotic H_2A, where A is anything, or H_2L^+, where HL is leucine.

The Acidic Form, H_2L^+

Easy stuff.

A salt such as leucine hydrochloride contains the protonated species H_2L^+, which can dissociate twice, as indicated in Reactions 10-3 and 10-4. Because $K_1 = 4.69 \times 10^{-3}$, H_2L^+ is a weak acid. HL is an even weaker acid, with $K_2 = 1.79 \times 10^{-10}$. It appears that H_2L^+ will dissociate only partly, and the resulting HL will hardly dissociate at all. For this reason, we make the (superb) approximation that a solution of H_2L^+ behaves as a *monoprotic* acid, with $K_a = K_1$.

With this approximation, the calculation of the pH of 0.050 0 M H_2L^+ is simple:

$$\underset{\substack{H_2L^+ \\ 0.050\,0 - x}}{\overset{\overset{\displaystyle R}{|}}{H_3NCHCO_2H}} \; \overset{K_1 = 4.69 \times 10^{-3}}{\rightleftharpoons} \; \underset{\substack{HL \\ x}}{\overset{\overset{\displaystyle R}{|}}{H_3NCHCO_2^-}} + \underset{x}{H^+}$$

H_2L^+ can be treated as monoprotic, with $K_a = K_1$.

$$\frac{x^2}{F - x} = K_1 \;\Rightarrow\; x = 1.31 \times 10^{-2}\,M$$

F is the formal concentration of H_2L^+ (=0.0500M in this example).

$$[HL] = x = 1.31 \times 10^{-2}\,M$$

$$[H^+] = x = 1.31 \times 10^{-2}\,M \;\Rightarrow\; pH = 1.88$$

$$[H_2L^+] = F - x = 3.69 \times 10^{-2}\,M$$

What is the concentration of L^- in the solution? We can find $[L^-]$ from the K_2 equation:

$$HL \overset{K_2}{\rightleftharpoons} L^- + H^+ \qquad K_2 = \frac{[H^+][L^-]}{[HL]} \;\Rightarrow\; [L^-] = \frac{K_2[HL]}{[H^+]}$$

$$[L^-] = \frac{(1.79 \times 10^{-10})(1.31 \times 10^{-2})}{(1.31 \times 10^{-2})} = 1.79 \times 10^{-10}\,M\;(=K_2)$$

Our approximation that the second dissociation of a diprotic acid is much less than the first dissociation is confirmed by this last result. The concentration of L^- is about eight orders of magnitude smaller than that of HL. As a source of protons, the dissociation of HL is negligible relative to the dissociation of H_2L^+. For most diprotic acids, K_1 is sufficiently larger than K_2 for this approximation to be valid. Even if K_2 were just 10 times less than K_1, the value of $[H^+]$ calculated by ignoring the second ionization would be in error by only 4%. The error in pH would be only 0.01 pH unit. In summary, *a solution of a diprotic acid behaves like a solution of a monoprotic acid, with $K_a = K_1$.*

Dissolved carbon dioxide is one of the most important diprotic acids in the earth's ecosystem. A remote sensor for monitoring CO_2 in the ocean is discussed in Box 10-1.

Box 10-1 *Informed Citizen*

Carbon Dioxide in the Air and Ocean

Carbon dioxide is the principal *greenhouse gas* in the atmosphere, with a significant role in regulating the temperature of earth's surface and, therefore, earth's climate. Earth absorbs sunlight and then radiates energy away by discharging infrared radiation to space. The balance between sunlight absorbed and radiation to space determines the surface temperature. A greenhouse gas is so named because it absorbs infrared radiation emitted from the ground and reradiates back to the ground. By intercepting earth's radiation, atmospheric carbon dioxide keeps our planet warmer than it would otherwise be in the absence of this gas.

Atmospheric carbon dioxide was constant at ~ 285 ppm ($285\,\mu L/L = 285\,\mu atm$) for a millennium (or more) prior to the industrial revolution. The burning of fossil fuel and the destruction of earth's forests since 1800 caused an exponential increase in CO_2 that threatens to alter earth's climate in your lifetime.

The ocean is a major reservoir for CO_2. Long-term monitoring of dissolved CO_2 is necessary to derive accurate models of the influence of the ocean on atmospheric CO_2. Dissolved CO_2 behaves as a diprotic acid

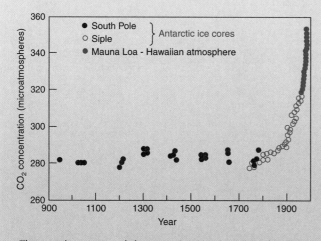

Thousand-year record documenting rise in atmospheric CO_2 concentration is derived from measurements of CO_2 trapped in ice cores in Antarctica and from direct atmospheric measurements. [From J. L. Sarmiento, *Chem. Eng. News* 31 May 1993, p. 30.]

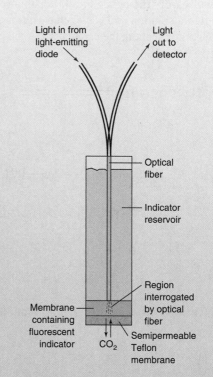

CO_2 sensor based on florescent indicator for remote measurements in the ocean. [From M. B. Tabacco, M. Uttamlal, M. McAllister, and D. R. Walt, *Anal. Chem.* **1999**, 71, 154.]

More easy stuff.

The Basic Form, L⁻

The fully basic species, L^-, would be found in a salt such as sodium leucinate, which could be prepared by treating leucine with an equimolar quantity of NaOH. Dissolving sodium leucinate in water gives a solution of L^-, the fully basic species. The K_b values for this dibasic anion are

$$L^- + H_2O \rightleftharpoons HL + OH^- \qquad K_{b1} = K_w/K_{a2} = 5.59 \times 10^{-5}$$

$$HL + H_2O \rightleftharpoons H_2L^+ + OH^- \qquad K_{b2} = K_w/K_{a1} = 2.13 \times 10^{-12}$$

(carbonic acid) whose equilibria were given in Demonstration 8-2. When the concentration of dissolved CO_2 goes up, the pH of the ocean goes down.

A remote sensor for monitoring dissolved CO_2 is shown at the left. The sensor contains a membrane soaked in a fluorescent indicator that emits yellow-green light with maximum intensity at a wavelength of 545 nm in its acidic form (HIn) and emits red light with a maximum at 625 nm in its basic form (In^-). When CO_2 from seawater diffuses across the thin, semipermeable Teflon membrane at the bottom of the sensor, it lowers the pH of the indicator solution. An optical fiber delivers light from a light-emitting diode down to the indicator in the membrane and collects fluorescent emission from the indicator. The ratio of emission intensities at the two wavelengths gives the relative concentrations of HIn and In^-. From the quotient $[HIn]/[In^-]$, the sensor can compute the pH and concentration of CO_2 in the seawater outside the sensor.

Results of a trial run for one month in the ocean near Woods Hole, Massachusetts, are shown in the graph. The oscillations arise from diurnal variations in CO_2 in the water due to metabolic activity of microorganisms. The general increase in CO_2 concentration was thought to arise from microbial growth on the sensor, a source of error that would need to be eliminated before this could be a reliable device.

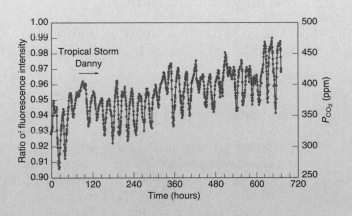

Ocean CO_2 measurements made by a remote sensor in August 1997. Concentration is expressed as the pressure of gaseous CO_2 (P_{CO_2} in ppm = μatm) in the atmosphere that would be in equilibrium with the measured concentration of dissolved CO_2. [From M. B. Tabacco, M. Uttamlal, M. McAllister, and D. R. Walt, *Anal. Chem.* **1999**, *71*, 154.]

K_{b1} tells us that L^- will not **hydrolyze** (react with water) very much to give HL. Furthermore, K_{b2} tells us that the resulting HL is such a weak base that hardly any further reaction to make H_2L^+ will occur.

We therefore treat L^- as a monobasic species, with $K_b = K_{b1}$. The results of this (fantastic) approximation can be outlined as follows:

Hydrolysis is the reaction of anything with water. Specifically, the reaction $L^- + H_2O \rightleftharpoons HL + OH^-$ is called hydrolysis.

$$\underset{\substack{L^- \\ 0.050\,0-x}}{H_2NCHCO_2^-} + H_2O \xrightleftharpoons{K_{b1}=5.59\times10^{-5}} \underset{\substack{HL \\ x}}{\overset{+}{H_3NCHCO_2^-}} + \underset{x}{OH^-}$$

(with R group above each structure)

L^- can be treated as monobasic with $K_b = K_{b1}$.

215

$$\frac{x^2}{F - x} = 5.59 \times 10^{-5} \quad \Rightarrow \quad x = [OH^-] = 1.64 \times 10^{-3} M$$

$$[HL] = x = 1.64 \times 10^{-3} M$$

$$[H^+] = K_w/[OH^-] = K_w/x = 6.08 \times 10^{-12} M \quad \Rightarrow \quad pH = 11.22$$

$$[L^-] = F - x = 4.84 \times 10^{-2} M$$

The concentration of H_2L^+ can be found from the K_{b2} equilibrium:

$$HL + H_2O \xrightleftharpoons{K_{b2}} H_2L^+ + OH^-$$

$$K_{b2} = \frac{[H_2L^+][OH^-]}{[HL]} = \frac{[H_2L^+]x}{x} = [H_2L^+]$$

We find that $[H_2L^+] = K_{b2} = 2.13 \times 10^{-12} M$, and the approximation that $[H_2L^+]$ is insignificant relative to [HL] is well justified. In summary, if there is any reasonable separation between K_1 and K_2 (and, therefore, between K_{b1} and K_{b2}), *the fully basic form of a diprotic acid can be treated as monobasic, with $K_b = K_{b1}$.*

The Intermediate Form, HL

A tougher problem.

A solution prepared from leucine, HL, is more complicated than one prepared from either H_2L^+ or L^-, because HL is both an acid and a base.

HL is both an acid and a base.

$$HL \rightleftharpoons H^+ + L^- \qquad\qquad K_a = K_2 = 1.79 \times 10^{-10} \qquad (10\text{-}7)$$

$$HL + H_2O \rightleftharpoons H_2L^+ + OH^- \qquad\qquad K_b = K_{b2} = 2.13 \times 10^{-12} \qquad (10\text{-}8)$$

A molecule that can both donate and accept a proton is said to be **amphiprotic.** The acid dissociation reaction (10-7) has a larger equilibrium constant than the base hydrolysis reaction (10-8), so we expect the solution of leucine to be acidic.

However, we cannot simply ignore Reaction 10-8, even when K_a and K_b differ by several orders of magnitude. Both reactions proceed to a nearly equal extent, because H^+ produced in Reaction 10-7 reacts with OH^- from Reaction 10-8, thereby driving Reaction 10-8 to the right.

The charge balance is discussed further in Section 11-3.

To treat this case correctly, we resort to writing a *charge balance,* which says that the sum of positive charges in solution must equal the sum of negative charges in solution. The procedure is applied to leucine, whose intermediate form (HL) has no net charge. However, the results apply to the intermediate form of *any* diprotic acid, regardless of its charge.

Our problem concerns 0.050 0 M leucine, in which both Reactions 10-7 and 10-8 can happen. The charge balance is

Charge balance for HL:

sum of positive charges =
       sum of negative charges

$$\underbrace{[H^+] + [H_2L^+]}_{\substack{\text{Sum of positive} \\ \text{charges}}} = \underbrace{[L^-] + [OH^-]}_{\substack{\text{Sum of negative} \\ \text{charges}}}$$

which can be rearranged to

$$[H_2L^+] - [L^-] + [H^+] - [OH^-] = 0 \qquad (10\text{-}9)$$

From the acid dissociation equilibria (Equations 10-3 and 10-4), we replace $[H_2L^+]$ with $[HL][H^+]/K_1$, and $[L^-]$ with $[HL]K_2/[H^+]$. Also, we can always write $[OH^-] = K_w/[H^+]$. Putting these expressions into Equation 10-9 gives

$$\frac{[HL][H^+]}{K_1} - \frac{[HL]K_2}{[H^+]} + [H^+] - \frac{K_w}{[H^+]} = 0$$

which can be solved for $[H^+]$. First multiply all terms by $[H^+]$:

$$\frac{[HL][H^+]^2}{K_1} - [HL]K_2 + [H^+]^2 - K_w = 0$$

Then factor out $[H^+]^2$ and rearrange:

$$[H^+]^2\left(\frac{[HL]}{K_1} + 1\right) = K_2[HL] + K_w$$

$$[H^+]^2 = \frac{K_2[HL] + K_w}{\dfrac{[HL]}{K_1} + 1}$$

Multiplying the numerator and denominator by K_1 and taking the square root of both sides give

$$[H^+] = \sqrt{\frac{K_1K_2[HL] + K_1K_w}{K_1 + [HL]}} \qquad (10\text{-}10)$$

We solved for $[H^+]$ in terms of known constants plus the single unknown, $[HL]$. Where do we proceed from here?

Luckily, a chemist gallops down from the mountain mists on her snow-white unicorn to provide the missing insight: "The major species will be HL, because it is both a weak acid and a weak base. Neither Reaction 10-7 nor Reaction 10-8 goes very far. For the concentration of HL in Equation 10-10, you can simply substitute the value 0.050 0 M."

The missing insight!

Taking the chemist's advice, we rewrite Equation 10-10 as follows:

$$[H^+] \approx \sqrt{\frac{K_1K_2F + K_1K_w}{K_1 + F}} \qquad (10\text{-}11)$$

where F is the formal concentration of HL ($=0.050\ 0M$). Equation 10-11 can be further simplified under most conditions. The first term in the numerator is almost always much greater than the second term, so the second term can be dropped:

$$[H^+] \approx \sqrt{\frac{K_1K_2F + \cancel{K_1K_w}}{K_1 + F}}$$

Then, if $K_1 \ll F$, the first term in the denominator can also be neglected.

$$[H^+] \approx \sqrt{\frac{K_1K_2F}{\cancel{K_1} + F}}$$

Canceling F in the numerator and denominator gives

$$[H^+] \approx \sqrt{K_1 K_2} = (K_1 K_2)^{1/2} \qquad (10\text{-}12)$$

Making use of the identity $\log(x^{1/2}) = \frac{1}{2}\log x$, we rewrite Equation 10-12 in the form

$$\log[H^+] \approx \tfrac{1}{2}\log(K_1 K_2)$$

Noting that $\log xy = \log x + \log y$, we can rewrite the equation once more:

$$\log[H^+] \approx \tfrac{1}{2}(\log K_1 + \log K_2)$$

To convert the left side of the preceding equation to $pH(=-\log[H^+])$ and the right side to $pK(=-\log K)$, multiply both sides by -1:

$$\underbrace{-\log[H^+]}_{pH} \approx \tfrac{1}{2}(\underbrace{-\log K_1}_{pK_1} \underbrace{-\log K_2}_{pK_2})$$

The pH of the intermediate form of a diprotic acid is close to midway between the two pK_a values and is almost independent of concentration.

Intermediate form of diprotic acid:

$$pH \approx \tfrac{1}{2}(pK_1 + pK_2) \qquad (10\text{-}13)$$

where K_1 and K_2 are the acid dissociation constants ($=K_{a1}$ and K_{a2}) of the diprotic acid.

Equation 10-13 is a good one to keep in your head. It says that *the pH of a solution of the intermediate form of a diprotic acid is approximately midway between pK_1 and pK_2, regardless of the formal concentration.*

For leucine, Equation 10-13 gives a pH of $\frac{1}{2}(2.329 + 9.747) = 6.038$, or $[H^+] = 10^{-pH} = 9.16 \times 10^{-7}$ M. The concentrations of H_2L^+ and L^- can be found from the K_1 and K_2 equilibria, using $[HL] = 0.050\,0$ M.

$$[H_2L^+] = \frac{[H^+][HL]}{K_1} = \frac{(9.16 \times 10^{-7})(0.050\,0)}{4.69 \times 10^{-3}} = 9.77 \times 10^{-6}\,M$$

$$[L^-] = \frac{K_2[HL]}{[H^+]} = \frac{(1.79 \times 10^{-10})(0.050\,0)}{9.16 \times 10^{-7}} = 9.77 \times 10^{-6}\,M$$

Was the approximation $[HL] \approx 0.050\,0$ M a good one? It certainly was, because $[H_2L^+]$ ($=9.77 \times 10^{-6}$ M) and $[L^-]$ ($=9.77 \times 10^{-6}$ M) are small in comparison with $[HL]$ ($\approx 0.050\,0$ M). Nearly all leucine remained in the form HL.

EXAMPLE   pH of the Intermediate Form of a Diprotic Acid

Potassium hydrogen phthalate, KHP, is a salt of the intermediate form of phthalic acid. Calculate the pH of 0.10 M KHP and 0.010 M KHP.

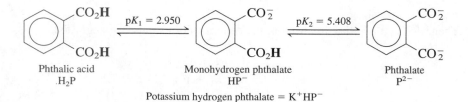

Potassium hydrogen phthalate = K^+HP^-

SOLUTION From Equation 10-13, the pH of potassium hydrogen phthalate is estimated as $\frac{1}{2}(pK_1 + pK_2) = 4.18$, regardless of concentration.

Summary of Diprotic Acid Calculations

The fully protonated species H_2A is treated as a monoprotic acid with acid dissociation constant K_1. The fully basic species A^{2-} is treated as a monoprotic base, with base association constant $K_{b1} = K_w/K_{a2}$. For the intermediate form HA^-, use the equation $pH \approx \frac{1}{2}(pK_1 + pK_2)$, where K_1 and K_2 are the acid dissociation constants for H_2A. The same considerations apply to a diprotic base ($B \rightarrow BH^+ \rightarrow BH_2^{2+}$): B is treated as monobasic; BH_2^{2+} is treated as monoprotic; and BH^+ is treated as an intermediate with $pH \approx \frac{1}{2}(pK_1 + pK_2)$, where K_1 and K_2 are the acid dissociation constants of BH_2^{2+}.

Diprotic systems
- Treat H_2A and BH_2^{2+} as monoprotic weak acids
- Treat A^{2-} and B as monoprotic weak bases
- Treat HA^- and BH^+ as intermediates: $pH \approx \frac{1}{2}(pK_1 + pK_2)$

Ask Yourself

10-B. Find the pH and the concentrations of H_2SO_3, HSO_3^-, and SO_3^{2-} in each of the following solutions: **(a)** 0.050 M H_2SO_3; **(b)** 0.050 M $NaHSO_3$; **(c)** 0.050 M Na_2SO_3.

10-3 Which Is the Principal Species?

We are often faced with the problem of identifying which species of acid, base, or intermediate is predominant under given conditions. For example, what is the principal form of benzoic acid in an aqueous solution at pH 8?

$$\langle\bigcirc\rangle\!-\!CO_2H \quad \begin{array}{l}\text{Benzoic acid}\\ pK_a = 4.20\end{array}$$

At $pH = pK_a$, $[A^-] = [HA]$ because

$$pH = pK_a + \log\left(\frac{[A^-]}{[HA]}\right)$$
$$= pK_a + \log 1 = pK_a.$$

The pK_a for benzoic acid is 4.20. This means that at pH 4.20 there is 1:1 mixture of benzoic acid (HA) and benzoate ion (A^-). At $pH = pK_a + 1$ (=5.20), the quotient $[A^-]/[HA]$ is 10:1. At $pH = pK_a + 2$ (=6.20), the quotient $[A^-]/[HA]$ is 100:1. As the pH increases, the quotient $[A^-]/[HA]$ increases still further.

For a monoprotic system, the basic species, A^-, is the predominant form when $pH > pK_a$. The acidic species, HA, is the predominant form when $pH < pK_a$. The predominant form of benzoic acid at pH 8 is the benzoate anion, $C_6H_5CO_2^-$.

pH	Major species
$< pK_a$	HA
$> pK_a$	A^-

EXAMPLE Principal Species—Which One and How Much?

What is the predominant form of ammonia in a solution at pH 7.0? Approximately what fraction is in this form?

SOLUTION In Appendix B, we find $pK_a = 9.24$ for the ammonium ion (NH_4^+, the conjugate acid of ammonia, NH_3). At $pH = 9.24$, $[NH_4^+] = [NH_3]$. Below

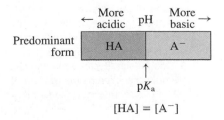

$$[HA] = [A^-]$$

pH 9.24, NH_4^+ will be the predominant form. Because pH = 7.0 is about 2 pH units below pK_a, the quotient $[NH_3]/[NH_4^+]$ will be around 1:100. Approximately 99% is in the form NH_4^+.

pH	Major species
$pH < pK_1$	H_2A
$pK_1 < pH < pK_2$	HA^-
$pH > pK_2$	A^{2-}

← More acidic   pH   More basic →

H_2A	HA^-	A^{2-}

pK_1

$[H_2A] = [HA^-]$

pK_2

$[HA^-] = [A^{2-}]$

For diprotic systems, the reasoning is the same, but there are two pK_a's. Consider fumaric acid, H_2A, with $pK_1 = 3.053$ and $pK_2 = 4.494$. At pH = pK_1, $[H_2A] = [HA^-]$. At pH = pK_2, $[HA^-] = [A^{2-}]$. The chart in the margin shows the major species in each pH region. At pH values below pK_1, H_2A is dominant. At pH values above pK_2, A^{2-} is dominant. At pH values between pK_1 and pK_2, HA^- is dominant. Figure 10-1 shows the fraction of each species as a function of pH.

The diagram below shows the major species for a triprotic system and introduces an important extension of what we learned in the previous section.

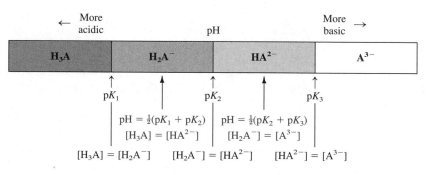

← More acidic   pH   More basic →

H_3A	H_2A^-	HA^{2-}	A^{3-}

pK_1 pK_2 pK_3

$pH = \frac{1}{2}(pK_1 + pK_2)$ $pH = \frac{1}{2}(pK_2 + pK_3)$
$[H_3A] = [HA^{2-}]$ $[H_2A^-] = [A^{3-}]$

$[H_3A] = [H_2A^-]$ $[H_2A^-] = [HA^{2-}]$ $[HA^{2-}] = [A^{3-}]$

Triprotic systems:

- Treat H_3A as monoprotic weak acid
- Treat A^{3-} as monoprotic weak base
- Treat H_2A^- as intermediate: $pH \approx \frac{1}{2}(pK_1 + pK_2)$
- Treat HA^{2-} as intermediate: $pH \approx \frac{1}{2}(pK_2 + pK_3)$

H_3A is the dominant species of a triprotic system in the most acidic solution at $pH < pK_1$. H_2A^- is dominant between pK_1 and pK_2. HA^{2-} is the major species between pK_2 and pK_3, and A^{3-} dominates in the most basic solution at $pH > pK_3$.

The diagram above shows that the pH of the first intermediate species, H_2A^-, is $\frac{1}{2}(pK_1 + pK_2)$. At this pH, the concentrations of H_3A and HA^{2-} are small and equal to each other. *The new feature of the diagram is that the pH of the second intermediate species, HA^{2-}, is $\frac{1}{2}(pK_2 + pK_3)$.* At this pH, the concentrations of H_2A^- and A^{3-} are small and equal.

FIGURE 10-1 Fractional composition diagram for fumaric acid (*trans*-butenedioic acid). α_i is the fraction of species i at each pH. At low pH, H_2A is the dominant form. Between pH = pK_1 and pH = pK_2, HA^- is dominant. Above pH = pK_2, A^{2-} dominates. Because pK_1 and pK_2 are not very different, the fraction of HA^- never gets very close to unity.

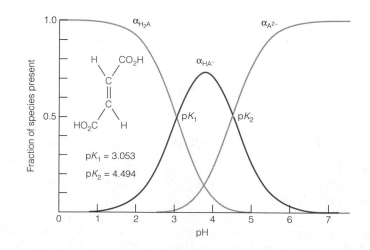

EXAMPLE   Principal Species in a Polyprotic System

The amino acid arginine has the following forms:

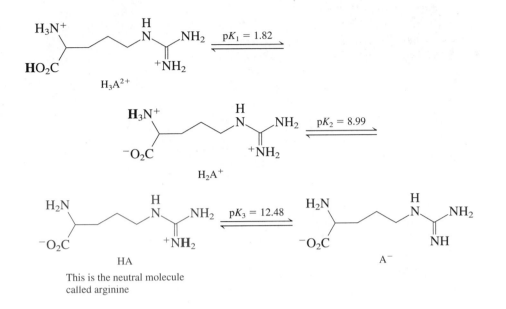

H_3A^{2+}

H_2A^+

HA

This is the neutral molecule
called arginine

A^-

Note that the ammonium group next to the carboxyl group at the left is more acidic than the substituent group at the right. What is the principal form of arginine at pH 10.0? Approximately what fraction is in this form? What is the second most abundant form at this pH?

SOLUTION We know that at pH = pK_2 = 8.99, $[H_2A^+] = [HA]$. At pH = pK_3 = 12.48, $[HA] = [A^-]$. At pH = 10.0, the major species is HA. Because pH 10.0 is about one pH unit higher than pK_2, we can say that $[HA]/[H_2A^+] \approx 10:1$. About 90% of arginine is in the form HA. The second most important species is H_2A^+, which makes up about 10% of the arginine.

EXAMPLE   More on Polyprotic Systems

In the pH range 1.82 to 8.99, H_2A^+ is the principal form of arginine. Which is the second most prominent species at pH 6.0? At pH 5.0?

SOLUTION We know that the pH of the pure intermediate (amphiprotic) species, H_2A^+, is

$$pH \text{ of } H_2A^+ \approx \tfrac{1}{2}(pK_1 + pK_2) = 5.40$$

Above pH 5.40 (and below pH = pK_2), HA is the second most important species. Below pH 5.40 (and above pH = pK_1), H_3A^{2+} is the second most important species.

Ask Yourself

10-C. (a) Draw the structure of the predominant form (principal species) of 1,3-dihydroxybenzene at pH 9.00 and at pH 11.00. What is the second most prominent species at each pH?

  (b) Cysteine is a triprotic system whose fully protonated form could be designated H_3C^+. Which form of cysteine is drawn below: H_3C^+, H_2C, HC^-, or C^{2-}? What would the pH of a 0.10 M solution of this form of cysteine be?

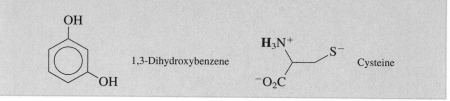

1,3-Dihydroxybenzene

Cysteine

10-4 Titrations in Polyprotic Systems

Figure 9-2 showed the titration curve for the monoprotic acid HA treated with OH^-. As a brief reminder, the pH at several critical points was computed as follows:

Initial solution: Has the pH of the weak acid HA
$V_e/2$: $pH = pK_a$ because $[HA] = [A^-]$
V_e: Has the pH of the conjugate base, A^-
Past V_e: pH is governed by concentration of excess OH^-

The slope of the titration curve is minimum at $V_e/2$ when $pH = pK_a$. The slope is maximum at the equivalence point.

  Before delving into titration curves for diprotic systems, you should realize that there are two buffer pairs derived from the acid H_2A. H_2A and HA^- constitute one buffer pair and HA^- and A^{2-} constitute a second pair. For the acid H_2A, there are *two* Henderson-Hasselbalch equations, both of which are *always* true. If you happen to know the concentrations $[H_2A]$ and $[HA^-]$, then use the pK_1 equation. If you know $[HA^-]$ and $[A^{2-}]$, use the pK_2 equation.

$$pH = pK_1 + \log\left(\frac{[HA^-]}{[H_2A]}\right) \qquad pH = pK_2 + \log\left(\frac{[A^{2-}]}{[HA^-]}\right)$$

Remember that pK_a in the Henderson-Hasselbalch equation always refers to the acid in the denominator.

  So now let us turn our attention to the titration of diprotic acids. Figure 10-2 shows calculated curves for 50.0 mL each of three different 0.020 0 M diprotic acids, H_2A, titrated with 0.100 M OH^-. For all three curves, H_2A has $pK_1 = 4.00$. In the lowest curve, pK_2 is 6.00. In the middle curve, pK_2 is 8.00; and in the upper curve, pK_2 is 10.00. The first equivalence volume (V_{e1}) occurs when the moles of added OH^- equal the moles of H_2A. The second equivalence volume (V_{e2}) is

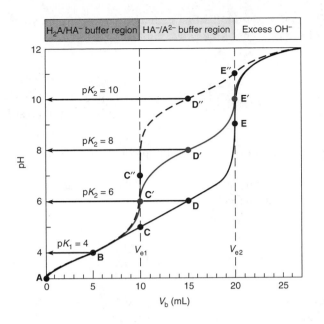

| H$_2$A/HA$^-$ buffer region | HA$^-$/A^{2-} buffer region | Excess OH$^-$ |

FIGURE 10-2   Calculated titration curves for three different diprotic acids, H$_2$A. For each curve, 50.0 mL of 0.020 0 M H$_2$A are titrated with 0.100 M NaOH. *Lowest curve:* $pK_1 = 4.00$ and $pK_2 = 6.00$. *Middle curve:* $pK_1 = 4.00$ and $pK_2 = 8.00$. *Upper curve:* $pK_1 = 4.00$ and $pK_2 = 10.00$.

always exactly twice as great as the first equivalence volume, because we must add the same amount of OH$^-$ to convert HA$^-$ to A^{2-}. Let's consider why pH varies as it does during these titrations.

$V_{e2} = 2V_{e1}$ (always!)

Point A has the same pH in all three cases. It is the pH of the acid H$_2$A, which is treated as a monoprotic acid with $pK_a = pK_1 = 4.00$ and formal concentration F.

$$\underset{\text{F}-x}{\text{HA}} \rightleftharpoons \underset{x}{\text{A}^-} + \underset{x}{\text{H}^+} \qquad \frac{x^2}{\text{F}-x} = K_1 \qquad (10\text{-}14)$$

Point A: weak acid H$_2$A

$$\frac{x^2}{0.0200-x} = 10^{-4.00} \quad \Rightarrow \quad x = 1.37 \times 10^{-3}\,\text{M} \quad \Rightarrow \quad \text{pH} = -\log(x) = 2.86$$

Point B, which is halfway to the first equivalence point, has the same pH in all three cases. It is the pH of a 1:1 mixture of H$_2$A:HA$^-$, which is treated as a monoprotic acid with $pK_a = pK_1 = 4.00$ and formal concentration F.

$$\text{pH} = pK_1 + \log\left(\frac{[\text{HA}^-]}{[\text{H}_2\text{A}]}\right) = pK_1 + \log 1 = pK_1 = 4.00 \qquad (10\text{-}15)$$

Point B: buffer containing H$_2$A + HA$^-$

Because all three acids have the same pK_1, the pH at this point is the same in all three cases.

Point C (and C' and C'') is the first equivalence point. H$_2$A has been converted to HA$^-$, the intermediate form of a diprotic acid. The pH is calculated from Equation 10-13:

$$\text{pH} \approx \tfrac{1}{2}(pK_1 + pK_2) = \begin{cases} 5.00 \text{ at C} \\ 6.00 \text{ at C'} \\ 7.00 \text{ at C''} \end{cases} \qquad (10\text{-}13)$$

Point C: intermediate form HA$^-$

The three acids have the same pK_1, but different values of pK_2. Therefore the pH at the first equivalence point is different in all three cases.

Point D is halfway from the first equivalence point to the second equivalence point. Half of the HA^- has been converted to A^{2-}. The pH is

Point D: buffer containing HA^- + A^{2-}

$$pH = pK_2 + \log\left(\frac{[A^{2-}]}{[HA^-]}\right) = pK_2 + \log 1 = pK_2 = \begin{cases} 6.00 \text{ at D} \\ 8.00 \text{ at D}' \\ 10.00 \text{ at D}'' \end{cases} \quad (10\text{-}16)$$

Look at Figure 10-2 and you will see that points D, D', and D'' come at pK_2 for each of the acids.

Point E is the second equivalence point. All of the acid has been converted into A^{2-}, a weak base with concentration F'. We find the pH by treating A^{2-} as a monoprotic base with $K_{b1} = K_w/K_{a2}$:

Point E: weak base A^{2-}

$$\underset{F-x}{A^{2-}} + H_2O \rightleftharpoons \underset{x}{HA^-} + \underset{x}{OH^-} \qquad \frac{x^2}{F'-x} = K_{b1} \quad (10\text{-}17)$$

The pH is different for the three acids, because K_{b1} is different for all three. Here is how we find the pH at the second equivalence point:

$$F' = [A^{2-}] = \frac{\text{mmol } A^{2-}}{\text{total mL}} = \frac{(0.020\,0\,M)(50.0\,mL)}{70.0\,mL} = 0.014\,3\,M$$

$$\frac{x^2}{0.014\,3 - x} = K_{b1} = \begin{cases} 10^{-8.00} \text{ at E} \\ 10^{-6.00} \text{ at E}' \\ 10^{-4.00} \text{ at E}'' \end{cases} \Rightarrow x = \begin{cases} 1.20 \times 10^{-5} \text{ at E} \\ 1.19 \times 10^{-4} \text{ at E}' = [OH^-] \\ 1.15 \times 10^{-3} \text{ at E}'' \end{cases}$$

$$\Rightarrow pH = -\log(K_w/x) = \begin{cases} 9.08 \text{ at E} \\ 10.08 \text{ at E}' \\ 11.06 \text{ at E}'' \end{cases}$$

Beyond V_{e2}, the pH is governed by the concentration of excess OH^-. pH rapidly converges to the same value for all three titrations.

You can see in Figure 10-2 that in a favorable case (the middle curve), we observe two obvious, steep equivalence points in the titration curve. When the pK values are too close together, or when the pK values are too low or too high, there may not be a distinct break at each equivalence point.

EXAMPLE   Titration of Sodium Carbonate

Let's reverse the process of Figure 10-2 and calculate the pH at points A–E in the titration of a diprotic base. Figure 10-3 shows the calculated titration curve for 50.0 mL of 0.020 0 M Na_2CO_3 treated with 0.100 M HCl. The first equivalence point is at 10.0 mL, and the second is 20.0 mL. Find the pH at points A–E.

$H_2CO_3 \overset{pK_1}{\rightleftharpoons} HCO_3^- \overset{pK_2}{\rightleftharpoons} CO_3^{2-}$
Carbonic   Carbonate
acid

$Na_2CO_3 \qquad pK_1 = 6.352 \qquad K_{a1} = 4.45 \times 10^{-7} \qquad K_{b1} = K_w/K_{a2} = 2.13 \times 10^{-4}$

$pK_2 = 10.329 \qquad K_{a2} = 4.69 \times 10^{-11} \qquad K_{b2} = K_w/K_{a1} = 2.25 \times 10^{-8}$

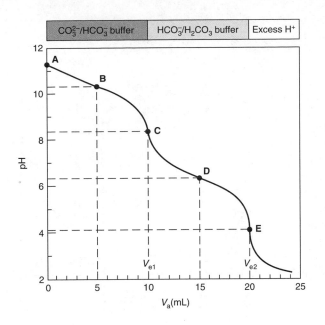

FIGURE 10-3 Calculated titration curve for 50.0 mL of 0.020 0 M Na_2CO_3 titrated with 0.100 M HCl.

SOLUTION

Point A. The initial point in the titration is just a solution of 0.020 0 M Na_2CO_3, which can be treated as a monoprotic base:

$$CO_3^{2-} + H_2O \overset{K_{b1}}{\rightleftharpoons} HCO_3^- + OH^- \qquad \frac{x^2}{0.020\ 0 - x} = K_{b1}$$

$$\underset{0.0200 - x}{} \qquad \underset{x}{} \quad \underset{x}{}$$

$$\Rightarrow \quad x = 1.96 \times 10^{-3} = [OH^-] \quad \Rightarrow \quad pH = -\log(K_w/x) = 11.29$$

Point B. Now we are halfway to the first equivalence point. Half of the carbonate has been converted to bicarbonate, so there is a 1:1 mixture of CO_3^{2-} and HCO_3^- —*Aha! A buffer!*

$$pH = pK_2 + \log\left(\frac{[CO_3^{2-}]}{[HCO_3^-]}\right) = pK_2 + \log 1 = pK_2 = 10.33$$

Use $pK_2(=pK_{a2})$ because the equilibrium
involves CO_3^{2-} and HCO_3^-

Point C. At the first equivalence point, we have a solution of HCO_3^-, the intermediate form of a diprotic acid. To a good approximation, the pH is independent of concentration and is given by

$$pH \approx \tfrac{1}{2}(pK_1 + pK_2) = \tfrac{1}{2}(6.352 + 10.329) = 8.34$$

Point D. We are halfway to the second equivalence point. Half of the bicarbonate has been converted to carbonic acid, so there is a 1:1 mixture of HCO_3^- and H_2CO_3 —*Aha! Another buffer!*

$$pH = pK_1 + \log\left(\frac{[HCO_3^-]}{[H_2CO_3]}\right) = pK_1 + \log 1 = pK_1 = 6.35$$

Use pK_1 because the equilibrium
involves HCO_3^- and H_2CO_3

Box 10-2 *Explanation*

What Is Isoelectric Focusing?

At its *isoelectric pH,* a protein has zero net charge and will therefore not migrate in an electric field. This effect is the basis of a very sensitive technique of protein separation called **isoelectric focusing.** A mixture of proteins is subjected to a strong electric field in a medium specifically designed to have a pH gradient. Positively charged molecules move toward the negative pole and negatively charged molecules move toward the positive pole. The proteins migrate in one direction or the other until they reach the point where the pH is the same as their isoelectric pH. At this point, they have no net charge and no longer move. Thus, each protein in the mixture is focused in one small region at its isoelectric pH.

An example of isoelectric focusing is shown in the figure. A mixture of proteins was applied to a polyacrylamide gel containing a mixture of polyprotic compounds called *ampholytes.* Several hundred volts were applied across the length of the gel. The ampholytes migrated until they formed a stable pH gradient (ranging from about pH 3 at one end of the gel to pH 10 at the other). Each protein migrated until it reached the zone with its isoelectric pH, at which point the protein had no net charge and ceased migrating. If a molecule diffuses out of its isoelectric region, it becomes charged and immediately migrates back to its isoelectric zone. When the proteins finished migrating, the electric field was removed and proteins were precipitated in place on the gel and stained with a dye to make them visible.

The stained gel is shown at the bottom of the figure. A spectrophotometric scan of the dye peaks is shown on the graph, and a profile of measured pH is also plotted. Each dark band of stained protein gives an absorbance peak.

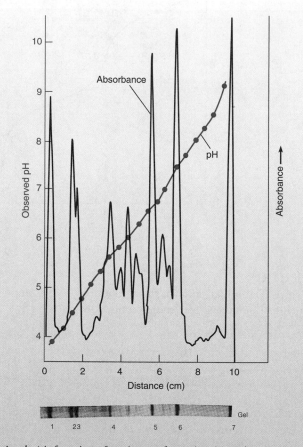

Isoelectric focusing of a mixture of proteins: (1) soybean trypsin inhibitor; (2) β-lactoglobulin A; (3) β-lactoglobulin B; (4) ovotransferrin; (5) horse myoglobin; (6) whale myoglobin; (7) cytochrome *c.* [Bio-Rad Laboratories, Hercules, CA.]

Point E. At the second equivalence point, all carbonate has been converted to carbonic acid, which has been diluted from its initial volume of 50.0 mL to a volume of 70.0 mL.

$$F' = [H_2CO_3] = \frac{\text{mmol } H_2CO_3}{\text{total mL}} = \frac{(0.020\ 0\text{M})(50.0\text{mL})}{70.0\text{mL}} = 0.014\ 3\text{M}$$

$$\begin{array}{ccc} H_2CO_3 & \stackrel{K_1}{\rightleftharpoons} & HCO_3^- + H^+ \\ F - x & & x \quad\ x \end{array} \qquad \frac{x^2}{0.014\ 3 - x} = K_1$$

$$\Rightarrow \quad x = 7.98 \times 10^{-5} = [H^+] \quad \Rightarrow \quad pH = -\log x = 4.10$$

Proteins Are Polyprotic Acids and Bases

Proteins are polymers made of amino acids:

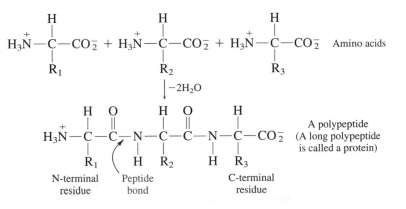

Proteins have biological functions such as structural support, catalysis of chemical reactions, immune response to foreign substances, transport of molecules across membranes, and control of genetic expression. The three-dimensional structure and function of a protein are determined by the sequence of amino acids from which the protein is made. Figure 10-4 shows the general shape of the protein myoglobin, whose function is to store O_2 in muscle cells. Of the 153 amino acids in sperm-whale myoglobin, 35 have basic side groups and 23 are acidic.

At high pH, most proteins have lost so many protons that they have a negative charge. At low pH, most proteins have gained so many protons that they have a positive charge. At some intermediate pH, called the *isoelectric pH* (or isoelectric point), each protein has exactly zero net charge. Box 10-2 explains how proteins can be separated from one another because of their different isoelectric points.

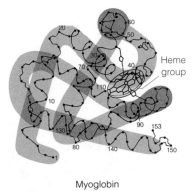

Myoglobin

FIGURE 10–4 Structure showing folding of the amino acid backbone of the protein myoglobin, which stores oxygen in muscle tissue. Substituents (R groups from Table 10-1) are omitted for clarity. The flat *heme* group at the right side of the protein contains an iron atom that can bind O_2, CO, and other small molecules. [From M. F. Perutz, "The Hemoglobin Molecule." Copyright ©1964 by Scientific American, Inc.]

Ask Yourself

10-D. Consider the titration of 50.0 mL of 0.050 0 M malonic acid with 0.100 M NaOH.

    (a) How many milliliters of titrant are required to reach each equivalence point?

    (b) Calculate the pH at $V_b = 0.0$, 12.5, 25.0, 37.5, 50.0, and 55.0 mL.

    (c) Put the points from **(b)** on a graph and sketch the titration curve.

Key Equations

Diprotic acid equilibria

$$H_2A \rightleftharpoons HA^- + H^+ \qquad K_{a1} \equiv K_1$$
$$HA^- \rightleftharpoons A^{2-} + H^+ \qquad K_{a2} \equiv K_2$$

Diprotic base equilibria

$$A^{2-} + H_2O \rightleftharpoons HA^- + OH^- \qquad K_{b1}$$
$$HA^- + H_2O \rightleftharpoons H_2A + OH^- \qquad K_{b2}$$

Relation between K_a and K_b

Monoprotic system $\qquad K_a K_b = K_w$

Diprotic system $\qquad K_{a1} K_{b2} = K_w$

$$K_{a2} K_{b1} = K_w$$

Triprotic system $\qquad K_{a1} K_{b3} = K_w$

$$K_{a2} K_{b2} = K_w$$

$$K_{a3} K_{b1} = K_w$$

pH of H_2A (or BH_2^{2+})

$$H_2A \xrightarrow{K_{a1}} H^+ + HA^- \qquad \frac{x^2}{F - x} = K_{a1}$$

(This gives $[H^+]$, $[HA^-]$, and $[H_2A]$. You can solve for $[A^{2-}]$ from the K_{a2} equilibrium.)

pH of HA^- (or diprotic BH^+)

$$pH \approx \frac{1}{2}(pK_1 + pK_2)$$

pH of A^{2-} (or diprotic B)

$$\underset{F-x}{A^{2-}} + H_2O \xrightarrow{K_{b1}} \underset{x}{HA^-} + \underset{x}{OH^-} \qquad \frac{x^2}{F-x} = K_{b1} = \frac{K_w}{K_{a2}}$$

(This gives $[OH^-]$, $[HA^-]$, and $[A^{2-}]$. You can find $[H^+]$ from the K_w equilibrium and $[H_2A]$ from the K_{a1} equilibrium.)

Diprotic buffer

$$pH = pK_1 + \log\left(\frac{[HA^-]}{[H_2A]}\right) \qquad pH = pK_2 + \log\left(\frac{[A^{2-}]}{[HA^-]}\right)$$

Both equations are always true and either can be used, depending on which set of concentrations you happen to know.

Titration of H_2A with OH^-

$V_b = 0$ Find pH of H_2A

$V_b = \frac{1}{2}V_{e1}$ $pH = pK_1$

$V_b = V_{e1}$ $pH \approx \frac{1}{2}(pK_1 + pK_2)$

$V_b = \frac{3}{2}V_{e1}$ $pH = pK_2$

$V_b = V_{e2}$ Find pH of A^{2-}

$V_b > V_{e2}$ Find concentration of excess OH^-

Titration of diprotic B with H^+

$V_a = 0$ Find pH of B

$V_a = \frac{1}{2}V_{e1}$ $pH = pK_{a2}$ (for BH_2^{2+})

$V_a = V_{e1}$ $pH \approx \frac{1}{2}(pK_{a1} + pK_{a2})$

$V_a = \frac{3}{2}V_{e1}$ $pH = pK_{a1}$ (for BH_2^{2+})

$V_a = V_{e2}$ Find pH of BH_2^{2+}

$V_a > V_{e2}$ Find concentration of excess H^+

Principal species

Monoprotic system

Diprotic system

Triprotic system

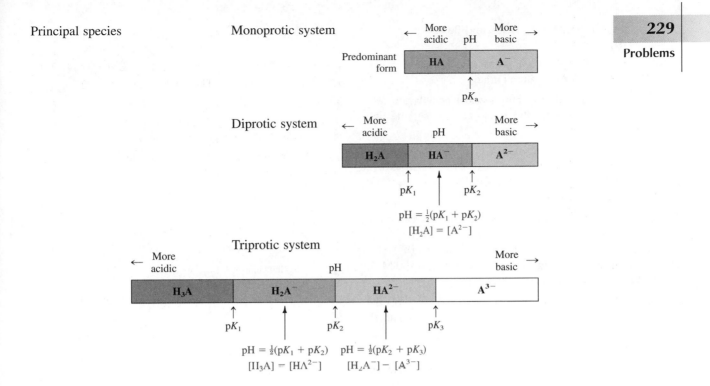

Important Terms

amino acid

amphiprotic

hydrolysis

isoelectric focusing

polyprotic acid

zwitterion

Problems

10-1. State what chemistry governs the pH at each point A through E in Figure 10-2.

10-2. Write the K_{a2} reaction of sulfuric acid (H_2SO_4) and the K_{b2} reaction of disodium oxalate $(Na_2C_2O_4)$ and find their numerical values.

10-3. The base association constants of phosphate are $K_{b1} = 0.014$, $K_{b2} = 1.58 \times 10^{-7}$, and $K_{b3} = 1.41 \times 10^{-12}$. From the K_b values, calculate K_{a1}, K_{a2}, and K_{a3} for H_3PO_4.

10-4. Write the general structure of an amino acid. Why do some amino acids in Table 10-1 have two pK values and others three?

10-5. Write the stepwise acid-base reactions for the following species in water. Write the correct symbol (e.g., K_{b1}) for the equilibrium constant for each reaction and find its numerical value.

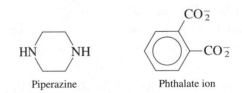

Piperazine

Phthalate ion

10-6. Write the K_{a2} reaction of proline and the K_{b2} reaction of the trisodium salt below.

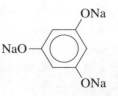

10-7. From the K_a values for citric acid in Appendix B, find K_{b1}, K_{b2}, and K_{b3} for trisodium citrate.

10-8. Write the chemical reactions whose equilibrium constants are K_{b1} and K_{b2} for the amino acid serine, and find their numerical values.

10-9. Abbreviating malonic acid, $CH_2(CO_2H)_2$, as H_2M, find the pH and concentrations of H_2M, HM^-, and M^{2-} in each of the following solutions: **(a)** 0.100 M H_2M; **(b)** 0.100 M NaHM; **(c)** 0.100 M Na_2M.

10-10. Consider the dibasic compound, B, that can form BH^+ and BH_2^{2+} with $K_{b1} = 1.00 \times 10^{-5}$ and $K_{b2} = 1.00 \times 10^{-9}$. Find the pH and concentrations of B, BH^+, and BH_2^{2+} in each of the following solutions: **(a)** 0.100 M B; **(b)** 0.100 M BH^+I^-; **(c)** 0.100 M $BH_2^{2+}(I^-)_2$.

10-11. Calculate the pH of a 0.300 M solution of the dibasic compound piperazine, which we will designate B. Calculate the concentration of each form of piperazine (B, BH^+, BH_2^{2+}).

10-12. Piperazine monohydrochloride is formed when one mole of HCl is added to the dibasic compound piperazine:

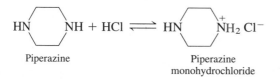

Piperazine   Piperazine
                   monohydrochloride

Find the pH of 0.150 M piperazine monohydrochloride and calculate the concentration of each form of piperazine in this solution.

10-13. Draw the structure of the amino acid glutamine and satisfy yourself that it is the intermediate form of a diprotic system. Find the pH of 0.050 M glutamine.

10-14. The acid HA has $pK_a = 7.00$.
(a) Which is the principal species, HA or A^-, at pH 6.00?
(b) Which is the principal species at pH 8.00?
(c) What is the quotient $[A^-]/[HA]$ at pH 7.00? at pH 6.00?

10-15. The acid H_2A has $pK_1 = 4.00$ and $pK_2 = 8.00$.
(a) At what pH is $[H_2A] = [HA^-]$?
(b) At what pH is $[HA^-] = [A^{2-}]$?
(c) Which is the principal species, H_2A, HA^-, or A^{2-} at pH 2.00?
(d) Which is the principal species at pH 6.00?
(e) Which is the principal species at pH 10.00?

10-16. The base B has $pK_b = 5.00$.
(a) What is the value of pK_a for the acid BH^+?
(b) At what pH is $[BH^+] = [B]$?

(c) Which is the principal species, B or BH^+, at pH 7.00?
(d) What is the quotient $[B]/[BH^+]$ at pH 12.00?

10-17. Draw the structures of the predominant forms of glutamic acid and tyrosine at pH 9.0 and pH 10.0. What is the second most abundant species at each pH?

10-18. Calculate the pH of a 0.10 M solution of each amino acid in the form drawn below:

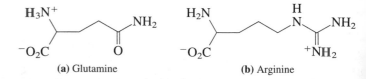

(a) Glutamine   (b) Arginine

10-19. Draw the structure of the predominant form of pyridoxal-5-phosphate at pH 7.00.

10-20. Find the pH and concentration of each species of arginine in 0.050 M arginine · HCl solution.

10-21. What is the charge of the predominant form of citric acid at pH 5.00?

10-22. A 100.0-mL aliquot of 0.100 M diprotic acid H_2A ($pK_1 = 4.00$, $pK_2 = 8.00$) was titrated with 1.00 M NaOH. At what volumes are the two equivalence points? Find the pH at the following volumes of base added and sketch a graph of pH versus V_b: V_b = 0, 5.0, 10.0, 15.0, 20.0, and 22.0 mL.

10-23. The dibasic compound B ($pK_{b1} = 4.00$, $pK_{b2} = 8.00$) was titrated with 1.00 M HCl. The initial solution of B was 0.100 M and had a volume of 100.0 mL. At what volumes are the two equivalence points? Find the pH at the following volumes of acid added and sketch a graph of pH versus V_a: V_a = 0, 5.0, 10.0, 15.0, 20.0, and 22.0 mL.

10-24. Select one indicator from Table 8-3 that would be useful for detecting each equivalence point shown in Figure 10-2 and listed below. State the color change you would observe in each case.
(a) Second equivalence point of the lowest curve
(b) First equivalence point of the upper curve
(c) First equivalence point of the middle curve
(d) Second equivalence point of the middle curve

10-25. Write two consecutive reactions that occur when 40.0 mL of 0.100 M piperazine are titrated with 0.100 M HCl and find the equivalence volumes. Find the pH at V_a = 0, 20.0, 40.0, 60.0, 80.0, and 100.0 mL and sketch the titration curve.

10-26. Write the chemical reactions (including structures of reactants and products) that occur when the amino acid histidine is titrated with perchloric acid. (Histidine has no net charge.) A solution containing 25.0 mL of 0.050 0 M histidine was titrated with 0.050 M HClO$_4$. List the equivalence volumes and calculate the pH at $V_a = 0$, 12.5, 25.0, and 50.0 mL.

10-27. An aqueous solution containing ~ 1 g of oxobutanedioic acid (FM 132.073) per 100 mL was titrated with 0.094 32 M NaOH to measure the acid molarity.

(a) Calculate the pH at the following volumes of added base: $\frac{1}{2}V_{e1}$, V_{e1}, $\frac{3}{2}V_{e1}$, V_{e2}, 1.05 V_{e2}, and sketch the titration curve.

(b) Which equivalence point would be best to use in this titration?

(c) You have the indicators erythrosine, ethyl orange, bromocresol green, bromothymol blue, thymolphthalein, and alizarin yellow. Which indicator will you use and what color change will you look for?

How Would You Do It?

10-28. *Finding the mean molecular mass of a polymer.*[1] Poly(ethylene glycol), abbreviated PEG in the scheme below is a polymer with $-OCH_2CH_2-$ repeat groups capped at each end by $-CH_2CH_2OH$ (alcohol) groups. The alcohols react with the reagent PMDA shown in the scheme to make Product 1. After the reaction is complete, water converts Product 1 into Product 2, which has six carboxylic acid groups.

Standardization of PMDA: A 10.00-mL solution of PMDA in a dry organic solvent was treated with water to hydrolyze it to Product 3. Titration of the carboxylic acid groups in Product 3 required 28.98 mL of 0.294 4 M NaOH to reach a phenolphthalein end point.

Analysis of PEG: A 220.0 mg quantity of PEG was treated with 10.00 mL of standard PMDA solution. After 30 min to complete the reaction, 30 mL of water were added to convert Product 1 to Product 2 and to convert unused PMDA to Product 3. The mixture of Products 2 and 3 required 24.41 mL of 0.294 4 M NaOH to reach a phenolphthalein end point. Find the mean molecular mass of PEG and the mean number of repeat groups, n, in the formula for PEG.

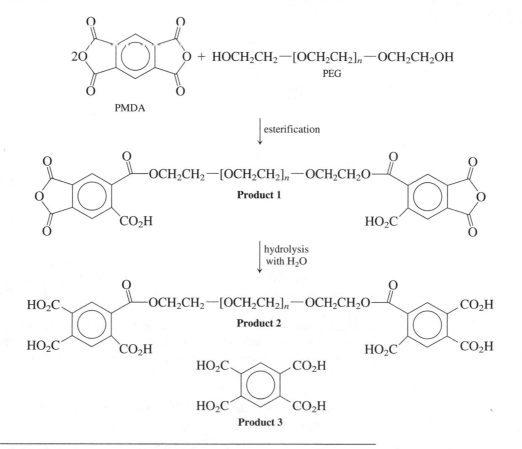

Notes and References

1. The actual procedure is slightly more complicated than described in this problem. Instructions for a student experiment based on the reaction scheme are given by K. R. Williams and U. R. Bernier, *J. Chem. Ed.* **1994,** *71,* 265.

Chemical Equilibrium in the Environment

C. Dalpra, Potomac River Basin Commission

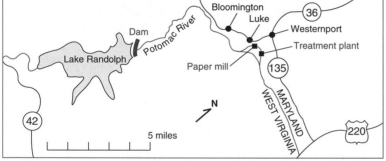

Paper mill on the Potomac River near Westernport, Maryland, neutralizes acid mine drainage in the water. Upstream of the mill, the river is acidic and lifeless; below the mill, the river teems with life.

*P*art of the North Branch of the Potomac River runs crystal clear through the scenic Appalachian Mountains, but it is lifeless—a victim of acid drainage from abandoned coal mines. As the river passes a paper mill and wastewater treatment plant near Westernport, Maryland, the pH rises from an acidic, lethal value of 4.5 to a neutral value of 7.2, at which fish and plants thrive. This fortunate accident comes about when the calcium carbonate by-product from papermaking exits the paper mill and reacts with massive quantities of carbon dioxide from bacterial respiration at the sewage treatment plant. The resulting soluble bicarbonate neutralizes the acidic river and restores life downstream of the plant.

$$CaCO_3(s) + CO_2(aq) + H_2O(l) \rightleftharpoons Ca^{2+}(aq) + 2HCO_3^-(aq)$$

Calcium carbonate $\qquad\qquad$ Dissolved calcium bicarbonate

$$HCO_3^-(aq) + H^+(aq) \xrightarrow{\text{neutralization}} CO_2(g)\uparrow + H_2O(l)$$

Acid from river

We call these two reactions *coupled equilibria.* Consumption of bicarbonate in the second reaction drives the first reaction to make more product.

Chapter **11**

A DEEPER LOOK AT CHEMICAL EQUILIBRIUM

We now pause to look more carefully at chemical equilibrium. This chapter is optional in that later chapters do not depend on it in any critical way. Some instructors consider the topics in this chapter to be fundamental to your understanding of chemistry and will ask you to take the time to savor every morsel. Others find it necessary to pass over this chapter and press onward so that there is more time to discuss instrumental methods of analysis that come later in the book.

11-1 | The Effect of Ionic Strength on Solubility of Salts

When slightly soluble lead(II) iodide dissolves in pure water, many species are formed.

$$PbI_3^-$$
0.01%

$$\Big\updownarrow I^-$$

$$PbI_2(s) \rightleftharpoons PbI_2(aq) \rightleftharpoons Pb^{2+} + 2I^-$$
0.8% 81%

$$\Big\updownarrow \qquad\qquad \Big\updownarrow H_2O$$

$$PbI^+ + I^- \qquad PbOH^+ + H^+$$
18% 0.3%

Composition was computed from the following equilibria with activity coefficients:[1]

$$PbI_2(s) \rightleftharpoons Pb^{2+} + 2I^-$$
$$K_{sp} = 7.9 \times 10^{-9}$$

$$Pb^{2+} + I^- \rightleftharpoons PbI^+$$
$$K \equiv \beta_1 = 1.0 \times 10^2$$

$$Pb^{2+} + 2I^- \rightleftharpoons PbI_2(aq)$$
$$K \equiv \beta_2 = 1.6 \times 10^3$$

$$Pb^{2+} + 3I^- \rightleftharpoons PbI_3^-$$
$$K \equiv \beta_3 = 7.9 \times 10^3$$

$$Pb^{2+} + H_2O \rightleftharpoons PbOH^+ + H^+$$
$$K_a = 2.5 \times 10^{-8}$$

233

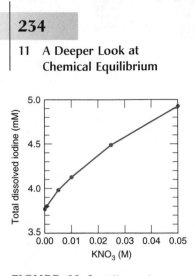

FIGURE 11-1 Effect of KNO_3 on the solubility of PbI_2. [Data from D. B. Green, G. Rechtsteiner, and A. Honodel, *J. Chem. Ed.* **1996**, *73*, 789.]

An anion is surrounded by more cations than anions. A cation is surrounded by more anions than cations.

Ion dissociation is increased by increasing the ionic strength of the solution.

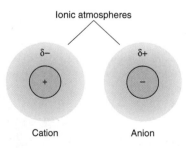

FIGURE 11-2 An ionic atmosphere, shown as a spherical cloud of charge $\delta+$ or $\delta-$, surrounds each ion in solution. The charge of the atmosphere is less than the charge of the central ion. The greater the ionic strength of the solution, the greater the charge in each ionic atmosphere.

Approximately 81% of the lead is found as Pb^{2+}, 18% is PbI^+, 0.8% is $PbI_2(aq)$, 0.3% is $PbOH^+$, and 0.01% is PbI_3^-. The *solubility product* is the equilibrium constant for the reaction $PbI_2(s) \rightleftharpoons Pb^{2+} + 2I^-$, which clearly tells only part of the story.

Now a funny thing happens when the "inert" salt, KNO_3, is added to the saturated PbI_2 solution. (By "inert," we mean that there is no chemical reaction between K^+ or NO_3^- with any of the lead iodide species.) As more KNO_3 is added, the total concentration of dissolved iodine increases, as shown in Figure 11-1. (Dissolved iodine includes free iodide and iodide attached to lead.) It turns out that adding any inert salt such as KNO_3 to a sparingly soluble salt such as PbI_2 increases the solubility of the sparingly soluble substance.

The Explanation

Why does solubility increase when salts are added to the solution? Consider one particular Pb^{2+} ion and one particular I^- ion in the solution. The I^- ion is surrounded by cations (K^+, Pb^{2+}) and anions (NO_3^-, I^-) in the solution. However, the typical anion has more cations than anions near it because cations are attracted but anions are repelled. These interactions create a region of net positive charge around any particular anion. We call this region the **ionic atmosphere** (Figure 11-2). Ions continually diffuse into and out of the ionic atmosphere. The net charge in the atmosphere, averaged over time, is less than the charge of the anion at the center. Similarly, an atmosphere of negative charge surrounds any cation in solution.

The ionic atmosphere *attenuates* (decreases) the attraction between ions in solution. The cation plus its negative atmosphere has less positive charge than the cation alone. The anion plus its ionic atmosphere has less negative charge than the anion alone. The net attraction between the cation with its ionic atmosphere and the anion with its ionic atmosphere is smaller than it would be between pure cation and anion in the absence of ionic atmospheres. *The higher the concentration of ions in a solution, the higher the charge in the ionic atmosphere. Each ion-plus-atmosphere contains less net charge and there is less attraction between any particular cation and anion.*

Increasing the concentration of ions in a solution therefore reduces the attraction between any particular Pb^{2+} ion and any I^- ion, relative to their attraction to each other in pure water. The effect is to reduce their tendency to come together, thereby increasing the solubility of PbI_2.

Increasing the concentration of ions in a solution promotes dissociation of ions. Thus, each of the following reactions is driven to the right if KNO_3 is added (Demonstration 11-1):

$$Fe(SCN)^{2+} \rightleftharpoons Fe^{3+} + SCN^-$$
$$\text{Thiocyanate}$$

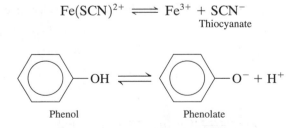

What Do We Mean by "Ionic Strength"?

Ionic strength, μ, is a measure of the total concentration of ions in solution. The more highly charged an ion, the more it is counted.

Effect of Ionic Strength on Ion Dissociation[2]

This experiment demonstrates the effect of ionic strength on the dissociation of the red iron(III) thiocyanate complex:

$$Fe(SCN)^{2+} \rightleftharpoons Fe^{3+} + SCN^-$$
$$\text{Red} \qquad \text{Pale yellow} \quad \text{Colorless}$$

Prepare a solution of 1 mM $FeCl_3$ by dissolving 0.27 g of $FeCl_3 \cdot 6H_2O$ in 1 L of water containing 3 drops of 15 M (concentrated) HNO_3. The acid slows the precipitation of $Fe(OH)_3$, which occurs in a few days and necessitates the preparation of fresh solution for this demonstration.

To demonstrate the effect of ionic strength on the dissociation reaction, mix 300 mL of the 1 mM $FeCl_3$ solution with 300 mL of 1.5 mM NH_4SCN or KSCN. Divide the pale red solution into two equal portions and add 12 g of KNO_3 to one of them to increase the ionic strength to 0.4 M. As the KNO_3 dissolves, the red $Fe(SCN)^{2+}$ complex dissociates and the color fades noticeably (Color Plate 6).

Add a few crystals of NH_4SCN or KSCN to either solution to drive the reaction toward formation of $Fe(SCN)^{2+}$, thereby intensifying the red color. This reaction demonstrates Le Châtelier's principle—adding a product creates more reactant.

Ionic strength:

$$\mu = \tfrac{1}{2}(c_1 z_1^2 + c_2 z_2^2 + \ldots) = \tfrac{1}{2}\sum_i c_i z_i^2 \qquad (11\text{-}1)$$

where c_i is the concentration of the ith species and z_i is its charge. The sum extends over *all* ions in solution.

EXAMPLE   Calculation of Ionic Strength

Find the ionic strength of (a) 0.10 M $NaNO_3$; (b) 0.010 M Na_2SO_4; and (c) 0.020 M KBr plus 0.010 M Na_2SO_4.

SOLUTION

(a) $\mu = \tfrac{1}{2}\{[Na^+] \cdot (+1)^2 + [NO_3^-] \cdot (-1)^2\}$
$= \tfrac{1}{2}\{0.10 \cdot 1 + 0.10 \cdot 1\} = 0.10\,M$

(b) $\mu = \tfrac{1}{2}\{[Na^+] \cdot (+1)^2 + [SO_4^{2-}] \cdot (-2)^2\}$
$= \tfrac{1}{2}\{(0.020 \cdot 1) + (0.010 \cdot 4)\} = 0.030\,M$

Note that $[Na^+] = 0.020\,M$ because there are two moles of Na^+ per mole of Na_2SO_4.

(c) $\mu = \tfrac{1}{2}\{[K^+] \cdot (+1)^2 + [Br^-] \cdot (-1)^2 + [Na^+] \cdot (+1)^2 + [SO_4^{2-}] \cdot (-2)^2\}$
$= \tfrac{1}{2}\{(0.020 \cdot 1) + (0.020 \cdot 1) + (0.020 \cdot 1) + (0.010 \cdot 4)\} = 0.050\,M$

Electrolyte	Molarity	Ionic strength
1:1	M	M
2:1	M	3M
3:1	M	6M
2:2	M	4M

$NaNO_3$ is called a 1:1 electrolyte because the cation and the anion both have a charge of 1. For 1:1 electrolytes, the ionic strength equals the molarity. For any other stoichiometry (such as the 2:1 electrolyte, Na_2SO_4), the ionic strength is greater than the molarity.

Ask Yourself

11-A. From the solubility product of PbI_2, calculate the expected concentration of dissolved iodine in a saturated solution of PbI_2. Why is your result different from the experimental observation in Figure 11-1, and why does the concentration of dissolved iodine increase with increasing KNO_3 concentration?

11-2   Activity Coefficients

Until now, we have written the equilibrium for the reaction $aA + bB \rightleftharpoons cC + dD$ in the form of $K = [C]^c[D]^d/[A]^a[B]^b$. This equilibrium constant does not account for any effect of ionic strength on the chemical reaction. To account for ionic strength, concentrations are replaced by **activities**:

Activity of C:

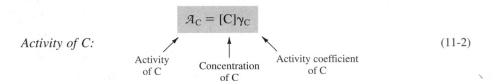

$$\mathcal{A}_C = [C]\gamma_C$$

(11-2)

Activity of C   Concentration of C   Activity coefficient of C

Do not confuse the terms *activity* and *activity coefficient*.

The activity of species C is its concentration multiplied by its **activity coefficient.** The activity coefficient depends on ionic strength. If there were no effect of ionic strength on the chemical reaction, the activity coefficient would be 1. The correct form of the equilibrium constant for the reaction $aA + bB \rightleftharpoons cC + dD$ is

This is the "real" equilibrium constant.

General form of equilibrium constant: $K = \dfrac{\mathcal{A}_C^c \mathcal{A}_D^d}{\mathcal{A}_A^a \mathcal{A}_B^b} = \dfrac{[C]^c\gamma_C^c[D]^d\gamma_D^d}{[A]^a\gamma_A^a[B]^b\gamma_B^b}$ (11-3)

For the reaction $PbI_2(s) \rightleftharpoons Pb^{2+} + 2I^-$, the equilibrium constant is

$$K_{sp} = \mathcal{A}_{Pb^{2+}}\mathcal{A}_{I^-}^2 = [Pb^{2+}]\gamma_{Pb^{2+}}[I^-]^2\gamma_{I^-}^2$$

(11-4)

If the concentrations of Pb^{2+} and I^- are to *increase* when a second salt is added to increase ionic strength, the activity coefficients must *decrease* with increasing ionic strength. Conversely, at low ionic strength, activity coefficients approach unity.

Activity Coefficients of Ions

Detailed consideration of the ionic atmosphere model leads to the **extended Debye-Hückel equation,** relating activity coefficients to ionic strength:

Extended Debye-Hückel equation: $\log\gamma = \dfrac{-0.51z^2\sqrt{\mu}}{1 + (\alpha\sqrt{\mu}/305)}$ (at 25°C)   (11-5)

1 pm (picometer) = 10^{-12} m

In Equation 11-5, γ is the activity coefficient of an ion of charge $\pm z$ and size α (picometers, pm) in an aqueous solution of ionic strength μ. The equation works fairly well for $\mu \leq 0.1\,M$.

The size α is the effective **hydrated radius** of the ion and its tightly bound sheath of water molecules. Small, highly charged ions bind water more tightly and have *larger* hydrated radii than do larger or less highly charged ions. F^-, for example, has a hydrated radius greater than that of I^- (Figure 11-3). Each anion attracts solvent molecules mainly by electrostatic interaction between the negative ion and the positive pole of the H_2O dipole:

Table 11-1 lists sizes (α) and activity coefficients (γ) of many ions. All ions of the same size and charge appear in the same group and have the same activity coefficients. For example, Ba^{2+} and succinate ion $[^-O_2CCH_2CH_2CO_2^-$ listed as $(CH_2CO_2^-)_2]$ each have a hydrated radius of 500 pm and are listed among the charge $=\pm2$ ions. In a solution with an ionic strength of 0.001 M, both of these ions have an activity coefficient of 0.868.

Effect of Ionic Strength, Ion Charge, and Ion Size on the Activity Coefficient

Over the range of ionic strengths from 0 to 0.1 M, we find that

1. As ionic strength increases, the activity coefficient decreases (Figure 11-4). The activity coefficient (γ) approaches unity as the ionic strength (μ) approaches 0.

2. As the charge of the ion increases, the departure of its activity coefficient from unity increases. Activity corrections are much more important for an ion with a charge of ±3 than for one with a charge of ±1 (Figure 11-4). Note that the activity coefficients in Table 11-1 depend on the magnitude of the charge but not on its sign.

3. The smaller the hydrated radius of the ion, the more important activity effects become.

EXAMPLE   Using Table 11-1

Find the activity coefficient of Mg^{2+} in a solution of 3.3 mM $Mg(NO_3)_2$.

SOLUTION The ionic strength is

$$\mu = \tfrac{1}{2}\{[Mg^{2+}]\cdot 2^2 + [NO_3^-]\cdot(-1)^2\}$$
$$= \tfrac{1}{2}\{(0.0033)\cdot 4 + (0.0066)\cdot 1\} = 0.010\,M$$

In Table 11-1, Mg^{2+} is listed under the charge ±2 and has a size of 800 pm. When $\mu = 0.010\,M$, $\gamma = 0.69$.

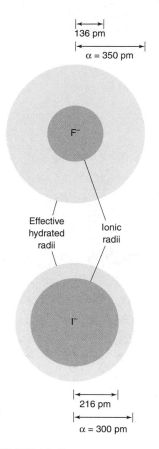

FIGURE 11-3 Ionic and hydrated radii of fluoride and iodide. The *smaller* F^- ion binds water molecules more tightly than does I^-, so F^- has the *larger* hydrated radius.

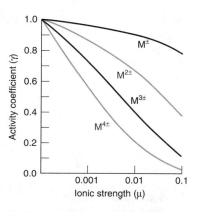

FIGURE 11-4 Activity coefficients for differently charged ions with a constant hydrated radius of 500 pm. At zero ionic strength, $\gamma = 1$. The greater the charge of the ion, the more rapidly γ decreases as ionic strength increases. Note that the abscissa is logarithmic.

TABLE 11-1 **Activity coefficients for aqueous solutions at 25°C**

Ion	Hydrated radius (α, pm)	Ionic strength (μ, M)				
		0.001	0.005	0.01	0.05	0.1
Charge = ± 1						
H^+	900	0.967	0.933	0.914	0.86	0.83
$(C_6H_5)_2CHCO_2^-$, $(C_3H_7)_4N^+$	800	0.966	0.931	0.912	0.85	0.82
$(O_2N)_3C_6H_2O^-$, $(C_3H_7)_3NH^+$, $CH_3OC_6H_4CO_2^-$	700	0.965	0.930	0.909	0.845	0.81
Li^+, $C_6H_5CO_2^-$, $HOC_6H_4CO_2^-$, $ClC_6H_4CO_2^-$, $C_6H_5CH_2CO_2^-$, $CH_2{=}CHCH_2CO_2^-$, $(CH_3)_2CHCH_2CO_2^-$, $(CH_3CH_2)_4N^+$, $(C_3H_7)_2NH_2^+$	600	0.965	0.929	0.907	0.835	0.80
$Cl_2CHCO_2^-$, $Cl_3CCO_2^-$, $(CH_3CH_2)_3NH^+$, $(C_3H_7)NH_3^+$	500	0.964	0.928	0.904	0.83	0.79
Na^+, $CdCl^+$, ClO_2^-, IO_3^-, HCO_3^-, $H_2PO_4^-$, HSO_3^-, $H_2AsO_4^-$, $Co(NH_3)_4(NO_2)_2^+$, $CH_3CO_2^-$, $ClCH_2CO_2^-$, $(CH_3)_4N^+$, $(CH_3CH_2)_2NH_2^+$, $H_2NCH_2CO_2^-$	450	0.964	0.928	0.902	0.82	0.775
$^+H_3NCH_2CO_2H$, $(CH_3)_3NH^+$, $CH_3CH_2NH_3^+$	400	0.964	0.927	0.901	0.815	0.77
OH^-, F^-, SCN^-, OCN^-, HS^-, ClO_3^-, ClO_4^-, BrO_3^-, IO_4^-, MnO_4^-, HCO_2^-, H_2citrate$^-$, $CH_3NH_3^+$, $(CH_3)_2NH_2^+$	350	0.964	0.926	0.900	0.81	0.76
K^+, Cl^-, Br^-, I^-, CN^-, NO_2^-, NO_3^-	300	0.964	0.925	0.899	0.805	0.755
Rb^+, Cs^+, NH_4^+, Tl^+, Ag^+	250	0.964	0.924	0.898	0.80	0.75
Charge = ± 2						
Mg^{2+}, Be^{2+}	800	0.872	0.755	0.69	0.52	0.45
$CH_2(CH_2CH_2CO_2^-)_2$, $(CH_2CH_2CH_2CO_2^-)_2$	700	0.872	0.755	0.685	0.50	0.425
Ca^{2+}, Cu^{2+}, Zn^{2+}, Sn^{2+}, Mn^{2+}, Fe^{2+}, Ni^{2+}, Co^{2+}, $C_6H_4(CO_2^-)_2$, $H_2C(CH_2CO_2^-)_2$, $(CH_2CH_2CO_2^-)_2$	600	0.870	0.749	0.675	0.485	0.405
Sr^{2+}, Ba^{2+}, Cd^{2+}, Hg^{2+}, S^{2-}, $S_2O_4^{2-}$, WO_4^{2-}, $H_2C(CO_2^-)_2$, $(CH_2CO_2^-)_2$, $(CHOHCO_2^-)_2$	500	0.868	0.744	0.67	0.465	0.38
Pb^{2+}, CO_3^{2-}, SO_3^{2-}, MoO_4^{2-}, $Co(NH_3)_5Cl^{2+}$, $Fe(CN)_5NO^{2-}$, $C_2O_4^{2-}$, Hcitrate^{2-}	450	0.867	0.742	0.665	0.455	0.37
Hg_2^{2+}, SO_4^{2-}, $S_2O_3^{2-}$, $S_2O_6^{2-}$, $S_2O_8^{2-}$, SeO_4^{2-}, CrO_4^{2-}, HPO_4^{2-}	400	0.867	0.740	0.660	0.445	0.355
Charge = ± 3						
Al^{3+}, Fe^{3+}, Cr^{3+}, Sc^{3+}, Y^{3+}, In^{3+}, lanthanides[a]	900	0.738	0.54	0.445	0.245	0.18
citrate^{3-}	500	0.728	0.51	0.405	0.18	0.115
PO_4^{3-}, $Fe(CN)_6^{3-}$, $Cr(NH)_6^{3+}$, $Co(NH_3)_6^{3+}$, $Co(NH_3)_5H_2O^{3+}$	400	0.725	0.505	0.395	0.16	0.095
Charge = ± 4						
Th^{4+}, Zr^{4+}, Ce^{4+}, Sn^{4+}	1 100	0.588	0.35	0.255	0.10	0.065
$Fe(CN)_6^{4-}$	500	0.57	0.31	0.20	0.048	0.021

a. Lanthanides are elements 57–71 in the periodic table.

SOURCE: J. Kielland, *J. Am. Chem. Soc.* **1937**, *59*, 1675.

How to Interpolate

If you need to find an activity coefficient for an ionic strength that is between values in Table 11-1, you can use Equation 11-5. Alternatively, in the absence of a spreadsheet, it is usually easier to interpolate than to use Equation 11-5. In *linear interpolation,* we assume that values between two entries of a table lie on a straight line. For example, consider a table in which $y = 0.67$ when $x = 10$ and $y = 0.83$ when $x = 20$. What is the value of y when $x = 16$?

Interpolation is the estimation of a number that lies *between* two values in a table. Estimating a number that lies *beyond* values in a table is called *extrapolation.*

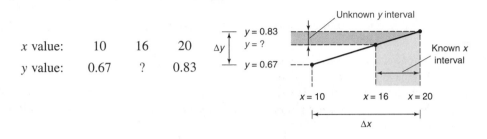

x value:	10	16	20
y value:	0.67	?	0.83

To interpolate the value of y, we can set up a proportion:

Interpolation:
$$\frac{\text{unknown } y \text{ interval}}{\Delta y} = \frac{\text{known } x \text{ interval}}{\Delta x} \qquad (11\text{-}6)$$

$$\frac{0.83 - y}{0.83 - 0.67} = \frac{20 - 16}{20 - 10}$$

$$\Rightarrow \quad y = 0.76_6$$

For $x = 16$, our estimate of y is 0.76_6.

This calculation is equivalent to saying:

"16 is 60% of the way from 10 to 20, so the y value will be 60% of the way from 0.67 to 0.83."

| EXAMPLE Interpolating Activity Coefficients

Calculate the activity coefficient of H^+ when $\mu = 0.025$ M.

SOLUTION H^+ is the first entry in Table 11-1.

	$\mu = 0.01$	0.025	0.05
H^+:	$\gamma = 0.914$	?	0.86

The linear interpolation is set up as follows:

$$\frac{\text{unknown } \gamma \text{ interval}}{\Delta \gamma} = \frac{\text{known } \mu \text{ interval}}{\Delta \mu} \qquad \frac{0.86 - \gamma}{0.86 - 0.914} = \frac{0.05 - 0.025}{0.05 - 0.01}$$

$$\Rightarrow \quad \gamma = 0.89_4$$

Another Solution A more accurate and slightly more tedious calculation uses Equation 11-5, with the hydrated radius $\alpha = 900$ pm listed for H^+ in Table 11-1:

$$\log \gamma_{H^+} = \frac{(-0.51)(1^2)\sqrt{0.025}}{1 + (900\sqrt{0.025}/305)} = -0.054_{98}$$

$$\gamma_{H^+} = 10^{-0.054_{98}} = 0.88_1$$

The difference between this calculated value and the interpolated value is less than 2%. Equation 11-5 is easy to implement in a spreadsheet.

Activity Coefficients of Nonionic Compounds

Neutral molecules, such as benzene and acetic acid, have no ionic atmosphere because they have no charge. To a good approximation, their activity coefficients are unity when the ionic strength is less than 0.1 M. In this text, we set $\gamma = 1$ for neutral molecules. That is, *the activity of a neutral molecule will be assumed to be equal to its concentration.*

For neutral species, $\mathcal{A}_C \approx [C]$.

For gases such as H_2, the activity is written

$$\mathcal{A}_{H_2} = P_{H_2}\gamma_{H_2}$$

where P_{H_2} is pressure in bars. The activity of a gas is called its *fugacity,* and the activity coefficient is called the *fugacity coefficient.* Deviation of gas behavior from the ideal gas law results in deviation of the fugacity coefficient from unity. For most gases at or below 1 bar, $\gamma \approx 1$. Therefore, for all gases, *we will set $\mathcal{A} = P$ (bar).*

For gases, $\mathcal{A} \approx P$ (bar).

High Ionic Strengths

At high ionic strength, γ increases with increasing μ.

The extended Debye-Hückel equation 11-5 predicts that the activity coefficient, γ, will decrease as ionic strength, μ, increases. In fact, above an ionic strength of approximately 1 M, activity coefficients of most ions *increase,* as shown for H^+ in $NaClO_4$ solutions in Figure 11-5. We should not be too surprised that activity coefficients in concentrated salt solutions are not the same as those in dilute aqueous solution. The "solvent" is no longer just H_2O but, rather, a mixture of H_2O and $NaClO_4$. Hereafter, we limit our attention to dilute aqueous solutions in which Equation 11-5 applies.

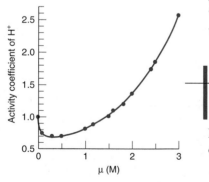

FIGURE 11-5 Activity coefficient of H^+ in solutions containing 0.010 0 M $HClO_4$ and various amounts of $NaClO_4$ [Data derived from L. Pezza, M. Molina, M. de Moraes, C. B. Melios, and J. O. Tognolli, *Talanta* **1996**, *43*, 1689.]

EXAMPLE   A Better Estimate of the Solubility of PbI_2

From the solubility product alone, you estimated in Ask Yourself 11-A that the concentration of dissolved iodine in a saturated solution of PbI_2 is 2.5 mM.

$$PbI_2(s) \xrightleftharpoons{K_{sp}} \underset{x}{Pb^{2+}} + \underset{2x}{2I^-}$$

$$x(2x)^2 = K_{sp} = 7.9 \times 10^{-9} \quad \Rightarrow \quad x = [Pb^{2+}] = 1.2_5 \times 10^{-3}\ M$$

$$2x = [I^-] = 2.5_0 \times 10^{-3}\ M$$

The observed concentration of dissolved iodine in the absence of KNO_3 in Figure 11-1 is 3.8 mM, which is 50% higher than the predicted concentration of I^- of

2.5 mM. The Pb^{2+} and I^- ions increase the ionic strength of the solution and therefore increase the solubility of PbI_2. Use activity coefficients to estimate the increased solubility.

SOLUTION The ionic strength of the solution is

$$\mu = \tfrac{1}{2}\{[Pb^{2+}] \cdot (+2)^2 + [I^-] \cdot (-1)^2\}$$
$$= \tfrac{1}{2}\{(0.001\,2_5 \cdot 4) + (0.002\,5_0 \cdot 1)\} = 0.003\,7_5\ M$$

If $\mu = 0.0037_5\ M$, interpolation in Table 11-1 tells us that the activity coefficients are $\gamma_{Pb^{2+}} = 0.781$ and $\gamma_{I^-} = 0.937$. A better estimate of the solubility of PbI_2 is obtained by using these activity coefficients in the solubility product expression (11-4):

$$K_{sp} = [Pb^{2+}]\gamma_{Pb^{2+}}[I^-]^2\gamma_{I^-}^2 = (x_2)(0.781)(2x_2)^2(0.937)^2$$
$$\Rightarrow\quad x_2 = [Pb^{2+}] = 1.4_2\ mM \text{ and } [I^-] = 2x_2 = 2.8_4\ mM$$

We wrote a subscript 2 in x_2 to indicate that it is our second approximation. The new concentrations of Pb^{2+} and I^- give a new estimate of the ionic strength, $\mu = 0.004\,2_6\ M$, which gives new activity coefficients: $\gamma_{Pb^{2+}} = 0.765$ and $\gamma_{I^-} = 0.932$. Repeating the solubility computation gives

$$K_{sp} = [Pb^{2+}]\gamma_{Pb^{2+}}[I^-]^2\gamma_{I^-}^2 = (x_3)(0.765)(2x_3)^2(0.932)^2$$
$$\Rightarrow\quad x_3 = [Pb^{2+}] = 1.4_4\ mM \text{ and } [I^-] = 2x_3 = 2.8_8\ mM$$

This third estimate is only slightly different from the second estimate. With activity coefficients, we estimate $[I^-] = 2.9\ mM$ instead of 2.5 mM calculated without activity coefficients. The remaining difference between 2.9 mM and the observed solubility of 3.8 mM is that we have not accounted for other species (PbI^+ and $PbI_2(aq)$) in the solution.

The Real Definition of pH

The pH measured by a pH electrode is not the negative logarithm of the hydrogen ion *concentration*. The ideal quantity that we measure with the pH electrode is the negative logarithm of the hydrogen ion *activity*.

Real definition of pH:
$$pH = -\log\mathcal{A}_{H^+} = -\log[H^+]\gamma_{H^+} \tag{11-7}$$

A pH electrode measures $-\log\mathcal{A}_{H^+}$.

Ask Yourself

11-B. Considering the equilibrium $H_2O \rightleftharpoons H^+ + OH^-$, for which $K_w = \mathcal{A}_{H^+}\mathcal{A}_{OH^-}$, what is the concentration of H^+ in pure water (in which the ionic strength is essentially 0) and in 0.1 M NaCl? What is the pH of each solution?

Difficult equilibrium problems can be tackled by writing all the relevant chemical equilibria plus two more equations: the balances of charge and mass. We now examine these two conditions.

Charge Balance

Solutions must have zero total charge.

The **charge balance** is an algebraic statement of electroneutrality: *The sum of the positive charges in solution equals the sum of the negative charges in solution.*

Suppose that a solution contains the following ionic species: H^+, OH^-, K^+, $H_2PO_4^-$, HPO_4^{2-}, and PO_4^{3-}. The charge balance is

$$[H^+] + [K^+] = [OH^-] + [H_2PO_4^-] + 2[HPO_4^{2-}] + 3[PO_4^{3-}] \tag{11-8}$$

$$\underbrace{}_{\text{Total positive charge}} \qquad \underbrace{\phantom{[OH^-] + [H_2PO_4^-] + 2[HPO_4^{2-}] + 3[PO_4^{3-}]}}_{\text{Total negative charge}}$$

The coefficient of each term in the charge balance equals the magnitude of the charge of each ion.

This statement says that the total charge contributed by H^+ and K^+ equals the magnitude of the charge contributed by all of the anions on the right side of the equation. *The coefficient in front of each species always equals the magnitude of the charge on the ion.* This statement is true because a mole of, say, PO_4^{3-} contributes three moles of negative charge. If $[PO_4^{3-}] = 0.01\,M$, the negative charge is $3[PO_4^{3-}] = 3(0.01) = 0.03$ M.

Equation 11-8 appears unbalanced to many people. "The right side of the equation has much more charge than the left side!" you might think. But you would be wrong.

For example, consider a solution prepared by weighing out 0.025 0 mol of KH_2PO_4 plus 0.030 0 mol of KOH and diluting to 1.00 L. The concentrations of the species at equilibrium are

$$[H^+] = 5.1 \times 10^{-12}\,M \qquad\qquad [H_2PO_4^-] = 1.3 \times 10^{-6}\,M$$
$$[K^+] = 0.0550\,M \qquad\qquad [HPO_4^{2-}] = 0.0220\,M$$
$$[OH^-] = 0.002\,0\,M \qquad\qquad [PO_4^{3-}] = 0.0030\,M$$

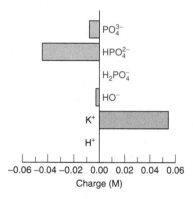

Are the charges balanced? Yes, indeed. Plugging into Equation 11-8, we find

$$[H^+] + [K^+] = [OH^-] + [H_2PO_4^-] + 2[HPO_4^{2-}] + 3[PO_4^{3-}]$$
$$5.1 \times 10^{-12} + 0.0550 = 0.0020 + 1.3 \times 10^{-6} + 2(0.0220) + 3(0.0030)$$
$$0.0550 = 0.0550$$

FIGURE 11-6   Charge contributed by each ion in 1.00 L of solution containing 0.025 0 mol KH_2PO_4 plus 0.030 0 mol KOH. The total positive charge equals the total negative charge.

The total positive charge is 0.055 0 M; the total negative charge is also 0.055 0 M (Figure 11-6). Charges must balance in every solution. Otherwise a beaker with excess positive charge would glide across the lab bench and smash into another beaker with excess negative charge.

The general form of the charge balance for any solution is

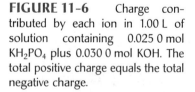

$\sum[\text{positive charges}] =$
$\qquad \sum[\text{negative charges}]$

Activity coefficients do not appear in the charge balance. The charge contributed by 0.1 M H^+ is *exactly* 0.1 M. Think about this.

Charge balance: $n_1[C_1] + n_2[C_2] + \ldots = m_1[A_1] + m_2[A_2] + \ldots$ \qquad (11-9) where [C] is the concentration of a cation, n is the charge of the cation, [A] is the concentration of an anion, and m is the magnitude of the charge of the anion.

EXAMPLE Writing a Charge Balance

Write the charge balance for a solution of lead(II) iodide containing the species Pb^{2+}, I^-, PbI^+, $PbI_2(aq)$, PbI_3^-, $PbOH^+$, H_2O, H^+, and OH^-.

SOLUTION The species $PbI_2(aq)$ and H_2O contribute no charge, so the charge balance is

$$2[Pb^{2+}] + [PbI^+] + [PbOH^+] + [H^+] = [I^-] + [PbI_3^-] + [OH^-]$$

Mass Balance

The **mass balance,** also called the *material balance,* is a statement of the conservation of matter. The mass balance states that *the quantity of all species in a solution containing a particular atom (or group of atoms) must equal the amount of that atom (or group) delivered to the solution.* Let's look at some examples.

Suppose that a solution is prepared by dissolving 0.050 mol of acetic acid in water to give a total volume of 1.00 L. The acetic acid partially dissociates into acetate:

$$\underset{\text{Acetic acid}}{CH_3CO_2H} \rightleftharpoons \underset{\text{Acetate}}{CH_3CO_2^-} + H^+$$

The mass balance states that the quantity of dissociated and undissociated acetic acid in the solution must equal the amount of acetic acid put into the solution.

Mass balance for acetic acid in water:
$$0.050\,M = \underset{\substack{\text{Undissociated} \\ \text{product}}}{[CH_3CO_2H]} + \underset{\substack{\text{Dissociated} \\ \text{product}}}{[CH_3CO_2^-]}$$
What we put into the solution

When a compound dissociates in several ways, the mass balance must include all the products. Phosphoric acid (H_3PO_4), for example, dissociates to $H_2PO_4^-$, HPO_4^{2-}, and PO_4^{3-}. The mass balance for a solution prepared by dissolving 0.025 0 mol of H_3PO_4 in 1.00 L is

$$0.0250\,M = [H_3PO_4] + [H_2PO_4^-] + [HPO_4^{2-}] + [PO_4^{3-}]$$

Now consider a solution prepared by dissolving PbI_2 in water to give Pb^{2+}, I^-, PbI^+, $PbI_2(aq)$, PbI_3^-, $PbOH^+$, H_2O, H^+, and OH^-. The source of all these species is PbI_2, so there must be two I atoms for every Pb atom in the solution. The mass balance is therefore

$$\underbrace{2\{[Pb^{2+}] + [PbI^+] + [PbOH^+] + [PbI_3^-]\}}_{2 \times \text{total concentration of Pb atoms}} =$$

$$\underbrace{[I^-] + [PbI^+] + 2[PbI_2(aq)] + 3[PbI_3^-]}_{\text{Total concentration of I atoms}} \quad (11\text{-}10)$$

The mass balance is a statement of the conservation of matter. It really refers to conservation of atoms, not to mass.

Activity coefficients do not appear in the mass balance. The concentration of each species counts exactly the number of atoms of that species.

We do not know the total concentration of either side of the equation because we do not know how much PbI_2 dissolved. However, we do know that there must be two I atoms for every Pb atom in the solution.

On the right side of Equation 11-10, there is a 2 in front of $[PbI_2(aq)]$ because each mole of $PbI_2(aq)$ contains two moles of I atoms. There is a 3 in front of $[PbI_3^-]$ because each mole of PbI_3^- contains three moles of I atoms.

Ask Yourself

11-C. Consider a buffer solution prepared by mixing 5.00 mmol of $Na_2C_2O_4$ (sodium oxalate) with 2.50 mmol of HCl in 0.100 L.

   (a) List all the chemical species in the solution. Oxalate can accept 1 or 2 protons.
   (b) What is the charge balance for the solution?
   (c) What mass balances can you write for the solution?
   (d) Decide which species are negligible and simplify the expressions in (b) and (c).

11-4   Systematic Treatment of Equilibrium

Now that we have learned about the charge and mass balances, we are ready for the systematic treatment of equilibrium. The general prescription follows these steps:

Step 1. Write the *pertinent reactions.*

Step 2. Write the *charge balance* equation.

Step 3. Write *mass balance* equations. There may be more than one.

Step 4. Write the *equilibrium constant* for each chemical reaction. This step is the only one in which activity coefficients enter.

Step 5. *Count the equations and unknowns.* At this point, you should have as many equations as unknowns (chemical concentrations). If not, you must either find more equilibria or fix some concentrations at known values.

Step 6. By hook or by crook, *solve* for all the unknowns.

Steps 1 and 6 are the heart of the problem. Guessing what chemical equilibria exist in a given solution requires a fair degree of chemical intuition. In this text, you will usually be given help with step 1. Unless we know all the relevant equilibria, it is not possible to calculate the composition of a solution correctly. Because we do not know all the chemical reactions, we undoubtedly oversimplify many equilibrium problems.

Step 6 is likely to be your biggest challenge. With n equations involving n unknowns, the problem can always be solved, at least in principle. In the simplest cases, you can do this by hand; but for most problems, approximations are made or a spreadsheet is employed.

A Simple Example: The pH of $10^{-8}M$ KOH

Here is a trick question we can now consider. When asked "What is the pH of 1.0×10^{-8} M KOH?" your first response might be $[OH^-] = 1.0 \times 10^{-8}$ M, so

$[H^+] = 1.0 \times 10^{-6}$ M, so the pH is 6.00. But this cannot be so, because adding base to a neutral solution could not possibly make it acidic. So let's see how the systematic treatment of equilibrium works in this case.

Step 1. *Pertinent reactions:* The only one is $H_2O \rightleftharpoons H^+ + OH^-$, which exists in every aqueous solution.

Step 2. *Charge balance:* The ions are K^+, H^+, and OH^-, so $[K^+] + [H^+] = [OH^-]$.

Step 3. *Mass balance:* You might be tempted to write $[K^+] = [OH^-]$, but this is false because OH^- comes from both KOH and H_2O. For every mole of K^+, one mole of OH^- is introduced into the solution. We also know that for every mole of H^+ from H_2O, one mole of OH^- is introduced. Therefore one mass balance is $[OH^-] = [K^+] + [H^+]$, which is the same as the charge balance in this particularly simple example. A second mass balance is $[K^+] = 1.00 \times 10^{-8}$ M.

Step 4. *Equilibrium constants:* The only one is $K_w = [H^+]\gamma_{H^+}[OH^-]\gamma_{OH^-}$.

Step 5. *Count equations and unknowns:* At this point, you should have as many equations as unknowns (chemical species). There are three unknowns, $[K^+]$, $[H^+]$, and $[OH^-]$, and three equations:

If you have fewer equations than unknowns, look for another mass balance or a chemical equilibrium that you overlooked.

Charge balance: $[K^+] + [H^+] = [OH^-]$
Mass balance: $[K^+] = 1.0 \times 10^{-8}$ M
Equilibrium constant: $K_w = [H^+]\gamma_{H^+}[OH^-]\gamma_{OH^-}$

Step 6. *Solve:* The ionic strength must be very low in this solution (someplace near 10^{-7} M), so it is safe to say that the activity coefficients are 1.00. Substituting 1.0×10^{-8} M for $[K^+]$ and $[OH^-] = K_w/[H^+]$ into the charge balance gives

$$[1.0 \times 10^{-8}] + [H^+] = K_w/[H^+]$$

Multiplying both sides by $[H^+]$ gives a quadratic equation

$$[1.0 \times 10^{-8}][H^+] + [H^+]^2 = K_w$$
$$[H^+]^2 + [1.0 \times 10^{-8}][H^+] - K_w = 0$$

whose two solutions are $[H^+] = 9.6 \times 10^{-8}$ and -1.1×10^{-7} M. Rejecting the negative solution (because the concentration cannot be negative), we find

$$pH = -\log[H^+]\gamma_{H^+} = -\log[9.6 \times 10^{-8}](1.00) = 7.02$$

It should not be too surprising that the pH is close to 7 and very slightly basic.

Coupled Equilibria: Solubility of CaF_2

The opening of this chapter gave an example of *coupled equilibria* in which calcium carbonate dissolves and the resulting bicarbonate reacts with H^+. The second reaction drives the first reaction forward.

Now we look at a similar case in which CaF_2 dissolves in water:

$$CaF_2(s) \xrightarrow{K_{sp} = 3.9 \times 10^{-11}} Ca^{2+} + 2F^- \qquad (11\text{-}11)$$

If we were very smart, we might also write the reaction

$$Ca^{2+} + OH^- \overset{K = 20}{\rightleftharpoons} CaOH^+$$

It turns out that this reaction is important only at high pH.

The fluoride ion can then react with water to give HF(aq):

$$F^- + H_2O \overset{K_b = 1.5 \times 10^{-11}}{\rightleftharpoons} HF + OH^- \qquad (11\text{-}12)$$

Also, for every aqueous solution, we can write

$$H_2O \overset{K_w}{\rightleftharpoons} H^+ + OH^- \qquad (11\text{-}13)$$

If Reaction 11-12 occurs, then the solubility of CaF_2 is greater than that predicted by the solubility product because that F^- produced in Reaction 11-11 is consumed in Reaction 11-12. According to Le Châtelier's principle, Reaction 11-11 will be driven to the right. The systematic treatment of equilibrium allows us to find the net effect of all three reactions.

Step 1. *Pertinent reactions:* The three reactions are (11-11) to (11-13).

Step 2. *Charge balance:* $[H^+] + 2[Ca^{2+}] = [OH^-] + [F^-]$ $\qquad (11\text{-}14)$

Step 3. *Mass balance:* If all fluoride remained in the form F^-, we could write $[F^-] = 2[Ca^{2+}]$ from the stoichiometry of Reaction 11-11. But some F^- reacts to give HF. The total moles of fluorine atoms is equal to the sum of F^- plus HF, and the mass balance is

$$\underbrace{[F^-] + [HF]}_{\substack{\text{Total concentration} \\ \text{of fluorine atoms}}} = 2[Ca^{2+}] \qquad (11\text{-}15)$$

Step 4. *Equilibrium constants:*

$$K_{sp} = [Ca^{2+}]\gamma_{Ca^{2+}}[F^-]^2\gamma_{F^-}^2 = 3.9 \times 10^{-11} \qquad (11\text{-}16)$$

$$K_b = \frac{[HF]\gamma_{HF}[OH^-]\gamma_{OH^-}}{[F^-]\gamma_{F^-}} = 1.5 \times 10^{-11} \qquad (11\text{-}17)$$

$$K_w = [H^+]\gamma_{H^+}[OH^-]\gamma_{OH^-} = 1.0 \times 10^{-14} \qquad (11\text{-}18)$$

For simplicity, we are generally going to ignore the activity coefficients.

Although we wrote activity coefficients in the equilibrium equations, we are not so masochistic as to use them in the rest of the problem. At this point, we are going to explicitly ignore the activity coefficients, which is equivalent to saying that they are unity. There will be some inaccuracy in the results, but you could go back after the calculation and compute the ionic strength and the activity coefficients and find a better approximation for the solution if you had to.

Step 5. *Count equations and unknowns:* There are five equations (11-14 through 11-18) and five unknowns: $[H^+]$, $[OH^-]$, $[Ca^{2+}]$, $[F^-]$, and $[HF]$.

Step 6. *Solve:* This is no simple matter for these five equations. Instead, let us ask a simpler question: What will be the concentrations of $[Ca^{2+}]$, $[F^-]$, and $[HF]$ if the pH is *fixed* at the value 3.00 by adding a buffer?

Once we know that $[H^+] = 1.0 \times 10^{-3}$ M, there is a straightforward procedure for solving the equations. From Equation 11-18, we know that $[OH^-] = K_w/[H^+] = 1.0 \times 10^{-11}$ M. Putting this value of $[OH^-]$ into Equation 11-17 gives

$$\frac{[HF]}{[F^-]} = \frac{K_b}{[OH^-]} = \frac{1.5 \times 10^{-11}}{1.0 \times 10^{-11}} = 1.5$$

$$\Rightarrow \quad [HF] = 1.5[F^-]$$

Substituting this expression for [HF] into the mass balance (Equation 11-15) gives

$$[F^-] + [HF] = 2[Ca^{2+}] \qquad (11\text{-}15)$$
$$[F^-] + 1.5[F^-] = 2[Ca^{2+}]$$
$$[F^-] = 0.80[Ca^{2+}]$$

Finally, we use this expression for $[F^-]$ in the solubility product (Equation 11-16):

$$[Ca^{2+}][F^-]^2 = K_{sp}$$
$$[Ca^{2+}](0.80[Ca^{2+}])^2 = K_{sp}$$
$$[Ca^{2+}] = \left(\frac{K_{sp}}{0.80^2}\right)^{1/3} = 3.9 \times 10^{-4}\ M$$

Challenge Use the concentration of Ca^{2+} calculated here to show that the concentrations $[F^-]$ and $[HF]$ are $3.1 \times 10^{-4}\ M$ and $4.7 \times 10^{-4}\ M$, respectively.

You should realize that the charge balance equation (11-14) is no longer valid if the pH is fixed by external means. To adjust the pH, an ionic compound must necessarily have been added to the solution. Equation 11-14 is therefore incomplete, because it omits those ions. However, we did not use Equation 11-14 to solve the problem because we omitted $[H^+]$ as a variable when we fixed the pH.

If we had selected a pH other than 3.00, we would have found a different set of concentrations because of the coupling of Reactions 11-11 and 11-12. Figure 11-7 shows the pH dependence of the concentrations of Ca^{2+}, F^-, and HF. At high pH, there is very little HF, so $[F^-] \approx 2[Ca^{2+}]$. At low pH, there is very little F^-, so $[HF] \approx 2[Ca^{2+}]$. The concentration of Ca^{2+} increases at low pH, because Reaction 11-11 is drawn to the right by the reaction of F^- with H_2O to make HF in Reaction 11-12.

In general, many minerals are more soluble at low pH because the anions react with acid.[3] Box 11-1 describes an environmental consequence of solubility in acids.

Fixing the pH invalidates the original charge balance because we added unspecified ions to the solution to fix the pH. There exists a new charge balance, but we do not know enough to write an equation for it.

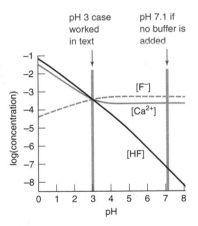

FIGURE 11-7 pH dependence of the concentrations of Ca^{2+}, F^-, and HF in a saturated solution of CaF_2. As the pH is lowered, H^+ reacts with F^- to make HF, and the concentration of Ca^{2+} increases. Note the logarithmic ordinate.

Ask Yourself

11-D. **(a)** Ignoring activity coefficients, find the concentrations of Ag^+, CN^-, and HCN in a saturated solution of AgCN, whose pH is fixed at 9.00. Consider the equilibria:

$$AgCN(s) \rightleftharpoons Ag^+ + CN^- \qquad K_{sp} = 2.2 \times 10^{-16}$$
$$CN^- + H_2O \rightleftharpoons HCN(aq) + OH^- \qquad K_b = 1.6 \times 10^{-5}$$

(b) What would be the mass balance if the following equilibria also occur?

$$Ag^+ + CN^- \rightleftharpoons AgCN(aq) \qquad AgCN(aq) + CN^- \rightleftharpoons Ag(CN)_2^-$$
$$Ag^+ + H_2O \rightleftharpoons AgOH(aq) + H^+$$

Box 11-1 *Informed Citizen*

**Aluminum Mobilization from Minerals
by Acid Rain**

Aluminum is the third most abundant element on earth (after oxygen and silicon), but it is tightly locked into insoluble minerals such as kaolinite $(Al_2(OH)_4Si_2O_5)$ and bauxite $(AlOOH)$. Acid rain from human activities is a recent change in the history of the earth and it is introducing soluble forms of aluminum (and lead and

mercury) into the environment.[4] The graph shows that at a pH below 5, aluminum is mobilized from minerals and its concentration in lake water rises rapidly. At a concentration of $130\,\mu g/L$, aluminum kills fish. In humans, high concentrations of aluminum cause dementia, softening of bones, and anemia.

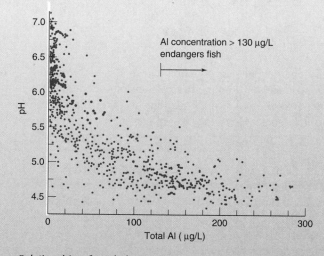

Relationship of total aluminum (including dissolved and suspended species) in 1 000 Norwegian lakes as a function of the pH of the lake water. The more acidic the water, the greater the aluminum concentration. [From G. Howells, *Acid Rain and Acid Waters*, 2nd ed. (Hertfordshire, Ellis Horwood, 1995).]

11-5 Fractional Composition Equations

As a final topic in our study of chemical equilibrium, we derive expressions for the fractions of a weak acid, HA, in each form (HA and A^-). These equations were already used in Section 9-7 for an acid-base titration spreadsheet. In Equation 7-15, we defined the *fraction of dissociation* as

$$\text{fraction of HA in the form A}^- \equiv \alpha_{A^-} = \frac{[A^-]}{[A^-] + [HA]} \qquad (11\text{-}19)$$

Similarly, we define the fraction in the form HA as

$$\text{fraction of HA in the form HA} \equiv \alpha_{HA} = \frac{[HA]}{[A^-] + [HA]} \qquad (11\text{-}20)$$

Consider an acid with formal concentration F:

$$HA \xrightleftharpoons{K_a} H^+ + A^- \qquad K_a = \frac{[H^+][A^-]}{[HA]}$$

The mass balance is simply

$$F = [HA] + [A^-]$$

Rearranging the mass balance gives $[A^-] = F - [HA]$, which can be plugged into the K_a equilibrium to give

$$K_a = \frac{[H^+](F - [HA])}{[HA]}$$

or, with a little algebra,

$$[HA] = \frac{[H^+]F}{[H^+] + K_a} \qquad (11\text{-}21)$$

Rearranging Equation 11-21 gives the fraction α_{HA}:

Fraction in the form HA: $\qquad \alpha_{HA} = \frac{[HA]}{F} = \frac{[H^+]}{[H^+] + K_a} \qquad (11\text{-}22)$

If we substitute $[HA] = F - [A^-]$ into the K_a equation, we could rearrange and solve for the fraction α_{A^-}:

Fraction in the form A⁻: $\qquad \alpha_{A^-} = \frac{[A^-]}{F} = \frac{K_a}{[H^+] + K_a} \qquad (11\text{-}23)$

Figure 11-8 shows α_{HA} and α_{A^-} for a system with $pK_a = 5.00$. At low pH, almost all the acid is in the form HA. At high pH, almost everything is in the form A^-. HA is the predominant species when $pH < pK_a$. A^- is the predominant species when $pH > pK_a$. Figure 10-1 is an analogous diagram of species in a diprotic system. At low pH, α_{H_2A} approaches 1; and at high pH, $\alpha_{A^{2-}}$ approaches 1. At intermediate pH, α_{HA^-} is the largest fraction.

If you were dealing with the conjugate pair BH^+ and B instead of HA and A^-, Equation 11-22 gives the fraction in the form BH^+ and Equation 11-23 gives the fraction in the form B. In this case, K_a is the acid dissociation constant for BH^+ (which is K_w/K_b).

α_{HA} = fraction of species in the form HA

α_{A^-} = fraction of species in the form A^-

$\alpha_{HA} + \alpha_{A^-} = 1$

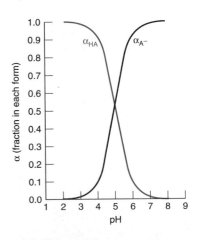

FIGURE 11-8 Fractional composition diagram of a monoprotic system with $pK_a = 5.00$. Below pH 5, HA is the dominant form; whereas, above pH 5, A^- dominates. Figure 10-1 showed the analogous plot for a diprotic system.

EXAMPLE   **Fractional Composition for an Acid**

pK_a for benzoic acid (HA) is 4.20. Find the concentration of A^- at pH 5.31 if the formal concentration of HA is 0.021 3 M.

SOLUTION At pH = 5.31, $[H^+] = 10^{-5.31} = 4.9 \times 10^{-6}$ M.

$$\alpha_{A^-} = \frac{K_a}{[H^+] + K_a} = \frac{6.3 \times 10^{-5}}{(4.9 \times 10^{-6}) + (6.3 \times 10^{-5})} = 0.92_8$$

From Equation 11-23, we can say

$$[A^-] = \alpha_{A^-}F = (0.92_8)(0.0213) = 0.020 \text{ M}$$

EXAMPLE Fractional Composition for a Base

K_a for the ammonium ion, NH_4^+, is 5.70×10^{-10} ($pK_a = 9.244$). Find the fraction in the form BH^+ at pH 10.38.

SOLUTION At pH = 10.38, $[H^+] = 10^{-10.38} = 4.1_7 \times 10^{-11}$ M. Using Equation 11-22, with BH^+ in place of HA, we find

$$\alpha_{BH^+} = \frac{[H^+]}{[H^+] + K_a} = \frac{4.1_7 \times 10^{-11}}{(4.1_7 \times 10^{-11}) + (5.70 \times 10^{-10})} = 0.068$$

Ask Yourself

11-E. The acid HA has $pK_a = 3.00$. Find the fraction in the form HA and the fraction in the form A^- at pH = 2.00, 3.00, and 4.00. Compute the quotient $[HA]/[A^-]$ at each pH.

Key Equations

Activity	$\mathcal{A}_C = [C]\gamma_C$ $\mathcal{A}$ = activity; γ = activity coefficient
Equilibrium constant	For the reaction $aA + bB \rightleftharpoons cC + dD$,

$$K = \frac{\mathcal{A}_C^c \mathcal{A}_D^d}{\mathcal{A}_A^a \mathcal{A}_B^b} = \frac{[C]^c\gamma_C^c[D]^d\gamma_D^d}{[A]^a\gamma_A^a[B]^b\gamma_B^b}$$

Ionic strength

$$\mu = \frac{1}{2}(c_1z_1^2 + c_2z_2^2 + \ldots) = \frac{1}{2}\sum_i c_iz_i^2$$

c = concentration; z = charge

Extended Debye-Hückel equation

$$\log\gamma = \frac{-0.51z^2\sqrt{\mu}}{(1 + (\alpha\sqrt{\mu}/305)}$$

z = charge; μ = ionic strength; α = hydrated radius

Linear interpolation

$$\frac{\text{unknown } y \text{ interval}}{\Delta y} = \frac{\text{known } x \text{ interval}}{\Delta x}$$

pH

$$pH = -\log\mathcal{A}_{H^+} = -\log[H^+]\gamma_{H^+}$$

Charge balance

positive charge in solution = negative charge in solution

Mass balance | Quantity of all species in a solution containing a particular atom (or group of atoms) must equal the amount of that atom (or group) delivered to the solution.

Systematic treatment of equilibrium

1. Pertinent reactions   4. Equilibrium constants
2. Charge balance   5. Count equations/unknowns
3. Mass balance   6. Solve

Fraction of HA in acidic form

$$\alpha_{HA} = \frac{[HA]}{F} = \frac{[H^+]}{[H^+] + K_a} = \alpha_{BH^+} = \frac{[BH^+]}{F}$$

Fraction of HA in basic form

$$\alpha_{A^-} = \frac{[A^-]}{F} = \frac{K_a}{[H^+] + K_a} = \alpha_B = \frac{[B]}{F}$$

Important Terms

activity
activity coefficient
charge balance

extended Debye-Hückel equation
hydrated radius
interpolation

ionic atmosphere
ionic strength
mass balance

Problems

11-1. What is an ionic atmosphere?

11-2. Explain why the solubility of an ionic compound increases as the ionic strength of the solution increases (at least up to ~0.5 M).

11-3. The figure shows the quotient of concentrations $[CH_3CO_2^-][H^+]/[CH_3CO_2H]$ for the dissociation of acetic acid as a function of the concentration of KCl added to the solution. Explain the shape of the curve.

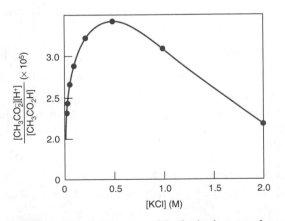

11-4. Which statements are true? In the ionic strength range 0–0.1 M, activity coefficients decrease with **(a)** increasing ionic strength; **(b)** increasing ionic charge; **(c)** decreasing hydrated radius.

11-5. Explain the following observations:

(a) Mg^{2+} has a greater hydrated radius than Ba^{2+}.

(b) Hydrated radii decrease in the order $Sn^{4+} > In^{3+} > Cd^{2+} > Rb^+$.

(c) H^+ has a hydrated radius of 900 pm, whereas that of OH^- is just 350 pm.

11-6. State in words the meaning of the charge balance equation.

11-7. State the meaning of the mass balance equation.

11-8. Why does the solubility of a salt of a basic anion increase with decreasing pH? Write chemical reactions for the minerals galena (PbS) and cerussite ($PbCO_3$) to explain how acid rain mobilizes trace quantities of toxic metallic elements from relatively inert forms into the environment, where the metals can be taken up by plants and animals. Why are the minerals kaolinite and bauxite in Box 11-1 more soluble in acidic solution than in neutral solution?

11-9. Assuming complete dissociation of the salts, calculate the ionic strength of **(a)** 0.2 mM KNO_3; **(b)** 0.2 mM Cs_2CrO_4; **(c)** 0.2 mM $MgCl_2$ plus 0.3 mM $AlCl_3$.

11-10. Find the activity coefficient of each ion at the indicated ionic strength:

(a) SO_4^{2-} ($\mu = 0.01\,M$)

(b) Sc^{3+} ($\mu = 0.005\,M$)

(c) Eu^{3+} $(\mu = 0.1\,M)$
(d) $(CH_3CH_2)_3NH^+$ $(\mu = 0.05\,M)$

11-11. Find the activity (not the activity coefficient) of the $(C_3H_7)_4N^+$ (tetrapropylammonium) ion in a solution containing 0.005 0 M $(C_3H_7)_4N^+Br^-$ plus 0.005 0 M $(CH_3)_4N^+Cl^-$.

11-12. Interpolate in Table 11-1 to find the activity coefficient of H^+ when $\mu = 0.030\,M$.

11-13. Calculate the activity coefficient of Zn^{2+} when $\mu = 0.083\,M$ by using **(a)** Equation 11-5; **(b)** linear interpolation with Table 11-1.

11-14. Using activities, find the concentration of Ag^+ in a saturated solution of AgSCN in **(a)** 0.060 M KNO_3; **(b)** 0.060 M KSCN.

11-15. Find the activity coefficient of H^+ in a solution containing 0.010 M HCl plus 0.040 M $KClO_4$. What is the pH of the solution?

11-16. Using activities, calculate the pH and concentration of H^+ in pure water containing 0.050 M LiBr at 25°C.

11-17. Using activities, calculate the pH of a solution containing 0.010 M NaOH plus 0.012 0 M $LiNO_3$. What would be the pH if you neglected activities?

11-18. Using activities, find the concentration of OH^- in a solution of 0.075 M $NaClO_4$ saturated with $Mn(OH)_2$. What is the pH of this solution?

11-19. Using activities, find the concentration of Ba^{2+} in a 0.100 M $(CH_3)_4N^+IO_3^-$ solution saturated with $Ba(IO_3)_2$.

11-20. Using activities, calculate $[Pb^{2+}]$ in a saturated solution of PbF_2 in water.

11-21. (a) Using activities and K_{sp} for $CaSO_4$, calculate the concentration of dissolved Ca^{2+} in a saturated aqueous solution of $CaSO_4$.
(b) The observed total concentration of dissolved calcium is 15–19 mM. Explain.

11-22. Write a charge balance for a solution containing H^+, OH^-, Ca^{2+}, HCO_3^-, CO_3^{2-}, $Ca(HCO_3)^+$, $Ca(OH)^+$, K^+, and ClO_4^-.

11-23. Write a charge balance for a solution of H_2SO_4 in water if the H_2SO_4 ionizes to HSO_4^- and SO_4^{2-}.

11-24. Write the charge balance for an aqueous solution of arsenic acid, H_3AsO_4, in which the acid can dissociate to $H_2AsO_4^-$, $HAsO_4^{2-}$, and AsO_4^{3-}. Look up the structure of arsenic acid in Appendix B and write the structure of $HAsO_4^{2-}$.

11-25. (a) Suppose that $MgBr_2$ dissolves to give Mg^{2+} and Br^-. Write a charge balance equation for this aqueous solution.

(b) What is the charge balance if, in addition to Mg^{2+} and Br^-, $MgBr^+$ is formed?

11-26. For a 0.1 M aqueous solution of sodium acetate, $Na^+CH_3CO_2^-$, one mass balance is simply $[Na^+] = 0.1$ M. Write a mass balance involving acetate.

11-27. Suppose that $MgBr_2$ dissolves to give Mg^{2+} and Br^-.
(a) Write the mass balance for Mg^{2+} for 0.20 M $MgBr_2$.
(b) Write a mass balance for Br^- for 0.20 M $MgBr_2$.
Now suppose that $MgBr^+$ is formed in addition to Mg^{2+} and Br^-.
(c) Write a mass balance for Mg^{2+} for 0.20 M $MgBr_2$.
(d) Write a mass balance for Br^- for 0.20 M $MgBr_2$.

11-28. (a) Write the mass balance for CaF_2 in water if the reactions are $CaF_2(s) \rightleftharpoons Ca^{2+} + 2F^-$ and $F^- + H^+ \rightleftharpoons HF(aq)$
(b) Write a mass balance for CaF_2 in water if, in addition to the previous reactions, the following reaction occurs: $HF(aq) + F^- \rightleftharpoons HF_2^-$.

11-29. Write a mass balance for an aqueous solution of $Ca_3(PO_4)_2$, if the aqueous species are Ca^{2+}, PO_4^{3-}, HPO_4^{2-}, $H_2PO_4^-$, and H_3PO_4.

11-30. Consider the dissolution of the compound X_2Y_3, which gives $X_2Y_2^{2+}$, X_2Y^{4+}, $X_2Y_3(aq)$, and Y^{2-}. Use the mass balance to find an expression for $[Y^{2-}]$ in terms of the other concentrations. Simplify your answer as much as possible.

11-31. Ignoring activity coefficients, calculate the concentration of each ion in a solution of $4.0 \times 10^{-8}\,M\ Mg(OH)_2$, which is completely dissociated to Mg^{2+} and OH^-.

11-32. Consider a saturated solution of $R_3NH^+Br^-$, where R is an organic group. Find the solubility (mol/L) of $R_3NH^+Br^-$ in a solution maintained at pH 9.50.

$$R_3NH^+Br^-(s) \rightleftharpoons R_3NH^+ + Br^- \quad K_{sp} = 4.0 \times 10^{-8}$$
$$R_3NH^+ \rightleftharpoons R_3N + H^+ \quad K_a = 2.3 \times 10^{-9}$$

11-33. (a) Ignoring activity coefficients, find the concentrations of Ag^+, CN^-, and HCN in a saturated solution of AgCN whose pH is somehow fixed at 9.00. Consider the following equilibria:

$$AgCN(s) \rightleftharpoons Ag^+ + CN^- \quad K_{sp} = 2.2 \times 10^{-16}$$
$$CN^- + H_2O \rightleftharpoons HCN(aq) + OH^- \quad K_b = 1.6 \times 10^{-5}$$

(b) *Activity problem.* Use activity coefficients to answer **(a)**. Assume that the ionic strength is fixed at 0.10 M by addition of an inert salt. When using activities, the statement that the pH is 9.00 means that $-\log[H^+]\gamma_{H^+} = 9.00$.

11-34. (a) How many moles of PbO will dissolve in a 1.00-L solution if the pH is fixed at 10.50? Consider the equilibrium

$$PbO(s) + H_2O \rightleftharpoons Pb^{2+} + 2OH^- \qquad K = 5.0 \times 10^{-16}$$

(b) Answer the same question asked in (a), but also consider the reaction

$$Pb^{2+} + H_2O \rightleftharpoons PbOH^+ + H^+ \qquad K_a = 2.5 \times 10^{-8}$$

(c) *Activity problem.* Answer (a) by using activity coefficients, assuming the ionic strength is fixed at 0.050 M.

11-35. Find the fraction of pyridine (B) in the form B and the fraction in the form BH$^+$ at pH = (a) 4.00; (b) 5.00; (c) 6.00.

11-36. Find the fraction of 1-naphthoic acid in the form HA and the fraction in the form A$^-$ at pH = (a) 2.00; (b) 3.00; (c) 3.50.

11-37. The base B has $pK_b = 4.00$. Find the fraction in the form B and the fraction in the form BH$^+$ at pH = (a) 9.00; (b) 10.00; (c) 10.30.

11-38. Create a spreadsheet using Equations 11-22 and 11-23 to compute the concentrations of HA and A$^-$ in a 0.200 M solution of hydroxybenzene as a function of pH from pH 2 to pH 12. Use graphics software to plot the curves in a beautifully labeled figure.

How Would You Do It?

11-39. Figure 11-1 shows student data for the total concentration of dissolved iodine in solutions saturated with $PbI_2(s)$ in the presence of added KNO_3. Dissolved iodine is present as iodide ion (I$^-$) or as iodide attached to Pb^{2+}. Dissolved iodine was measured by adding nitrite to convert iodide to iodine (I$_2$):

$$2I^- + 2NO_2^- + 4H^+ \longrightarrow I_2(aq) + 2NO + 2H_2O$$
Iodide   Nitrite   Iodine   Nitric oxide
(colorless)   (orange-brown)
$$K = 5 \times 10^{15}$$

Only the product I$_2$ is colored, so it can be measured by its absorption of visible light. To collect the data in Figure 11-1, excess $PbI_2(s)$ was shaken with solutions containing various concentrations of KNO_3. The solutions were then centrifuged and the clear supernatant liquid was removed by pipet for analysis by reaction with nitrite.

By assuming that the only dissolved species were Pb^{2+} and I$^-$, and taking activity coefficients into account, the people who measured the concentration of dissolved iodine calculated that the solubility product for PbI$_2$ is 1.64×10^{-8}. This number is considerably higher than the value 7.9×10^{-9} in Appendix A because no account was made for species such as PbI$^+$, $PbI_2(aq)$, PbI$_3^-$, and PbOH$^+$.

(a) Write the chemical reaction that produces PbOH$^+$ and propose an experiment to measure the concentration of this species, using a pH meter.

(b) I can't think of a way to distinguish the species PbI$^+$ from I$^-$ by measuring total dissolved iodine. Can you propose a different kind of experiment to measure the equilibrium constant for formation of PbI$^+$?

Notes and References

1. The primary source of equilibrium constants for serious work is the computer database compiled by R. M. Smith, A. E. Martell, and R. J. Motekaitis, *NIST Critical Stability Constants of Metal Complexes Database 46,* (Gaithersburg, MD: National Institute of Standards & Technology, 1995). Compilations of equilibrium constants are found in L. G. Sillén and A.E. Martell, *Stability Constants of Metal-Ion Complexes* (London: Chemical Society, Special Publications No. 17 and 25, 1964 and 1971); and A. E. Martell and R. M. Smith, *Critical Stability Constants* (New York: Plenum Press, 1974–1989), a multivolume collection.

2. D. R. Driscol, *J. Chem. Ed.* **1979,** *56,* 603. See also R. W. Ramette, *J. Chem. Ed.* **1963,** *40,* 252.

3. For a general spreadsheet approach to computing the solubility of salts in which the anion and cation can react with water, see J. L. Guiñón, J. Garcia-Antón, and V. Pérez-Herranz, *J. Chem. Ed.* **1999,** *76,* 1157.

4. R. B. Martin, *Acc. Chem. Res.* **1994,** *27,* 204.

Further Reading

W. B. Guenther, *Unified Equilibrium Calculations* (New York: Wiley, 1991).

J. N. Butler, *Ionic Equilibrium: Solubility and pH Calculations* (New York: Wiley, 1998).

A. Martell and R. Motekaitis, *Determination and Use of Stability Constants* (New York: VCH Publishers, 1992).

M. Meloun, *Computation of Solution Equilibria* (New York: Wiley, 1988).

An Antibiotic Chelate Captures Its Prey

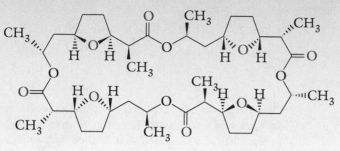

Structure of nonactin with ligand atoms highlighted

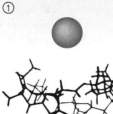

① ② ③

④ ⑤ ⑥

Nonactin molecule envelops a K⁺ ion. [From T. J. Marrone and K. M. Merz, Jr., *J. Am. Chem. Soc.* **1992,** *114,* 7542.] The term "chelate" is derived from the great claw, or *chela* (from the Greek *chēlē*), of the lobster.[1] Frame 3 at the right is reminiscent of a lobster reaching out with its claw to grab the metal ion.

A as a nerve impulse passes along an axon, the *hydrophilic* ("water-loving") ions, K^+ and Na^+, must cross the *hydrophobic* ("water-hating") cell membrane. Carrier molecules with *polar* interiors and *nonpolar* exteriors perform this function. (Polar compounds have positive and negative regions that attract neighboring molecules by electrostatic forces. Nonpolar compounds do not have charged regions and are soluble inside the nonpolar cell membrane.)

A class of antibiotics called *ionophores,* including nonactin, gramicidin, and nigericin, alter the permeability of bacterial cells to metal ions and thus disrupt their metabolism. Ionophores are said to be *chelates* (pronounced KEE-lates), which means that they bind metal ions through more than one atom of the ionophore. One molecule of nonactin locks onto a K^+ ion through eight oxygen atoms. The sequence in the figure shows how the conformation of nonactin changes as it wraps itself around a K^+ ion. The outside of the K^+-nonactin complex is mainly hydrocarbon, so the complex easily passes through the nonpolar cell membrane.

Chapter 12

EDTA TITRATIONS

*E*DTA is a merciful abbreviation for *ethylenediaminetetraacetic acid,* a compound that can be used to titrate most metal ions by forming strong 1:1 complexes (Figure 12-1). In addition to its place in chemical analysis, EDTA plays a larger role as a strong metal-binding agent in industrial processes and in household products such as soaps, cleaning agents, and food additives that prevent metal-catalyzed oxidation of food. EDTA is an emerging player in environmental chemistry.[2] For example, the majority of nickel discharged into San Francisco Bay, which passes unscathed through wastewater treatment plants, is nickel-EDTA. A significant fraction of the iron, lead, copper, and zinc flowing into the bay are also EDTA complexes.

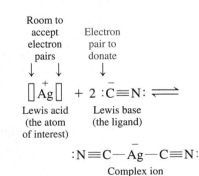

Lewis acid: electron pair acceptor
Lewis base: electron pair donor

12-1 Metal–Chelate Complexes

An atom or group of atoms bound to whatever atom you are interested in is called a **ligand.** A ligand with a pair of electrons to share can bind to a metal ion that can accept a pair of electrons. Electron pair acceptors are called **Lewis acids** and electron pair donors are called **Lewis bases.** Cyanide is said to be a **monodentate** ("one-toothed") **ligand** because it binds to a metal ion through only one atom (the carbon atom).

255

Box 12-1 *Informed Citizen*

Chelation Therapy and Thalassemia

Oxygen is carried in the human circulatory system bound to iron in the protein hemoglobin, which consists of two pairs of subunits, designated α and β. β-Thalassemia major is a genetic disease in which β subunits are not synthesized in adequate quantities. Children afflicted with this disease survive only with frequent transfusions of normal red blood cells. However, the child accumulates 4–8 g of iron per year from hemoglobin in the transfused cells. Our bodies have no mechanism for excreting such large quantities of iron, and most patients die by age 20 from the toxic effects of iron overload. One mechanism by which iron is so toxic is that it catalyzes the formation of hydroxyl radical (HO·), a powerful and destructive oxidant.

To enhance excretion of iron, intensive chelation therapy is used. The most successful drug is *desferrioxamine B*, a powerful Fe^{3+}-chelator produced by the microbe *Streptomyces pilosus*. The formation constant for binding Fe^{3+} to form the iron complex, ferrioxamine B, is $10^{30.6}$. The illustration below shows the structure of ferrioxamine B and a graph of its life-prolonging effects.

Used in conjunction with ascorbic acid—vitamin C, a reducing agent that reduces Fe^{3+} to the more soluble Fe^{2+}—desferrioxamine clears several grams of iron per year from an overloaded patient. The ferrioxamine complex is excreted in the urine.

Desferrioxamine reduces the incidence of heart and liver disease in thalassemia patients and maintains approximate iron balance. In patients for whom desferrioxamine effectively controls iron overload, there is a 91% rate of cardiac disease-free survival after 15 years of chelation therapy.[3] Among the negative effects of desferrioxamine is that too high a dose stunts the growth of children.

Desferrioxamine is expensive and must be taken by continuous injection. It is not absorbed through the intestine. Over more than two decades, many potent iron chelators have been tested to find an effective one that can be taken orally, but none has proved to be as efficient or free of side effects as desferrioxamine. In the long term, bone marrow transplants or gene therapy might cure the disease.

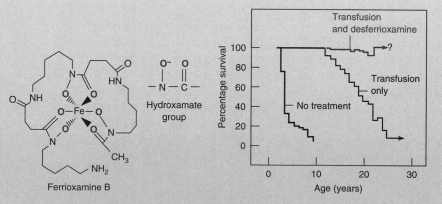

Structure of the iron complex ferrioxamine B and graph showing success of transfusions and transfusions plus chelation therapy. [From P. S. Dobbin and R. C. Hider, *Chem. Br.* **1990**, *26*, 565.]

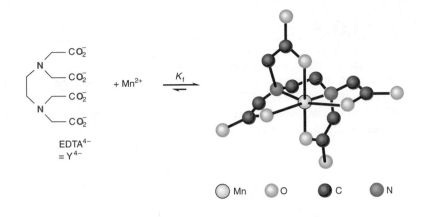

FIGURE 12-1 EDTA forms strong
1:1 complexes with most metal ions,
binding through four oxygen and two
nitrogen atoms. The six-coordinate
structure of Mn^{2+}-EDTA is found in
crystals of $KMnEDTA \cdot 2H_2O$. [From
J. Stein, J. P. Fackler, Jr., G. J. McClune,
J. A. Fee, and L. T. Chan, *Inorg. Chem.*
1979, *18,* 3511.]

A **multidentate ligand** (also called a *polydentate ligand,* a **chelating ligand,** or just
a *chelate*), binds to a metal ion through more than one ligand atom. EDTA in Figure 12-1 is a chelating ligand.

Most transition metal ions bind six ligand atoms. An important *tetradentate*
ligand is adenosine triphosphate (ATP), which binds to divalent metal ions (such as Mg^{2+},
Mn^{2+}, Co^{2+}, and Ni^{2+}) through four of their six coordination positions (Figure 12-2).
The fifth and sixth positions are occupied by water molecules. The biologically active
form of ATP is generally the Mg^{2+} complex.

The aminocarboxylic acids in Figure 12-3 are synthetic chelating agents whose
nitrogen and carboxylate oxygen atoms can lose protons and bind to metal ions. The
chelating agents in Figure 12-3 form strong 1:1 complexes with all metal ions,
except univalent ions such as Li^+, Na^+, and K^+. *The stoichiometry is 1:1 regardless
of the charge on the ion.* A titration based on complex formation is called a **complexometric titration.** Box 12-1 describes an important use of chelating agents in medicine.

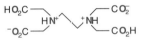

EDTA
Ethylenediaminetetraacetic acid
(also called ethylenedinitrilotetraacetic
acid)

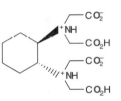

DCTA
trans-1,2-Diaminocyclohexanetetraacetic
acid

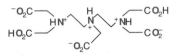

DTPA
Diethylenetriaminepentaacetic acid

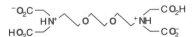

EGTA
bis-(Aminoethyl)glycolether-*N,N,N',N'*-
tetraacetic acid

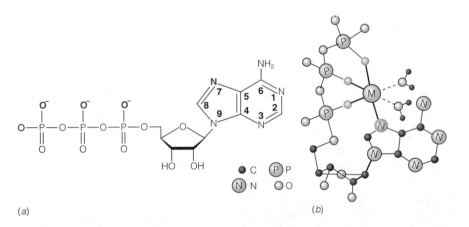

FIGURE 12-2 (*a*) Structure of adenosine triphosphate (ATP), with ligand atoms shown
in **color**. (*b*) Possible structure of a metal-ATP complex, with four bonds to ATP and two
bonds to H_2O ligands.

FIGURE 12-3 Analytically useful synthetic chelating agents that
form strong 1:1 complexes with
most metal ions.

Ask Yourself

12-A. What is the difference between a monodentate and a multidentate lig-
and? Is a chelating ligand monodentate or multidentate?

12-2 EDTA

EDTA is, by far, the most widely used chelator in analytical chemistry. By direct titration or through an indirect sequence of reactions, virtually every element of the periodic table can be analyzed with EDTA.

EDTA is a hexaprotic system, designated H_6Y^{2+}. The highlighted acidic hydrogen atoms are the ones that are lost upon metal-complex formation:

$$\begin{array}{ll}
\text{HO}_2\text{CCH}_2 \qquad\qquad \text{CH}_2\text{CO}_2\text{H} & pK_1 = 0.0 \ (\text{CO}_2\text{H}) \\
\overset{+}{\text{HN}}\text{CH}_2\text{CH}_2\overset{+}{\text{NH}} & pK_2 = 1.5 \ (\text{CO}_2\text{H}) \\
\text{HO}_2\text{CCH}_2 \qquad\qquad \text{CH}_2\text{CO}_2\text{H} & pK_3 = 2.0 \ (\text{CO}_2\text{H}) \\
\qquad\qquad\qquad H_6Y^{2+} & pK_4 = 2.66 \ (\text{CO}_2\text{H}) \\
& pK_5 = 6.16 \ (\text{NH}^+) \\
& pK_6 = 10.24 \ (\text{NH}^+)
\end{array}$$

The first four pK values apply to carboxyl protons, and the last two are for the ammonium protons. Below a pH of 10.24, most EDTA is protonated and is not in the form Y^{4-} that binds to metal ions (Figure 12-1).

Neutral EDTA is tetraprotic, with the formula H_4Y. A common reagent is the disodium salt, $Na_2H_2Y \cdot 2H_2O$, which attains the dihydrate composition upon heating at 80°C.

Box 12-2 describes the notation used for formation constants.

The equilibrium constant for the reaction of a metal with a ligand is called the **formation constant, K_f,** or the *stability constant:*

Formation constant: $\quad M^{n+} + Y^{4-} \rightleftharpoons MY^{n-4} \qquad K_f = \dfrac{[MY^{n-4}]}{[M^{n+}][Y^{4-}]}$ (12-1)

Table 12-1 shows that formation constants for EDTA complexes are large and tend to be larger for more positively charged metal ions. Note that K_f is defined for reaction of the species Y^{4-} with the metal ion. At low pH, most EDTA is in one of its protonated forms, not Y^{4-}.

A metal-EDTA complex becomes unstable at low pH because H^+ competes with the metal ion for the EDTA. At too high a pH, the EDTA complex is unstable because OH^- competes with EDTA for the metal ion and may precipitate the metal hydroxide or form unreactive hydroxide complexes. Figure 12-4 (see page 260) shows the pH ranges over which some common metal ions can be titrated. Pb^{2+}, for example, reacts "quantitatively" with EDTA between pH 3 and pH 12. Between pH 9 and 12, it is necessary to use an **auxiliary complexing agent,** which forms a weak complex with Pb^{2+} and keeps it in solution (Color Plate 7). The auxiliary complexing agent is displaced by EDTA during the titration. Auxiliary complexing agents, such as ammonia, tartrate, citrate, or triethanolamine, prevent metal ions from precipitating in the absence of EDTA. The titration of Pb^{2+} is carried out at pH 10 in the presence of tartrate, which complexes the metal ion and does not allow $Pb(OH)_2$ to precipitate. The lead-tartrate complex must be less stable than the lead-EDTA complex, or the titration would not be feasible.

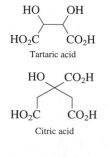

Tartaric acid

Citric acid

$N(CH_2CH_2OH)_3$
Triethanolamine

Notation for Formation Constants

Formation constants are the equilibrium constants for complex ion formation. The **stepwise formation constants,** designated K_i, are defined as follows:

$$M + X \xrightleftharpoons{K_1} MX \qquad K_1 = [MX]/[M][X]$$

$$MX + X \xrightleftharpoons{K_2} MX_2 \qquad K_2 = [MX_2]/[MX][X]$$

$$MX_{n-1} + X \xrightleftharpoons{K_n} MX_n \qquad K_n = [MX_n]/[MX_{n-1}][X]$$

where M is a metal ion and X is a ligand. The **overall,** or **cumulative, formation constants** are denoted β_i:

$$M + 2X \xrightleftharpoons{\beta_2} MX_2 \qquad \beta_2 = [MX_2]/[M][X]^2$$

$$M + nX \xrightleftharpoons{\beta_n} MX_n \qquad \beta_n = [MX_n]/[M][X]^n$$

A useful relationship is that $\beta_n = K_1 K_2 \cdots K_n$. Page 233 shows cumulative formation constants for lead-iodide complexes used to calculate the composition of a saturated solution of PbI_2.

Figure 12-4 shows the experimentally determined pH ranges in which various *metal ion indicators* (discussed in the next section) are useful for finding the end points. The figure also provides a strategy for the selective titration of one ion in

TABLE 12-1 Formation constants for metal-EDTA complexes

Ion	log K_f	Ion	log K_f	Ion	log K_f
Li^+	2.79	Fe^{2+}	14.32	Pd^{2+}	18.5[a]
Na^+	1.66	Co^{2+}	16.31	Zn^{2+}	16.50
K^+	0.8	Ni^{2+}	18.62	Cd^{2+}	16.46
Be^{2+}	9.2	Cu^{2+}	18.80	Hg^{2+}	21.7
Mg^{2+}	8.79	Ti^{3+}	21.3[a]	Sn^{2+}	18.3
Ca^{2+}	10.69	V^{3+}	26.0	Pb^{2+}	18.04
Sr^{2+}	8.73	Cr^{3+}	23.4	Al^{3+}	16.3
Ba^{2+}	7.86	Mn^{3+}	25.3[a]	Ga^{3+}	20.3
Ra^{2+}	7.1	Fe^{3+}	25.1	In^{3+}	25.0
Sc^{3+}	23.1	Co^{3+}	41.4[a]	Tl^{3+}	37.8
Y^{3+}	18.09	Zr^{4+}	29.5	Bi^{3+}	27.8
La^{3+}	15.50	VO^{2+}	18.8	Ce^{3+}	15.98
V^{2+}	12.7	VO_2^+	15.55	Gd^{3+}	17.37
Cr^{2+}	13.6	Ag^+	7.32	Th^{4+}	23.2
Mn^{2+}	13.87	Tl^+	6.54	U^{4+}	25.8

Note: The stability constant is the equilibrium constant for the reaction $M^{n+} + Y^{4-} \rightleftharpoons MY^{n-4}$. Values in table generally apply at 20°C and ionic strength 0.1 M.

a. 25°C.

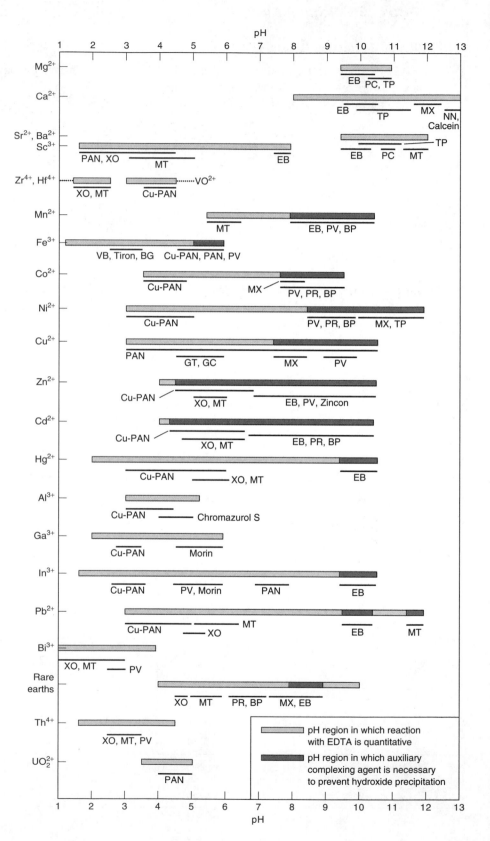

FIGURE 12-4 Guide to EDTA titrations of some common metals. Light color shows pH region in which reaction with EDTA is quantitative. Dark color shows pH region in which auxiliary complexing agent such as ammonia is required to prevent metal from precipitating. [Adapted from K. Ueno, *J. Chem. Ed.* **1965,** *42,* 432.]

Abbreviations for indicators:
BG, Bindschedler's green leuco base
BP, Bromopyrogallol red
Cu-PAN, PAN plus Cu-EDTA
EB, Eriochrome black T
GC, Glycinecresol red
GT, Glycinethymol blue
MT, Methylthymol blue
MX, Murexide
NN, Patton & Reeder's dye
PAN, Pyridylazonaphthol
PC, *o*-Cresolphthalein complexone
PR, Pyrogallol red
PV, Pyrocatechol violet
TP, Thymolphthalein complexone
VB, Variamine blue B base
XO, Xylenol orange

the presence of another. For example, a solution containing both Fe^{3+} and Ca^{2+} could be titrated with EDTA at pH 4. At this pH, Fe^{3+} is titrated without interference from Ca^{2+}.

Ask Yourself

12-B. (a) Write the reaction whose equilibrium constant is the formation constant for EDTA complex formation and write the algebraic form of K_f.
   (b) Why is EDTA complex formation less complete at lower pH?
   (c) What is the purpose of an auxiliary complexing agent?
   (d) The following diagram is analogous to those in Section 10-3 for polyprotic acids. It shows the pH at which each species of EDTA is predominant. Fill in the pH at each arrow.

H_6Y^{2+}	H_5Y^+	H_4Y	H_3Y^-	H_2Y^{2-}	HY^{3-}	Y^{4-}

pH: 1.5   8.20

12-3   Metal Ion Indicators

A **metal ion indicator** is a compound whose color changes when it binds to a metal ion. Two common indicators are shown in Table 12-2. *For an indicator to be useful, it must bind metal less strongly than EDTA does.*

A typical analysis is illustrated by the titration of Mg^{2+} with EDTA, using Eriochrome black T as the indicator:

$$\text{MgIn} + \text{EDTA} \longrightarrow \text{MgEDTA} + \text{In} \qquad (12\text{-}2)$$
$$\text{(red)} \quad \text{(colorless)} \qquad \text{(colorless)} \quad \text{(blue)}$$

> The indicator must release its metal to EDTA.

At the start of the experiment, a small amount of indicator (In) is added to the colorless solution of Mg^{2+} to form a red complex. As EDTA is added, it reacts first with free, colorless Mg^{2+}. When free Mg^{2+} is used up, the last EDTA added before the equivalence point displaces indicator from red MgIn complex. The change from the red of MgIn to the blue of unbound In signals the end point of the titration (Demonstration 12-1).

Most metal ion indicators are also acid-base indicators. Because the color of free indicator is pH dependent, most indicators can be used only in certain pH ranges. For example, xylenol orange (pronounced ZY-leen-ol) in Table 12-2 changes from yellow to red when it binds to a metal ion at pH 5.5. This is an easy color change to observe. At pH 7.5, the change is from violet to red and is rather difficult to see.

For an indicator to be useful in an EDTA titration, the indicator must give up its metal ion to EDTA. If metal does not freely dissociate from the indicator, the

TABLE 12-2 **Some metal ion indicators**

Name	Structure	pK_a	Color of free indicator	Color of metal ion complex
Eriochrome black T	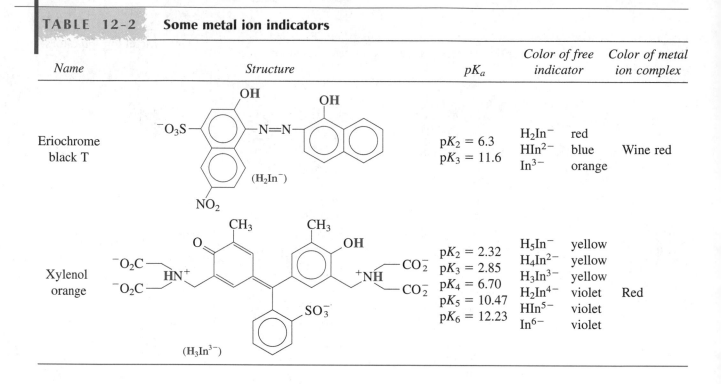	$pK_2 = 6.3$ $pK_3 = 11.6$	H_2In^- red HIn^{2-} blue In^{3-} orange	Wine red
Xylenol orange		$pK_2 = 2.32$ $pK_3 = 2.85$ $pK_4 = 6.70$ $pK_5 = 10.47$ $pK_6 = 12.23$	H_5In^- yellow H_4In^{2-} yellow H_3In^{3-} yellow H_2In^{4-} violet HIn^{5-} violet In^{6-} violet	Red

Question What will the color change be when the back titration is performed?

metal is said to **block** the indicator. Eriochrome black T is blocked by Cu^{2+}, Ni^{2+}, Co^{2+}, Cr^{3+}, Fe^{3+}, and Al^{3+}. It cannot be used for the direct titration of any of these metals. It can be used for a back titration, however. For example, excess standard EDTA can be added to Cu^{2+}. Then indicator is added and excess EDTA is back-titrated with Mg^{2+}.

🫙 *Demonstration 12-1*

Metal Ion Indicator Color Changes

This demonstration illustrates the color change associated with Reaction 12-2 and shows how a second dye can be added to a solution to produce a more easily detected color change.

Stock solutions
Eriochrome black T: Dissolve 0.1 g of the solid in 7.5 mL of triethanolamine plus 2.5 mL of absolute ethanol.
Methyl red: Dissolve 0.02 g in 60 mL of ethanol; then add 40 mL of water.
Buffer: Add 142 mL of concentrated (14.5 M) aqueous ammonia to 17.5 g of ammonium chloride and dilute to 250 mL with water.

$MgCl_2$: 0.05 M
EDTA: 0.05 M $Na_2H_2EDTA \cdot 2H_2O$

Prepare a solution containing 25 mL of $MgCl_2$, 5 mL of buffer, and 300 mL of water. Add 6 drops of Eriochrome black T indicator and titrate with EDTA. Note the color change from wine red to pale blue at the end point (Color Plate 8*a*).

For some people, the change of indicator color is not so easy to see. Adding 3 mL of methyl red (or other yellow dyes) produces an orange color prior to the end point and a green color after it. This sequence of colors is shown in Color Plate 8*b*.

Ask Yourself

12-C. (a) Explain why the change from red to blue in Reaction 12-2 occurs suddenly at the equivalence point instead of gradually throughout the entire titration.

(b) EDTA buffered to pH 5 was titrated with standard Pb^{2+}, using xylenol orange as indicator (Table 12-2).

(i) Which is the principal species of the indicator at pH 5?
(ii) What color was observed before the equivalence point?
(iii) What color was observed after the equivalence point?
(iv) What would the color change be if the titration were conducted at pH 8 instead of pH 5?

12-4 EDTA Titration Techniques

EDTA can be used directly or indirectly to analyze most elements of the periodic table. In this section, we discuss several important techniques.

Direct Titration

In a **direct titration,** analyte is titrated with standard EDTA. The analyte is buffered to an appropriate pH, at which the reaction with EDTA is essentially complete and the free indicator has a color distinctly different from that of the metal-indicator complex. An auxiliary complexing agent might be required to maintain the metal ion in solution in the absence of EDTA.

Back Titration

In a **back titration,** a known excess of EDTA is added to the analyte. Excess EDTA is then titrated with a standard solution of metal ion. A back titration is necessary if the analyte precipitates in the absence of EDTA, if analyte reacts too slowly with EDTA, or if analyte blocks the indicator. The metal used in the back titration must not displace analyte from EDTA.

EXAMPLE   A Back Titration

Ni^{2+} can be analyzed by a back titration with standard Zn^{2+} at pH 5.5 and xylenol orange indicator. A solution containing 25.00 mL of Ni^{2+} in dilute HCl was treated with 25.00 mL of 0.052 83 M Na_2EDTA. The solution was neutralized with NaOH, and the pH was adjusted to 5.5 with acetate buffer. The solution turned yellow when a few drops of indicator were added. Titration with 0.022 99 M Zn^{2+} required 17.61 mL to reach the red end point. What was the molarity of Ni^{2+} in the unknown?

SOLUTION The unknown was treated with 25.00 mL of 0.052 83 M EDTA, which contains (25.00 mL)(0.052 83 M) = 1.320 8 mmol of EDTA. Back titration required $(17.61 \text{ mL})(0.022\ 99 \text{ M}) = 0.404\ 9$ mmol of Zn^{2+}. Because 1 mol of EDTA reacts with 1 mol of any metal ion, there must have been

$$1.320\ 8 \text{ mmol EDTA} - 0.404\ 9 \text{ mmol } Zn^{2+} = 0.915\ 9 \text{ mmol } Ni^{2+}$$

The concentration of Ni^{2+} is $0.915\ 9$ mmol$/25.00$ mL $= 0.036\ 64$ M.

An EDTA back titration can prevent precipitation of analyte. For example, $Al(OH)_3$ precipitates at pH 7 in the absence of EDTA. An acidic solution of Al^{3+} can be treated with excess EDTA, adjusted to pH 7 with sodium acetate, and boiled to ensure complete complexation. The Al^{3+}-EDTA complex is stable at pH 7. The solution is then cooled; Eriochrome black T indicator is added; and back titration with standard Zn^{2+} is performed.

Displacement Titration

For some metal ions, there is no satisfactory indicator, but a **displacement titration** may be feasible. In this procedure, the analyte usually is treated with excess $Mg(EDTA)^{2-}$ to displace Mg^{2+}, which is later titrated with standard EDTA.

Challenge Calculate the equilibrium constant for Reaction 12-3 if $M^{n+} = Hg^{2+}$. Why is $Mg(EDTA)^{2-}$ used for a displacement titration?

$$M^{n+} + MgY^{2-} \longrightarrow MY^{n-4} + Mg^{2+} \qquad (12\text{-}3)$$

Hg^{2+} is determined in this manner. The formation constant of $Hg(EDTA)^{2-}$ must be greater than the formation constant of $Mg(EDTA)^{2-}$, or else the displacement of Mg^{2+} from $Mg(EDTA)^{2-}$ does not occur.

There is no suitable indicator for Ag^+. However, Ag^+ will displace Ni^{2+} from the tetracyanonickelate(II) ion:

$$2Ag^+ + Ni(CN)_4^{2-} \longrightarrow 2Ag(CN)_2^- + Ni^{2+}$$

The liberated Ni^{2+} can then be titrated with EDTA to find out how much Ag^+ was added.

Indirect Titration

Anions that precipitate metal ions can be analyzed with EDTA by **indirect titration.** For example, sulfate can be analyzed by precipitation with excess Ba^{2+} at pH 1. The $BaSO_4(s)$ is filtered, washed, and boiled with excess EDTA at pH 10 to bring Ba^{2+} back into solution as $Ba(EDTA)^{2-}$. The excess EDTA is back-titrated with Mg^{2+}.

Alternatively, an anion can be precipitated with excess metal ion. The precipitate is filtered and washed, and the excess metal ion in the filtrate is titrated with EDTA. CO_3^{2-}, CrO_4^{2-}, S^{2-}, and SO_4^{2-} can be determined in this manner.

What Is Hard Water?

Hardness refers to the total concentration of alkaline earth ions in water. Because the concentrations of Ca^{2+} and Mg^{2+} are usually much greater than the concentrations of other Group 2 ions, hardness can be equated to $[Ca^{2+}] + [Mg^{2+}]$. Hardness is commonly expressed as the equivalent number of milligrams of $CaCO_3$ per liter. Thus, if $[Ca^{2+}] + [Mg^{2+}] = 1$ mM, we would say that the hardness is 100 mg $CaCO_3$ per liter because 100 mg $CaCO_3 = 1$ mmol $CaCO_3$. Water whose hardness is less than 60 mg $CaCO_3$ per liter is considered to be "soft."

Hard water reacts with soap to form insoluble curds:

$$Ca^{2+} + 2RSO_3^- \longrightarrow Ca(RSO_3)_2(s) \qquad (A)$$

Soap   Precipitate

Enough soap to consume the Ca^{2+} and Mg^{2+} must be used before the soap is useful for cleaning. Hard water is not thought to be unhealthy. Hardness is beneficial in irrigation water because the alkaline earth ions tend to *flocculate* (cause to aggregate) *colloidal* particles in soil and thereby increase the permeability of the soil to water. Colloids are soluble particles that are 1–100 nm in diameter (see Demonstration 6-1). Such small particles tend to plug the paths by which water can drain through soil.

To measure hardness, water is treated with ascorbic acid to reduce Fe^{3+} to Fe^{2+} and with cyanide to mask Fe^{2+}, Cu^+, and several other minor metal ions. Titration with EDTA at pH 10 in ammonia buffer gives $[Ca^{2+}] + [Mg^{2+}]$. $[Ca^{2+}]$ can be determined separately if the titration is carried out at pH 13 without ammonia. At this pH, $Mg(OH)_2$ precipitates and is inaccessible to the EDTA.

Insoluble carbonates are converted to soluble bicarbonates by excess carbon dioxide:

$$CaCO_3(s) + CO_2 + H_2O \longrightarrow Ca(HCO_3)_2(aq)$$

Calcium carbonate   Calcium bicarbonate

(B)

Heating reverses Reaction B to form a solid scale of $CaCO_3$ that clogs boiler pipes. The fraction of hardness due to $Ca(HCO_3)_2(aq)$ is called *temporary hardness* because this calcium is lost (by precipitation of $CaCO_3$) upon heating. Hardness arising from other salts (mainly dissolved $CaSO_4$) is called *permanent hardness* because it is not removed by heating.

Masking

A **masking agent** is a reagent that protects some component of the analyte from reaction with EDTA. For example, Mg^{2+} in a mixture of Mg^{2+} and Al^{3+} can be titrated by first masking the Al^{3+} with F^-, thereby leaving only the Mg^{2+} to react with EDTA.

Cyanide is a masking agent that forms complexes with Cd^{2+}, Zn^{2+}, Hg^{2+}, Co^{2+}, Cu^+, Ag^+, Ni^{2+}, Pd^{2+}, Pt^{2+}, Fe^{2+}, and Fe^{3+}, but not with Mg^{2+}, Ca^{2+}, Mn^{2+}, or Pb^{2+}. When CN^- is added to a solution containing Cd^{2+} and Pb^{2+}, only the Pb^{2+} can react with EDTA. (***Caution:*** Cyanide forms toxic gaseous HCN below pH 11. Cyanide solutions should be made strongly basic and handled in a hood.) Fluoride masks Al^{3+}, Fe^{3+}, Ti^{4+}, and Be^{2+}. (***Caution:*** HF formed by F^- in acidic solution is extremely hazardous and should not contact skin or eyes. It may not be immediately painful, but the affected area should be flooded with water for 5 min and then treated with 2.5 wt % calcium gluconate gel that you have on hand *before* the accident. First aiders must wear rubber gloves to protect themselves. Damage from HF exposure can continue for several days after exposure. Exposure of 2% of your body to concentrated HF can kill you.[4]) Triethanolamine masks Al^{3+}, Fe^{3+}, and Mn^{2+}; and 2,3-dimercaptopropanol masks Bi^{3+}, Cd^{2+}, Cu^{2+}, Hg^{2+}, and Pb^{2+}. Selectivity afforded by masking and pH control allows individual components of complex mixtures to be analyzed by EDTA titration.

Masking prevents one element from interfering in the analysis of another element. Box 12-3 describes an important application of masking.

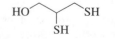

2,3-Dimercaptopropanol

12-5   The pH-Dependent Metal-EDTA Equilibrium

This is the point in the chapter where those who want to spend more time on instrumental methods of analysis might want to move to a new chapter. The remainder of this chapter deals with equilibrium calculations required to understand the shape of EDTA titration curves.

Fractional Composition of EDTA Solutions

The fraction of EDTA in each of its protonated forms is plotted in Figure 12-5. The fraction, α, is defined as in Section 11-5 for any weak acid. For example, $\alpha_{Y^{4-}}$ is

Fraction of EDTA in the form Y^{4-}:

$$\alpha_{Y^{4-}} = \frac{[Y^{4-}]}{[H_6Y^{2+}] + [H_5Y^+] + [H_4Y] + [H_3Y^-] + [H_2Y^{2-}] + [HY^{3-}] + [Y^{4-}]}$$

$$\alpha_{Y^{4-}} = \frac{[Y^{4-}]}{[EDTA]} \tag{12-4}$$

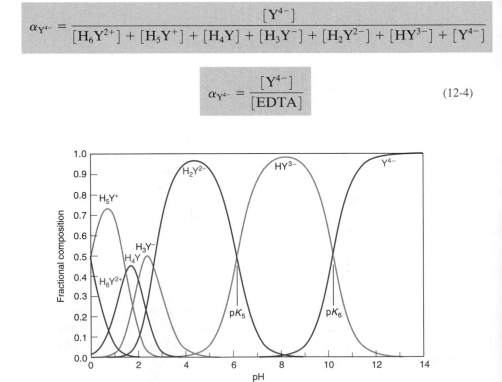

FIGURE 12-5 Fraction of EDTA in each of its protonated forms as a function of pH. This figure should remind you of Figure 10-1, a diagram for a diprotic acid.

where [EDTA] is the total concentration of all *free* EDTA species in the solution. By "free," we mean EDTA not complexed to metal ions. Following a derivation similar to the one in Section 11-5, it can be shown that $\alpha_{Y^{4-}}$ is given by

$$\alpha_{Y^{4-}} = \frac{K_1 K_2 K_3 K_4 K_5 K_6}{\{[H^+]^6 + [H^+]^5 K_1 + [H^+]^4 K_1 K_2 + [H^+]^3 K_1 K_2 K_3 + [H^+]^2 K_1 K_2 K_3 K_4 \atop + [H^+] K_1 K_2 K_3 K_4 K_5 + K_1 K_2 K_3 K_4 K_5 K_6\}} \qquad (12\text{-}5)$$

Table 12-3 gives values for $\alpha_{Y^{4-}}$ as a function of pH.

EXAMPLE   What Does $\alpha_{Y^{4-}}$ Mean?

The fraction of all free EDTA in the form Y^{4-} is called $\alpha_{Y^{4-}}$. At pH 6.00 at a formal concentration of 0.10 M, the composition of an EDTA solution is

$[H_6 Y^{2+}] = 8.4 \times 10^{-20}$ M $[H_5 Y^+] = 8.4 \times 10^{-14}$ M $[H_4 Y] = 2.7 \times 10^{-9}$ M
$[H_3 Y^-] = 2.7 \times 10^{-5}$ M $[H_2 Y^{2-}] = 0.059$ M $[HY^{3-}] = 0.041$ M
$$[Y^{4-}] = 2.3 \times 10^{-6} \text{ M}$$

Find $\alpha_{Y^{4-}}$.

SOLUTION $\alpha_{Y^{4-}}$ is the fraction in the form Y^{4-}:

$$\alpha_{Y^{4-}} = \frac{[Y^{4-}]}{[H_6 Y^{2+}] + [H_5 Y^+] + [H_4 Y] + [H_3 Y^-] + [H_2 Y^{2-}] + [HY^{3-}] + [Y^{4-}]}$$

$$= \frac{[2.3 \times 10^{-6}]}{[8.4 \times 10^{-20}] + [8.4 \times 10^{-14}] + [2.7 \times 10^{-9}] + [2.7 \times 10^{-5}]}{ + [0.059] + [0.041] + [2.3 \times 10^{-6}]}$$

$$= 2.3 \times 10^{-5}$$

Conditional Formation Constant

The formation constant in Equation 12-1 describes the reaction between Y^{4-} and a metal ion. As you can see in Figure 12-5, most of the EDTA is not Y^{4-} below pH = pK_6 = 10.24. The species HY^{3-}, $H_2 Y^{2-}$, and so on, predominate at lower pH. It is convenient to express the fraction of free EDTA in the form Y^{4-} by rearranging Equation 12-4 to give

$$[Y^{4-}] = \alpha_{Y^{4-}}[EDTA] \qquad (12\text{-}6)$$

where [EDTA] refers to the total concentration of all EDTA species not bound to metal ion.

The equilibrium constant for Reaction 12-1 can now be rewritten as

$$K_f = \frac{[MY^{n-4}]}{[M^{n+}][Y^{4-}]} = \frac{[MY^{n-4}]}{[M^{n+}]\alpha_{Y^{4-}}[EDTA]}$$

**TABLE 12-3
Values of $\alpha_{Y^{4-}}$ for EDTA at 20°C and μ = 0.10 M**

pH	$\alpha_{Y^{4-}}$
0	1.3×10^{-23}
1	1.9×10^{-18}
2	3.3×10^{-14}
3	2.6×10^{-11}
4	3.8×10^{-9}
5	3.7×10^{-7}
6	2.3×10^{-5}
7	5.0×10^{-4}
8	5.6×10^{-3}
9	5.4×10^{-2}
10	0.36
11	0.85
12	0.98
13	1.00
14	1.00

Challenge Use Figure 12-5 to decide which three species should have the highest concentrations at pH 6. See whether you agree with the numbers given in the example. Which species would have the greatest concentration at pH 7?

Equation 12-1 should not be interpreted to mean that Y^{4-} is the only species that reacts with M^{n+}. It simply says that the equilibrium constant is expressed in terms of the concentration of Y^{4-}.

The product $\alpha_{Y^{4-}}$ [EDTA] in the denominator accounts for the fact that only some of the free EDTA is in the form Y^{4-}.

If the pH is fixed by a buffer, then $\alpha_{Y^{4-}}$ is a constant that can be combined with K_f:

Conditional formation constant: $$K_f' = \alpha_{Y^{4-}} K_f = \frac{[MY^{n-4}]}{[M^{n+}][EDTA]} \qquad (12\text{-}7)$$

The number $K_f' = \alpha_{Y^{4-}} K_f$ is called the **conditional formation constant,** or the *effective formation constant*. It describes the formation of MY^{n-4} at any particular pH.

The conditional formation constant allows us to look at EDTA complex formation as if the uncomplexed EDTA were all in one form:

$$M^{n+} + EDTA \rightleftharpoons MY^{n-4} \qquad K_f' = \alpha_{Y^{4-}} K_f$$

With the conditional formation constant, we can treat EDTA complex formation as if all the free EDTA were in one form.

At any given pH, we can find $\alpha_{Y^{4-}}$ and evaluate K_f'.

EXAMPLE Using the Conditional Formation Constant

The formation constant in Table 12-1 for FeY^- is $10^{25.1} = 1.3 \times 10^{25}$. Calculate the concentrations of free Fe^{3+} in a solution of 0.10 M FeY^- at pH 4.00 and pH 1.00.

SOLUTION The complex formation reaction is

$$Fe^{3+} + EDTA \rightleftharpoons FeY^- \qquad K_f' = \alpha_{Y^{4-}} K_f$$

where EDTA on the left side of the equation refers to all forms of unbound EDTA ($= Y^{4-}$, HY^{3-}, H_2Y^{2-}, H_3Y^-, etc.). Using $\alpha_{Y^{4-}}$ from Table 12-3, we find

$$\text{at pH 4.00:} \quad K_f' = (3.8 \times 10^{-9})(1.3 \times 10^{25}) = 4.9 \times 10^{16}$$
$$\text{at pH 1.00:} \quad K_f' = (1.9 \times 10^{-18})(1.3 \times 10^{25}) = 2.5 \times 10^{7}$$

Because dissociation of FeY^- must produce equal quantities of Fe^{3+} and EDTA, we can write

	Fe^{3+}	+	EDTA	$\rightleftharpoons$	FeY^-
Initial concentration (M)	0		0		0.10
Final concentration (M)	x		x		$0.10 - x$

$$\frac{[FeY^-]}{[Fe^{3+}][EDTA]} = \frac{0.10 - x}{x^2} = K_f' = 4.9 \times 10^{16} \qquad \text{at pH 4.00}$$
$$= 2.5 \times 10^{7} \qquad \text{at pH 1.00}$$

Solving for x, we find $[Fe^{3+}] = x = 1.4 \times 10^{-9}$ M at pH 4.00 and 6.4×10^{-5} M at pH 1.00. *Using the conditional formation constant at a fixed pH, we treat the dissociated EDTA as if it were a single species.*

We see that a metal-EDTA complex becomes less stable at lower pH. For a titration reaction to be effective, it must go "to completion"; which means that the equilibrium constant is large—the analyte and titrant are essentially completely reacted at the equivalence point. Figure 12-6 shows how pH affects the titration of Ca^{2+} with EDTA. Below pH $\approx$ 8, the break at the end point is not sharp enough to allow accurate determination. At lower pH, the inflection point disappears because the conditional formation constant for CaY^{2-} is too small.

Ask Yourself

12-E. The equivalence point in the titration of Ca^{2+} with EDTA is the same as a solution of pure CaY^{2-}. Suppose that the formal concentration $[CaY^{2-}] = 0.010$ M at the equivalence point in Figure 12-6. Let's see how complete the reaction is at low and high pH.

    (a) Find the concentration of free Ca^{2+} at pH 5.00 at the equivalence point.

    (b) What is the fraction of bound Ca^{2+}, which is $[CaY^{2-}]/([CaY^{2-}] + [Ca^{2+}])$?

    (c) Find the concentration of free Ca^{2+} at pH 9.00 and the fraction of bound Ca^{2+}.

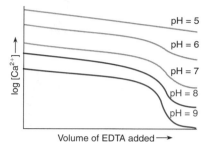

FIGURE 12-6   Titration of Ca^{2+} with EDTA as a function of pH. The experimental ordinate is the voltage difference between two electrodes (mercury and calomel) immersed in the titration solution. This voltage is a measure of $\log[Ca^{2+}]$. [From C. N. Reilley and R. W. Schmid, *Anal. Chem.* **1958**, *30*, 947.]

12-6   EDTA Titration Curves

Now we calculate the concentration of free metal ion during the course of the titration of metal with EDTA. The titration reaction is

$$M^{n+} + EDTA \rightleftharpoons MY^{n-4} \qquad K'_f = \alpha_{Y^{4-}} K_f \qquad (12-8)$$

K'_f is the effective formation constant at the fixed pH of the solution.

If K'_f is large, we can consider the reaction to be complete at each point in the titration.

    The titration curve is a graph of pM ($\equiv -\log[M]$) versus the volume of added EDTA. The curve is analogous to plotting pH versus volume of titrant in an acid-base titration. There are three natural regions of the titration curve in Figure 12-7.

Region 1: Before the Equivalence Point
In this region, there is excess M^{n+} left in solution after the EDTA has been consumed. The concentration of free metal ion is equal to the concentration of excess, unreacted M^{n+}. The dissociation of MY^{n-4} is negligible.

Region 2: At the Equivalence Point
There is exactly as much EDTA as metal in the solution. We can treat the solution as if it had been made by dissolving pure MY^{n-4}. Some free M^{n+} is generated by the slight dissociation of MY^{n-4}:

$$MY^{n-4} \rightleftharpoons M^{n+} + EDTA$$

In this reaction, EDTA refers to the total concentration of free EDTA in all of its forms. At the equivalence point, $[M^{n+}] = [EDTA]$.

Region 3: After the Equivalence Point
Now there is excess EDTA, and virtually all the metal ion is in the form MY^{n-4}. The concentration of free EDTA can be equated to the concentration of excess EDTA added after the equivalence point.

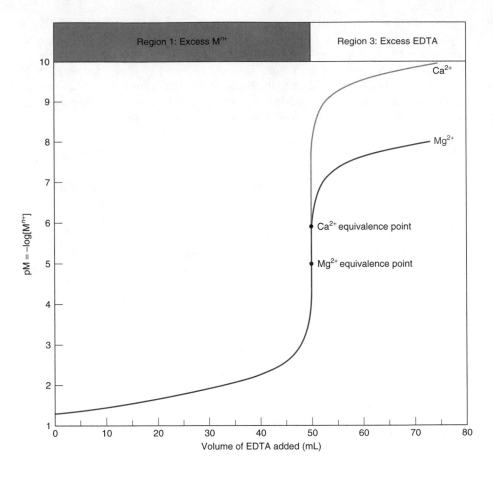

FIGURE 12-7 Three regions in an EDTA titration illustrated for reaction of 50.0 mL of 0.050 0 M M^{n+} (M = Mg or Ca) with 0.050 0 M EDTA at pH 10. The concentration of free M^{n+} decreases as the titration proceeds.

Titration Calculations

Let's calculate the shape of the titration curve for the reaction of 50.0 mL of 0.050 0 M Mg^{2+} (buffered to pH 10.00) with 0.050 0 M EDTA:

$$Mg^{2+} + EDTA \longrightarrow MgY^{2-}$$
$$K_f' = \alpha_{Y^{4-}}K_f = (0.36)(6.2 \times 10^8) = 2.2 \times 10^8$$

$\alpha_{Y^{4-}}$ comes from Table 12-3.

The equivalence volume is 50.0 mL. Because K_f' is large, it is reasonable to say that the reaction goes to completion with each addition of titrant. We want to make a graph in which pMg^{2+} ($= -\log[Mg^{2+}]$) is plotted versus milliliters of added EDTA.

Equivalence volume = 50.0 mL

Region 1: Before the Equivalence Point

Before the equivalence point, there is excess unreacted M^{n+}.

Consider the addition of 5.00 mL of EDTA. Because the equivalence point is 50.0 mL, one-tenth of the Mg^{2+} will be consumed and nine-tenths remains.

$$\text{initial mmol } Mg^{2+} = (0.050\ 0 \text{ M } Mg^{2+})(50.0 \text{ mL}) = 2.50 \text{ mmol}$$
$$\text{mmol remaining} = (0.900)(2.50 \text{ mmol}) = 2.25 \text{ mmol}$$
$$[Mg^{2+}] = \frac{2.25 \text{ mmol}}{55.0 \text{ mL}} = 0.040\ 9 \text{ M} \implies pMg^{2+} = -\log[Mg^{2+}] = 1.39$$

In a similar manner, we could calculate pMg^{2+} for any volume of EDTA less than 50.0 mL.

Region 2: At the Equivalence Point

Virtually all the metal is in the form MgY^{2-}. We began with 2.50 mmol Mg^{2+}, which is now close to 2.50 mmol MgY^{2-} in a volume of $50.0 + 50.0 = 100.0$ mL.

$$[MgY^{2-}] = \frac{2.50 \text{ mmol}}{100.0 \text{ mL}} = 0.025\ 0 \text{ M}$$

The concentration of free Mg^{2+} is small and unknown. We can write

	Mg^{2+}	$+$	EDTA	$\rightleftharpoons$	MgY^{2-}
Initial concentration (M)	—		—		0.025 0
Final concentration (M)	x		x		$0.025\ 0 - x$

At the equivalence point, the major species is MY^{n-4}, in equilibrium with small, equal amounts of free M^{n+} and EDTA.

$$\frac{[MgY^{2-}]}{[Mg^{2+}][EDTA]} = K_f' = 2.2 \times 10^8$$

$$\frac{0.025\ 0 - x}{x^2} = 2.2 \times 10^8 \implies x = 1.07 \times 10^{-5} \text{ M}$$

$$pMg^{2+} = -\log x = 4.97$$

[EDTA] refers to the total concentration of all forms of EDTA not bound to metal.

Region 3: After the Equivalence Point

In this region, virtually all the metal is in the form MgY^{2-}, and there is excess, unreacted EDTA. The concentrations of MgY^{2-} and excess EDTA are easily calculated. For example, when we have added 51.00 mL of EDTA, there is 1.00 mL of excess EDTA $= (0.050\ 0 \text{ M})(1.00 \text{ mL}) = 0.050\ 0$ mmol.

After the equivalence point, virtually all the metal is present as MY^{n-4}. There is a known excess of EDTA present. A small amount of free M^{n+} exists in equilibrium with the MY^{n-4} and EDTA.

$$[EDTA] = \frac{0.050\ 0 \text{ mmol}}{101.0 \text{ mL}} = 0.000\ 495 \text{ M}$$

$$[MgY^{2-}] = \frac{2.50 \text{ mmol}}{101.0 \text{ mL}} = 0.024\ 8 \text{ M}$$

The concentration of Mg^{2+} is governed by

$$\frac{[MgY^{2-}]}{[Mg^{2+}][EDTA]} = K_f' = 2.2 \times 10^8$$

$$\frac{[0.024\ 8]}{[Mg^{2+}](0.000\ 495)} = 2.2 \times 10^8$$

$$[Mg^{2+}] = 2.3 \times 10^{-7} \text{ M} \implies pMg^{2+} = 6.64$$

The same sort of calculation can be used for any volume past the equivalence point.

The Titration Curve

The calculated titration curves for Mg^{2+} and for Ca^{2+} are shown in Figure 12-7. Both show a distinct break at the equivalence point, where the slope is greatest. The break is greater for Ca^{2+} than for Mg^{2+} because the formation constant for CaY^{2-} is greater than K_f for MgY^{2-}. Notice the strong analogy between

The lower the pH, the less distinct the end point.

Figure 12-7 and acid-base titration curves. The greater the metal-EDTA formation constant, the more pronounced the break at the equivalence point. The stronger the acid (HA), the greater the break at the equivalence point in the titration with OH^-.

The completeness of reaction (and hence the sharpness of the equivalence point) is determined by the conditional formation constant, $\alpha_{Y^{4-}} K_f$, which is pH dependent. Because $\alpha_{Y^{4-}}$ decreases as the pH is lowered, the pH is an important variable determining whether a titration is feasible. The end point is more distinct at high pH. However, the pH must not be so high that metal hydroxide precipitates. The effect of pH on the titration of Ca^{2+} was shown in Figure 12-6.

Beware

The calculation we just did was oversimplified because we neglected any other chemistry of M^{n+}, such as formation of MOH^+, $M(OH)_2(aq)$, $M(OH)_2(s)$, $M(OH)_3^-$, etc. To the extent that these species are present, they decrease the concentration of available M^{n+} and decrease the sharpness of the titration curve that we computed. Mg^{2+} is normally titrated in ammonia buffer at pH 10 in which $Mg(NH_3)^{2+}$ is also present. The accurate calculation of metal-EDTA titration curves requires full knowledge of the chemistry of the metal with water and any other ligands present in the solution.

Ask Yourself

12-F. What would be pCa^{2+} ($= -\log[Ca^{2+}]$) in the titration in Figure 12-7 at the following volumes of EDTA: 5.00, 50.00, and 51.00 mL? See that your answers agree with the figure.

Key Equations

Formation constant

$$M^{n+} + Y^{4-} \rightleftharpoons MY^{n-4} \qquad K_f = \frac{[MY^{n-4}]}{[M^{n+}][Y^{4-}]}$$

Fraction of EDTA as Y^{4-}

$$\alpha_{Y^{4-}} = \frac{[Y^{4-}]}{[EDTA]}$$

$[EDTA]$ = total concentration of EDTA not bound to metal

Conditional formation constant

$$K_f' = \alpha_{Y^{4-}} K_f = \frac{[MY^{n-4}]}{[M^{n+}][EDTA]}$$

Titration calculations

Before V_e, there is a known excess of M^{n+}

$$pM = -\log[M]$$

At V_e: Some M^{n+} is generated by the dissociation of MY^{n-4}

$$\underset{x}{M^{n+}} + \underset{x}{EDTA} \overset{K_f'}{\rightleftharpoons} \underset{F-x}{MY^{n-4}}$$

After V_e, [EDTA] and $[MY^{n-4}]$ are known

$$M^{n+} + EDTA \underset{}{\overset{K_f'}{\rightleftharpoons}} MY^{n-4}$$

x Known   Known

Important Terms

auxiliary complexing agent	direct titration	masking agent
back titration	displacement titration	metal ion indicator
blocking	formation constant	monodentate ligand
chelating ligand	indirect titration	multidentate ligand
complexometric titration	Lewis acid	overall formation constant
conditional formation constant	Lewis base	stepwise formation constant
cumulative formation constant	ligand	

Problems

12-1. How many milliliters of 0.050 0 M EDTA are required to react with 50.0 mL of **(a)** 0.010 0 M Ca^{2+} or **(b)** 0.010 0 M Al^{3+}?

12-2. Give three circumstances in which an EDTA back titration might be necessary.

12-3. Describe what is done in a displacement titration and give an example.

12-4. Give an example of the use of a masking agent.

12-5. What is meant by water hardness? Explain the difference between temporary and permanent hardness.

12-6. State the purpose of an auxiliary complexing agent and give an example of its use.

12-7. Draw a reasonable structure for a complex between Fe^{3+} and nitrilotriacetic acid, $N(CH_2CO_2H)_3$.

12-8. A 25.00-mL sample containing Fe^{3+} was treated with 10.00 mL of 0.036 7 M EDTA to complex all the Fe^{3+} and leave excess EDTA in solution. The excess EDTA was then back-titrated, requiring 2.37 mL of 0.046 1 M Mg^{2+}. What was the concentration of Fe^{3+} in the original solution?

12-9. A 50.0-mL solution containing Ni^{2+} and Zn^{2+} was treated with 25.0 mL of 0.045 2 M EDTA to bind all the metal. The excess unreacted EDTA required 12.4 mL of 0.012 3 M Mg^{2+} for complete reaction. An excess of the reagent 2,3-dimercapto-1-propanol was then added to displace the EDTA from zinc. Another 29.2 mL of Mg^{2+} was required for reac-

tion with the liberated EDTA. Calculate the molarities of Ni^{2+} and Zn^{2+} in the original solution.

12-10. Sulfide ion was determined by indirect titration with EDTA. To a solution containing 25.00 mL of 0.043 32 M $Cu(ClO_4)_2$ plus 15 mL of 1 M acetate buffer (pH 4.5) was added 25.00 mL of unknown sulfide solution with vigorous stirring. The CuS precipitate was filtered and washed with hot water. Ammonia was added to the filtrate (which contains excess Cu^{2+}) until the blue color of $Cu(NH_3)_4^{2+}$ was observed. Titration with 0.039 27 M EDTA required 12.11 mL to reach the end point with the indicator murexide. Find the molarity of sulfide in the unknown.

12-11. The potassium ion in a 250.0 (± 0.1)-mL water sample was precipitated with sodium tetraphenylborate:

$$K^+ + (C_6H_5)_4B^- \longrightarrow KB(C_6H_5)_4(s)$$

The precipitate was filtered, washed, and dissolved in an organic solvent. Treatment of the organic solution with an excess of Hg^{2+}-EDTA then gave the following reaction:

$$4HgY^{2-} + (C_6H_5)_4B^- + 4H_2O \longrightarrow H_3BO_3 + 4C_6H_5Hg^+ +$$
$$4HY^{3-} + OH^-$$

The liberated EDTA was titrated with 28.73 (± 0.03) mL of 0.043 7 (± 0.000 1) M Zn^{2+}. Find the concentration (and uncertainty) of K^+ in the original sample.

12-12. A 25.00-mL sample of unknown containing Fe^{3+} and Cu^{2+} required 16.06 mL of 0.050 83 M EDTA for complete titration. A 50.00-mL sample of the unknown was treated with NH_4F to protect the Fe^{3+}. Then the Cu^{2+} was reduced and masked by addition of thiourea. Upon addition of 25.00 mL of 0.050 83 M EDTA, the Fe^{3+} was liberated from its fluoride complex and formed an EDTA complex. The excess EDTA required 19.77 mL of 0.018 83 M Pb^{2+} to reach an end point, using xylenol orange. Find $[Cu^{2+}]$ in the unknown.

12-13. Cyanide recovered from the refining of gold ore can be determined indirectly by EDTA titration. A known excess of Ni^{2+} is added to the cyanide to form tetracyanonickelate(II):

$$4CN^- + Ni^{2+} \longrightarrow Ni(CN)_4^{2-}$$

When the excess Ni^{2+} is titrated with standard EDTA, $Ni(CN)_4^{2-}$ does not react. In a cyanide analysis, 12.7 mL of cyanide solution was treated with 25.0 mL of standard solution containing excess Ni^{2+} to form tetracyanonickelate. The excess Ni^{2+} required 10.1 mL of 0.013 0 M EDTA for complete reaction. In a separate experiment, 39.3 mL of 0.013 0 M EDTA was required to react with 30.0 mL of the standard Ni^{2+} solution. Calculate the molarity of CN^- in the 12.7-mL sample of unknown.

12-14. A mixture of Mn^{2+}, Mg^{2+}, and Zn^{2+} was analyzed as follows: The 25.00-mL sample was treated with 0.25 g of $NH_3OH^+Cl^-$ (hydroxylammonium chloride, a reducing agent that maintains manganese in the $+2$ state), 10 mL of ammonia buffer (pH 10), and a few drops of Eriochrome black T indicator and then diluted to 100 mL. It was warmed to 40°C and titrated with 39.98 mL of 0.045 00 M EDTA to the blue end point. Then 2.5 g of NaF were added to displace Mg^{2+} from its EDTA complex. The liberated EDTA required 10.26 mL of standard 0.020 65 M Mn^{2+} for complete titration. After this second end point was reached, 5 mL of 15 wt % aqueous KCN was added to displace Zn^{2+} from its EDTA complex. This time the liberated EDTA required 15.47 mL of standard 0.020 65 M Mn^{2+}. Calculate the number of milligrams of each metal $(Mn^{2+}, Zn^{2+},$ and $Mg^{2+})$ in the 25.00-mL sample of unknown.

12-15. The sulfur content of insoluble sulfides that do not readily dissolve in acid can be measured by oxidation with Br_2 to SO_4^{2-}.[5] Metal ions are then replaced with H^+ by an ion-exchange column (Chapter 22), and sulfate is precipitated as $BaSO_4$ with a known excess of $BaCl_2$. The excess Ba^{2+} is then titrated with EDTA to determine how much was present. (To make the indicator end point clearer, a small, known quantity of Zn^{2+} is also added. The EDTA titrates both the Ba^{2+} and Zn^{2+}.) Knowing the excess Ba^{2+}, we can calculate

how much sulfur was in the original material. To analyze the mineral sphalerite (ZnS, FM 97.46), 5.89 mg of powdered solid were suspended in a mixture of carbon tetrachloride and water containing 1.5 mmol Br_2. After 1 h at 20° and 2 h at 50°C, the powder dissolved and the solvent and excess Br_2 were removed by heating. The residue was dissolved in 3 mL water and passed through an ion-exchange column to replace Zn^{2+} with H^+. Then 5.000 mL of 0.014 63 M $BaCl_2$ were added to precipitate all sulfate as $BaSO_4$. After adding 1.000 mL of 0.010 00 M $ZnCl_2$ and 3 mL of ammonia buffer, pH 10, the excess Ba^{2+} and Zn^{2+} required 2.39 mL of 0.009 63 M EDTA to reach the Eriochrome black T end point. Find the weight percent of sulfur in the sphalerite. What is the theoretical value?

12-16. State (in words) what $\alpha_{Y^{4-}}$ means. Calculate $\alpha_{Y^{4-}}$ for EDTA at **(a)** pH 3.50 and **(b)** pH 10.50.

12-17. The cumulative formation constants for the reaction of Co^{2+} with ammonia are $\log \beta_1 = 1.99$, $\log \beta_2 = 3.50$, $\log \beta_3 = 4.43$, $\log \beta_4 = 5.07$, $\log \beta_5 = 5.13$, and $\log \beta_6 = 4.39$.

(a) Write the chemical reaction whose equilibrium constant is β_4.

(b) Write the reaction whose stepwise formation constant is K_4 and find its numerical value.

12-18. (a) Find the conditional formation constant for $Mg(EDTA)^{2-}$ at pH 9.00.

(b) Find the concentration of free Mg^{2+} in 0.050 M $Na_2[Mg(EDTA)]$ at pH 9.00.

12-19. The ion M^{n+} (100.0 mL of 0.050 0 M metal ion buffered to pH 9.00) was titrated with 0.050 0 M EDTA.

(a) What is the equivalence volume, V_e, in milliliters?

(b) Calculate the concentration of M^{n+} at $V = \frac{1}{2}V_e$.

(c) What fraction $(\alpha_{Y^{4-}})$ of free EDTA is in the form Y^{4-} at pH 9.00?

(d) The formation constant (K_f) is $10^{12.00}$. Calculate the value of the conditional formation constant $K_f' (= \alpha_{Y^{4-}} K_f)$.

(e) Calculate the concentration of M^{n+} at $V = V_e$.

(f) What is the concentration of M^{n+} at $V = 1.100V_e$?

12-20. Consider the titration of 25.0 mL of 0.020 0 M $MnSO_4$ with 0.010 0 M EDTA in a solution buffered to pH 6.00 Calculate pMn^{2+} at the following volumes of added EDTA and sketch the titration curve:

(a) 0 mL	**(d)** 49.0 mL	**(g)** 50.1 mL
(b) 20.0 mL	**(e)** 49.9 mL	**(h)** 55.0 mL
(c) 40.0 mL	**(f)** 50.0 mL	**(i)** 60.0 mL

12-21. Using the same volumes used in the previous problem, calculate pCa^{2+} for the titration of 25.00 mL of 0.020 00 M EDTA with 0.010 00 M $CaSO_4$ at pH 10.00 and sketch the titration curve.

12-22. Explain the analogies between the titration of a metal with EDTA and the titration of a weak acid (HA) with OH^-. Make comparisons in all three regions of the titration curve.

12-23. *Metal ion buffers.* By analogy to a hydrogen ion buffer, a metal ion buffer tends to maintain a particular metal ion concentration in solution. A mixture of the acid HA and its conjugate base A^- is a hydrogen ion buffer that maintains a pH defined by the equation $K_a = [A^-][H^+]/[HA]$. A mixture of CaY^{2-} and Y^{4-} serves as a Ca^{2+} buffer governed by the equation $1/K_f' = [EDTA][Ca^{2+}]/[CaY^{2-}]$. How many grams of $Na_2H_2EDTA \cdot 2H_2O$ (FM 372.23) should be mixed with 1.95 g of $Ca(NO_3)_2 \cdot 2H_2O$ (FM 200.12) in a 500-mL volumetric flask to give a buffer with $pCa^{2+} = 9.00$ at pH 9.00?

How Would You Do It?

12-24. Consider the titration curve for Mg^{2+} in Figure 12-7. The curve was calculated under the simplifying assumption that all the magnesium not bound to EDTA is free Mg^{2+}. In fact, at pH 10 in a buffer containing 1 M NH_3, approximately 63% of the magnesium not bound to EDTA is $Mg(NH_3)^{2+}$ and 4% is $MgOH^+$. Both NH_3 and OH^- are readily displaced from Mg^{2+} by EDTA during the course of the titration.

(a) *Sketch* how the titration curve in Figure 12-7 would be different prior to the equivalence point if two-thirds of the magnesium not bound to EDTA is bound to the ligands NH_3 and OH^-.

(b) Consider a point that is 10% past the equivalence point. We know that almost all magnesium is bound to EDTA at this point. We also know that there is 10% excess unbounded EDTA at this point. Is the concentration of free Mg^{2+} the same or different from what we calculated in Figure 12-7? Will the titration curve be the same or different from Figure 12-7 past the equivalence point?

Notes and References

1. D. T. Haworth, *J. Chem. Ed.* **1998,** *75,* 47.

2. W. D. Bedsworth and D. L. Sedlak, *Environ. Sci. Technol.* **1999,** *33,* 926.

3. N. F. Olivieri and G. M. Brittenham, *Blood,* **1997,** *89,* 739.

4. *Chem. Eng. News,* 13 September 1999, p. 40.

5. T. Darjaa, K. Yamada, N. Sato, T. Fujino, and Y. Waseda, *Fresenius J. Anal. Chem.* **1998,** *361,* 442.

Further Reading

A. E. Martell and R. D. Hancock, *Metal Complexes in Aqueous Solution* (New York: Plenum Press, 1996).

G. Schwarzenbach and H. Flaschka, *Complexometric Titrations,* H. M. N. H. Irving, trans. (London: Methuen, 1969).

A. Ringbom, *Complexation in Analytical Chemistry* (New York: Wiley, 1963).

Redox Barriers to Remediate Underground Contamination

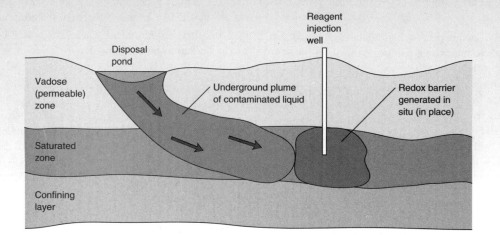

Large-scale, underground redox chemistry is proving to be a valuable tool to remediate plumes of polluted water and prevent them from leaking into aquifers and rivers. The diagram above shows a plume of contaminant migrating underground beneath a toxic waste disposal pond on the surface. A permeable, reactive barrier is created ahead of the migrating plume by inject-ing sodium dithionite ($Na_2S_2O_4$) through a well.[1] Dithionite reacts with Fe(III) in the soil to produce a porous zone containing particles of Fe(0).

When a plume containing the carcinogen chromate (CrO_4^{2-}) and chlori-nated solvents, such as trichloroethene, reaches the reactive barrier, both of the pollutants are reduced to far less hazardous species:

$$CrO_4^{2-} + Fe(0) + 4H_2O \longrightarrow \underbrace{Cr(OH)_3(s) + Fe(OH)_3(s)}_{\text{Mixed hydroxide precipitate}} + 2OH^-$$

Dissolved   **Iron**
chromate   **particles**

$$ClHC{=}CCl_2 \xrightarrow{\text{Fe(0) particles}} H_2C{=}CH_2 + HC{\equiv}CH + Cl^-$$

Trichloroethene   **Ethene**   **Ethyne**
(trichloroethylene)   **(ethylene)** **(acetylene)**

This scheme is employed at Hanford, Washington, to prevent a waste plume from reaching the Columbia River. At a U.S. Coast Guard station in North Carolina, underground treatment of a waste plume with buried iron filings reduced chromate from 4 mg/L to < 0.01 mg/L and eliminated 95% of chlo-rinated hydrocarbons.[2]

Chapter 13

ELECTRODE POTENTIALS

Measurements of the acidity of rainwater, the fuel-air mixture in an automobile engine, and the concentration of gases and electrolytes in your bloodstream are all made with electrical sensors. This chapter provides the foundation needed to discuss some of the most common electrochemical sensors in the next chapter.

13-1 Redox Chemistry and Electricity

In a **redox reaction,** electrons are transferred from one species to another. A molecule is said to be **oxidized** when it *loses electrons.* It is **reduced** when it *gains electrons.* An **oxidizing agent,** also called an **oxidant,** takes electrons from another substance and becomes reduced. A **reducing agent,** also called a **reductant,** gives electrons to another substance and is oxidized in the process. In the reaction

$$\underset{\substack{\text{Oxidizing} \\ \text{agent}}}{Fe^{3+}} + \underset{\substack{\text{Reducing} \\ \text{agent}}}{V^{2+}} \longrightarrow Fe^{2+} + V^{3+}$$

Oxidation: loss of electrons
Reduction: gain of electrons
Oxidizing agent: takes electrons
Reducing agent: gives electrons

Fe^{3+} is the oxidizing agent because it takes an electron from V^{2+}. V^{2+} is the reducing agent because it gives an electron to Fe^{3+}. Fe^{3+} is reduced (its oxidation state goes down from $+3$ to $+2$), and V^{2+} is oxidized (its oxidation state goes up from $+2$ to $+3$) as the reaction proceeds from left to right. Appendix D discusses oxidation numbers and balancing redox equations. You should be able to write a balanced redox reaction as the sum of two *half-reactions,* one involving oxidation and the other involving reduction.

277

Michael Faraday (1791–1867) was a self-educated English "natural philosopher" (the old term for "scientist") who discovered the basic law of electrochemistry: The extent of an electrochemical reaction is proportional to the electric charge that passes through the cell. More important, he discovered many fundamental laws of electromagnetism. He gave us the electric motor, electric generator, and electric transformer, as well as the terms *ion, cation, anion, electrode, cathode, anode,* and *electrolyte.* His gift for lecturing is best remembered from his Christmas season lecture demonstrations for children at the Royal Institution. Faraday "took great delight in talking to [children], and easily won their confidence.... They felt as if he belonged to them; and indeed he sometimes, in his joyous enthusiasm, appeared like an inspired child."[3]

Faraday constant:

$F \approx 9.649 \times 10^4$ C/mol

$1 A = 1$ C/s

$1 \text{ ampere} = 1 \dfrac{\text{coulomb}}{\text{second}}$

The *quantity* of electrons flowing from a reaction is proportional to the quantity of analyte that reacts.

Chemistry and Electricity

Electric charge (q) is measured in **coulombs (C).** The magnitude of the charge of a single electron (or proton) is 1.602×10^{-19} C. A mole of electrons therefore has a charge of $(1.602 \times 10^{-19} \text{ C})(6.022 \times 10^{23}/\text{mol}) = 9.649 \times 10^4$ C/mol, which is called the **Faraday constant, F.**

Relation between charge and moles:

$$\underset{\text{coulombs}}{q} = \underset{\text{moles}}{n} \cdot \underset{\dfrac{\text{coulombs}}{\text{mole}}}{F} \tag{13-1}$$

In the reaction $Fe^{3+} + V^{2+} \longrightarrow Fe^{2+} + V^{3+}$, one electron is transferred to oxidize one atom of V^{2+} and to reduce one atom of Fe^{3+}. If we know how many moles of electrons are transferred from V^{2+} to Fe^{3+}, then we know how many moles of product have been formed.

EXAMPLE   Relating Coulombs to Quantity of Reaction

If 5.585 g of Fe^{3+} were reduced in the reaction $Fe^{3+} + V^{2+} \longrightarrow Fe^{2+} + V^{3+}$, how many coulombs of charge must have been transferred from V^{2+} to Fe^{3+}?

SOLUTION First, we find that 5.585 g of Fe^{3+} equal 0.100 0 mol of Fe^{3+}. Because each Fe^{3+} ion requires one electron, 0.100 0 mol of electrons must have been transferred. Equation 13-1 relates coulombs of charge to moles of electrons:

$$q = nF = (0.100\ 0 \text{ mol e}^-)\left(9.649 \times 10^4 \frac{C}{\text{mol e}^-}\right) = 9.649 \times 10^3 \text{ C}$$

Electric Current Is Proportional to the Rate of a Redox Reaction

Electric **current** (I) is the quantity of charge flowing each second past a point in an electric circuit. The unit of current is the **ampere** (A), which represents a flow of one coulomb per second.

Consider Figure 13-1, in which electrons flow into a Pt wire dipped into a solution in which the reduction of Sn^{4+} to Sn^{2+} occurs:

$$Sn^{4+} + 2e^- \longrightarrow Sn^{2+}$$

The Pt wire is an **electrode**—a device to conduct electrons into or out of the chemicals involved in the redox reaction. Platinum is an *inert* electrode; it does not participate in the reaction except as a conductor of electrons. We call a molecule that can donate or accept electrons at an electrode an **electroactive species.** The rate at which electrons flow into the electrode is a measure of the rate of reduction of Sn^{4+}.

Suppose that Sn^{4+} is reduced to Sn^{2+} at a constant rate of 4.24 mmol/h in Figure 13-1. How much current flows into the solution?

SOLUTION *Two* electrons are required to reduce *one* Sn^{4+} ion to Sn^{2+}. If Sn^{4+} is reacting at a rate of 4.24 mmol/h, electrons flow at a rate of $2(4.24) = 8.48$ mmol/h, which corresponds to

$$\frac{8.48 \text{ mmol/h}}{3\,600 \text{ s/h}} = 2.356 \times 10^{-3} \text{ mmol/s} = 2.356 \times 10^{-6} \text{ mol/s}$$

To find the current, we use the Faraday constant to convert moles of electrons per second to coulombs per second:

$$\text{current} = \frac{\text{coulombs}}{\text{second}} = \frac{\text{moles}}{\text{second}} \cdot \frac{\text{coulombs}}{\text{mole}}$$

$$= \left(2.356 \times 10^{-6} \frac{\text{mol}}{\text{s}} \right)\left(9.649 \times 10^4 \frac{\text{C}}{\text{mol}} \right) = 0.227 \text{ C/s} = 0.227 \text{ A}$$

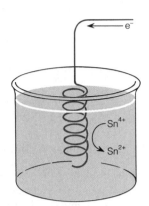

FIGURE 13-1   Electrons flowing into a coil of Pt wire at which Sn^{4+} ions in solution are reduced to Sn^{2+} ions. This process could not happen by itself because there is no complete circuit. If Sn^{4+} is to be reduced at this Pt electrode, some other species must be oxidized at some other place.

Voltage and Electrical Work

Because of their negative charge, electrons are attracted to positively charged regions and repelled from negatively charged regions. If electrons are attracted from one point to another, they can do useful work along the way. If we want to force electrons into a region from which they are repelled, we must do work on the electrons to push them along. Work has the dimensions of energy, whose units are *joules* (J).

The difference in **electric potential** between two points measures the work that is needed (or can be done) when electrons move from one point to another. The greater the potential difference between points A and B, the more work can be done (or must be done) when electrons travel from point A to point B. Potential difference is measured in **volts** (V).

A good analogy for understanding current and potential is to think of water flowing through a garden hose. Electric current is the quantity of electric charge flowing per second through a wire. Electric current is analogous to the volume of water flowing per second through the hose. Electric potential is a measure of the force pushing on the electrons. The greater the force, the more current flows. Electric potential is analogous to the pressure on the water in the hose. The greater the pressure, the faster the water flows.

When a charge, *q*, moves through a potential difference, *E*, the work done is

It costs energy to move like charges toward each other. Energy is released when opposite charges move toward each other.

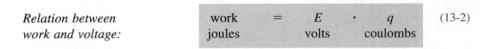

Relation between	work	=	*E*	·	*q*	(13-2)
work and voltage:	joules		volts		coulombs	

One joule of energy is gained or lost when one coulomb of charge moves through a potential difference of one volt. Equation 13-2 tells us that the dimensions of volts are J/C.

1 volt = 1 J/C

The garden hose analogy for work is this: Suppose that one end of a hose is raised 1 m above the other end and a volume of 1 L of water flows through the hose. The flowing water could go through a mechanical device to do a certain amount of work. If one end of the hose is raised 2 m above the other, the amount of work that can be done by the falling water is twice as great. The difference in elevation between the ends of the hose is analogous to electric potential difference, and the volume of water is analogous to electric charge. The greater the electric potential difference between two points in a circuit, the more work can be done by the charge flowing between those two points.

Demonstration 13-1

Electrochemical Writing

Approximately 7% of electric power in the United States goes into electrolytic chemical production. The electrolysis apparatus shown here consists of a sheet of aluminum foil taped or cemented to a glass or wooden surface. Any size will work, but an area about 15 cm on a side is convenient for a classroom demonstration. On the metal foil is taped (at one edge only) a sandwich consisting of filter paper, writing paper, and another sheet of filter paper. A stylus is prepared from a length of copper wire (18 gauge or thicker) looped at the end and passed through a length of glass tubing.

Prepare a fresh solution from 1.6 g of KI, 20 mL of water, 5 mL of 1 wt % starch solution, and 5 mL of phenolphthalein indicator solution. (If the solution darkens after standing for several days, decolorize it by adding a few drops of dilute $Na_2S_2O_3$.) Soak the three layers of paper with the KI-starch-phenolphthalein solution. Connect the stylus and foil to a 12-V DC power source and write on the paper with the stylus.

When the stylus is the cathode, pink color appears from the reaction of OH^- with phenolphthalein:

cathode: $$H_2O + e^- \longrightarrow \tfrac{1}{2}H_2(g) + OH^-$$
$$OH^- + \text{phenolphthalein} \longrightarrow \text{pink color}$$

When the polarity is reversed and the stylus is the anode, a black (very dark blue) color appears from the reaction of the newly generated I_2 with starch:

anode: $$I^- \longrightarrow \tfrac{1}{2}I_2 + e^-$$
$$I_2 + \text{starch} \longrightarrow \text{dark blue complex}$$

Pick up the top sheet of filter paper and the typing paper, and you will discover that the writing appears in the opposite color on the bottom sheet of filter paper. This sequence is shown in Color Plate 9. Color Plate 10 shows the same oxidation reaction in a solution where you can clearly see the process in action.

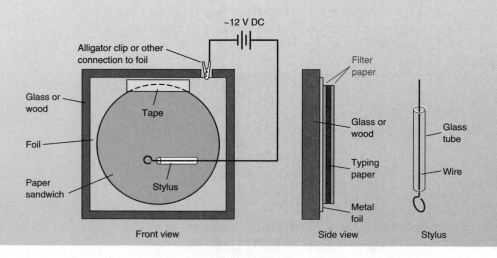

~12 V DC

Alligator clip or other connection to foil

Glass or wood

Tape

Foil

Paper sandwich

Stylus

Front view

Filter paper

Glass or wood

Typing paper

Metal foil

Side view

Glass tube

Wire

Stylus

EXAMPLE Electrical Work

How much work can be done when 2.36 mmol of electrons move "downhill" through a potential difference of 1.05 V? By "downhill," we mean that the flow of current is energetically favorable.

SOLUTION To use Equation 13-2, first convert moles of electrons to coulombs of charge:

$$q = nF = (2.36 \times 10^{-3} \text{ mol})(9.649 \times 10^4 \text{ C/mol}) = 2.277 \times 10^2 \text{ C}$$

The work that can be done by the moving electrons is

$$\text{work} = E \cdot q = (1.05 \text{ V})(2.277 \times 10^2 \text{ C}) = 239 \text{ J}$$

Electrolysis is a chemical reaction in which we apply a voltage to drive a redox reaction that would not otherwise occur. For example, electrolysis is used to make aluminum metal from Al^{3+} and to make chlorine gas (Cl_2) from Cl^- in seawater. Demonstration 13-1 is a great classroom illustration of electrolysis.

Ask Yourself

13-A. Consider the redox reaction

$$I_2 + 2\overset{O^-}{\underset{O}{O=S-S^-}} \rightleftharpoons 2I^- + O=\overset{O^-}{\underset{O}{S}}-S-S-\overset{O^-}{\underset{O}{S}}=O$$

Iodine   Thiosulfate   Iodide   Tetrathionate
FM 112.13

    (a) Identify the oxidizing agent on the left side of the reaction and write a balanced oxidation half-reaction.
    (b) Identify the reducing agent of the left side of the reaction and write a balanced reduction half-reaction.
    (c) How many coulombs of charge are passed from reductant to oxidant when 1.00 g of thiosulfate reacts?
    (d) If the rate of reaction is 1.00 g of thiosulfate consumed per minute, what current (in amperes) flows from reductant to oxidant?
    (e) If the charge in (c) flows "downhill" through a potential difference of 0.200 V, how much work (in joules) can be done by the electric current?

Electrolytic production of aluminum by the Hall-Heroult process consumes 4.5% of the electrical output of the United States! Al^{3+} in a molten solution of Al_2O_3 and cryolite (Na_3AlF_6) is reduced to aluminum metal at the cathode of a cell that typically draws 250 000 A. This process was invented by **Charles Hall** in 1886 when he was 22 years old, just after graduating from Oberlin College. [Photo courtesy of Alcoa Co., Pittsburgh, PA.]

A *spontaneous reaction* is one in which it is energetically favorable for reactants to be converted to products. Energy released from the chemicals is available as electrical energy.

For more on salt bridges, see Demonstration 13-2.

13-2 Galvanic Cells

In a **galvanic cell**, a *spontaneous* chemical reaction generates electricity. To accomplish this, one reagent must be oxidized and another must be reduced. The two cannot be in contact, or electrons would flow directly from the reducing agent to the oxidizing agent without going through the external circuit. Therefore, the oxidizing and reducing agents are physically separated, and electrons flow through a wire to get from one reactant to the other.

A Cell in Action

Figure 13-2 shows a galvanic cell consisting of two *half-cells* connected by a **salt bridge** through which ions can migrate to maintain electroneutrality in each vessel. The left half-cell has a metallic zinc electrode dipped into aqueous $ZnCl_2$. The right half-cell has a copper electrode immersed in aqueous $CuSO_4$. The salt bridge is filled with a gel containing saturated aqueous KCl. The electrodes are connected by a *potentiometer* (a voltmeter) to measure the voltage difference between the two half-cells.

The chemical reactions occurring in this cell are

reduction half-reaction:	$Cu^{2+}(aq) + 2e^- \rightleftharpoons Cu(s)$
oxidation half-reaction:	$Zn(s) \rightleftharpoons Zn^{2+}(aq) + 2e^-$
net reaction:	$Cu^{2+}(aq) + Zn(s) \rightleftharpoons Cu(s) + Zn^{2+}(aq)$

The two half-reactions are always written with equal numbers of electrons so that their sum includes no free electrons.

Oxidation of $Zn(s)$ at the left side of Figure 13-2 produces $Zn^{2+}(aq)$. Electrons from the Zn metal flow through the potentiometer into the copper electrode, where $Cu^{2+}(aq)$ is reduced to $Cu(s)$. If there were no salt bridge, the left half-cell would

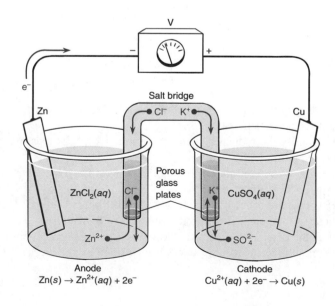

FIGURE 13-2 A galvanic cell consisting of two half-cells and a salt bridge through which ions can diffuse to maintain electroneutrality on each side.

The Human Salt Bridge

A salt bridge is an ionic medium through which ions diffuse to maintain electroneutrality in each compartment of an electrochemical cell. One way to prepare a salt bridge is to heat 3 g of agar (the stuff used to grow bacteria in a Petri dish) with 30 g of KCl in 100 mL of water until a clear solution is obtained. Pour the solution into a U-tube and allow it to gel. Store the bridge in saturated aqueous KCl.

To conduct this demonstration, set up the galvanic cell in Figure 13-2 with 0.1 M $ZnCl_2$ on the left side and 0.1 M $CuSO_4$ on the right side. You can use a voltmeter or a pH meter to measure voltage. If you use a pH meter, the positive terminal is the connection for the glass electrode and the negative terminal is the reference electrode connection.

You should be able to write the two half-reactions for this cell and use the Nernst equations (13-6 and 13-7) to calculate the theoretical voltage. Measure the voltage with a conventional salt bridge. Then replace the salt bridge with one made of filter paper freshly soaked in NaCl solution and measure the voltage again. Finally, replace the filter paper with two fingers of the same hand and measure the voltage again. Your body is really just a bag of salt housed in a *semipermeable membrane* (skin) through which ions can diffuse. Small differences in voltage observed when the salt bridge is replaced can be attributed to the junction potential discussed in Section 14-2.

Challenge One hundred and eighty students at Virginia Polytechnic Institute and State University made a salt bridge by holding hands.[4] (Their electrical resistance was lowered by a factor of 100 by wetting everyone's hands.) Can your school beat this record?

soon build up positive charge (from excess Zn^{2+}) and the right half-cell would build up negative charge (by depletion of Cu^{2+}). In an instant, the charge buildup would oppose the driving force for the chemical reaction and the process would cease.

To maintain electroneutrality, Zn^{2+} on the left side diffuses into the salt bridge and Cl^- from the salt bridge diffuses into the left half-cell. In the right half-cell, SO_4^{2-} diffuses into the salt bridge and K^+ diffuses out of the salt bridge. The result is that positive and negative charges in each half-cell remain exactly balanced.

Chemists define the electrode at which *reduction* occurs as the **cathode.** The **anode** is the electrode at which *oxidation* occurs. In Figure 13-2, Cu is the cathode because reduction takes place at its surface ($Cu^{2+} + 2e^- \rightarrow Cu$) and Zn is the anode because it is oxidized ($Zn \rightarrow Zn^{2+} + 2e^-$).

cathode $\leftrightarrow$ reduction

anode $\leftrightarrow$ oxidation

Line Notation

We often use a notation employing just two symbols to describe electrochemical cells:

| phase boundary || salt bridge

The cell in Figure 13-2 is represented by the *line diagram*

$$Zn(s) \mid ZnCl_2(aq) \parallel CuSO_4(aq) \mid Cu(s)$$

Each phase boundary is indicated by a vertical line. The electrodes are shown at the extreme left- and right-hand sides of the diagram. The contents of the salt bridge are not specified.

EXAMPLE   Interpreting Line Diagrams of Cells

Write a line diagram for the cell in Figure 13-3. Write an anode reaction (an oxidation) for the left half-cell, a cathode reaction (a reduction) for the right half-cell, and the net cell reaction.

SOLUTION The right half-cell contains two solid phases (Ag and AgCl) and an aqueous phase. The left half-cell contains one solid phase and one aqueous phase. The line diagram is

$$Cd(s) \mid Cd(NO_3)_2(aq) \parallel KCl(aq) \mid AgCl(s) \mid Ag(s)$$

$Cd(NO_3)_2$ is a strong electrolyte, so it is completely dissociated to Cd^{2+} and NO_3^-; and KCl is dissociated into K^+ and Cl^-. In the left half-cell, we find cadmium in the oxidation states 0 and +2. In the right half-cell, we find silver in the 0 and +1 states. The Ag(I) is solid AgCl adhering to the Ag electrode. The electrode reactions are

anode (oxidation): $Cd(s) \rightleftharpoons Cd^{2+}(aq) + 2e^-$

cathode (reduction): $AgCl(s) + e^- \rightleftharpoons Ag(s) + Cl^-(aq)$

net reaction: $Cd(s) + AgCl(s) \rightleftharpoons Cd^{2+}(aq) + Ag(s) + Cl^-(aq) + e^-$

Oops! We didn't write both half-reactions with the same number of electrons, so the net reaction still has electrons in it. This is not allowed, so we double the coefficients of the second half-reaction:

anode (oxidation): $Cd(s) \rightleftharpoons Cd^{2+}(aq) + 2e^-$

cathode (reduction): $2AgCl(s) + 2e^- \rightleftharpoons 2Ag(s) + 2Cl^-(aq)$

net reaction: $Cd(s) + 2AgCl(s) \rightleftharpoons Cd^{2+}(aq) + 2Ag(s) + 2Cl^-(aq)$

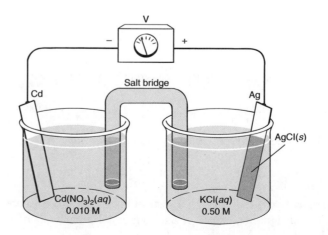

FIGURE 13-3 Another galvanic cell.

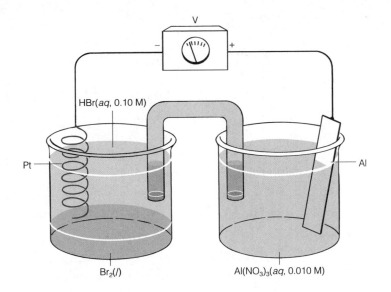

FIGURE 13-4 Cell for Ask Yourself 13-B.

Ask Yourself

13-B. (a) What is the line notation for the cell in Figure 13-4?

(b) Draw a picture of the following cell. Write an oxidation half-reaction for the left half-cell, a reduction half-reaction for the right half-cell, and a balanced net reaction.

$$Au(s)\,|\,Fe(CN)_6^{4-}(aq),\,Fe(CN)_6^{3-}(aq)\,\|\,Ag(S_2O_3)_2^{3-}(aq),\,S_2O_3^{2-}(aq)\,|\,Ag(s)$$

13-3 Standard Potentials

The voltage measured in the experiment in Figure 13-2 is the difference in electric potential between the Cu electrode on the right and the Zn electrode on the left. The greater the voltage, the more favorable is the net cell reaction and the more work can be done by electrons flowing from one side to the other (Equation 13-2). The potentiometer terminals are labeled + and −. The voltage is positive when electrons flow into the negative terminal, as in Figure 13-2. If electrons flow the other way, the voltage is negative. We will draw every cell with the negative terminal of the potentiometer at the left.

To predict the voltage when different half-cells are connected to each other, the **standard reduction potential** ($E°$) for each half-cell is measured by an experiment shown in idealized form in Figure 13-5. The half-reaction of interest in this diagram is

$$Ag^+ + e^- \rightleftharpoons Ag(s) \tag{13-3}$$

occurring in the right-hand half-cell, which is connected to the *positive* terminal of the potentiometer. The term *standard* means that species are solids or liquids or their concentrations are 1 M or their pressures are 1 bar. We call these conditions the *standard states* of the reactants and products.

We oversimplify the standard half-cell in this text. The word *standard* really means that all *activities* are 1. Although activity (Section 11-2) is not the same as concentration, we will consider *standard* to mean that concentrations are 1 M or 1 bar for simplicity.

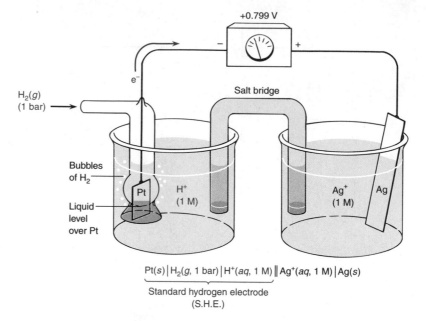

+0.799 V

H₂(g)
(1 bar)

Salt bridge

e⁻

Bubbles
of H₂

Liquid
level
over Pt

Pt

H⁺
(1 M)

Ag⁺
(1 M)

Ag

$$Pt(s)\,|\,H_2(g,\,1\,bar)\,|\,H^+(aq,\,1\,M)\,\|\,Ag^+(aq,\,1\,M)\,|\,Ag(s)$$

Standard hydrogen electrode
(S.H.E.)

FIGURE 13-5 Setup used to measure the standard reduction potential ($E°$) for the half-reaction $Ag^+ + e^- \rightleftharpoons Ag(s)$. The left half-cell is called the standard hydrogen electrode (S.H.E.).

Question What is the pH of the standard hydrogen electrode?

We will write all half-reactions as *reductions*. By convention, $E° = 0$ for S.H.E.

The left half-cell, connected to the *negative* terminal of the potentiometer, is called the **standard hydrogen electrode (S.H.E.).** It consists of a catalytic Pt surface in contact with an acidic solution in which $[H^+] = 1$ M. A stream of $H_2(g,\,1\,bar)$ is bubbled through the electrode. The reaction at the surface of the Pt electrode is

$$H^+(aq,\,1\,M) + e^- \rightleftharpoons \tfrac{1}{2}H_2(g,\,1\,bar) \tag{13-4}$$

We *assign* a potential of 0 to the standard hydrogen electrode. The voltage measured in Figure 13-5 can therefore be *assigned* to Reaction 13-3, which occurs in the right half-cell. The measured value $E° = +0.799$ V is the standard reduction potential for Reaction 13-3. The positive sign tells us that electrons flow from left to right through the meter.

The line notation for the cell in Figure 13-5 is

$$Pt(s)\,|\,H_2(g,\,1\,bar),\,H^+(aq,\,1\,M)\,\|\,Ag^+(aq,\,1\,M)\,|\,Ag(s)$$

which is abbreviated

$$S.H.E.\,\|\,Ag^+(aq,\,1\,M)\,|\,Ag(s)$$

Measured voltage =
   right-hand electrode potential −
   left-hand electrode potential

The standard reduction potential is really the *difference* between the standard potential of the reaction of interest and the potential of the S.H.E., which we have arbitrarily set to 0.

If we wanted to measure the standard potential of the half-reaction

$$Cd^{2+} + 2e^- \rightleftharpoons Cd(s) \tag{13-5}$$

we would construct the cell

$$S.H.E.\,\|\,Cd^{2+}(aq,\,1\,M)\,|\,Cd(s)$$

with the cadmium half-cell *at the right*. In this case, we would observe a *negative* voltage of -0.402 V. The negative sign means that electrons flow from Cd to Pt, a direction opposite that of the cell in Figure 13-5.

Challenge Draw a picture of the cell S.H.E.$\|Cd^{2+}(aq, 1\ M)\,|\,Cd(s)$ and show the direction of electron flow.

What the Standard Potential Means

Table 13-1 lists a few reduction half-reactions in order of decreasing $E°$ value. The more positive $E°$, the more energetically favorable is the half-reaction. The strongest oxidizing agents are the reactants at the upper left side of the table because they have the strongest tendency to accept electrons. $F_2(g)$ is the strongest oxidizing agent in the table. Conversely, F^- is the weakest reducing agent, because it has the least tendency to give up electrons to make F_2. The strongest reducing agents in Table 13-1 are at the bottom right side. $Li(s)$ and $K(s)$ are very strong reducing agents.

Formal Potential

Appendix C lists many standard reduction potentials. Sometimes, multiple potentials are listed for one reaction, as for the $AgCl(s)\,|\,Ag(s)$ half-reaction:

$$AgCl(s) + e^- \rightleftharpoons Ag(s) + Cl^- \qquad \begin{cases} 0.222\ V \\ 0.197\ V\ \text{saturated KCl} \end{cases}$$

Standard potential: 0.222 V

Formal potential for saturated KCl: 0.197 V

The value 0.222 V is the standard potential that would be measured in the cell

$$\text{S.H.E.} \parallel Cl^-(aq, 1\ M)\,|\,AgCl(s)\,|\,Ag(s)$$

TABLE 13-1   **Ordered redox potentials**

Oxidizing agent	Reducing agent	$E°$ (V)
$F_2(g) + 2e^- \rightleftharpoons 2F^-$		2.890
$O_3(g) + 2H^+ + 2e^- \rightleftharpoons O_2(g) + H_2O$		2.075
$\vdots$		
$MnO_4^- + 8H^+ + 5e^- \rightleftharpoons Mn^{2+} + 4H_2O$		1.507
$\vdots$		
$Ag^+ + e^- \rightleftharpoons Ag(s)$		0.799
$\vdots$		
$Cu^{2+} + 2e^- \rightleftharpoons Cu(s)$		0.339
$\vdots$		
$2H^+ + 2e^- \rightleftharpoons H_2(g)$		0.000
$\vdots$		
$Cd^{2+} + 2e^- \rightleftharpoons Cd(s)$		-0.402
$\vdots$		
$K^+ + e^- \rightleftharpoons K(s)$		-2.936
$Li^+ + e^- \rightleftharpoons Li(s)$		-3.040

Oxidizing power increases ↑   Reducing power increases ↓

The value 0.197 V is measured in a cell containing saturated KCl solution instead of 1 M Cl⁻:

$$\text{S.H.E.} \parallel \text{KCl } (aq, \text{ saturated}) \mid \text{AgCl}(s) \mid \text{Ag}(s)$$

The potential for a cell containing a specified concentration of reagent other than 1 M is called the **formal potential.**

Ask Yourself

13-C. Draw a line diagram and a picture of the cell used to measure the standard potential of the reaction $Fe^{3+} + e^- \rightleftharpoons Fe^{2+}$, which comes to equilibrium at the surface of a Pt electrode. Use Appendix C to find the cell voltage. From the sign of the voltage, show the direction of electron flow in your picture.

13-4   The Nernst Equation

Le Châtelier's principle tells us that if we increase reactant concentrations, we drive a reaction to the right. Increasing the product concentrations drives a reaction to the left. The net driving force for a reaction is expressed by the **Nernst equation,** whose two terms include the driving force under standard conditions ($E°$, which applies when concentrations are 1 M or 1 bar) and a term that shows the dependence on concentrations.

Nernst Equation for a Half-Reaction

For the half-reaction

$$a\text{A} + n\text{e}^- \rightleftharpoons b\text{B}$$

the Nernst equation giving the half-cell potential, E, is

In the Nernst equation:

- E and $E°$ are measured in volts
- solute concentration = mol/L
- gas concentration = bar
- solid, liquid, solvent omitted

Nernst equation: $\quad E = E° - \dfrac{0.059\ 16}{n} \log\left(\dfrac{[\text{B}]^b}{[\text{A}]^a}\right) \quad$ (at 25° C)   (13-6)

where $E°$ is the standard reduction potential that applies when $[\text{A}] = [\text{B}] = 1$ M.

The logarithmic term in the Nernst equation is the *reaction quotient,* $Q\ (= [\text{B}]^b/[\text{A}]^a)$. Q has the same form as the equilibrium constant, but the concentrations need not be at their equilibrium values. Concentrations of solutes are expressed as moles per liter and concentrations of gases are expressed as pressures in bars. Pure solids, pure liquids, and solvents are omitted from Q. When all concentrations are 1 M and all pressures are 1 bar, $Q = 1$ and log $Q = 0$, thus giving $E = E°$.

EXAMPLE   **Writing the Nernst Equation for a Half-Reaction**

Let's write the Nernst equation for the reduction of white phosphorus to phosphine gas:

$$\tfrac{1}{4}P_4(s, \text{white}) + 3H^+ + 3e^- \rightleftharpoons PH_3(g) \qquad E° = -0.046 \text{ V}$$

Phosphine

SOLUTION We omit solids from the reaction quotient, and the concentration of phosphine gas is expressed as its pressure in bar (P_{PH_3}):

$$E = -0.046 - \frac{0.059\ 16}{3} \log\left(\frac{P_{PH_3}}{[H^+]^3}\right)$$

EXAMPLE   **Multiplication of a Half-Reaction**

If you multiply a half-reaction by any factor, the value of E° does not change. However, the factor n before the log term and the exponents in the reaction quotient do change. Write the Nernst equation for the reaction in the previous example, multiplied by 2:

$$\tfrac{1}{2}P_4(s, \text{white}) + 6H^+ + 6e^- \rightleftharpoons 2PH_3(g) \qquad E° = -0.046 \text{ V}$$

SOLUTION

$$E = -0.046 - \frac{0.059\ 16}{6} \log\left(\frac{P_{PH_3}^2}{[H^+]^6}\right)$$

$E°$ remains at -0.046 V, as in the previous example. However, the factor in front of the log term and the exponents in the log term have changed.

Nernst Equation for a Complete Reaction

Consider the cell in Figure 13-3 in which the negative terminal of the potentiometer is connected to the Cd electrode and the positive terminal is connected to the Ag electrode. The voltage, *E,* is the difference between the potentials of the two electrodes:

Nernst equation for a complete cell: $E = E_+ - E_-$ (13-7)

where E_+ is the potential of the half-cell attached to the positive terminal of the potentiometer and E_- is the potential of the half-cell attached to the negative terminal. The potential of each half-reaction (*written as a reduction*) is governed by the Nernst equation 13-6.

Both half-reactions are written as *reductions* when we use Equation 13-7.

Here is a procedure for writing a net cell reaction and finding its voltage:

Step 1. Write *reduction* half-reactions for both half-cells and find $E°$ for each in Appendix C. Multiply the half-reactions as necessary so that they each contain the same number of electrons. When you multiply a reaction by any number, *do not* multiply $E°$.

Step 2. Write a Nernst equation for the half-reaction in the right half-cell, which is attached to the positive terminal of the potentiometer. This is E_+.

Step 3. Write a Nernst equation for the half-reaction in the left half-cell, which is attached to the negative terminal of the potentiometer. This is E_-.

Step 4. Find the net cell voltage by subtraction: $E = E_+ - E_-$.

Step 5. To write a balanced net cell reaction, subtract the left half-reaction from the right half-reaction. (*This operation is equivalent to reversing the left half-reaction and adding.*)

If the net cell voltage, $E (= E_+ - E_-)$, is positive, then the net cell reaction is spontaneous in the forward direction. If the net cell voltage is negative, then the reaction is spontaneous in the reverse direction.

EXAMPLE   Nernst Equation for a Complete Reaction

Find the voltage of the cell in Figure 13-3 if the right half-cell contains 0.50 M KCl(*aq*) and the left half-cell contains 0.010 M Cd(NO$_3$)$_2$(*aq*). Write the net cell reaction and state whether it is spontaneous in the forward or reverse direction.

SOLUTION

Step 1. right half-cell: $2AgCl(s) + 2e^- \rightleftharpoons 2Ag(s) + 2Cl^-$ $E_+° = 0.222$ V

left half-cell: $Cd^{2+} + 2e^- \rightleftharpoons Cd(s)$ $E_-° = -0.402$ V

Step 2. Nernst equation for right half-cell:

Pure solids, pure liquids, and solvents are omitted from Q.

$$E_+ = E_+° - \frac{0.059\ 16}{2} \log([Cl^-]^2) \tag{13-8}$$

$$= 0.222 - \frac{0.059\ 16}{2} \log([0.50]^2) = 0.240 \text{ V}$$

Step 3. Nernst equation for left half-cell:

$$E_- = E_-° - \frac{0.059\ 16}{2} \log\left(\frac{1}{[Cd^{2+}]}\right) = -0.402 - \frac{0.059\ 16}{2} \log\left(\frac{1}{[0.010]}\right)$$

$$= -0.461 \text{ V}$$

Step 4. cell voltage: $E = E_+ - E_- = 0.240 - (-0.461) = 0.701$ V

Step 5. net cell reaction:

$$2AgCl(s) + 2e^- \rightleftharpoons 2Ag(s) + 2Cl^-$$

$$- \quad Cd^{2+} + 2e^- \rightleftharpoons Cd(s)$$

$$Cd(s) + 2AgCl(s) \rightleftharpoons Cd^{2+} + 2Ag(s) + 2Cl^-$$

Subtracting a reaction is the same as *reversing the reaction and adding.*

Because the voltage is positive, the net reaction is spontaneous in the forward direction. $Cd(s)$ is oxidized to Cd^{2+} and $AgCl(s)$ is reduced to $Ag(s)$. Electrons flow from the left electrode to the right electrode.

What if you had written the Nernst equation for the right half-cell with just one electron instead of two: $AgCl(s) + e^- \rightleftharpoons Ag(s) + Cl^-$? Try this and you will discover that the half-cell potential is unchanged. *Neither $E°$ nor E changes when you multiply a reaction.*

Different Descriptions of the Same Reaction

We know that the right half-cell in Figure 13-3 must contain some $Ag^+(aq)$ in equilibrium with $AgCl(s)$. Suppose that instead of writing the reaction $2AgCl(s) + 2e^- \rightleftharpoons 2Ag(s) + 2Cl^-$, a different, less handsome, author wrote the reaction

$$2Ag^+(aq) + 2e^- \rightleftharpoons 2Ag(s) \qquad E_+° = 0.799 \text{ V}$$

$$E_+ = 0.799 - \frac{0.059\,16}{2} \log\left(\frac{1}{[Ag^+]^2}\right) \qquad (13\text{-}9)$$

Both descriptions of the right half-cell are equally valid. In both cases, Ag(I) is reduced to Ag(0).

If the two descriptions are equal, then they should predict the same voltage. To use Equation 13-9, you must know the concentration of Ag^+ in the right half-cell, which is not obvious. But you are clever and cunning and realize that you can find $[Ag^+]$ from the solubility product for AgCl and the known concentration of Cl^- (which is 0.50 M).

$$K_{sp} = [Ag^+][Cl^-] \Rightarrow [Ag^+] = \frac{K_{sp}}{[Cl^-]} = \frac{1.8 \times 10^{-10}}{0.50} = 3.6 \times 10^{-10} \text{ M}$$

Putting this concentration into Equation 13-9 gives

$$E_+ = 0.799 - \frac{0.059\,16}{2} \log\left(\frac{1}{(3.6 \times 10^{-10})^2}\right) = 0.240 \text{ V}$$

The cell voltage is a measured, experimental quantity that cannot depend on how we write the reaction!

which is exactly the same voltage computed in Equation 13-8! The two choices of half-reaction give the same voltage because they describe the same cell.

Advice for Finding Relevant Half-Reactions

When faced with a cell drawing or line diagram, the first step is to write reduction reactions for each half-cell. To do this, *look for elements in two oxidation states.* For the cell

$$Pb(s) \mid PbF_2(s) \mid F^-(aq) \parallel Cu^{2+}(aq) \mid Cu(s)$$

we see lead in the oxidation states 0 in Pb(s) and +2 in $PbF_2(s)$, and copper in the oxidation states 0 in Cu(s) and +2 in Cu^{2+}. Thus, the half-reactions are

right half-cell: $\qquad\qquad Cu^{2+} + 2e^- \rightleftharpoons Cu(s)$

left half-cell: $\qquad\qquad PbF_2(s) + 2e^- \rightleftharpoons Pb(s) + 2F^-$ (13-10)

You might have chosen to write the lead half-reaction as

left half-cell: $\qquad\qquad Pb^{2+} + 2e^- \rightleftharpoons Pb(s)$ (13-11)

because you know that if $PbF_2(s)$ is present, there must be some Pb^{2+} in the solution. Reactions 13-10 and 13-11 are equally valid descriptions of the cell, and each should predict the same cell voltage. Your choice of reaction depends on whether the F^- or Pb^{2+} concentration is more easily known to you.

Don't write a reaction such as $F_2(g) + 2e^- \rightleftharpoons 2F^-$, because $F_2(g)$ is not shown in the line diagram of the cell. $F_2(g)$ is neither a reactant nor a product.

Ask Yourself

13-D. **(a)** Arsine (AsH_3) is a poisonous gas used to make gallium arsenide for diode lasers. Find E for the half-reaction $As(s) + 3H^+ + 3e^- \rightleftharpoons AsH_3(g)$ if pH = 3.00 and $P_{AsH_3} = 0.010\ 0$ bar.
   (b) The cell in Demonstration 13-2 (and Figure 13-2) can be written

$$Zn(s) \mid Zn^{2+}(0.1\ M) \parallel Cu^{2+}(0.1\ M) \mid Cu(s)$$

Write a reduction half-reaction for each half-cell and use the Nernst equation to predict the cell voltage.

13-5 $E°$ and the Equilibrium Constant

A galvanic cell produces electricity because the cell reaction is not at equilibrium. If the cell runs long enough, reactants are consumed until the reaction comes to equilibrium, and the cell voltage, E, reaches 0. This is what happens to a battery when it runs down.

At equilibrium, E (not $E°$) = 0.

If $E_+°$ is the standard reduction potential for the right half-cell and $E_-°$ is the standard reduction potential for the left half-cell, $E°$ for the net cell reaction is

Remember that $E°$ is the potential when all reactants and products are present in their standard state (1 M, 1 bar, pure solid, pure liquid).

E° for net cell reaction: $\qquad\qquad E° = E_+° - E_-°$ (13-12)

Now let's relate $E°$ to the equilibrium constant for the net cell reaction.

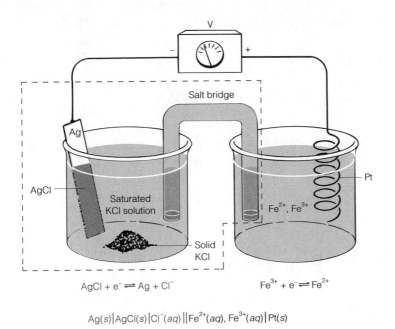

AgCl + e⁻ ⇌ Ag + Cl⁻   Fe³⁺ + e⁻ ⇌ Fe²⁺

$Ag(s)|AgCl(s)|Cl^-(aq)\|Fe^{2+}(aq), Fe^{3+}(aq)|Pt(s)$

FIGURE 13-6 Cell used to illustrate the relation of $E°$ to the equilibrium constant. The dashed line encloses the part of the cell that we call the *reference electrode* in Section 13-6.

As an example, consider the cell in Figure 13-6. The left half-cell contains a silver electrode coated with solid AgCl and dipped into saturated aqueous KCl. The right half-cell has a platinum wire dipped into a solution containing Fe^{2+} and Fe^{3+}.

The two half-reactions are

right: $Fe^{3+} + e^- \rightleftharpoons Fe^{2+}$ $E_+^\circ = 0.771$ V   (13-13)

left: $AgCl(s) + e^- \rightleftharpoons Ag(s) + Cl^-$ $E_-^\circ = 0.222$ V   (13-14)

E_+° and E_+ refer to the half-cell connected to the positive terminal of the potentiometer.

E_-° and E_- refer to the half-cell connected to the negative terminal.

The net cell reaction is found by subtracting the left half-reaction from the right half-reaction:

net reaction: $Fe^{3+} + Ag(s) + Cl^- \rightleftharpoons Fe^{2+} + AgCl(s)$

$$E° = E_+^\circ - E_-^\circ = 0.771 - 0.222 = 0.549 \text{ V} \qquad (13\text{-}15)$$

The two electrode potentials are given by the Nernst equation 13-6:

$$E_+ = E_+^\circ - \frac{0.059\ 16}{n} \log\left(\frac{[Fe^{2+}]}{[Fe^{3+}]}\right) \qquad (13\text{-}16)$$

$$E_- = E_-^\circ - \frac{0.059\ 16}{n} \log\left([Cl^-]\right) \qquad (13\text{-}17)$$

where $n = 1$ for Reactions 13-13 and 13-14. The cell voltage is the difference $E_+ - E_-$:

$$E = E_+ - E_- = \left\{ E_+^\circ - \frac{0.059\ 16}{n} \log\left(\frac{[Fe^{2+}]}{[Fe^{3+}]}\right) \right\} - \left\{ E_-^\circ - \frac{0.059\ 16}{n} \log([Cl^-]) \right\}$$

$$= (E_+^\circ - E_-^\circ) - \left\{ \frac{0.059\ 16}{n} \log\left(\frac{[Fe^{2+}]}{[Fe^{3+}]}\right) - \frac{0.059\ 16}{n} \log([Cl^-]) \right\} \quad (13\text{-}18)$$

To combine logarithms, we use the equality $\log x - \log y = \log(x/y)$:

$$E = \underbrace{(E_+^\circ - E_-^\circ)}_{E^\circ \text{ for net reaction}} - \frac{0.059\,16}{n} \log \left(\underbrace{\frac{[Fe^{2+}]}{[Fe^{3+}][Cl^-]}}_{\substack{\text{Reaction quotient } (Q) \\ \text{for net reaction}}} \right) = E^\circ - \frac{0.059\,16}{n} \log Q \qquad (13\text{-}19)$$

Algebra of logarithms:

$$\log x + \log y = \log(xy)$$
$$\log x - \log y = \log(x/y)$$

Equation 13-19 is true at any time. *In the special case when the cell is at equilibrium, E = 0 and Q = K, the equilibrium constant.* Therefore,

$$0 = E^\circ - \frac{0.059\,16}{n} \log K \Rightarrow E^\circ = \frac{0.059\,16}{n} \log K \qquad (13\text{-}20)$$

To go from Equation 13-20 to 13-21:

$$\frac{0.059\,16}{n} \log K = E^\circ$$

$$\log K = \frac{nE^\circ}{0.059\,16}$$

$$10^{\log K} = 10^{nE^\circ/0.059\,16}$$

$$K = 10^{nE^\circ/0.059\,16}$$

Finding K from E°: $\qquad K = 10^{nE^\circ/0.059\,16} \qquad$ (at 25°C) $\qquad (13\text{-}21)$

Equation 13-21 gives the equilibrium constant for a net cell reaction from E°.

A positive value of E° means that $K > 1$ in Equation 13-21 and a negative value means that $K < 1$. A reaction is spontaneous under standard conditions (i.e., when all concentrations of reactants and products are 1 M or 1 bar) if E° is positive. Biochemists use a different potential, called $E^{\circ\prime}$, which is described in Box 13-1.

EXAMPLE Using E° to Find the Equilibrium Constant

Find the equilibrium constant for the reaction $Fe^{3+} + Ag(s) + Cl^- \rightleftharpoons Fe^{2+} + AgCl(s)$, which is the net cell reaction in Figure 13-6.

SOLUTION Equation 13-15 states that $E^\circ = 0.549$ V. The equilibrium constant is computed with Equation 13-21, using $n = 1$, because one electron is transferred in each half-reaction:

$$K = 10^{nE^\circ/0.059\,16} = 10^{(1)(0.549)/(0.059\,16)} = 1.9 \times 10^9$$

Significant figures in logarithms and exponents were discussed in Section 3.2.

K has two significant figures because E° has three digits. One digit of E° is used for the exponent (9), and the other two are left for the multiplier (1.9).

Ask Yourself

13-E. (a) Write the half-reactions for Figure 13-2. Calculate E° and the equilibrium constant for the net cell reaction.

   (b) The solubility product reaction of AgBr is $AgBr(s) \rightleftharpoons Ag^+ + Br^-$. Use the reactions $Ag^+ + e^- \rightleftharpoons Ag(s)$ and $AgBr(s) + e^- \rightleftharpoons Ag(s) + Br^-$ to compute the solubility product of AgBr. Compare your answer to that in Appendix A.

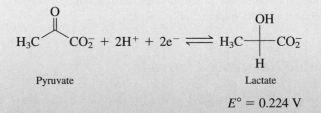

Box 13-1 Explanation

Why Biochemists Use $E°'$

Redox reactions are essential for life. For example, the enzyme-catalyzed reduction of pyruvate to lactate is a step in the anaerobic fermentation of sugar by bacteria:

H_3C (Pyruvate) $CO_2^- + 2H^+ + 2e^- \rightleftharpoons H_3C$ (Lactate) CO_2^-

$$E° = 0.224 \text{ V}$$

The same reaction causes lactate to build up in your muscles and makes you feel fatigued during intense exercise, when the flow of oxygen cannot keep pace with your requirement for energy.

$E°$ applies when the concentrations of reactants and products are 1 M. For a reaction involving H^+, $E°$ applies when the pH is 0 (because $[H^+] = 1$ M and log 1 = 0). Biochemists studying the energetics of fermentation or respiration are more interested in reduction potentials that apply near physiologic pH, not at pH 0. Therefore, biochemists use a formal potential designated $E°'$, which applies at pH 7.

$E°$ applies at pH 0
$E°'$ applies at pH 7

For the conversion of pyruvate to lactate, the reactant and product are carboxylic acids at pH 0 and carboxylate anions at pH 7. The value of $E°'$ at pH 7 is -0.190 V, which is quite different from $E° = +0.224$ V at pH 0.

13-6 Reference Electrodes

Imagine a solution containing an electroactive species whose concentration we wish to measure. We construct a half-cell by inserting an electrode (such as a Pt wire) into the solution to transfer electrons to or from the species of interest. Because this electrode responds directly to the analyte, it is called the **indicator electrode.** The potential of the indicator electrode is E_+. We then connect this half-cell to a second half-cell via a salt bridge. The second half-cell has a fixed composition that provides a known, constant potential, E_-. Because the second half-cell has a constant potential, it is called a **reference electrode.** The cell voltage $(E = E_+ - E_-)$ is the difference between the variable potential that reflects changes in the analyte concentration and the constant reference potential.

Suppose you have a solution containing Fe^{2+} and Fe^{3+}. If you are clever, you can make this solution part of a cell whose voltage tells you the relative concentrations of these species—that is, the value of $[Fe^{2+}]/[Fe^{3+}]$. Figure 13-6 shows one way to do this. A Pt wire is inserted as an indicator electrode through which Fe^{3+} can receive electrons or Fe^{2+} can lose electrons. The left half-cell completes the galvanic cell and has a known, constant potential.

The two half-reactions were given in Reactions 13-13 and 13-14, and the two electrode potentials were given in Equations 13-16 and 13-17. The cell voltage is the difference $E_+ - E_-$:

$$E = E_+ - E_- = \left\{ \underbrace{E_+°}_{\text{Constant}} - \underbrace{0.059\ 16 \log\left(\frac{[Fe^{2+}]}{[Fe^{3+}]}\right)}_{\text{The variable of interest}} \right\} -$$

$$\{ \underbrace{E_-°}_{\text{Constant}} - \underbrace{0.059\ 16 \log([Cl^-])}_{\substack{\text{Constant concentration} \\ \text{in left half cell}}} \} \quad (13\text{-}22)$$

Indicator electrode: responds to analyte concentration

Reference electrode: maintains a fixed (reference) potential

The cell voltage in Figure 13-6 responds only to changes in the quotient $[Fe^{2+}]/[Fe^{3+}]$. Everything else is constant.

295

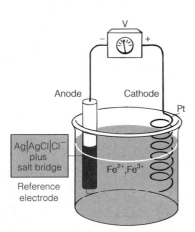

FIGURE 13-7 Another view of Figure 13-6. The contents of the dashed box in Figure 13-6 are now considered to be a reference electrode dipped into the analyte solution.

Because $[Cl^-]$ in the left half-cell is constant (fixed by the solubility of KCl, with which the solution is saturated), the cell voltage changes only when the quotient $[Fe^{2+}]/[Fe^{3+}]$ changes.

The half-cell on the left in Figure 13-6 can be thought of as a *reference electrode*. We can picture the cell and salt bridge enclosed by the dashed line as a single unit dipped into the analyte solution, as shown in Figure 13-7. The Pt wire is the indicator electrode, whose potential responds to changes in the quotient $[Fe^{2+}]/[Fe^{3+}]$. The reference electrode completes the redox reaction and provides a *constant potential* to the left side of the potentiometer. Changes in the cell voltage can be assigned to changes in the quotient $[Fe^{2+}]/[Fe^{3+}]$.

Silver-Silver Chloride Reference Electrode

The half-cell enclosed by the dashed line in Figure 13-6 is called a **silver-silver chloride electrode.** Figure 13-8 shows how the half-cell is reconstructed as a thin, glass-enclosed electrode that can be dipped into the analyte solution in Figure 13-7. The porous plug at the base of the electrode functions as a salt bridge. It allows electrical contact between solutions inside and outside the electrode with minimal physical mixing.

The standard reduction potential for AgCl | Ag is +0.222 V at 25°C. If the cell is saturated with KCl, the potential is +0.197 V. This is the value we will use for all problems involving the AgCl | Ag reference electrode. The advantage of a saturated KCl solution is that the concentration of chloride does not change if some of the liquid evaporates.

$$Ag\ |\ AgCl\ electrode:\qquad AgCl(s) + e^- \rightleftharpoons Ag(s) + Cl^- \qquad E° = +0.222\ V$$
$$E(saturated\ KCl) = +0.197\ V$$

EXAMPLE Using a Reference Electrode

Calculate the cell voltage in Figure 13-7 if the reference electrode is a saturated silver-silver chloride electrode and $[Fe^{2+}]/[Fe^{3+}] = 10$.

SOLUTION We use Equation 13-22, noting that $E_- = 0.197$ V for a saturated silver-silver chloride electrode:

$$E = E_+ - E_- = \left\{ \underbrace{E_+^\circ}_{0.771\ V} - 0.059\ 16 \log \left(\underbrace{\frac{[Fe^{2+}]}{[Fe^{3+}]}}_{10} \right) \right\} - \underbrace{0.197}_{\substack{Reference \\ electrode\ voltage}}$$

$$E = \{0.712\} - 0.197 = 0.515\ V$$

- Wire lead
- Air inlet to allow electrolyte to drain slowly through porous plug
- Ag wire bent into a loop
- Aqueous solution saturated with KCl and AgCl
- AgCl paste
- Solid KCl plus some AgCl
- Porous plug for contact with external solution (salt bridge)

FIGURE 13-8 Silver-silver chloride reference electrode.

Calomel Reference Electrode

The *calomel electrode* in Figure 13-9 is based on the reaction

Calomel electrode: $\frac{1}{2}Hg_2Cl_2(s) + e^- \rightleftharpoons Hg(l) + Cl^-$ $E° = +0.268$ V

$\qquad$ Mercury(I) chloride $\qquad\qquad$ $E(\text{saturated KCl}) = +0.241$ V
$\qquad\qquad$ (calomel)

If the cell is saturated with KCl, it is called a **saturated calomel electrode** and the cell potential is $+0.241$ V at 25°C. This electrode is encountered so frequently that it is abbreviated **S.C.E.**

Voltage Conversions Between Different Reference Scales

It is sometimes necessary to convert potentials between different reference scales. If an electrode has a potential of -0.461 V with respect to a calomel electrode, what is the potential with respect to a silver-silver chloride electrode? What would be the potential with respect to the standard hydrogen electrode?

$\qquad$ To answer these questions, Figure 13-10 shows the positions of the calomel and silver-silver chloride electrodes with respect to the standard hydrogen electrode. Point A, which is -0.461 V from S.C.E., is -0.417 V from the silver-silver chloride electrode and -0.220 V from S.H.E. What about point B, whose potential is $+0.033$ V with respect to silver-silver chloride? Its position is -0.011 V from S.C.E. and $+0.230$ V with respect to S.H.E. By keeping this diagram in mind, you can convert potentials from one scale to another.

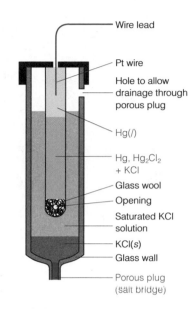

FIGURE 13-9   Saturated calomel electrode (S.C.E.).

Ask Yourself

13-F. **(a)** Find the concentration ratio $[Fe^{2+}]/[Fe^{3+}]$ for the cell in Figure 13-7 if the measured voltage is 0.703 V.

$\qquad$ **(b)** Convert the potentials listed below. The Ag | AgCl and calomel reference electrodes are saturated with KCl.

$\qquad\qquad$ **(i)**   0.523 V versus S.H.E. = ? vs Ag | AgCl

$\qquad\qquad$ **(ii)** 0.222 V versus S.C.E. = ? vs S.H.E.

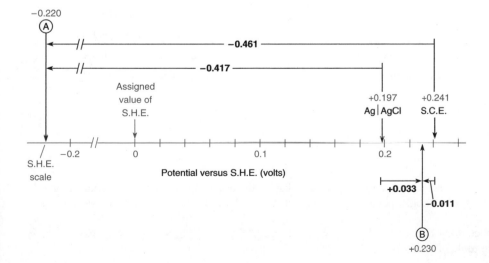

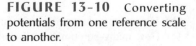

FIGURE 13-10   Converting potentials from one reference scale to another.

Key Equations

Definitions	Oxidizing agent—takes electrons Reducing agent—gives electrons Anode—where oxidation occurs Cathode—where reduction occurs
Standard potential	Measured by the cell S.H.E. ‖ half-reaction of interest, where S.H.E. is the standard hydrogen electrode and all reagents in the right half-cell are in their standard state ($= 1$ M, 1 bar, pure solid, or pure liquid)
Nernst equation	For the half-reaction $aA + ne^- \rightleftharpoons bB$ $$E = E° - \frac{0.059\ 16}{n} \log\left(\frac{[B]^b}{[A]^a}\right) \quad \text{(at 25°C)}$$
Voltage of complete cell	$E = E_+ - E_-$ E_+ = voltage of electrode connected to + terminal of meter E_- = voltage of electrode connected to − terminal of meter
$E°$ for net cell reaction	$E° = E°_+ - E°_-$ $E°_+$ = standard potential for half-reaction in cell connected to + terminal of meter $E°_-$ = standard potential for half-reaction in cell connected to − terminal of meter
Finding K from $E°$	$K = 10^{nE°/0.059\ 16}$ (n = number of e$^-$ in half-reaction)

Important Terms

ampere	formal potential	reductant
anode	galvanic cell	reduction
cathode	indicator electrode	reference electrode
coulomb	Nernst equation	salt bridge
current	oxidant	saturated calomel electrode (S.C.E.)
electric potential	oxidation	silver-silver chloride electrode
electroactive species	oxidizing agent	standard hydrogen electrode (S.H.E.)
electrode	redox reaction	standard reduction potential
electrolysis	reducing agent	volt
Faraday constant		

Problems

13-1. (a) Explain the difference between electric charge (q, coulombs), electric current (I, amperes), and electric potential (E, volts).

(b) How many electrons are in one coulomb?

(c) How many coulombs are in one mole of charge?

13-2. Identify the oxidizing and reducing agents among the reactants below and write a balanced half-reaction for each.

$$2S_2O_4^{2-} + TeO_3^{2-} + 2OH^- \rightleftharpoons 4SO_3^{2-} + Te(s) + H_2O$$
Dithionite   Tellurite   Sulfite

(b) How many coulombs of charge are passed from reductant to oxidant when 1.00 g of Te is deposited?

(c) If Te is created at a rate of 1.00 g/h, how much current is flowing?

13-3. Hydrogen ions can carry out useful work in a living cell (such as the synthesis of the molecule ATP that provides energy for chemical synthesis) when they pass from a region of high potential to a region of lower potential. How many joules of work can be done when 1.00 μmol of H^+ crosses a membrane and goes from a potential of $+0.075$ V to a potential of -0.090 V (i.e., through a potential difference of 0.165 V)?

13-4. The basal rate of consumption of O_2 by a 70-kg human is 16 mol O_2/day. This O_2 oxidizes food and is reduced to H_2O, thereby providing energy for the organism:

$$O_2 + 4H^+ + 4e^- \rightleftharpoons 2H_2O$$

(a) To what current (in amperes $= C/s$) does this respiration rate correspond? (Current is defined by the flow of electrons from food to O_2.)

(b) If the electrons flow from reduced nicotinamide adenine dinucleotide (NADH) to O_2, they experience a potential drop of 1.1 V. How many joules of work can be done by 16 mol O_2?

13-5. Draw a picture of the following cell.

$$Pt(s) \mid Hg(l) \mid Hg_2Cl_2(s) \mid KCl(aq) \parallel ZnCl_2(aq) \mid Zn(s)$$

Write an oxidation for the left half-cell and a reduction for the right half-cell.

13-6. Redraw the cell in Figure 13-5, showing KNO_3 in the salt bridge. Noting the direction of electron flow through the circuit, show what happens to each reactant and product in each half-cell and show the direction of motion of each ion in each half-cell and in the salt bridge. Show the reaction at each electrode. When you finish this, you should understand what is happening in the cell.

13-7. Suppose that the concentrations of NaF and KCl were each 0.10 M in the cell

$$Pb(s) \mid PbF_2(s) \mid F^-(aq) \parallel Cl^-(aq) \mid AgCl(s) \mid Ag(s)$$

(a) Using the half-reactions $2AgCl(s) + 2e^- \rightleftharpoons 2Ag(s) + 2Cl^-$ and $PbF_2(s) + 2e^- \rightleftharpoons Pb(s) + 2F^-$, calculate the cell voltage.

(b) Now calculate the cell voltage by using the reactions $2Ag^+ + 2e^- \rightleftharpoons 2Ag(s)$ and $Pb^{2+} + 2e^- \rightleftharpoons Pb(s)$ and K_{sp} for AgCl and PbF$_2$ (see Appendix A).

13-8. Consider a circuit in which the left half-cell was prepared by dipping a Pt wire in a beaker containing an equimolar mixture of Cr^{2+} and Cr^{3+}. The right half-cell contained a Tl rod immersed in 1.00 M TlClO$_4$.

(a) Use line notation to describe this cell.

(b) Calculate the cell voltage.

(c) Write the spontaneous net cell reaction.

(d) When the two electrodes are connected by a salt bridge and a wire, which terminal (Pt or Tl) will be the anode?

13-9. Consider the cell

$$Pt(s) \mid H_2(g, 0.100 \text{ bar}) \mid H^+(aq, pH = 2.54) \parallel$$
$$Cl^-(aq, 0.200 \text{ M}) \mid Hg_2Cl_2(s) \mid Hg(l) \mid Pt(s)$$

(a) Write a reduction reaction and Nernst equation for each half-cell. For the Hg$_2$Cl$_2$ half-reaction, $E° = 0.268$ V.

(b) Find E for the net cell reaction and state whether reduction will occur at the left- or right-hand electrode.

13-10. **(a)** Calculate the cell voltage (E, not $E°$), and state the direction in which electrons will flow through the potentiometer in Figure 13-4. Write the spontaneous net cell reaction.

(b) The left half-cell was loaded with 14.3 mL of $Br_2(l)$ (density $= 3.12$ g/mL). The aluminum electrode contains 12.0 g of Al. Which element, Br_2 or Al, is the limiting reagent in this cell? (That is, which reagent will be used up first?)

(c) If the cell is somehow operated under conditions in which it produces a constant voltage of 1.50 V, how much electrical work will have been done when 0.231 mL of $Br_2(l)$ has been consumed?

(d) If the current is 2.89×10^{-4} A, at what rate (grams per second) is Al(s) dissolving?

13-11. Calculate $E°$ and K for each reaction:

(a) $Cu(s) + Cu^{2+} \rightleftharpoons 2Cu^+$

(b) $2F_2(g) + H_2O \rightleftharpoons F_2O(g) + 2H^+ + 2F^-$

13-12. From the half-reactions below, calculate the solubility product of $Mg(OH)_2$.

$$Mg^{2+} + 2e^- \rightleftharpoons Mg(s) \qquad E° = -2.360 \text{ V}$$
$$Mg(OH)_2(s) + 2e^- \rightleftharpoons Mg(s) + 2OH^- \qquad E° = -2.690 \text{ V}$$

13-13. The cell in Ask Yourself 13-B(b) contains 1.3 mM Fe(CN)$_6^{4-}$, 4.9 mM Fe(CN)$_6^{3-}$, 1.8 mM Ag(S$_2$O$_3$)$_2^{3-}$, and 55 mM S$_2$O$_3^{2-}$.

(a) Find K for the net cell reaction.

(b) Find the cell voltage and state whether Ag(s) is oxidized or reduced by the spontaneous cell reaction.

13-14. From the standard potentials for reduction of $Br_2(aq)$ and $Br_2(l)$ in Appendix C, calculate the solubility of Br_2 in water at 25°C. Express your answer as g/L.

13-15. A solution contains 0.010 0 M IO_3^-, 0.010 0 M I^-, 1.00×10^{-4} M I_3^-, and pH 6.00 buffer. Consider the reactions

$$2IO_3^- + I^- + 12H^+ + 10e^- \rightleftharpoons I_3^- + 6H_2O$$
$$E° = 1.210 \text{ V}$$
$$I_3^- + 2e^- \rightleftharpoons 3I^- \qquad E° = 0.535 \text{ V}$$

(a) Write a balanced net reaction that can occur in this solution.

(b) Calculate $E°$ and K for the reaction.

(c) Calculate E for the conditions given above.

(d) At what pH would the concentrations of IO_3^-, I^-, and I_3^- listed above be in equilibrium?

13-16. (a) Write the half-reactions for the silver-silver chloride and calomel reference electrodes.

(b) Predict the voltage for the cell below.

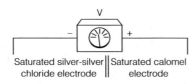

Saturated silver-silver chloride electrode ‖ Saturated calomel electrode

13-17. Convert the potentials listed below. The Ag | AgCl and calomel reference electrodes are saturated with KCl.

(a) -0.111 V versus Ag | AgCl = ? versus S.H.E.

(b) 0.023 V versus Ag | AgCl = ? versus S.C.E.

(c) -0.023 V versus calomel = ? versus Ag | AgCl

13-18. Suppose that the silver-silver chloride electrode in Figure 13-7 is replaced by a saturated calomel electrode. Calculate the cell voltage if $[Fe^{2+}]/[Fe^{3+}] = 2.5 \times 10^{-3}$.

13-19. The formal reduction potential for $Fe^{3+} + e^- \rightleftharpoons Fe^{2+}$ in 1 M $HClO_4$ is 0.73 V. The formal reduction potential for the complex LFe(III) + e⁻ ⇌ LFe(II) (where L is the chelator desferrioxamine B in Box 12-1) is -0.48 V.[5] What do these potentials tell you about the relative stability of the complexes LFe(III) and LFe(II)?

How Would You Do It?

13-20. A *concentration cell* has the same half-reaction in both half-cells, but the concentration of a reactant or product is different in the half-cells. Consider the cell

$$Ag(s) \mid Ag^+(aq, c_l) \parallel Ag^+(aq, c_r) \mid Ag(s)$$

where c_l is the concentration of Ag^+ in the left half-cell and c_r is the concentration of Ag^+ in the right half-cell.

(a) Write the Nernst equation for the cell, showing how the cell voltage depends on the concentrations c_l and c_r. What is the expected voltage when $c_l = c_r$?

(b) The *formation constant* is the equilibrium constant for the reaction of a metal with a ligand: $M + L \rightleftharpoons ML$. Propose a method for measuring formation constants of silver, using a concentration cell.[6] Make up a hypothetical example to predict what voltage would be observed.

13-21. One measure of the capability of a battery is how much electricity it can produce per kilogram of reactants. The quantity of electricity could be measured in coulombs, but it is customarily measured in ampere · hours, where 1 A · h provides 1 A for 1 h. Thus if 0.5 kg of reactants can produce 3 A · h, the storage capacity would be 3 A · h/0.5 kg = 6 A · h/kg. Compare the capabilities of a conventional lead-acid car battery to that of a hydrogen-oxygen fuel cell in terms of A · h/kg.

lead-acid battery: $Pb + PbO_2 + 2H_2SO_4 \longrightarrow$
$$2PbSO_4 + 2H_2O$$
(FM of reactants =
$$207.2 + 239.2 + 2 \times 98.079 = 642.6)$$

hydrogen-oxygen fuel cell: $2H_2 + O_2 \longrightarrow 2H_2O$
(FM of reactants = 36.031)

Notes and References

1. K. S. Betts, *Environ. Sci. Technol.* **1998**, *32*, 495A.

2. http://www.epa.gov/ada/eliz.html; K. R. Waybrant, D. W. Blowes, and C. J. Ptacek, *Environ. Sci. Technol.* **1998**, *32*, 1972; R. W. Powell and R. W. Puls, *Environ. Sci. Technol.* **1997**, *31*, 2244; A. L. Roberts, L. A. Totten, W. A. Arnold, D. R. Burris, and T. J. Campbell, *Environ. Sci. Technol.* **1996**, *30*, 2654. See also S. G. Benner, D. W. Blowes, W. D. Gould, R. B. Herbert, Jr., and C. J. Ptacek, *Environ. Sci. Technol.* **1999**, *33*, 2793.

3. The quotation about Faraday was made by Lady Pollock and cited in J. Kendall, *Great Discoveries by Young Chemists* (New York: Thomas Y. Crowell Co., 1953, p. 63.)

4. L. P. Silverman and B. B. Bunn, *J. Chem. Ed.* **1992,** *69,* 309.

5. I. Spasojević, S. K. Armstrong, T. J. Brickman, and A. L. Crumbliss, *Inorg. Chem.* **1999,** *38,* 449.

6. For a student experiment using concentration cells to measure formation constants, see M. L. Thompson and L. J. Kateley, *J. Chem. Ed.* **1999,** *76,* 95.

Further Reading

A. Hamnett, C. H. Hamann, and W. Vielstich, *Electrochemistry* (New York: Wiley, 1998).

Measuring Ions in a Beating Heart

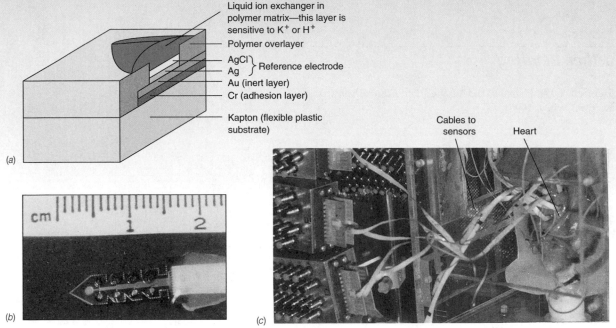

(a) Structure of each ion-sensing dot in the array.

(b) Flexible sensor array that can be inserted inside a beating heart.

(c) Multiple sensor arrays inserted into a beating pig heart monitor simultaneous changes in [K⁺] and [H⁺] when blood is denied to certain regions of the heart. [From V. V. Cosofret, M. Erdösy, T. A. Johnson, R. P. Buck, R. B. Ash, and M. R. Neuman, *Anal. Chem.* **1995,** *67,* 1647.]

(d) Changes in extracellular pH, lactate, and K⁺ when blood flow to the heart is interrupted at time 0. [Data from S. A. M. Marzouk, S. Ufer, R. P. Buck, T. A. Johnson, L. A. Dunlap, and W. E. Cascio, *Anal. Chem.* **1998,** *70,* 5054.]

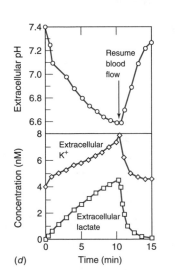

Myocardial ischemia is a condition in which blood flow to part of the heart muscle is interrupted. To study the chemical changes in such an event, flexible, miniature, ion-selective electrodes were designed to be inserted into vessels of a beating heart. In the device shown in panel *b*, four dots are sensors for **K⁺** and four detect **H⁺**. The dot at the front tip is a **Ag|AgCl reference electrode. The sensors provide a linear response to pH over the range 4–12 and to K⁺ from 10^{-1} to 10^{-5} M. Fabrication required photolithographic techniques from the microelectronics industry and polymer know-how to produce a flexible device. When blood flow is interrupted at time 0 in the graphs at the left, muscle metabolism switches from aerobic to anaerobic, increasing the concentrations of H⁺, lactate, and K⁺ outside the cells.**

Chapter 14

ELECTRODE MEASUREMENTS

A critically ill patient is wheeled into the emergency room, and the doctor needs blood chemistry information quickly to help her make a diagnosis and begin treatment. Every analyte in Table 14-1, which is part of the critical care profile of blood chemistry, can be measured by electrochemical means. Ion-selective electrodes, which we study in this chapter, are the method of choice for Na^+, K^+, Cl^-, pH, and P_{CO_2}. The "Chem 7" test constitutes up to 70% of tests performed in the hospital lab. It measures Na^+, K^+, Cl^-, total CO_2, glucose, urea, and creatinine, four of which are analyzed with ion-selective electrodes. The use of voltage measurements to extract chemical information is called **potentiometry.**

14-1 The Silver Indicator Electrode

In the last chapter, we learned that the voltage of an electrochemical cell is related to the concentrations of species in the cell. We saw that some cells could be divided into a *reference electrode* that provides a constant electric potential and an *indicator electrode* whose potential varies in response to analyte concentration.

Chemically inert platinum, gold, and carbon indicator electrodes are frequently used to conduct electrons to or from species in solution. In contrast to chemically inert elements, silver readily participates in the reaction $Ag^+ + e^- \rightleftharpoons Ag(s)$.

Figure 14-1 shows how a silver electrode can be used in conjunction with a calomel reference electrode to measure $[Ag^+]$ during the titration of halide ions by Ag^+ (as shown in Figures 5-4 and 5-5). The reaction at the silver indicator electrode is

$$Ag^+ + e^- \rightleftharpoons Ag(s) \qquad E^\circ_+ = 0.799 \text{ V}$$

TABLE 14-1
Critical care profile

Function	Analyte
Conduction	K^+, Ca^{2+}
Contraction	Ca^{2+}, Mg^{2+}
Energy level	Glucose, P_{O_2}, lactate, hematocrit
Ventilation	P_{O_2}, P_{CO_2}
Perfusion	Lactate, $SO_2\%$, hematocrit
Acid-base	pH, P_{CO_2}, HCO_3^-
Osmolality	Na^+, glucose
Electrolyte balance	Na^+, K^+, Ca^{2+}, Mg^{2+}
Renal function	Blood urea nitrogen, creatinine

SOURCE: C. C. Young, *J. Chem. Ed.* **1997,** *74,* 177.

Demonstration 14-1 uses a Pt indicator electrode to monitor a pretty amazing chemical reaction.

303

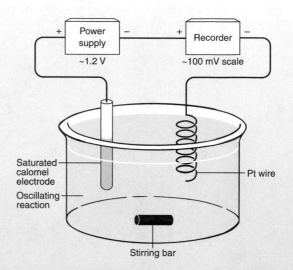

Demonstration 14-1

Potentiometry with an Oscillating Reaction

Principles of potentiometry are illustrated in a fascinating manner by *oscillating reactions* in which chemical concentrations oscillate between high and low values. An example is the Belousov-Zhabotinskii reaction:

$$3CH_2(CO_2H)_2 + 2BrO_3^- + 2H^+ \xrightarrow{\text{Ce}^{3+/4+} \text{ catalyst}}$$

Malonic acid   Bromate

$$2BrCH(CO_2H)_2 + 3CO_2 + 4H_2O$$

Bromomalonic acid

During this reaction, the quotient $[Ce^{3+}]/[Ce^{4+}]$ oscillates by a factor of 10 to 100. When the Ce^{4+} concentration is high, the solution is yellow. When Ce^{3+} predominates, the solution is colorless.

To start the show, combine the following solutions in a 300-mL beaker:

160 mL of 1.5 M H_2SO_4
40 mL of 2 M malonic acid
30 mL of 0.5 M $NaBrO_3$ (or saturated $KBrO_3$)
4 mL of saturated ceric ammonium sulfate,
     $Ce(SO_4)_2 \cdot 2(NH_4)_2SO_4 \cdot 2H_2O$

After an induction period of 5 to 10 min with magnetic stirring, you can initiate oscillations by adding 1 mL of ceric ammonium sulfate solution. The reaction is somewhat temperamental and may need more Ce^{4+} over a 5-min period to initiate oscillations.

The $[Ce^{3+}]/[Ce^{4+}]$ ratio is monitored by Pt and calomel electrodes. You should be able to write the cell reactions and a Nernst equation for this experiment.

Apparatus used to monitor relative concentrations of Ce^{3+} and Ce^{4+} in an oscillating reaction. [George Rossman, California Institute of Technology.]

In place of a potentiometer (a pH meter), we use a chart recorder to obtain a permanent record of the oscillations. The potential oscillates over a range of ~100 mV centered near ~1.2 V, so we offset the cell voltage by ~1.2 V with any available power supply. Trace *a* shows what is usually observed. The potential changes rapidly during the abrupt colorless-to-yellow transition and more gradually during the gentle yellow-to-colorless transition. Trace *b* shows two different cycles superimposed in the same solution.

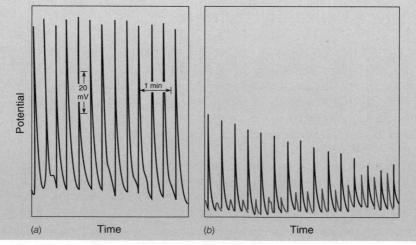

and the reference half-cell reaction is

$$Hg_2Cl_2(s) + 2e^- \rightleftharpoons 2Hg(l) + 2Cl^- \qquad E_- = 0.241 \text{ V}$$

The reference half-cell potential (E_-, not E_-°) is constant at 0.241 V because $[Cl^-]$ is fixed by the concentration of saturated KCl. The Nernst equation for the entire cell is therefore

$$E = E_+ - E_- = \underbrace{\left\{ 0.799 - 0.059\ 16 \log\left(\frac{1}{[Ag^+]} \right) \right\}}_{\substack{\text{Potential of Ag|Ag}^+ \\ \text{indicator electrode}}} - \underbrace{\{0.241\}}_{\substack{\uparrow \\ \text{Constant potential of} \\ \text{S.C.E. reference electrode}}}$$

E_- = reference electrode potential with actual concentrations in the reference cell

E_-° = standard potential of reference half-reaction when all species are in their standard states (pure solid, pure liquid, 1 M, or 1 bar)

Noting that $\log(1/[Ag^+]) = -\log[Ag^+]$, we rewrite the expression above as

$$E = 0.558 + 0.059\ 16 \log[Ag^+] \qquad (14\text{-}1)$$

The voltage changes by 0.059 16 V (at 25°C) for each factor-of-10 change in $[Ag^+]$.

The experiment in Figure 5-4 used a silver indicator electrode and a *glass* reference electrode. The glass electrode responds to the pH of the solution, which is held constant by a buffer. Therefore, the glass electrode remains at a constant potential.

Titration of a Halide Ion with Ag⁺

Let's consider how the concentration of Ag^+ varies during the titration of I^- with Ag^+, as shown in Figure 5-5b. The titration reaction is

$$Ag^+ + I^- \longrightarrow AgI(s) \qquad K = \frac{1}{K_{sp}} = \frac{1}{8.3 \times 10^{-17}}$$

If you were monitoring the reaction with silver and calomel electrodes, you could use Equation 14-1 to compute the expected voltage at each point in the titration.

At any point prior to the equivalence point (V_e), there is a known excess of I^- from which we can calculate $[Ag^+]$:

Before V_e: $\qquad K_{sp} = [Ag^+][I^-] \Rightarrow [Ag^+] = K_{sp}/[I^-] \qquad (14\text{-}2)$

At the equivalence point, the quantity of Ag^+ added is exactly equal to the I^- that was originally present. We can imagine that $AgI(s)$ is made stoichiometrically and a little bit redissolves:

At V_e: $\qquad K_{sp} = [Ag^+][I^-] \Rightarrow [Ag^+] = [I^-] = \sqrt{K_{sp}} \qquad (14\text{-}3)$

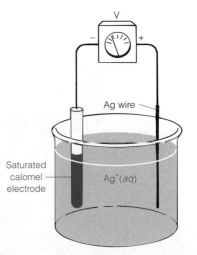

FIGURE 14-1 Use of silver and calomel electrodes to measure the concentration of Ag^+ in a solution. The calomel electrode has a double junction, like that in Figure 14-3. The outer compartment of the electrode is filled with KNO_3, so there is no direct contact between the KCl solution in the inner compartment and Ag^+ in the beaker.

Beyond the equivalence point, the quantity of excess Ag^+ added from the buret is known, and the concentration is just

$$After\ V_e: \qquad [Ag^+] = \frac{\text{moles of excess } Ag^+}{\text{total volume of solution}} \qquad (14\text{-}4)$$

EXAMPLE   Potentiometric Precipitation Titration

When adding Ag^+ to I^-:
- Before V_e, there is a known excess of I^-: $[Ag^+] = K_{sp}/[I^-]$
- At V_e, $[Ag^+] = [I^-] = \sqrt{K_{sp}}$
- After V_e, there is a known excess of Ag^+

A 20.00-mL solution containing 0.100 4 M KI was titrated with 0.084 5 M $AgNO_3$, using the cell in Figure 14-1. Calculate the voltage at volumes $V_{Ag^+} = 15.00$, V_e, and 25.00 mL.

SOLUTION The titration reaction is $Ag^+ + I^- \longrightarrow AgI(s)$, and the equivalence volume is

$$\underbrace{(V_e(mL))(0.084\ 5\ M)}_{\text{mmol } Ag^+} = \underbrace{(20.00\ mL)(0.100\ 4\ M)}_{\text{mmol } I^-} \quad \Rightarrow \quad V_e = 23.76\ mL$$

15.00 mL: We began with $(20.00\ mL)(0.100\ 4\ M) = 2.008$ mmol I^- and added $(15.00\ mL)(0.084\ 5\ M) = 1.268$ mmol Ag^+. The concentration of unreacted I^- is

$$[I^-] = \frac{(2.008 - 1.268)\ \text{mmol}}{(20.00 + 15.00)\ mL} = 0.021\ 1\ M$$

The concentration of Ag^+ in equilibrium with the solid AgI is therefore

$$[Ag^+] = \frac{K_{sp}}{[I^-]} = \frac{8.3 \times 10^{-17}}{0.021\ 1} = 3.9 \times 10^{-15}\ M$$

The cell voltage is computed with Equation 14-1:

$$E = 0.558 + 0.059\ 16\ \log(3.9 \times 10^{-15}) = -0.294\ V$$

At V_e: Equation 14-3 tells us that $[Ag^+] = \sqrt{K_{sp}} = 9.1 \times 10^{-9}\ M$, so

$$E = 0.558 + 0.059\ 16\ \log(9.1 \times 10^{-9}) = 0.082\ V$$

25.00 mL: Now there is an excess of $25.00 - 23.76 = 1.24$ mL of 0.084 5 M $AgNO_3$ in a total volume of 45.00 mL.

$$[Ag^+] = \frac{(1.24\ mL)(0.084\ 5\ M)}{45.00\ mL} = 2.33 \times 10^{-3}\ M$$

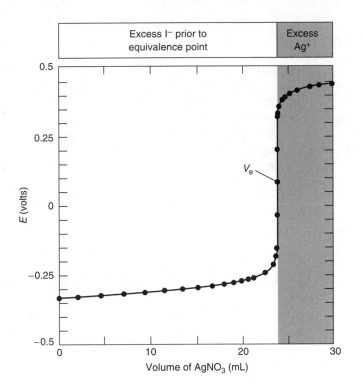

FIGURE 14-2 Calculated titration curve for the addition of 0.084 5 M Ag^+ to 20.00 mL of 0.100 4 M I^-, using the cell in Figure 14-1.

and the cell voltage is

$$E = 0.0558 + 0.059\ 16\ \log(2.33 \times 10^{-3}) = 0.402\ \text{V}$$

The voltage in Figure 14-2 barely changes prior to the equivalence point because the concentration of Ag^+ is very low and relatively constant until I^- is used up. When I^- has been consumed, $[Ag^+]$ suddenly increases and so does the voltage. Figure 14-2 is upside down relative to curve *b* in Figure 5-5. The reason is that, in Figure 14-1, the indicator electrode is connected to the *positive* terminal of the potentiometer. In Figure 5-4, the indicator electrode is connected to the *negative* terminal because the glass pH electrode fits into the positive terminal of the meter. In addition to their opposite polarities, the voltages in Figures 14-2 and 5-5 are different because each experiment uses a different reference electrode.

Double-Junction Reference Electrode

If you tried to titrate I^- with Ag^+ by using the cell in Figure 14-1, KCl solution would slowly leak into the titration beaker from the porous plug at the base of the reference electrode (Figure 13-9). Cl^- introduces a titration error because it consumes Ag^+. The *double-junction reference electrode* in Figure 14-3 prevents the inner electrolyte solution from leaking directly into the titration vessel.

FIGURE 14-3 Double-junction reference electrode has an inner electrode identical to those in Figures 13-8 and 13-9. The outer compartment is filled with any desired electrolyte, such as KNO_3, that is compatible with the titration solution. KCl electrolyte from the inner electrode slowly leaks into the outer electrode, so the outer electrolyte should be changed periodically. [Fisher Scientific, Pittsburgh, PA.]

Ask Yourself

14-A. Consider the titration of 40.0 mL of 0.050 0 M NaCl with 0.200 M $AgNO_3$, using the cell in Figure 14-1. The equivalence volume is $V_e = 10.0$ mL.

(a) Prior to V_e, there is a known excess of Cl^-. Find $[Cl^-]$ at the following volumes of added silver: $V_{Ag^+} = 0.10, 2.50, 5.00, 7.50,$ and 9.90 mL. From $[Cl^-]$, use K_{sp} for AgCl to find $[Ag^+]$ at each volume.

(b) Find $[Cl^-]$ and $[Ag^+]$ at $V_{Ag^+} = V_e = 10.00$ mL.

(c) After V_e, there is a known excess of Ag^+. Find $[Ag^+]$ at $V_{Ag^+} = 10.10$ and 12.00 mL.

(d) Find the cell voltage at each volume in (a)–(c) and make a graph of the titration curve.

14-2   What Is a Junction Potential?

$$E_{observed} = E_{cell} + E_{junction}$$

Because the junction potential is usually unknown, E_{cell} is uncertain.

When two dissimilar electrolyte solutions are placed in contact, a voltage difference called the **junction potential** develops at the interface. This small, unknown voltage (usually a few millivolts) exists at each end of a salt bridge connecting two half-cells. *The junction potential puts a fundamental limitation on the accuracy of direct potentiometric measurements,* because we usually do not know the contribution of the junction to the measured voltage.

To see why a junction potential occurs, consider a solution containing NaCl in contact with distilled water (Figure 14-4). The Na^+ and Cl^- ions diffuse from the NaCl solution into the water phase. However, Cl^- ion has a greater *mobility* than Na^+. That is, Cl^- diffuses faster than Na^+. As a result, a region rich in Cl^-, with excess negative charge, develops at the front. Behind it is a positively charged region depleted of Cl^-. The result is an electric potential difference at the junction of the NaCl and H_2O phases.

Mobilities of ions are shown in Table 14-2, and several junction potentials are listed in Table 14-3. Because K^+ and Cl^- have similar mobilities, junction potentials of a KCl salt bridge are slight. This is why saturated KCl is used in salt bridges.

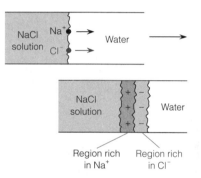

FIGURE 14-4 Development of the junction potential caused by unequal mobilities of Na^+ and Cl^-.

TABLE 14-3   Liquid junction potentials at 25°C

Junction	Potential (mV)
0.1 M NaCl│0.1 M KCl	−6.4
0.1 M NaCl│3.5 M KCl	−0.2
1 M NaCl│3.5 M KCl	−1.9
0.1 M HCl│0.1 M KCl	+27
0.1 M HCl│3.5 M KCl	+3.1

Note: A positive sign means that the right side of the junction becomes positive with respect to the left side.

TABLE 14-2   Mobilities of ions in water at 25°C

Ion	Mobility $[m^2/(s \cdot V)]^a$	Ion	Mobility $[m^2/(s \cdot V)]^a$
H^+	36.30×10^{-8}	OH^-	20.50×10^{-8}
K^+	7.62×10^{-8}	SO_4^{2-}	8.27×10^{-8}
NH_4^+	7.61×10^{-8}	Br^-	8.13×10^{-8}
La^{3+}	7.21×10^{-8}	I^-	7.96×10^{-8}
Ba^{2+}	6.59×10^{-8}	Cl^-	7.91×10^{-8}
Ag^+	6.42×10^{-8}	NO_3^-	7.40×10^{-8}
Ca^{2+}	6.12×10^{-8}	ClO_4^-	7.05×10^{-8}
Cu^{2+}	5.56×10^{-8}	F^-	5.70×10^{-8}
Na^+	5.19×10^{-8}	$CH_3CO_2^-$	4.24×10^{-8}
Li^+	4.01×10^{-8}		

a. The mobility of an ion is the velocity achieved in an electric field of 1 V/m. Mobility = velocity/field. The units of mobility are therefore (m/s)/(V/m) = $m^2/(s \cdot V)$.

Direct versus Relative Potentiometric Measurements

In a *direct potentiometric measurement,* we use an electrode such as a silver wire to measure $[Ag^+]$ or a pH electrode to measure $[H^+]$ or a calcium ion-selective electrode to measure $[Ca^{2+}]$. There is inherent inaccuracy in most direct potentiometric measurements because there is usually a liquid-liquid junction with an unknown voltage difference making the intended indicator electrode potential uncertain. For example, Figure 14-5 shows a 4% standard deviation among 14 measurements by direct potentiometry. Part of the variation could be attributed to differences in the indicator (ion-selective) electrodes and part could be from varying liquid junction potentials.

By contrast, in *relative potentiometric measurements,* changes in the potential observed during a titration, such as Figure 5-5, are relatively precise and permit an end point to be identified with little uncertainty. Measuring the *absolute* Ag^+ concentration by direct potentiometry is inherently inaccurate, but measuring *changes* in Ag^+ can be done accurately and precisely.

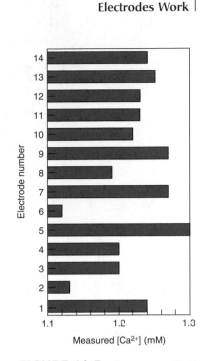

FIGURE 14-5 Response of 14 different Ca^{2+} ion-selective electrodes to identical human blood serum samples. The mean value is 1.22 ± 0.05 mM. [From M. Umemoto, W. Tani, K. Kuwa, and Y. Ujihira, *Anal. Chem.* **1994,** *66,* 352A.]

Ask Yourself

14-B. A 0.1 M NaCl solution is placed in contact with a 0.1 M NaNO₃ solution. The concentration of Na^+ is the same on both sides of the junction, so

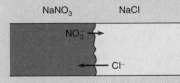

there is no net diffusion of Na^+ from one side to the other. The mobility of Cl^- is greater than that of NO_3^-, so Cl^- diffuses away from the NaCl side faster than NO_3^- diffuses away from NaNO₃. Which side of the junction will become positive and which will become negative? Explain your reasoning.

14-3 How Ion-Selective Electrodes Work

An **ion-selective electrode** responds preferentially to one species in a solution. Differences in concentration of the selected ion inside and outside the electrode produce a voltage difference across the membrane.

To understand how ion-selective electrodes work, imagine a membrane separating two solutions of $CaCl_2$ (Figure 14-6). The membrane contains a molecule that binds and transports Ca^{2+}, but *not* Cl^-. Ca^{2+} can enter and cross the membrane, but Cl^- cannot. Initially the potential difference across the membrane is 0, because both solutions are neutral (Figure 14-6a). After some time, there is a net diffusion of Ca^{2+} from the right (high concentration) to the left (low concentration).

As Ca^{2+} migrates from right to left, positive charge builds up on the left side of the membrane (Figure 14-6b). Eventually the excess positive charge, which repels Ca^{2+}, prevents further migration of Ca^{2+} to the positively charged side. At equilibrium, the potential difference across the membrane is

Mechanism of ion-selective electrode:

1. A specific ion crosses the membrane, creating charge imbalance.

2. Charge buildup opposes further movement of ion.

3. Result: Potential difference across membrane is related to difference in concentrations of specific ion on either side.

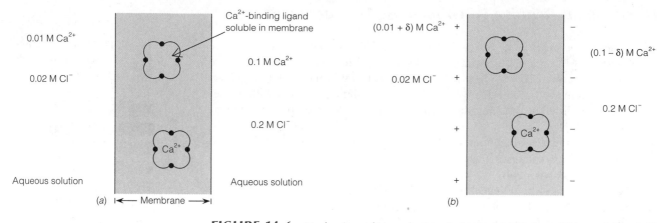

FIGURE 14-6 Mechanism of ion-selective electrode. (*a*) Initial conditions prior to Ca^{2+} migration across membrane. (*b*) After δ moles of Ca^{2+} per liter have crossed the membrane, giving the left-hand solution a charge of $+2\delta$ mol/L and the right-hand solution a charge of -2δ mol/L.

The factor in Equation 14-5 is 0.054 20 V at 0°C and 0.061 54 V at 37°C. Equation 14-5 should really be written with activities (Section 11-2) instead of concentrations.

Electric potential difference due to concentration difference:

$$E = \left(\frac{0.059\,16}{n}\right) \log\left(\frac{[Ca^{2+}]_{\text{right}}}{[Ca^{2+}]_{\text{left}}}\right) \quad \text{(volts at 25°C)} \quad \text{(14-5)}$$

where n is the charge on the species of interest. Because the charge of Ca^{2+} is $n = 2$, a potential difference of $0.059\,16/2 = 0.029\,58$ V is expected for every factor-of-10 difference in concentration of Ca^{2+} across the membrane. No charge buildup results from the Cl^- because Cl^- has no way to cross the membrane.

The key to designing an ion-selective electrode is to fabricate a membrane across which only the desired species can migrate. The approach is to select soluble molecules or solid crystals that selectively interact with the desired analyte. No membrane is perfectly selective, so there is always some interference from unintended species.

Two Classes of Indicator Electrodes

Metal electrodes such as silver or platinum function as surfaces on which redox reactions can occur:

equilibrium on a silver electrode: $\quad Ag^+ + e^- \rightleftharpoons Ag(s)$

equilibrium on a platinum electrode: $\quad Fe(CN)_6^{3+} + e^- \rightleftharpoons Fe(CN)_6^{2+}$

Metal electrode: surface on which redox reaction occurs

Ion-selective electrode: selectively permeable to one ion—no redox chemistry

Ion-selective electrodes such as a calcium electrode or a glass pH electrode have selectively permeable membranes that ideally allow just one kind of ion to pass through. *There is no redox chemistry associated with the ion-selective electrode.* The potential difference across the membrane is a result of its selective permeability to just one ion.

Ask Yourself

14-C. Why doesn't Ca^{2+} continue to diffuse across the membrane in Figure 14-6 until the concentration is the same on both sides?

14-4 pH Measurement with a Glass Electrode

The most widely employed ion-selective electrode is the **glass electrode** for measuring pH. A pH electrode responds selectively to H^+, building up a potential difference of 0.059 16 V for every factor-of-10 difference in $[H^+]$ across the electrode membrane. A factor-of-10 difference in $[H^+]$ is one pH unit, so a change of, say, 4.00 pH units leads to a change in electrode potential of $4.00 \times 0.059\ 16 = 0.237$ V.

A typical **combination electrode,** incorporating both glass and reference electrodes in one body, is shown in Figure 14-7. The line diagram of this cell is

Glass membrane

$$\underbrace{Ag(s)\ |\ AgCl(s)\ |\ Cl^-(aq)}_{\substack{\text{Outer reference}\\ \text{electrode}}}\ \|\ \underbrace{H^+(aq,\ \text{outside})}_{\substack{H^+\ \text{outside}\\ \text{glass electrode}\\ \text{(analyte solution)}}}\ \vdots\ \underbrace{H^+(aq,\ \text{inside}),\ Cl^-(aq)}_{\substack{H^+\ \text{inside}\\ \text{glass}\\ \text{electrode}}}\ |\ \underbrace{AgCl(s)\ |\ Ag(s)}_{\substack{\text{Inner reference}\\ \text{electrode}}}$$

The pH-sensitive part of the electrode is the thin glass membrane in the shape of a bulb at the bottom of the electrodes in Figures 14-7 and 14-8.[1]

The glass membrane at the bottom of the pH electrode consists of an irregular network of SiO_4 tetrahedra through which Na^+ ions move sluggishly. Studies with

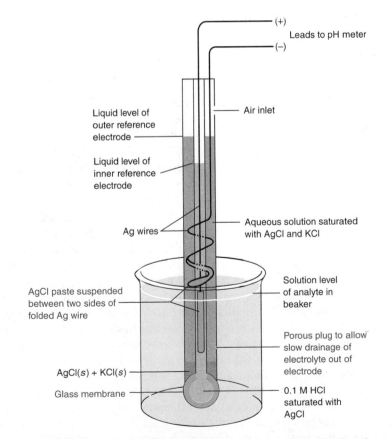

(+)
Leads to pH meter
(−)

Air inlet

Liquid level of outer reference electrode

Liquid level of inner reference electrode

Ag wires

Aqueous solution saturated with AgCl and KCl

AgCl paste suspended between two sides of folded Ag wire

Solution level of analyte in beaker

Porous plug to allow slow drainage of electrolyte out of electrode

AgCl(s) + KCl(s)

Glass membrane

0.1 M HCl saturated with AgCl

FIGURE 14–7 Glass combination electrode with a silver–silver chloride reference electrode. The glass electrode is immersed in a solution of unknown pH so that the porous plug on the lower right is below the surface of the liquid. The two Ag electrodes measure the voltage across the glass membrane.

(a)   (b)

Porous junction

FIGURE 14–8 (a) Glass-body combination electrode with pH–sensitive glass bulb at the bottom. The circle at the side of the electrode near the bottom is the porous junction salt bridge to the reference electrode compartment. Two silver wires coated with AgCl are visible inside the electrode. (b) Polymer body surrounds glass electrode to protect the delicate bulb. [Fisher Scientific, Pittsburgh, PA.]

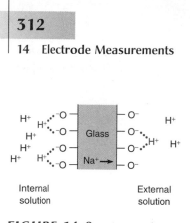

FIGURE 14-9 Ion-exchange equilibria on the inner and outer surfaces of the glass membrane. The pH of the internal solution is fixed. As the pH of the external solution (the sample) changes, the electric potential difference across the glass membrane changes.

A pH electrode *must* be calibrated before use. It should be calibrated about every 2 h during sustained use. Ideally, calibration standards should bracket the pH of the unknown.

Don't leave a glass electrode out of water (or in a nonaqueous solvent) longer than necessary.

tritium (the radioactive isotope ^{3}H) show that H^+ does *not* diffuse through the membrane. The glass surface contains exposed $—O^-$ groups that can bind H^+ from the solutions on either side of the membrane (Figure 14-9). H^+ equilibrates with the glass surface, thereby giving the side of the membrane exposed to the higher concentration of H^+ the more positive charge. Na^+ ions that are already in the glass migrate across the membrane from the positive side to the negative side, so the potential changes by 0.059 16 V for a unit change in pH. Because the electrical resistance of the glass membrane is high, little current actually flows across it.

The potential difference between the inner and outer silver-silver chloride electrodes in Figure 14-7 depends on $[Cl^-]$ in each electrode compartment and on the potential difference across the glass membrane. Because $[Cl^-]$ is fixed, and because $[H^+]$ is constant inside the glass electrode, the only variable is the pH of the analyte solution outside the glass membrane.

Real glass electrodes are described by the equation

Response of glass electrode:

$$E = \text{constant} + \beta(0.059\ 16)\Delta pH \qquad \text{(at 25°C)} \qquad (14\text{-}6)$$

where ΔpH is the difference in pH between the analyte solution and the solution inside the glass bulb. The factor β, which is ideally 1, is typically 0.98–1.00. The constant term, called the *asymmetry potential*, arises because no two sides of a real object are identical, so a small voltage exists even if the pH is the same on both sides of the membrane. We correct for asymmetry by calibrating the electrode in solutions of known pH.

Calibrating a Glass Electrode

A pH electrode should be calibrated with two (or more) standard buffers selected so that the pH of the unknown lies within the range of the standards. Before you calibrate the electrode, wash it with distilled water and gently *blot* it dry with a tissue. Don't *wipe* it, because this action might produce a static charge on the glass. Dip the electrode in a standard buffer whose pH is near 7 and allow the electrode to equilibrate for at least a minute. For better results, solutions should be stirred during testing. Adjust the meter reading (usually with a knob labeled "Calibrate") to indicate the pH of the standard buffer. Wash the electrode, blot it dry, and immerse it in the second standard buffer. If the electrode response were perfect, the voltage would change by 0.059 16 V per pH unit at 25°C. The actual change may be slightly less, so these two measurements establish the value of β in Equation 14-6. The pH of the second buffer is set on the meter with a knob labeled "Slope" or "Temperature." Finally, dip the electrode in the unknown and read the pH on the meter.

Store the glass electrode in aqueous solution to prevent dehydration of the glass. Ideally, the solution should be similar to that inside the reference compartment of the electrode. Distilled water is not a good storage medium. If the electrode has dried out, it should be reconditioned in aqueous solution for several hours. If the electrode is to be used above pH 9, soak it in a high-pH buffer.

If electrode response becomes sluggish, or if the electrode cannot be calibrated properly, try soaking it in 6 M HCl, followed by water. As a last resort, dip the electrode in 20 wt % aqueous ammonium bifluoride, NH_4HF_2, (in a plastic beaker) for 1 min. This reagent dissolves a little of the glass, exposing fresh surface. Wash the electrode with water and try calibrating it again. *Ammonium bifluoride must not contact your skin, because it produces HF burns.* (See page 265 for precautions with HF.)

To use a glass electrode intelligently, you should understand its limitations:

1. *Standards.* A pH measurement cannot be more accurate than our standards, which are typically ±0.01–0.02 pH units.

2. *Junction potential.* A junction potential exists at the porous plug near the bottom of the electrode in Figure 14-7. If the ionic composition of the analyte solution is different from that of the standard buffer, the junction potential will change *even if the pH of the two solutions is the same.* This factor gives an uncertainty of at least ~0.01 pH unit. Box 14-1 describes how junction potentials affect the measurement of the pH of rainwater.

Box 14-1 *Informed Citizen*

Systematic Error in Rainwater pH Measurement: The Effect of Junction Potential

The opening of Chapter 7 shows the pH of rainfall over North America. Acidity in rainfall is partly a result of human activities and is slowly changing the nature of many ecosystems as we know them. Monitoring the pH of rainwater is a critical component of programs to reduce the production of acid rain.

To identify and correct systematic errors in the measurement of pH of rainwater, eight samples were provided to each of 17 laboratories, along with explicit instructions for how to conduct the measurements. Each lab used two standard buffers to calibrate its pH meter.

The figure below shows typical results for the pH of rainwater. The average of the 17 measurements is given by the horizontal line at pH 4.14 and the letters s, t, u, v, w, x, y, and z identify the types of pH electrodes used for the measurements. Laboratories using electrode types s and w had relatively large systematic errors. The type s electrode was a combination electrode (Figure 14-7) containing a reference electrode liquid

junction with an exceptionally large area. Electrode type w had a reference electrode filled with a gel.

It was hypothesized that variability in the liquid junction potential (Section 14-2) led to the variability of the pH measurements. Standard buffers used for pH meter calibration typically have ion concentrations of ~0.05 M, whereas rainwater samples have ion concentrations two or more orders of magnitude lower. To test the hypothesis that junction potential caused systematic errors, a pure HCl solution with a concentration near 2×10^{-4} M was used as a pH calibration standard in place of high ionic strength buffers. The following data were obtained, with good results from all but the first lab. The standard deviation of all 17 measurements was reduced from 0.077 pH unit with the standard buffer to 0.029 pH unit with the HCl standard. It was concluded that junction potential is the cause of most of the variability between labs and that a low ionic strength standard is appropriate for rainwater pH measurements.

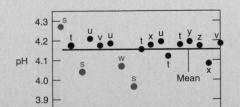

pH of rainwater from identical samples measured at 17 different labs, which all used the same standard calibration buffers. Letters designate different types of pH electrodes.

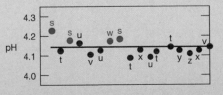

Rainwater pH measured after using low ionic strength HCl for calibration. [W. F. Koch, G. Marinenko, and R. C. Paule, *J. Res. National Bureau of Standards* **1986**, *91*, 23.]

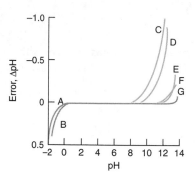

FIGURE 14-10 Acid and alkaline errors of some glass electrodes. A: Corning 015, H_2SO_4. B: Corning 015, HCl. C: Corning 015, 1 M Na^+. D: Beckman–GP, 1 M Na^+. E: L & N Black Dot, 1 M Na^+. F: Beckman Type E, 1 M Na^+. G: Ross electrode.[2] [From R. G. Bates, *Determination of pH: Theory and Practice*, 2nd ed. (New York: Wiley, 1973). Ross electrode data are from Orion, *Ross pH Electrode Instruction Manual*.]

Challenge Show that the potential of the glass electrode changes by 1.3 mV when the analyte H^+ concentration changes by 5.0%. Because 59 mV ≈ 1 pH unit, 1.3 mV = 0.02 pH unit.

Moral: A small uncertainty in voltage (1.3 mV) or pH (0.02 units) corresponds to a large uncertainty (5%) in H^+ concentration. Similar uncertainties arise in other potentiometric measurements.

3. *Junction potential drift.* Most combination electrodes have a silver-silver chloride reference electrode containing saturated KCl solution. More than 350 mg of silver per liter dissolve in the KCl (mainly as $AgCl_4^{3-}$ and $AgCl_3^{2-}$). In the porous plug (the salt bridge) near the bottom of the electrode in Figure 14-7, KCl is diluted and AgCl precipitates in the plug. If the analyte solution contains a reducing agent, Ag(s) can also precipitate in the plug. Both effects change the junction potential, causing slow drift of the pH reading. You can compensate for this error by recalibrating the electrode every 2 h.

4. *Sodium error.* When $[H^+]$ is very low and $[Na^+]$ is high, the electrode responds to Na^+ as if Na^+ were H^+, and the apparent pH is lower than the true pH. This is called the *alkaline error,* or *sodium error* (Figure 14-10).

5. *Acid error.* In strong acid, the measured pH is higher than the actual pH, for reasons that are not well understood (Figure 14-10).

6. *Equilibration time.* The electrode needs time to equilibrate with each solution. In a well-buffered solution, with adequate stirring, equilibration takes just seconds. In a poorly buffered solution near the equivalence point of a titration, it could take minutes.

7. *Hydration.* A dry electrode requires several hours of soaking in aqueous solution before it responds to H^+ correctly.

8. *Temperature.* A pH meter should be calibrated at the same temperature at which the measurement will be made. You cannot calibrate your equipment at one temperature and make accurate measurements at a second temperature.

Errors 1 and 2 limit the accuracy of pH measurements with the glass electrode to ±0.02 pH unit, at best. Measurement of pH differences *between* solutions can be accurate to about ±0.002 pH unit, but knowledge of the true pH will still be at least an order of magnitude more uncertain. An uncertainty of ±0.02 pH unit corresponds to an uncertainty of ±5 % in $[H^+]$.

Solid-State pH Sensors

There are pH sensors that do not depend on a fragile glass membrane. The *field effect transistor* in Figure 14-11 is a tiny semiconductor device whose surface binds H^+ from the medium in which the transistor is immersed. The higher the concentration of H^+ in the external medium, the more positively charged is the transistor's surface. The surface charge regulates the flow of current through the transistor, which therefore behaves as a pH sensor.

Ask Yourself

14-D. (a) List the sources of error associated with pH measurement made with the glass electrode.

(b) When the difference in pH across the membrane of a glass electrode at 25°C is 4.63 pH units, how much voltage is generated by the pH gradient? Assume that the constant β in Equation 14-6 is 1.00.

(c) Why do glass pH electrodes tend to indicate a pH lower than the actual pH in strongly basic solution?

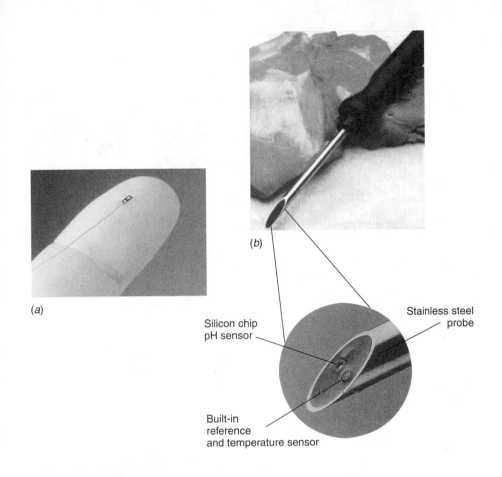

(a)

(b)

Silicon chip
pH sensor

Stainless steel
probe

Built-in
reference
and temperature sensor

FIGURE 14-11 (a) pH–sensitive field effect transistor. (b) Rugged field effect transistor mounted in a steel shaft can be inserted into meat, poultry, or other damp solids to measure pH. [Sentron, Gig Harbor, WA and IQ Scientific Instruments, San Diego, CA.]

14-5 Ion-Selective Electrodes

The glass pH electrode is an example of a *solid-state ion-selective electrode* whose operation depends on (1) an ion-exchange reaction of H^+ between the glass surface and the analyte solution and (2) transport of an ion (Na^+) across the glass membrane. We now examine several very useful ion-selective electrodes.

Solid-State Electrodes

The ion-sensitive component of a fluoride **solid-state ion-selective electrode** is a crystal of LaF_3 doped with EuF_2 (Figure 14-12a). *Doping* is the intentional addition of a small amount of an impurity (EuF_2 in this case) into the solid crystal (LaF_3). The inner surface of the crystal is exposed to filler solution containing a constant concentration of F^-. The outer surface is exposed to a variable concentration of F^- in the unknown. F^- from solution is selectively adsorbed on each surface of the crystal. Doping LaF_3 with EuF_2 creates anion vacancies that allow F^- to jump from one site to the next. Unlike H^+ in a pH electrode, F^- itself migrates across the LaF_3 crystal, as shown in Figure 14-12b.

The response of the electrode to F^- is given by

Response of F^- electrode: $E = \text{constant} - \beta(0.059\ 16)\log[F^-]_{\text{outside}}$ (14-7)

Electrode response really depends on $\log\left(\dfrac{[F^-]_{\text{outside}}}{[F^-]_{\text{inside}}}\right)$. The constant value of $[F^-]_{\text{inside}}$ is incorporated into the constant term in Equation 14-7.

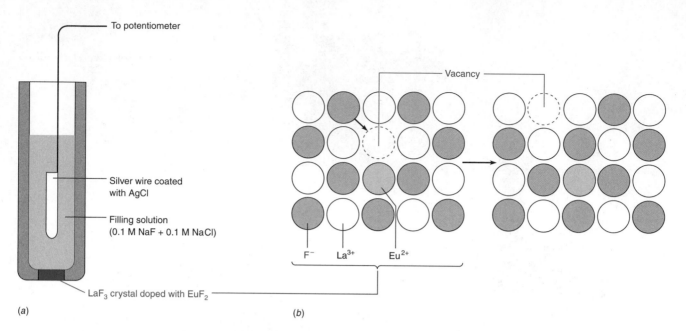

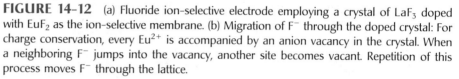

(a)

(b)

FIGURE 14-12 (a) Fluoride ion–selective electrode employing a crystal of LaF_3 doped with EuF_2 as the ion-selective membrane. (b) Migration of F^- through the doped crystal: For charge conservation, every Eu^{2+} is accompanied by an anion vacancy in the crystal. When a neighboring F^- jumps into the vacancy, another site becomes vacant. Repetition of this process moves F^- through the lattice.

where $[F^-]_{outside}$ is the concentration of F^- in the analyte solution and β is close to 1.00. The electrode response is close to 59 mV per decade over a F^- concentration range from about 10^{-6} M to 1 M. The electrode is more responsive to F^- than to most other ions by a factor greater than 1 000. (Response to OH^- is one-tenth as great as the response to F^-.) At low pH, F^- is converted to HF ($pK_a = 3.17$), to which the electrode is insensitive. The F^- electrode is used to monitor and control the fluoridation of municipal water supplies. Fluoride is added to drinking water to help prevent tooth decay. Several other solid-state ion-selective electrodes are listed in Table 14-4.

TABLE 14-4		**Solid-state ion-selective electrodes**		
Ion	*Concentration range (M)*	*Membrane crystal[a]*	*pH range*	*Interfering species*
F^-	10^{-6}–1	LaF_3	5–8	OH^-
Cl^-	10^{-4}–1	AgCl	2–11	$CN^-, S^{2-}, I^-, S_2O_3^{2-}, Br^-$
Br^-	10^{-5}–1	AgBr	2–12	CN^-, S^{2-}, I^-
I^-	10^{-6}–1	AgI	3–12	S^{2-}
CN^-	10^{-6}–10^{-2}	AgI	11–13	S^{2-}, I^-
S^{2-}	10^{-5}–1	Ag_2S	13–14	

a. Electrodes containing silver-based crystals such as Ag_2S should be stored in the dark and protected from light during use to prevent light-induced chemical degradation.

EXAMPLE   **Calibration Curve for an Ion-Selective Electrode**

A fluoride electrode immersed in standard solutions gave the following potentials:

$[F^-]$ (M)	$\log[F^-]$	E (mV vs. S.C.E.)
1.00×10^{-5}	5.00	100.0
1.00×10^{-4}	4.00	41.4
1.00×10^{-3}	3.00	-17.0
1.00×10^{-2}	2.00	-75.4

(a) What potential is expected if $[F^-] = 5.00 \times 10^{-5}$ M? **(b)** What concentration of F^- will give a potential of 0.0V?

SOLUTION **(a)** Our strategy is to fit the calibration data with Equation 14-7 and then to substitute the concentration of F^- into the equation to find the potential:

$$E = \underbrace{\text{constant}}_{\substack{y \quad \text{Intercept}}} - \underbrace{m}_{\text{Slope}} \cdot \underbrace{\log[F^-]}_{x}$$

Using the method of least squares in Section 4-6, we plot E versus $\log[F^-]$ to find a straight line with a slope of -58.46 mV and an intercept of -192.4 mV (Figure 14-13). Setting $[F^-] = 5.00 \times 10^{-5}$ M gives

$$E = -192.4 - 58.46 \log[5.00 \times 10^{-5}] = 59.0 \text{ mV}$$

(b) If $E = 0.0$ mV, we can solve for the concentration of $[F^-]$:

$$0.0 = -192.4 - 58.46 \log[F^-] \implies [F^-] = 5.1 \times 10^{-4} \text{ M}$$

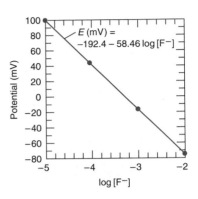

FIGURE 14-13   Calibration curve for fluoride ion–selective electrode.

Liquid-Based Ion-Selective Electrodes

The principle of a **liquid-based ion-selective electrode** was described in Figure 14-6. Figure 14-14 shows a Ca^{2+} ion-selective electrode, at the base of which is a membrane saturated with a liquid ion exchanger (a Ca^{2+}-binding compound dissolved in organic solvent). Ca^{2+} is selectively transported across the membrane to establish a voltage related to the difference in concentrations of Ca^{2+} between the sample and internal solution:

Response of Ca^{2+}
electrode: $$E = \text{constant} + \beta \left(\frac{0.059\ 16}{2} \right) \log[Ca^{2+}]_{\text{outside}} \qquad (14\text{-}8)$$

where β is close to 1.00. Equations 14-8 and 14-7 have different signs before the log term because one involves an anion and the other a cation. The charge of the calcium ion requires a factor of 2 in the denominator before the logarithm.

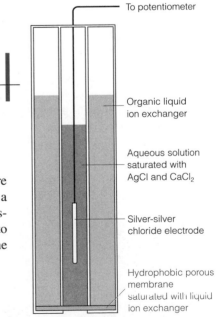

FIGURE 14-14   Schematic diagram of a Ca^{2+} ion-selective electrode based on a liquid ion exchanger. Figure 14-5 showed the variability of results among different Ca^{2+} ion-selective electrodes.

The most serious interference with the Ca^{2+} electrode comes from Zn^{2+}, Fe^{2+}, Pb^{2+}, and Cu^{2+}, but high concentrations of Sr^{2+}, Mg^{2+}, Ba^{2+}, and Na^+ also interfere. Interference from H^+ is substantial below pH 4. pH and the ionic strength of standards and unknowns must generally be held constant when using ion-selective electrodes.

An interesting liquid-based ion-selective electrode being developed for clinical use is the heparin electrode whose response is shown in Figure 14-15. Heparin is administered during surgery to prevent blood clotting, but the dose must be limited to avoid uncontrolled bleeding. Heparin is highly negatively charged with numerous carboxylate ($-CO_2^-$) and sulfonate ($-SO_3^-$) groups. The heparin-sensitive electrode contains positively charged ammonium groups with long hydrocarbon tails to anchor them in the electrode membrane. The ammonium groups bind to the negative heparin molecule. The electrode allows a surgeon to maintain heparin at the right level during an operation.

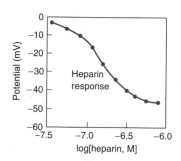

FIGURE 14-15 Response of an ion-selective electrode used to monitor the concentration of the anticoagulant drug heparin in a patient during surgery. [From M. E. Meyerhoff, B. Fu, E. Bakker, J.-H. Yun, and V. C. Yang, *Anal. Chem.* **1996**, *68*, 168A.]

Selectivity Coefficient

No electrode responds exclusively to one kind of ion, but the glass electrode is among the most selective. A high-pH glass electrode responds to Na^+ only when $[H^+] \leq 10^{-12}$ M and $[Na^+] \geq 10^{-2}$ M (Figure 14-10).

If an electrode that measures ion X also responds to ion Y, the **selectivity coefficient** is defined as

Selectivity coefficient:
$$k_{X,Y} = \frac{\text{response to Y}}{\text{response to X}} \tag{14-9}$$

Ideally, the selectivity coefficient should be very small ($k \ll 1$).

The behavior of ion-selective electrodes is described approximately by the equation

Equation 14-10 describes the response of an electrode to its primary ion, X, and to interfering ions, Y.

Response of ion-selective electrode:
$$E = \text{constant} \pm \beta \left(\frac{0.059\ 16}{n_X} \right) \log \left\{ [X] + \sum_Y (k_{X,Y}[Y]^{n_X/n_Y}) \right\} \tag{14-10}$$

where [X] is the concentration of the ion intended to be measured and [Y] is the concentration of an interfering species. The magnitude of the charge of each species is n_X or n_Y, and $k_{X,Y}$ is the selectivity coefficient. If the ion-selective electrode is connected to the positive terminal of the potentiometer, the sign before the log term is positive if X is a cation and negative if X is an anion. The value of β is near 1.00.

EXAMPLE   Using the Selectivity Coefficient

A Ca^{2+} ion-selective electrode has a selectivity coefficient $k_{Ca^{2+},Mg^{2+}} = 0.01$. What will the change in electrode potential be when 1 mM Mg^{2+} is added to 0.1 mM Ca^{2+}?

SOLUTION Using Equation 14-10 with $\beta = 1.00$, n_X = charge of Ca^{2+} = +2, and n_Y = charge of Mg^{2+} = +2, the potential without Mg^{2+} is

$$E = \text{constant} + \left(\frac{0.059\ 16}{2}\right)\log[10^{-4}] = \text{constant} - 118.3\ \text{mV}$$

Addition of 1 mM Mg^{2+} gives an electrode potential of

$$E = \text{constant} + \left(\frac{0.059\ 16}{2}\right)\log[10^{-4} + (10^{-2})(10^{-3})^{2/2}] = \text{constant} - 117.1\ \text{mV}$$

The change is $-117.1 - (-118.3) = 1.2$ mV. If you did not know that Mg^{2+} were present, this voltage would lead you to believe that $[Ca^{2+}]$ is 10% higher than it actually is.

The response of this electrode to Mg^{2+} is only 1% as great as the response to Ca^{2+}. The selectivity coefficient $k_{Ca^{2+},Fe^{2+}} = 0.80$, which means that the response to Fe^{2+} is 80% as great as the response to Ca^{2+}.

Compound Electrodes

A **compound electrode** is a conventional electrode surrounded by a membrane that isolates (or generates) the analyte to which the electrode responds. The CO_2 gas-sensing electrode in Figure 14-16 is an ordinary glass pH electrode surrounded by electrolyte solution enclosed in a semipermeable membrane made of rubber, Teflon, or polyethylene.[3] A silver-silver chloride reference electrode is immersed in the electrolyte solution. When CO_2 diffuses through the semipermeable membrane, it lowers the pH in the electrolyte compartment. The response of the glass electrode to the change in pH is measured.

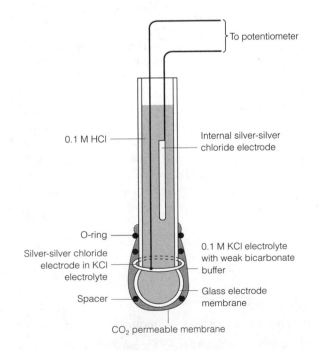

To potentiometer

0.1 M HCl

Internal silver-silver chloride electrode

O-ring

Silver-silver chloride electrode in KCl electrolyte

0.1 M KCl electrolyte with weak bicarbonate buffer

Spacer

Glass electrode membrane

CO_2 permeable membrane

FIGURE 14-16 A CO_2 gas-sensing electrode.

Other acidic or basic gases, including NH_3, SO_2, H_2S, NO_x (nitrogen oxides), and HN_3 (hydrazoic acid), can be detected in the same manner. These electrodes can be used to measure gases *in the gas phase* or dissolved in solution. Some ingenious compound electrodes contain a conventional electrode coated with an enzyme that catalyzes a reaction of the analyte. The product of the reaction is detected by the electrode. Compound electrodes based on enzymes are among the most selective because enzymes tend to be extremely specific in their reactivity with just the species of interest.

Linda A. Hughes

Ask Yourself

14-E. Box 5-1 discussed nitrogen species found in a saltwater aquarium. Now we consider the measurement of ammonia in the fishtank, using an ammonia-selective compound electrode. The procedure is to mix 100.0 mL of unknown or standard with 1.0 mL of 10 M NaOH and then to measure NH_3 with an electrode. The purpose of NaOH is to raise the pH above 11, so that ammonia is in the form NH_3, not NH_4^+. (In a more rigorous procedure, EDTA is added prior to NaOH to bind metal ions and displace NH_3 from the metals.)

    (a) A series of standards gave the readings below. Prepare a calibration curve of potential (mV) versus log(nitrogen concentration in ppm) and determine the equation of the straight line by the method of least squares. (Calibration is done in terms of nitrogen in the original standards. There is no dilution factor to consider from the NaOH.)

NH_3 nitrogen concentration (ppm)	log[N]	Electrode potential (mV vs. S.C.E.)
0.100	−1.000	72
0.500	−0.301	42
1.000	0.000	25

    (b) Two students measured NH_3 in the aquarium and observed values of 106 and 115 mV. What NH_3 nitrogen concentration (in ppm) should be reported by each student?

    (c) Artificial seawater for the aquarium is prepared by adding a commercial seawater salt mix to the correct volume of distilled water. There is an unhealthy level of NH_4Cl impurity in the salt mix, so the instructions call for several hours of aerating freshly prepared seawater to remove $NH_3(g)$ before adding the water to a tank containing live fish. A student measured the concentration of NH_3 in freshly prepared seawater prior to aeration and observed a potential of 56 mV. What is the concentration of NH_3 in the freshly prepared seawater?

Key Equations

Voltage of complete cell

$$E = E_+ - E_- \quad \text{(repeated from Chapter 13)}$$

$E_+ = $ voltage of electrode connected to + terminal of meter

$E_- = $ voltage of electrode connected to − terminal of meter

Titration of X^- with M^+

Before V_e: $[M^+] = K_{sp}/[X^-]$

At V_e: $[M^+] = [X^-] = \sqrt{K_{sp}}$

After V_e: $[M^+] = \dfrac{\text{mol excess } M^+}{\text{total volume}}$

Response of glass pH electrode

$E = \text{constant} + \beta(0.059\ 16)\Delta pH$

$\Delta pH = (\text{analyte pH}) - (\text{pH of internal solution})$

$\beta\ (\approx 1.00)$ is measured with standard buffers

$\text{constant} = $ asymmetry potential (measured by calibration)

Ion-selective electrode response

$E = \text{constant} \pm \beta\left(\dfrac{0.059\ 16}{n_X}\right)\log\left\{[X] + \sum_{Y} \left(k_{X,Y}[Y]^{n_X/n_Y}\right)\right\}$

$X = $ analyte ion with charge n_X

Use $+$ sign if X is cation and $-$ sign if X is anion

$Y = $ interfering ion with charge n_Y

$k_{X,Y} = $ selectivity coefficient

Important Terms

combination electrode

compound electrode

glass electrode

ion-selective electrode

junction potential

liquid-based ion-selective electrode

potentiometry

selectivity coefficient

solid-state ion-selective electrode

Problems

14-1. A cell was prepared by dipping a Cu wire and a saturated $Ag|AgCl$ electrode into 0.10 M $CuSO_4$ solution. The Cu wire was attached to the positive terminal of a potentiometer and the reference electrode was attached to the negative terminal.

(a) Write a half-reaction for the Cu electrode.

(b) Write the Nernst equation for the Cu electrode.

(c) Calculate the cell voltage.

14-2. Pt and saturated calomel electrodes are dipped into a solution containing 0.002 17 M $Br_2(aq)$ and 0.234 M Br^-.

(a) Write the reaction that occurs at Pt and find the half-cell potential E_+.

(b) Find the net cell voltage, E.

14-3. A 50.0-mL solution of 0.100 M NaSCN was titrated with 0.200 M $AgNO_3$ in the cell in Figure 14-1. Find $[Ag^+]$ and E at $V_{Ag^+} = 0.1, 10.0, 25.0,$ and 30.0 mL and sketch the titration curve.

14-4. A 10.0-mL solution of 0.050 0 M $AgNO_3$ was titrated with 0.025 0 M NaBr in the cell in Figure 14-1. Find the cell voltage at $V_{Br^-} = 0.1, 10.0, 20.0,$ and 30.0 mL and sketch the titration curve.

14-5. Which side of the liquid junction 0.1 M $KNO_3|0.1$ M NaCl will be negative? Explain your answer.

14-6. If electrode C in Figure 14-10 is placed in a solution of pH 11.0, what will the pH reading be?

14-7. Suppose that the $Ag|AgCl$ outer electrode in Figure 14-7 is filled with 0.1 M NaCl instead of saturated KCl. Suppose that the electrode is calibrated in a dilute buffer containing 0.1 M KCl at pH 6.54 at 25°C. The electrode is then dipped in a second buffer *at the same pH* and same temperature, but containing 3.5 M KCl.

(a) Use Table 14-3 to estimate the change in junction potential and how much the indicated pH will change.

(b) Suppose that a change in junction potential causes the apparent pH to change from 6.54 to 6.60. By what percentage does $[H^+]$ appear to change?

14-8. Why is measuring $[H^+]$ with a pH electrode somewhat inaccurate, whereas locating the end point in an acid-base titration with a pH electrode can be very accurate?

14-9. Explain the principle of operation of ion-selective electrodes. How does a compound electrode differ from a simple ion-selective electrode?

14-10. What does the selectivity coefficient tell us? Is it better to have a large or a small selectivity coefficient?

14-11. By how many volts will the potential of a Mg^{2+} ion-selective electrode change if the electrode is transferred from 1.00×10^{-4} M $MgCl_2$ to 1.00×10^{-3} M $MgCl_2$?

14-12. When measured with a F^- ion-selective electrode with a Nernstian response at 25°C, the potential due to F^- in unfluoridated groundwater in Foxboro, Massachusetts was 40.0 mV more positive than the potential of tap water in Providence, Rhode Island. Providence maintains its fluoridated water at the recommended level of 1.00 ± 0.05 mg F^-/L. What is the concentration of F^- in mg/L in groundwater in Foxboro? (Disregard the uncertainty.)

14-13. A cyanide ion-selective electrode obeys the equation

$$E = \text{constant} - (0.059\ 16) \log[CN^-]$$

The electrode potential was -0.230 V when the electrode was immersed in 1.00×10^{-3} M NaCN.

(a) Evaluate the constant in the equation above.

(b) Using the result from **(a)**, find the concentration of CN^- if $E = -0.300$ V.

14-14. The selectivity coefficient, $k_{Li^+,Ca^{2+}}$, for a lithium electrode is 5×10^{-5}. When this electrode is placed in 3.44×10^{-4} M Li^+ solution, the potential is -0.333 V versus S.C.E. What would the potential be if Ca^{2+} were added to give 0.100 M Ca^{2+}?

14-15. An ammonia gas-sensing electrode gave the following calibration points when all solutions contained 1 M NaOH:

NH_3 (M)	E (mV)	NH_3 (M)	E (mV)
1.00×10^{-5}	268.0	5.00×10^{-4}	368.0
5.00×10^{-5}	310.0	1.00×10^{-3}	386.4
1.00×10^{-4}	326.8	5.00×10^{-3}	427.6

A dry food sample weighing 312.4 mg was digested by the Kjeldahl procedure (Section 9-6) to convert all the nitrogen to NH_4^+. The digestion solution was diluted to 1.00 L, and 20.0 mL was transferred to a 100-mL volumetric flask. The 20.0-mL aliquot was treated with 10.0 mL of 10.0 M NaOH plus enough NaI to complex the Hg catalyst from the digestion, and diluted to 100.0 mL. When measured with the ammonia electrode, this solution gave a reading of 339.3 mV.

(a) From the calibration data, find $[NH_3]$ in the 100-mL solution.

(b) Calculate the wt % of nitrogen in the food sample.

14-16. The selectivities of a lithium ion-selective electrode are indicated in the diagram. Which alkali metal (Group 1) ion causes the most interference?

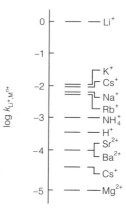

14-17. (a) Write an expression analogous to Equation 14-8 for the response of a La^{3+}-selective electrode to La^{3+} ion.

(b) If $\beta \approx 1.00$, by how many millivolts will the potential change when the electrode is removed from 1.00×10^{-4} M $LaClO_4$ and placed in 1.00×10^{-3} M $LaClO_4$?

(c) By how many millivolts will the potential of the electrode change if the electrode is removed from 2.36×10^{-4} M $LaClO_4$ and placed in 4.44×10^{-3} M $LaClO_4$?

(d) The electrode potential is $+100$ mV in 1.00×10^{-4} M $LaClO_4$ and the selectivity coefficient $k_{La^{3+},Fe^{3+}}$ is $\frac{1}{200}$. What will the potential be when 0.010 M Fe^{3+} is added?

14-18. The following data were obtained when a Ca^{2+} ion-selective electrode was immersed in a series of standard solutions.

Ca^{2+} (M)	E (mV)
3.38×10^{-5}	-74.8
3.38×10^{-4}	-46.4
3.38×10^{-3}	-18.7
3.38×10^{-2}	$+10.0$
3.38×10^{-1}	$+37.7$

(a) Calculate the slope and the y-intercept (and their standard deviations) of the best straight line through the points, using your least-squares spreadsheet from Chapter 4.

(b) Calculate the concentration of a sample that gave a reading of -22.5 mV.

(c) Your spreadsheet gives the uncertainty in $\log[Ca^{2+}]$. Using the upper and lower limits for $\log[Ca^{2+}]$, express the Ca^{2+} concentration as $[Ca^{2+}] = x \pm y$.

14-19. Fourteen ion-selective electrodes were used to measure Ca^{2+} in the same solution with the following results: $[Ca^{2+}]$ = 1.24, 1.13, 1.20, 1.20, 1.30, 1.12, 1.27, 1.19, 1.27, 1.22, 1.23, 1.23, 1.25, 1.24 mM. Find the 95% confidence interval for the mean. If the actual concentration is known to be 1.19 mM, are the results of the ion-selective electrode within experimental error of the known value of the 95% confidence level?

How Would You Do It?

14-20. The graph at the right shows the effect of pH on the response of a liquid-based nitrite (NO_2^-) ion-selective electrode. Ideally, the response would be flat—dependent on $[NO_2^-]$ but not on pH.

(a) Nitrite is the conjugate base of nitrous acid. Why do the curves rise at low pH?

(b) Why do the curves fall at high pH?

(c) What is the optimum pH for using this electrode?

(d) Measure points on the graph at the optimum pH and construct a curve of mV versus $\log [NO_2^-]$. What is the lower concentration limit for linear response?

14-21. Consider the cell: $Ag(s)\,|\,Ag^+(aq, c_1)\,\|\,Ag^+(aq, c_r)\,|$ $Ag(s)$, where c_1 is the concentration of Ag^+ in the left half-cell and c_r is the concentration of Ag^+ in the right half-cell.

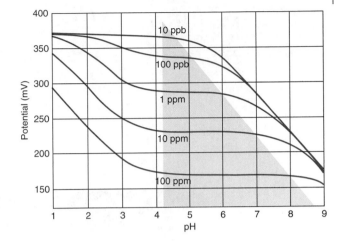

Response of nitrite ion-selective electrode. [From S. J. West and X. Wen, *Am. Environ. Lab.* September 1997, 15.]

When both cells contain 0.010 0 M $AgNO_3$, the measured voltage is very close to 0. When the solution in the right-hand cell is replaced by 15.0 mL of 0.020 0 M $AgNO_3$ plus 15.0 mL of 0.200 M NH_3, the voltage changes to -0.289 V. Under these conditions, nearly all the silver in the right half-cell is $Ag(NH_3)_2^+$. From the measured voltage, can you find the formation constant (called β_2) for the reaction $Ag^+ + 2NH_3 \rightleftharpoons Ag(NH_3)_2^+$?

Notes and References

1. You can build your own glass electrode from a glass Christmas tree ornament in a very instructive experiment. See R. T. da Rocha, I. G. R. Gutz, and C. L. do Lago, *J. Chem. Ed.* **1995**, *72*, 1135.

2. The reference electrode in the Ross combination electrode is $Pt\,|\,I_2, I^-$. This electrode is claimed to give improved precision

and accuracy over conventional pH electrodes [R. C. Metcalf, *Analyst* **1987**, *112*, 1573].

3. You can build your own CO_2 compound electrode. See S. Kocmur, E. Cortón, L. Haim, G. Locascio, and L. Galagosky, *J. Chem. Ed.* **1999**, *76*, 1253.

Further Reading

For the history of ion-selective electrodes, see M. S. Frant, *J. Chem. Ed.* **1997**, *74*, 159; J. Ruzicka, *J. Chem. Ed.* **1997**, *74*, 167; T. S. Light, *J. Chem. Ed.* **1997**, *74*, 171; and C. C. Young, *J. Chem. Ed.* **1997**, *74*, 177.

High-Temperature Superconductors

Permanent magnet levitates above superconducting disk cooled in a pool of liquid nitrogen. Redox titrations are crucial in measuring the chemical composition of a superconductor. [Photo courtesy D. Cornelius and T. Vanderah, Michelson Laboratory.]

*S*uperconductors are materials that lose all electric resistance when cooled below a critical temperature. Prior to 1987, all known superconductors required cooling to temperatures near that of liquid helium (4 K), a process that is costly and impractical for all but a few applications. In 1987, a giant step was taken when "high-temperature" superconductors that retain their superconductivity above the boiling point of liquid nitrogen (77 K) were discovered.

The most startling characteristic of a superconductor is magnetic levitation, which is shown here. When a magnetic field is applied to a superconductor, current flows in the outer skin of the material such that the applied magnetic field is exactly canceled by the induced magnetic field, and the net field inside the specimen is 0. Expulsion of a magnetic field from a superconductor is called the *Meissner effect.*

A prototypical high-temperature superconductor is yttrium barium copper oxide, $YBa_2Cu_3O_7$, in which two-thirds of the copper is in the +2 oxidation state and one-third occurs in the unusual +3 state. Another example is $Bi_2Sr_2(Ca_{0.8}Y_{0.2})Cu_2O_{8.295}$, in which the average oxidation state of copper is +2.105 and the average oxidation state of bismuth is +3.090 (which is formally a mixture of Bi^{3+} and Bi^{5+}). The most reliable means to unravel these complex formulas is through a redox titration described in the last problem of this chapter.

Chapter 15

REDOX TITRATIONS

A **redox titration** is based on an oxidation-reduction reaction between analyte and titrant. Common oxidizing agents in analytical chemistry include iodine (I_2), permanganate (MnO_4^-), cerium(IV), and dichromate (CrO_7^{2-}). Reducing agents such as Fe^{2+} (ferrous ion) and Sn^{2+} (stannous ion) are less commonly encountered because solutions of most reducing agents need protection from air to avoid reaction with O_2. We begin with the theory of redox titrations and redox indicators and then proceed to iodine titrations that are widely used for chemical analysis.

Box 15-1 explains how dichromate is used to measure *chemical oxygen demand* in environmental analysis.

15-1 Theory of Redox Titrations

Consider the titration of iron(II) with standard cerium(IV), the course of which could be monitored potentiometrically as shown in Figure 15-1. The titration reaction is

titration reaction:
$$Ce^{4+} + Fe^{2+} \longrightarrow Ce^{3+} + Fe^{3+} \qquad (15\text{-}1)$$

Ceric   Ferrous   Cerous   Ferric

(titrant)   (analyte)

for which $K \approx 10^{16}$ in 1 M $HClO_4$. Each mole of ceric ion oxidizes 1 mol of ferrous ion rapidly and quantitatively. The titration reaction creates a mixture of Ce^{4+}, Ce^{3+}, Fe^{2+}, and Fe^{3+} in the beaker in Figure 15-1.

The *titration reaction* is $Ce^{4+} + Fe^{2+} \rightarrow Ce^{3+} + Fe^{3+}$. It goes to completion after each addition of titrant. The equilibrium constant is given by Equation 13-21: $K = 10^{nE^\circ/0.059\,16}$ at 25°C.

325

Environmental Carbon Analysis and Oxygen Demand

Industrial waste streams are partially characterized and regulated on the basis of their carbon content or oxygen demand. *Total carbon* (TC) is defined by the amount of CO_2 evolved when a sample is completely oxidized:

total carbon analysis: all carbon $\xrightarrow[\text{catalyst}]{O_2/900°C}$ CO_2

Total carbon includes dissolved organic material (called *total organic carbon,* TOC) and dissolved CO_3^{2-} and HCO_3^- (called *inorganic carbon,* IC). By definition, TC = TOC + IC.

To distinguish TOC from IC, the pH of a fresh sample is lowered below 2 to convert CO_3^{2-} and HCO_3^- to CO_2, which is purged from the solution with N_2 prior to combustion analysis. Because IC has been removed, this procedure measures TOC only. The IC content is the difference between the two experiments. TOC is widely used to determine compliance with discharge laws.

Total oxygen demand (TOD) tells us how much O_2 is required for complete combustion of pollutants in a waste stream. A volume of N_2 containing a known quantity of O_2 is mixed with the sample and complete combustion is carried out. The remaining O_2 is measured by a potentiometric sensor. This measurement is sensitive to the oxidation states of species in the waste stream. For example, urea $((NH_2)_2C{=}O)$ consumes five times as much O_2 as formic acid (HCO_2H) does. Species such as NH_3 and H_2S also contribute to TOD.

Pollutants can be oxidized by refluxing with dichromate $(Cr_2O_7^{2-})$. *Chemical oxygen demand* (COD) is defined as the O_2 that is chemically equivalent to the $Cr_2O_7^{2-}$ consumed in this process. Each $Cr_2O_7^{2-}$ consumes $6e^-$ (to make $2\ Cr^{3+}$) and each O_2 consumes $4e^-$ (to make $2\ H_2O$). Therefore, 1 mol of $Cr_2O_7^{2-}$ is chemically equivalent to 1.5 mol of O_2 for this computation. COD analysis is carried out by refluxing polluted water for 2 h with excess standard $Cr_2O_7^{2-}$ in H_2SO_4 solution containing Ag^+ catalyst. Unreacted $Cr_2O_7^{2-}$ is then measured by titration with standard Fe^{2+} or by spectrophotometry. Many permits for industrial operations are defined in terms of COD analysis of the waste streams.

Biochemical oxygen demand (BOD) is defined as the O_2 required for biochemical degradation of organic materials by microorganisms. The procedure calls for incubating a sealed container of wastewater with no extra air space for 5 days at 20°C in the dark while microbes metabolize organic compounds in the waste. The O_2 dissolved in the solution is measured before and after the incubation. The difference is BOD.[1] BOD also measures species such as HS^- and Fe^{2+} that may be in the water. Inhibitors are added to prevent oxidation of nitrogen species such as NH_3.

To follow the course of the reaction, we place a pair of electrodes into the titration vessel. At the *calomel reference electrode,* the reaction is

reference half-reaction: $2Hg(l) + 2Cl^- \rightleftharpoons Hg_2Cl_2(s) + 2e^-$

At the *Pt indicator electrode,* there are *two* reactions that come to equilibrium:

Equilibria 15-2 and 15-3 are both established at the Pt electrode.

indicator half-reaction: $Fe^{3+} + e^- \rightleftharpoons Fe^{2+}$ $E° = 0.767$ V   (15-2)

indicator half-reaction: $Ce^{4+} + e^- \rightleftharpoons Ce^{3+}$ $E° = 1.70$ V   (15-3)

The potentials cited here are the formal potentials that apply in 1 M $HClO_4$.

The cell reaction can be described in either of two ways:

cell reaction: $2Fe^{3+} + 2Hg(l) + 2Cl^- \rightleftharpoons 2Fe^{2+} + Hg_2Cl_2(s)$ (15-4)

cell reaction: $2Ce^{4+} + 2Hg(l) + 2Cl^- \rightleftharpoons 2Ce^{3+} + Hg_2Cl_2(s)$ (15-5)

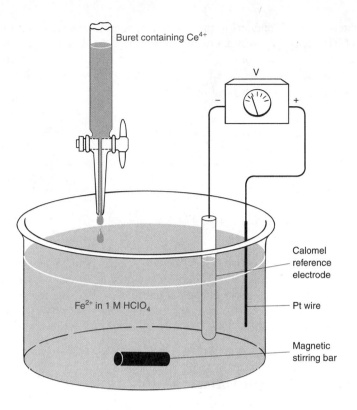
Buret containing Ce^{4+}

V

− +

Calomel
reference
electrode

Fe^{2+} in 1 M $HClO_4$

Pt wire

Magnetic
stirring bar

FIGURE 15-1 Apparatus for potentiometric titration of Fe^{2+} with Ce^{4+}.

The cell reactions are not the same as the titration reaction (15-1). The potentiometer has no interest in how the Ce^{4+}, Ce^{3+}, Fe^{3+}, and Fe^{2+} happened to get into the beaker. What it does care about is that electrons want to flow from the anode to the cathode through the meter. If the solution has come to equilibrium, the potential driving Reactions 15-4 and 15-5 must be the same. **We may describe the cell with either Reaction 15-4 or Reaction 15-5, or both, however we please.**

The titration reaction goes to completion. The cell reactions proceed to a negligible extent because the potentiometer allows negligible current to flow. The potentiometer *measures* the concentrations of species in solution but does not *change* them.

To reiterate, the physical picture of the titration is this: Ce^{4+} is added from the buret to create a mixture of Ce^{4+}, Ce^{3+}, Fe^{3+}, and Fe^{2+} in the beaker. Because the equilibrium constant for Reaction 15-1 is large, the titration reaction goes to "completion" after each addition of Ce^{4+}. The potentiometer measures the voltage driving electrons from the reference electrode, through the meter, and out at the Pt electrode. That is, **the circuit measures the potential for reduction of Fe^{3+} or Ce^{4+} at the Pt surface by electrons from the $Hg\,|\,Hg_2Cl_2$ couple in the reference electrode.** The titration reaction, on the other hand, is an oxidation of Fe^{2+} and a reduction of Ce^{4+}. The titration reaction produces a certain mixture of Ce^{4+}, Ce^{3+}, Fe^{3+}, and Fe^{2+}. The circuit measures the potential for reduction of Ce^{4+} and Fe^{3+} by Hg. **The titration reaction goes to completion. The cell reaction is negligible. The cell is being used to measure concentrations, not to change them.**

We now set out to calculate how the cell voltage changes as Fe^{2+} is titrated with Ce^{4+}. The titration curve has three regions.

Region 1: Before the Equivalence Point

As each aliquot of Ce^{4+} is added, titration reaction 15-1 consumes the Ce^{4+} and creates an equal number of moles of Ce^{3+} and Fe^{3+}. Prior to the equivalence point,

E_+ is the potential of the half-cell connected to the positive terminal of the potentiometer in Figure 15-1. E_- refers to the half-cell connected to the negative terminal.

excess unreacted Fe^{2+} remains in the solution. Therefore, we can find the concentrations of Fe^{2+} and Fe^{3+} without difficulty. On the other hand, we cannot find the concentration of Ce^{4+} without solving a fancy little equilibrium problem. Because the amounts of Fe^{2+} and Fe^{3+} are both known, it is *convenient* to calculate the cell voltage with Reaction 15-2 instead of 15-3.

$$E = E_+ - E_- = \left[0.767 - 0.059\ 16 \log \left(\frac{[Fe^{2+}]}{[Fe^{3+}]} \right) \right] - 0.241$$

$$\uparrow \qquad\qquad\qquad\qquad\qquad\qquad\qquad\qquad \uparrow$$

Formal potential for   Potential of
Fe^{3+} reduction in   saturated calomel
1 M $HClO_4$ electrode

$$E = 0.526 - 0.059\ 16 \log \left(\frac{[Fe^{2+}]}{[Fe^{3+}]} \right) \tag{15-6}$$

For Reaction 15-2, $E_+ = E°$ for the $Fe^{3+} | Fe^{2+}$ couple when $V = \frac{1}{2}V_e$.

One special point is reached before the equivalence point. When the volume of titrant is one-half of the amount required to reach the equivalence point ($V = \frac{1}{2}V_e$), the concentrations of Fe^{3+} and Fe^{2+} are equal. In this case, the log term is 0, and $E_+ = E°$ for the $Fe^{3+} | Fe^{2+}$ couple. *The point at which $V = \frac{1}{2}V_e$ is analogous to the point at which $pH = pK_a$ when $V = \frac{1}{2}V_e$ in an acid-base titration.*

Region 2: At the Equivalence Point

Exactly enough Ce^{4+} has been added to react with all the Fe^{2+}. Virtually all cerium is in the form Ce^{3+}, and virtually all iron is in the form Fe^{3+}. Tiny amounts of Ce^{4+} and Fe^{2+} are present at equilibrium. From the stoichiometry of Reaction 15-1, we can say that

$$[Ce^{3+}] = [Fe^{3+}] \tag{15-7}$$

and

$$[Ce^{4+}] = [Fe^{2+}] \tag{15-8}$$

To understand why Equations 15-7 and 15-8 are true, imagine that *all* the cerium and the iron have been converted to Ce^{3+} and Fe^{3+}. Because we are at the equivalence point, $[Ce^{3+}] = [Fe^{3+}]$. Now let Reaction 15-1 come to equilibrium:

$$Fe^{3+} + Ce^{3+} \rightleftharpoons Fe^{2+} + Ce^{4+} \text{ (reverse of Reaction 15-1)}$$

If a little bit of Fe^{3+} goes back to Fe^{2+}, an equal number of moles of Ce^{4+} must be made. So $[Ce^{4+}] = [Fe^{2+}]$.

At any time, Reactions 15-2 and 15-3 are *both* in equilibrium at the Pt electrode. At the equivalence point, it is *convenient* to use both reactions to find the cell voltage. The Nernst equations are

At the equivalence point, we use both Reactions 15-2 and 15-3 to calculate the cell voltage. This is strictly a matter of algebraic convenience.

$$E_+ = 0.767 - 0.059\ 16 \log \left(\frac{[Fe^{2+}]}{[Fe^{3+}]} \right) \tag{15-9}$$

$$E_+ = 1.70 - 0.059\ 16 \log \left(\frac{[Ce^{3+}]}{[Ce^{4+}]} \right) \tag{15-10}$$

Here is where we stand: Each equation above is a statement of algebraic truth. But neither one alone allows us to find E_+ because we do not know exactly what tiny concentrations of Fe^{2+} and Ce^{4+} are present. It is possible to solve the four simultaneous equations 15-7 through 15-10, by first *adding* Equations 15-9 and 15-10. The logic and beauty of this algebraic manipulation will become apparent momentarily.

Adding Equations 15-9 and 15-10 gives

$$2E_+ = 0.767 + 1.70 - 0.059\,16\log\left(\frac{[Fe^{2+}]}{[Fe^{3+}]}\right) - 0.059\,16\log\left(\frac{[Ce^{3+}]}{[Ce^{4+}]}\right)$$

$$2E_+ = 2.46_7 - 0.059\,16\log\left(\frac{[Fe^{2+}][Ce^{3+}]}{[Fe^{3+}][Ce^{4+}]}\right)$$

$$\log a + \log b = \log ab$$

But, because $[Ce^{3+}] = [Fe^{3+}]$ and $[Ce^{4+}] = [Fe^{2+}]$ at the equivalence point, the ratio of concentrations in the log term is unity. Therefore, the logarithm is 0 and

$$2E_+ = 2.46_7\text{ V} \Rightarrow E_+ = 1.23\text{ V}$$

E_+ is just the average of the standard potentials for the two half-reactions at the Pt electrode. The cell voltage is

$$E = E_+ - E(\text{calomel}) = 1.23 - 0.241 = 0.99\text{ V} \qquad (15\text{-}11)$$

In this particular titration, the equivalence point voltage is independent of the con centrations and volumes of the reactants.

Region 3: After the Equivalence Point

Now virtually all iron atoms are Fe^{3+}. The moles of Ce^{3+} equal the moles of Fe^{3+}, and there is a known excess of unreacted Ce^{4+}. Because we know both $[Ce^{3+}]$ and $[Ce^{4+}]$, it is *convenient* to use Reaction 15-3 to describe the chemistry at the Pt electrode:

$$E = E_+ - E(\text{calomel}) = \left[1.70 - 0.059\,16\log\left(\frac{[Ce^{3+}]}{[Ce^{4+}]}\right)\right] - 0.241 \qquad (15\text{-}12)$$

After the equivalence point, it is convenient to use Reaction 15-3 because we can easily calculate the concentrations of Ce^{3+} and Ce^{4+}. It is not convenient to use Reaction 15-2, because we do not know the concentration of Fe^{2+}, which has been "used up."

At the special point when $V = 2V_e$, $[Ce^{3+}] = [Ce^{4+}]$ and $E_+ = E°(Ce^{4+}|Ce^{3+}) = 1.70$ V.

Before the equivalence point, the voltage is fairly steady near the value $E = E_+ - E(\text{calomel}) \approx E°(Fe^{3+} | Fe^{2+}) - 0.241$ V $= 0.53$ V. After the equivalence point, the voltage levels off near $E \approx E°(Ce^{4+} | Ce^{3+}) - 0.241$ V $= 1.46$ V. At the equivalence point, there is a rapid rise in voltage.

EXAMPLE   Potentiometric Redox Titration

Suppose that we titrate 100.0 mL of 0.050 0 M Fe^{2+} with 0.100 M Ce^{4+}, using the cell in Figure 15-1. The equivalence point occurs when $V_{Ce^{4+}} = 50.0$ mL, because the Ce^{4+} is twice as concentrated as the Fe^{2+}. Calculate the cell voltage at 36.0, 50.0, and 63.0 mL.

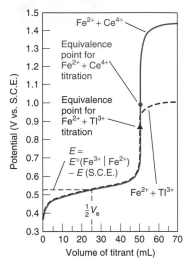

FIGURE 15-2 Solid line: theoretical curve for titration of 100.0 mL of 0.050 0 M Fe^{2+} with 0.100 M Ce^{4+} in 1 M $HClO_4$. You cannot calculate the potential for zero titrant, but you can start at a small volume such as 0.1 mL. Dashed line: theoretical curve for titration of 100.0 mL of 0.050 0 M Fe^{2+} with 0.050 0 M Tl^{3+} in 1 M $HClO_4$.

The shape of the curve in Figure 15-2 is essentially independent of the concentrations of analyte and titrant. The solid curve is symmetric near V_e because the stoichiometry is 1:1.

You would not choose a weak acid to titrate a weak base because the break at V_e would not be very large.

SOLUTION

At 36.0 mL: This is 36.0/50.0 of the way to the equivalence point. Therefore, 36.0/50.0 of the iron is in the form Fe^{3+} and 14.0/50.0 is in the form Fe^{2+}. Putting $[Fe^{2+}]/[Fe^{3+}] = 14.0/36.0$ into Equation 15-6 gives $E = 0.550$ V.

At 50.0 mL: Equation 15-11 tells us that the cell voltage at the equivalence point is 0.99 V, regardless of the concentrations of reagents for this particular titration.

At 63.0 mL: The first 50.0 mL of cerium has been converted to Ce^{3+}. Because 13.0 mL of excess Ce^{4+} has been added, $[Ce^{3+}]/[Ce^{4+}] = 50.0/13.0$ in Equation 15-12, and $E = 1.424$ V.

Shapes of Redox Titration Curves

The preceding calculations allow us to plot the solid titration curve for Reaction 15-1 in Figure 15-2, which shows the potential as a function of the volume of added titrant. The equivalence point is marked by a steep rise in the voltage. The calculated value of E_+ at $\frac{1}{2}V_e$ is the formal potential of the $Fe^{3+} \mid Fe^{2+}$ couple, because the quotient $[Fe^{2+}]/[Fe^{3+}]$ is unity at this point. The calculated voltage at any point in this titration depends only on the *ratio* of reactants; their *concentrations* do not figure in any calculations in this example. We expect, therefore, that the curve in Figure 15-2 will be independent of dilution. We should observe the same curve if both reactants were diluted by a factor of 10.

The voltage at zero titrant volume cannot be calculated because we do not know how much Fe^{3+} is present. If $[Fe^{3+}] = 0$, the voltage calculated with Equation 15-9 would be $-\infty$. In fact, there must be some Fe^{3+} in each reagent, either as an impurity or from oxidation of Fe^{2+} by atmospheric oxygen. In any case, the voltage could not be lower than that needed to reduce the solvent ($H_2O + e^- \rightarrow \frac{1}{2}H_2 + OH^-$).

For Reaction 15-1, the titration curve in Figure 15-2 is symmetric near the equivalence point because the reaction stoichiometry is 1:1. For oxidation of Fe(II) by Tl(III),

$$2Fe^{2+} + Tl^{3+} \longrightarrow 2Fe^{3+} + Tl^+ \qquad (15-13)$$

the dashed curve in Figure 15-2 is not symmetric about the equivalence point because the stoichiometry of reactants is 2:1, not 1:1. Still, the curve is so steep near the equivalence point that negligible error is introduced if the center of the steepest portion is taken as the end point. Demonstration 15-1 provides an example of an asymmetric titration curve whose shape also depends on the pH of the reaction medium.

The change in voltage near the equivalence point for the dashed curve in Figure 15-2 is smaller than the voltage change for the solid curve because Tl^{3+} is a weaker oxidizing agent than Ce^{4+}. Clearest results are achieved with the strongest oxidizing and reducing agents. The same rule applies to acid-base titrations where strong-acid or strong-base titrants give the sharpest break at the equivalence point.

Ask Yourself

15-A. A 20.0-mL solution of 0.005 00 M Sn^{2+} in 1 M HCl was titrated with 0.020 0 M Ce^{4+} to give Sn^{4+} and Ce^{3+}. What is the potential (versus S.C.E.) at the following volumes of Ce^{4+}: 0.100, 1.00, 5.00, 9.50, 10.00, 10.10, and 12.00 mL? Sketch the titration curve.

Potentiometric Titration of Fe^{2+} with MnO_4^-

The titration of Fe^{2+} with $KMnO_4$ nicely illustrates principles of potentiometric titrations.

$$MnO_4^- + 5Fe^{2+} + 8H^+ \longrightarrow$$
Titrant   Analyte

$$Mn^{2+} + 5Fe^{3+} + 4H_2O \quad \text{(A)}$$

Dissolve 0.60 g of $Fe(NH_4)_2(SO_4)_2 \cdot 6H_2O$ (FM 392.13; 1.5 mmol) in 400 mL of 1 M H_2SO_4. Titrate the well-stirred solution with 0.02 M $KMnO_4$ (~15 mL is required to reach the equivalence point), using Pt and calomel electrodes with a pH meter as a potentiometer. The reference socket of the pH meter is the negative input terminal. Before starting the titration, calibrate the meter by connecting the two input sockets directly to each other with a wire and setting the millivolt scale of the meter to 0.

The demonstration is much more meaningful if you calculate some points on the theoretical titration curve before performing the experiment. Then compare the theoretical and experimental results. Also note the coincidence of the potentiometric and visual end points.

Question Potassium permanganate is purple, and all the other species in this titration are colorless (or very faintly colored). What color change is expected at the equivalence point?

To calculate points on the theoretical titration curve, we use the following half-reactions:

$$Fe^{3+} + e^- \rightleftharpoons Fe^{2+}$$

$$E° = 0.68 \text{ V in 1 M } H_2SO_4 \quad \text{(B)}$$

$$MnO_4^- + 8H^+ + 5e^- \longrightarrow Mn^{2+} + 4H_2O$$

$$E° = 1.507 \text{ V} \quad \text{(C)}$$

Prior to the equivalence point, calculations are similar to those in Section 15-1 for the titration of Fe^{2+} by Ce^{4+}, but using $E° = 0.68$ V. After the equivalence point, you can find the potential by using Reaction C. For example, suppose that you titrate 0.400 L of 3.75 mM Fe^{2+} with 0.020 0 M $KMnO_4$. From the stoichiometry of Reaction A, the equivalence point is $V_e = 15.0$ mL. When you have added 17.0 mL of $KMnO_4$, the concentrations of species in Reaction C are $[Mn^{2+}] = 0.719$ mM, $[MnO_4^-] = 0.095\ 9$ mM, and $[H^+] = 0.959$ M. The cell potential is

$$E = E_+ - E(\text{calomel})$$

$$= \left[1.507 - \frac{0.059\ 16}{5} \log \left(\frac{[Mn^{2+}]}{[MnO_4^-][H^+]^8} \right) \right] - 0.241$$

$$= \left[1.507 - \frac{0.059\ 16}{5} \log \left(\frac{7.19 \times 10^{-4}}{(9.59 \times 10^{-5})(0.959)^8} \right) \right]$$

$$- 0.241 = 1.254 \text{ V}$$

To calculate the potential at the equivalence point, we add the Nernst equations for Reactions B and C, as we did for the cerium and iron reactions in Section 15-1. Before doing so, however, multiply the permanganate equation by 5 so that we can add the log terms:

$$E_+ = 0.68 - 0.059\ 16 \log \left(\frac{[Fe^{2+}]}{[Fe^{3+}]} \right)$$

$$5E_+ = 5 \left[1.507 - \frac{0.059\ 16}{5} \log \left(\frac{[Mn^{2+}]}{[MnO_4^-][H^+]^8} \right) \right]$$

Now we can add the two equations to get

$$6E_+ = 8.215 - 0.059\ 16 \log \left(\frac{[Mn^{2+}][Fe^{2+}]}{[MnO_4^-][Fe^{3+}][H^+]^8} \right)$$
$$\text{(D)}$$

But the stoichiometry of the titration reaction A tells us that at the equivalence point $[Fe^{3+}] = 5[Mn^{2+}]$ and $[Fe^{2+}] = 5[MnO_4^-]$. Substituting these values into Equation D gives

$$6E_+ = 8.215 - 0.059\ 16 \log \left(\frac{[Mn^{2+}](5[MnO_4^-])}{[MnO_4^-](5[Mn^{2+}])[H^+]^8} \right)$$

$$= 8.215 - 0.059\ 16 \log \left(\frac{1}{[H^+]^8} \right) \quad \text{(E)}$$

Inserting the concentration of $[H^+]$, which is $(400/415)(1.00 \text{ M}) = 0.964$ M, we find

$$6E_+ = 8.215 - 0.059\ 16 \log \left(\frac{1}{(0.964)^8} \right) \Rightarrow$$

$$E_+ = 1.368 \text{ V}$$

The predicted cell voltage at V_e is $E = E_+ - E(\text{calomel}) = 1.368 - 0.241 = 1.127$ V.

15-2   Redox Indicators

An indicator may be used to detect the end point of a redox titration, just as an indicator may be used in an acid-base titration. A **redox indicator** changes color when it goes from its oxidized to its reduced state. One common indicator is ferroin, whose color change is from pale blue (almost colorless) to red.

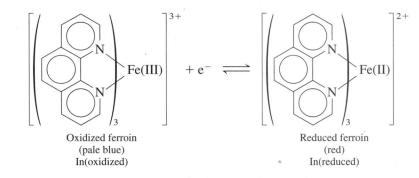

Oxidized ferroin
(pale blue)
In(oxidized)

Reduced ferroin
(red)
In(reduced)

To predict the potential range over which the indicator color will change, we first write a Nernst equation for the indicator.

$$\text{In(oxidized)} + ne^- \rightleftharpoons \text{In(reduced)}$$

$$E = E^\circ - \frac{0.059\ 16}{n} \log\left(\frac{[\text{In(reduced)}]}{[\text{In(oxidized)}]}\right)$$

As with acid-base indicators, the color of In(reduced) will be observed when

$$\frac{[\text{In(reduced)}]}{[\text{In(oxidized)}]} \gtrsim \frac{10}{1}$$

and the color of In(oxidized) will be observed when

$$\frac{[\text{In(reduced)}]}{[\text{In(oxidized)}]} \lesssim \frac{1}{10}$$

Putting these quotients into the Nernst equation for the indicator tells us that the color change will occur over the range

A redox indicator changes color over a range of $\pm(59/n)$ mV, centered at E° for the indicator. n is the number of electrons in the indicator half-reaction.

Redox indicator color change range:

$$E = \left(E^\circ \pm \frac{0.059\ 16}{n}\right) \text{volts} \qquad (15\text{-}14)$$

For ferroin, with $E^\circ = 1.147$ V (Table 15-1), we expect the color change to occur in the approximate range 1.088 V to 1.206 V with respect to the standard hydrogen electrode. If a saturated calomel electrode is used as the reference instead, the indicator transition range is

Figure 13-10 will help you understand Equation 15-15.

The indicator transition range should overlap the steep part of the titration curve.

$$\begin{pmatrix} \text{Indicator transition} \\ \text{range versus calomel} \\ \text{electrode (S.C.E.)} \end{pmatrix} = \begin{pmatrix} \text{Transition range} \\ \text{versus standard hydrogen} \\ \text{electrode (S.H.E.)} \end{pmatrix} - E(\text{calomel}) \qquad (15\text{-}15)$$

$$= (1.088 \text{ to } 1.206) - (0.241)$$

$$= 0.847 \text{ to } 0.965 \text{ V (versus S.C.E.)}$$

TABLE 15-1   **Redox indicators**

	Color		
Indicator	Reduced	Oxidized	$E°$
Phenosafranine	Colorless	Red	0.28
Indigo tetrasulfonate	Colorless	Blue	0.36
Methylene blue	Colorless	Blue	0.53
Diphenylamine	Colorless	Violet	0.75
4′-Ethoxy-2,4-diaminoazobenzene	Red	Yellow	0.76
Diphenylamine sulfonic acid	Colorless	Red-violet	0.85
Diphenylbenzidine sulfonic acid	Colorless	Violet	0.87
Tris(2,2′-bipyridine)iron	Red	Pale blue	1.120
Tris(1,10-phenanthroline)iron (ferroin)	Red	Pale blue	1.147
Tris(5-nitro-1,10-phenanthroline)iron	Red-violet	Pale blue	1.25
Tris(2,2′-bipyridine)ruthenium	Yellow	Pale blue	1.29

Ferroin would therefore be a useful indicator for the solid curve in Figure 15-2.

The larger the difference in standard potential between titrant and analyte, the sharper the break in the titration curve at the equivalence point. A redox titration is usually feasible if the difference between analyte and titrant is $\gtrsim 0.2$ V. However, the end point of such a titration is not very sharp and is best detected potentiometrically. If the difference in formal potentials is $\gtrsim 0.4$ V, then a redox indicator usually gives a satisfactory end point.

Ask Yourself

15-B. What would be the best redox indicator in Table 15-1 for the titration of $Fe(CN)_6^{4-}$ with Tl^{3+} in 1 M HCl? (*Hint:* The potential at the equivalence point must be between the potentials for each redox couple.) What color change would you look for?

15-3 Titrations Involving Iodine

Convenient redox titrations (Tables 15-2 and 15-3) are available for many analytes with iodine (I_2, a mild oxidizing agent) or iodide (I^-, a mild reducing agent).

Iodine as oxidizing agent:	$I_2(aq) + 2e^- \longrightarrow 2I^-$	(15-16)
Iodide as reducing agent:	$2I^- \longrightarrow I_2(aq) + 2e^-$	(15-17)

For example, vitamin C in foods and the compositions of superconductors (shown at the opening of the chapter) can be measured with iodine. When a reducing analyte is titrated with iodine (I_2), the method is called *iodimetry. Iodometry* is the

TABLE 15-2   Iodimetric titrations: Titrations with standard iodine (actually I_3^-)

Species analyzed	Oxidation reaction	Notes
SO_2	$SO_2 + H_2O \rightleftharpoons H_2SO_3$ $H_2SO_3 + H_2O \rightleftharpoons SO_4^{2-} + 4H^+ + 2e^-$	Add SO_2 (or H_2SO_3 or HSO_3^- or SO_3^{2-}) to excess standard I_3^- in dilute acid and back-titrate unreacted I_3^- with standard thiosulfate.
H_2S	$H_2S \rightleftharpoons S(s) + 2H^+ + 2e^-$	Add H_2S to excess I_3^- in 1 M HCl and back-titrate with thiosulfate.
$Zn^{2+}, Cd^{2+}, Hg^{2+}, Pb^{2+}$	$M^{2+} + H_2S \longrightarrow MS(s) + 2H^+$ $MS(s) \rightleftharpoons M^{2+} + S + 2e^-$	Precipitate and wash metal sulfide. Dissolve in 3 M HCl with excess standard I_3^- and back-titrate with thiosulfate.
Cysteine, glutathione, mercaptoethanol	$2RSH \rightleftharpoons RSSR + 2H^+ + 2e^-$	Titrate the sulfhydryl compound at pH 4–5 with I_3^-.
$H_2C{=}O$	$H_2CO + 3OH^- \rightleftharpoons HCO_2^- + 2H_2O + 2e^-$	Add excess I_3^- plus NaOH to the unknown. After 5 min, add HCl and back-titrate with thiosulfate.
Glucose (and other reducing sugars)	$\overset{\displaystyle O}{\overset{\displaystyle \|}{R C H}} + 3OH^- \rightleftharpoons RCO_2^- + 2H_2O + 2e^-$	Add excess I_3^- plus NaOH to the sample. After 5 min, add HCl and back-titrate with thiosulfate.

titration of iodine produced when an oxidizing analyte is added to excess I^-. The iodine is usually titrated with standard thiosulfate solution.

I_2 is only slightly soluble in water (1.3×10^{-3} M at 20°C), but its solubility is enhanced by complexation with iodide:

$$\underset{\text{Iodine}}{I_2(aq)} + \underset{\text{Iodide}}{I^-} \rightleftharpoons \underset{\text{Triiodide}}{I_3^-} \qquad K = 7 \times 10^2 \tag{15-18}$$

A typical 0.05 M solution of I_3^- for titrations is prepared by dissolving 0.12 mol of KI plus 0.05 mol of I_2 in 1 L of water. When we speak of using "iodine" as a titrant, we almost always mean I_2 plus excess I^-. For simplicity, we often write reactions in terms of I_2 rather than I_3^-, but you should realize that a mole of I_2 is equivalent to a mole of I_3^- through Reaction 15-18.

Starch Indicator

Starch is the indicator of choice for iodine because it forms an intense blue complex with iodine. The active fraction of starch is amylose, a polymer of the sugar α-D-glucose (Figure 15-3). The polymer coils into a helix, inside of which chains of I_6 (made from $3I_2$) fit and form an intense blue color.

$$\cdots [I{-}I{-}I{-}I{-}I{-}I] \cdots [I{-}I{-}I{-}I{-}I{-}I] \cdots$$

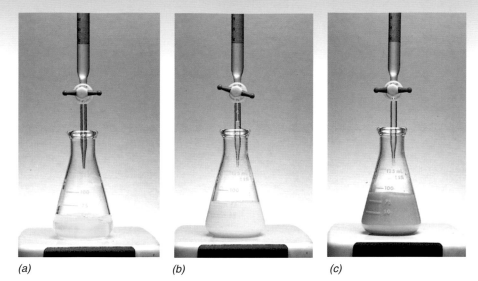

COLOR PLATE 1
Fajans Titration of Cl⁻ with AgNO₃ Using Dichlorofluorescein (Demonstration 5-1)
(a) Indicator before beginning titration.
(b) AgCl precipitate before end point.
(c) Indicator adsorbed on precipitate after end point.

(a)   *(b)*   *(c)*

(a)   *(b)*   *(c)*

COLOR PLATE 2   Colloids and Dialysis (Demonstration 6-1)
(a) Ordinary aqueous Fe(III) (right) and colloidal Fe(III) (left). *(b)* Dialysis bags containing colloidal Fe(III) (left) and a solution of Cu(II) (right) immediately after placement in flasks of water. *(c)* After 24 h of dialysis, the Cu(II) has diffused out and is dispersed uniformly between the bag and the flask, but the colloidal Fe(III) remains inside the bag.

COLOR PLATE 3
HCl Fountain
(Demonstration 7-1)
(a) Basic indicator solution in beaker. *(b)* Indicator is drawn into flask and changes to acidic color.
(c) Solution levels at end of experiment.

(a)   *(b)*   *(c)*

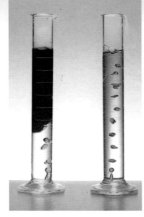
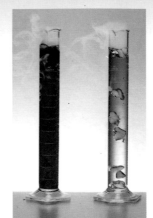

(a) (b) (c)

(d) (e)

COLOR PLATE 4 **Indicators and Acidity of CO₂**
(Demonstration 8-2) *(a)* Cylinder before adding Dry
Ice. Ethanol indicator solutions of phenolphthalein (left)
and bromothymol blue (right) have not yet mixed with
entire cylinder. *(b)* Adding Dry Ice causes bubbling and
mixing. *(c)* Further mixing. *(d)* Phenolphthalein changes
to its colorless acidic form. Color of bromothymol blue
is due to mixture of acidic and basic forms. *(e)* After
addition of HCl and stirring of right–hand cylinder, bubbles
of CO₂ can be seen leaving solution, and indicator
changes completely to its acidic color.

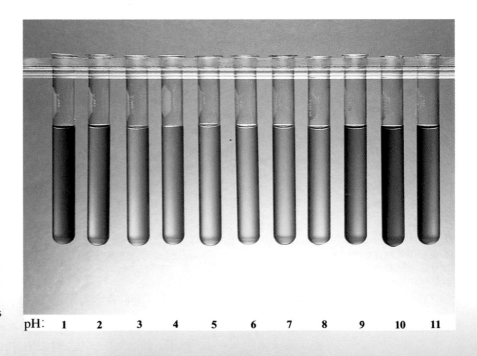

COLOR PLATE 5
Thymol Blue (Section 8-6)
Acid–base indicator thymol
blue between pH 1 (left)
and 11 (right). The pK values
are 1.7 and 8.9.

pH: **1** **2** **3** **4** **5** **6** **7** **8** **9** **10** **11**

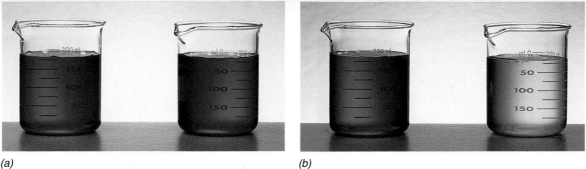

(a) (b)

COLOR PLATE 6   Effect of Ionic Strength on Ionic Dissociation (Demonstration 11-1)
(a) Two beakers containing identical solutions with $FeSCN^{2+}$, Fe^{3+}, and SCN^-. *(b)* Color change when KNO_3 is added to the right-hand beaker.

COLOR PLATE 7   Titration of Cu(II) with EDTA, Using Auxiliary Complexing Agent (Section 12-2)
0.02 M $CuSO_4$ before titration (left). Color of Cu(II)–ammonia complex after adding ammonia buffer, pH 10 (center). End-point color when all ammonia ligands have been displaced by EDTA (right). In this case, the purpose of the auxiliary complexing agent is not to keep the metal ion in solution but to accentuate the color change at the equivalence point.

(a)   (b)

COLOR PLATE 8   Titration of Mg^{2+} by EDTA Using Eriochrome Black T Indicator (Demonstration 12-1) *(a)* Before (left), near (center), and after (right) equivalence point.
(b) Same titration with methyl red added as inert dye to alter colors.

COLOR PLATE 14 Grating Dispersion
(Section 18-1) Visible spectrum produced by grating inside spectrophotometer.

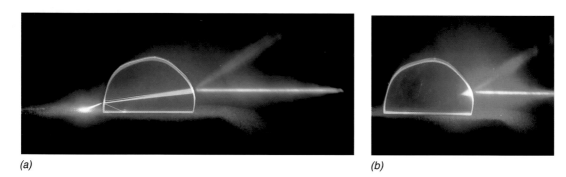

(a)   *(b)*

COLOR PLATE 15 Transmission, Reflection, Refraction, and Absorption of Light (Section 18-1)
(a) Blue-green laser is directed into a semicircular crystal of yttrium aluminum garnet containing a small amount of Er^{3+}, which emits yellow light when it absorbs the laser light. Light entering the crystal from the right is refracted (bent) and partially reflected at the right-hand surface of the crystal. The laser beam appears yellow inside the crystal because of luminescence from Er^{3+}. As it exits the crystal at the left side, the laser beam is refracted again, and partially reflected back into the crystal. *(b)* Same experiment, but with blue light instead of blue-green light. The blue light is absorbed by Er^{3+} and does not penetrate very far into the crystal. [Courtesy M. D. Seltzer, M. Johnson, and D. O'Connor, Michelson Laboratory, China Lake, CA.]

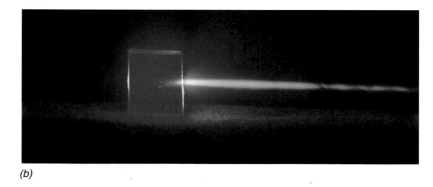

(a)

(b)

COLOR PLATE 16   Luminescence (Section 18-4) *(a)* Green crystal of yttrium aluminum garnet containing a small amount of Cr^{3+}. *(b)* When irradiated with high-intensity blue light from a laser at the right side, Cr^{3+} absorbs blue light and emits lower energy red light. When the laser is removed, the crystal appears green again. [Courtesy M. D. Seltzer, M. Johnson, and D. O'Connor, Michelson Laboratory, China Lake, CA.]

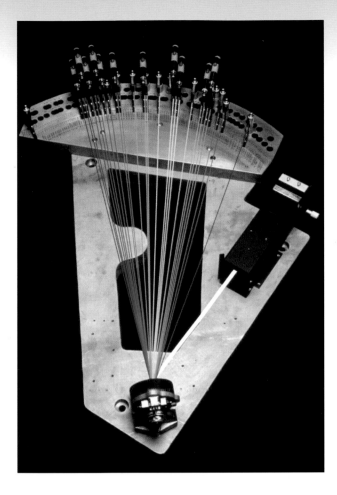

COLOR PLATE 17 Polychromator for Inductively Coupled Plasma Atomic Emission Spectrometer (Section 19-4)
"White" light emitted by a complex sample in the plasma enters the polychromator at the right and is dispersed into its component wavelengths by a grating at the bottom of the diagram. Each different emission wavelength (shown schematically by colored lines) is diffracted at a different angle and directed to a different detector on the focal curve. Each detector therefore responds to one preselected element in the sample and all elements are measured simultaneously. There is one detector for each element to be observed and the element is selected by the location of the detector on the focal curve. [Courtesy TJA Solutions, Franklin, MA.]

COLOR PLATE 18 Thin-Layer Chromatography (Section 20-1)
(a) Solvent ascends past mixture of dyes near bottom of flat plate coated with solid adsorbent.
(b) Separation achieved after solvent has ascended most of the way up the plate.

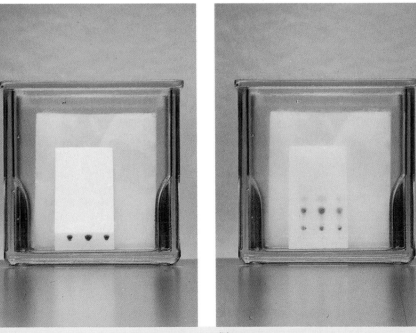

(a)   *(b)*

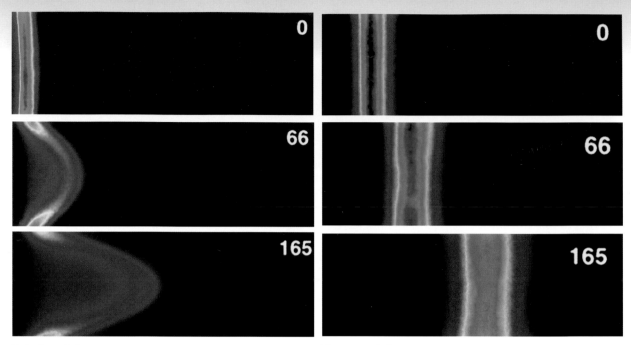

Hydrodynamic flow →
  100-μm-diameter capillary

Electroosmotic flow →
  75-μm-diameter capillary

COLOR PLATE 19 Observed Velocity Profiles for Hydrodynamic and Electroosmotic Flow (Section 22-6) A fluorescent dye was imaged inside a capillary tube at times of 0, 66, and 165 ms after initiating flow. The highest concentration of dye is represented by blue and the lowest concentration is red in these images in which different colors are assigned to different fluorescence intensities. [From P. H. Paul, M. G. Garguilo, and D. J. Rakestraw, *Anal. Chem.* **1998**, *70*, 2459.]

COLOR PLATE 20 Titration of VO^{2+} with Potassium Permanganate (Experiment 23-12) Blue VO^{2+} solution prior to titration (left). Mixture of blue VO^{2+} and yellow VO_2^+ observed during titration (center). Dark color of MnO_4^- at end point (right).

TABLE 15-3 Iodometric titrations: Titrations of iodine (actually I_3^-) produced by analyte

Species analyzed	Reaction	Notes
HOCl	$HOCl + H^+ + 3I^- \rightleftharpoons Cl^- + I_3^- + H_2O$	Reaction in 0.5 M H_2SO_4.
Br_2	$Br_2 + 3I^- \rightleftharpoons 2Br^- + I_3^-$	Reaction in dilute acid.
IO_3^-	$2IO_3^- + 16I^- + 12H^+ \rightleftharpoons 6I_3^- + 6H_2O$	Reaction in 0.5 M HCl.
IO_4^-	$2IO_4^- + 22I^- + 16H^+ \rightleftharpoons 8I_3^- + 8H_2O$	Reaction in 0.5 M HCl.
O_2	$O_2 + 4Mn(OH)_2 + 2H_2O \rightleftharpoons 4Mn(OH)_3$ $2Mn(OH)_3 + 6H^+ + 6I^- \rightleftharpoons$ $\qquad\qquad 2Mn^{2+} + 2I_3^- + 6H_2O$	The sample is treated with Mn^{2+}, NaOH, and KI. After 1 min, it is acidified with H_2SO_4, and the I_3^- is titrated.
H_2O_2	$H_2O_2 + 3I^- + 2H^+ \rightleftharpoons I_3^- + 2H_2O$	Reaction in 1 M H_2SO_4 with NH_4MoO_3 catalyst.
O_3	$O_3 + 3I^- + 2H^+ \rightleftharpoons O_2 + I_3^- + H_2O$	O_3 is passed through neutral 2 wt % KI solution. Add H_2SO_4 and titrate.
NO_2^-	$2HNO_2 + 2H^+ + 3I^- \rightleftharpoons 2NO + I_3^- + 2H_2O$	The nitric oxide is removed (by bubbling CO_2 generated in situ) prior to titration of I_3^-.
$S_2O_8^{2-}$	$S_2O_8^{2-} + 3I^- \rightleftharpoons 2SO_4^{2-} + I_3^-$	Reaction in neutral solution. Then acidify and titrate.
Cu^{2+}	$2Cu^{2+} + 5I^- \rightleftharpoons 2CuI(s) + I_3^-$	NH_4HF_2 is used as a buffer.
MnO_4^-	$2MnO_4^- + 16H^+ + 15I^- \rightleftharpoons$ $\qquad\qquad 2Mn^{2+} + 5I_3^- + 8H_2O$	Reaction in 0.1 M HCl.
MnO_2	$MnO_2(s) + 4H^+ + 3I^- \rightleftharpoons$ $\qquad\qquad Mn^{2+} + I_3^- + 2H_2O$	Reaction in 0.5 M H_3PO_4 or HCl.

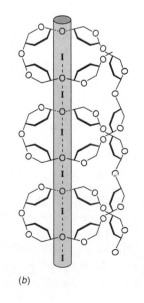

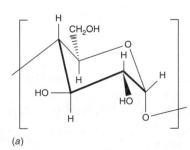

(a)

(b)

FIGURE 15-3 (a) Structure of the repeating unit of the sugar amylose found in starch. (b) Schematic structure of the starch-iodine complex. The sugar chain forms a helix around nearly linear I_6 chains. [From R. C. Teitelbaum, S. L. Ruby, and T. J. Marks, *J. Am. Chem. Soc.* **1980**, *102*, 3332.]

An alternative to the use of starch is the addition of a few milliliters of *p*-xylene to the vigorously stirred titration vessel. After each addition of reagent near the end point, stop stirring long enough to examine the color of the xylene. I_2 is 400 times more soluble in xylene than in water, and its color is readily detected in the xylene.

There is a significant vapor pressure of toxic I_2 above solid I_2 and aqueous I_3^-. Vessels containing I_2 or I_3^- should be sealed and kept in a fume hood. Waste solutions of I_3^- should not be dumped into a sink in the open lab.

Disproportionation means that an element in one oxidation state reacts to give the same element in both higher and lower oxidation states.

In a solution with no other colored species, it is possible to see the color of $\sim 5 \times 10^{-6}$ M I_3^-. With a starch indicator, the limit of detection is extended by about a factor of 10.

Starch is readily biodegraded, so either it should be freshly dissolved or the solution should contain a preservative, such as HgI_2 or thymol. A hydrolysis product of starch is glucose, which is a reducing agent. A partially hydrolyzed solution of starch could thus be a source of error in a redox titration.

In iodimetry (titration *with* I_3^-), starch can be added at the beginning of the titration. The first drop of excess I_3^- after the equivalence point causes the solution to turn dark blue. In iodometry (titration *of* I_3^-), I_3^- is present throughout the reaction up to the equivalence point. *Starch should not be added to such a reaction until immediately before the equivalence point* (as detected visually, by fading of the I_3^- (Color Plate 11)). Otherwise some iodine tends to remain bound to starch particles after the equivalence point is reached.

Preparation and Standardization of I_3^- Solutions

Triiodide (I_3^-) is prepared by dissolving solid I_2 in excess KI. I_2 is seldom used as a primary standard because some sublimes (evaporates) during weighing. Instead, an approximate amount is rapidly weighed, and the solution of I_3^- is standardized with a pure sample of the intended analyte or with As_4O_6 or $Na_2S_2O_3$.

Acidic solutions of I_3^- are unstable because the excess I^- is slowly oxidized by air:

$$4I^- + O_2 + 4H^+ \longrightarrow 2I_2 + 2H_2O$$

In neutral solutions, oxidation is insignificant in the absence of heat, light, and metal ions. Above pH 11, iodine disproportionates to hypoiodous acid (HOI), iodate (IO_3^-), and iodide.

Iodine can be standardized by reaction with primary standard grade arsenic(III) oxide, As_4O_6 (Figure 15-4), at pH 7–8 in bicarbonate buffer:

$$\underset{\text{Arsenic(III) oxide}}{As_4O_6(s)} \; + \; 6H_2O \rightleftharpoons \underset{\text{Arsenious acid}}{4H_3AsO_3}$$

$$H_3AsO_3 + I_2 + H_2O \rightleftharpoons H_3AsO_4 + 2I^- + 2H^+ \qquad (15\text{-}19)$$

Standard iodine (actually I_3^-) is made by adding a weighed quantity of pure potassium iodate to a small excess of KI. Addition of excess strong acid (to give pH $\approx$ 1) produces I_3^-:

Preparing standard I_3^-:

$$IO_3^- + 8I^- + 6H^+ \rightleftharpoons 3I_3^- + 3H_2O \qquad (15\text{-}20)$$

Freshly acidified iodate plus iodide can be used to standardize thiosulfate. The I_3^- reagent must be used immediately, for it is soon oxidized by air. The only disadvantage of KIO_3 is its low formula mass relative to the number of electrons it accepts. The small quantity of KIO_3 leads to a larger than desirable relative weighing error in preparing solutions.

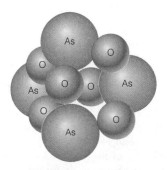

FIGURE 15-4 As_4O_6 consists of an As_4 tetrahedron with a bridging oxygen atom on each edge.

Use of Sodium Thiosulfate

Sodium thiosulfate is the almost universal titrant for iodine. In neutral or acidic solution ($pH < 9$), iodine oxidizes thiosulfate cleanly to tetrathionate:

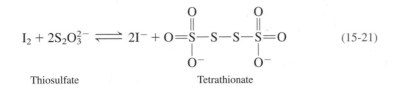

$$I_2 + 2S_2O_3^{2-} \rightleftharpoons 2I^- + O=S-S-S-S=O \quad (15\text{-}21)$$

Thiosulfate   Tetrathionate

One mol of I_2 in Reaction 15-21 is equivalent to one mol of I_3^- in Reaction 15-20. I_2 and I_3^- are interchangeable via the equilibrium $I_2 + I^- \rightleftharpoons I_3^-$.

The common form of thiosulfate, $Na_2S_2O_3 \cdot 5H_2O$, is not pure enough to be a primary standard. Instead, thiosulfate is standardized by reaction with a fresh solution of I_3^- prepared from KIO_3 plus KI or a solution of I_3^- standardized with As_4O_6.

KIO_3 is a primary standard for the generation of I_3^-.

A stable solution of $Na_2S_2O_3$ is prepared by dissolving the reagent in high-quality, freshly boiled distilled water. Dissolved CO_2 promotes disproportionation of $S_2O_3^{2-}$:

$$S_2O_3^{2-} + H^+ \rightleftharpoons HSO_3^- + S(s) \quad (15\text{-}22)$$
$$\text{Bisulfite} \quad \text{Sulfur}$$

and metal ions catalyze atmospheric oxidation of thiosulfate. Thiosulfate solutions are stored in the dark with 0.1 g of Na_2CO_3 per liter to maintain optimum pH. Three drops of chloroform should be added to a thiosulfate solution to prevent bacterial growth. Although an acidic solution of thiosulfate is unstable, the reagent can be used to titrate iodine in acid because Reaction 15-21 is faster than Reaction 15-22.

Ask Yourself

15-C. (a) Potassium iodate solution was prepared by dissolving 1.022 g of KIO_3 (FM 214.00) in a 500-mL volumetric flask. Then 50.00 mL of the solution were pipetted into a flask and treated with excess KI (2 g) and acid (10 mL of 0.5 M H_2SO_4) to drive Reaction 15-20 to completion. How many moles of I_3^- are created by the reaction?

(b) The triiodide from part **(a)** required 37.66 mL of sodium thiosulfate solution for Reaction 15-21. What is the concentration of the sodium thiosulfate solution?

(c) A 1.223-g sample of solid containing ascorbic acid and inert ingredients was dissolved in dilute H_2SO_4 and treated with 2 g of KI and 50.00 mL of potassium iodate solution from part **(a)**. After Reaction 15-20 and the reaction with ascorbic acid in Box 15-2 went to completion, the excess, unreacted triiodide required 14.22 mL of sodium thiosulfate solution from part **(b)** for complete titration. Find the moles of ascorbic acid and weight percent of ascorbic acid (FM 176.13) in the unknown .

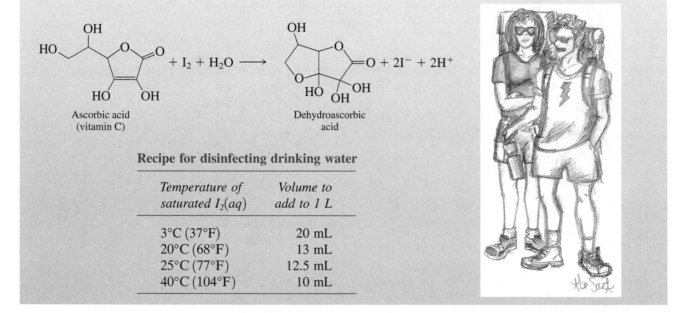

Box 15-2 *Informed Citizen*

Disinfecting Drinking Water with Iodine

Many hikers use iodine to disinfect water from streams and lakes to make it safe to drink. Iodine is more effective than filter pumps, which remove bacteria but not viruses, because viruses are small enough to pass through the filter. Iodine kills everything in the water.

When I hike, I carry a 60-mL glass bottle of water containing a few large crystals of solid iodine and a Teflon-lined cap. The crystals keep the solution saturated with I_2. I use the cap to measure out liquid from this bottle and add it to a 1-L bottle of water from a stream or lake. The required volume of saturated aqueous I_2 is shown in the table below. For example, I use 4 caps of iodine solution when the air temperature

is near 20°C to deliver approximately 13 mL of disinfectant to my 1-L water bottle. It is important to use just the supernatant liquid, not the crystals of solid iodine, because too much iodine is harmful to humans. After allowing 30 min for the iodine to kill any critters, the water is safe to drink. Every time iodine solution is used, I refill the small bottle with water so that saturated aqueous I_2 is available at the next water stop.

Vitamin C, a reducing agent present in many foods, reacts rapidly with I_2. Beverages such as the orange drink Tang are loaded with vitamin C. *Don't put Tang or other beverages in the disinfected stream water until the 30-min waiting period is over.* If you add beverage too soon, it consumes the I_2 before the water is disinfected.

Ascorbic acid (vitamin C) $+ I_2 + H_2O \longrightarrow$ Dehydroascorbic acid $=O + 2I^- + 2H^+$

Recipe for disinfecting drinking water

Temperature of saturated $I_2(aq)$	Volume to add to 1 L
3°C (37°F)	20 mL
20°C (68°F)	13 mL
25°C (77°F)	12.5 mL
40°C (104°F)	10 mL

Key Equations

Redox titration potential calculations	Cell voltage $= E_+ - E$(reference electrode), where E_+ is the indicator electrode potential

Prior to equivalence point: Analyte is in excess; use analyte Nernst equation to find indicator electrode potential.

At equivalence point: Add analyte and titrant Nernst equations (with equal number of electrons) and use stoichiometry to cancel many terms; if necessary, use known concentrations to evaluate log term.

Past equivalence point: Titrant in excess; use titrant Nernst equation to find indicator electrode potential.

Redox indicator color change range

$$E = \left(E° \pm \frac{0.059\ 16}{n} \right) \text{volts}$$

n = number of electrons in indicator half-reaction

Important Terms

redox indicator   redox titration

Problems

15-1. What is the difference between the *titration reaction* and the *cell reactions* in a potentiometric titration?

15-2. Consider the titration of Fe^{2+} with Ce^{4+} in Figure 15-2.

(a) Write a balanced titration reaction.

(b) Write two different half-reactions for the indicator electrode.

(c) Write two different Nernst equations for the net cell reaction.

(d) Calculate E at the following volumes of Ce^{4+}: 10.0, 25.0, 49.0, 50.0, 51.0, 60.0, and 100.0 mL. Compare your results with Figure 15-2.

15-3. Consider the titration of 100.0 mL of 0.010 0 M Ce^{4+} in 1 M $HClO_4$ by 0.040 0 M Cu^+ to give Ce^{3+} and Cu^{2+}, using Pt and saturated Ag | AgCl electrodes to find the end point.

(a) Write a balanced titration reaction.

(b) Write two different half-reactions for the indicator electrode.

(c) Write two different Nernst equations for the net cell reaction.

(d) Calculate E at the following volumes of Cu^+: 1.00, 12.5, 24.5, 25.0, 25.5, 30.0, and 50.0 mL. Sketch the titration curve.

15-4. Consider the titration of 25.0 mL of 0.010 0 M Sn^{2+} by 0.050 0 M Tl^{3+} in 1 M HCl, using Pt and saturated calomel electrodes to find the end point.

(a) Write a balanced titration reaction.

(b) Write two different half-reactions for the indicator electrode.

(c) Write two different Nernst equations for the net cell reaction.

(d) Calculate E at the following volumes of Tl^{3+}: 1.00, 2.50, 4.90, 5.00, 5.10, and 10.0 mL. Sketch the titration curve.

15-5. Compute the titration curve for Demonstration 15-1, in which 400.0 mL of 3.75 mM Fe^{2+} are titrated with 20.0 mM MnO_4^- at a *fixed pH* of 0.00 in 1 M H_2SO_4. Calculate the potential at titrant volumes of 1.0, 7.5, 14.0, 15.0, 16.0, and 30.0 mL and sketch the titration curve.

15-6. Consider the titration of 50.0 mL of 1.00 mM I^- with 5.00 mM $Br_2(aq)$ to give $I_2(aq)$ and Br^-. Calculate the potential (versus S.C.E.) at the following volumes of Br_2: 0.100, 2.50, 4.99, 5.01, and 6.00 mL. Sketch the titration curve.

15-7. Ascorbic acid (0.010 0 M) (structure in Box 15-2) was added to 10.0 mL of 0.020 0 M Fe^{3+} in a solution buffered to pH 0.30, and the potential was monitored with Pt and saturated Ag | AgCl electrodes.

$$\text{dehydroascorbic acid} + 2H^+ + 2e^- \rightleftharpoons$$
$$\text{ascorbic acid} + H_2O \quad E° = 0.390\ \text{V}$$

(a) Write a balanced equation for the titration reaction.

(b) Write two balanced equations to describe the cell reactions. This question is asking for two complete reactions, not two half-reactions.

(c) Using $E° = 0.767$ V for the Fe^{3+} | Fe^{2+} couple, calculate the cell voltage when 5.0, 10.0, and 15.0 mL of ascorbic acid have been added. (*Hint:* Whenever $[H^+]$ appears in a Nernst equation, use the numerical value $10^{-pH} = 10^{-0.30}$.)

15-8. Select indicators from Table 15-1 that would be suitable for finding the two end points in Figure 15-2. What color changes would be observed?

15-9. Would tris(2,2′-bipyridine) iron be a useful indicator for the titration of Sn^{2+} in 1 M HCl with $Mn(EDTA)^-$?

(*Hint:* The potential at the equivalence point must be between the potentials for each redox couple.)

15-10. Why is iodine almost always used in a solution containing excess I^-?

15-11. A solution of I_2 was standardized by titrating freshly dissolved arsenic(III) oxide (As_4O_6, FM 395.68). The titration of 25.00 mL of a solution prepared by dissolving 0.366 3 g of As_4O_6 in a volume of 100.0 mL required 31.77 mL of I_2.

(a) Calculate the molarity of the I_2 solution.

(b) Does it matter whether starch indicator is added at the beginning or near the end point in this titration? Why?

15-12. Ozone (O_3) is a colorless gas with a pungent odor. It can be generated by passing a high-voltage electric spark through air. O_3 can be analyzed by its stoichiometric reaction with I^- in neutral solution:

$$O_3 + 2I^- + H_2O \longrightarrow O_2 + I_2 + 2OH^-$$

(The reaction must be carried out in neutral solution. In acidic solution, more I_2 is made than the reaction above indicates.) A 1.00-L bulb of air containing O_3 produced by an electric spark was treated with 25 mL of 2 M KI, shaken well, and left closed for 30 min so that all O_3 would react. The aqueous solution was then drained from the bulb, acidified with 2 mL of 1 M H_2SO_4, and required 29.33 mL of 0.050 44 M $S_2O_3^{2-}$ for titration of the I_2.

(a) What color would you expect the KI solution to be before and after reaction with O_3?

(b) Calculate the mass of O_3 in the 1.00-L bulb.

(c) Does it matter whether starch indicator is added at the beginning or near the end point in this titration? Why?

15-13. A 128.6-mg sample of a protein (FM 58 600) was treated with 2.000 mL of 0.048 7 M $NaIO_4$ to react with all the serine and threonine residues.

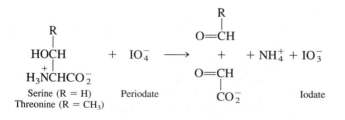

Serine (R = H)   Periodate
Threonine (R = CH₃)

The solution was then treated with excess iodide to convert the unreacted periodate to iodine (actually, I_3^-):

$$IO_4^- + 2I^- + H_2O \rightleftharpoons IO_3^- + I_2 + 2OH^-$$

Titration of the iodine required 823 μL of 0.098 8 M thiosulfate.

(a) Calculate the number of serine plus threonine residues per molecule of protein. Round your answer to the nearest integer.

(b) How many milligrams of As_4O_6 (FM 395.68) would be required to react with the I_3^- liberated in this experiment?

15-14. The Kjeldahl analysis in Section 9-6 is used to measure the nitrogen content of organic compounds, which are digested in boiling sulfuric acid to decompose to ammonia, which, in turn, is distilled into standard acid. The remaining acid is then back-titrated with base. Kjeldahl himself had difficulty discerning by lamplight in 1880 the methyl red indicator end point in the back titration. He could have refrained from working at night, but instead he chose to complete the analysis differently. After distilling the ammonia into standard sulfuric acid, he added a mixture of KIO_3 and KI to the acid. The liberated iodine was then titrated with thiosulfate, using starch for easy end point detection—even by lamplight. Explain how the thiosulfate titration is related to the nitrogen content of the unknown. Derive a relationship between moles of NH_3 liberated in the digestion and moles of thiosulfate required for titration of iodine.

15-15. From the following reduction potentials,

$$I_2(s) + 2e^- \rightleftharpoons 2I^- \qquad E° = 0.535 \text{ V}$$
$$I_2(aq) + 2e^- \rightleftharpoons 2I^- \qquad E° = 0.620 \text{ V}$$
$$I_3^- + 2e^- \rightleftharpoons 3I^- \qquad E° = 0.535 \text{ V}$$

(a) Calculate the equilibrium constant for the reaction $I_2(aq) + I^- \rightleftharpoons I_3^-$.

(b) Calculate the equilibrium constant for the reaction $I_2(s) + I^- \rightleftharpoons I_3^-$.

(c) Calculate the solubility (g/L) of I_2 in water.

How Would You Do It?

15-16. Ozone (O_3) in smog is formed by the action of solar ultraviolet light on organic vapors plus nitric oxide (NO) in the air. An O_3 level of 100 to 200 ppb (nL per liter of air) for 1 hr creates a "1st stage smog alert" and is considered "unhealthful." A level above 200 ppb, which defines a "2nd stage smog alert" is "very unhealthful."

(a) The ideal gas law tells us that $PV = nRT$, where P is pressure (bar), V is volume (L), n is moles, R is the gas constant ($8.314\ 472 \times 10^{-2}$ L·bar/(mol·K)), and T is temperature (K). If the pressure of air in a flask is 1 bar, the partial pressure of a 1 ppb component is 10^{-9} bar. Find the number of moles of O_3 in a liter of air if the concentration of O_3 is 200 ppb and the temperature is 300 K.

(b) Is it feasible to use the iodometric procedure of Problem 15-12 to measure O_3 at a level of 200 ppb in smog? State your reason.

15-17. *Iodometric analysis of a superconductor.* An analysis was carried out to find the effective copper oxidation state, and therefore the number of oxygen atoms, in the superconductor $YBa_2Cu_3O_{7-z}$, where z ranges from 0 to 0.5. Common oxidation states of yttrium and barium are Y^{3+} and Ba^{2+}, and common states of copper are Cu^{2+} and Cu^+. If copper were Cu^{2+}, the formula of the superconductor would be $(Y^{3+})(Ba^{2+})_2(Cu^{2+})_3(O^{2-})_{6.5}$, with a cation charge of $+13$ and an anion charge of -13. The composition $YBa_2Cu_3O_7$ formally requires Cu^{3+}, which is rather rare. $YBa_2Cu_3O_7$ can be thought of as $(Y^{3+})(Ba^{2+})_2(Cu^{2+})_2(Cu^{3+})(O^{2-})_7$, with a cation charge of $+14$ and an anion charge of -14.

An iodometric analysis of $YBa_2Cu_3O_x$ involves two experiments. In *Experiment 1,* $YBa_2Cu_3O_x$ is dissolved in dilute acid, in which Cu^{3+} is converted to Cu^{2+}. For simplicity, we write the equations for the formula $YBa_2Cu_3O_7$, but you could balance these equations for $x \neq 7$:

$$YBa_2Cu_3O_7 + 13H^+ \longrightarrow$$
$$Y^{3+} + 2Ba^{2+} + 3Cu^{2+} + \tfrac{13}{2}H_2O + \tfrac{1}{4}O_2 \quad \text{(A)}$$

The total copper content is measured by treatment with iodide:

$$3Cu^{2+} + 6I^- \longrightarrow 3CuI(s) + \tfrac{3}{2}I_2 \quad \text{(B)}$$

and titration of the liberated I_2 with standard thiosulfate by Reaction 15-21. Each mole of Cu in $YBa_2Cu_3O_7$ is equivalent to 1 mol of $S_2O_3^{2-}$ in Experiment 1.

In *Experiment 2,* $YBa_2Cu_3O_x$ is dissolved in dilute acid containing I^-. Each mole of Cu^{3+} produces 1 mol of I_2, and each mole of Cu^{2+} produces 0.5 mol of I_2:

$$Cu^{3+} + 3I^- \longrightarrow CuI(s) + I_2 \quad \text{(C)}$$

$$Cu^{2+} + 2I^- \longrightarrow CuI(s) + \tfrac{1}{2}I_2 \quad \text{(D)}$$

The moles of thiosulfate required in Experiment 1 equal the total moles of Cu in the superconductor. The difference in thiosulfate required between Experiments 2 and 1 gives the Cu^{3+} content.

(a) In Experiment 1, 1.00 g of superconductor required 4.55 mmol of $S_2O_3^{2-}$. In Experiment 2, 1.00 g of superconductor required 5.68 mmol of $S_2O_3^{2-}$. What is the value of z in the formula $YBa_2Cu_3O_{7-z}$ (FM $666.246 - 15.999\,4\,z$)?

(b) *Propagation of uncertainty.* In several replications of Experiment 1, the thiosulfate required was 4.55 (±0.10) mmol of $S_2O_3^{2-}$ per gram of $YBa_2Cu_3O_{7-z}$. In Experiment 2, the thiosulfate required was 5.68 (±0.05) mmol of $S_2O_3^{2-}$ per gram. Find the uncertainty of x in the formula $YBa_2Cu_3O_x$.

Notes and References

1. Biochemical oxygen demand (BOD) and chemical oxygen demand (COD) procedures are described in *Standard Methods for the Examination of Wastewater,* 18th ed. (Washington, DC: American Public Health Association, 1992), which is the standard reference for water analysis.

A Biosensor for Personal Glucose Monitoring

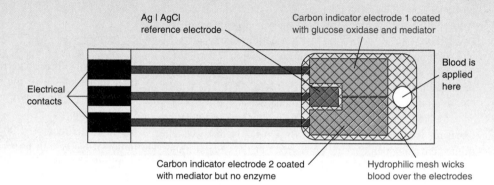

Ag | AgCl reference electrode

Carbon indicator electrode 1 coated with glucose oxidase and mediator

Blood is applied here

Electrical contacts

Carbon indicator electrode 2 coated with mediator but no enzyme

Hydrophilic mesh wicks blood over the electrodes

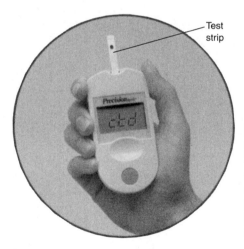

Test strip

Personal glucose monitor is used by diabetics to measure blood sugar level. Diagram shows principal components of disposable test strip to which a drop of blood is applied. [Courtesy Abbott Laboratories MediSense Products, Bedford, MA.]

A *biosensor* is an analytical device that uses a biological component such as an enzyme or antibody for highly specific sensing of one substance. Many people with diabetes must monitor their blood sugar (glucose) level several times a day to control their disease through diet and insulin injections. The photograph shows a home glucose monitor featuring a disposable test strip to which as little as 4 μL of blood is applied for each measurement. This biosensor uses the enzyme glucose oxidase to catalyze the oxidation of glucose. The electrodes measure an oxidation product. Section 16-2 explains how the sensor works.

Chapter 16

INSTRUMENTAL METHODS
IN ELECTROCHEMISTRY

We now introduce a variety of electrochemical methods used in chemical analysis. These techniques are used in applications as diverse as home glucose monitors and chromatography detectors.

16-1 Electrogravimetric and Coulometric Analysis

Electrolysis is a chemical reaction in which we apply a voltage to drive a redox reaction that would not otherwise occur. An **electroactive species** is one that can be oxidized or reduced at an electrode.

Electrogravimetric Analysis

One of the oldest electrolytic methods in quantitative analysis is **electrogravimetric analysis,** in which the analyte is plated out on an electrode and weighed. For example, an excellent procedure for the measurement of copper is to pass a current through a solution of a copper salt to deposit all of the copper on the cathode:

$$Cu^{2+}(aq) + 2e^- \longrightarrow Cu(s, \text{deposited on the cathode}) \qquad (16-1)$$

The increase in mass of the cathode tells us how much copper was present in the solution.

Figure 16-1 shows how this experiment might be done. Analyte is typically deposited on a carefully cleaned, chemically inert Pt gauze cathode with a large surface area.

343

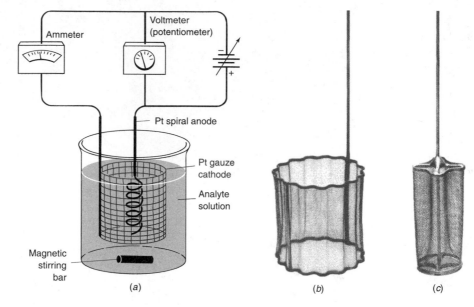

FIGURE 16-1 (a) Electrogravimetric analysis. Analyte is deposited
on the large Pt gauze electrode. If
analyte is to be oxidized, rather than
reduced, the polarity of the power
supply is reversed so that deposition
always occurs on the large electrode.
(b) Outer Pt gauze electrode. (c) Optional inner Pt gauze electrode
designed to be spun by a motor in
place of magnetic stirring.

Tests for completion of the
deposition:

1. Disappearance of color
2. Deposition on freshly exposed
   electrode surface
3. Qualitative test for analyte in
   solution

How do you know when electrolysis is complete? One way is to observe the
disappearance of color in a solution from which a colored species such as Cu^{2+} is
removed. Another way is to expose most, but not all, of the surface of the cathode
to the solution during electrolysis. To test whether the reaction is complete, raise
the beaker or add water so that fresh surface of the cathode is exposed to the solution. After an additional period of electrolysis (15 min, say), see whether the newly
exposed electrode surface has a deposit. If it does, repeat the procedure. If not, the
electrolysis is done. A third method is to remove a small sample of solution and
perform a qualitative test for analyte.

Electrogravimetric analysis would be simple if it merely involved a single
analyte species in an otherwise inert solution. In practice, there may be other electroactive species that interfere. Water is electroactive, decomposing to $H_2 + \frac{1}{2}O_2$ at
sufficiently high voltage. Gas bubbles at the electrode interfere with deposition of
solids. Because of these complications, control of electrode potential is important
for successful analysis.

Coulometric Analysis

In **coulometry,** electrons participating in a chemical reaction are counted to learn
how much analyte reacted. For example, hydrogen sulfide (H_2S) can be measured
by its reaction with I_2 generated at an anode:

$$\text{anode generates } I_2: \quad 2I^- \longrightarrow I_2 + 2e^- \tag{16-2}$$

$$\text{reaction in solution:} \quad I_2 + H_2S \longrightarrow S(s) + 2H^+ + 2I^- \tag{16-3}$$

We measure the electric current and the time required to generate enough I_2 in
Reaction 16-2 to reach the equivalence point of Reaction 16-3. From the current
and time, we calculate how many electrons participated in Reaction 16-2 and therefore how many moles of H_2S were involved in Reaction 16-3. A way to find the end
point in this example would be to have some starch in the solution. As long as I_2 is
consumed rapidly by H_2S, the solution remains colorless. After the equivalence point,
the solution turns blue because excess I_2 accumulates.

Review of Section 13-1:

Electric charge is measured in
coulombs (C).

Electric current (charge per
unit time) is measured in
amperes (A).

1 A = 1 C/s

Faraday constant relates coulombs
to moles:

$F \approx 96\ 485$ C/mol

$$\underset{\substack{\text{Coulombs}}}{q} = \underset{\substack{\text{Moles of}\\\text{electrons}}}{n} \cdot \underset{\substack{\text{C/mol}}}{F}$$

Find the moles of H_2S in an unknown if the end point in Reaction 16-3 came after a current of 0.058 2 A flowed for 184 s in Reaction 16-2.

SOLUTION The quantity of charge in Reaction 16-2 was $(0.058\ 2\ C/s)(184\ s) = 10.7_1\ C$. We use the Faraday constant to convert coulombs to moles of electrons:

$$n = mol\ e^- = \frac{q}{F} = \frac{10.7_1\ C}{96\ 485\ C/mol} = 1.11_0 \times 10^{-4}\ mol$$

Because 2 electrons in Reaction 16-2 correspond to 1 mol of H_2S in Reaction 16-3, $\frac{1}{2}(1.11_0 \times 10^{-4}\ mol) = 5.55 \times 10^{-5}\ mol\ H_2S$ must have been in the unknown.

Ask Yourself

16-A. The apparatus in Figure 16-2 was used in a coulometric titration of an unknown acid.

  (a) What chemical reactions produce OH^- and H^+?

  (b) If a current of 89.2 mA for 666 s was required to reach the end point in the titration of 5.00 mL of an unknown acid, HA, what was the molarity of HA?

  (c) How might you measure the end point of the reaction?

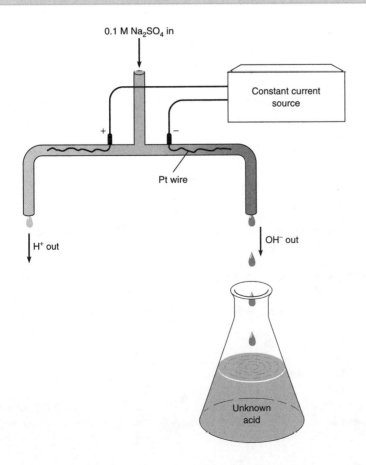

FIGURE 16-2 Using coulometrically generated base to titrate an unknown acid.

In amperometry, we measure an electric current that is proportional to the concentration of a species in solution. In *coulometry,* we measure the total number of electrons (= current × time) that flow during a chemical reaction.

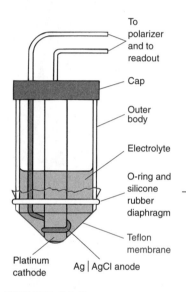

To polarizer and to readout

Cap

Outer body

Electrolyte

O-ring and silicone rubber diaphragm

Teflon membrane

Platinum cathode

Ag | AgCl anode

FIGURE 16-3   A Clark oxygen electrode. [From D. T. Sawyer, A. Sobkowiak, and J. L. Roberts, Jr., *Electrochemistry for Chemists,* 2nd ed. (New York: Wiley, 1995).]

Enzyme: A protein that catalyzes a biochemical reaction. The enzyme increases the rate of reaction by many orders of magnitude.

16-2   Amperometry

In **amperometry,** we measure the electric current between a pair of electrodes that are driving an electrolysis reaction. One of the reactants is the intended analyte, and the measured current is proportional to the concentration of analyte.

A common amperometric method is the measurement of dissolved O_2 with the **Clark electrode** in Figure 16-3.[1] This device has a Pt cathode and a Ag|AgCl anode enclosed by a thin Teflon membrane. Oxygen diffuses through the membrane and equilibrates between the external solution and the internal KCl solution in a few seconds. If the Pt is held at −0.6 V with respect to the Ag|AgCl electrode, dissolved O_2 is reduced to H_2O.

$$\text{cathode:}\quad O_2 + 4H^+ + 4e^- \longrightarrow 2H_2O \qquad (16\text{-}4)$$

$$\text{anode:}\quad 4Ag + 4Cl^- \longrightarrow 4AgCl + 4e^-$$

The electric current is proportional to the concentration of dissolved O_2. A Clark electrode is calibrated by placing it in solutions of known oxygen concentration and a graph of current versus oxygen concentration is constructed.

A Clark electrode can fit into the tip of a surgical catheter to measure O_2 in the umbilical artery of a newborn child to detect respiratory distress. The sensor responds within 20 to 50 s to administration of O_2 for breathing or to mechanical ventilation of the lungs.

EXAMPLE   A Digression on Henry's Law

Henry's law is the observation that in dilute solutions the concentration of a gaseous species dissolved in the liquid is proportional to the pressure of that species in the gas phase. For oxygen dissolved in water at 25°C, Henry's law takes the form

$$[O_2(aq)] = (0.001\ 26\ \text{M/bar}) \times P_{O_2}\ (\text{bar})$$

where P_{O_2} is expressed in bar and $[O_2(aq)]$ is in mol/L. Clark electrodes are often calibrated in terms of P_{O_2} rather than $[O_2(aq)]$, because P_{O_2} is easier to measure. For example, an electrode might be calibrated in solutions bubbled with pure nitrogen ($P_{O_2} = 0$), dry air ($P_{O_2} \approx 0.212$ bar), and pure oxygen at 1 atm ($P_{O_2} \approx 1.01$ bar). If a Clark electrode gives a reading of "0.100 bar," what is the molarity of $O_2(aq)$?

SOLUTION   Henry's law tells us that

$$[O_2(aq)] = 0.001\ 26 \times P_{O_2} = (0.001\ 26\ \text{M/bar}) \times (0.100\ \text{bar}) = 1.26 \times 10^{-4}\ \text{M}$$

Glucose Monitors

The blood glucose sensor at the opening of this chapter is probably the most widely used **biosensor**—a device that uses a biological component such as an *enzyme* or *antibody* for highly selective response to one analyte. The disposable test strip shown at the opening of the chapter has two carbon indicator electrodes and a Ag|AgCl reference electrode. As little as 4 μL of blood applied in the circular opening at the right of the figure is wicked over all three electrodes by a thin *hydrophilic* ("water-loving") mesh. A 20-s measurement begins when the liquid reaches the reference electrode.

Indicator electrode 1 is coated with the enzyme glucose oxidase and a *mediator*, which we describe soon. The enzyme is a protein that catalyzes the reaction of glucose with oxygen:

Antibody: A protein that binds to a specific target molecule called an *antigen.* Foreign cells that infect your body are marked by antibodies and destroyed by *lysis* (bursting them open with fluid) or gobbled up by macrophage cells.

Reaction in coating above indicator electrode 1:

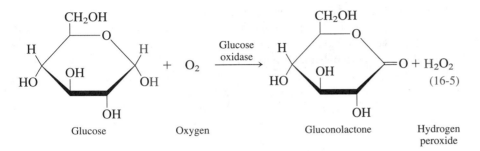

(16-5)

Glucose   Oxygen   Gluconolactone   Hydrogen peroxide

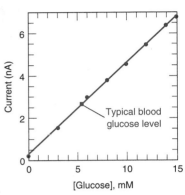

FIGURE 16–4 Response of an amperometric glucose electrode with dissolved O_2 concentration corresponding to an oxygen pressure of 0.027 bar, which is 20% lower than the typical concentration in subcutaneous tissue. [Data from S.-K. Jung and G. W. Wilson, *Anal. Chem.* **1996**, *68*, 591.]

In the absence of enzyme, the rate of Reaction 16-5 is negligible.

Early glucose monitors measured H_2O_2 from Reaction 16-5 by oxidation at a single indicator electrode, which was held at +0.6 V versus Ag|AgCl:

Reaction at indicator electrode 1: $H_2O_2 \longrightarrow O_2 + 2H^+ + 2e^-$ (16-6)

The current is proportional to the concentration of H_2O_2, which, in turn, is proportional to the glucose concentration in blood (Figure 16-4).

One problem with early glucose monitors is that their response depended on the concentration of O_2 in the enzyme layer, because O_2 participates in Reaction 16-5. If the O_2 concentration is low, the monitor responds as though the glucose concentration were low.

A good way to reduce O_2 dependence is to incorporate into the enzyme layer a species that substitutes for O_2 in Reaction 16-5. A substance that transports electrons between the analyte (glucose, in this case) and the electrode is called a **mediator.** Ferricinium salts serve this purpose nicely:

A *mediator* transports electrons between analyte and the working electrode. The mediator undergoes no net reaction itself.

Reaction in coating above indicator electrode 1:

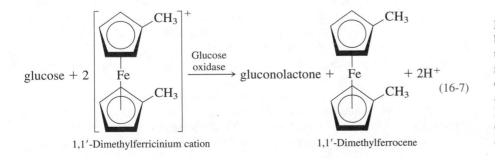

(16-7)

1,1′-Dimethylferricinium cation   1,1′-Dimethylferrocene

Ferrocene contains flat five-membered aromatic carbon rings, similar to benzene. Each ring formally carries one negative charge, so the oxidation state of iron is +2. The iron atom sits between the two flat rings. Because of its shape, this type of molecule is called a *sandwich complex.*

347

The ferricinium mediator lowers the required working electrode potential from 0.6 V to 0.2 V versus Ag|AgCl, thereby improving the stability of the glucose sensor and eliminating some interference by other species in the blood.

The mediator consumed in Reaction 16-7 is then regenerated at the indicator electrode:

Reaction at indicator electrode 1:

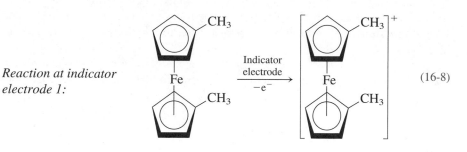

$$(16\text{-}8)$$

The current at the indicator electrode is proportional to the concentration of ferrocene, which, in turn, is proportional to the concentration of glucose in the blood.

One problem with glucose monitors is that other species found in blood can be oxidized at the same potential required to oxidize the mediator in Reaction 16-8. Interfering species include ascorbic acid (vitamin C), uric acid, and acetaminophen (Tylenol®). To correct for this interference, the test strip at the opening of this chapter has a second indicator electrode coated with mediator, *but not with glucose oxidase.* Interfering species that are reduced at electrode 1 are also reduced at electrode 2. The current due to glucose is the current at electrode 1 minus the current at electrode 2 (both measured with respect to the reference electrode). Now you see why the test strip has three electrodes.

A major challenge is to manufacture glucose monitors in such a reproducible manner that they do not require calibration. A user expects to add a drop of blood to the test strip and get a reliable reading without first constructing a calibration curve from known concentrations of glucose in blood. Each lot of test strips must be highly reproducible and calibrated at the factory.

Cells with Three Electrodes

All the cells we have discussed so far are based on two electrodes: an indicator electrode and a reference electrode. Current is measured between the two electrodes. The apparent exception—the glucose monitor at the opening of the chapter—has two indicator electrodes. Current due to analyte plus interfering species is measured between indicator electrode 1 and the reference electrode. Current due to interfering species alone is measured between indicator electrode 2 and the reference electrode. The difference between the two currents is the current due to analyte.

Reference electrode: Provides fixed reference potential with negligible current flow.

Working electrode: Analyte reacts here. Voltage is measured between working and reference electrodes.

Auxiliary electrode: Other half of the electrochemistry occurs here. Current flows between working and auxiliary electrodes.

Potentiostat: Controls potential difference between working and reference electrodes.

For many electrochemical techniques, an entirely different kind of cell with three electrodes is required for fine control of the electrochemistry. The cell in Figure 16-5 features a conventional **reference electrode** (such as calomel or silver-silver chloride), a **working electrode** at which the reaction of interest occurs, and an **auxiliary electrode** that is the current-carrying partner of the working electrode. The working electrode is equivalent to the indicator electrode of two-electrode cells. The auxiliary electrode is something new that we have not encountered before. *Current flows between the working and auxiliary electrodes. Voltage is measured between the working and reference electrodes.*

The voltage between the working and reference electrodes is precisely controlled by a device called a **potentiostat.** Virtually no current flows through the reference electrode; it simply establishes a fixed reference potential with which to measure the working electrode potential. Current flows between the working and auxiliary

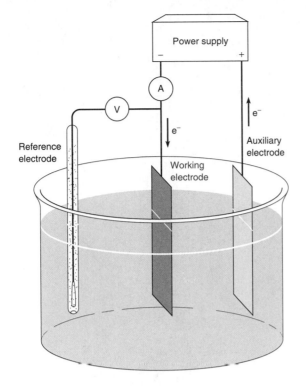

FIGURE 16-5 Controlled-potential electrolysis with a three-electrode cell. Voltage is measured between the working and reference electrodes. Current is measured between the working and auxiliary electrodes. Negligible current flows through the reference electrode. Ⓥ is a voltmeter (potentiometer) and Ⓐ is an ammeter.

electrodes. The potential of the auxiliary electrode varies with time in an uncontrolled manner in response to changing concentrations and current flow in an electrolysis cell. It is beyond the scope of this text to explain why the electrode potential varies. Suffice it to say, however, that, in a two-electrode cell, the potential of the working electrode can drift as the reaction proceeds. As it drifts, reactions other than the intended analytical reaction can occur. In a three-electrode cell, the potentiostat maintains the working electrode at the desired potential, while the auxiliary electrode potential drifts out of our control.

Figure 16-5 shows reduction of analyte at the working electrode, which is therefore the cathode in this figure. In other cases, the working electrode could be the anode. The working electrode is an indicator electrode at which the reaction of analyte occurs.

Amperometric Detector for Chromatography

Figure 0-4 provided an example of chromatography used to separate caffeine from theobromine in a chemical analysis. Absorption of light and electrochemical reactions are two common means to detect analytes as they emerge from the column. Sugars in beverages can be measured by separating them by anion-exchange chromatography (described in Chapter 22) and detecting them with an electrode as they emerge. The —OH groups of sugars such as glucose partially dissociate to —O⁻ anions in 0.1 M NaOH. Anions are separated from one another when they pass through a column packed with particles having fixed positive charges.

The amperometric detector in Figure 16-6 features a Cu working electrode over which the liquid from the column flows. The Ag|AgCl reference electrode and a stainless steel auxiliary electrode are further downstream at the upper left of the diagram. The working electrode is poised by a potentiostat at a potential of +0.55 V

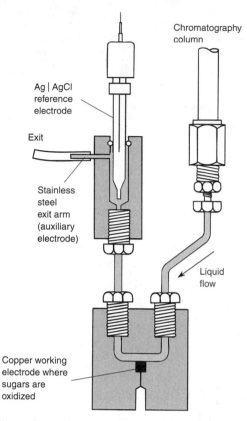

FIGURE 16-6   Electrochemical detector measures sugars emerging from a chromatography column by using amperometry. Sugars are oxidized at the copper electrode and water is reduced at the stainless steel exit arm. [Adapted from Bioanalytical Systems, West Lafayette, IN.]

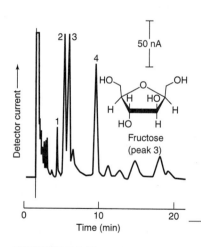

versus $Ag|AgCl$. As sugars emerge from the column, they are oxidized at the Cu surface. Reduction of water $(H_2O + e^- \rightarrow \frac{1}{2}H_2 + OH^-)$ occurs at the auxiliary electrode. The electric current flowing between the working and auxiliary electrodes is proportional to the concentration of each sugar exiting the column. Figure 16-7 shows the chromatogram, which is a trace of detector current versus time as different sugars emerge from the chromatography column. Table 16-1 shows the sugar contents in various beverages measured by this method.

FIGURE 16-7   Anion-exchange chromatogram of Bud Dry beer diluted by a factor of 100 with water and filtered through a 0.45-μm membrane to remove particles. Column stationary phase is CarboPac PA1 and the mobile phase is 0.1 M NaOH. Labeled peaks are the sugars (1) arabinose, (2) glucose, (3) fructose, and (4) lactose. [From P. Luo, M. Z. Luo, and R. P. Baldwin, *J. Chem. Ed.* **1993**, *70*, 679.]

TABLE 16-1	**Partial list of sugars in beverages**			
	Sugar concentration (g/L)			
Brand	*Glucose*	*Fructose*	*Lactose*	*Maltose*
Budweiser	0.54	0.26	0.84	2.05
Bud Dry	0.14	0.29	0.46	—
Coca Cola	45.1	68.4	—	1.04
Pepsi	44.0	42.9	—	1.06
Diet Pepsi	0.03	0.01	—	—

SOURCE: P. Luo, M. Z. Luo, and R. P. Baldwin, *J. Chem. Ed.* **1993**, *70*, 679.

16-3   Voltammetry

In **voltammetry,** current is measured while the voltage between two electrodes is varied. (In amperometry, we held the voltage fixed during the measurement of current.) Consider the apparatus in Figure 16-8, which is used to measure the concentration of vitamin C (ascorbic acid) in fruit drinks in Experiment 23-14. Oxidation of the analyte takes place at the exposed tip of the graphite working electrode:

Working electrode:

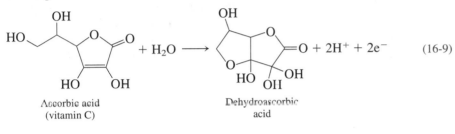

$$+ H_2O \longrightarrow \quad \quad =O + 2H^+ + 2e^- \quad (16\text{-}9)$$

Ascorbic acid
(vitamin C)

Dehydroascorbic
acid

Graphite electrodes were chosen because they are inexpensive. The working electrode has a very small exposed tip to decrease distortion of the electrochemical signal from electrical resistance of the solution and capacitance of the electrode.

and reduction of H^+ occurs at the auxiliary electrode:

Auxiliary electrode: $\quad 2H^+ + 2e^- \longrightarrow H_2(g)$

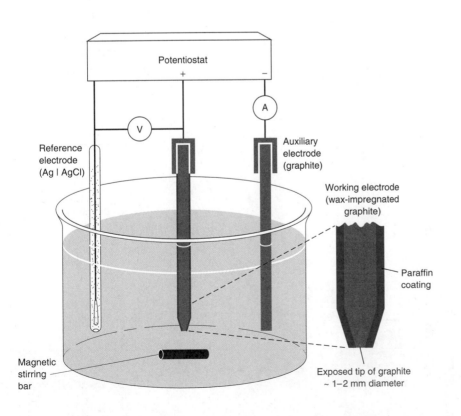

FIGURE 16-8 Three-electrode cell for voltammetric measurement of vitamin C in fruit drinks. The voltage between the working and reference electrodes is measured by the voltmeter Ⓥ, and the current between the working and auxiliary electrodes is measured by the ammeter Ⓐ. The potentiostat varies the voltage in a chosen manner.

Conditioning is repeated before each measurement (including each standard addition) to obtain a clean, fairly reproducible electrode surface.

We measure current between the working and auxiliary electrodes as the potential of the working electrode is varied with respect to the reference electrode.

To record the **voltammogram** (the graph of current versus potential) of orange juice in Figure 16-9, the working electrode was first held at a potential of −1.5 V (versus Ag|AgCl) for 2 min while the solution was stirred. This *conditioning* reduces and removes organic material from the tip of the electrode. The potential was then changed to −0.4 V and stirring continued for 30 s while bubbles of gas were dislodged from the electrode by gentle tapping. Stirring was then discontinued for 30 s so that the solution would be calm for the measurement. Finally, the voltage was scanned from −0.4 V to +1.2 V at a rate of +33 mV/s to record the solid trace in Figure 16-9.

What happens as the voltage is scanned? At −0.4 V, there is no significant reaction and little current flows. At a potential near +0.2 V in Figure 16-9, ascorbic acid begins to be oxidized at the tip of the working electrode and the current rises. Beyond ~+0.8 V, ascorbic acid in the vicinity of the electrode tip is depleted by the electrochemical reaction. The current falls slightly because analyte cannot diffuse fast enough to the electrode to maintain the peak reaction rate.

The peak current is proportional to the concentration of ascorbic acid in the orange juice. We measure the peak current at the arrow in Figure 16-9 from the extrapolated baseline. Any species in the juice that is oxidized near +0.8 V will give a false high result in the analysis. We do not yet know the proportionality constant between the current and the ascorbic acid concentration. To complete the measurement, we make several *standard additions* of known quantities of ascorbic acid, shown by the dashed curves in Figure 16-9.

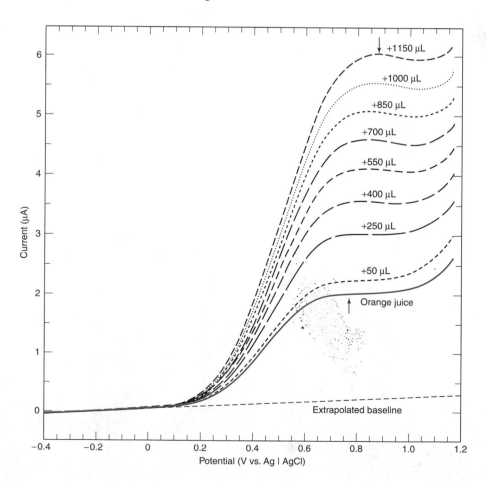

FIGURE 16-9 *Solid line:* Voltammogram of 50 mL of orange juice. *Dashed lines:* Standard additions of 0.279 M ascorbic acid in 0.029 M HNO_3. Voltage was scanned at +33 mV/s, using the apparatus in Figure 16-8. The peak position changes slightly as standard is added because the solution becomes more acidic. Peak positions are marked by arrows in the lowest and highest curves.

In the method of **standard addition,** a known quantity of analyte is added to a specimen and the increase in signal is measured. The relative increase in signal allows us to infer how much analyte was in the original specimen. The key assumption is that signal is proportional to the concentration of analyte.

Standard addition is especially appropriate when the sample *matrix* is complex or unknown. **Matrix** refers to everything in the unknown, other than analyte. A matrix such as orange juice has unknown constituents that you could not incorporate into standard solutions for a calibration curve. We add *small* volumes of concentrated standard to the unknown so that we do not change the matrix very much. An assumption in standard addition is that the matrix has the same effect on added standard as it has on the original analyte in the unknown.

Suppose that a sample with unknown initial concentration $[X]_i$ gives a signal I_X, where I might be the peak current in voltammetry, or absorbance in spectrophotometry, or peak area in chromatography. Then a known concentration of standard S (a known concentration of analyte) is added to the sample and a signal I_{S+X} is observed. Because signal is proportional to analyte concentration, we can say that

Standard additions are most appropriate when the sample matrix is complex and difficult to reproduce in standard solutions.

Bear in mind that the species X and S are the same.

$$\frac{\text{concentration of analyte in unknown}}{\text{concentration of analyte plus standard in mixture}} = \frac{\text{signal from unknown}}{\text{signal from mixture}}$$

Standard addition equation:
$$\frac{[X]_i}{[X]_f + [S]_f} = \frac{I_X}{I_{S+X}} \qquad (16\text{-}10)$$

where $[X]_f$ is the final concentration of unknown analyte after adding the standard and $[S]_f$ is the final concentration of standard after addition to the unknown. If we began with an initial volume V_0 of unknown and added the volume V_s of standard with initial concentration $[S]_i$, the total volume is $V = V_0 + V_s$ and the concentrations in Equation 16-10 are

$$[X]_f = [X]_i\left(\frac{V_0}{V}\right) \qquad [S]_f = [S]_i\left(\frac{V_s}{V}\right) \qquad (16\text{-}11)$$

EXAMPLE   Standard Addition

A 50.0-mL sample of orange juice gave a peak current of 1.78 μA in Figure 16-9. A standard addition of 0.400 mL of 0.279 M ascorbic acid increased the peak current to 3.35 μA. Find the concentration of ascorbic acid in the orange juice.

SOLUTION If the initial concentration of ascorbic acid in the juice is $[X]_i$, the concentration after dilution of 50.0 mL of juice with 0.400 mL of standard is

$$\text{final concentration of analyte} = [X]_f = [X]_i\left(\frac{V_0}{V}\right) = [X]_i\left(\frac{50.0}{50.4}\right)$$

The final concentration of the added standard after addition to the orange juice is

$$[S]_f = [S]_i \left(\frac{V_s}{V}\right) = [0.279 \text{ M}] \left(\frac{0.400}{50.4}\right) = 2.21_4 \text{ mM}$$

The standard addition equation 16-10 therefore becomes

$$\frac{[X]_i}{[X]_f + [S]_f} = \frac{[X]_i}{\left(\frac{50.0}{50.4}\right)[X]_i + 2.21_4 \text{ mM}} = \frac{1.78 \ \mu\text{A}}{3.35 \ \mu\text{A}} \quad \Rightarrow \quad [X]_i = 2.49 \text{ mM}$$

Graphical Procedure for Standard Addition

A more accurate procedure is to make a series of standard additions that increase the original signal by a factor of 1.5 to 3 and use all the results together. If you were to take the expressions for $[X]_f$ and $[S]_f$ from Equations 16-11, plug them into the standard addition equation 16-10, and do a little rearranging, you would find

Equation to plot for standard addition:

$$\underbrace{I_{S+X}\left(\frac{V}{V_0}\right)}_{\substack{\text{Function to plot} \\ \text{on } y\text{-axis}}} = I_X + \frac{I_X}{[X]_i}\underbrace{[S]_i\left(\frac{V_s}{V_0}\right)}_{\substack{\text{Function to plot} \\ \text{on } x\text{-axis}}} \tag{16-12}$$

A graph of I_{S+X} (V/V_0) (the corrected current) on the y-axis versus the function $[S]_i(V_s/V_0)$ on the x-axis should be a straight line. The intercept on the x-axis is the initial concentration of unknown, $[X]_i$.

Data from Figure 16-9 are set out in the spreadsheet in Figure 16-10. The current in column D is the observed peak current minus the baseline current at the peak position. Equation 16-12 is plotted in Figure 16-11. From the x-intercept, we conclude that the concentration of ascorbic acid in the original orange juice was

FIGURE 16-10   Spreadsheet with data from Figure 16-9 for graphing Equation 16-12.

	A	B	C	D	E
1	Vitamin C standard addition experiment				
2	Add 0.279 M ascorbic acid to 50.0 mL of orange juice				
3					
4		Vs =			
5	Vo (mL) =	mL ascorbic	x-axis function	I(s+x) =	y-axis function
6	50	acid added	Si*Vs/Vo	signal (μA)	I(s+x)*Vs/Vo
7	[S]i (mM) =	0.000	0.000	1.78	1.780
8	279	0.050	0.279	2.00	2.002
9		0.250	1.395	2.81	2.824
10		0.400	2.232	3.35	3.377
11		0.550	3.069	3.88	3.923
12		0.700	3.906	4.37	4.431
13		0.850	4.743	4.86	4.943
14		1.000	5.580	5.33	5.437
15		1.150	6.417	5.82	5.954
16					
17	C7 = $A$8*B7/$A$6		E7 = D7*($A$6+B7)/$A$6		

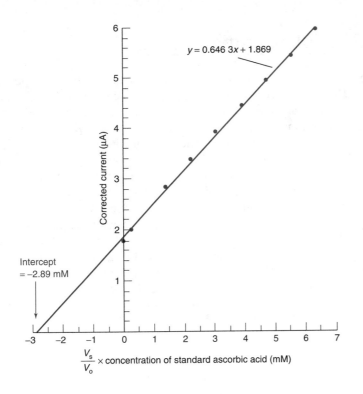

$$y = 0.646\,3x + 1.869$$

Intercept
= −2.89 mM

$\dfrac{V_s}{V_o}$ × concentration of standard ascorbic acid (mM)

FIGURE 16-11   Graphical treatment of the method of standard addition using Equation 16-12.

$[X]_i = 2.89$ mM. In the preceding example, we found $[X]_i = 2.49$ mM from a single standard addition. The 14% difference between the two results is experimental error attributable to using one point instead of all the points.

Often, all solutions in a standard addition experiment are made up to the same total volume. In this case, $V = V_0$ and $[S]_i(V_s/V_0) = [S]_i(V_s/V) = [S]_f$. The left side of Equation 16-12 is just I_{S+X} and the function on the right side is just $[S]_f$. *If all solutions are made up to the same final volume, we plot current versus* $[S]_f$.

If all solutions are made up to the same final volume, plot I_{S+X} versus $[S]_f$.

Ask Yourself

16-C. Successive standard additions of 1.00 mL of 25.0 mM ascorbic acid were made to 50.0 mL of orange juice in an experiment similar to the one shown in Figures 16-8 and 16-9. Prepare a graph similar to Figure 16-11 to find the concentration of ascorbic acid in the orange juice.

Total volume of added standard (mL)	Peak current (μA)
0	1.66
1.00	2.03
2.00	2.39
3.00	2.79
4.00	3.16
5.00	3.51

Polarography was invented in 1922 by Jaroslav Heyrovský, who received the Nobel Prize in 1959.

Potential limit (versus S.C.E.) for electrodes in 1 M H_2SO_4:

Pt	-0.2 to $+0.9$ V
Au	-0.3 to $+8$ V
Glassy carbon	-0.8 to $+1.1$ V
Hg	-1.3 to $+0.1$ V

In the presence of 1 M Cl^-, Hg is oxidized even more easily (near 0 V) by virtue of the reaction $Hg(l) + 4Cl^- \rightarrow HgCl_4^{2-} + 2e^-$.

16-5 Polarography

Polarography is a voltammetry experiment conducted with a *dropping-mercury electrode.* The cell in Figure 16-12 has a dropping-mercury working electrode, a Pt auxiliary electrode, and a calomel reference electrode. An electronically controlled dispenser suspends one drop of mercury from the tip of a glass capillary tube immersed in the analyte solution. An electrical measurement is made in ~1 s, the drop is released, and a fresh drop is suspended from the capillary for the next measurement. There is always a fresh, reproducible metal surface for each measurement.

Mercury is particularly useful for reduction processes. With other working electrodes, such as Pt, Au, or carbon, H^+ is readily reduced to H_2 at modest negative potentials. The high current from this reaction obscures the signal from reduction of analyte. Reduction of H^+ is difficult at a Hg surface and requires much more negative potentials. Conversely, Hg has little useful range for oxidations, because Hg itself is oxidized to Hg^{2+} at modest positive potentials. Therefore, a dropping mercury electrode is usually used to reduce analytes; and Pt, Au, or carbon are used to oxidize analytes such as vitamin C in Reaction 16-9.

The Polarogram

To record the voltammogram of vitamin C in Figure 16-9, the potential applied to the working electrode was varied at a constant rate from -0.4 V to $+1.2$ V. We call this voltage profile a *linear voltage ramp* (Figure 16-13a).

One of many ways to conduct a polarography experiment is with a *staircase voltage ramp* (Figure 16-13b). When each new drop of Hg is dispensed, the potential is made more negative by 4 mV. After waiting almost 1 s, the current is measured during the last 17 ms of the life of each Hg drop. The **polarogram** in Figure 16-14a is a graph of current versus voltage when Cd^{2+} is the analyte. The chemistry at the working electrode is

Reaction at working electrode: $Cd^{2+} + 2e^- \longrightarrow Cd(dissolved\ in\ Hg)$ (16-13)

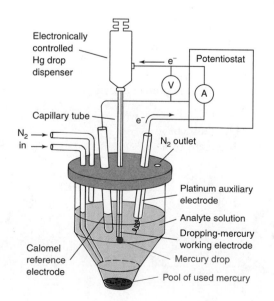

FIGURE 16-12 A cell for polarography.

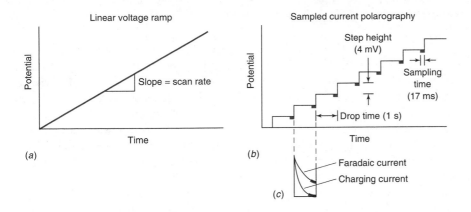

FIGURE 16-13 Voltage profiles for voltammetry: (*a*) Linear voltage ramp used in vitamin C experiment; (*b*) Staircase profile for *sampled current* polarography. Inset (*c*) shows how faradaic and charging currents decay after each potential step.

The product Cd(0) is dissolved in the liquid Hg drop. A solution of anything in Hg is called an **amalgam.** We call the curve in Figure 16-14a a *sampled current polarogram* because the current is only measured at the end of each drop life.

The curve in Figure 16-14a is called a **polarographic wave.** The potential at which half the maximum current is reached is called the **half-wave potential** ($E_{1/2}$) in Figure 16-14a. The constant current in the plateau region is called the **diffusion current** because it is limited by the rate of diffusion of analyte to the electrode. *For quantitative analysis, diffusion current is proportional to the concentration of analyte.* Diffusion current is measured from the baseline recorded without analyte in Figure 16-14b. The small **residual current** of the baseline in the absence of analyte is due mainly to reduction of impurities in the solution and on the surface of the electrodes. At sufficiently negative potential (-1.2 V in Figure 16-14), current increases rapidly as reduction of H^+ to H_2 in the aqueous solution commences.

For quantitative analysis, we require that the peak current (the diffusion current) be governed by the rate at which analyte can diffuse to the electrode. Other mechanisms by which analyte can reach the electrode are convection and electrostatic attraction. We minimize convection by using an unstirred solution. Electrostatic attraction is decreased by using a high concentration of inert ions (called *supporting electrolyte*), such as 1 M HCl in Figure 16-14.

Oxygen must be absent because O_2 gives two polarographic waves when it is reduced first to H_2O_2 and then to H_2O. In Figure 16-12, N_2 is bubbled through the analyte solution for 10 min to remove O_2. Then the liquid is maintained under a blanket of flowing N_2 in the gas phase to keep O_2 out. The liquid should not be purged with N_2 during a measurement because we do not want convection of analyte to the electrode.

For the first 50 years of polarography, current was measured continuously as Hg flowed from an open capillary tube. Each drop grew until it fell off and was replaced by a new drop. The current oscillated from a low value when the drop was small to a high value when the drop was big. Polarograms in the older literature have large oscillations superimposed on the wave in Figure 16-14a.

$E_{1/2}$ is characteristic of a particular analyte in a particular medium. Different analytes can be distinguished from each other by their half-wave potentials.

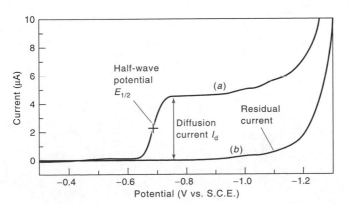

FIGURE 16-14 Sampled current polarogram of (*a*) 5 mM Cd^{2+} in 1 M HCl and (*b*) 1 M HCl alone.

Faradaic and Charging Currents

Faradaic current: due to redox reaction at the electrode

Charging current: due to electrostatic attraction or repulsion of ions in solution and electrons in the electrode

By waiting after each potential step before measuring the current, we observe significant faradaic current from the redox reaction with little interference from the charging current.

The current we seek to measure in voltammetry is **faradaic current** due to reduction (or oxidation) of analyte at the working electrode. In Figure 16-14a, faradaic current is from reduction of Cd^{2+} at the Hg electrode. Another current, called **charging current** (or *capacitor current*), interferes with every measurement. To step the working electrode to a more negative potential, electrons are forced into the electrode from the potentiostat. In response, cations in solution flow toward the electrode, and anions flow away from the electrode. This flow of ions and electrons, called the *charging current,* is not from redox reactions. We try to minimize charging current because it obscures the faradaic current. The charging current usually controls the detection limit in polarography or voltammetry.

Figure 16-13c shows the behavior of faradaic and charging currents after each potential step in Figure 16-13b. Faradaic current decays because analyte cannot diffuse to the electrode fast enough to sustain the high reaction rate. Charging current decays even faster because ions near the electrode redistribute themselves rapidly. By waiting 1 s after each potential step, the faradaic current is still significant and the charging current is very small.

Square Wave Voltammetry

The optimum height of the square wave, E_p in Figure 16-15, is $50/n$ mV, where n is the number of electrons in the half-reaction. For Reaction 16-13, $n = 2$ and $E_p = 25$ mV.

Advantages of square wave voltammetry:

- increased signal
- derivative (peak) shape permits better resolution of neighboring signals
- faster measurement

The most efficient voltage profile for polarography or voltammetry, called **square wave voltammetry,** uses the waveform in Figure 16-15, which consists of a square wave superimposed on a staircase.[2] During each cathodic pulse in Figure 16-15, there is a rush of analyte to be reduced at the electrode surface. During the anodic pulse, analyte that was just reduced is reoxidized. The square wave polarogram in Figure 16-16 is the *difference* in current between the intervals 1 and 2 in Figure 16-15. Electrons flow from the electrode to analyte at point 1 and in the reverse direction at point 2. Because the two currents have opposite signs, their difference is larger than either current alone. By taking a difference, the shape of the square wave polarogram in Figure 16-16 is essentially the derivative of the sampled current polarogram.

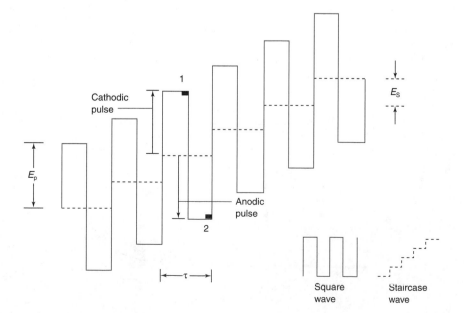

FIGURE 16–15 Waveform for square wave voltammetry. Typical parameters are pulse height (E_p) = 25 mV, step height (E_s) = 10 mV, and pulse period (τ) = 5 ms. Current is measured in regions 1 and 2.

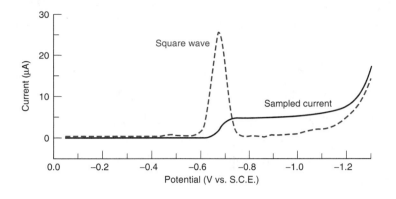

FIGURE 16-16 Comparison of polarograms of 5 mM Cd^{2+} in 1 M HCl. The following operating parameters are defined in Figures 16-13b and 16-15. Sampled current: drop time = 1 s, step height = 4 mV, sampling time = 17 ms. Square wave: drop time = 1 s, step height (E_s) = 4 mV, pulse period (τ) = 67 ms, pulse height (E_p) = 25 mV, sampling time = 17 ms.

The signal in square voltammetry is increased relative to a sampled current voltammogram, and the wave becomes peak shaped. The detection limit is reduced from $\sim 10^{-5}$ M for sampled current polarography to $\sim 10^{-7}$ M in square wave polarography. Because it is easier to resolve neighboring peaks than neighboring waves, square wave polarography can resolve species whose half-wave potentials differ by ~ 0.05 V, whereas the potentials must differ by ~ 0.2 V to be resolved in sampled current polarography. Square wave voltammetry is much faster than other voltammetric techniques. The square wave polarogram in Figure 16-16 was recorded in one-fifteenth of the time required for the sampled current polarogram. In principle, the shorter the pulse period, τ, in Figure 16-15, the greater the current that will be observed. In practice, a pulse period of 50 ms is a practical lower limit for common equipment.

Stripping Analysis

In **stripping analysis,** analyte from a dilute solution is first concentrated into a single drop of Hg (or onto a solid electrode) by electroreduction.[3] Analyte is then *stripped* from the electrode by making the potential more positive, thereby oxidizing it back into solution. Current measured during oxidation is proportional to the quantity of analyte that was initially deposited. Figure 16-17 shows an anodic stripping polarogram of traces of Cd, Pb, and Cu from honey. Field-portable instruments are used for stripping analysis of Cr, Cd, Zn, Cu, Pb, and Hg in contaminated soil.[4]

Stripping is the most sensitive polarographic technique because analyte is concentrated from a dilute solution. The longer the period of concentration, the more sensitive is the analysis. Only a fraction of analyte from the solution is deposited, so deposition must be done for a reproducible time (such as 5 min) with reproducible stirring. Detection limits are $\sim 10^{-10}$ M.

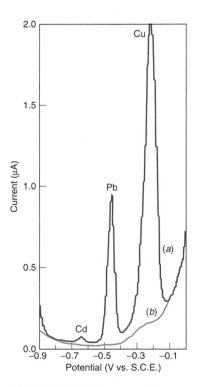

FIGURE 16-17 (*a*) Anodic stripping voltammogram of honey dissolved in water and acidified to pH 1.2 with HCl. Cd, Pb, and Cu were reduced from solution into a thin film of Hg for 5 min at -1.4 V (versus S.C.E.) prior to recording the voltammogram. (*b*) Voltammogram obtained without 5-min reduction step. The concentrations of Cd and Pb in the honey were 7 and 27 ng/g (ppb), respectively. The precision of the analysis was 2–4%. [From Y. Li, F. Wahdat, and R. Neeb, *Fresenius J. Anal. Chem.* **1995,** *351,* 678.]

Ask Yourself

16-D. (a) What is the difference between faradaic and charging current?

(b) Why is it desirable to wait 1 s after a potential pulse before recording the current in voltammetry?

(c) What are the advantages of square wave polarography over sampled current polarography?

(d) Explain what is done in anodic stripping voltammetry. Why is stripping the most sensitive polarographic technique?

Key Equations

Standard addition

$$\frac{[X]_i}{[X]_f + [S]_f} = \frac{I_X}{I_{S+X}}$$

$[X]_i$ = concentration of analyte in initial unknown

$[X]_f$ = concentration of analyte after standard addition

$[S]_f$ = concentration of standard after addition to unknown

Standard addition graph

You should be able to graph Equation 16-12 to interpret a standard addition experiment.

Important Terms

amalgam	electrogravimetric analysis	potentiostat
amperometry	electrolysis	reference electrode
auxiliary electrode	faradaic current	residual current
biosensor	half-wave potential	square wave voltammetry
charging current	matrix	standard addition
Clark electrode	mediator	stripping analysis
coulometry	polarogram	voltammetry
diffusion current	polarographic wave	voltammogram
electroactive species	polarography	working electrode

Problems

16-1. (a) State the general idea behind electrogravimetric analysis.

(b) How can you know when an electrogravimetric deposition is complete?

16-2. How do the measurements of current and time in Reaction 16-2 allow us to measure the quantity of H_2S in Reaction 16-3?

16-3. In the diagram below, ──o is the symbol for the working electrode, ──┤ is the auxiliary electrode, and ──▶ is the reference electrode. Which voltage, V_1 or V_2, is held constant in an electrolysis with three electrodes?

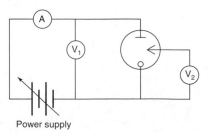

Power supply

16-4. A 50.0-mL aliquot of unknown Cu(II) solution was

exhaustively electrolyzed to deposit all copper on the cathode. The mass of the cathode was 15.327 g prior to electrolysis and 16.414 g after electrolysis. Find the molarity of Cu(II) in the unknown.

16-5. A solution containing 0.402 49 g of $CoCl_2 \cdot xH_2O$ (a solid with an unknown number of waters of hydration) was exhaustively electrolyzed to deposit 0.099 37 g of metallic cobalt on a platinum cathode by the reaction $Co^{2+} + 2e^- \rightarrow Co(s)$. Calculate the number of moles of water per mole of cobalt in the reagent.

16-6. Ions that react with Ag^+ can be determined electrogravimetrically by deposition on a silver anode: $Ag(s) + X^- \rightarrow AgX(s) + e^-$. What will be the final mass of a silver anode used to electrolyze 75.00 mL of 0.023 80 M KSCN if the initial mass of the anode is 12.463 8 g?

16-7. A 0.326 8-g unknown containing inert material plus lead lactate, $Pb(CH_3CHOHCO_2)_2$ (FM 385.3), was electrolyzed to produce 0.111 1 g of PbO_2 (FM 239.2). Was the PbO_2 deposited at the anode or at the cathode? Find the weight percent of lead lactate in the unknown.

16-8. $H_2S(aq)$ is analyzed by titration with coulometrically generated I_2 in Reactions 16-2 and 16-3. To 50.00 mL of unknown H_2S sample were added 4 g of KI. Electrolysis required 812 s at 52.6 mA. Find the concentration of H_2S ($\mu g/mL$) in the sample.

16-9. A 1.00-L electrolysis cell initially containing 0.025 0 M Mn^{2+} and another metal ion, M^{3+}, is fitted with Mn and Pt electrodes. The reactions are

$$Mn(s) \longrightarrow Mn^{2+} + 2e^-$$
$$M^{3+} + 3e^- \longrightarrow M(s)$$

(a) Is the Mn electrode the anode or the cathode?

(b) A constant current of 2.60 A was passed through the cell for 18.0 min, causing 0.504 g of the metal M to plate out on the Pt electrode. What is the atomic mass of M?

(c) What will the concentration of Mn^{2+} in the cell be at the end of the experiment?

16-10. The sensitivity of a coulometer is governed by the delivery of its minimum current for its minimum time. Suppose that 5 mA can be delivered for 0.1 s.

(a) How many moles of electrons are delivered at 5 mA for 0.1 s?

(b) How many milliliters of a 0.01 M solution of a two-electron reducing agent are required to deliver the same number of electrons?

16-11. The electrolysis cell shown below was run at a constant current of 0.021 96 A. On one side, 49.22 mL of H_2 were produced (at 303 K and 0.996 bar); on the other side, Cu metal was oxidized to Cu^{2+}.

(a) How many moles of H_2 were produced? (See Problem 15-16 for the ideal gas law.)

(b) If 47.36 mL of EDTA were required to titrate the Cu^{2+} produced by the electrolysis, what was the molarity of the EDTA?

(c) For how many hours was the electrolysis run?

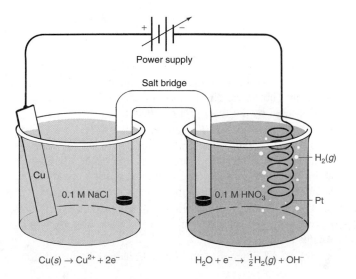

$$Cu(s) \rightarrow Cu^{2+} + 2e^-$$

$$H_2O + e^- \rightarrow \tfrac{1}{2}H_2(g) + OH^-$$

16-12. A mixture of trichloroacetate and dichloroacetate can be analyzed by selective reduction in a solution containing 2 M KCl, 2.5 M NH_3, and 1 M NH_4Cl. At a mercury cathode potential of -0.90 V (versus S.C.E.), only trichloroacetate is reduced:

$$Cl_3CCO_2^- + H_2O + 2e^- \longrightarrow Cl_2CHCO_2^- + OH^- + Cl^-$$

At a potential of -1.65 V, dichloroacetate reacts:

$$Cl_2CHCO_2^- + H_2O + 2e^- \longrightarrow ClCH_2CO_2^- + OH^- + Cl^-$$

A hygroscopic mixture of trichloroacetic acid (FM 163.386) and dichloroacetic acid (FM 128.943) containing an unknown quantity of water weighed 0.721 g. Upon controlled potential electrolysis, 224 C passed at -0.90 V, and 758 C were required to complete the electrolysis at -1.65 V. Calculate the weight percent of each acid in the mixture.

16-13. Chlorine has been used for decades to disinfect drinking water. An undesirable side effect of this treatment is the reaction of chlorine with organic impurities to create organochlorine compounds, some of which could be toxic. Monitoring total organic halide (designated TOX) is now required for many water providers. A standard procedure for TOX is to pass water through activated charcoal that adsorbs organic compounds. Then the charcoal is combusted to liberate hydrogen halides:

$$\text{organic halide (RX)} \xrightarrow{O_2/800°C} CO_2 + H_2O + HX$$

The HX is absorbed into aqueous solution and measured by automatic coulometric titration with a silver anode:

$$X^-(aq) + Ag(s) \longrightarrow AgX(s) + e^-$$

When 1.00 L of drinking water was analyzed, a current of 4.23 mA was required for 387 s. A blank prepared by oxidizing charcoal required 6 s at 4.23 mA. Express the TOX of the drinking water as μmol halogen/L. If all halogen is chlorine, express the TOX as μg Cl/L.

16-14. *Propagation of uncertainty.* In an extremely accurate measurement of the Faraday constant, a pure silver anode was oxidized to Ag^+ with a constant current of 0.203 639 0 ($\pm0.000\ 000\ 4$) A for 18 000.075 (±0.010) s to give a mass loss of 4.097 900 ($\pm0.000\ 003$) g from the anode. Given that the atomic mass of Ag is 107.868 2 ($\pm0.000\ 2$), find the value of the Faraday constant and its uncertainty.

16-15. (a) How does a Clark electrode measure the concentration of dissolved O_2?

(b) What does it mean when we say that the concentration of dissolved O_2 is "0.20 bar"? What is the actual molarity of O_2?

16-16. (a) What are the advantages of a dropping Hg electrode in polarography? Why is polarography used mainly to study reductions rather than oxidations?

(b) What is the difference between faradaic and charging current? How does sampled current polarography reduce the contribution of charging current to the polarogram?

16-17. Suppose that a peak current of 3.9 μA was observed in the oxidation of 50 mL of 2.4 mM ascorbic acid in the experiment in Figures 16-8 and 16-9. Suppose that this much current flowed for 10 min during the course of several measurements. From the current and time, calculate what fraction of the ascorbic acid is oxidized at the electrode. Is it fair to say that the ascorbic acid concentration is nearly constant during the measurements?

16-18. A 50.0-mL sample of orange juice gave a peak current of 2.02 μA in an experiment similar to the one in Figures 16-8 and 16-9. A standard addition of 1.00 mL of 29.4 mM ascorbic acid increased the peak current to 3.79 μA. Find the concentration of ascorbic acid in the orange juice.

16-19. An unknown gave a polarographic signal of 10.0 μA. When 1.00 mL of 0.050 0 M standard was added to 100.0 mL of unknown, the signal increased to 14.0 μA. Find the concentration of the original unknown.

16-20. *Calibration curve and error estimate.* The following polarographic diffusion currents were measured at -0.6 V for $CuSO_4$ in 2 M NH_4Cl/2M NH_3. Use the method of least squares to estimate the molarity (and its uncertainty) of an unknown solution giving $I_d = 15.6$ μA.

$[Cu^{2+}]$ (mM)	I_d (μA)	$[Cu^{2+}]$ (mM)	I_d (μA)
0.039 3	0.256	0.990	6.37
0.078 0	0.520	1.97	13.00
0.158 5	1.058	3.83	25.0
0.489	3.06	8.43	55.8

16-21. The drug Librium gives a polarographic wave with $E_{1/2} = -0.265$ V (versus S.C.E.) in 0.05 M H_2SO_4. A 50.0-mL sample containing Librium gave a wave height of 0.37 μA. When 2.00 mL of 3.00 mM Librium in 0.05 M H_2SO_4 were added to the sample, the wave height increased to 0.80 μA. Find the molarity of Librium in the unknown.

16-22. A polarogram of reagent-grade methanol is shown at the top of the next column, along with that of a standard made by adding an additional 0.001 00 wt % acetone, 0.001 00 wt % acetaldehyde, and 0.001 00 wt % formaldehyde to reagent-grade methanol. Estimate the weight percent of acetone in reagent-grade methanol.

16-23. The figure on the next page shows standard additions of Cu^{2+} to acidified tap water measured by anodic stripping voltammetry at an Ir electrode. The unknown and all standard additions were made up to the same final volume.

(a) What chemical reaction occurs during the concentration stage of the analysis?

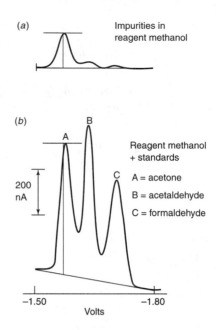

Problem 16-22: Polarogram of reagent-grade methanol and methanol containing added standards. [From D. B. Palladino, *Am. Lab.* August 1992, p. 56.] Solutions were prepared by diluting 25.0 mL of methanol (or methanol plus standards) up to 100.0 mL with water containing buffer and hydrazine sulfate. The latter reacts with carbonyl compounds to form hydrazones, which are the electroactive species in this analysis.

$$(CH_3)_2C{=}O + H_2N{-}NH_2 \longrightarrow$$
Acetone · · · · · · Hydrazine

$$(CH_3)_2C{=}N{-}NH_2 \xrightarrow{2e^- + 2H^+}$$
Acetone hydrazone

$$(CH_3)_2CH{-}NH{-}NH_2$$

(b) What chemical reaction occurs during the stripping stage of the analysis?

(c) Find the concentration of Cu^{2+} in the tap water.

How Would You Do It?

16-24. Nitrite (NO_2^-) is a potential carcinogen for humans, but it is also an important food preservative in foods such as bacon and hot dogs. Nitrite and nitrate (NO_3^-) in foods can be measured by the following procedure:

1. A 10-g food sample is blended with 82 mL of 0.07 M NaOH, transferred to a 200-mL volumetric flask, and heated on a steam bath for 1 h. Then 10 mL of 0.42 M $ZnSO_4$ are added; and after an additional 10 min of heating, the mixture is cooled and diluted to 200 mL.

2. After the solids settle, an aliquot of supernatant solution is mixed with 200 mg of charcoal to remove some organic solutes and filtered.

3. A 5.00-mL aliquot of the filtered solution is placed in a

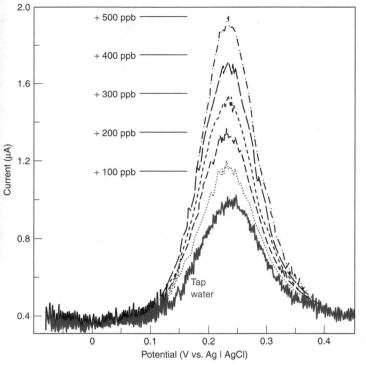

Problem 16-23: Anodic stripping voltammograms of tap water and five standard additions of 100 ppb Cu^{2+}. [From M. A. Nolan and S. P. Kounaves, *Anal. Chem.* **1999**, *71*, 3567.]

test tube containing 5 mL of 9 M H_2SO_4 and 150 μL of 85 wt % NaBr, which reacts with nitrite, but not nitrate:

$$HNO_2 + Br^- \longrightarrow NO(g) + Br_2$$

4. The NO is transported by bubbling with purified N_2 that is subsequently passed through a trapping solution to convert NO into diphenylnitrosamine:

$$NO + (C_6H_5)_2NH \xrightarrow{H^+ \text{ and catalyst}} (C_6H_5)_2N-N{=}O$$
$$\text{Diphenylamine} \qquad\qquad\qquad \text{Diphenylnitrosamine}$$

5. The acidic diphenylnitrosamine is analyzed by amperometry at -0.66 V (versus Ag|AgCl) to measure the NO liberated from the food sample:

$$(C_6H_5)_2N-N{=}OH^+ + 4H^+ + 4e^- \longrightarrow$$
$$(C_6H_5)_2N-NH_3^+ + H_2O$$

6. To measure nitrate, an additional 6 mL of 18 M H_2SO_4 are added to the sample tube in step 3. The acid promotes reduction of HNO_3 to NO, which is then purged with N_2 and trapped and analyzed as in steps 4 and 5.

$$HNO_3 + Br^- \xrightarrow{\text{strong acid}} NO(g) + Br_2$$

A 10.0-g bacon sample gave a current of 8.9 μA in step 5, which increased to 23.2 μA in step 6. (Step 6 measures the sum of signals from nitrite and nitrate, not just the signal from nitrate.) In a second experiment, the 5.00-mL sample in step 3 was spiked with a standard addition of 5.00 μg of NO_2^- ion. The analysis was repeated to give currents of 14.6 μA in step 5 and 28.9 μA in step 6.

(a) Find the nitrite content of the bacon, expressed as micrograms per gram of bacon.

(b) From the ratio of signals due to nitrate and nitrite in the first experiment, find the micrograms of nitrate per gram of bacon.

Notes and References

1. Construction of an oxygen electrode is described by J. E. Brunet, J. I. Gardiazabal, and R. Schrebler, *J. Chem. Ed.* **1983**, *60*, 677.

2. For an excellent account of square wave voltammetry, see J. G. Osteryoung and R. A. Osteryoung, *Anal. Chem.* **1985**, *57*, 101A.

3. J. Wang, *Stripping Analysis: Principles, Instrumentation and Applications* (Deerfield Beach, FL: VCH Publishers, 1984).

4. K. B. Olsen, J. Wang, R. Setladji, and J. Lu, *Environ. Sci. Tech.* **1994**, *28*, 2074; D. F. Faust and J. Y. Gui, *Am. Environ. Lab.*, April 1997, p. 17; K. White and A. Clapp, *Am. Environ. Lab.*, October 1996, p. 19.

Further Reading

A. J. Cunningham, *Introduction to Bioanalytical Sensors* (New York: Wiley, 1998).

D. Diamond, *Principles of Chemical and Biological Sensors* (New York: Wiley, 1998).

G. Ramsay, *Commercial Biosensors: Applications to Clinical, Bioprocess, and Environmental Samples* (New York: Wiley, 1998).

B. Eggins, *Biosensors* (Chichester, UK: Wiley Teubner, 1996).

M. Alvarez-Icaza and U. Bilitewski, "Mass Production of Biosensors," *Anal. Chem.* **1993**, *65*, 525A.

The Ozone Hole

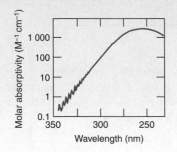

Spectrum of ozone, showing maximum absorption of ultraviolet radiation at a wavelength near 260 nm. [From R. P. Wayne, *Chemistry of Atmospheres* (Oxford: Clarendon Press, 1991).]

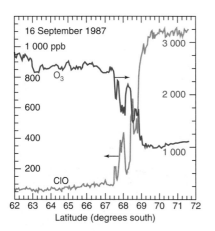

Spectroscopically measured concentrations of O_3 and ClO (measured in ppb = nL/L) in the stratosphere near the South Pole in 1987. Where ClO increases, O_3 decreases, in agreement with the reaction sequence on this page. [From J. G. Anderson, W. H. Brune, and M. H. Proffitt, *J. Geophys. Res.* **1989**, *94D*, 11465.]

*O*zone, formed at altitudes of 20 to 40 km by the action of solar ultraviolet radiation ($h\nu$) on O_2, absorbs the ultraviolet radiation that causes sunburns and skin cancer:

$$O_2 \xrightarrow{h\nu} 2O \qquad O + O_2 \longrightarrow O_3$$
$$\text{Ozone}$$

In 1985, the British Antarctic Survey reported that ozone over Antarctica had decreased by 50% in early spring, relative to levels observed in the preceding 20 years. This "ozone hole" occurs only in early spring and is gradually becoming worse.

The current explanation of this phenomenon begins with chlorofluorocarbons such as Freon-12 (CCl_2F_2) from refrigerants. These long-lived compounds diffuse to the stratosphere, where they catalyze ozone decomposition:

(1) $\quad CCl_2F_2 \xrightarrow{h\nu} CClF_2 + Cl \qquad$ Photochemical Cl formation

(2) $\quad Cl + O_3 \longrightarrow ClO + O_2$

(3) $\quad O_3 \xrightarrow{h\nu} O + O_2$

(4) $\quad O + ClO \longrightarrow Cl + O_2$

Net reaction of (2)–(4):
Catalytic O_3 destruction
$2O_3 \longrightarrow 3O_2$

Cl produced in step 4 goes back to destroy another ozone molecule in step 2. A single Cl atom in this chain reaction can destroy $> 10^5$ molecules of O_3. The chain is terminated when Cl or ClO reacts with hydrocarbons or NO_2 to form HCl or $ClONO_2$.

Stratospheric clouds catalyze the reaction of HCl with $ClONO_2$ to form Cl_2, which is split by sunlight into Cl atoms to initiate O_3 destruction:

$$HCl + ClONO_2 \xrightarrow[\text{polar clouds}]{\text{Surface of}} Cl_2 + HNO_3 \qquad Cl_2 \xrightarrow{h\nu} 2Cl$$

The clouds require winter cold to form. Only when the sun is rising at the South Pole in September and October, and the clouds are still present, are conditions right for O_3 destruction.

To protect life from ultraviolet radiation, international treaties now ban or phase out chlorofluorocarbons. However, so much of these compounds has already been released, and so much more remains in use in your house and mine, that ozone depletion is expected to become more severe before slowly returning to historic values late in the twenty-first century.

Chapter 17

Let There Be Light

A bsorption and emission of *electromagnetic radiation* (a fancy term for light) are key physical characteristics of molecules used in quantitative and qualitative analysis. This chapter discusses basic aspects of **spectrophotometry**—the use of electromagnetic radiation to measure chemical concentrations—and the next chapter provides further detail on instrumentation and applications of spectrophotometry.

17-1   Properties of Light

Light can be described both as waves and as particles. Light waves consist of perpendicular, oscillating electric and magnetic fields (Figure 17-1). The **wavelength,** λ, is the crest-to-crest distance between waves. The **frequency,** ν, is the number of complete oscillations that the wave makes each second. The unit of frequency is *reciprocal seconds,* s^{-1}. One oscillation per second is also called one **hertz** (Hz). A frequency of 10^6 s^{-1} is therefore said to be 10^6 Hz, or one *megahertz* (MHz). The product of frequency times wavelength is c, the speed of light (2.998×10^8 m/s in vacuum):

Relation between frequency and wavelength:
$$\nu\lambda = c$$
(17-1)

Following the discovery of the Antarctic ozone "hole" in 1985, atmospheric chemist Susan Solomon led the first expedition in 1986 specifically intended to make chemical measurements of the Antarctic atmosphere by using high-altitude balloons and ground-based spectroscopy. The expedition discovered that ozone depletion occurred after polar sunrise and that the concentration of chemically active chlorine in the stratosphere was 100 times greater than had been predicted from gasphase chemistry. Solomon's group identified chlorine as the culprit in ozone destruction and polar stratospheric clouds as the catalytic surface for the release of so much chlorine.

365

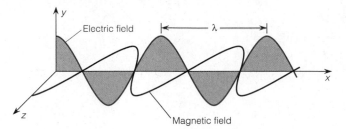

FIGURE 17-1 *Plane-polarized* electromagnetic radiation of wavelength λ, propagating along the *x*-axis. The electric field oscillates in the *xy*-plane and the magnetic field oscillates in the *xz*-plane. Ordinary, unpolarized light has electric and magnetic field components in all planes.

EXAMPLE   Relating Wavelength and Frequency

What is the wavelength of radiation in your microwave oven, whose frequency is 2.45 GHz?

SOLUTION First recognize that 2.45 GHz means 2.45×10^9 Hz $= 2.45 \times 10^9\, s^{-1}$. From Equation 17-1, we write

$$\lambda = \frac{c}{\nu} = \frac{2.998 \times 10^8 \text{ m/s}}{2.45 \times 10^9 \text{ s}^{-1}} = 0.122 \text{ m}$$

Light can also be thought of as particles called **photons.** The energy, *E,* of a photon is proportional to its frequency:

Relation between energy and frequency:
$$E = h\nu \qquad (17\text{-}2)$$

Physical constants are listed inside the book cover.

where *h* is *Planck's constant* ($= 6.626 \times 10^{-34}$ J · s).
    Combining Equations 17-1 and 17-2, we can write

$$E = h\frac{c}{\lambda} = hc\frac{1}{\lambda} = hc\tilde{\nu} \qquad (17\text{-}3)$$

Energy increases if
• frequency (ν) increases
• wavelength (λ) decreases
• wavenumber ($\tilde{\nu}$) increases

where $\tilde{\nu}$ ($= 1/\lambda$) is called the **wavenumber.** Energy is inversely proportional to wavelength and directly proportional to wavenumber. Red light, with a wavelength longer than that of blue light, is less energetic than blue light. The SI unit for wavenumber is m^{-1}. However, the most common unit of wavenumber is cm^{-1}, read "reciprocal centimeters" or "wavenumbers." Wavenumber units are most common in infrared spectroscopy.
    Regions of the **electromagnetic spectrum** are shown in Figure 17-2. Visible light—the kind our eyes detect—represents only a small fraction of the electromagnetic spectrum.
    The lowest energy state of an atom or a molecule is called the **ground state.** When the atom or molecule absorbs a photon, the energy of the atom or molecule

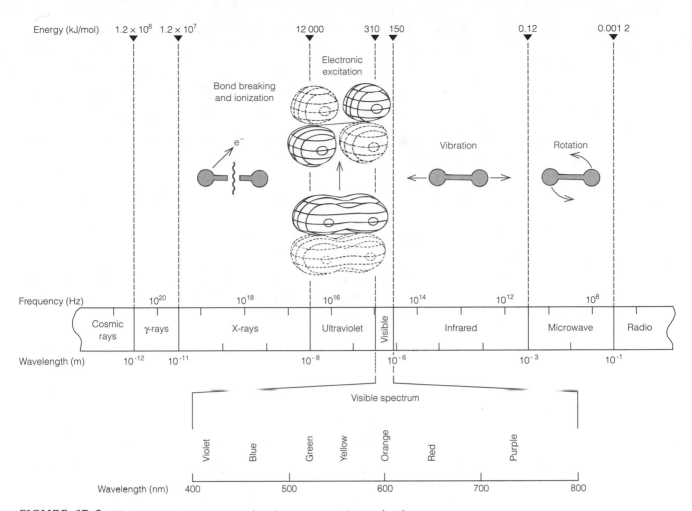

FIGURE 17-2 Electromagnetic spectrum, showing representative molecular processes that occur when radiation in each region is absorbed. The visible spectrum spans the wavelength range 380 to 780 nanometers (1 nm = 10^{-9} m).

increases and we say that the atom or molecule is promoted to an **excited state** (Figure 17-3). If the atom or molecule emits a photon, the energy of the atom or molecule decreases. Figure 17-2 indicates that microwave radiation stimulates molecules to rotate faster. A microwave oven heats food by increasing the rotational energy of water in the food. Infrared radiation excites vibrations of molecules. Visible light and ultraviolet radiation promote electrons from lower energy states to higher energy states. (A molecule does not absorb visible light unless the molecule is colored.) X-rays and short-wavelength ultraviolet radiation are harmful because they break chemical bonds and ionize molecules, which is why you should minimize your exposure to medical X-rays.

EXAMPLE Photon Energies

By how many joules is the energy of a molecule increased when it absorbs **(a)** visible light with a wavelength of 500 nm or **(b)** infrared radiation with a wavenumber of 1 251 cm^{-1}?

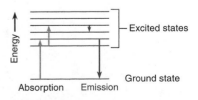

FIGURE 17-3 Absorption of light increases the energy of a molecule. Emission of light decreases its energy.

SOLUTION **(a)** The visible wavelength is $500 \text{ nm} = 500 \times 10^{-9} \text{ m}$.

$$E = h\nu = h\frac{c}{\lambda}$$

$$= (6.626 \times 10^{-34} \text{ J} \cdot \text{s})\left(\frac{2.998 \times 10^8 \text{ m/s}}{500 \times 10^{-9} \text{ m}}\right) = 3.97 \times 10^{-19} \text{ J}$$

This is the energy of one photon absorbed by one molecule. If a mole of molecules absorbed a mole of photons, the energy increase is

$$E = \left(3.97 \times 10^{-19} \frac{\text{J}}{\text{molecule}}\right)\left(6.022 \times 10^{23} \frac{\text{molecules}}{\text{mol}}\right) = 2.39 \times 10^5 \frac{\text{J}}{\text{mol}}$$

$$\left(2.39 \times 10^5 \frac{\text{J}}{\text{mol}}\right)\left(\frac{1 \text{ kJ}}{1\,000 \text{ J}}\right) = 239 \frac{\text{kJ}}{\text{mol}}$$

(b) When given the wavenumber, we use Equation 17-3. However, we must remember to convert the wavenumber unit cm^{-1} to the unit m^{-1} with the conversion factor 100 cm/m. The energy of one photon is

$$E = hc\tilde{\nu} = (6.626 \times 10^{-34} \text{ J} \cdot \text{s})\left(2.998 \times 10^8 \frac{\text{m}}{\text{s}}\right)(1\,251 \text{ cm}^{-1})\underbrace{\left(100 \frac{\text{cm}}{\text{m}}\right)}_{\text{Conversion of cm}^{-1} \text{ to m}^{-1}}$$

$$= 2.485 \times 10^{-20} \text{ J}$$

Multiplying by Avogadro's number, we find this photon energy corresponds to 14.97 kJ/mol, which falls in the infrared region and excites molecular vibrations.

Ask Yourself

17-A. What is the frequency (Hz), wavenumber (cm^{-1}), and energy (kJ/mol) of light with a wavelength of **(a)** 100 nm; **(b)** 500 nm; **(c)** 10 μm; and **(d)** 1 cm? In which spectral region does each kind of radiation lie and what molecular process occurs when the radiation is absorbed?

17-2   Absorption of Light

A **spectrophotometer** is an instrument that measures transmission of light through a substance. If light is absorbed, the *radiant power* of the light beam decreases. Radiant power, P, is the energy per second per unit area of the beam. Light of one wavelength is said to be **monochromatic** ("one color"). In Figure 17-4, light passes through a *monochromator,* a device that selects one wavelength. Light of this wave-

FIGURE 17-4 Schematic diagram of a single-beam spectrophotometric experiment.

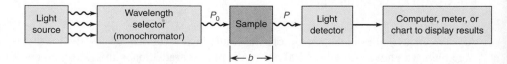

length, with radiant power P_0, strikes a sample of length b. The radiant power of the beam emerging from the other side of the sample is P. Some of the light may be absorbed by the sample, so $P \leq P_0$.

Transmittance, Absorbance, and Beer's Law

Transmittance, T, is the fraction of incident light that passes through a sample.

Transmittance:
$$T = \frac{P}{P_0}$$
(17-4)

Transmittance lies in the range 0 to 1. If no light is absorbed, the transmittance is 1. If all light is absorbed, the transmittance is 0. *Percent transmittance* ($100T$) ranges from 0% to 100%. A transmittance of 30% means that 70% of the light does not pass through the sample.

The most useful quantity for chemical analysis is **absorbance,** A, defined as

Absorbance:
$$A = \log\left(\frac{P_0}{P}\right) = -\log\left(\frac{P}{P_0}\right) = -\log T$$
(17-5)

Of course, you remember that
$$\log\left(\frac{1}{x}\right) = -\log x.$$

When no light is absorbed, $P = P_0$ and $A = 0$. If 90% of the light is absorbed, 10% is transmitted and $P = P_0/10$. This ratio gives $A = 1$. If 1% of the light is transmitted, $A = 2$.

P/P_0	% T	A
1	100	0
0.1	10	1
0.01	1	2

EXAMPLE   Absorbance and Transmittance

What absorbance corresponds to 99% transmittance? To 0.10% transmittance?

SOLUTION Use the definition of absorbance in Equation 17-5.

99% *T*: $A = -\log T = -\log 0.99 = 0.004\,4$
0.10% *T*: $A = -\log T = -\log 0.001\,0 = 3.0$

The higher the absorbance, the less light is transmitted through a sample.

The reason why absorbance is so important is that *absorbance is proportional to the concentration of light-absorbing molecules in the sample,* as given by **Beer's law:**

Beer's law:
$$A = \varepsilon b c$$
(17-6)

Absorbance (A) is dimensionless. Concentration (c) has units of moles per liter (M), and pathlength (b) is commonly expressed in centimeters. The quantity ε (epsilon) is a constant called the **molar absorptivity.** It has the units $M^{-1}\,cm^{-1}$ because the product $\varepsilon b c$ must be dimensionless. Molar absorptivity tells how much light is absorbed at a particular wavelength.

Box 17-1 gives a physical picture of Beer's law that could be the basis for a classroom exercise.

Color Plate 12 shows standard solutions of an iron compound. You can see that the color intensity increases as the concentration increases. Absorbance is a measure of the color. The more intense the color, the greater the absorbance.

EXAMPLE   Absorbance, Transmittance, and Beer's Law

Find the absorbance and transmittance of a 0.002 40 M solution of a substance with a molar absorptivity of (**a**) 1.00×10^2 or (**b**) $2.00 \times 10^2 \ M^{-1} \ cm^{-1}$ in a cell with a 2.00-cm pathlength.

SOLUTION Beer's law tells us the absorbance:

(**a**) $A = \varepsilon bc = (100 \ M^{-1} \ cm^{-1})(2.00 \ cm)(0.002 \ 40 \ M) = 0.480$

(**b**) $A = (200 \ M^{-1} \ cm^{-1})(2.00 \ cm)(0.002 \ 40 \ M) = 0.960$

Box 17-1 *Explanation*

Discovering Beer's Law[1]

Each photon passing through a solution has a certain probability of striking a light-absorbing molecule and being absorbed. Let's model this process by thinking of an inclined plane with holes representing absorbing molecules. The number of molecules is equal to the number of holes and the pathlength is equal to the length of the plane. Suppose that 1 000 small balls, representing 1 000 photons, are rolled down the incline. Whenever a ball drops through a hole, we consider it to have been "absorbed" by a molecule.

Let the plane be divided into 10 equal intervals and let the probability that a ball will fall through a hole in the first interval be 1/10. Of the 1 000 balls entering the first interval, one-tenth—100 balls—are absorbed (dropping through the holes) and 900 pass into the second interval. Of the 900 balls entering the second interval, one-tenth—90 balls—are absorbed and 810 proceed to the third interval. Of these 810 balls, 81 are absorbed and 729 proceed to the fourth interval. The table summarizes the action.

Transmittance is defined as

$$T = \frac{\text{number of surviving balls}}{\text{initial number of balls} \ (= 1 \ 000)}$$

Interval	Photons absorbed	Photons trans- mitted	Trans- mittance (P/P_0)
0		1 000	1.000
1	100	900	0.900
2	90	810	0.810
3	81	729	0.729
4	73	656	0.656
5	66	590	0.590
6	59	531	0.531
7	53	478	0.478
8	48	430	0.430
9	43	387	0.387
10	39	348	0.348

Balls represent photons

Holes represent molecules that can absorb photons

Inclined plane model for photon absorption.

Doubling the molar absorptivity doubles the absorbance. Doubling the concentration would also double the absorbance.

Transmittance is obtained from Equation 17-5 by raising 10 to the power on each side of the equation:

$$\log T = -A$$
$$\underbrace{10^{\log T}}_{} = 10^{-A}$$
$$10^{\log T} \text{ is the same as } T$$

(a) $T = 10^{-0.480} = 0.331 = 33.1\%$

(b) $T = 10^{-0.960} = 0.110 = 11.0\%$

Transmittance is not linearly related to molar absorptivity (or concentration). Doubling the molar absorptivity or the concentration does not decrease transmittance by a factor of 2.

To evaluate $10^{-0.480}$ on your calculator, use y^x or *antilog*. If you use y^x, $y = 10$ and $x = -0.480$. If you use *antilog*, find the antilog of -0.480. Be sure you can show that $10^{-0.480} = 0.331$.

Graph *a* shows that a plot of transmittance versus interval number (which is analogous to plotting transmittance versus pathlength in a spectrophotometric experiment) is not linear. However, the plot of $-\log$ (transmittance) versus interval number in graph b is linear and passes through the origin. Graph *b* shows that $\log T$ is proportional to pathlength.

To investigate how transmittance depends on the concentration of absorbing molecules, you could do the same mental experiment with a different number of holes in the inclined plane. For example, try setting up a table to show what happens if the probability of absorption in each interval is 1/20 instead of 1/10. This change corresponds to decreasing the concentration of absorbing molecules to half of its initial value. You will discover that a graph of $-\log T$ versus interval number has a slope equal to one-half that in graph b. That is, $-\log T$ is proportional to concentration as well as to pathlength. So, we have just shown that $-\log T$ is proportional to both concentration and pathlength. Defining absorbance as $-\log T$ gives us the essential terms in Beer's law:

$$A \equiv -\log T \propto \text{concentration} \times \text{pathlength}$$

This symbol means "is proportional to."

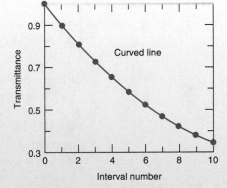

(a) Transmittance versus interval is analogous to T versus pathlength.

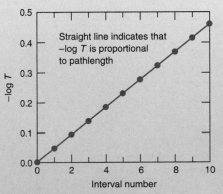

(b) $-\log T$ versus interval is analogous to $-\log T$ versus pathlength.

EXAMPLE   Finding Concentration from the Absorbance

Gaseous ozone has a molar absorptivity of 2 700 M^{-1} cm^{-1} at the absorption peak near 260 nm in the spectrum at the beginning of this chapter. Find the concentration of ozone (mol/L) in air if a sample has an absorbance of 0.23 in a 10.0-cm cell. Air has negligible absorbance at 260 nm.

SOLUTION We rearrange Beer's law to solve for concentration:

$$c = \frac{A}{\varepsilon b} = \frac{0.23}{(2\ 700\ M^{-1}\ cm^{-1})(10.0\ cm)} = 8.5 \times 10^{-6}\ M$$

Absorption Spectra and Color

The plural of "spectrum" is "spectra."

An **absorption spectrum** is a graph showing how A (or ε) varies with wavelength (or frequency or wavenumber). The opening highlight of this chapter shows the ultraviolet absorption spectrum of ozone, which peaks near 260 nm. Figure 17-5 shows the absorption spectrum of a typical sunscreen lotion, which absorbs harmful solar radiation below about 350 nm. Demonstration 17-1 illustrates the meaning of an absorption spectrum.

EXAMPLE   How Effective Is Sunscreen?

What fraction of ultraviolet radiation is transmitted through the sunscreen in Figure 17-5 at the peak absorbance near 300 nm?

SOLUTION From the spectrum in Figure 17-5, the absorbance at 300 nm is approximately 0.35. Therefore the transmittance is $T = 10^{-A} = 10^{-0.35} = 0.45 = 45\%$. Just over half the ultraviolet radiation (55%) is absorbed by sunscreen and does not reach your skin.

FIGURE 17-5   Absorption spectrum of typical sunscreen lotion shows absorbance versus wavelength in the ultraviolet region. Sunscreen was thinly coated onto a transparent window to make this measurement. [From D. W. Daniel, *J. Chem. Ed.* **1994**, *71*, 83.] Sunscreen makers refer to the region 400–320 nm as UV-A and 320–280 nm as UV-B. Problem 17-27 gives more data on sunscreen.

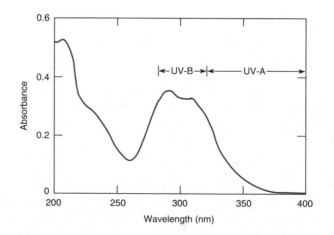

Absorption Spectra[2]

The spectrum of visible light can be projected on a screen in a darkened room in the following manner: Four layers of plastic diffraction grating† are mounted on a cardboard frame that has a square hole large enough to cover the lens of an overhead projector. This assembly is taped over the projector lens facing the screen. An opaque cardboard surface with two 1×3 cm slits is placed on the working surface of the projector.

When the lamp is turned on, the white image of each slit is projected on the center of the screen. A visible spectrum appears on either side of each image. When a beaker of colored solution is placed over one slit, you can see color projected on the screen where the white image previously appeared. The spectrum beside the colored image loses its intensity in regions where the colored solution absorbs light.

Color Plate 13a shows the spectrum of white light and the absorption spectra of three different colored solutions. We see that potassium dichromate, which appears orange or yellow, absorbs blue wavelengths. Bromophenol blue absorbs yellow and orange wavelengths and appears blue to our eyes. The absorption of phenolphthalein is located near the center of the visible spectrum. For comparison, the spectra of these three solutions recorded with a spectrophotometer are shown in Color Plate 13b.

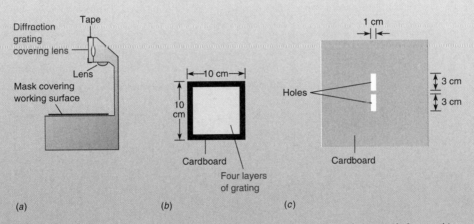

(a) Overhead projector. (b) Diffraction grating mounted on cardboard. (c) Mask for working surface.

†Diffraction grating is available from Edmund Scientific Co., 5975 Edscorp Building, Barrington, NJ 08007, catalog no. 40,267.

White light contains all the colors of the rainbow. Any substance that absorbs visible light appears colored when white light is transmitted through it or reflected from it. The substance absorbs certain wavelengths of the white light, and our eyes detect the wavelengths that are not absorbed. A rough guide to colors is given in Table 17-1. The observed color is called the *complement* of the absorbed color. As an example, bromophenol blue in Color Plate 13 has a visible absorbance maximum at 591 nm, and its observed color is blue.

The color of a substance is the complement of the color of the light that it absorbs.

TABLE 17-1	Colors of visible light	
Wavelength of maximum absorption (nm)	*Color absorbed*	*Color observed*
380–420	Violet	Green-yellow
420–440	Violet-blue	Yellow
440–470	Blue	Orange
470–500	Blue-green	Red
500–520	Green	Purple
520–550	Yellow-green	Violet
550–580	Yellow	Violet-blue
580–620	Orange	Blue
620–680	Red	Blue-green
680–780	Purple	Green

Ask Yourself

17-B. (a) What is the absorbance of a 2.33×10^{-4} M solution of a compound with a molar absorptivity of 1.05×10^3 M^{-1} cm^{-1} in a 1.00-cm cell?

(b) What is the transmittance of the solution in **(a)**?

(c) Find A and $\% T$ when the pathlength is doubled to 2.00 cm.

(d) Find A and $\% T$ when the pathlength is 1.00 cm but the concentration is doubled.

(e) What would be the absorbance in **(a)** for a different compound with twice as great a molar absorptivity ($\varepsilon = 2.10 \times 10^3$ M^{-1} cm^{-1})? The concentration and pathlength are unchanged from **(a)**.

17-3   Practical Matters

For chemical analysis, transmittance is converted to the more useful quantity, absorbance: $A = -\log T$.

The minimum requirements for a spectrophotometer were shown in Figure 17-4. The instrument measures the fraction of incident light (the transmittance) that passes through a sample to the detector. The sample is usually contained in a cell called a **cuvet** (Figure 17-6), which has flat fused silica or quartz faces. Fused silica (a glass made of SiO_2) and quartz (crystalline SiO_2) both transmit visible and ultraviolet radiation. Glass and plastic absorb ultraviolet radiation, so glass or clear plastic can

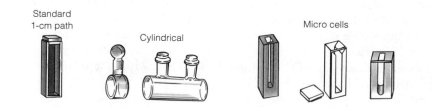

FIGURE 17-6 Common cuvets for ultraviolet and visible measurements. [A. H. Thomas Co., Philadelphia, PA.]

be used only for measurements at visible wavelengths. Infrared cells are typically made of sodium chloride or potassium bromide crystals.

The instrument represented in Figure 17-4 is called a *single-beam spectrophotometer* because it has only one beam of light. We do not measure the incident radiant power, P_0, directly. Rather, the radiant power of light passing through a reference cuvet containing pure solvent is *defined* as P_0 in Equation 17-4. This cuvet is then removed and repaced by an identical one containing sample. The radiant power of light striking the detector is then taken as P in Equation 17-4, thus allowing T or A to be determined. The reference cuvet, containing pure solvent, compensates for reflection, scattering, or absorption of light by the cuvet and solvent. The radiant power of light striking the detector would not be the same if the reference cuvet were removed from the beam. A double-beam spectrophotometer housing both a sample cuvet and a reference cuvet is described in Section 18-1.

Radiant power passing through
   cuvet filled with solvent $\equiv P_0$

Radiant power passing through
   cuvet filled with sample $\equiv P$

Transmittance $= P/P_0$

Good Operating Techniques

Cuvets should be handled with a tissue to avoid putting fingerprints on the cuvet faces and must be kept scrupulously clean. Fingerprints or contamination from previous samples can scatter or absorb light. Wash the cuvet and rinse it with distilled water as soon as you are finished using it. Let it dry upside down so that water drains out and no water marks are left on the walls. All vessels should be covered to protect them from dust, which scatters light and therefore makes it look like the absorbance of the sample has increased. Another reason to cover a cuvet is to prevent evaporation of the sample.

Any slight mismatch between sample and reference cuvets leads to systematic errors in spectrophotometry. Always use a *matched set* of cuvets, manufactured as carefully as possible to have identical pathlength. Place each cuvet in the spectrophotometer as reproducibly as possible. One side of the cuvet should be marked in a manner allowing you to be sure that the cuvet is always oriented the same way. Slight misplacement of the cuvet in its holder, or turning a flat cuvet around by 180°, or rotation of a circular cuvet, all lead to random errors in absorbance.

Spectrophotometry is most accurate at intermediate levels of absorbance ($A \approx 0.4$–0.9), as shown in Figure 17-7. If too little light gets through the sample (high absorbance), the intensity is hard to measure. If too much light gets through (low absorbance), it is hard to distinguish the transmittance of the sample from that of the reference. It is therefore desirable to adjust the concentration of the sample so that its absorbance falls in the range 0.4–0.9.

For spectrophotometric analysis, measurements are made at a wavelength (λ_{max}) corresponding to a peak in the absorbance spectrum. This wavelength gives the greatest sensitivity—maximum response for a given concentration of analyte. Errors due to wavelength drift and the finite bandwidth of wavelengths selected by the monochromator are minimized because the spectrum varies least with wavelength at the absorbance maximum.

In measuring a spectrum, it is routine to first record a baseline with pure solvent or a reagent blank in *both* cuvets. In principle, the baseline absorbance should be 0. However, small mismatches between the two cuvets and instrumental imperfections lead to small positive or negative baseline absorbance. The absorbance of the sample is then recorded and the absorbance of the baseline is subtracted from that of the sample to obtain true absorbance.

Do not touch the clear faces of a cuvet with your fingers. Keep the cuvet scrupulously clean.

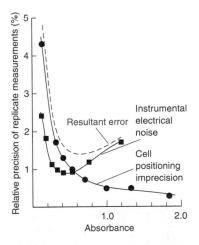

FIGURE 17-7 Errors in spectrophotometry due to instrumental electrical noise and cell positioning imprecision. Best precision is generally observed at intermediate absorbance ($A \approx 0.4$–0.9). [Data from L. D. Rothman, S. R. Crouch, and J. D. Ingle, Jr., *Anal. Chem.* **1975**, *47*, 1226.]

Ask Yourself

17-C. (a) What precautions should you take in handling a cuvet and placing it in the spectrophotometer?

    (b) Why is it most accurate to measure absorbances in the range $A = 0.4$–0.9?

17-4   Using Beer's Law

Spectrophotometric analysis with visible radiation is called *colorimetric* analysis.

For a compound to be analyzed by spectrophotometry, it must absorb electromagnetic radiation, and this absorption should be distinguishable from that of other species in the sample. Biochemists assay proteins in the ultraviolet region at 280 nm because the aromatic amino acids tyrosine, phenylalanine, and tryptophan (Table 10-1) have maximum absorbance near 280 nm. Other common solutes such as salts, buffers, and carbohydrates have little or no absorbance at this wavelength. In this section, we use Beer's law for a simple analysis and then discuss the measurement of nitrite in an aquarium.

EXAMPLE   Measuring Benzene in Hexane

Benzene
C_6H_6

(a) The solvent hexane has negligible ultraviolet absorbance above a wavelength of 200 nm. A solution prepared by dissolving 25.8 mg of benzene (C_6H_6, FM 78.114) in hexane and diluting to 250.0 mL has an absorption peak at 256 nm, with an absorbance of 0.266 in a 1.000-cm cell. Find the molar absorptivity of benzene at this wavelength.

Beer's law:

$$A = \varepsilon bc$$

A = absorbance (dimensionless)

ε = molar absorptivity ($M^{-1}\ cm^{-1}$)

b = pathlength (cm)

c = concentration (M)

ε has funny units so that the product εbc will be dimensionless.

SOLUTION The concentration of benzene is

$$[C_6H_6] = \frac{(0.025\ 8\ \text{g})/(78.114\ \text{g/mol})}{0.250\ 0\ \text{L}} = 1.32_1 \times 10^{-3}\ M$$

We find the molar absorptivity from Beer's law:

$$\text{molar absorptivity} = \varepsilon = \frac{A}{bc} = \frac{0.266}{(1.000\ \text{cm})(1.32_1 \times 10^{-3}\ M)} = 201._3\ M^{-1}\ cm^{-1}$$

(b) A sample of hexane contaminated with benzene has an absorbance of 0.070 at 256 nm in a cell with a 5.000-cm pathlength. Find the concentration of benzene.

SOLUTION Use the molar absorptivity from **(a)** in Beer's law:

$$[C_6H_6] = \frac{A}{\varepsilon b} = \frac{0.070}{(201._3\ M^{-1}\ cm^{-1})(5.000\ \text{cm})} = 6.95_3 \times 10^{-5}\ M$$

Using a Standard Curve to Measure Nitrite

Box 5-1 showed that nitrogen compounds derived from animals and plants are broken down to ammonia by heterotrophic bacteria. Ammonia is oxidized first to

Linda A. Hughes

nitrite (NO_2^-) and then to nitrate (NO_3^-) by nitrifying bacteria. In Section 5-3, we saw how a permanganate titration was used to standardize a nitrite stock solution. The nitrite solution is used here to prepare standards for a spectrophotometric analysis of nitrite in aquarium water.

The aquarium nitrite analysis is based on a reaction whose colored product has an absorbance maximum at 543 nm (Figure 17-8):

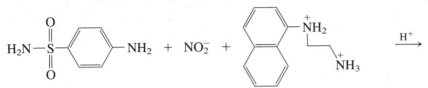

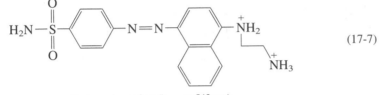

Sulfanilamide   Nitrite N-(1-Naphthyl)ethylenediammonium ion

Red-purple product (λ_{max} = 543 nm)   (17-7)

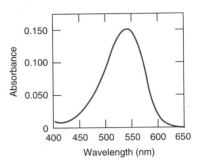

FIGURE 17-8 Spectrum of the red–purple product of Reaction 17-7, beginning with a standard nitrite solution containing 0.915 ppm nitrogen. [Kenneth Hughes, Georgia Institute of Technology.]

For quantitative analysis, we can prepare a **standard curve** (also called a *calibration curve* in Section 4-5) in which absorbance at 543 nm is plotted against nitrite concentration in a series of standards (Figure 17-9).

The general procedure for measuring nitrite is to add color-forming reagent to an unknown or standard, wait 10 min for the reaction to be completed, and measure the absorbance. A *reagent blank* is prepared with nitrite-free, artificial seawater in place of unknown or standards. *The absorbance of the blank is subtracted from the absorbance of all other samples prior to any calculations.* The purpose of the blank is to subtract absorbance at 543 nm arising from starting materials or impurities. Here are the details:

The saltwater aquarium is filled with artificial seawater made by adding water to a mixture of salts.

Reagents:

1. *Color-forming reagent* is prepared by mixing 1.0 g of sulfanilamide, 0.10 g of *N*-(1-naphthyl)ethylenediamine dihydrochloride, and 10 mL of 85 wt % phosphoric acid and diluting to 100 mL. Store the solution in a dark bottle in the refrigerator to prevent thermal and photochemical degradation.

2. *Standard nitrite* (~0.02 M) is prepared by dissolving $NaNO_2$ in water and standardizing (measuring the concentration) by the back titration described in Section 5-3. Dilute the concentrated standard with artificial seawater (containing no nitrite) to prepare standards containing 0.5–3 ppm nitrite nitrogen.

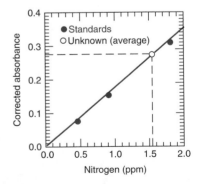

FIGURE 17-9 Calibration curve for nitrite analysis, using corrected absorbance values from Table 17-2.

General procedure:

For each analysis, dilute 10.00 mL of standard or unknown up to 100.0 mL with water. Then place 50.00 mL of the diluted solution in a flask and add 2.00 mL of color-forming reagent. After 10 min, measure the absorbance in a cuvet with a 1.000-cm pathlength.

(a) Construct a *standard curve* from known nitrite solutions. Prepare a *reagent blank* by carrying artificial seawater through the same steps as a standard.

(b) Analyze duplicate samples of *unknown* aquarium water that has been filtered prior to dilution to remove suspended solids. Several trial dilutions may be required before the aquarium water is dilute enough to have a nitrite concentration that falls within the calibration range.

The unknown should always be adjusted to fall within the calibration range, because you have not verified a linear response outside the calibration range.

Table 17-2 and Figure 17-9 show typical results. The equation of the calibration line in Figure 17-9, determined by the method of least squares, is

$$\text{absorbance} = 0.176\ 9\ [\text{ppm}] + 0.001\ 5 \tag{17-8}$$

where [ppm] represents micrograms of nitrite nitrogen per milliliter. In principle, the intercept should be 0, but we will use the observed intercept (0.001 5) in our calculations. By plugging the average absorbance of unknown into Equation 17-8, we can solve for the concentration of nitrite (ppm) in the unknown.

EXAMPLE **Using the Standard Curve**

From the data in Table 17-2, find the molarity of nitrite in the aquarium.

SOLUTION The average corrected absorbance of unknowns in Table 17-2 is 0.276. Substituting this value into Equation 17-8 gives ppm of nitrite nitrogen in the aquarium:

$$0.276 = 0.176\ 9\ [\text{ppm}] + 0.001\ 5$$

$$[\text{ppm}] = \frac{0.276 - 0.001\ 5}{0.176\ 9} = 1.55\ \text{ppm} = 1.55\ \frac{\mu g\ N}{mL}$$

To find the molarity of nitrite nitrogen, we first find the mass of nitrogen in a liter, which is

$$1.55 \times 10^{-6}\ \frac{g\ N}{mL} \times 1\ 000\ \frac{mL}{L} = 1.55 \times 10^{-3}\ \frac{g\ N}{L}$$

Then we convert mass of nitrogen into moles of nitrogen:

$$[\text{nitrite nitrogen}] = \frac{1.55 \times 10^{-3}\ g\ N/L}{14.007\ g\ N/mol} = 1.11 \times 10^{-4}\ M$$

Because one mole of nitrite (NO_2^-) contains one mole of nitrogen, the concentration of nitrite is also 1.11×10^{-4} M.

TABLE 17-2 **Aquarium nitrite analysis**

Sample	Absorbance at 543 nm in 1.000-cm cuvet	Corrected absorbance (blank subtracted)
Blank	0.003	—
Standards		
0.457 5 ppm	0.085	0.082
0.915 0 ppm	0.167	0.164
1.830 ppm	0.328	0.325
Unknown	0.281	0.278
Unknown	0.277	0.274

EXAMPLE   Preparing Nitrite Standards

How would you prepare a nitrite standard containing approximately 2 ppm nitrite nitrogen from a concentrated standard containing 0.018 74 M $NaNO_2$?

SOLUTION First let's find out how many ppm of nitrogen are in 0.018 74 M $NaNO_2$. Because one mole of nitrite contains one mole of nitrogen, the concentration of nitrogen in the concentrated standard is 0.018 74 M. The mass of nitrogen in 1 mL is

$$\frac{\text{g of N}}{\text{mL}} = \left(0.018\ 74\ \frac{\text{mol}}{\text{L}}\right)\left(14.007\ \frac{\text{g}}{\text{mol}}\right)\left(0.001\ \frac{\text{L}}{\text{mL}}\right) = 2.625 \times 10^{-4}\ \frac{\text{g}}{\text{mL}}$$

Assuming that 1.00 mL of solution has a mass of 1.00 g, we use the definition of parts per million to convert the mass of nitrogen to ppm:

$$\text{ppm} = \frac{\text{g N}}{\text{g solution}} \times 10^6 = \frac{2.625 \times 10^{-4}\ \text{g N}}{1.00\ \text{g solution}} \times 10^6 = 262.5\ \text{ppm}$$

Dilution formula 1-5:

$$\boxed{M_{\text{conc}} \cdot V_{\text{conc}} = M_{\text{dil}} \cdot V_{\text{dil}}}$$

Use any units you like for M and V as long as you use the same units on both sides of the equation. M could be ppm and V could be mL.

To prepare a standard containing ~2 ppm N, you could dilute the concentrated standard by a factor of 100 to give a 2.625 ppm N. This dilution could be done by pipeting 10.00 mL of concentrated standard into a 1-L volumetric flask and diluting to the mark.

Ask Yourself

17-D. You have been sent to India to investigate the occurrence of goiter disease attributed to iodine deficiency. As part of your investigation, you must make field measurements of traces of iodide (I^-) in groundwater. The procedure is to oxidize I^- to I_2 and convert the I_2 into an intensely colored complex with the dye brilliant green in the organic solvent toluene.

(a) A 3.15×10^{-6} M solution of the colored complex exhibited an absorbance of 0.267 at 635 nm in a 1.000-cm cuvet. A blank solution made from distilled water in place of groundwater had an absorbance of 0.019. Find the molar absorptivity of the colored complex.

(b) The absorbance of an unknown solution prepared from groundwater was 0.175. Subtract the blank absorbance from the unknown absorbance and use Beer's law to find the concentration of the unknown.

Key Equations

Frequency-wavelength relation	$\nu\lambda = c$
	ν = frequency λ = wavelength c = speed of light
Wavenumber	$\tilde{\nu} = 1/\lambda$
Photon energy	$E = h\nu = hc/\lambda = hc\tilde{\nu}$
	h = Planck's constant

Transmittance	$T = P/P_0$
	P_0 = radiant intensity of light incident on sample
	P = radiant intensity of light emerging from the sample
Absorbance	$A = - \log T$
Beer's law	$A = \varepsilon bc$
	ε = molar absorptivity of the absorbing species
	b = pathlength
	c = concentration of absorbing species

Important Terms

absorbance	frequency	spectrophotometer
absorption spectrum	ground state	spectrophotometry
Beer's law	hertz	standard curve
cuvet	molar absorptivity	transmittance
electromagnetic spectrum	monochromatic light	wavelength
excited state	photon	wavenumber

Problems

17-1. (a) When you double the frequency of electromagnetic radiation, you _____ the energy.

(b) When you double the wavelength, you _____ the energy.

(c) When you double the wavenumber, you _____ the energy.

17-2. How much energy (J) is carried by one photon of **(a)** red light with $\lambda = 650$ nm? **(b)** violet light with $\lambda = 400$ nm? After finding the energy of one photon of each wavelength, express the energy of a mole of each type of photon in kJ/mol.

17-3. What color would you expect for light transmitted through a solution with an absorption maximum at **(a)** 450; **(b)** 550; **(c)** 650 nm?

17-4. State the difference between transmittance, absorbance, and molar absorptivity. Which one is proportional to concentration?

17-5. An absorption spectrum is a graph of _____ or _____ versus _____.

17-6. Why does a compound which has a visible absorption maximum at 480 nm (blue-green) appear to be red?

17-7. What color would you expect to observe for a solution with a visible absorbance maximum at 562 nm?

17-8. Calculate the frequency (Hz), wavenumber (cm^{-1}), and energy (J/photon and kJ per mole of photons) of **(a)**

ultraviolet light with a wavelength of 250 nm and **(b)** infrared light with a wavelength of 2.50 μm.

17-9. Convert transmittance (T) to absorbance (A):

T:	0.99	0.90	0.50	0.10	0.010	0.001 0	0.000 10
A:			1.0				

17-10. The absorbance of a 2.31×10^{-5} M solution is 0.822 at a wavelength of 266 nm in a 1.00-cm cell. Calculate the molar absorptivity at 266 nm.

17-11. The iron-transport protein in your blood is called transferrin. When its two iron-binding sites do not contain metal ions, the protein is called apotransferrin.

(a) Apotransferrin has a molar absorptivity of 8.83×10^4 M^{-1} cm^{-1} at 280 nm. Find the concentration of apotransferrin in water if the absorbance is 0.244 in a 0.100 cm cell.

(b) The formula mass of apotransferrin is 81 000. Express the concentration from **(a)** in g/L.

17-12. A 15.0-mg sample of a compound with a formula mass of 384.63 was dissolved in a 5-mL volumetric flask. A 1.00-mL aliquot was withdrawn, placed in a 10-mL volumetric flask, and diluted to the mark.

(a) Find the concentration of sample in the 5-mL flask.

(b) Find the concentration in the 10-mL flask.

(c) The 10-mL sample was placed in a 0.500-cm cuvet and gave an absorbance of 0.634 at 495 nm. Find the molar absorptivity at 495 nm.

17-13. (a) In Figure 17-5, measure the peak absorbance of sunscreen near 215 nm.

(b) What fraction of ultraviolet radiation is transmitted through the sunscreen near 215 nm?

17-14. (a) What value of absorbance corresponds to 45.0% T?

(b) When the concentration of a solution is doubled, the *absorbance* is doubled. If a 0.010 0 M solution exhibits 45.0% T at some wavelength, what will be the percent transmittance for a 0.020 0 M solution of the same substance?

17-15. A 0.267-g quantity of a compound with a formula mass of 337.69 was dissolved in 100.0 mL of ethanol. Then 2.000 mL were withdrawn and diluted to 100.0 mL. The spectrum of this solution exhibited a maximum absorbance of 0.728 at 438 nm in a 2.000-cm cell. Find the molar absorptivity of the compound.

17-16. Transmittance of vapor from the solid compound, pyrazine, was measured at a wavelength of 266 nm in a 3.00-cm cell at 298 K.

Pressure (μbar)	Transmittance (%)
4.3	83.8
11.4	61.6
20.0	39.6
30.3	24.4
60.7	5.76
99.7	0.857
134.5	0.147

Data from M. A. Muyskens and E. T. Sevy, *J. Chem. Ed.* **1997**, *74*, 1138.

(a) Use the ideal gas law (Problem 15-16) to convert pressure to concentration in mol/L. Convert transmittance to absorbance.

(b) Prepare a graph of absorbance versus concentration to see whether the data conform to Beer's law. Find the molar absorptivity from the slope of the graph.

17-17. (a) A 3.96×10^{-4} M solution of compound A exhibited an absorbance of 0.624 at 238 nm in a 1.000-cm cuvet. A blank solution containing only solvent had an absorbance of 0.029 at the same wavelength. Find the molar absorptivity of compound A.

(b) The absorbance of an unknown solution of compound A in the same solvent and cuvet was 0.375 at 238 nm. Subtract the blank absorbance from the unknown absorbance and use Beer's law to find the concentration of the unknown.

(c) A concentrated solution of a compound A in the same solvent was diluted from an initial volume of 2.00 mL to a final volume of 25.00 mL and then had an absorbance of 0.733. What is the concentration of A in the 25.00-mL solution?

(d) Considering the dilution from 2.00 mL up to 25.00 mL, what was the concentration of A in the 2.00-mL solution in (c)?

17-18. A compound with a formula mass of 292.16 was dissolved in a 5-mL volumetric flask. A 1.00-mL aliquot was withdrawn, placed in a 10-mL volumetric flask, and diluted to the mark. The absorbance measured at 340 nm was 0.427 in a 1.000-cm cuvet. The molar absorptivity for this compound at 340 nm is $\varepsilon_{340} = 6\ 130\ \text{M}^{-1}\ \text{cm}^{-1}$.

(a) Calculate the concentration of compound in the cuvet.

(b) What was the concentration of compound in the 5-mL flask?

(c) How many milligrams of compound were used to make the 5-mL solution?

17-19. When I was a boy, Uncle Wilbur let me watch as he analyzed the iron content of runoff from his banana ranch. A 25.0-mL sample was acidified with nitric acid and treated with excess KSCN to form a red complex. (KSCN itself is colorless.) The solution was then diluted to 100.0 mL and put in a variable-pathlength cell. For comparison, a 10.0-mL reference sample of 6.80×10^{-4} M Fe^{3+} was treated with HNO_3 and KSCN and diluted to 50.0 mL. The reference was placed in a cell with a 1.00-cm light path. The runoff sample exhibited the same absorbance as the reference when the pathlength of the runoff cell was 2.48 cm. What was the concentration of iron in Uncle Wilbur's runoff?

17-20. A nitrite analysis was conducted according to the procedure in Section 17-4, giving the data in the table. Fill in the corrected absorbance, which is the measured absorbance minus the average blank absorbance (0.023). Construct a calibration line to find the **(a)** ppm of nitrite nitrogen and **(b)** molar concentration of nitrite in the aquarium. Use the average blank and the average unknown absorbances.

Sample	Absorbance	Corrected absorbance
Blank	0.022	—
Blank	0.024	—
Standards:		
0.538 ppm	0.121	0.098
1.076 ppm	0.219	
2.152 ppm	0.413	
3.228 ppm	0.600	
4.034 ppm	0.755	
Unknown	0.333	
Unknown	0.339	
Unknown	0.338	

17-21. Use the method of least squares in Chapter 4 to find the *uncertainty* in nitrite nitrogen concentration (ppm) in the previous problem.

17-22. Starting with 0.015 83 M $NaNO_2$ solution, explain how you would prepare standards containing *approximately* 0.5, 1, 2, and 3 ppm nitrogen (1 ppm = 1 μg/mL). You may use any volumetric flasks and transfer pipets in Tables 2-2 and 2-3. What would be the exact concentrations of the standards prepared by your method?

17-23. Ammonia (NH_3) is determined spectrophotometrically by reaction with phenol in the presence of hypochlorite (OCl^-):

$$\text{phenol} + \text{ammonia} \xrightarrow{\ OCl^-\ } \text{blue product}$$

phenol   ammonia   blue product
Colorless   Colorless $\lambda_{max} = 625$ nm

1. A 4.37-mg sample of protein was chemically digested to convert its nitrogen to ammonia and then diluted to 100.0 mL.

2. Then 10.0 mL of the solution were placed in a 50-mL volumetric flask and treated with 5 mL of phenol solution plus 2 mL of sodium hypochlorite solution. The sample was diluted to 50.0 mL, and the absorbance at 625 nm was measured in a 1.00-cm cuvet after 30 min.

3. A standard solution was prepared from 0.010 0 g of NH_4Cl (FM 53.49) dissolved in 1.00 L of water. A 10.0-mL aliquot of this standard was placed in a 50-mL volumetric flask and analyzed in the same manner as the unknown.

4. A reagent blank was prepared by using distilled water in place of unknown.

Sample	Absorbance at 625 nm
Blank	0.140
Standard	0.308
Unknown	0.592

(a) From step 3, calculate the molar absorptivity of the blue product.

(b) Using the molar absorptivity, find the concentration of ammonia in step 2.

(c) From your answer to **(b)**, find the concentration of ammonia in the 100-mL solution in step 1.

(d) Find the weight percent of nitrogen in the protein.

17-24. Cu^+ reacts with neocuproine to form (neocuproine)$_2$-Cu^+, with an absorption maximum at 454 nm. Neocuproine is particularly useful because it reacts with few other metals. The copper complex is soluble in isoamyl alcohol, an organic solvent that does not dissolve appreciably in water. When isoamyl alcohol is added to water, a two-layered mixture results, with the denser water layer at the bottom. If (neocuproine)$_2Cu^+$ is present, virtually all of it goes into the organic phase. For the purpose of this problem, assume that no isoamyl alcohol dissolves in water and that all the colored complex will be in the organic phase. Suppose that the procedure in the diagram below is carried out.

1. A rock containing copper is pulverized, and all metals are extracted from it with strong acid. The acidic solution is neutralized with base and made up to 250.0 mL in flask A.

2. Next 10.00 mL of the solution are transferred to flask B and treated with 10.00 mL of a reducing agent to reduce Cu^{2+} to Cu^+. Then 10.00 mL of buffer are added to bring the pH to a value suitable for complex formation with neocuproine.

3. After that, 15.00 mL of this solution are withdrawn and placed in flask C. To the flask are added 10.00 mL of an aqueous solution containing neocuproine and 20.00 mL of isoamyl alcohol. After shaking well and allowing the phases to separate, all (neocuproine)$_2Cu^+$ is in the organic phase.

4. A few milliliters of the upper layer are withdrawn, and the absorbance at 454 nm is measured in a 1.00-cm cell. A blank carried through the same procedure gave an absorbance of 0.056.

(a) Suppose that the rock contained 1.00 mg of Cu. What will the concentration of copper (moles per liter) in the isoamyl alcohol phase be?

(b) If the molar absorptivity of (neocuproine)$_2Cu^+$ is 7.90×10^3 M^{-1} cm^{-1}, what will the observed absorbance be? Remember that a blank carried through the same procedure gave an absorbance of 0.056.

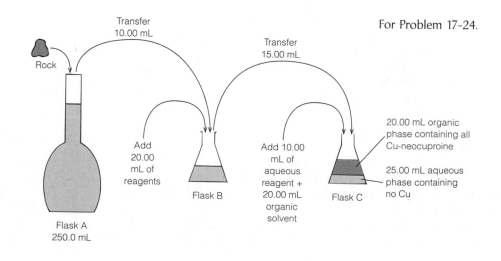

Transfer
10.00 mL

Transfer
15.00 mL

For Problem 17-24.

Rock

Add
20.00
mL of
reagents

Add 10.00
mL of
aqueous
reagent +
20.00 mL
organic
solvent

Flask B

Flask C

20.00 mL organic
phase containing all
Cu-neocuproine

25.00 mL aqueous
phase containing
no Cu

Flask A
250.0 mL

(c) A rock is analyzed and found to give a final absorbance of 0.874 (uncorrected for the blank). How many milligrams of copper are in the rock?

17-25. Spectrophotometric analysis of phosphate can be performed by the following procedure.

Standard solutions:
A. KH_2PO_4 (potassium dihydrogen phosphate, FM 136.09): 81.37 mg dissolved in 500.0 mL H_2O
B. $Na_2MoO_4 \cdot 2H_2O$ (sodium molybdate): 1.25 g in 50 mL of 5 M H_2SO_4
C. $H_3NNH_3^{2+}SO_4^{2-}$ (hydrazine sulfate): 0.15 g in 100 mL H_2O

Procedure:
Place the sample (either an unknown or the standard phosphate solution, A) in a 5-mL volumetric flask, and add 0.500 mL of B and 0.200 mL of C. Dilute to almost 5 mL with water, and heat at 100°C for 10 min to form a blue product ($H_3PO_4(MoO_3)_{12}$, 12-molybdophosphoric acid). Cool the flask to room temperature, dilute to the mark with water, mix well, and measure the absorbance at 830 nm in a 1.00-cm cell.

(a) When 0.140 mL of solution A was analyzed, an absorbance of 0.829 was recorded. A blank carried through the same procedure gave an absorbance of 0.017. Find the molar absorptivity of blue product.

(b) A solution of the phosphate-containing iron-storage protein ferritin was analyzed by this procedure. The unknown contained 1.35 mg of ferritin, which was digested in a total volume of 1.00 mL to release phosphate from the protein. Then 0.300 mL of this solution was analyzed by the procedure above and found to give an absorbance of 0.836. A blank carried through this procedure gave an absorbance of 0.038. Find the weight percent of phosphorus in the ferritin.

17-26. Starting with an ammonia solution known to contain 28.6 wt % NH_3, explain how to prepare ammonia standards containing exactly 1.00, 2.00, 4.00, and 8.00 ppm nitrogen (1 ppm = 1 $\mu g/mL$) for a spectrophotometric calibration curve. You will need to use a known *mass* of concentrated reagent, and you may use any flasks and pipets from Tables 2-2 and 2-3.

How Would You Do It?

17-27. *Sunscreens.* Ultraviolet radiation in the wavelength range 280–320 nm, called UV-B, is principally responsible for sunburns. UV-A radiation (320–400 nm) is less dangerous. UV-C radiation (100–280) is largely absorbed by the atmosphere and does not reach us. The "sun protection factor" on sunscreen creams is defined as[3]

$$SPF = \frac{\text{time to burn with sunscreen at 2 mg/cm}^2}{\text{time to burn without sunscreen}}$$

where "time to burn" is the length of time needed to become sunburned, which is a highly variable biological response to sun. If sunscreen has an SPF rating of 2, we expect that a person could be exposed to the sun twice as long as a person without sunscreen before getting burned. That is, we might expect that a sunscreen with SPF = 2 would have 50% transmittance if applied in a dose of 2 mg/cm^2.

The table shows the absorbance of various Banana Boat® sunscreens. A 2.00-g quantity of each sunscreen was suspended in 0.200 L of the solvent 2-propanol, mixed and heated to dissolve the active ingredient of the sunscreen, and centrifuged to remove undissolved solids. The supernatant solution containing soluble components was diluted by a factor of 100 with 2-propanol and the absorbance measured in a 1.00-cm cell.

SPF	Absorbance at 310 nm
2	0.15
8	0.65
15	0.85
30	1.14
50+	1.27

Data from J. R. Abney and B. A. Scalettar, *J. Chem. Ed.* **1998,** *75,* 757.

(a) From the definition of SPF, suggest a mathematical relationship between SPF and transmittance and between SPF and absorbance.

(b) Prepare a graph to test the proposed relationship and comment on the results.

(c) From the experimental data, find the transmittance of sunscreen with SPF = 2 spread on skin at a dose of 2 mg/cm^2.

Notes and References

1. R. W. Ricci, M. A. Ditzler, and L. P. Nestor, *J. Chem. Ed.* **1994,** *71,* 983.

2. D. H. Alman and F. W. Billmeyer, Jr., *J. Chem. Ed.* **1976,** *53,* 166.

3. C. Walters, A. Keeney, C. T. Wigal, C. R. Johnston, and R. D. Cornelius, *J. Chem. Ed.* **1997,** *74,* 99.

Remote Sensing of Airborne Bacteria

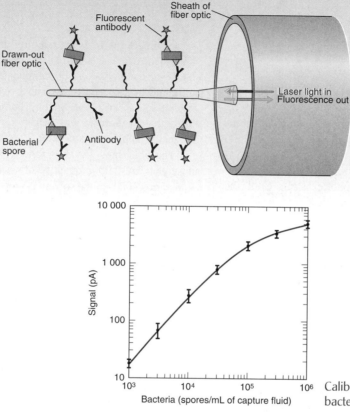

Optical fiber coated with antibodies to detect the spores of a specific bacterium. [Adapted from F. S. Ligler, G. P. Anderson, P. T. Davidson, R. J. Foch, J. T. Ives, K. D. King, G. Page, D. A. Stenger, and J. P. Whelan, *Environ. Sci. Technol.* **1998**, *32*, 2461.]

Calibration curve for airborne bacterial spores.

A *biosensor* combines electronic or optical components with biological molecules such as antibodies or enzymes for the specific sensing of one analyte. A battery-powered device based on an optical fiber coated with antibodies was developed for detection of airborne bacterial spores to warn of a biological warfare attack over a battlefield. The antibodies recognize specific bacterial spores. The unit is flown in a remote-controlled airplane with a wingspan of 3.6 m.

A tube on the airplane takes >100 L of air per minute into a plastic chamber in which the air forms a cyclone that contacts a film of water on the walls. Bacterial spores from the air are transferred into the liquid, which is pumped over the optical fiber. If the specific target spores are present, they bind to the antibodies on the optical fiber. After washing away unbound substances, a second antibody with a fluorescent label is then pumped over the fiber. If spores are present, the second antibody binds to the spores on the surface of the fiber.

Laser light in the fiber excites fluorescent antibodies attached to the spore stuck on the surface of the fiber. Fluorescence captured by the fiber travels back out to an electronic detector, whose output is radioed to a computer on the ground.

Chapter *18*

SPECTROPHOTOMETRY: INSTRUMENTS AND APPLICATIONS

*I*n this chapter, we describe the components of a spectrophotometer, some of the physical processes that occur when light is absorbed by molecules, and a few important applications of spectrophotometry in analytical chemistry. New analytical instruments for medicine and biology such as the one on the facing page are being developed by combining sensitive optical methods with biologically specific recognition elements.

18-1   The Spectrophotometer

The minimum requirements for a *single-beam spectrophotometer* were shown in Figure 17-4. *Polychromatic light* from a lamp passes through a *monochromator* that separates different wavelengths from one another and selects one narrow band of wavelengths to pass through the sample. Transmittance is P/P_0, where P_0 is the radiant power reaching the detector when the sample cell contains a blank solution with no analyte and P is the power reaching the detector when analyte is present in the sample cell. A single-beam spectrophotometer is somewhat inconvenient because two different samples must be placed alternately in the beam. Error occurs if the source intensity or detector response drifts between the two measurements.

The *double-beam spectrophotometer* in Figure 18-1 features a rotating mirror (the *beam chopper*), which alternately directs light through the sample or reference

Polychromatic light contains many wavelengths (literally, "many colors").

A refresher:

$$\text{transmittance} = T = \frac{P}{P_0}$$

$$\text{absorbance} = -\log T$$

385

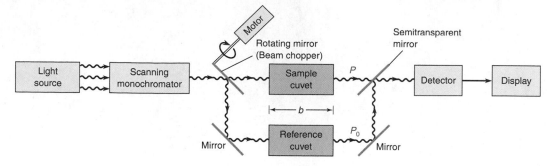

FIGURE 18-1 Schematic diagram of a double-beam scanning spectrophotometer. The incident beam is passed alternately through the sample and reference cuvets by the rotating beam chopper.

cell several times per second. Radiant power emerging from the reference cell, which is filled with a blank solution or pure solvent, is P_0. Light emerging from the sample cell has power P. By measuring P and P_0 many times per second, the instrument compensates for drift in the source intensity or detector response. Figure 18-2 shows a double-beam instrument and the layout of its components. Let's examine the principal components.

Light Source

The two lamps in the upper part of Figure 18-2b provide visible or ultraviolet radiation. An ordinary *tungsten lamp,* whose filament glows at a temperature near 3 000 K, produces radiation in the visible and near-infrared regions at wavelengths of 320 to 2 500 nm (Figure 18-3). For ultraviolet spectroscopy, we normally employ a *deuterium arc lamp* in which an electric discharge (a spark) dissociates D_2 molecules, which then emit ultraviolet radiation from 200 to 400 nm (Figure 18-3). Typically, a change is made between the deuterium and tungsten lamps when passing through 360 nm, so that the source giving the most radiation is always employed. Other sources of visible and ultraviolet radiation are electric discharge (spark) lamps filled with mercury vapor or xenon. Infrared radiation (5 000 to 200 cm^{-1}) is commonly obtained from a silicon carbide rod called a *globar,* electrically heated to near 1 500 K.

Lasers are extremely bright sources of light at just one or a few wavelengths. Lasers do not provide a wide range of wavelengths, but they are used in devices such as the bacterial spore monitor at the opening of this chapter that operate at a single wavelength.

Monochromator

A **monochromator** disperses light into its component wavelengths and selects a narrow band of wavelengths to pass through the sample. The monochromator in Figure 18-2b consists of entrance and exit slits, mirrors, and a *grating* to disperse the light. *Prisms* were used in older instruments to disperse light.

A **grating** has a series of closely ruled lines. When light is reflected from or transmitted through the grating, each line behaves as a separate source of radiation. Different wavelengths are reflected or transmitted at different angles from the

Ultraviolet radiation is harmful to the naked eye. Do not view an ultraviolet source without protection.

Question What wavelengths, in nm and μm, correspond to 5 000 and 200 cm^{-1}?

386

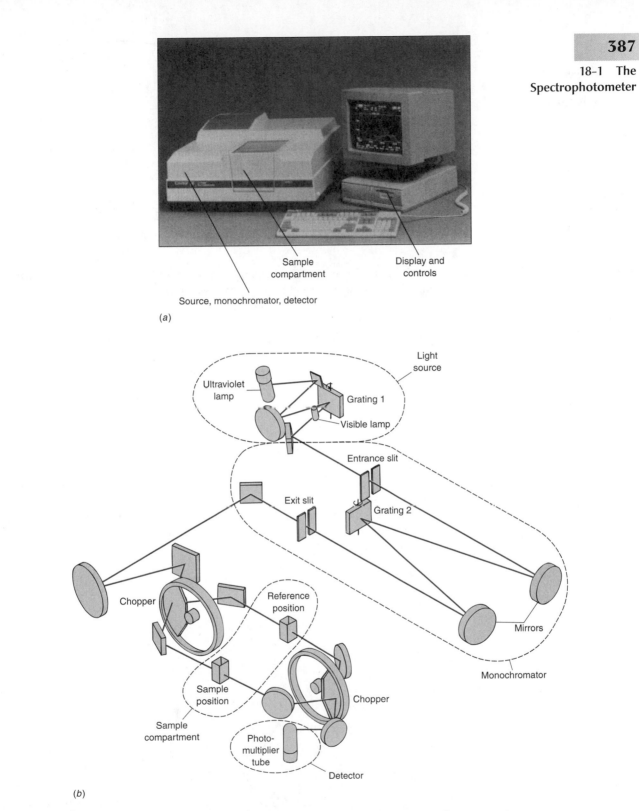

(a)

(b)

FIGURE 18-2 (a) Varian Cary 3E Ultraviolet-Visible Spectrophotometer. (b) Diagram of the optical train. [From Varian Australia Pty. Ltd., Victoria, Australia.]

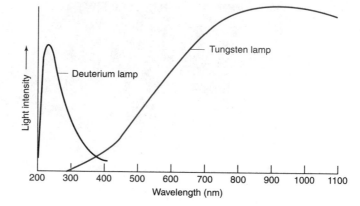

FIGURE 18-3 Intensities of a tungsten filament at 3 200 K and a deuterium arc lamp.

Grating: optical element with closely spaced lines

Diffraction: bending of light by a grating

Refraction: bending of light by a lens or prism

grating (Color Plate 14). The bending of light rays by a grating is called **diffraction.** (In contrast, the bending of light rays by a prism or lens, which is called *refraction,* is shown in Color Plate 15.)

In the grating monochromator in Figure 18-4, *polychromatic* radiation from the entrance slit is *collimated* (made into a beam of parallel rays) by a concave mirror. These rays fall on a reflection grating, whereupon different wavelengths are diffracted at different angles. The light strikes a second concave mirror, which focuses each wavelength at a different point. The grating directs a narrow band of wavelengths to the exit slit. Rotation of the grating allows different wavelengths to pass through the exit slit.

Diffraction at a reflection grating is shown in Figure 18-5. The closely spaced parallel grooves of the grating have a repeat distance *d.* When light is reflected from the grating, each groove behaves as a source of radiation. If adjacent light rays are in phase, they reinforce one another. If they are out of phase, they cancel one another (Figure 18-6).

Consider the two rays in Figure 18-5. Constructive interference occurs if the difference in pathlength $(a - b)$ traveled by the two rays is an integer multiple of the wavelength

$$n\lambda = a - b \tag{18-1}$$

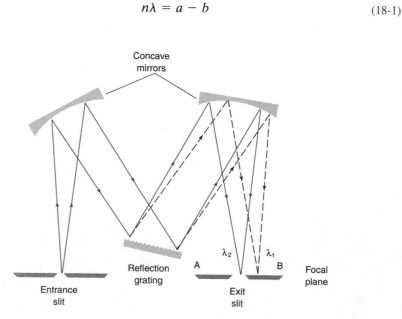

FIGURE 18-4   Czerny-Turner grating monochromator.

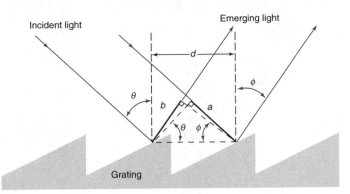

Incident light

Emerging light

Grating

FIGURE 18-5   Principle of a reflection grating.

where the diffraction order, n, is ± 1, ± 2, ± 3, ± 4, ... The maximum for which $n = \pm 1$ is called *first-order diffraction.* When $n = 2$, we have *second-order diffraction,* and so on.

From the geometry in Figure 18-5, we see that $a = d \sin \theta$ and $b = d \sin \phi$. Therefore, the condition for constructive interference is

Grating equation:
$$n\lambda = d(\sin \theta - \sin \phi) \qquad (18\text{-}2)$$

For each incident angle θ, there are reflection angles ϕ at which a given wavelength will produce maximum constructive interference.

In general, first-order diffraction of one wavelength overlaps higher-order diffraction of another wavelength. Therefore *filters* that reject many wavelengths are used to select a desired wavelength while rejecting other wavelengths at the same diffraction angle. High-quality spectrophotometers use several gratings with different line spacings optimized for different wavelengths. Spectrophotometers with two monochromators in series (a *double monochromator*) reduce unwanted radiation by orders of magnitude.

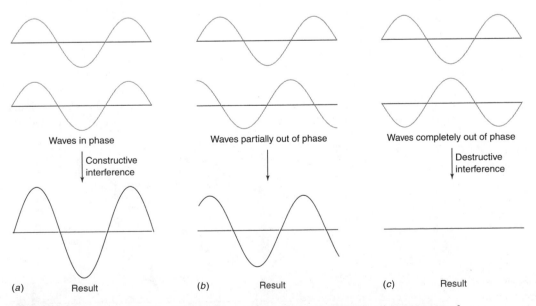

Waves in phase

Constructive interference

Waves partially out of phase

Waves completely out of phase

Destructive interference

(a)   Result

(b)   Result

(c)   Result

FIGURE 18-6   Interference of adjacent waves that are (a) 0°, (b) 90°, and (c) 180° out of phase.

18 Spectrophotometry: Instruments and Applications

FIGURE 18-7 Choosing wavelength and monochromator bandwidth. For quantitative analysis, use a wavelength of maximum absorbance so that small errors in the wavelength do not change the absorbance very much. Choose a monochromator bandwidth (controlled by the exit slit width in Figure 18-4) small enough so that it does not distort the band shape, but not so small that the spectrum is too noisy. Widening the slit width distorts the spectrum. In the lowest trace, a monochromator bandwidth that is 1/5 of the width of the sharp absorption bands (measured at half the peak height) avoids distortion. [Courtesy M. D. Seltzer, Michelson Laboratory, China Lake, CA.]

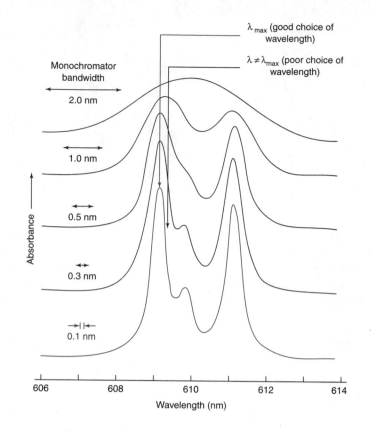

Trade-off between resolution and signal: The narrower the exit slit, the greater the ability to resolve closely spaced peaks and the noisier the spectrum.

Decreasing the exit slit width in Figure 18-4 decreases the selected bandwidth and decreases the energy reaching the detector. Thus, *resolution of closely spaced bands, which requires a narrow slit width, can be achieved at the expense of decreased signal-to-noise ratio.* For quantitative analysis, a monochromator bandwidth that is $\lesssim 1/5$ of the width of the absorption band is reasonable (Figure 18-7).

Detector

Dectector response is a function of wavelength of incident light.

A detector produces an electric signal when it is struck by photons. Figure 18-8 shows that detector response depends on the wavelength of the incident photons. In a single-beam spectrophotometer, the 100% transmittance control must be readjusted each time the wavelength is changed because the maximum possible detector signal depends on the wavelength. Subsequent readings are scaled to the 100% reading.

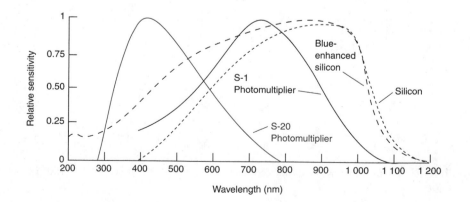

FIGURE 18-8 Detector response. Each curve is normalized to a maximum value of 1. [Courtesy Barr Associates, Inc., Westford, MA.]

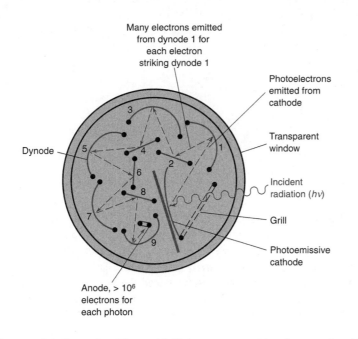

Many electrons emitted
from dynode 1 for
each electron
striking dynode 1

Photoelectrons
emitted from
cathode

Transparent
window

Dynode

Incident
radiation ($h\nu$)

Grill

Photoemissive
cathode

Anode, $> 10^6$
electrons for
each photon

FIGURE 18-9   *Photomultiplier tube with nine dynodes. Amplification occurs at each dynode, which is approximately 90 volts more positive than the previous dynode.*

A **photomultiplier tube** (Figure 18-9) is a very sensitive detector in which electrons emitted from a photosensitive cathode strike a second surface, called a *dynode*, which is positive with respect to the cathode. Electrons strike the dynode with more than their original kinetic energy. Each energetic electron knocks more than one electron from the dynode. These new electrons are accelerated toward a second dynode, which is more positive than the first dynode. Upon striking the second dynode, even more electrons are knocked off and accelerated toward a third dynode. This process is repeated so that more than 10^6 electrons are finally collected for each photon striking the cathode. Extremely low light intensities are translated into measurable electric signals.

Photodiode Array Spectrophotometer

Dispersive spectrophotometers described so far scan through a spectrum one wavelength at a time. A *diode array spectrophotometer* records the entire spectrum at once. The entire spectrum of a compound emerging from a chromatography column can be recorded in a fraction of a second by a photodiode array spectrophotometer. At the heart of rapid spectroscopy is a **photodiode array** such as the one in Figure 18-10, which contains 1 024 individual semiconductor detector elements (diodes) in a row.

A *dispersive* spectrophotometer spreads light from the source into its component wavelengths and then measures the absorption of one narrow band of wavelengths at a time.

A typical photodiode array responds to visible and ultraviolet radiation, with a response curve similar to that of the blue-enhanced silicon in Figure 18-8.

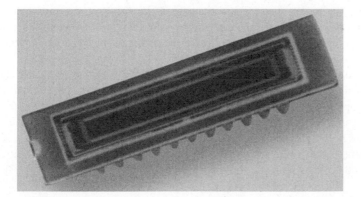

FIGURE 18-10   *Photodiode array with 1 024 elements, each 25 μm wide and 2.5 mm high. The entire chip is 5 cm long. [Courtesy Oriel Corporation, Stratford, CT.]*

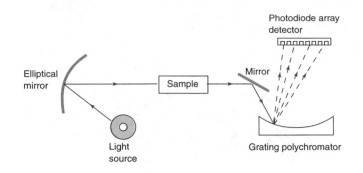

Attributes of diode array spectrophotometer:

- speed ($\sim$1 s per spectrum)
- excellent wavelength repeatability (because the grating does not rotate)
- simultaneous measurements at multiple wavelengths
- relatively insensitive to errors from stray light
- relatively poor resolution (1 to 3 nm)

In the diode array spectrophotometer in Figure 18-11, *white light* (with all wavelengths) passes through the sample. The beam then enters a **polychromator,** which disperses light into its component wavelengths and directs the light to the diode array. *A different wavelength band strikes each diode.* The resolution, which is typically 1 to 3 nm, depends on how closely spaced the diodes are and how much dispersion is produced by the polychromator. By comparison, high-quality dispersive spectrophotometers can resolve features that are 0.1 nm apart. The diode array spectrophotometer is faster than the dispersive spectrophotometer because the array measures all wavelengths at once, instead of one at a time. A diode array spectrophotometer is usually a single-beam instrument, subject to absorbance errors from drift in the source intensity and detector response between calibrations.

Ask Yourself

18-A. Explain what each component in the optical train in Figure 18-2(b) does, beginning with the lamp and ending with the detector.

18-2 Analysis of a Mixture

When there is more than one absorbing species in a solution, *the absorbance at a particular wavelength is the sum of absorbances from all species at that wavelength:*

Absorbance is additive.

Absorbance of a mixture:
$$A = \varepsilon_X b[X] + \varepsilon_Y b[Y] + \varepsilon_Z b[Z] + \cdots \qquad (18\text{-}3)$$

where ε is the molar absorptivity of each species (X, Y, Z, etc.) and b is the pathlength (Figure 18-1). If we measure the spectra of the pure components in a separate experiment, we can mathematically disassemble the spectrum of the mixture into those of its components.

Figure 18-12 shows spectra of titanium and vanadium complexes and an unknown mixture of the two. Let's denote the titanium complex by X and the vanadium complex by Y. To analyze a mixture, we usually choose wavelengths of maximum absorption for the individual components. Accuracy is improved if compound Y absorbs weakly at the maximum for compound X and if compound X absorbs weakly at the maximum for compound Y. In Figure 18-12, the two spectra overlap badly, so there will be some loss of accuracy.

We normally choose wavelengths at absorbance maxima. The absorbance of the mixture should not be too small or too great, so that uncertainty in absorbance is small (Figure 17-7).

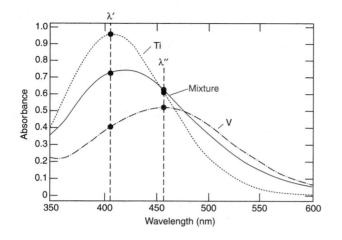

FIGURE 18-12 Visible spectra of hydrogen peroxide complexes of Ti(IV) (1.32 mM), V(V) (1.89 mM), and an unknown mixture containing both ions. All solutions contain 0.5 wt % H_2O_2 and ~0.01 M H_2SO_4 in a 1.00-cm-pathlength cell. [From M. Blanco, H. Iturriaga, S. Maspoch, and P. Tarín, *J. Chem. Ed.* **1989,** *66,* 178. Consult this article for a more accurate method to find the composition of the mixture by using more than two wavelengths.]

Choosing wavelengths λ' and λ'' as the absorbance maxima in Figure 18-12, we can write a Beer's law expression for each wavelength:

$$A' = \varepsilon'_X b[X] + \varepsilon'_Y b[Y] \qquad\qquad A'' = \varepsilon''_X b[X] + \varepsilon''_Y b[Y] \qquad (18\text{-}4)$$

Solving Equations 18-4 for [X] and [Y], we find

Analysis of a mixture when spectra are resolved:

$$[X] = \frac{1}{D}(A'\varepsilon''_Y - A''\varepsilon'_Y)$$

$$[Y] = \frac{1}{D}(A''\varepsilon'_X - A'\varepsilon''_X) \qquad (18\text{-}5)$$

where $D = b(\varepsilon'_X\varepsilon''_Y - \varepsilon'_Y\varepsilon''_X)$. To analyze the mixture, we measure absorbances at two wavelengths and must know ε at each wavelength for each compound.

EXAMPLE Analysis of a Mixture with Equations 18-5

The molar absorptivities of X (the Ti complex) and Y (the V complex) in Figure 18-12 were measured with pure samples of each:

	ε (M^{-1} cm^{-1})	
λ (nm)	X	Y
$\lambda' \equiv 406$	$\varepsilon'_X = 720$	$\varepsilon'_Y = 212$
$\lambda'' \equiv 457$	$\varepsilon''_X = 479$	$\varepsilon''_Y = 274$

A mixture of X and Y in a 1.00-cm cell had an absorbance of $A' = 0.722$ at 406 nm and $A'' = 0.641$ at 457 nm. Find the concentrations of X and Y in the mixture.

SOLUTION Using Equations 18-5 and setting $b = 1.00$ cm, we find

$$D = b(\varepsilon_X' \varepsilon_Y'' - \varepsilon_Y' \varepsilon_X'') = (1.00)[(720)(274) - (212)(479)] = 9.57_3 \times 10^4$$

$$[X] = \frac{1}{D}(A' \varepsilon_Y'' - A'' \varepsilon_Y') = \frac{(0.722)(274) - (0.641)(212)}{9.57_3 \times 10^4} = 6.47 \times 10^{-4} \text{ M}$$

$$[Y] = \frac{1}{D}(A'' \varepsilon_X' - A' \varepsilon_X'') = \frac{(0.641)(720) - (0.722)(479)}{9.57_3 \times 10^4} = 1.21 \times 10^{-3} \text{ M}$$

Isosbestic Points

If a species X is converted to species Y during the course of a chemical reaction, a spectrum of the mixture has a very obvious and characteristic behavior, shown in Figure 18-13. If the spectra of pure X and pure Y cross each other at some wavelength, then every spectrum recorded during this chemical reaction will cross at that same point, called an **isosbestic point.** *The observation of an isosbestic point during a chemical reaction is good evidence that only two principal species are present.*

Consider methyl red, which is shown in the margin. This acid-base indicator changes between red (HIn) and yellow (In⁻) near pH 5.1. Because the spectra of HIn and In⁻ (at the same concentration) happen to cross at 465 nm in Figure 18-13, all spectra cross at this point. (If the spectra of HIn and In⁻ crossed at several points, each would be an isosbestic point.)

To see why there is an isosbestic point, we write an equation for the absorbance of the solution at 465 nm:

$$A^{465} = \varepsilon_{\text{HIn}}^{465} b[\text{HIn}] + \varepsilon_{\text{In}^-}^{465} b[\text{In}^-] \tag{18-6}$$

But the spectra of pure HIn and pure In⁻ (at the same concentration) cross at 465 nm, so $\varepsilon_{\text{HIn}}^{465}$ must be equal to $\varepsilon_{\text{In}^-}^{465}$. Setting $\varepsilon_{\text{HIn}}^{465} = \varepsilon_{\text{In}^-}^{465} = \varepsilon^{465}$, we rewrite Equation 18-6 in the form

$$A^{465} = \varepsilon^{465} b \left([\text{HIn}] + [\text{In}^-]\right) \tag{18-7}$$

In Figure 18-13, all solutions contain the same total concentration of methyl red (= [HIn] + [In⁻]). Only the pH varies. Therefore, the sum of concentrations in Equation 18-7 is constant, and there is an isosbestic point because A^{465} is *constant.*

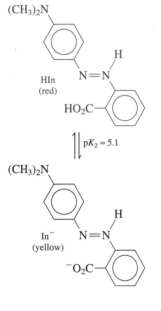

HIn
(red)

pK₂ ≈ 5.1

In⁻
(yellow)

An isosbestic point occurs when $\varepsilon_X = \varepsilon_Y$ and $[X] + [Y]$ is constant.

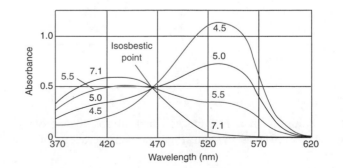

FIGURE 18–13 Absorption spectrum of 3.7×10^{-4} M methyl red as a function of pH between pH 4.5 and 7.1. [From E. J. King, *Acid-Base Equilibria* (Oxford: Pergamon Press, 1965).]

Ask Yourself

18-B. Consider the example beneath Equations 18-5. A different mixture of these same compounds X and Y in a *0.100-cm cell* had an absorbance of 0.233 at 406 nm and 0.200 at 457 nm. Find the concentrations of X and Y in this new mixture.

18-3   Spectrophotometric Titrations

In a **spectrophotometric titration,** we monitor changes in absorption or emission of electromagnetic radiation to detect the end point. We now consider an example from biochemistry.

Iron for biosynthesis is transported through the bloodstream by the protein *transferrin* (Figure 18-14). A solution of transferrin can be titrated with iron to measure its iron binding capacity. Transferrin without iron, called *apotransferrin,* is colorless. Each protein molecule with a formula mass of 81 000 binds two Fe^{3+} ions. When the iron binds to the protein, a red color with an absorbance maximum at 465 nm develops. The intensity of the red color allows us to follow the course of the titration of an unknown amount of apotransferrin with a standard solution of Fe^{3+}.

$$\text{apotransferrin} + 2Fe^{3+} \longrightarrow (Fe^{3+})_2\text{transferrin} \qquad (18\text{-}8)$$
$$\underset{\text{Colorless}}{} \qquad\qquad\qquad \underset{\text{Red}}{}$$

Figure 18-15 shows the titration of 2.000 mL of apotransferrin with 1.79×10^{-3} M ferric nitrilotriacetate. As iron is added to the protein, red color develops and the absorbance increases. When the protein is saturated with iron, no more iron binds to the protein and the curve levels off. The end point is the extrapolated intersection of the two straight lines at 203 μL. Absorbance rises slowly after the equivalence point because ferric nitrilotriacetate has some absorbance at 465 nm.

To construct the graph in Figure 18-15, we must account for the volume change as titrant is added. Each point plotted on the graph represents the absorbance that

Ferric nitrilotriacetate is used because Fe^{3+} precipitates as $Fe(OH)_3$ in neutral solution. Nitrilotriacetate binds Fe^{3+} through four **bold** atoms:

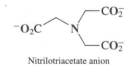

Nitrilotriacetate anion

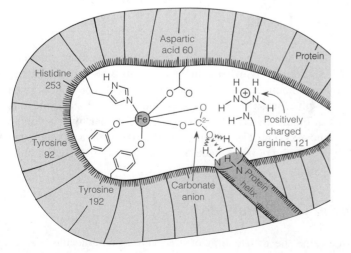

FIGURE 18-14 Each of the two iron-binding sites of transferrin is located in a cleft of the protein. Fe^{3+} binds to one nitrogen atom from the amino acid histidine and three oxygen atoms from tyrosine and aspartic acid. Two more binding sites of the metal are occupied by oxygen atoms from a carbonate anion (CO_3^{2-}), which is anchored in place by electrostatic interaction to the positively charged amino acid arginine and by hydrogen bonding to part of the protein helix. When transferrin is taken up by a cell, it is brought into a vesicle (Box 1-1) whose pH is lowered to 5.5. H^+ then reacts with the carbonate ligand to make HCO_3^- and H_2CO_3, thereby releasing Fe^{3+} from the protein. [Adapted from E. N. Baker, B. F. Anderson, H. M. Baker, M. Haridas, G. E. Norris, S. V. Rumball, and C. A. Smith, *Pure Appl. Chem.* **1990,** *62,* 1067.]

would be observed *if the solution had not been diluted from its original volume of 2.000 mL.*

$$\text{corrected absorbance} = \left(\frac{\text{total volume}}{\text{initial volume}}\right)(\text{observed absorbance}) \qquad (18\text{-}9)$$

EXAMPLE   Correcting Absorbance for the Effect of Dilution

The absorbance measured after adding 125 μL (= 0.125 mL) of ferric nitrilotriacetate to 2.000 mL of apotransferrin was 0.260. Calculate the corrected absorbance that should be plotted in Figure 18-15.

SOLUTION The total volume was $2.000 + 0.125 = 2.125$ mL. If the volume had been 2.000 mL, the absorbance would have been greater than 0.260 by a factor of 2.125/2.000.

$$\text{corrected absorbance} = \left(\frac{2.125 \text{ mL}}{2.000 \text{ mL}}\right)(0.260) = 0.276$$

The absorbance plotted in the graph is 0.276.

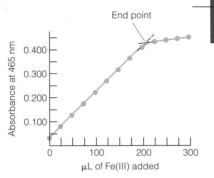

FIGURE 18-15   Spectrophotometric titration of apotransferrin with ferric nitrilotriacetate. Absorbance is corrected for dilution. The initial absorbance of the solution, before iron is added, is due to a colored impurity.

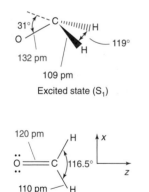

Excited state (S_1)

Ground state (S_0)

FIGURE 18-16   Geometry of formaldehyde in its ground state (S_0) and lowest excited singlet state (S_1).

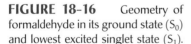
In *sigma* orbitals, electrons are localized between atoms. In *pi* orbitals, electrons are concentrated on either side of the plane of the formaldehyde molecule.

Ask Yourself

18-C. A 2.00-mL solution of apotransferrin, titrated as in Figure 18-15, required 163 μL of 1.43 mM ferric nitrilotriacetate to reach the end point.
    (a) How many moles of Fe^{3+} were required to reach the end point?
    (b) Each apotransferrin molecule binds two Fe^{3+} ions. Find the concentration of apotransferrin in the 2.00-mL solution.
    (c) Why does the slope in Figure 18-15 change abruptly at the equivalence point?

18-4   What Happens When a Molecule Absorbs Light?

When a molecule absorbs a photon, the molecule is promoted to a more energetic *excited state* (Figure 17-3). Conversely, when a molecule emits a photon, the energy of the molecule falls by an amount equal to the energy of the photon that is given off. As an example, we will discuss formaldehyde, whose ground state and one excited state are shown in Figure 18-16. The ground state is planar, with a double bond between carbon and oxygen. The double bond consists of a sigma bond between carbon and oxygen and a pi bond made from the $2p_y$ (out-of-plane) atomic orbitals of carbon and oxygen.

Electronic States of Formaldehyde

Molecular orbitals describe the distribution of electrons in a molecule, just as *atomic orbitals* describe the distribution of electrons in an atom. In Figure 18-17,

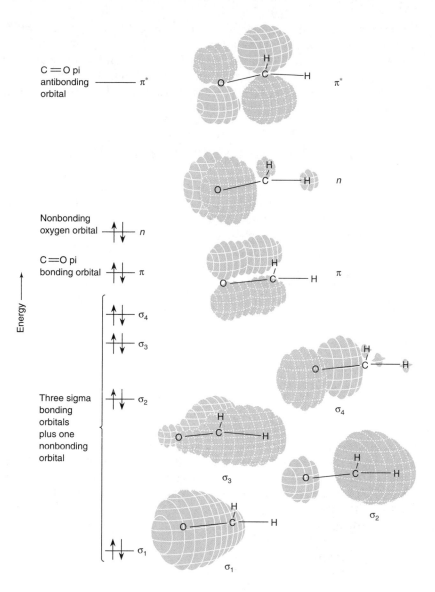

C=O pi antibonding orbital ——— π^*

Nonbonding oxygen orbital n

C=O pi bonding orbital π

σ_4
σ_3

Three sigma bonding orbitals plus one nonbonding orbital

σ_2

σ_1

Energy

FIGURE 18-17 Molecular orbital diagram of formaldehyde, showing energy levels and orbital shapes. The coordinate system of the molecule was shown in the previous figure. [From W. L. Jorgensen and L. Salem, *The Organic Chemist's Book of Orbitals* (New York: Academic Press, 1973).]

four low-lying orbitals of formaldehyde, labeled σ_1 through σ_4, are each occupied by a pair of electrons with opposite spins (spin quantum numbers $= +\frac{1}{2}$ and $-\frac{1}{2}$ represented by ↑ and ↓). At higher energy is a pi bonding orbital (π), made of the p_y atomic orbitals of carbon and oxygen. The highest-energy occupied orbital is a nonbonding orbital (n), composed principally of the oxygen $2p_x$ atomic orbital. The lowest-energy unoccupied orbital is a pi antibonding orbital (π^*). An electron in this orbital produces repulsion, rather than attraction, between the carbon and oxygen atoms.

In an **electronic transition,** an electron moves from one orbital to another. The lowest-energy electronic transition of formaldehyde promotes a nonbonding (n) electron to the antibonding pi orbital (π^*). There are actually two possible transitions, depending on the spin quantum numbers in the excited state. The state in which the spins are opposed in Figure 18-18 is called a **singlet state.** If the spins are parallel, the excited state is a **triplet state.**

The lowest-energy excited singlet and triplet states are called S_1 and T_1. In general, T_1 has lower energy than S_1. In formaldehyde, the transition $n \rightarrow \pi^*(T_1)$

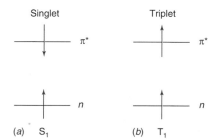

Singlet   Triplet

π^* π^*

n n

(a) S_1 (b) T_1

FIGURE 18-18 Two possible electronic states arising from an $n \rightarrow \pi^*$ transition. (a) Excited singlet state, S_1. (b) Excited triplet state, T_1.

The terms *singlet* and *triplet* are used because a triplet state splits into three slightly different energy levels in a magnetic field, but a singlet state is not split.

Remember, the shorter the wavelength, the greater the energy.

C—O stretching is reduced from 1 746 cm^{-1} in the S_0 state to 1 183 cm^{-1} in the S_1 state because the strength of the C—O bond decreases when the antibonding π^* orbital is populated.

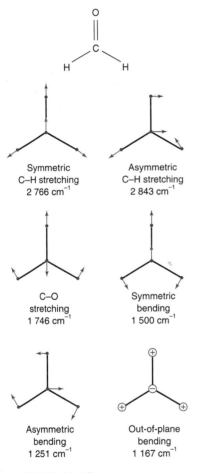

Symmetric C–H stretching 2 766 cm^{-1}

Asymmetric C–H stretching 2 843 cm^{-1}

C–O stretching 1 746 cm^{-1}

Symmetric bending 1 500 cm^{-1}

Asymmetric bending 1 251 cm^{-1}

Out-of-plane bending 1 167 cm^{-1}

FIGURE 18-19 The six kinds of vibrations of formaldehyde.

Internal conversion is a radiationless transition between states with the same spin quantum numbers (e.g., $S_1 \rightarrow S_0$).

Intersystem crossing is a radiationless transition between states with different spin quantum numbers (e.g., $T_1 \rightarrow S_0$).

requires absorption of visible light with a wavelength of 397 nm. The $n \rightarrow \pi^*(S_1)$ transition takes 355-nm ultraviolet radiation.

Although formaldehyde is planar in its ground state (S_0), it is pyramidal in both the S_1 (Figure 18-16) and T_1 excited states. Promotion of a nonbonding electron to an antibonding C—O orbital weakens and lengthens the C—O bond and changes the molecular geometry.

Vibrational and Rotational States of Formaldehyde

Infrared and microwave radiation are not energetic enough to induce electronic transitions, but they can change the vibrational or rotational motion of the molecule. The six kinds of vibrations of formaldehyde are shown in Figure 18-19. When formaldehyde absorbs an infrared photon with a wavenumber of 1 746 cm^{-1}, for example, C—O stretching is stimulated: Oscillations of the atoms increase in amplitude and the energy of the molecule increases.

Rotational energies of a molecule are even smaller than vibrational energies. Absorption of microwave radiation increases the rotational speed of a molecule.

Combined Electronic, Vibrational, and Rotational Transitions

In general, when a molecule absorbs light of sufficient energy to cause an electronic transition, **vibrational** and **rotational transitions**—changes in the vibrational and rotational states—occur as well. Formaldehyde can absorb one photon with just the right energy to (1) promote the molecule from S_0 to the S_1 electronic state; (2) increase the vibrational energy from the ground vibrational state of S_0 to an excited vibrational state of S_1; and (3) change from one rotational state of S_0 to a different rotational state of S_1. Electronic absorption bands are usually very broad (as in Figures 17-5 and 17-8) because many different vibrational and rotational levels are excited at slightly different energies.

What Happens to Absorbed Energy?

Suppose that absorption of a photon promotes a molecule from the ground electronic state, S_0, to a vibrationally and rotationally excited level of the excited electronic state S_1 (Figure 18-20, process A). Usually, the first event after absorption is *vibrational relaxation* to the lowest vibrational level of S_1. In this process, labeled R_1 in Figure 18-20, energy is lost to other molecules (solvent, for example) through collisions. The net effect is to convert part of the energy of the absorbed photon into heat spread through the entire medium.

From S_1, the molecule could enter a highly excited vibrational level of S_0 having the same energy as S_1. This process is called *internal conversion* (IC). Then the molecule can relax back to the ground vibrational state, transferring energy to neighboring molecules through collisions. If a molecule follows the path A-R_1-IC-R_2 in Figure 18-20, the entire energy of the photon will have been converted to heat.

Alternatively, the molecule could cross from S_1 into an excited vibrational level of T_1. Such an event is known as *intersystem crossing* (ISC). Following relaxation R_3, the molecule finds itself at the lowest vibrational level of T_1. From here, the molecule might undergo a second intersystem crossing to S_0, followed by the relaxation R_4, which liberates heat.

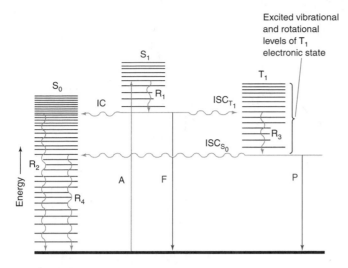

FIGURE 18-20 Physical processes that can occur after a molecule absorbs an ultraviolet or visible photon. S_0 is the ground electronic state of the molecule. S_1 and T_1 are the lowest excited singlet and triplet states, respectively. Straight arrows represent processes involving photons, and wavy arrows are radiationless transitions. A, absorption; F, fluorescence; P, phosphorescence; IC, internal conversion; ISC, intersystem crossing; R, vibrational relaxation.

Alternatively a molecule could relax from S_1 or T_1 to S_0 by emitting a photon. The transition $S_1 \rightarrow S_0$ is called **fluorescence** (Demonstration 18-1), and $T_1 \rightarrow S_0$ is called **phosphorescence**. (Fluorescence and phosphorescence can terminate in any of the vibrational levels of S_0, not just the ground state shown in Figure 18-20.) The rates of internal conversion, intersystem crossing, fluorescence, and phosphorescence

Demonstration 18-1

In Which Your Class Really Shines[1]

A fluorescent lamp is a glass tube filled with mercury vapor; the inner walls are coated with a *phosphor* (luminescent substance) consisting of a calcium halophosphate ($Ca_5(PO_4)_3F_{1-x}Cl_x$) doped with Mn^{2+} and Sb^{3+}. (*Doping* means adding an intentional impurity, called a *dopant*.) The mercury atoms, promoted to an excited state by electric current passing through the lamp, emit mostly ultraviolet radiation at 254 and 185 nm. This radiation is absorbed by the Sb^{3+}, and some of the energy is passed on to Mn^{2+}. Sb^{3+} emits blue light and Mn^{2+} emits yellow light, with the combined emission appearing white. The emission spectrum is shown at the right. Fluorescent lamps are important energy-saving devices because they are more efficient than incandescent lamps in the conversion of electricity to light.

White fabrics are sometimes made "whiter" by treatment with a fluorescent dye. Turn on an ultraviolet lamp in a darkened classroom and illuminate some people

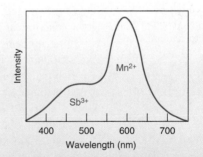

standing at the front of the room. (*The victims should not look directly at the lamp,* because ultraviolet light is harmful to eyes.) You will discover a surprising amount of emission from white fabrics, including shirts, pants, shoelaces, and unmentionables. You may also be surprised to see fluorescence from teeth and from recently bruised areas of skin that show no surface damage.

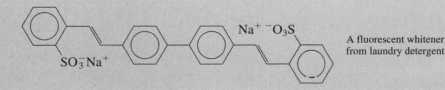

A fluorescent whitener from laundry detergent

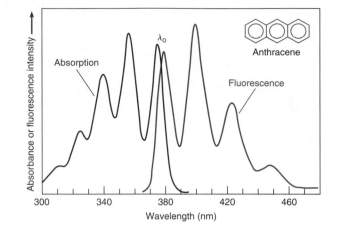

FIGURE 18-21 Spectrum of anthracene shows typical approximate mirror image relationship between absorption and fluorescence. Fluorescence comes at lower energy (longer wavelength) than absorption. [From C. M. Byron and T. C. Werner, *J. Chem. Ed.* **1991**, *68*, 433.]

Fluorescence: emission of a photon caused by a transition between states with the same spin quantum numbers (e.g., $S_1 \rightarrow S_0$).

Phosphorescence: emission of a photon from a transition between states with different spin quantum numbers (e.g., $T_1 \rightarrow S_0$).

An example of emission at lower energy (longer wavelength) than absorption is seen in Color Plate 16, in which blue light absorbed by a crystal gives rise to *red* emission.

depend on the solvent and conditions such as temperature and pressure. We see in Figure 18-20 that phosphorescence occurs at lower energy (longer wavelength) than fluorescence.

Fluorescence and phosphorescence are relatively rare. Molecules generally decay from the excited state through collisions—not by emitting light. The *lifetime* of fluorescence is always very short (10^{-8} to 10^{-4} s). The lifetime of phosphorescence is much longer (10^{-4} to 10^2 s). Phosphorescence is even rarer than fluorescence, because a molecule in the T_1 state has a good chance of collisional deactivation before phosphorescence can occur.

Figure 18-21 compares absorption and fluorescence spectra of anthracene. Fluorescence comes at lower energy and is roughly the mirror image of absorption. To understand the mirror image relationship, consider the energy levels in Figure 18-22. In the absorption spectrum, wavelength λ_0 corresponds to a transition from the ground vibrational level of S_0 to the lowest vibrational level of S_1. Absorption maxima at higher energy (shorter wavelength) correspond to the $S_0 \rightarrow S_1$ transition accompanied by absorption of one or more quanta of vibrational energy. In polar solvents, vibrational structure is often broadened beyond recognition, and only a broad envelope of absorption is observed. In Figure 18-21, the solvent is cyclohexane, which is nonpolar, and the vibrational structure is easily seen.

Following absorption, the vibrationally excited S_1 molecule relaxes back to the lowest vibrational level of S_1 prior to emitting any radiation. Emission from S_1 can go to any of the vibrational levels of S_0 in Figure 18-22. The highest-energy transition comes at wavelength λ_0, with a series of peaks following at longer wavelength. The absorption and emission spectra will have an approximate mirror image relationship if the spacings between vibrational levels are roughly equal and if the transition probabilities are similar.

Another possible consequence of the absorption of light is the breaking of chemical bonds. **Photochemistry** is a chemical reaction initiated by absorption of light (as in the reaction $O_2 \xrightarrow{h\nu} 2O$ in the upper atmosphere at the opening of Chapter 17). Some chemical reactions (not initiated by light) release energy in the form of light, which is called **chemiluminescence**. The light from a firefly or a light stick[2] is chemiluminescence.

Ask Yourself

18-D. (a) What is the difference between electronic, vibrational, and rotational transitions?

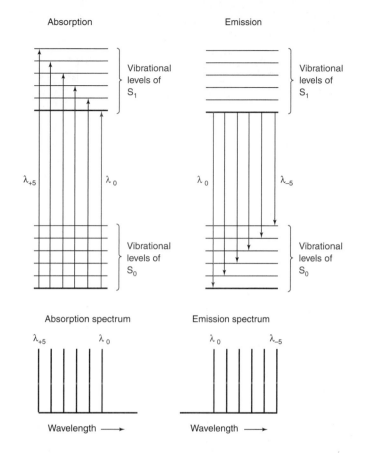

Absorption   Emission

Absorption spectrum   Emission spectrum

FIGURE 18-22   Energy-level diagram showing why structure is seen in the absorption and emission spectra, and why the spectra are roughly mirror images of each other. In absorption, wavelength λ_0 comes at lowest energy, and λ_{+5} is at highest energy. In emission, wavelength λ_0 comes at highest energy, and λ_{-5} is at lowest energy.

(b) How can the energy of an absorbed photon be released without emission of light?

(c) What processes lead to fluorescence and phosphorescence? Which comes at higher energy? Which is faster?

(d) Why does fluorescence tend to be the mirror image of absorption?

(e) What is the difference between photochemistry and chemiluminescence?

18-5   Luminescence in Analytical Chemistry

Luminescence is any emission of electromagnetic radiation and includes fluorescence, phosphorescence, and other possible processes. Luminescence is measured in Figure 18-23 by exciting a sample at a wavelength it absorbs and observing emission at a different wavelength.

Luminescence is useful in quantitative analysis because, over some concentration range, the intensity of luminescence (I) is proportional to the concentration of the emitting species (c):

*Relation of emission
intensity to concentration:*

$$I = kP_0c \qquad (18\text{-}10)$$

Two ways to increase luminescence are to increase the concentration of the emitting compound and to increase the incident radiant power.

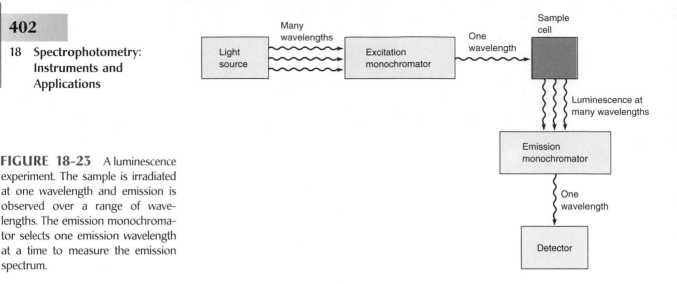

FIGURE 18-23 A luminescence experiment. The sample is irradiated at one wavelength and emission is observed over a range of wavelengths. The emission monochromator selects one emission wavelength at a time to measure the emission spectrum.

Luminescence is more sensitive than absorbance for detecting very low concentrations of analyte.

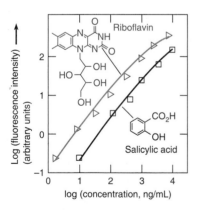

FIGURE 18-24 Calibration curves demonstrating a nearly linear relation between fluorescence intensity and concentration for riboflavin (vitamin B_2) and salicylic acid. A log-log scale is used because the data span several orders of magnitude. [Data from B. T. Jones, B. W. Smith, M. B. Leong, M. A. Mignardi, and J. D. Winefordner, *J. Chem. Ed.* **1989**, *66*, 357.]

where P_0 is the incident radiant power and k is a constant. Figure 18-24 shows calibration curves illustrating Equation 18-10.

Luminescence is more sensitive than absorption. Imagine yourself in a stadium at night with the lights off, but each of the 50 000 raving fans is holding a lighted candle. If 500 people blow out their candles, you will hardly notice the difference. Now imagine that the stadium is completely dark, and then 500 people light their candles. The change would be dramatic. The first case is analogous to changing transmittance from 100% to 99%. It is hard to measure such a small change because the 50 000-candle background is so bright. The second case is analogous to observing luminescence from 1% of the molecules in a sample. Against the black background, luminescence is easy to detect.

Immunoassays

An important application of luminescence is in **immunoassays,** which employ antibodies to detect analyte.[3,4] An **antibody** is a protein produced by the immune system of an animal in response to a foreign molecule, which is called an **antigen.** An antibody specifically recognizes and binds to the antigen that stimulated its synthesis.

Figure 18-25 illustrates the principle of an *enzyme-linked immunosorbent assay,* abbreviated ELISA in biochemical literature. Antibody 1, which is specific for the analyte of interest (the antigen), is bound to a polymer support. In steps 1 and 2, analyte is incubated with the polymer-bound antibody to form the antibody-antigen complex. The fraction of antibody sites that bind analyte is proportional to the concentration of analyte in the unknown. The surface is then washed to remove unbound substances. In steps 3 and 4, the antibody-antigen complex is treated with antibody 2, which recognizes a different region of the analyte. An enzyme that will be used later was covalently attached to antibody 2 (prior to step 3). Again, excess unbound substances are washed away.

Figure 18-26 shows two ways in which the enzyme attached to antibody 2 is used for quantitative analysis. In Figure 18-26a, the enzyme transforms a colorless reactant into a colored product. Because one enzyme molecule catalyzes the same reaction many times, many molecules of colored product are created for each

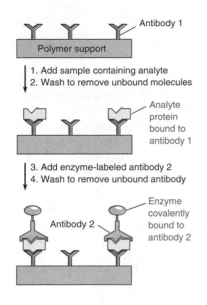

1. Add sample containing analyte
2. Wash to remove unbound molecules

Antibody 1

Polymer support

Analyte protein bound to antibody 1

3. Add enzyme-labeled antibody 2
4. Wash to remove unbound antibody

Antibody 2

Enzyme covalently bound to antibody 2

FIGURE 18-25 Enzyme-linked immunosorbent assay. Antibody 1, which is specific for the analyte of interest, is bound to a polymer support and treated with unknown. After washing away excess, unbound molecules, the analyte remains bound to antibody 1. The bound analyte is then treated with antibody 2, which recognizes a different site on the analyte and to which an enzyme is covalently attached. After washing away unbound material, each molecule of analyte is coupled to an enzyme, which will be used in Figure 18-26.

molecule of antigen. The enzyme thereby *amplifies* the signal in chemical analysis. The higher the concentration of analyte in the unknown, the more enzyme is bound and the greater the extent of the enzyme-catalyzed reaction. In Figure 18-26b, the enzyme converts a nonfluorescent reactant into a fluorescent product. Enzyme-linked immunosorbent assays are sensitive to <1 ng of analyte. Pregnancy tests are based on the immunoassay of a placental protein in urine. Box 18-1 shows how immunoassays are used in field-portable environmental analyses.

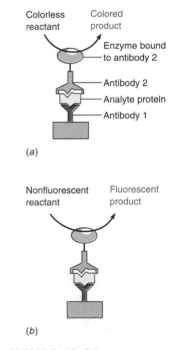

Colorless reactant   Colored product

Enzyme bound to antibody 2

Antibody 2

Analyte protein

Antibody 1

(a)

Nonfluorescent reactant   Fluorescent product

(b)

FIGURE 18-26 Enzyme bound to antibody 2 can catalyze reactions that produce (a) colored or (b) fluorescent products. Each molecule of analyte bound in the immunoassay leads to many molecules of colored or fluorescent product that are easily measured.

Ask Yourself

18-E. Red tide is a phenomenon in which the sea becomes red and mass fish kills may occur. Chemiluminescence can be used to detect the early stages of red tide caused by the plankton *Chattonella marina*, which secretes superoxide ion (O_2^-). The procedure uses a flow cell with a spiral mixing chamber to combine unknown seawater with a chemiluminescent probe designated MCLA at pH 9.7. When O_2^- reacts with MCLA, the product emits light that is detected with a photomultiplier tube. The table below shows the detector signal as a function of known cell concentrations of *C. marina* in seawater. Prepare a calibration curve and find the concentration of *C. marina* in a seawater sample that produces a detector signal of 15.9 units.

Cells/mL in seawater	Detector signal (arbitrary units)	Cells/mL in seawater	Detector signal (arbitrary units)
188	2.2	2 336	16.5
376	4.6	3 087	20.6
738	6.0	3 973	26.6
1 154	9.1	4 913	31.5
1 544	11.6		

Box 18-1 *Informed Citizen*

Immunoassays in Environmental Analysis[5]

Immunoassays are becoming available for screening and analysis of environmental samples in the field. An advantage of screening in the field is that uncontaminated regions that require no further attention are readily identified. Assays have been developed to monitor pesticides, industrial chemicals, and microbial toxins at the parts-per-trillion to parts-per-million levels in groundwater, soil, and food. In some cases, the immunoassay field test is 20–40 times less expensive than a chromatographic analysis in the laboratory. Immunoassays require less than a milliliter of sample and can be completed in 2–3 h in the field. Chromatographic analyses might require up to 2 days, because analyte must first be extracted or concentrated from liter-quantity samples to obtain a sufficient concentration.

The diagram shows how an assay works. In step 1, antibody for the intended analyte is adsorbed to the bottom of a microtiter well, which is a depression in a plate having as many as 96 wells for simultaneous analyses. In step 2, known volumes of sample and standard solution containing enzyme-labeled analyte are added to the well. In step 3, analyte in the sample competes with enzyme-labeled analyte for binding sites on the antibody. This is the key step: *The greater the concentration of analyte in the unknown sample, the more it will bind to antibody and the less enzyme-labeled analyte will bind.* After an incubation period, unbound sample and standard are washed away. In step 4, a *chromogenic* substrate is added to the well. This is a colorless substance that reacts in the presence of enzyme to make a colored product. In step 5, colored product is measured visually by comparison to standards or quantitatively with a spectrophotometer. The greater the concentration of analyte in the unknown sample, the lighter the color in step 5.

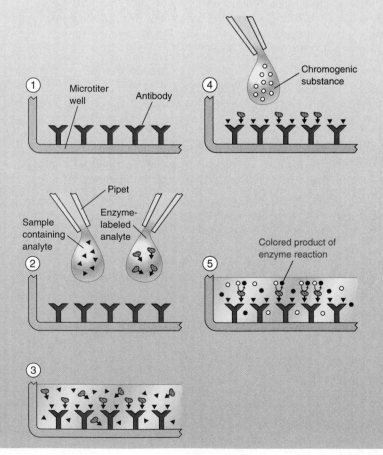

Key Equations

Absorption of a mixture of species X, Y, Z, etc.

$$A = \varepsilon_X b[X] + \varepsilon_Y b[Y] + \varepsilon_Z b[Z] + \cdots$$

A = absorbance at wavelength λ

ε_i = molar absorptivity of species i at wavelength λ

b = pathlength

You should be able to use Equations 18-5 to analyze the spectrum of a mixture

Fluorescence intensity

$$I = kP_0 c$$

I = fluorescence intensity

k = constant

P_0 = radiant power of incident radiation

c = concentration of fluorescing species

Important Terms

antibody	isosbestic point	photomultiplier tube
antigen	luminescence	polychromator
chemiluminescence	molecular orbital	rotational transition
diffraction	monochromator	singlet state
electronic transition	phosphorescence	spectrophotometric titration
fluorescence	photochemistry	triplet state
grating	photodiode array	vibrational transition
immunoassay		

Problems

18-1. State the differences between single- and double-beam spectrophotometers and explain how each measures the transmittance of a sample. What source of error in a single-beam instrument is absent in the double-beam instrument?

18-2. Would you use a tungsten or a deuterium lamp as a source of 300-nm radiation?

18-3. What are the advantages and disadvantages of decreasing monochromator slit width?

18-4. Consider a reflection grating operating with an incident angle of 40° in Figure 18-5.

(a) How many lines per centimeter should be etched in the grating if the first-order diffraction angle for 600 nm (visible) light is to be 30°?

(b) Answer the same question for $1\,000\ \text{cm}^{-1}$ (infrared) light.

18-5. Why is a photomultiplier such a sensitive photodetector?

18-6. What characteristic makes a photodiode array spectrophotometer suitable for measuring the spectrum of a compound as it emerges from a chromatography column and a dispersive spectrophotometer not suitable? What is the disadvantage of the photodiode array spectrophotometer?

18-7. When are isosbestic points observed and why?

18-8. Explain how signal amplification is achieved in enzyme-linked immunosorbent assays.

18-9. Molar absorptivities of compounds X and Y were measured with pure samples of each:

	ε ($\text{M}^{-1}\ \text{cm}^{-1}$)	
λ (nm)	X	Y
$\lambda' \equiv 272$	$\varepsilon_X' = 16\,440$	$\varepsilon_Y' = 3\,870$
$\lambda'' \equiv 327$	$\varepsilon_X'' = 3\,990$	$\varepsilon_Y'' = 6\,420$

A mixture of X and Y in a 1.000-cm cell had an absorbance of $A' = 0.957$ at 272 nm and $A'' = 0.559$ at 327 nm. Find the concentrations of X and Y in the mixture.

18-10. *Spreadsheet for simultaneous equations.* Write a spreadsheet for the analysis of a mixture, using Equations 18-5. The input will be the sample pathlength, the observed absorbances at two wavelengths, and the molar absorptivities of the two pure compounds at two wavelengths. The output will be the concentration of each component of the mixture. Test your spreadsheet with numbers from Problem 18-9.

18-11. Ultraviolet absorbance for 1.00×10^{-4} M MnO_4^-, 1.00×10^{-4} M $Cr_2O_7^{2-}$, and an unknown mixture of both (all in a 1.000-cm cell) are given below. Find the concentration of each species in the mixture.

Wavelength (nm)	MnO_4^- standard	$Cr_2O_7^{2-}$ standard	Mixture
266	0.042	0.410	0.766
320	0.168	0.158	0.422

18-12. Transferrin is the iron-transport protein in blood. It has a molecular mass of 81 000 and carries two Fe^{3+} ions. Desferrioxamine B (Box 12-1) is a potent iron chelator used to treat patients with iron overload. It has a molecular mass of about 650 and can bind one Fe^{3+}. Desferrioxamine can take iron from many sites within the body and is excreted (with its iron) through the kidneys. The molar absorptivities of these compounds (saturated with iron) at two wavelengths are given in the table. Both compounds are colorless (no visible absorption) in the absence of iron.

	ε (M^{-1} cm^{-1})	
λ (nm)	Transferrin	Desferrioxamine
428	3 540	2 730
470	4 170	2 290

(a) A solution of transferrin exhibits an absorbance of 0.463 at 470 nm in a 1.000-cm cell. Calculate the concentration of transferrin in milligrams per milliliter and the concentration of iron in micrograms per milliliter.

(b) A short time after adding desferrioxamine (which dilutes the sample), the absorbance at 470 nm was 0.424, and the

absorbance at 428 nm was 0.401. Calculate the fraction of iron in transferrin. Remember that transferrin binds two Fe^{3+} ions and desferrioxamine binds only one. If you wrote one, use the spreadsheet from Problem 18-10 to work this problem.

18-13. *Finding pK_a by spectrophotometry.* Consider an indicator, HIn, which dissociates according to the equation.

$$HIn \xrightleftharpoons{K_{HIn}} H^+ + In^-$$

The molar absorptivity, ε, is 2 080 M^{-1} cm^{-1} for HIn and 14 200 M^{-1} cm^{-1} for In$^-$, at a wavelength of 440 nm.

(a) Use Beer's law to write an expression for the absorbance at 440 nm of a solution in a 1.00-cm cuvet containing HIn at a concentration [HIn] and In$^-$ at a concentration [In$^-$].

(b) A 1.84×10^{-4} M solution of HIn is adjusted to pH 6.23 to give a mixture of HIn and In$^-$ with a total concentration of 1.84×10^{-4} M. The measured absorbance of the solution at 440 nm is 0.868. Using your expression from **(a)** and the fact that $[HIn] + [In^-] = 1.84 \times 10^{-4}$ M, calculate pK_{HIn} for this indicator.

18-14. *Graphical method to find pK_a by spectrophotometry.* This method requires a series of solutions containing a compound of unknown but constant concentration at different pH values. The figure shows that we choose a wavelength at which one of the species, say In$^-$, has maximum absorbance (A_{In^-}) and HIn has a different absorbance (A_{HIn}).

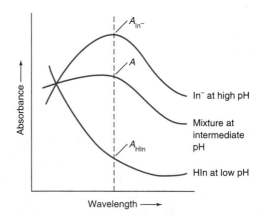

At intermediate pH, the absorbance (A) is between the two extremes. Let the total concentration be $c_o = [HIn] + [In^-]$. At high pH, the absorbance is $A_{In^-} = \varepsilon_{In^-}bc_o$, and at low pH, the absorbance is $A_{HIn} = \varepsilon_{HIn}bc_o$, where ε is molar absorptivity and b is pathlength. At intermediate pH, both species are present and the absorbance is $A = \varepsilon_{HIn}b[HIn] + \varepsilon_{In^-}b[In^-]$. You can combine these expressions to show that

$$\frac{[In^-]}{[HIn]} = \frac{A - A_{HIn}}{A_{In^-} - A}$$

Placing this expression into the Henderson-Hasselbalch equation 8-1 gives

$$pH = pK_{HIn} + \log\left(\frac{[In^-]}{[HIn]}\right) \Rightarrow \log\left(\frac{A - A_{HIn}}{A_{In^-} - A}\right) = pH - pK_{HIn}$$

If we have several solutions of intermediate pH, a graph of $\log[(A - A_{HIn})/(A_{In^-} - A)]$ versus pH should be a straight line with a slope of 1 that crosses the x-axis at pK_{HIn}.

pH	Absorbance
~2	$0.006 \equiv A_{HIn}$
3.35	0.170
3.65	0.287
3.94	0.411
4.30	0.562
4.64	0.670
~12	$0.818 \equiv A_{In^-}$

Data at 590 nm for HIn = bromophenol blue from G. S. Patterson, *J. Chem. Ed.* **1999,** *76,* 395.

Prepare a graph of $\log[(A - A_{HIn})/(A_{In^-} - A)]$ versus pH and find the slope and intercept by the method of least squares. Find pK_{HIn}.

18-15. Infrared spectra are customarily recorded on a transmittance scale so that weak and strong bands can be displayed on the same scale. The region near $2\,000\text{ cm}^{-1}$ in the infrared spectra of compounds A and B is shown in the figure at the top of the next column. Note that absorption corresponds to a downward peak on this scale. The spectra were recorded from a 0.010 0 M solution of each, in cells with 0.005 00-cm pathlengths. A mixture of A and B in a 0.005 00-cm cell gave a transmittance of 34.0% at $2\,022\text{ cm}^{-1}$ and 38.3% at $1\,993\text{ cm}^{-1}$. Find the concentrations of A and B. If it is available, use the spreadsheet from Problem 18-10 to work this problem.

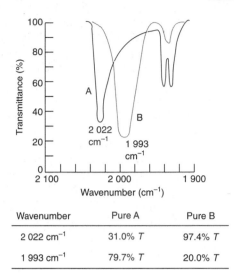

Wavenumber	Pure A	Pure B
$2\,022\text{ cm}^{-1}$	31.0% T	97.4% T
$1\,993\text{ cm}^{-1}$	79.7% T	20.0% T

18-16. The iron-binding site of transferrin in Figure 18-14 can accommodate certain other metal ions besides Fe^{3+} and certain other anions besides CO_3^{2-}. Data are given below for the titration of transferrin (3.57 mg in 2.00 mL) with 6.64 mM Ga^{3+} solution in the presence of the anion oxalate, $C_2O_4^{2-}$, and in the absence of a suitable anion. Prepare a graph similar to Figure 18-15, showing both sets of data. Indicate the theoretical equivalence point for the binding of two Ga^{3+} ions per molecule of protein and the observed end point. How many Ga^{3+} ions are bound to transferrin in the presence and in the absence of oxalate?

Titration in presence of $C_2O_4^{2-}$		Titration in absence of anion	
Total μL Ga^{3+} added	Absorbance at 241 nm	Total μL Ga^{3+} added	Absorbance at 241 nm
0.0	0.044	0.0	0.000
2.0	0.143	2.0	0.007
4.0	0.222	6.0	0.012
6.0	0.306	10.0	0.019
8.0	0.381	14.0	0.024
10.0	0.452	18.0	0.030
12.0	0.508	22.0	0.035
14.0	0.541	26.0	0.037
16.0	0.558		
18.0	0.562		
21.0	0.569		
24.0	0.576		

18-17. The metal-binding compound semi-xylenol orange is yellow at pH 5.9 but turns red (λ_{max} = 490 nm) when it reacts with Pb^{2+}. A 2.025-mL sample of semi-xylenol orange was titrated with 7.515×10^{-4} M $Pb(NO_3)_2$, with the following results:

Total μL Pb^{2+} added	Absorbance at 490 nm in 1-cm cell	Total μL Pb^{2+} added	Absorbance at 490 nm in 1-cm cell
0.0	0.227	42.0	0.425
6.0	0.256	48.0	0.445
12.0	0.286	54.0	0.448
18.0	0.316	60.0	0.449
24.0	0.345	70.0	0.450
30.0	0.370	80.0	0.447
36.0	0.399		

Make a graph of corrected absorbance versus microliters of Pb^{2+} added. Corrected absorbance is what would be observed if the volume were not changed from its initial value of 2.025 mL. Assuming that the reaction of semi-xylenol orange with Pb^{2+} has a 1:1 stoichiometry, find the molarity of semi-xylenol orange in the original solution.

18-18. Here is a procedure for detecting vapors of trinitrotoluene (TNT) explosive:

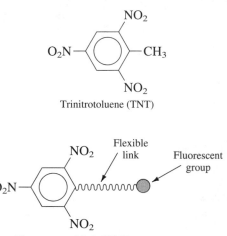

Trinitrotoluene (TNT)

Fluorescence-labeled TNT

1. A column containing covalently bound antibodies against TNT was prepared. Fluorescence-labeled TNT was passed through this column to saturate all the antibodies with labeled TNT. The column was washed with excess solvent until no fluorescence was detected at the outlet.

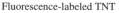

Problem 18-19: Titration of xylenol orange with VO^{2+} at pH 6.0. [From D. C. Harris and M. H. Gelb, *Biochim. Biophys. Acta* **1980**, *623*, 1.]

2. Air containing TNT vapors was drawn through a sampling device that concentrated traces of the gaseous vapors into distilled water.

3. Aliquots of water from the sampler were injected into the antibody column. The fluorescence intensity of the liquid exiting the column was proportional to the concentration of unlabeled TNT in the water from the sampler over the range 20 to 1 200 ng/mL.

Draw pictures showing the state of the column in steps 1 and 3 and explain how this scheme works.

How Would You Do It?

18-19. The metal ion indicator xylenol orange (Table 12-2) is yellow at pH 6 ($\lambda_{max} = 439$ nm). The spectral changes that occur as VO^{2+} (vanadyl ion) is added to the indicator at pH 6 are shown in the figure on page 408. The mole ratio VO^{2+}/xylenol orange at each point is shown in the table below. Suggest a sequence of chemical reactions to explain the spectral changes, especially the isosbestic points at 457 and 528 nm.

Trace	Mole ratio	Trace	Mole ratio	Trace	Mole ratio
0	0	7	0.70	12	1.3
1	0.10	8	0.80	13	1.5
2	0.20	9	0.90	14	2.0
3	0.30	10	1.0	15	3.1
4	0.40	11	1.1	16	4.1
5	0.50				
6	0.60				

Notes and References

1. J. A. DeLuca, *J. Chem. Ed.* **1980**, *57,* 541.

2. C. Salter, K. Range, and G. Salter, *J. Chem. Ed.* **1999**, *76,* 84; E. Wilson, *Chem. Eng. News,* 18 January 1999, p. 65.

3. R. S. Yalow, *J. Chem. Ed.* **1999**, *76,* 767; R. P. Ekins, *J. Chem. Ed.* **1999**, *76,* 769; E. F. Ullman, *J. Chem. Ed.* **1999**, *76,* 781; E. Strauss, *J. Chem. Ed.* **1999**, *76,* 788; D. J. Holme and H. Peck, *Analytical Biochemistry,* 3rd ed. (New York: Addison Wesley Longman, 1998).

4. For student immunoassay experiments, see G. L. Anderson and L. A. McNellis, *J. Chem. Ed.* **1998**, *75,* 1275; and L. A. Inda, P. Razquín, F. Lampreave, M. A. Alava, and M. Calvo, *J. Chem. Ed.* **1998**, *75,* 1618.

5. J. M. Van Emon and V. Lopez-Avila, *Anal. Chem.* **1992**, *64,* 79A.

Measuring Calcium in Single Cells

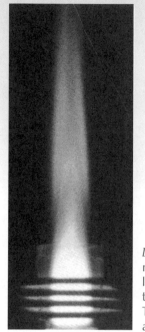

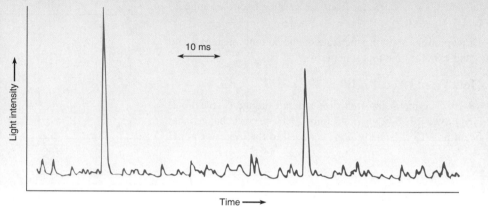

Left: Inductively coupled plasma torch decomposes molecules into atoms that emit specific wavelengths of light. Above: Bursts of light recorded from calcium when two individual animal cells enter the flame. [From T. Nomizu, S. Kaneco, T. Tanaka, D. Ito, H. Kawaguchi, and B. L. Vallee, Anal. Chem. **1994**, 66, 3000.]

An inductively coupled plasma is an extremely hot flame that decomposes molecules into excited atoms, which then emit characteristic frequencies of light. Emission intensity is proportional to the quantity of analyte atoms over many orders of magnitude. The photograph shows an inductively coupled plasma consisting of argon ions and electrons whose energy is derived by absorption of radio-frequency energy from the water-cooled electrical coil at the base of the torch. When a suspension of animal cells is injected into the flame, the calcium from each cell is vaporized all at once and a single burst of light is observed from each cell. The area of each spike in the graph is proportional to the mass of calcium in each cell. Representative results for three different cell types are given in the table.

Cell type	Ca content (pg/cell)	Cell diameter (μm)	[Ca] in cell (mM)
Mouse fibroblast	0.06 ± 0.03	10–15	0.8–2.8
Human pancreas	0.16 ± 0.04	15–20	0.9–2.2
Human endothelium	0.27 ± 0.04	15–20	1.6–3.7

ATOMIC SPECTROSCOPY

A*tomic spectroscopy* is a principal tool for measuring metallic elements at parts per million (and lower) levels in industrial and environmental laboratories. When this workhorse technique is automated with mechanical sample-changing devices, each instrument can turn out hundreds of analyses per day.

Parts per million (ppm) means micrograms of solute per gram of solution. Because the density of dilute aqueous solutions is close to 1.00 g/mL, ppm usually refers to μg/mL. A concentration of 1 ppm Fe corresponds to 1 μg Fe/mL $\approx$ 2×10^{-5} M.

19-1   What Is Atomic Spectroscopy?

In the atomic spectroscopy measurement outlined in Figure 19-1, a liquid sample is *aspirated* (sucked) through a plastic tube into a flame that is hot enough to break molecules apart into atoms. The concentration of an element in the flame is measured either by its absorption or emission of radiation. For **atomic absorption spectroscopy,** radiation of the correct frequency is passed through the flame (Figure 19-2) and the intensity of transmitted radiation is measured. For **atomic emission spectroscopy,** no lamp is required. Radiation is emitted by hot atoms whose electrons have been promoted to excited states in the flame. For both experiments in Figure 19-2, a monochromator selects the wavelength of radiation that will be viewed by the detector. It is routine to measure analyte concentrations at the parts per million level with a precision of 2%. To analyze major constituents of an unknown, the sample must be diluted to reduce concentrations to the parts per million level.

A typical spectrum of a molecule in solution, such as Figure 17-8, has bands that are ~100 nm in width. In contrast, the spectrum of gaseous atoms in a flame

Atomic spectroscopy:
- absorption (requires a lamp with light whose frequency is absorbed by atoms)
- emission (luminescence from excited atoms—no lamp required)

411

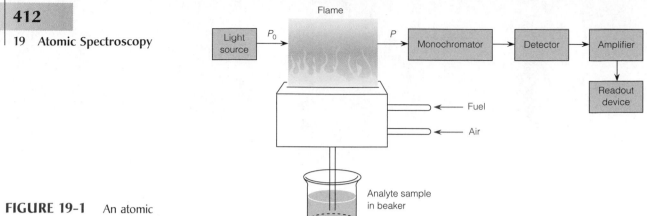

FIGURE 19-1 An atomic absorption experiment.

has extremely sharp lines with widths of 10^{-3} to 10^{-2} nm (Figure 19-3). Because the lines are so sharp, there is usually little overlap between the spectra of different elements in the same sample. The lack of overlap allows some instruments to measure over 60 elements in a sample simultaneously.

Ask Yourself

19-A. What is the difference between atomic absorption and atomic emission spectroscopy?

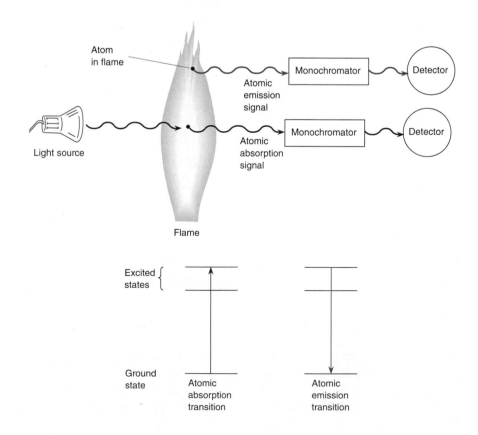

FIGURE 19-2 Absorption and emission of light by atoms in a flame. In atomic absorption, atoms absorb light from the lamp and unabsorbed light reaches the detector. In atomic emission, light is emitted by excited atoms in the flame.

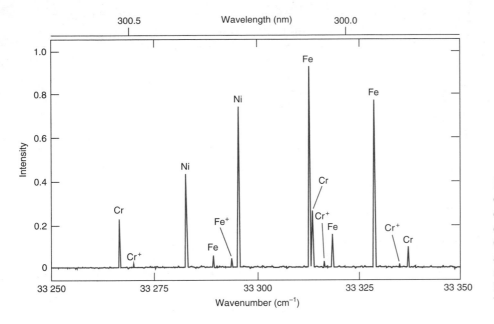

FIGURE 19-3 A tiny portion of the spectrum of a steel hollow-cathode lamp, showing sharp lines characteristic of gaseous Fe, Ni, and Cr atoms and weak lines from Cr^+ and Fe^+ ions. The resolution is 0.001 nm, which is about half of the true linewidths of the signals. [From A. P. Thorne, *Anal. Chem.* **1991,** *63,* 57A.]

19-2 Atomization: Flames, Furnaces, and Plasmas

Atomization is the process of breaking analyte into gaseous atoms, which then are measured by their absorption or emission of radiation. Older atomic absorption spectrometers—and the apparatus generally found in student laboratories—use a combustion flame to decompose analyte into atoms, as in Figure 19-1. Modern instruments employ an inductively coupled argon plasma or an electrically heated graphite tube for atomization. The combustion flame and the graphite furnace are used for both atomic absorption and atomic emission measurements. The plasma is so hot that many atoms are in excited states and their emission is readily observed. Plasmas are used almost exclusively for atomic emission measurements. The newest technique, which is becoming widespread in industrial, environmental, and research laboratories, is to atomize the sample with a plasma and measure the concentration of ions in the plasma with a *mass spectrometer*. This method of quantitative analysis does not involve electromagnetic radiation and is not a form of spectroscopy.

Atomization method	Quantitation method
flame	absorption or emission
graphite furnace	absorption or emission
plasma	emission or mass spectrometry

Flame

Most flame spectrometers use a *premix burner,* such as that in Figure 19-4, in which the sample, oxidant, and fuel are mixed before being introduced into the flame. Sample solution is drawn in by the rapid flow of oxidant and breaks into a fine mist when it leaves the tip of the *nebulizer* and strikes a glass bead. The formation of small droplets is termed *nebulization*. The mist flows past a series of baffles to promote further mixing and block large droplets of liquid (which flow out to the drain). A fine mist containing about 5% of the initial sample reaches the flame.

After solvent evaporates in the flame, the remaining sample vaporizes and decomposes to atoms. Many metal atoms (M) form oxides (MO) and hydroxides (MOH)

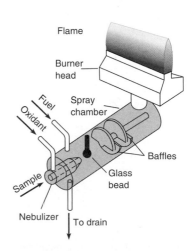

FIGURE 19-4 Premix burner with a pneumatic nebulizer. The slot in the burner head is typically 10 cm long and 0.5 mm wide.

TABLE 19-1   **Maximum flame temperatures**

Fuel	Oxidant	Temperature (K)
Acetylene	Air	2 400–2 700
Acetylene	Nitrous oxide	2 900–3 100
Acetylene	Oxygen	3 300–3 400
Hydrogen	Air	2 300–2 400
Hydrogen	Oxygen	2 800–3 000
Cyanogen	Oxygen	4 800

as they rise through the flame. Molecules do not have the same spectra as atoms, so the atomic signal is lowered. If the flame is relatively rich in fuel (a "rich" flame), excess carbon species tend to reduce MO and MOH back to M and thereby increase sensitivity. The opposite of a rich flame is a "lean" flame, which has excess oxidant and is hotter. We choose a lean or rich flame to provide optimum conditions for different elements.

The most common fuel-oxidant combination is acetylene and air, which produces a flame temperature of 2 400–2 700 K (Table 19-1). When a hotter flame is required to vaporize *refractory* elements (those with high boiling points), acetylene and nitrous oxide is usually the mixture of choice. The height above the burner head at which maximum atomic absorption or emission is observed depends on the element being measured, as well as the flow rates of sample, fuel, and oxidant. Each of these parameters must be optimized for a given analysis.

Furnace

Furnaces offer increased sensitivity and require less sample than a flame.

Graphite is a form of carbon. At high temperature in air, the oxidation reaction is $C(s) + O_2 \rightarrow CO_2$.

The electrically heated **graphite furnace** in Figure 19-5a provides greater sensitivity than a flame and requires less sample. A 1- to 100-μL sample is injected into the oven through the hole at the center. The light beam travels through windows at each end of the tube. The maximum recommended temperature for a graphite furnace is 2 550°C for not more than 7 s. Surrounding the graphite with an atmosphere of Ar helps prevent oxidation.

The graphite furnace has a high sensitivity because it confines atoms in the optical path for several seconds. In flame spectroscopy, the sample is diluted during nebulization, and its residence time in the optical path is only a fraction of a second. Flames require a sample volume of at least 10 mL, because sample is constantly flowing into the flame. The graphite furnace requires only tens of microliters. In an extreme case, when only nanoliters of kidney tubular fluid were available, a method was devised to reproducibly deliver 0.1 nL to a furnace for analysis of Na and K. Precision with a furnace is rarely better than 5–10% with manual sample injection, but automated injection improves reproducibility.

The operator must determine reasonable time and temperature for each stage of the analysis. Once a program is established, it can be applied to similar samples.

Sample is injected onto the *L'vov platform* inside the furnace in Figure 19-5a. Analyte does not vaporize until the furnace wall has reached constant temperature (Figure 19-5b). If sample were injected directly onto the inside wall of the furnace, atomization would occur while the wall is heating up at a rate of 2 000 K/s (Figure 19-5b) and the signal would be much less reproducible than with the platform.

A skilled operator must determine heating conditions for three or more steps to properly atomize a sample. To analyze Fe in the iron-storage protein ferritin, 10 μL of sample containing ~0.1 ppm Fe are injected into the cold graphite

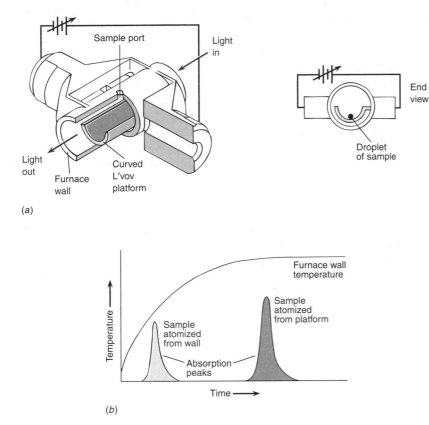

(a)

(b)

FIGURE 19-5 (a) Electrically heated graphite furnace for atomic spectroscopy. Sample is injected through the port at the top and light travels from end to end. The L'vov platform inside the furnace is uniformly heated by radiation from the outer wall. The platform is attached to the wall by one small connection hidden from view. [Courtesy Perkin-Elmer Corp., Norwalk, CT.] (b) Heating profile comparing analyte evaporation from wall and from platform.

furnace. The furnace is programmed to *dry* the sample at 125°C for 20 s to remove solvent. Drying is followed by 60 s of *charring* (also called *pyrolysis*) at 1 400°C to destroy organic matter, which creates smoke that would interfere with the optical measurement. *Atomization* is then carried out at 2 100°C for 10 s, during which the absorbance reaches a maximum and then decreases as Fe evaporates from the furnace. The time-integrated absorbance is taken as the analytical signal. Finally, the furnace is heated to 2 500°C for 3 s to vaporize any residue.

The temperature needed to char the sample **matrix** (the medium containing the analyte) may vaporize the analyte. *Matrix modifiers* can retard evaporation of analyte until the matrix has charred away. For example, $Mg(NO_3)_2$ matrix modifier prevents premature evaporation of Al analyte. At high temperature, $Mg(NO_3)_2$ produces gaseous MgO. Aluminum is converted to $Al_2O_3(s)$ during heating. At elevated temperature, Al_2O_3 decomposes to Al and O, and the Al evaporates. Evaporation of Al is retarded as long as MgO is present, by virtue of the reaction

$$3MgO(g) + 2Al(s) \rightleftharpoons 3Mg(g) + Al_2O_3(s)$$

When all the MgO is gone, Al_2O_3 finally decomposes and evaporates. The $Mg(NO_3)_2$ matrix modifier prevents Al from evaporating until a high temperature is reached.

Inductively Coupled Plasma

The **inductively coupled plasma** shown at the opening of this chapter and in Figure 19-6 reaches a much higher temperature than combustion flames. Its high temperature and stability eliminate many problems encountered with conventional flames. The plasma's disadvantage is its expense to purchase and to operate.

The plasma is energized by a radio-frequency induction coil wrapped around the quartz torch. High-purity argon gas plus analyte aerosol is fed through the torch. After a spark from a Tesla coil ionizes the Ar gas, free electrons accelerated by the radio-frequency field heat the gas to 6 000 to 10 000 K by colliding with atoms.

In the pneumatic nebulizer in Figure 19-4, liquid sample is sucked in by the flow of gas and breaks into small drops when it strikes the glass bead. The concentration of analyte needed for adequate signal (Table 19-2) is reduced by an order of magnitude with the *ultrasonic nebulizer* in Figure 19-7, in which liquid sample is directed onto a quartz crystal oscillating at 1 MHz. The vibrating crystal creates a fine aerosol that is carried by a stream of Ar through a heated tube, where solvent evaporates. The stream then passes through a cool zone in which solvent condenses and is removed. Analyte reaches the plasma as an aerosol of dry, solid particles. Plasma energy is not wasted in evaporating solvent, so more energy is available for atomization. Also, a larger fraction of the original sample reaches the flame than with a conventional nebulizer.

The quartz crystal is like the one that keeps time in your computer or wrist watch. It vibrates at a precisely reproducible frequency when stimulated by an oscillating electric field.

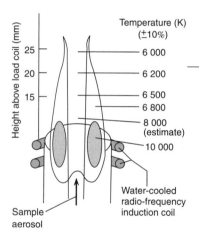

FIGURE 19-6 Temperature profile of a typical inductively coupled plasma used in analytical spectroscopy. [From V. A. Fassel, *Anal. Chem.* **1979**, *51*, 1290A.]

TABLE 19-2 **Comparison of detection limits for Ni⁺ ion at 231 nm**

Technique[a]	Detection limits for different instruments (ng/mL)
ICP-atomic emission (pneumatic nebulizer)	3–50
ICP-atomic emission (ultrasonic nebulizer)	0.3–4
Graphite furnace-atomic absorption	0.02–0.06
ICP-mass spectrometry	0.001–0.2

a. ICP, inductively coupled plasma.
SOURCE: J.-M. Mermet and E. Poussel, *Appl. Spectros.* **1995**, *49*, 12A.

Ask Yourself

19-B. Why can we detect smaller samples with lower concentrations with a furnace than with a flame or a plasma?

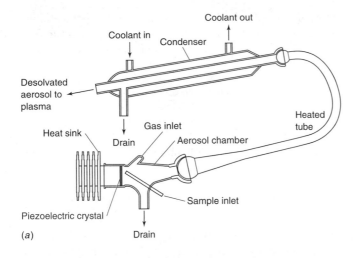

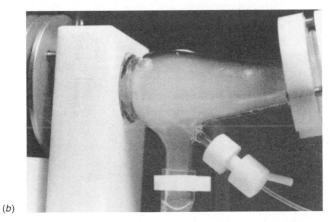

FIGURE 19-7 (a) Diagram of an ultrasonic nebulizer that lowers the detection limit for atomic spectroscopy for most elements by an order of magnitude. (b) Mist created when sample is sprayed against vibrating crystal. [Courtesy Cetac Technologies, Omaha, NE.]

19-3 How Temperature Affects Atomic Spectroscopy

Temperature determines the degree to which a sample breaks down to atoms and the extent to which an atom is found in its ground, excited, or ionized states. Each effect influences the strength of the observed signal.

The Boltzmann Distribution

Consider a molecule with two energy levels (Figure 19-8) separated by energy ΔE. Call the lower level E_0 and the upper level E^*. In general, an atom (or molecule) may have more than one state available at a given energy level. In Figure 19-8, we show three states at E^* and two at E_0. The number of states at each energy level is called the *degeneracy* of the level. We will call the degeneracies g_0 and g^*.

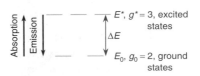

FIGURE 19-8 Two energy levels with different degeneracies. Ground-state atoms can absorb light to be promoted to the excited state. Excited-state atoms can emit light to return to the ground state.

The **Boltzmann distribution** describes the relative populations of different states at thermal equilibrium. If equilibrium exists (which is not true in all parts of a flame), the relative population (N^*/N_0) of any two states is

Boltzmann distribution:
$$\frac{N^*}{N_0} = \left(\frac{g^*}{g_0}\right)e^{-\Delta E/kT} \qquad (19\text{-}1)$$

where T is temperature (K) and k is Boltzmann's constant (1.381×10^{-23} J/K).

The Effect of Temperature on the Excited-State Population

The lowest excited state of a sodium atom lies 3.371×10^{-19} J/atom above the ground state. The degeneracy of the excited state is 2, whereas that of the ground state is 1. Let's calculate the fraction of sodium atoms in the excited state in an acetylene-air flame at 2 600 K:

$$\frac{N^*}{N_0} = \left(\frac{2}{1}\right)e^{-(3.371 \times 10^{-19}\text{ J})/[(1.381 \times 10^{-23}\text{ J/K})(2\,600\text{ K})]} = 0.000\,167$$

Fewer than 0.02% of the atoms are in the excited state.

How would the fraction of atoms in the excited state change if the temperature were 2 610 K instead?

$$\frac{N^*}{N_0} = \left(\frac{2}{1}\right)e^{-(3.371 \times 10^{-19}\text{ J})/[(1.381 \times 10^{-23}\text{ J/K})(2\,610\text{ K})]} = 0.000\,174$$

A 10-K temperature rise changes the excited-state population by 4% in this example.

The fraction of atoms in the excited state is still less than 0.02%, but that fraction has increased by $[(1.74 - 1.67)/1.67] \times 100 = 4\%$.

The Effect of Temperature on Absorption and Emission

We see that 99.98% of the sodium atoms are in their ground state at 2 600 K. *Varying the temperature by 10 K hardly affects the ground-state population and would not noticeably affect the signal in an atomic absorption experiment.*

How would emission intensity be affected by a 10-K rise in temperature? In Figure 19-8, we see that absorption arises from ground-state atoms, but emission arises from excited-state atoms. Emission intensity is proportional to the population of the excited state. *Because the excited-state population changes by 4% when the temperature rises 10 K, the emission intensity rises by 4%.* It is critical in atomic *emission* spectroscopy that the flame be very stable, or the emission intensity will vary significantly. In atomic *absorption* spectroscopy, flame temperature variation is not as critical.

The magnitude of atomic absorption is not as sensitive to temperature as the intensity of atomic emission, which is exponentially sensitive to temperature.

The inductively coupled plasma is almost always used for emission, not absorption, because it is so hot that a substantial fraction of atoms and ions are excited. Table 19-3 compares excited-state populations for a flame at 2 500 K and a plasma at 6 000 K. Although the fraction of excited atoms is small, each atom emits many photons per second because it is rapidly promoted back to the excited state by collisions.

Wavelength separation of states (nm)	Energy separation of states (J)	Excited-state fraction $(N^*/N_0)^a$	
		2 500 K	6 000 K
250	7.95×10^{-19}	1.0×10^{-10}	6.8×10^{-5}
500	3.97×10^{-19}	1.0×10^{-5}	8.3×10^{-3}
750	2.65×10^{-19}	4.6×10^{-4}	4.1×10^{-2}

a. Based on the equation $N^*/N_0 = (g^*/g_0)e^{-\Delta E/KT}$ in which $g^* = g_0 = 1$.

Ask Yourself

19-C. (a) A ground-state atom absorbs light of wavelength 400 nm to be promoted to an excited state. Find the energy difference (in joules) between the two states.

(b) If both states have a degeneracy of $g_0 = g^* = 1$, find the fraction of excited-state atoms (N^*/N_0) at thermal equilibrium at 2 500 K.

19-4   Instrumentation

Requirements for atomic absorption were shown in Figure 19-1. Principal differences between atomic and solution spectroscopy lie in the light source, the sample container (the flame, furnace, or plasma), and the need to subtract background emission from the observed signal.

The Linewidth Problem

For absorbance to be proportional to analyte concentration, the linewidth of radiation being measured must be substantially narrower than the linewidth of the absorbing atoms. Atomic absorption lines are very sharp, with an inherent width of $\sim 10^{-4}$ nm.

Two mechanisms broaden the lines in atomic spectroscopy. One is the *Doppler effect,* in which an atom moving toward the lamp samples the oscillating electromagnetic wave more frequently than one moving away from the lamp (Figure 19-9). That is, an atom moving toward the source "sees" higher frequency light than

The linewidth of the source must be narrower than the linewidth of the atomic vapor for Beer's law (Section 17-2) to be obeyed. The terms "linewidth" and "bandwidth" are used interchangeably, but "lines" are narrower than "bands."

Doppler and pressure effects broaden the atomic lines by 1–2 orders of magnitude relative to their inherent linewidths.

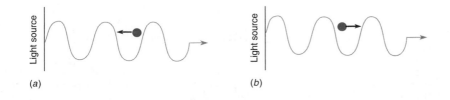

(a)   (b)

FIGURE 19-9 The Doppler effect. A molecule moving (a) toward the radiation source "feels" the electromagnetic field oscillate more often than one moving (b) away from the source.

419

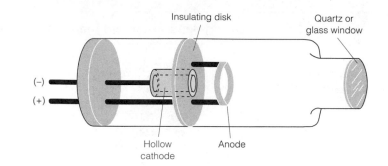

FIGURE 19-10 A hollow-cathode lamp.

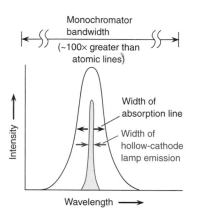

FIGURE 19-11 Relative line-widths of hollow-cathode emission, atomic absorption, and a monochromator. The linewidth from the hollow cathode is narrowest because the gas temperature in the lamp is lower than a flame temperature (so there is less Doppler broadening) and the pressure in the lamp is lower than the pressure in a flame (so there is less pressure broadening).

Background signal arises from absorption, emission, or scatter by everything in the sample besides analyte (the *matrix*), and from absorption, emission, or scatter by the flame, the plasma, or the furnace.

that encountered by one moving away. Linewidth is also affected by *pressure broadening* from collisions between atoms. Colliding atoms absorb a broader range of frequencies than do isolated atoms. Broadening is proportional to pressure. The Doppler effect and pressure broadening are similar in magnitude and yield linewidths of 10^{-3} to 10^{-2} nm in atomic spectroscopy.

Hollow-Cathode Lamps

To produce narrow lines of the correct frequency for atomic absorption, we use a **hollow-cathode lamp** containing a vapor of the same element as that being analyzed. The lamp in Figure 19-10 is filled with Ne or Ar at a pressure of 130–700 Pa. A high voltage between the anode and cathode ionizes the gas and accelerates cations toward the cathode. Ions striking the cathode "sputter" atoms from the metallic cathode into the gas phase. Free metal atoms are excited by collisions with high-energy electrons and then emit photons to return to the ground state. This atomic radiation shown in Figure 19-3 has the same frequency as that absorbed by atoms in the flame or furnace. The linewidth in Figure 19-11 is sufficiently narrow for Beer's law to hold. *A lamp with different cathode material is required for each element.*

Background Correction

Atomic spectroscopy requires **background correction** to distinguish analyte signal from absorption, emission, and optical scattering by the sample matrix, the flame, plasma, or white-hot graphite furnace. For example, Figure 19-12 shows the

FIGURE 19-12 Graphite furnace absorption spectrum of bronze dissolved in HNO_3. Notice the high, constant background absorbance of 0.3 in this narrow region of the spectrum. [From B. T. Jones, B. W. Smith, and J. D. Winefordner, *Anal. Chem.* **1989**, *61*, 1670.]

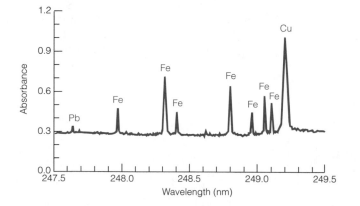

absorption spectrum of Fe, Cu, and Pb in a graphite furnace. Sharp atomic signals with maximum absorbance near 1.0 are superimposed on a flat background absorbance of 0.3. If we did not measure the background absorbance, significant errors would result. Background correction is most critical for graphite furnaces, which tend to be filled with smoke from the charring step. Optical scatter from smoke must somehow be distinguished from optical absorption by analyte.

For atomic absorption, *Zeeman background correction* (pronounced ZAY-man) is most commonly found in modern instruments for research and industrial laboratories. When a strong magnetic field is applied parallel to the light path through a flame or furnace, the absorption (or emission) lines of the atoms are split into three components (Figure 19-13). Two are shifted to slightly lower and higher wavelengths. The third component is unshifted but has the wrong electromagnetic polarization to absorb light from the lamp. Therefore analyte absorption decreases markedly in the presence of the magnetic field. For Zeeman background correction, the magnetic field is pulsed on and off. Absorption from the sample plus background is observed when the field is off, and absorption from the background alone is observed when the field is on. The difference between the two absorbances is due to analyte.

For atomic emission from an inductively coupled plasma, a popular form of background correction called *spectrum shifting* is shown in Figure 19-14. There is a quartz refractor plate located at the entrance to the monochromator. When light from the plasma strikes the plate at normal incidence, it is not refracted and the emission reaching the detector is centered at wavelength A—the peak of the analyte signal. If the refractor plate is rotated very slightly in one direction, the wavelength reaching the detector is shifted to B. If the plate is rotated slightly in the other direction, the wavelength reaching the detector is shifted to C. Emission intensity at A relative to the baseline drawn from B to C is the desired analytical signal.

Multielement Analysis with the Inductively Coupled Plasma

Inductively coupled plasma atomic emission spectroscopy is more versatile than atomic absorption with a flame or furnace because no lamps are required. Each element in the sample emits light at many characteristic frequencies. One way to measure many elements in a single sample is to scan the monochromator (Figure 18-4) to direct one wavelength at a time to the detector. This process requires time to integrate the signal at each wavelength (typically 1 to 10 s at each wavelength) and to rotate the monochromator from one wavelength to another. Tens of milliliters of sample may be aspirated into the plasma in the time required to measure several elements.

Alternatively, if you know in advance what elements you intend to measure, you can order a spectrometer with multiple fixed detectors—one at each wavelength of interest. Color Plate 17 shows "white" emission from a plasma directed at a grating that splits the light into its component wavelengths. An individual detector is located at each desired diffraction angle to measure emission from one element in the plasma. By this means, all elements of interest are measured simultaneously in 1 to 10 s. Errors are reduced because there are no moving parts. The price for this extremely convenient measurement is the cost of many detectors and the inflexibility of being limited to preselected analytes when you purchase the instrument.

Inductively coupled plasma-mass spectrometry (ICP-MS) is rapidly gaining ground as a method of choice in inorganic analysis. Instead of measuring emission from excited atoms and ions in the plasma, ions from the plasma are fed directly

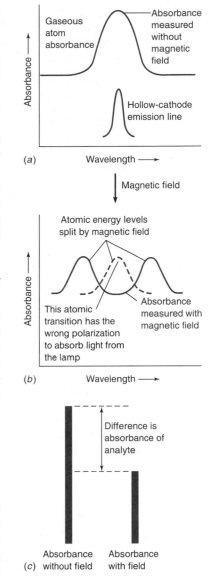

FIGURE 19-13 Principle of Zeeman background correction for atomic absorption spectroscopy. (*a*) In the absence of a magnetic field, we observe the sum of the absorbances of analyte and background. (*b*) In the presence of a magnetic field, the analyte absorbance is split away from the hollow cathode wavelength and the absorbance is due to background only. (*c*) The desired signal is the difference between those observed without and with the magnetic field.

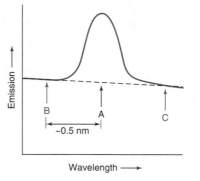

FIGURE 19-14 Correcting for background in atomic emission by spectrum shifting. Analyte emission is measured at wavelength A and background is measured at the slightly shifted wavelengths B and C. The height at A is measured relative to the baseline drawn between B and C.

Some comparisons:

Flame atomic absorption

- lowest cost equipment
- different lamp required for each element
- poor sensitivity

Furnace atomic absorption

- expensive equipment
- different lamp required for each element
- high background signals
- very high sensitivity

Plasma emission

- expensive equipment
- no lamps
- low background and low interference
- moderate sensitivity

ICP-MS

- very expensive equipment
- no lamps
- least background and interference
- highest sensitivity

into a mass spectrometer (Box 20-2) that separates ions according to their mass and measures the number of ions at each mass. ICP-MS provides lower detection limits than atomic spectroscopy (Table 19-2) and automatic multielement analysis because the entire range of atomic masses is electronically scanned. An integration time of 1–3 s is required for each atomic mass of interest.

Detection Limits

The **detection limit** is the concentration of an element that gives a signal equal to two times the peak-to-peak noise level of the baseline (Figure 19-15).[1] The baseline noise level is measured while a blank sample is aspirated into the flame.

Detection limits for a furnace are typically 100 times lower than those of a flame (Figure 19-16), because the sample is confined in a small volume for a relatively long time in the furnace. Detection limits for the plasma are intermediate and can be

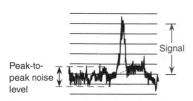

FIGURE 19-15 Measurement of peak–to-peak noise level and signal level. The signal is measured from its base at the midpoint of the noise component along the slightly slanted baseline. This sample exhibits a signal-to-noise ratio of 2.4.

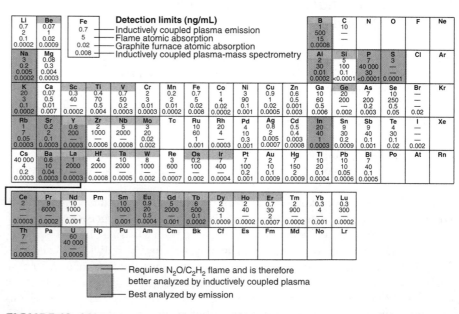

FIGURE 19-16 Flame, furnace, inductively coupled plasma emission, and inductively coupled plasma–mass spectrometry detection limits (ng/mL = ppb) with instruments from one manufacturer. Reliable quantitative analysis usually requires concentrations 10–100 times greater than the detection limit. [Flame, furnace, and ICP emission limits are for instruments from one manufacturer cited in R. J. Gill, *Am. Lab.* November 1993, 24F. ICP-MS limits are from T. T. Nham, *Am. Lab.* August 1998, 17A.]

Box 19-1 *Informed Citizen*

Atomic Emission Spectroscopy in the Clinical Laboratory

Before the advent of atomic spectroscopy, laborious gravimetric procedures were required to measure Na^+ and K^+ in serum and urine. Today these measurements are routine in clinical analysis.

The typical flame photometer shown in the figure operates on air and natural gas. Optical filters isolate the strong emission lines of Na, K, Li, Ca, and Ba. Full-scale sensitivity for Na and K corresponds to 3 μg/mL; 50 μg/mL for Ca. The digital readout is calibrated by aspirating standard samples through the narrow tube immersed in the beaker. Lithium is often used as an *internal standard* (Section 20-5) for sodium and potassium.

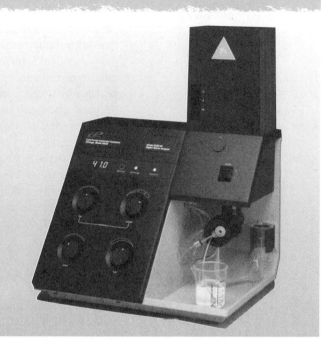

Digital flame photometer. [Courtesy Cole–Parmer Instrument Co., Chicago, IL.]

improved by a factor of 3 to 10 by viewing emission along the axis of the plasma, instead of perpendicular to the axis. The axial view increases the effective pathlength. Inductively coupled plasma-mass spectrometry has lower detection limits in Figure 19-16 than any of the atomic spectroscopies. Figure 19-17 compares the linear operating ranges for ICP-MS and ICP-atomic emission. The linear range is the concentration interval in which the measured signal is proportional to analyte concentration. ICP-MS achieves a lower detection range than any atomic spectroscopy.

Detection limits vary among instruments. The *ultrasonic nebulizer* in Figure 19-7 lowers the detection limit for some elements by a factor of 10.

Box 19-1 shows an application of atomic spectroscopy in clinical analysis.

Ask Yourself

19-D. How is background correction accomplished in atomic absorption and atomic emission spectroscopy?

19-5   Interference

Interference is any effect that changes the signal when analyte concentration remains unchanged. Interference can be corrected by counteracting the source of interference or by preparing standards that exhibit the same interference.

Types of Interference

Spectral interference occurs when analyte signal overlaps signals from other species in the sample or signals due to the flame or furnace. Interference from the flame is subtracted by background correction. The best means of dealing with overlap

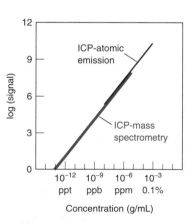

FIGURE 19-17 Approximate comparison of linear operating ranges for inductively coupled plasma with mass spectrometry or atomic emission detection. [Adapted from Thermo Jarrell Ash instrument brochure, Franklin, MA.]

423

Types of interference:

- spectral: unwanted signals overlap analyte signal
- chemical: chemical reactions decrease the concentration of analyte atoms
- ionization: ionization of analyte atoms decreases the concentration of neutral atoms

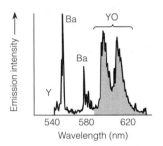

FIGURE 19-18 Emission from a plasma produced by laser irradiation of the high-temperature superconductor $YBa_2Cu_3O_7$. The solid is vaporized by the laser, and excited atoms and molecules in the gas phase emit light at their characteristic wavelengths. [From W. A. Weimer, *Appl. Phys. Lett.* **1988**, *52*, 2171.]

between lines of different elements in the sample is to choose another wavelength for analysis. The spectrum of a molecule is much broader than that of an atom, so spectral interference can occur at many wavelengths. Figure 19-18 shows an example of a plasma containing Y and Ba atoms in addition to YO molecules. Elements that form stable oxides in the flame commonly give spectral interference.

Chemical interference is caused by any substance that decreases the extent of atomization of analyte. For example, SO_4^{2-} and PO_4^{3-} hinder the atomization of Ca^{2+}, perhaps by forming nonvolatile salts. *Releasing agents* can be added to a sample to decrease chemical interference. EDTA and 8-hydroxyquinoline protect Ca^{2+} from the interfering effects of SO_4^{2-} and PO_4^{3-}. La^{3+} is also a releasing agent, apparently because it preferentially reacts with PO_4^{3-} and frees the Ca^{2+}. A fuel-rich flame reduces oxidized analyte species that would otherwise hinder atomization. Higher flame temperatures eliminate many kinds of chemical interference.

Ionization interference is a problem in the analysis of alkali metals, which have the lowest ionization potentials. For any element, we can write a gas-phase ionization reaction:

$$M(g) \rightleftharpoons M^+(g) + e^-(g) \qquad K = \frac{[M^+][e^-]}{[M]} \qquad (19\text{-}2)$$

At 2 450 K and a pressure of 0.1 Pa, Na is 5% ionized. With its lower ionization potential, K is 33% ionized under the same conditions. Ionized atoms have energy levels different from those of neutral atoms, so the desired signal is decreased.

An *ionization suppressor* is an element added to a sample to decrease the ionization of analyte. For example, 1 mg/mL of CsCl is added to the sample for the analysis of potassium, because cesium is more easily ionized than potassium. By producing a high concentration of electrons in the flame, ionization of Cs reverses Reaction 19-2 for K. This reversal is an example of Le Châtelier's principle.

The *method of standard addition,* described in Section 16-4, compensates for many types of interference by adding known quantities of analyte to the unknown in its complex matrix. For example, Figure 19-19 shows the analysis of strontium in aquarium water by standard addition. The slope of the standard addition curve is 0.018 8 absorbance units/ppm. If, instead, Sr is added to distilled water, the slope is 0.030 8 absorbance units/ppm. That is, in distilled water, the absorbance increases 0.030 8/0.018 8 = 1.64 times more than it does in aquarium water for each addition of standard Sr. We attribute the lower response in aquarium water to interference by other species that are present. The negative *x*-intercept of the standard

FIGURE 19-19 Atomic absorption calibration curve for Sr added to distilled water and standard addition of Sr to aquarium water. All solutions are made up to a constant volume, so the ordinate is the concentration of added Sr. [Data from L. D. Gilles de Pelichy, C. Adams, and E. T. Smith, *J. Chem. Ed.* **1997**, *74*, 1192.]

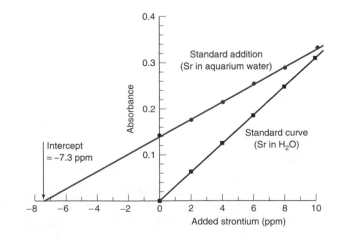

addition curve, 7.3 ppm, is a reliable measure of Sr in the aquarium. If we had just measured the atomic absorption of Sr in aquarium water and used the calibration curve for distilled water, we would have overestimated the Sr concentration by 64%.

Virtues of the Inductively Coupled Plasma

An Ar plasma eliminates common interferences. The plasma is twice as hot as a conventional flame, and the residence time of analyte in the plasma is about twice as long. Atomization is more complete, and the signal is correspondingly enhanced. Formation of analyte oxides and hydroxides is negligible. The plasma is relatively free of background radiation. Figure 19-20 shows plasma emission calibration curves that are linear over nearly five orders of magnitude. For atomic absorption in flames and furnaces, the linear range spans only two orders of magnitude.

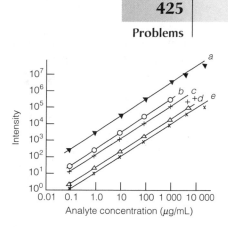

FIGURE 19-20 Analytical calibration curves for emission from (a) Ba$^+$, (b) Cu, (c) Na, (d) Fe, and (e) Ba in an inductively coupled plasma. Note that both axes are logarithmic. [From R. N. Savage and G. M. Hieftje, *Anal. Chem.* **1979**, *51*, 408.]

Ask Yourself

19-E. What is meant by **(a)** spectral, **(b)** chemical, and **(c)** ionization interference?

Key Equation

Boltzmann distribution

$$\frac{N^*}{N_0} = \left(\frac{g^*}{g_0}\right)e^{-\Delta E/kT}$$

N^* = population of excited state

N_0 = population of ground state

ΔE = energy difference between excited and ground states

g^* = number of states with energy E^*

g_0 = number of states with energy E_0

Important Terms

atomic absorption spectroscopy

atomic emission spectroscopy

atomization

background correction

Boltzmann distribution

chemical interference

detection limit

graphite furnace

hollow-cathode lamp

inductively coupled plasma

ionization interference

matrix

spectral interference

Problems

19-1. Compare the advantages and disadvantages of furnaces and flames in atomic absorption spectroscopy.

19-2. Compare the advantages and disadvantages of the inductively coupled plasma and flames in atomic spectroscopy.

19-3. In which technique, atomic absorption or atomic emission, is flame temperature stability more critical? Why?

19-4. Spectrum shifting of the monochromator (Figure 19-14) is used for background correction in atomic emission.

Why can't we use the same technique for background correction in atomic absorption?

19-5. What is the purpose of a matrix modifier in atomic spectroscopy?

19-6. The first excited state of Ca is reached by absorption of 422.7-nm light.

(a) What is the energy difference (J) between the ground and excited states? (*Hint:* See Section 17-1.)

(b) The degeneracies are $g^*/g_0 = 3$ for Ca. What is the ratio N^*/N_0 at 2 500 K?

(c) By what percentage will the fraction in **(b)** be changed by a 15-K rise in temperature?

(d) Find the ratio N^*/N_0 at 6 000 K.

19-7. The first excited state of Cu is reached by absorption of 327-nm radiation.

(a) What is the energy difference (J) between the ground and excited states?

(b) The ratio of degeneracies is $g^*/g_0 = 3$ for Cu. What is the ratio N^*/N_0 at 2 400 K?

(c) By what percentage will the fraction in **(b)** be changed by a 15-K rise in temperature?

(d) What will the ratio N^*/N_0 be at 6 000 K?

19-8. The atomic absorption signal shown in the figure below was obtained with 0.048 5 μg Fe/mL in a graphite furnace. Measure the signal height and the peak-to-peak noise level. The detection limit is the concentration of Fe that gives a signal equal to two times the peak-to-peak noise level. Estimate the detection limit for Fe.

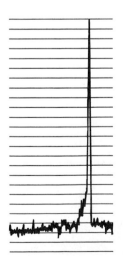

19-9. *Standard curve.* (Review Section 17-4 on standard curves.) A series of potassium standards gave the following emission intensities at 404.3 nm. Construct a standard curve and find the concentration of potassium in the unknown.

Sample (μg K/mL)	Relative emission
Blank	0
5.00	124
10.0	243
20.0	486
30.0	712
Unknown	417

19-10. *Standard addition.* (See Section 16-4 for the method of standard addition.) Suppose that 5.00 mL of serum containing an unknown concentration of potassium, $[K^+]_i$, gave an atomic emission signal of 3.00 mV in the flame photometer in Box 19-1. After adding 1.00 mL of 30.0 mM K^+ standard and diluting the mixture to 10.00 mL, the emission signal increased to 4.00 mV.

(a) Find the concentration of added standard, $[S]_f$, in the mixture.

(b) The initial 5.00-mL sample was diluted to 10.00 mL. Therefore the serum potassium concentration decreased from $[K^+]_i$ to $[K^+]_f = \frac{1}{2}[K^+]_i$. Use Equation 16-10 to find the original K^+ content of the serum, $[K^+]_i$.

19-11. *Standard addition.* A blood serum sample containing Na^+ gave an emission signal of 4.27 mV in a flame photometer. A small volume of concentrated standard Na^+ was then added to increase the Na^+ concentration by 0.104 M, without significantly diluting the sample. This "spiked" serum sample gave a signal of 7.98 mV in atomic emission. Find the original concentration of Na^+ in the serum.

19-12. *Standard addition.* An unknown sample of Cu^{2+} gave an absorbance of 0.262 in an atomic absorption analysis. Then 1.00 mL of solution containing 100.0 ppm ($= \mu$g/mL) Cu^{2+} was mixed with 95.0 mL of unknown, and the mixture was diluted to 100.0 mL in a volumetric flask. The absorbance of the new solution was 0.500. Find the original concentration of Cu^{2+} in the unknown.

19-13. *Standard addition.* An unknown containing element X was mixed with aliquots of a standard solution of element X for atomic absorption spectroscopy. The standard solution contained 1 000.0 μg of X per milliliter.

Volume of unknown (mL)	Volume of standard (mL)	Total volume (mL)	Absorbance
10.00	0	100.0	0.163
10.00	1.00	100.0	0.240
10.00	2.00	100.0	0.319
10.00	3.00	100.0	0.402
10.00	4.00	100.0	0.478

(a) Calculate the concentration (μg X/mL) of added standard in each solution.

(b) Prepare a graph similar to Figure 16-11 to determine the concentration of X in the unknown.

19-14. *Standard addition.* Li^+ was determined by atomic emission, using standard additions. From the data below, prepare a graph similar to Figure 16-11 to find the concentration of Li in pure unknown. The Li standard contained 1.62 μg Li/mL.

Unknown (mL)	Standard (mL)	Final volume (mL)	Emission intensity (arbitrary units)
10.00	0.00	100.0	309
10.00	5.00	100.0	452
10.00	10.00	100.0	600
10.00	15.00	100.0	765
10.00	20.00	100.0	906

19-15. *Internal standard.* (See Section 20-5 for a discussion of internal standards.) A solution was prepared by mixing 10.00 mL of unknown (X) with 5.00 mL of standard (S) containing 8.24 μg S/mL and diluting to 50.0 mL. The measured signal quotient was (signal due to X/signal due to S) = 1.69.

(a) In a separate experiment, it was found that, for equal concentrations of X and S, the signal due to X was 0.93 times as intense as the signal due to S. Find the concentration of X in the unknown.

(b) In a third experiment, it was found that, for the concentration of X equal to 3.42 times the concentration of S, the signal due to X was 0.93 times as intense as the signal due to S. Find the concentration of X in the unknown.

19-16. *Internal standard.* (See Section 20-5 for a discussion of internal standards.) Mn was used as an internal standard for measuring Fe by atomic absorption. A standard mixture containing 2.00 μg Mn/mL and 2.50 μg Fe/mL gave a quotient (Fe signal/Mn signal) = 1.05. A mixture with a volume of 6.00 mL was prepared by mixing 5.00 mL of unknown Fe solution with 1.00 mL containing 13.5 μg Mn/mL. The absorbance of this mixture at the Mn wavelength was 0.128, and the absorbance at the Fe wavelength was 0.185. Find the molarity of the unknown Fe solution.

How Would You Do It?

19-17. The measurement of Li in brine (salt water) is used by geochemists to help determine the origin of this fluid in oil fields. Flame atomic emission and absorption of Li are subject to interference by scattering, ionization, and overlapping spectral emission from other elements. Atomic absorption analysis of replicate samples of a marine sediment gave the results in the table below.[2]

(a) Suggest a reason for the increasing apparent Li concentration in samples 1 to 3.

(b) Why do samples 4 to 6 give an almost constant result?

(c) What value would you recommend for the real concentration of Li in the sample?

Sample and treatment	Li found (μg/g)	Analytical method	Flame type
1. None	25.1	Standard curve	air/C_2H_2
2. Dilute to 1/10 with H_2O	64.8	Standard curve	air/C_2H_2
3. Dilute to 1/10 with H_2O	82.5	Standard addition	air/C_2H_2
4. None	77.3	Standard curve	N_2O/C_2H_2
5. Dilute to 1/10 with H_2O	79.6	Standard curve	N_2O/C_2H_2
6. Dilute to 1/10 with H_2O	80.4	Standard addition	N_2O/C_2H_2

Notes and References

1. Detection limit is really the lowest concentration that can be "reliably" detected. For various statistical definitions of detection limit, see G. L. Long and J. D. Winefordner, *Anal. Chem.* **1983,** *55,* 712A; W. R. Porter, *Anal. Chem.* **1983,** *55,* 1290A; and J. R. Burdge, D. L. MacTaggart, and S. O. Farwell, *J. Chem. Ed.* **1999,** *76,* 434.

2. B. Baraj, L. F. H. Niencheski, R. D. Trapaga, R. G. França, V. Cocoli, and D. Robinson, *Fresenius J. Anal. Chem.* **1999,** *364,* 678.

Katia and Dante

Courageous Katia Krafft samples vapors from a volcano. [Photo by Maurice Krafft, from F. Press and R. Siever, *Understanding Earth* (New York: W. H. Freeman, 1994), p. 107.]

Robot Dante was equipped with a gas chromatograph to sample emissions from volcanic Mt. Erebus in Antarctica. [Philip R. Kyle, New Mexico Tech.]

Over billions of years, volcanoes released gases that created today's atmosphere and condensed into the earth's oceans. Generally 70–95% of volcanic gas is H_2O, followed by smaller volumes of CO_2 and SO_2, and traces of N_2, H_2, CO, sulfur, Cl_2, HF, HCl, and H_2SO_4. To reduce wear and tear on brave scientists who have ventured near a volcano to collect gases, the eight-legged robot Dante was designed to rappel 250 m into the sheer crater of Mt. Erebus in Antarctica in 1992 to sample volcanic emissions with a gas chromatograph and return information via a fiber-optic cable. Dante also had a quartz crystal microbalance (page 34) to measure aerosols, an infrared thermometer, and a γ-ray spectrometer to measure K, Th, and U in the crater's soil. Dante failed when its optical fiber broke after just 6 m of descent into the crater. In 1994, an improved Dante II successfully probed the volcano Mt. Spurr in Alaska.

The volcanologists Katia and Maurice Krafft died in the eruption of Mt. Unzen in Japan in 1991.

Chapter *20*

PRINCIPLES OF
CHROMATOGRAPHY

*C*hromatography is the most powerful tool in an analytical chemist's arsenal for separating and measuring the components of a complex mixture. In conjunction with mass spectrometry and infrared spectroscopy, chromatography can identify the components as well. This chapter discusses principles of chromatography, and the next two chapters take up specific methods of separation.

Chromatography is widely used for

Quantitative analysis: How much of a component is present?

Qualitative analysis: What is the identity of the component?

20-1   What Is Chromatography?

Chromatography is a process in which we separate compounds from one another by passing a mixture through a column that retains some compounds longer than others. In Figure 20-1, a solution containing compounds A and B is placed on top of a column previously packed with solid particles and filled with solvent. When the outlet is opened, A and B flow down into the column and are washed through with fresh solvent applied to the top of the column. If solute A is more strongly *adsorbed* than solute B on the solid particles, then A spends a smaller fraction of the time free in solution. Solute A moves down the column more slowly than B and emerges at the bottom after B.

Adsorption means sticking to the surface of the solid particles. Color Plate 18 illustrates thin-layer chromatography, which is a form of adsorption chromatography.

429

FIGURE 20-1 Separation by chromatography. Solute A has a greater affinity than solute B for the stationary phase, so A remains on the column longer than B. The term *chromatography* is derived from experiments in 1903 by M. Tswett, who separated plant pigments with a column containing solid $CaCO_3$ particles (the stationary phase) washed by hydrocarbon solvent (the mobile phase). The separation of colored bands led to the name *chromatography,* from the Greek word *chromatos,* meaning "color."

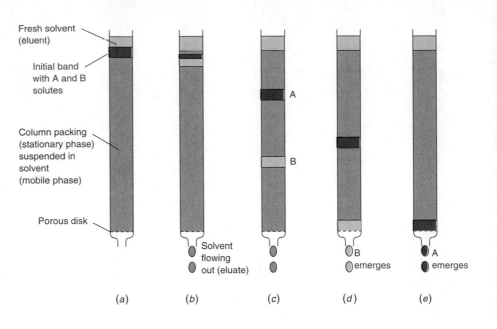

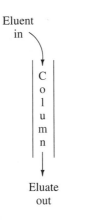

Eluent
in

Column

Eluate
out

For their pioneering work on liquid-liquid partition chromatography in 1941, A. J. P. Martin and R. L. M. Synge received the Nobel Prize in 1952.

B. A. Adams and E. L. Holmes developed the first synthetic ion-exchange resins in 1935. *Resins* are relatively hard, amorphous (noncrystalline) organic solids. *Gels* are relatively soft.

In molecular exclusion, large molecules pass through the column *faster* than small molecules.

The **mobile phase** (solvent moving through the column) in chromatography is either a liquid or a gas. The **stationary phase** (the substance that stays fixed inside the column) is either a solid or a viscous (syrupy) liquid bound to solid particles or to the inside wall of a hollow capillary column. Partitioning of solutes between the mobile and stationary phases gives rise to separation. In **gas chromatography,** the mobile phase is a gas; and in **liquid chromatography,** the mobile phase is a liquid.

Fluid entering the column is called **eluent.** Fluid exiting the column is called **eluate.** The process of passing liquid or gas through a chromatography column is called **elution.**

Chromatography can be classified by the type of interaction of the solute with the stationary phase, as shown in Figure 20-2.

Adsorption chromatography uses a solid stationary phase and a liquid or gaseous mobile phase. Solute is adsorbed on the surface of the solid particles.

Partition chromatography involves a thin liquid stationary phase coated on the surface of a solid support. Solute equilibrates between the stationary liquid and the mobile phase.

Ion-exchange chromatography features ionic groups such as $—SO_3^-$ or $—N(CH_3)_3^+$ covalently attached to the stationary solid phase, which is usually a *resin.* Solute ions are attracted to the stationary phase by electrostatic forces. The mobile phase is a liquid.

Molecular exclusion chromatography (also called *gel filtration, gel permeation,* or *molecular sieve* chromatography) separates molecules by size, with larger molecules passing through most quickly. Unlike other forms of chromatography, there is no attractive interaction between the "stationary phase" and the solute. The stationary phase has pores small enough to exclude large molecules, but not small ones. Large molecules stream past without entering the pores. Small molecules take longer to pass through the column because they enter the pores and therefore must flow through a larger volume before leaving the column.

Affinity chromatography, the most selective kind of chromatography, employs specific interactions between one kind of solute molecule and a second molecule that is covalently attached (immobilized) to the stationary phase. For example, the immobilized molecule might be an antibody to a particular protein. When a mixture containing a thousand different proteins is passed through the column, only that one protein binds to the antibody on the column. After washing all other proteins from the column, the desired protein is dislodged by changing the pH or ionic strength.

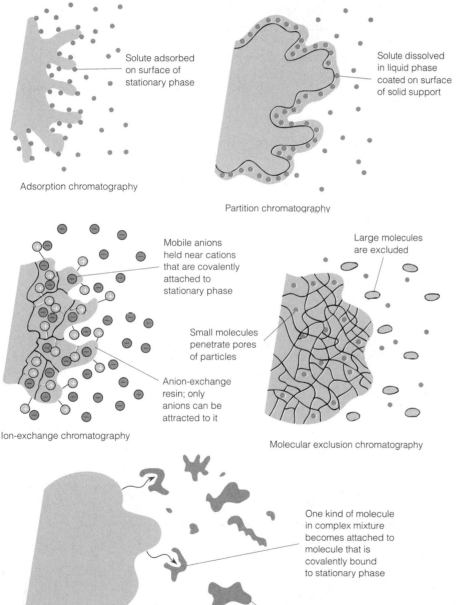

Solute adsorbed on surface of stationary phase

Adsorption chromatography

Solute dissolved in liquid phase coated on surface of solid support

Partition chromatography

Mobile anions held near cations that are covalently attached to stationary phase

Small molecules penetrate pores of particles

Anion-exchange resin; only anions can be attracted to it

Ion-exchange chromatography

Large molecules are excluded

Molecular exclusion chromatography

One kind of molecule in complex mixture becomes attached to molecule that is covalently bound to stationary phase

All other molecules simply wash through

Affinity chromatography

FIGURE 20-2 Types of chromatography.

20-A. Match the terms with their definitions:
   1. Adsorption chromatography
   2. Partition chromatography
   3. Ion-exchange chromatography
   4. Molecular exclusion chromatography
   5. Affinity chromatography

   A. Mobile-phase ions attracted to stationary phase ions.
   B. Solute attracted to specific groups attached to stationary phase.
   C. Solute equilibrates between mobile phase and surface of stationary phase.
   D. Solute equilibrates between mobile phase and stationary liquid film.
   E. Solute penetrates voids in stationary phase. Largest solutes eluted first.

20-2   How We Describe a Chromatogram

Detectors discussed in the next chapter respond to solutes as they exit the chromatography column. A **chromatogram** shows detector response as a function of time (or elution volume) in a chromatography experiment (Figure 20-3). Each peak corresponds to a different substance eluted from the column. The **retention time,** t_r, is the time needed after injection for an individual solute to reach the detector.

Theoretical Plates

An ideal chromatographic peak has a Gaussian shape, like that in Figure 20-3. If the height of the peak is h, the *width at half-height, $w_{1/2}$,* is measured at $\frac{1}{2}h$. For a Gaussian peak, $w_{1/2}$ is equal to 2.35σ, where σ is the standard deviation of the peak. The width of the peak at the baseline, as shown in Figure 20-3, is 4σ.

FIGURE 20-3 Schematic gas chromatogram, showing measurement of retention time (t_r) and width at half-height ($w_{1/2}$). The width at the base (w) is found by drawing tangents to the steepest parts of the Gaussian curve and extrapolating down to the baseline. The standard deviation of the Gaussian curve is σ. In gas chromatography, a small volume of CH_4 injected with the 0.1- to 2-μL sample is usually the first component to be eluted.

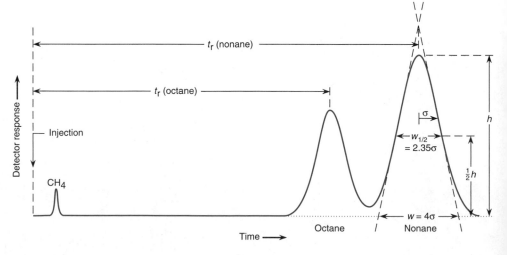

In days of old, distillation was the most powerful means for separating volatile compounds from one another. A distillation column was divided into sections (*plates*) in which liquid and vapor equilibrated with each other. The more plates on a column, the more equilibration steps, and the better the separation between compounds with different boiling points.

Nomenclature from distillation has carried over into chromatography. We speak of a chromatography column as if it were divided into discrete sections (called **theoretical plates**) in which a solute molecule equilibrates between the mobile and stationary phases. The retention of a compound on a column can be described by the number of theoretical equilibration steps that occur between injection and elution. The more equilibration steps (the more theoretical plates), the narrower the bandwidth when the compound emerges.

For any peak in Figure 20-3, the number of theoretical plates is computed by measuring the retention time and the width at half-height:

Even though we speak of discrete theoretical plates, chromatography is really a continuous process. Theoretical plates are an imaginary way to describe the process.

Number of plates on column:
$$N = \frac{5.55 t_r^2}{w_{1/2}^2}$$
(20-1)

Retention and width can be measured in units of time or volume (such as milliliters of eluate). Both t_r and $w_{1/2}$ must be expressed in the same units in Equation 20-1.

If a column is divided into N theoretical plates (only in our minds), then the **plate height,** H, is the length of one plate. H is calculated by dividing the length of the column (L) by the number of theoretical plates on the column:

Plate height:
$$H = L/N$$
(20-2)

The smaller the plate height, the narrower the peaks. The ability of a column to separate components of a mixture is improved by decreasing plate height. An efficient column has more theoretical plates than an inefficient column. Different solutes behave as if the column has somewhat different plate heights, because different compounds equilibrate between the mobile and stationary phases at different rates. Plate heights are in the neighborhood of ~0.1 to 1 mm in gas chromatography, ~10 μm in high-performance liquid chromatography, and <1 μm in capillary electrophoresis.

You can test for degradation of a column by injecting a test compound periodically and looking for peak asymmetry and changes in the number of plates.

small plate height $\Rightarrow$
narrow peaks $\Rightarrow$
better separations

EXAMPLE   Measuring Plates

A solute with a retention time of 407 s has a width at half-height of 8.1 s on a column 12.2 m long. Find the number of plates and the plate height.

SOLUTION number of plates $= N = \dfrac{5.55 t_r^2}{w_{1/2}^2} = \dfrac{5.55 \cdot 407^2}{8.1^2} = 1.40 \times 10^4$

plate height $= H = \dfrac{L}{N} = \dfrac{12.2 \text{ m}}{1.40 \times 10^4} = 0.87 \text{ mm}$

Resolution

The **resolution** between neighboring peaks is equal to the peak separation (Δt_r) divided by the average peak width (w_{av}, measured at the base, as in Figure 20-3):

Resolution:
$$\text{resolution} = \frac{\Delta t_r}{w_{av}} \qquad (20\text{-}3)$$

resolution $\propto \sqrt{\text{column length}}$

(The symbol $\propto$ means "is proportional to.")

The better the resolution, the more complete is the separation between neighboring peaks. Figure 20-4 shows that when two peaks have a resolution of 0.50, the overlap of area is 16%. When the resolution is 1.00, the overlap is 2.3%. If you double the length of an ideal chromatography column, you will improve resolution by $\sqrt{2}$.

Scaling Up a Separation

Analytical chromatography: small-scale analysis

Preparative chromatography: large-scale separation

Analytical chromatography is conducted on a small scale to separate, identify, and measure components of a mixture. *Preparative* chromatography is carried out on a large scale to isolate a significant quantity of one or more components of a mixture. Analytical chromatography typically uses long, thin columns to obtain good resolution. Preparative chromatography usually employs short, fat columns that can handle larger quantities of material but give inferior resolution. (Long, fat columns may be prohibitively expensive to buy or operate.)

If you have developed a chromatographic procedure to separate 2 mg of a mixture on a column with a diameter of 1.0 cm, what size column should you use to separate 20 mg of the mixture? The most straightforward way to scale up is to maintain the same column length and increase the cross-sectional area to maintain a constant ratio of unknown to column volume. The cross-sectional area is proportional to the square of the column radius (r), so

Scaling equation:
$$\frac{\text{large load (g)}}{\text{small load (g)}} = \left(\frac{\text{large column radius}}{\text{small column radius}}\right)^2 \qquad (20\text{-}4)$$

$$\frac{20\ \text{mg}}{2\ \text{mg}} = \left(\frac{\text{large column radius}}{0.50\ \text{cm}}\right)^2$$

$$\text{large column radius} = 1.58\ \text{cm}$$

A column with a diameter near 3 cm would be appropriate.

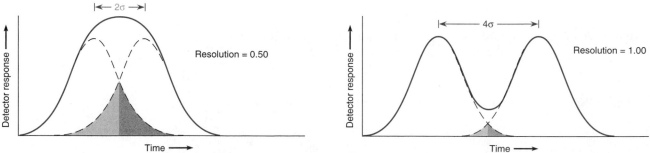

FIGURE 20-4 Resolution of Gaussian peaks of equal area and amplitude. Dashed lines show individual peaks; solid lines are the sum of two peaks. Overlapping area is shaded.

To reproduce the conditions of the smaller column in the larger column, the *volume flow rate* (mL/min) should be increased in proportion to the cross-sectional area of the column. If the area of the large column is 10 times greater than that of the small column, the volume flow rate should be 10 times greater. If the small sample is dissolved in a volume V before it is applied to the small column, the large sample should be dissolved in a volume $10V$.

Rules for scaling up without reducing separation:

- cross-sectional area of column $\propto$ mass of analyte
- keep column length constant
- volume flow rate $\propto$ cross-sectional area of column
- sample volume $\propto$ mass of analyte

Ask Yourself

20-B. (a) Use a ruler to measure retention times and widths at the base (w) of octane and nonane in Figure 20-3 to the nearest 0.1 mm.

(b) Calculate the number of theoretical plates for octane and nonane.

(c) If the column is 1.00 m long, find the plate height for octane and nonane.

(d) Use your measurements from **(a)** to compute the resolution between octane and nonane.

(e) Suppose that the sample size for Figure 20-3 was 3.0 mg and the column dimensions were 4.0 mm diameter $\times$ 1.00 m length and the flow rate was 7.0 mL/min. What size column and what flow rate should be used to obtain the same quality of separation of 27.0 mg of sample?

20-3   Why Do Bands Spread?

If a solute is applied to a column in an ideal manner as an infinitesimally thin band, the band will broaden as it moves through a chromatography column (Figure 20-5). Broadening occurs because of diffusion, slow equilibration of solute between the mobile and stationary phases, and irregular flow paths on the column.

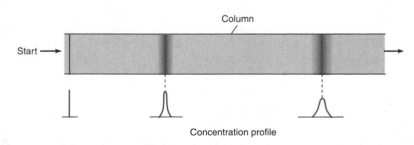

FIGURE 20-5 Broadening of an initially sharp band of solute as it moves through a chromatography column.

Bands Diffuse

An infinitely narrow band of solute that is stationary inside the column slowly broadens because solute molecules diffuse away from the center of the band in both directions. This inescapable process, called *longitudinal diffusion,* begins at the moment solute is injected into the column. In chromatography, the further a band has traveled, the more time it has had to diffuse and the broader it becomes (Figure 20-6).

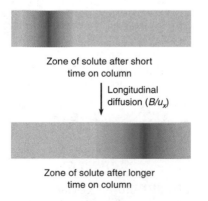

FIGURE 20-6 In longitudinal diffusion, solute continuously diffuses away from the concentrated zone at the center of its band. The greater the flow rate, the less time is spent on the column and the less longitudinal diffusion occurs.

The flow rate of importance in Equations 20-5 to 20-7 is *linear* flow rate (cm/min). This is the rate at which solvent goes past stationary phase. For a given column diameter, *volume* flow rate (mL/min) is proportional to linear flow rate.

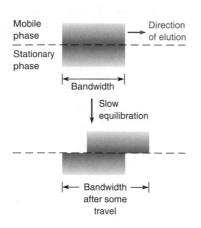

FIGURE 20-7 Solute requires a finite time to equilibrate between the mobile and stationary phases. If equilibration is slow, solute in the stationary phase tends to lag behind that in the mobile phase, thereby causing the band to spread. The slower the flow rate, the less the zone broadening by this mechanism.

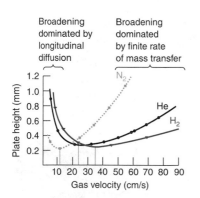

FIGURE 20-8 Optimum resolution (i.e., minimum plate height) occurs at an intermediate flow rate. Curves show measured plate height in gas chromatography of $n\text{-}C_{17}H_{36}$ at 175°C, using N_2, He, or H_2 mobile phase. [From R. R. Freeman, Ed., *High Resolution Gas Chromatography* (Palo Alto, CA: Hewlett Packard Co., 1981).]

The faster the flow rate, the less time a band spends in the column, and the less time there is for diffusion to occur. The faster the flow, the sharper the peaks. Broadening by longitudinal diffusion is inversely proportional to flow rate:

Broadening by longitudinal diffusion: $\qquad$ broadening $\propto \dfrac{1}{u}$ $\qquad$ (20-5)

where u is the flow rate, usually measured in mL/min.

Solute Requires Time to Equilibrate Between Phases

Imagine a solute that is distributed between the mobile and stationary phases at some moment in time at some position in a column with zero flow rate. Now set the flow into motion. If the solute cannot equilibrate rapidly enough between the phases, then solute in the stationary phase tends to lag behind solute in the mobile phase (Figure 20-7). This broadening due to the *finite rate of mass transfer* between phases becomes worse as the flow rate increases:

Broadening by finite rate of mass transfer: $\qquad$ broadening $\propto u$ $\qquad$ (20-6)

A Separation Has an Optimum Rate

When trying to separate closely spaced bands, we want to minimize band broadening during chromatography. If the bands broaden too much, we would not be able to resolve them from each other. Because broadening by longitudinal diffusion decreases with increasing flow rate (Equation 20-5), but broadening by the finite rate of mass transfer increases with increasing flow rate (Equation 20-6), there is some intermediate flow rate that gives minimum broadening and optimum resolution (Figure 20-8). Part of the science and art of chromatography is to experiment with conditions such as flow rate and solvent composition to obtain the best separation (or adequate separation) between components of a mixture.

The rate of mass transfer between phases increases with temperature. Raising the column temperature might improve the resolution or allow faster separations without reducing resolution.

Some Band Broadening Is Independent of Flow Rate

Some mechanisms of band broadening are independent of flow rate. Figure 20-9 shows a mechanism that is called *multiple paths* and occurs in any column packed with solid particles. Because some of the random flow paths are longer than others, solute molecules entering the column at the same time on the left are eluted at different times on the right.

Plate Height Equation

The **van Deemter equation** for plate height (H) as a function of flow rate (u) is the net result of the three band broadening mechanisms that we just discussed:

FIGURE 20-9 Band spreading from multiple flow paths. The smaller the stationary phase particles, the less serious is this problem. This process is absent in an open tubular column. [From H. M. McNair and E. J. Bonelli, *Basic Gas Chromatography* (Palo Alto, CA: Varian Instruments Division, 1968).]

van Deemter equation for plate height:

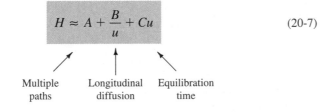

$$H \approx A + \frac{B}{u} + Cu \qquad (20\text{-}7)$$

Multiple paths Longitudinal diffusion Equilibration time

where *A*, *B*, and *C* are constants determined by the column, stationary phase, mobile phase, and temperature. Each of the curves in Figure 20-8 is described by Equation 20-7 with different values of *A*, *B*, and *C*.

Equation 20-7 describes the broadening of a band of solute as it passes through a chromatography column. If the band has some finite width when it is applied to the column, the eluted band emerging from the other end will be broader than we predict with the van Deemter equation. Bands can broaden outside the column if there is too much tubing to flow through or if the detector has too large a volume. Best results are obtained in chromatography if the lengths and diameters of all tubing outside the column are kept to a minimum. Also, there should be no void spaces inside the column where mixing can occur.

Open Tubular Columns

In contrast to a **packed column** that is filled with solid particles coated with stationary phase, an **open tubular column** is a hollow capillary whose inner wall is coated with a thin layer of stationary phase (Figure 20-10). In an open tubular column, there is no broadening from multiple paths, because there are no particles of stationary phase in the flow path. Therefore a given length of open tubular column generally gives better resolution than the same length of packed column. We say that the open tubular column has more theoretical plates (or a smaller plate height) than the packed column.

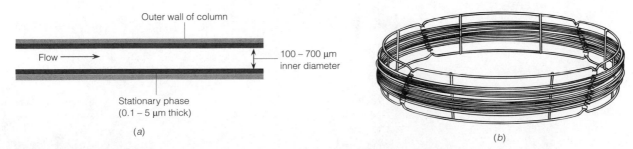

Outer wall of column

Flow

100 – 700 µm inner diameter

Stationary phase (0.1 – 5 µm thick)

(a)

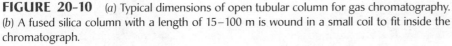

(b)

FIGURE 20-10 (a) Typical dimensions of open tubular column for gas chromatography. (b) A fused silica column with a length of 15–100 m is wound in a small coil to fit inside the chromatograph.

Box 20-1 *Explanation*

Polarity

Polar compounds have positive and negative regions that attract neighboring molecules by electrostatic forces.

Polar molecules attract each other by electrostatic forces. Positive attracts negative.

Polarity arises because different atoms have different electronegativity—different abilities to attract electrons from the chemical bonds. For example, oxygen is more electronegative than carbon. In acetone, the oxygen attracts electrons from the $C=O$ double bond and takes on a partial negative charge, designated $\delta-$. This shift leaves a partial positive charge ($\delta+$) on the neighboring carbon.

In contrast to acetone, hexane is considered to be a **nonpolar compound** because there is little charge separation within a hexane molecule.

Water is a very polar molecule that forms hydrogen bonds between the electronegative oxygen atom in one molecule and the electropositive hydrogen atom of another molecule

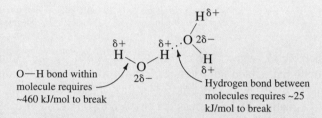

O—H bond within molecule requires ~460 kJ/mol to break

Hydrogen bond between molecules requires ~25 kJ/mol to break

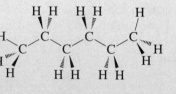

Acetone   Hexane

Ionic compounds usually dissolve in water. In general, polar organic compounds tend to be most soluble in polar solvents and least soluble in nonpolar solvents. Nonpolar compounds tend to be most soluble in nonpolar solvents. "Like dissolves like."

Typical nonpolar and weakly polar compounds		Typical polar compounds	
⌇⌇⌇⌇⌇	octane (C_8H_{18})	CH_3OH	methanol
⬡	benzene	CH_3CH_2OH	ethanol
⬡—CH_3	toluene	$CHCl_3$	chloroform
Cl—$C(Cl)_3$	carbon tetrachloride	$CH_3-C(=O)OH$	acetic acid
⌇O⌇	diethyl ether ($C_4H_{10}O$)	$CH_3C\equiv N$	acetonitrile

438

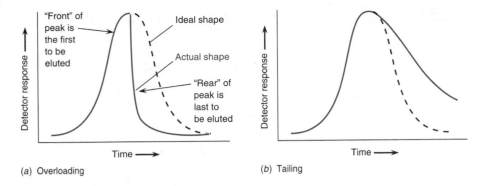

(a) Overloading   (b) Tailing

FIGURE 20-11 (a) Overloading gives rise to a chromatographic band with an ordinary front and an abruptly cutoff rear. (b) Tailing is a peak shape with a normal front and a grossly elongated rear.

A packed gas chromatography column has greater resistance to gas flow than does an open tubular column. Therefore an open tubular column can be made much longer than a packed column with the same operating pressure. Because of its resistance to gas flow, a packed gas chromatography column is usually just 2–3 m in length, whereas an open tubular column can be 100 m in length. The greater length and smaller plate height of the open tubular column provide much better resolution than a packed column does.

An open tubular column cannot handle as much solute as a packed column, because there is less stationary phase in the open tubular column. Therefore open tubular columns are useful for analytical separations but not for preparative separations.

Open tubular columns give better separations than packed columns because

- there is no multiple path band broadening ($A = 0$ in van Deemter equation)
- the open tubular column can be much longer

Sometimes Bands Have Funny Shapes

When a column is overloaded by too much solute in one band, the band emerges from the column with a gradually rising front and an abruptly cutoff back (Figure 20-11a). The reason for this behavior is that a compound is most soluble in itself. As the concentration rises from 0 at the front of the band to some high value inside the band, *overloading* can occur. When overloaded, the solute is so soluble in the concentrated part of the band that little solute trails behind the concentrated region.

Tailing is an asymmetric peak shape in which the trailing part of the band is elongated (Figure 20-11b). Such a peak occurs when there are strongly polar, highly adsorptive sites (such as exposed — OH groups) in the stationary phase that retain solute more strongly than other sites. To reduce tailing, we use a chemical treatment called *silanization* to convert polar — OH groups to nonpolar — $OSi(CH_3)_3$ groups.

Box 20-1 discusses *polarity*.

Ask Yourself

20-C. (a) Why is longitudinal diffusion a more serious problem in gas chromatography than in liquid chromatography? (Think about this one; the answer is not in the text.)

    (b) (i) In Figure 20-8, what is the optimal flow rate for best separation of solutes with He mobile phase?

        (ii) Why does plate height increase at high flow rate?

        (iii) Why does plate height increase at low flow rate?

    (c) Why does an open tubular gas chromatography column give better resolution than the same length of packed column?

    (d) Why is it desirable to use a very long open tubular column? What difference between packed and open tubular columns allows much longer open tubular columns to be used?

Two compounds might have the same retention time on a particular column. It is unlikely that they will have the same retention time on several columns with different stationary phases.

Chromatography is powerful for both qualitative and quantitative analysis. The simplest way to identify a chromatographic peak is to compare its retention time with that of an authentic sample of the suspected compound. The most reliable way to do this is by **co-chromatography,** in which authentic sample is added to the unknown. If the added compound is identical to one component of the unknown, then the relative size of that one peak will increase. Identification is only tentative when carried out with one column, but it is nearly conclusive when carried out with several different kinds of columns.

For qualitative analysis, each chromatographic peak can be directed into a mass spectrometer (Box 20-2) or infrared spectrophotometer to record a spectrum as the substance is eluted from the column. The spectrum can be identified by comparison with a library of spectra stored in a computer. Figure 20-12a shows a chromatogram from a student experiment in which 1-bromobutane was synthesized from 1-butanol:

$$CH_3CH_2CH_2CH_2OH + Br^- \longrightarrow CH_3CH_2CH_2CH_2Br + OH^- \qquad (20\text{-}8)$$

1-Butanol   1-Bromobutane

Peaks in the chromatogram are identified by recording the mass spectrum of each as it is eluted. Figure 20-12b shows the spectrum of peak 1 and identifies it as starting material by comparison with a spectral library of known compounds. Figure 20-12c shows that peak 2 is product. The progress of the reaction can be followed by chromatography, and additional products (if any) could be identified. Box 20-3 shows how gas chromatography can be used for qualitative identification of flavor components of candy and gum.

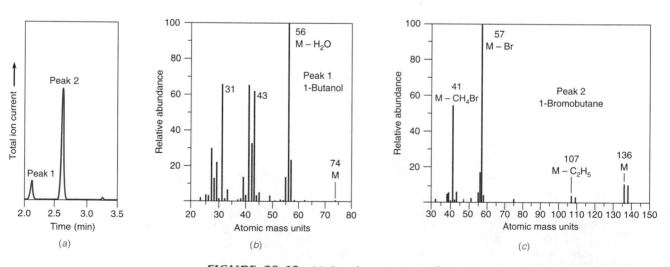

FIGURE 20-12 (*a*) Gas chromatogram of a student sample from Reaction 20-8. The detector is the mass spectrometer set to record the total current from all ions of all masses. Dichloromethane solvent eluted prior to 2 min is not shown. (*b*) Mass spectrum of peak 1, which is identical to the spectrum of 1-butanol. (*c*) Mass spectrum of peak 2, which is identical to the spectrum of 1-bromobutane. In (*b*) and (*c*), M stands for the parent ion. [From A. Illies, P. B. Shevlin, G. Childers, M. Peschke, and J. Tsai, *J. Chem. Ed.* **1995**, *72*, 717.]

Box 20-2 *Explanation*

What Is Mass Spectrometry?

A **mass spectrometer** is a device that ionizes molecules in the gas phase, accelerates the ions, and separates them according to their mass. The output of the instrument, called a *mass spectrum,* is a graph showing the relative abundance of each charged species versus m/z, where m is the mass of the ion and z is the number of charges it carries (see Figure 20-12b and 20-12c). When molecules are ionized by electron impact, as in the diagram below, most species carry a single charge, so m/z is numerically equal to the mass. Some ionization techniques that we will not discuss do create ions with multiple charges. Proteins separated by liquid chromatography and detected in a mass spectrometer commonly carry multiple charges.

The diagram shows a quadrupole mass spectrometer connected to the exit of a gas chromatograph to record a spectrum of each peak as it is eluted. The chromatography column passes through a heated, gas-tight connector into the ionization chamber of the mass spectrometer, which is pumped rapidly to maintain a good vacuum. Electrons from a hot filament are accelerated through 70 V and collide with gaseous eluate. Some molecules are ionized by collisions with the electrons to make cations that have enough energy to decompose into fragments.

An ionized molecule (designated M in Figure 20-12b and 20-12c) and its ionized fragments are accelerated through 15 V and enter the quadrupole mass separator. The separator consists of four parallel metal cylinders to which are applied both a constant voltage and a radio frequency alternating voltage. The electric field causes gaseous cations to travel in complex trajectories as they migrate from the ionization chamber toward the ion detector. The electric field allows cations with only one particular mass (resonant ions) to reach the detector. Cations with all other masses (nonresonant ions) collide with the cylinders and are lost before they reach the detector. By rapidly varying the applied voltages, ions of different mass are selected to generate the mass spectra in Figure 20-12. Quadrupole mass separators are scanned fast enough to record two to eight spectra per second, covering 800 mass units.

We can understand the pattern of fragments of 1-bromobutane in Figure 20-12c without too much difficulty. The peak at 136 mass units is the unfragmented ion called the *parent ion* (M). Its formula is $C_4H_9{}^{79}Br^+$. The peak at 138 mass units with almost equal intensity has the formula $C_4H_9{}^{81}Br^+$. Natural bromine consists of 50.5% ^{79}Br and 49.5% ^{81}Br, which explains why the peak at 136 mass units is slightly taller than the peak at 138 mass units. Small peaks at 107 and 109 mass units have lost C_2H_5 fragments. They are labeled M − C_2H_5 and have the formula $C_2H_4Br^+$. The peak at 57 mass units is $C_4H_9^+$, which has no bromine and therefore has no isotopic partner at 59 mass units. The peak at 41 mass units is $C_3H_5^+$.

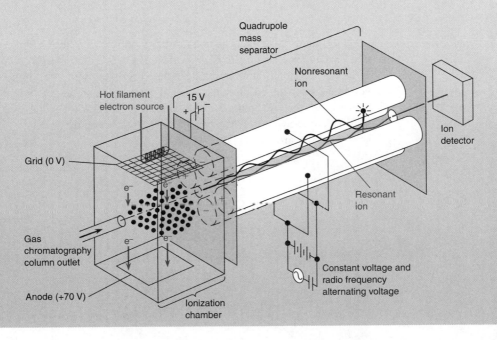

Quadrupole mass separator

Nonresonant ion

Hot filament electron source

15 V

Ion detector

Grid (0 V)

Resonant ion

Gas chromatography column outlet

Constant voltage and radio frequency alternating voltage

Anode (+70 V)

Ionization chamber

Box 20-3 *Informed Citizen*

Volatile Flavor Components of Candy and Gum

Students at Indiana State University identify components of candy and gum by using gas chromatography with a mass spectrometer as a detector. The extremely simple sampling technique for this qualitative analysis is called *headspace analysis*. A crushed piece of hard candy or a piece of gum is placed in a vial and allowed to stand for a few minutes so that *volatile* compounds (compounds with a high vapor pressure) can evaporate and fill the gas phase (the *headspace*) with vapor. A 5-μL sample of the vapor phase is then drawn into a syringe and injected into a gas chromatograph. Peaks are identified by comparison of their retention times and

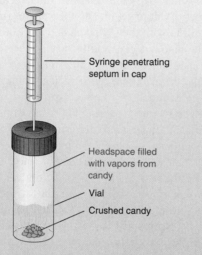

Headspace sampling. A syringe is inserted through a rubber *septum* (a rubber disk) in the cap of the vial to withdraw gas for analysis.

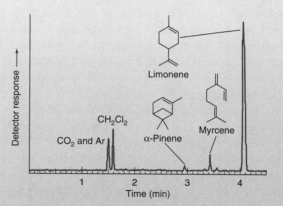

Chromatogram of headspace vapors from Orange Life Savers®. The mass spectral detector measures ions above 34 atomic mass units. CO_2 and Ar are from air and CH_2Cl_2 is the solvent used to clean the syringe. [From R. A. Kjonaas, J. L. Soller, and L. A. McCoy, *J. Chem. Ed.* **1997**, *74*, 1104.]

For quantitative analysis, we normally choose conditions under which *the area of a chromatographic peak is proportional to the quantity of that component.* Peak areas are measured automatically if the chromatograph is computer controlled. If you must make a measurement by hand, a triangle can be drawn with two sides tangent to the steepest points on each side of the peak (as in Figure 20-3). When the third side is drawn along the baseline, the area of the triangle ($= \frac{1}{2}$(base $\times$ height)) is 96% of the area of a Gaussian peak.

Ask Yourself

20-D. The mass spectrum of 1-bromobutane in Figure 20-12c has a pair of nearly equal-intensity peaks at 136 and 138 mass units and another pair at 107 and 109. The peaks at 57 and 41 have no similar equal-intensity partners. Read Box 20-2 and explain why this is so.

Volatile components from candy and gum

Sample	Major components	Minor components
Orange Life Savers®	limonene	α-pinene myrcene
Lemon Life Savers®	limonene	α-pinene β-pinene γ-terpinene
Wild Cherry Life Savers®	benzaldehyde benzyl alcohol	ethyl acetate
Wrigley's Spearmint Gum®	carvone limonene	cineole menthone menthol
Wrigley's Big Red Gum®	cinnamaldehyde	

Data from R. A. Kjonaas, J. L. Soller, and L. A. McCoy, _J. Chem. Ed._ **1997**, _74_, 1104.

mass spectra to those of known compounds. The table lists major volatile components identified by students.

If you look up some of these constituents of foods, you might be rather surprised. For example, according to the _Merck Index_ (9th ed.), α-pinene irritates skin and mucous membranes, causes skin eruption, gastrointestinal irritation, delirium, ataxia, kidney damage, and coma. Inhalation causes palpitation, dizziness, nervous disturbances, chest pain, bronchitis, nephritis, and benign skin tumors from chronic contact. A fatal dose is 180 g orally. Obviously, sucking Life Savers does not have these effects on you. The body can handle small quantities of some chemicals that would be disastrous in a high dose.

20-5 Internal Standards

An **internal standard** is a known amount of a compound, different from analyte, that is added to the unknown. Signal from analyte is compared with signal from the internal standard to find out how much analyte is present.

Internal standards are especially useful for analyses in which the quantity of sample analyzed or the instrument response varies slightly from run to run for reasons that are difficult to control. For example, gas or liquid flow rates that vary by a few percent in chromatography could change the detector response. A calibration curve is only accurate for the one set of conditions under which it was obtained. However, the _relative_ response of the detector to the analyte and standard is usually constant over a wide range of conditions. If signal from the standard increases by 8.4% because of a change in solvent flow rate, signal from the analyte usually increases by 8.4% also. As long as the concentration of standard is known, the correct concentration of analyte can be derived. Internal standards are widely used in chromatography because the small sample injected into the chromatograph is not very reproducible, so a calibration curve would not be accurate.

Internal standards are also desirable when sample loss can occur during sample preparation steps prior to analysis. If a known quantity of standard is added to

An _internal standard_ is different from the analyte. In _standard addition_ (Section 16-4), the standard is the same substance as the analyte.

the unknown prior to any manipulations, the ratio of standard to analyte remains constant because the same fraction of each is lost in any operation.

To use an internal standard, we prepare a known mixture of standard and analyte and measure the relative response of the detector to the two species. In Figure 20-13, the area under each peak is proportional to the concentration of each compound injected into the column. However, the detector generally has a different response to each component. For example, if both the analyte (X) and internal standard (S) have concentrations of 10.0 mM, the area under the analyte peak might be 2.30 times greater than the area under the standard peak. We say that the **response factor,** F, is 2.30 times greater for X than for S.

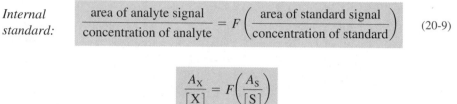

$$
\textit{Internal} \atop \textit{standard:} \qquad \frac{\text{area of analyte signal}}{\text{concentration of analyte}} = F\left(\frac{\text{area of standard signal}}{\text{concentration of standard}}\right) \qquad (20\text{-}9)
$$

$$
\frac{A_X}{[X]} = F\left(\frac{A_S}{[S]}\right)
$$

Linear response means that peak area is proportional to analyte concentration. For narrow peaks, height is often substituted for area.

[X] and [S] are the concentrations of analyte and standard *after they have been mixed together.* Equation 20-9 is predicated on linear response to both the analyte and the standard.

EXAMPLE   Using an Internal Standard

In a preliminary experiment, a solution containing 0.083 7 M X and 0.066 6 M S gave peak areas of $A_X = 423$ and $A_S = 347$. (Areas are measured in arbitrary units by the instrument's computer.) To analyze the unknown, 10.0 mL of 0.146 M S were added to 10.0 mL of unknown, and the mixture was diluted to 25.0 mL in a volumetric flask. This mixture gave the chromatogram in Figure 20-13, with peak areas $A_X = 553$ and $A_S = 582$. Find the concentration of X in the unknown.

SOLUTION First use the standard mixture to find the response factor in Equation 20-9:

$$
\textit{Standard mixture:} \qquad \frac{A_X}{[X]} = F\left(\frac{A_S}{[S]}\right)
$$

$$
\frac{423}{0.083\ 7} = F\left(\frac{347}{0.066\ 6}\right) \quad \Rightarrow \quad F = 0.970_0
$$

In the mixture of unknown plus standard, the concentration of S is

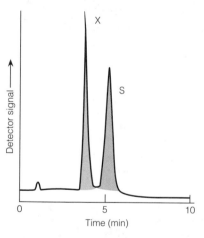

FIGURE 20-13 Chromatogram illustrating the use of an internal standard. A known amount of standard S is added to unknown X. From the areas of the peaks, we can tell how much X is in the unknown. To do this, we needed to measure the relative response to known amounts of each compound in a separate experiment.

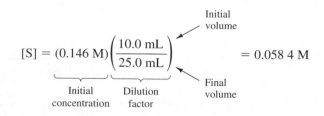

$$
[S] = (0.146\ \text{M})\left(\frac{10.0\ \text{mL}}{25.0\ \text{mL}}\right) = 0.058\ 4\ \text{M}
$$

Using the known response factor, we substitute back into Equation 20-9 to find the concentration of unknown in the mixture:

Unknown mixture:
$$\frac{A_X}{[X]} = F\left(\frac{A_S}{[S]}\right)$$

$$\frac{553}{[X]} = 0.970_0\left(\frac{582}{0.058\ 4}\right) \quad \Rightarrow \quad [X] = 0.057\ 2_1\ M$$

Because X was diluted from 10.0 to 25.0 mL when the mixture with S was prepared, the original concentration of X in the unknown was $(25.0/10.0) \times (0.057\ 2_1\ M) = 0.143\ M$.

Ask Yourself

20-E. A mixture of 52.4 nM iodoacetone (X) and 38.9 nM *p*-dichlorobenzene (S) gave the relative response (peak area of X)/(peak area of S) = 0.644/1.000. A second solution containing an unknown quantity of X plus 742 nM S had (peak area of X)/(peak area of S) = 1.093/1.000. Find [X] in the second solution.

Key Equations

Theoretical plates
$$N = \frac{5.55t_r^2}{w_{1/2}^2}$$

t_r = retention time of analyte
$w_{1/2}$ = peak width at half-height
} Both must be measured in the same units

Plate height
$$H = L/N \qquad (L = \text{length of column})$$

Resolution
$$\text{resolution} = \frac{\Delta t_r}{w_{av}}$$

Δt_r = difference in retention times between two peaks
w_{av} = average width of two peaks at baseline
(Δt_r and w_{av} must be measured in the same units.)

Scaling equation
$$\frac{\text{large load}}{\text{small load}} = \left(\frac{\text{large column radius}}{\text{small column radius}}\right)^2$$

van Deemter equation
$$H \approx A + \frac{B}{u} + Cu$$

H = plate height; u = flow rate
A = constant due to multiple flow paths
B = constant due to longitudinal diffusion of solute
C = constant due to equilibration time of solute between phases

Internal standard

$$\frac{\text{area of analyte signal}}{\text{concentration of analyte}} = F\left(\frac{\text{area of standard signal}}{\text{concentration of standard}}\right)$$

F = response factor measured in a separate experiment with
known concentrations of analyte and standard

Important Terms

adsorption chromatography	internal standard	partition chromatography
affinity chromatography	ion-exchange chromatography	plate height
chromatogram	liquid chromatography	polar compound
chromatography	mass spectrometer	resolution
co-chromatography	mobile phase	response factor
eluate	molecular exclusion chromatography	retention time
eluent	nonpolar compound	stationary phase
elution	open tubular column	theoretical plate
gas chromatography	packed column	van Deemter equation

Problems

20-1. What is the difference between eluent and eluate?

20-2. Which column is more efficient: plate height = 0.1 mm or 1 mm?

20-3. Explain why a chromatographic separation normally has an optimum flow rate that gives the best separation.

20-4. (a) Why is longitudinal diffusion a more serious problem in gas chromatography than in liquid chromatography?

(b) Suggest a reason why the optimum linear flow rate is much higher in gas chromatography than in liquid chromatography.

20-5. Why does silanization reduce tailing of chromatographic peaks?

20-6. What kind of information does a mass spectrometer detector give in gas chromatography that is useful for qualitative analysis? For quantitative analysis?

20-7. How is co-chromatography used in qualitative analysis? Why are several different types of columns necessary to make a convincing case for the identity of a compound?

20-8. (a) How many theoretical plates produce a chromatography peak eluting at 12.83 min with a width at half-height of 8.7 s?

(b) The length of the column is 15.8 cm. Find the plate height.

20-9. A gas chromatogram of a mixture of toluene and ethyl acetate is shown below.

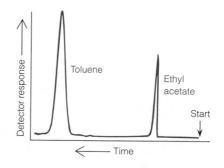

(a) Measure $w_{1/2}$ for each peak to the nearest 0.1 mm. When the thickness of the pen trace is significant relative to the length being measured, it is important to take the pen width into account. It is best to measure from the edge of one trace to the corresponding edge of the other trace, as shown here.

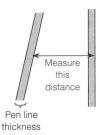

Measure this distance

Pen line thickness

(b) Find the number of theoretical plates and the plate height for each peak.

20-10. Two components of a 12-mg sample are adequately separated by chromatography through a column 1.5 cm in diameter and 25 cm long at a flow rate of 0.8 mL/min. What size column and what flow rate should be used to obtain similar separation with a 250-mg sample?

20-11. The chromatogram in the figure at the right has a peak for isooctane at 13.81 min. The column is 30.0 m long.

(a) Measure $w_{1/2}$ and find the number of theoretical plates for this peak.

(b) Find the plate height.

(c) Measure the width at the base, w, as defined in Figure 20-3. Figure 20-3 tells us that the expected ratio $w/w_{1/2}$ for a Gaussian peak is $4\sigma/2.35\sigma = 1.70$. Compare the measured ratio $w/w_{1/2}$ to the theoretical ratio.

20-12. Consider the peaks for heptane (14.56 min) and p-difluorobenzene (14.77 min) in the chromatogram for Problem 20-11. The column is 30.0 m long.

(a) Find the number of plates and the plate height for the two compounds.

(b) Measure w for each peak and find the resolution between the two peaks.

20-13. A column 3.00-cm in diameter and 32.6 cm long gives adequate resolution of a 72.4-mg mixture of unknowns, initially dissolved in 0.500 mL.

(a) If you wish to scale down to 10.0 mg of the same mixture with minimum use of chromatographic stationary phase and solvent, what length and diameter column would you use?

(b) In what volume would you dissolve the sample?

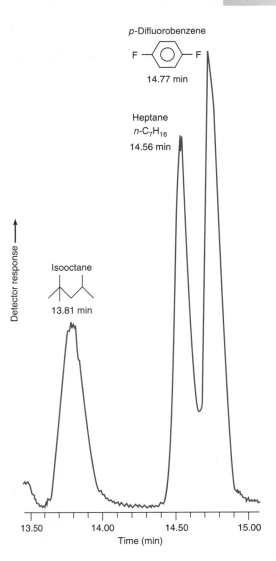

For Problem 20-11.

(c) If the flow rate in the large column is 1.85 mL/min, what should be the flow rate in the small column?

20-14. In Figure 20-8, flow rate is expressed as gas velocity in cm/s. The gas is flowing through an open tubular column with an inner diameter of 0.25 mm. What volume flow rate (mL/min) corresponds to a gas velocity of 50 cm/s? (Note that the volume of a cylinder is $\pi r^2 \times$ length, where r is the radius.)

20-15. *Internal standard.* A solution containing 3.47 mM X (analyte) and 1.72 mM S (standard) gave peak areas of 3 473 and 10 222, respectively, in a chromatographic analysis. Then

1.00 mL of 8.47 mM S was added to 5.00 mL of unknown X, and the mixed solution was diluted to 10.0 mL. This solution gave peak areas of 5 428 and 4 431 for X and S, respectively.

(a) Find the response factor for X relative to S in Equation 20-9.

(b) Find [S] (mM) in the 10.0 mL of mixed solution.

(c) Find [X] (mM) in the 10.0 mL of mixed solution.

(d) Find [X] in the original unknown.

20-16. *Internal standard.* A known mixture of compounds C and D gave the following chromatography results:

Compound	Concentration (μg/mL) in mixture	Peak area (cm^2)
C	236	4.42
D	337	5.52

A solution was prepared by mixing 1.23 mg of D in 5.00 mL with 10.00 mL of unknown containing just C and diluting to 25.00 mL. Peak areas of 3.33 and 2.22 cm^2 were observed for C and D, respectively. Find the concentration of C (μg/mL) in the unknown.

20-17. *Internal standard.* When 1.06 mmol of 1-pentanol and 1.53 mmol of 1-hexanol were separated by gas chromatography, they gave relative peak areas of 922 and 1 570 units, respectively. When 0.57 mmol of pentanol was added to an unknown containing hexanol, the relative chromatographic peak areas were 843:816 (pentanol:hexanol). How much hexanol did the unknown contain?

20-18. Suggest a molecular formula for the major peaks at 31, 41, 43, and 56 mass units in the mass spectrum of 1-butanol (C_4H_9OH) in Figure 20-12b.

20-19. **(a)** The plate height in a particular packed gas chromatography column is characterized by the van Deemter equation $H(\text{mm}) = A + B/u + Cu$, where $A = 1.50$ mm, $B = 25.0$ mm $\cdot$ mL/min, $C = 0.025\ 0$ mm $\cdot$ min/mL, and u is flow rate in mL/min. Construct a graph of plate height versus flow rate and find the optimum flow rate for minimum plate height.

(b) In the van Deemter equation, B is proportional to the rate of longitudinal diffusion. Predict whether the optimum flow rate would increase or decrease if the rate of longitudinal diffusion were doubled. To confirm your prediction, increase B to 50.0 mm $\cdot$ mL/min, construct a new graph, and find the optimum flow rate. Does the optimum plate height increase or decrease?

(c) The parameter C is inversely proportional to the rate of equilibration of solute between the mobile and stationary phase ($C \propto 1/\text{rate of equilibration}$). Predict whether the optimum flow rate would increase or decrease if the rate of equilibration between phases were doubled (i.e., $C = 0.012\ 5$ mm $\cdot$ min/mL). Does the optimum plate height increase or decrease?

20-20. (a) Using the van Deemter parameters in Problem 20-19(a), find the plate height for a flow rate of 20.0 mL/min.

(b) How many plates are on the column if the length is 2.00 m?

(c) What will be the width at half-height of a peak eluted in a time of 8.00 min?

How Would You Do It?

20-21. Caffeine can be measured by chromatography with $^{13}C_3$-caffeine as an internal standard.[1] $^{13}C_3$-caffeine has the same retention time as ordinary ^{12}C-caffeine.

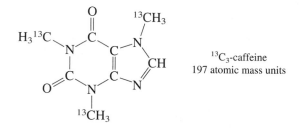

$^{13}C_3$-caffeine
197 atomic mass units

Caffeine can be extracted from aqueous solution by *solid-phase microextraction* (Section 21-4). In this procedure, a silica (SiO_2) fiber coated with a polymer is dipped into the liquid and solutes from the liquid distribute themselves between the polymer phase and the liquid. Then the fiber is withdrawn from the liquid and heated in the port of a gas chromatograph. Solutes evaporate from the polymer and are carried into the column. Suggest a procedure for the quantitative analysis of caffeine in coffee using solid-phase microextraction with a $^{13}C_3$-caffeine internal standard.

Notes and References

1. M. J. Yang, M. L. Orton, and J. Pawliszyn, *J. Chem. Ed.* **1997,** *74,* 1130.

Further Reading

J. C. Giddings, *Unified Separation Science* (New York: Wiley, 1991).

In Vivo Microdialysis for Measuring Drug Metabolism

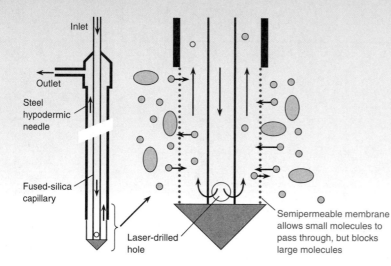

Microdialysis probe. Enlarged view of the lower end shows a semipermeable membrane that allows small molecules to pass in both directions but excludes large molecules.

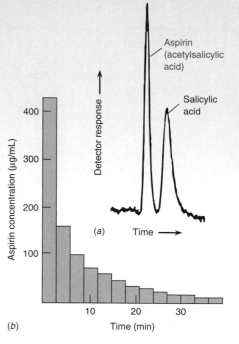

(a) Chromatogram of blood microdialysis sample drawn 5 min after intravenous injection of aspirin. (b) Aspirin concentration in blood versus time after injection of aspirin. [From K. L. Steele, D. O. Scott, and C. E. Lunte, *Anal. Chim. Acta* **1991**, *246*, 181.]

Dialysis is the process in which small molecules diffuse across a *semipermeable membrane* that has pore sizes large enough to pass small but not large molecules. A *microdialysis probe* has a semipermeable membrane attached to the shaft of a hypodermic needle, which can be inserted into an animal. Fluid is pumped through the probe from the inlet to the outlet. Small molecules from the animal diffuse into the probe and are rapidly transported to the outlet. Fluid exiting the probe (*dialysate*) can be analyzed by liquid chromatography.

An in vivo (inside the living organism) study of aspirin metabolism in a rat is shown above. Aspirin is converted to salicylic acid by enzymes in the bloodstream. To measure the conversion rate, the investigators injected aspirin into a rat and used chromatography to monitor dialysate from a microdialysis probe in a vein of the rat. If you simply were to withdraw blood for analysis, aspirin would continue to be metabolized by enzymes in the blood. Microdialysis separates the small aspirin molecule from large enzyme molecules.

Chapter *21*

GAS AND LIQUID CHROMATOGRAPHY

*H*igh-resolution capillary gas chromatography and high-performance liquid chromatography are workhorses of the modern analytical chemistry laboratory. This chapter describes the equipment and basic techniques of gas and liquid chromatography.

21-1   Gas Chromatography

In **gas chromatography,** a gaseous mobile phase transports a gaseous solute through a long, thin column containing stationary phase. Figure 21-1 shows the main components of a gas chromatograph. We begin the experiment by injecting a volatile liquid through a rubber *septum* (a thin disk) into a heated port, which vaporizes the sample. The sample is swept through the column by He, N_2, or H_2 *carrier gas,* and the separated solutes flow through a detector, whose response is displayed on a recorder or computer. The column must be hot enough to produce sufficient vapor pressure for each solute to be eluted in a reasonable time. The detector is maintained at a higher temperature than the column so that all solutes are gaseous. The injected sample size for a liquid is typically $0.1–2\ \mu L$ for analytical chromatography, whereas preparative columns can handle $20–1\ 000\ \mu L$. Gases can be introduced in volumes of $0.5–10$ mL by a gas-tight syringe or a gas-sampling valve.

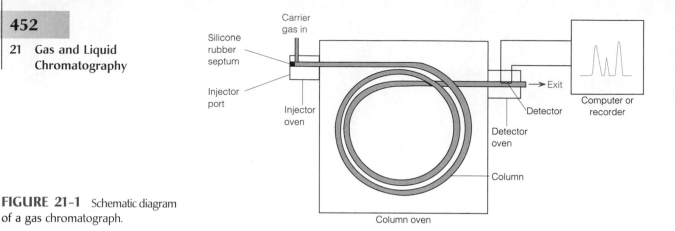

FIGURE 21-1 Schematic diagram
of a gas chromatograph.

Columns

Open tubular columns (see Figure 20-10) have a liquid or solid stationary phase
coated on the inside wall (Figure 21-2). For example, a solid, porous carbon stationary phase shown in Figure 21-3a can be used to separate gases found in beer,
as shown in Figure 21-3b.

Open tubular columns are usually made of fused silica (SiO_2). As the column
ages, stationary phase bakes off and exposes silanol groups ($Si-O-H$) on the
silica surface. The silanol groups strongly retain some polar compounds by hydrogen bonding, thereby causing *tailing* (Figure 20-11b) of chromatographic peaks. To
reduce the tendency of a stationary phase to bleed from the column at high temperature, the stationary phase can be *bonded* (chemically attached) to the silica
surface or *cross-linked* to itself with covalent chemical bridges.

Polarity was discussed in Box 20-1.

Liquid stationary phases in Table 21-1 have a range of polarities. The choice
of liquid phase for a given problem is based on the rule "like dissolves like." Nonpolar columns are usually best for nonpolar solutes, and polar columns are usually
best for polar solutes.

You can see the effects of column polarity on a separation in Figure 21-4. In
Figure 21-4a, 10 compounds are eluted nearly in order of increasing boiling point
from a nonpolar stationary phase: The higher the vapor pressure, the faster the compound is eluted. In Figure 21-4b, the strongly polar stationary phase strongly retains
the polar solutes. The three alcohols (with $-OH$ groups) are the last to be eluted,
following the three ketones (with $>C=O$ groups), which follow four alkanes

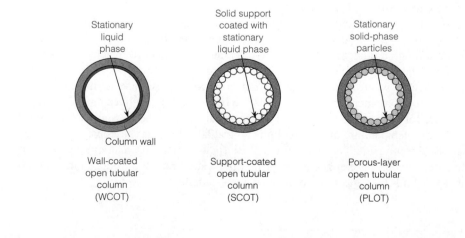

FIGURE 21-2 Cross-sectional
view of wall-coated, support-coated,
and porous-layer columns.

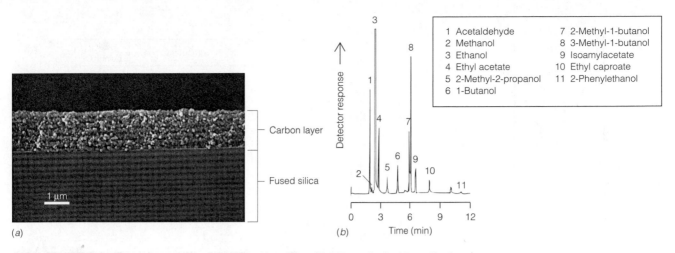

FIGURE 21-3 (a) Porous carbon stationary phase (2 μm thick) on the inside wall of a fused-silica open tubular capillary gas chromatography column. (b) Chromatogram of compounds in the headspace (gas phase) of a beer can, obtained with a porous carbon column (0.25 mm diameter × 30 m long) operated at 30°C for 2 min and then ramped up to 160°C at 20°/min. [Courtesy Alltech Associates, State College, PA.]

TABLE 21-1   Common stationary phases in capillary gas chromatography

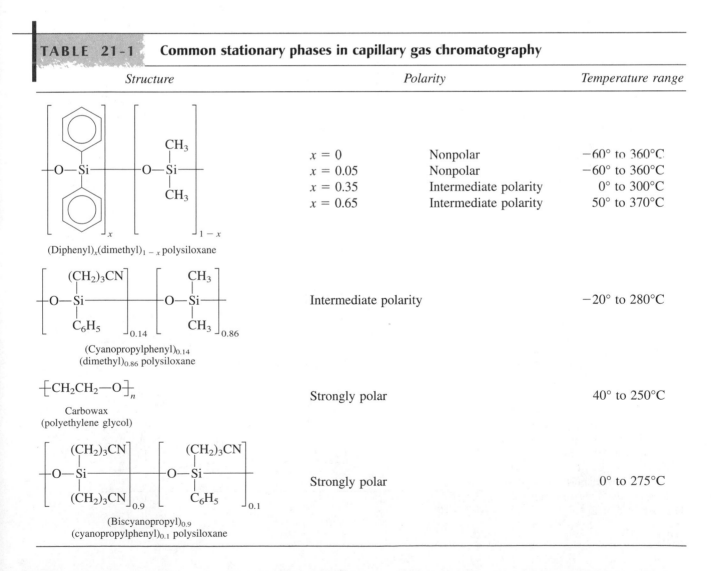

Structure	Polarity	Temperature range
(Diphenyl)$_x$(dimethyl)$_{1-x}$ polysiloxane	$x = 0$ Nonpolar	−60° to 360°C
	$x = 0.05$ Nonpolar	−60° to 360°C
	$x = 0.35$ Intermediate polarity	0° to 300°C
	$x = 0.65$ Intermediate polarity	50° to 370°C
(Cyanopropylphenyl)$_{0.14}$ (dimethyl)$_{0.86}$ polysiloxane	Intermediate polarity	−20° to 280°C
Carbowax (polyethylene glycol)	Strongly polar	40° to 250°C
(Biscyanopropyl)$_{0.9}$ (cyanopropylphenyl)$_{0.1}$ polysiloxane	Strongly polar	0° to 275°C

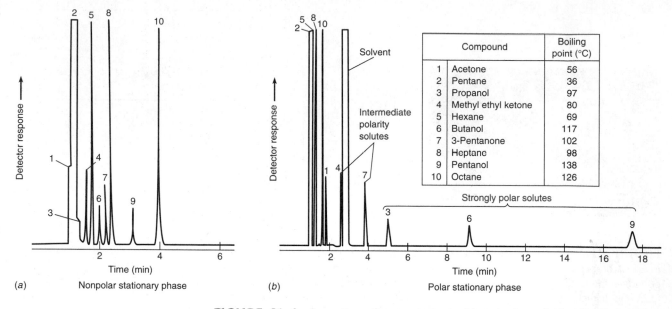

	Compound	Boiling point (°C)
1	Acetone	56
2	Pentane	36
3	Propanol	97
4	Methyl ethyl ketone	80
5	Hexane	69
6	Butanol	117
7	3-Pentanone	102
8	Heptane	98
9	Pentanol	138
10	Octane	126

(a) Nonpolar stationary phase

(b) Polar stationary phase

FIGURE 21-4 Separation of compounds on (a) nonpolar poly(dimethylsiloxane) and (b) strongly polar polyethylene glycol stationary phases (1 μm thick) in open tubular columns (0.32 mm diameter × 30 m long) at 70°C. [Courtesy Restek Co., Bellefonte, PA.]

Molecular sieves are also widely used to dry gases because sieves strongly retain water. Sieves are regenerated (freed of water) by heating to 300°C in vacuum.

(having only C—H bonds). Hydrogen bonding between solute and the stationary phase is probably the strongest force causing retention.

Common solid stationary phases include porous carbon (Figure 21-3a) and *molecular sieves*, which are inorganic materials with nanometer-size cavities that retain and separate small molecules such as H_2, O_2, N_2, CO_2, and CH_4. Figure 21-5

FIGURE 21-5 Gas chromatography with 5A molecular sieves. Upper chromatogram was obtained with a packed column (3.2 mm diameter × 4.6 m long) at 40°C, using 1 mL of sample containing 2 ppm (by volume) of each analyte in He. Lower chromatogram was obtained with an open tubular column (0.32 mm diameter × 30 m long) at 30°C, using 4 μL of the same sample. [From J. Madabushi, H. Cai, S. Steams, and W. Wentworth, *Am. Lab.* October 1995, p. 21.]

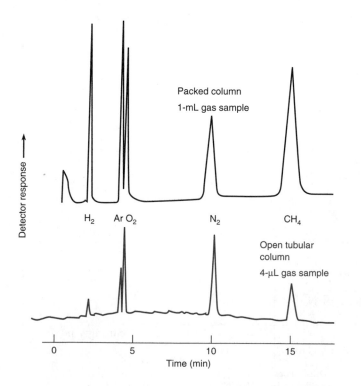

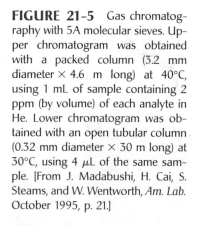

compares the separation of gases by molecular sieves in an open tubular column and a *packed column* filled with particles of the solid stationary phase. Open tubular columns typically give better separations (narrower peaks), but packed columns can handle larger samples. In Figure 21-5, the sample injected into the packed column was 250 times larger than the sample injected into the open tubular column.

Temperature Programming

If you increase the column oven temperature in Figure 21-1, solute vapor pressure increases and retention times decrease. To separate compounds with a wide range of boiling points or polarities, we usually raise the column temperature *during* the separation, a technique called **temperature programming.** Figure 21-6 shows the effect of temperature programming on the separation of nonpolar compounds with a range of boiling points from 69°C for C_6H_{14} to 356°C for $C_{21}H_{44}$. At a constant column temperature of 150°C, the lower boiling compounds emerge close together, and the higher boiling compounds may not be eluted from the column. If the temperature is programmed to increase from 50°C to 250°C, all compounds are eluted and the separation of peaks is fairly uniform. Even though 250°C is below the boiling point of some of the compounds in the mixture, these compounds have sufficient vapor pressure to be eluted at 250°C.

Raising column temperature
• decreases retention time
• sharpens peaks

We refer to constant-temperature conditions as *isothermal* conditions.

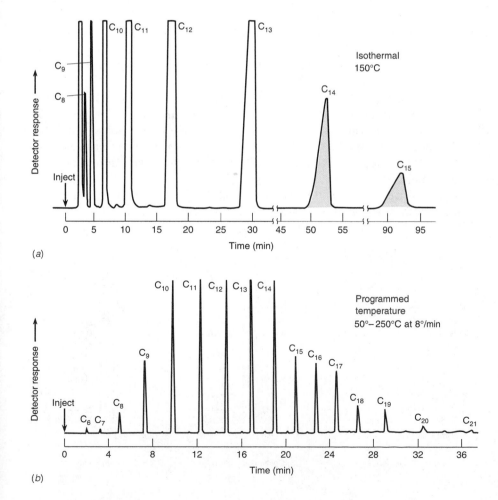

FIGURE 21-6 Comparison of (a) isothermal and (b) programmed temperature chromatography of linear alkanes through a packed column with a nonpolar stationary phase. Detector sensitivity is 16 times greater in (a) than in (b). [From H. M. McNair and E. J. Bonelli, *Basic Gas Chromatography* (Palo Alto, CA: Varian Instrument Division, 1968).]

Carrier Gas

Figure 20-8 showed that H_2 and He give better resolution (smaller plate height) than N_2 at high flow rate. The reason is that solutes diffuse more rapidly through H_2 and He and therefore equilibrate between the mobile and stationary phases more rapidly than they can in N_2. H_2 is explosive when mixed with air, so He is commonly used.

Sample Injection

Liquid samples are injected through a rubber septum into a heated glass port. Carrier gas sweeps the vaporized sample from the port into the chromatography column. A complete injection usually contains too much material for an open tubular capillary column. In **split injection** (Figure 21-7a), only 0.1–10% of the injected sample reaches the column. The remainder is blown out to waste. If the entire sample is not vaporized during injection, however, the higher boiling components will not be completely injected and there will be errors in quantitative analysis.

For quantitative analysis and for analysis of trace components of a mixture, **splitless injection** shown in Figure 21-7b is appropriate. (*Trace components* are those present at extremely low concentrations.) For this purpose, a dilute sample in a low-boiling solvent is injected when the column temperature is 40°C lower than the boiling point of the solvent. Solvent condenses at the beginning of the column and traps a thin band of solute. (Hence, this technique is called **solvent trapping.**) After purging additional vapors from the injection port, the column temperature is raised and chromatography is begun. In splitless injection, ~80% of the sample is applied to the column, and little fractionation (selective evaporation of components) occurs during injection.

Another technique called **cold trapping** can be used to focus high-boiling solutes at the beginning of a chromatography column. In this case, the column is initially 150° lower than the boiling points of solutes of interest. Solvent and low-boiling solutes are eluted rapidly, but high-boiling solutes condense in a narrow band at the start of the column. The column is later warmed to initiate chromatography for the components of interest.

For sensitive compounds that decompose above their boiling temperature, we use **on-column injection** of solution directly into the column (Figure 21-7c), without

Injection into open tubular columns:

Split: routine method

Splitless: best for quantitative analysis

On-column: best for thermally unstable solutes

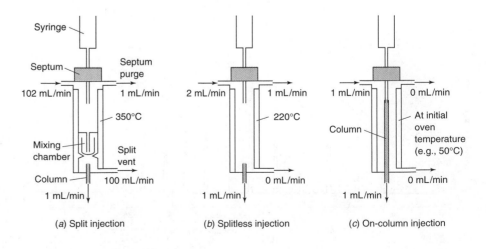

FIGURE 21-7   Injection port operation for (*a*) split, (*b*) splitless, and (*c*) on-column injection into an open tubular column. A slow flow of gas past the inside surface of the septum out to waste helps cool the rubber and prevents the entry of volatile species outgasing from the rubber into the chromatography column.

going through a hot port. Analytes are focused in a narrow band by solvent trapping or cold trapping. Warming the column initiates chromatography.

Open tubular columns narrower than 0.53 mm in diameter ("wide bore") are too narrow to accept the needle of common syringes and cannot be used for on-column injection. Narrower columns (typically 0.10–0.32 mm diameter) provide higher resolution (sharper peaks) than do wide-bore columns, but they have less capacity for analyte.

Thermal Conductivity Detector

Thermal conductivity measures the ability of a substance to transport heat from a hot region to a cold region. In the **thermal conductivity detector** in Figure 21-8, gas emerging from the chromatography column flows over a hot tungsten-rhenium filament. When solute emerges from the column, the thermal conductivity of the gas stream decreases, the filament gets hotter, its electrical resistance increases, and the voltage across the filament increases. The voltage change is the detector signal. Thermal conductivity detection is more sensitive at lower flow rates. To prevent overheating and oxidation of the filament, the detector should never be left on unless carrier gas is flowing.

The detector responds to *changes* in thermal conductivity, so the conductivities of solute and carrier gas should be as different as possible. Because H_2 and He have the highest thermal conductivity, these two are the carriers of choice for thermal conductivity detection. A thermal conductivity detector responds to every substance except the carrier gas.

Thermal conductivity detectors are not sensitive enough to detect the minute quantities of analyte eluted from open tubular columns smaller than 0.53 mm in diameter. For narrower columns, other detectors must be used. Thermal conductivity detectors are excellent for the larger quantities of solute injected into packed columns.

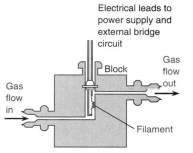

FIGURE 21-8 Thermal conductivity detector. [Courtesy Varian Associates, Palo Alto, CA.]

Flame Ionization Detector

In the **flame ionization detector** in Figure 21-9, eluate is burned in a mixture of H_2 and air. Carbon atoms (except carbonyl and carboxyl carbon atoms) produce CH radicals, which go on to produce CHO^+ ions in the flame:

$$CH + O \longrightarrow CHO^+ + e^-$$

CHO^+ cations are collected at the cathode above the flame. The electric current that flows between the anode and cathode is proportional to the number of CHO^+ ions. Only about 1 in 10^5 carbon atoms produces an ion, but ion production is strictly proportional to the number of susceptible carbon atoms entering the flame. The flame ionization detector is insensitive to O_2, CO_2, H_2O, and NH_3.

Response to organic compounds is directly proportional to solute mass over seven orders of magnitude. The detection limit is 100 times smaller than that of the thermal conductivity detector and is best with N_2 carrier gas. For open tubular columns, chromatography is conducted with H_2 or He at a low flow rate and N_2 *makeup gas* is added to the stream before it enters the detector. The makeup gas provides the higher flow rate needed by the detector and improves sensitivity.

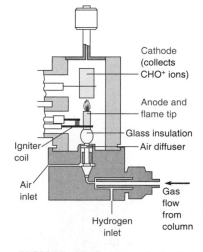

FIGURE 21-9 Flame ionization detector. [Courtesy Varian Associates, Palo Alto, CA.]

Electron capture detector:

$$^{63}\text{Ni} \longrightarrow \quad \beta^-$$
High-energy
electron

$$\beta^- + \text{N}_2 \longrightarrow \text{N}_2^+ + 2e^-$$
Collected at
anode

$$\text{analyte} + e^- \longrightarrow \text{analyte}^-$$
Too slow to
reach anode

Gas chromatography detectors:

Thermal conductivity: responds to
    everything

Flame ionization: responds to
    compounds with C—H

Electron capture: halogens,
    conjugated C=O,
    —C≡N, —NO$_2$

Flame photometer: P and S

Alkali flame: P and N

Sulfur chemiluminescence: S

Mass spectrometer: responds to
    everything

Electron Capture Detector

The **electron capture detector** is extremely sensitive to halogen-containing molecules, such as pesticides, but relatively insensitive to hydrocarbons, alcohols, and ketones. Carrier gas entering the detector is ionized by high-energy electrons ("β-rays") emitted from a foil containing radioactive ^{63}Ni. Electrons liberated from the gas are attracted to an anode, producing a small, steady current. When analyte molecules with a high electron affinity enter the detector, they capture some of the electrons and reduce the current. The decrease in electric current is the analytical signal. Electron capture detectors are sensitive to as little as 5 fg (femtograms, 10^{-15} g) of analyte eluted per second. The carrier gas is usually either N$_2$ or 5 vol % CH$_4$ in Ar. For open tubular columns, chromatography is conducted with H$_2$ or He at a low flow rate and N$_2$ *makeup gas* is added to the stream before it enters the detector. The makeup gas provides the higher flow rate needed by the detector and improves sensitivity.

Other Detectors

A *flame photometric detector* measures optical emission from phosphorus and sulfur compounds. When eluate passes through a H$_2$-air flame, excited sulfur- and phosphorus-containing species emit characteristic radiation, which is detected with a photomultiplier tube. The radiant emission is proportional to analyte concentration.

The *alkali flame detector*, also called a *nitrogen-phosphorus* detector, is a modified flame ionization detector that is selectively sensitive to phosphorus and nitrogen. It is especially important for drug analysis. When ions produced by these elements contact a Rb$_2$SO$_4$-containing glass bead at the burner tip, they create the electric current that is measured. N$_2$ carrier gas cannot be used for nitrogen-containing samples.

A *sulfur chemiluminescence detector* mixes the exhaust from a flame ionization detector with O$_3$ to form an excited state of SO$_2$ that emits light, which is detected. The *mass spectrometer* (see Box 20-2) is a sensitive detector that provides qualitative information about the structure of the analyte, as well as a quantitative measure.

Ask Yourself

21-A. **(a)** What is the advantage of temperature programming in gas chromatography?

(b) What is the advantage of an open tubular column over a packed column? Does a narrower or wider open tubular column provide higher resolution? What is the advantage of a packed column over an open tubular column?

(c) Why do H$_2$ and He allow more rapid linear flow rates in gas chromatography than does N$_2$, without loss of column efficiency (see Figure 20-8)?

(d) When would you use split, splitless, or on-column injection?

(e) To which kinds of analytes do the following detectors respond? **(i)** thermal conductivity, **(ii)** flame ionization, **(iii)** electron capture, **(iv)** flame photometric, **(v)** alkali flame, **(vi)** sulfur chemiluminescence, and **(viii)** mass spectrometer

Modern chromatography evolved from the experiment in Figure 20-1 in which sample is applied to the top of an open, gravity-feed column containing stationary phase. The next section describes high-performance liquid chromatography, which uses closed columns under high pressure and is the most common form of chromatography practiced today. However, open columns are used for preparative separations in biochemistry and chemical synthesis.

There is an art to pouring uniform columns, applying samples evenly, and obtaining symmetric elution bands. Stationary solid phase is normally poured into a column by first making a *slurry* (a mixture of solid and liquid) and pouring the slurry gently down the wall of the column. Try to avoid creating distinct layers, which form when some of the slurry is allowed to settle before more is poured in. Don't drain solvent below the top of the stationary phase, because air spaces and irregular flow patterns will be created. Solvent should be directed gently down the wall of the column. In *no* case should the solvent be allowed to dig a channel into the stationary phase. Maximum resolution demands a slow flow rate.

The Stationary Phase

For adsorption chromatography, *silica* ($SiO_2 \cdot xH_2O$, also called silicic acid) is a common stationary phase whose active adsorption sites are $Si-O-H$ (silanol) groups, which are slowly deactivated by adsorption of water from the air. Silica is activated by heating to 200°C to drive off the water. *Alumina* ($Al_2O_3 \cdot xH_2O$) is the other most common adsorbent. Preparative chromatography in the biochemistry lab is most often based on molecular exclusion and ion exchange, which are described in the next chapter.

Solvents

In adsorption chromatography, solvent competes with solute for adsorption sites on the stationary phase. *The relative abilities of different solvents to elute a given solute from the column are nearly independent of the nature of the solute.* Elution can be described as a displacement of solute from the adsorbent by solvent (Figure 21-10).

An *eluotropic series* ranks solvents by their relative abilities to displace solutes from a given adsorbent. **Eluent strength** in Table 21-2 is a measure of solvent adsorption energy, with the value for pentane defined as 0. The more polar the solvent, the greater its eluent strength. The greater the eluent strength, the more rapidly solutes will be eluted from the column.

A *gradient* (steady change) of eluent strength is used for many separations. First, weakly retained solutes are eluted with a solvent of low eluent strength. Then a second solvent is mixed with the first, either in discrete steps or continuously, to increase eluent strength and elute more strongly adsorbed solutes. A small amount of polar solvent markedly increases the eluent strength of a nonpolar solvent.

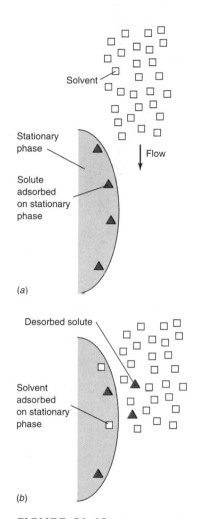

(a)

(b)

FIGURE 21-10 Solvent molecules compete with solute molecules for binding sites on the stationary phase. The more strongly the solvent binds to the stationary phase, the greater the *eluent strength* of the solvent.

Gradient elution in liquid chromatography is analogous to temperature programming in gas chromatography. Increased eluent strength is required to elute more strongly retained solutes.

Ask Yourself

21-B. Why are the relative eluent strengths of solvents in adsorption chromatography fairly independent of solute?

TABLE 21-2 Eluotropic series and ultraviolet cutoff wavelengths of solvents for adsorption chromatography on silica

Solvent	Eluent strength ($\varepsilon°$)	Ultraviolet cutoff (nm)
Pentane	0.00	190
Hexane	0.01	195
Heptane	0.01	200
Trichlorotrifluoroethane	0.02	231
Toluene	0.22	284
Chloroform	0.26	245
Dichloromethane	0.30	233
Diethyl ether	0.43	215
Ethyl acetate	0.48	256
Methyl t-butyl ether	0.48	210
Dioxane	0.51	215
Acetonitrile	0.52	190
Acetone	0.53	330
Tetrahydrofuran	0.53	212
2-Propanol	0.60	205
Methanol	0.70	205

Ultraviolet cutoff is the approximate minimum wavelength at which solutes can be detected above the strong ultraviolet absorbance of solvent. The ultraviolet cutoff for water is 190 nm.

SOURCES: L. R. Snyder in *High-Performance Liquid Chromatography* (C. Horváth, ed.), Vol. 3 (New York: Academic Press, 1983); *Burdick & Jackson Solvent Guide,* 3rd ed. (Muskegon, MI: Burdick & Jackson Laboratories, 1990).

FIGURE 21-11 Three-hundred-liter preparative chromatography column can purify a kilogram of material. [Courtesy Prochrom, Inc., Indianapolis, IN.]

Question According to the scaling rules in Section 20-2, if the column diameter is increased from 4 mm to 400 mm and all else remains the same, how much larger can the sample be to achieve the same resolution?

Decreased particle size increases resolution but requires high pressure to obtain a reasonable flow rate.

21-3 High-Performance Liquid Chromatography

High-performance liquid chromatography (HPLC) uses high pressure to force eluent through a closed column packed with micron-size particles that provide exquisite separations. The industrial preparative column in Figure 21-11 can handle up to 1 kg of sample. The analytical HPLC equipment in Figure 21-12 uses columns with diameters of 1–5 mm and lengths of 5–30 cm, yielding 50 000 to 100 000 plates per meter. Essential components include a solvent delivery system, a sample injection valve, a detector, and a computer to display results.

Figures 21-13 and 21-14 show that resolution increases when the stationary phase particle size decreases. Notice how much sharper the peaks become in Figure 21-13 and how a decomposition product is resolved from the slow-moving component. The penalty for using fine particles is resistance to solvent flow. Pressures of ~70–400 bar are required for flow rates of 0.5–5 mL/min. Smaller particles of stationary phase improve resolution by allowing solute to diffuse faster between the

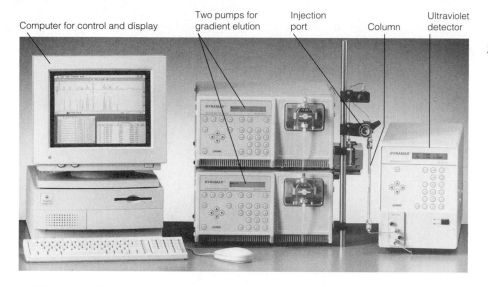

Computer for control and display   Two pumps for gradient elution   Injection port   Column   Ultraviolet detector

FIGURE 21-12   Typical laboratory equipment for high-performance liquid chromatography (HPLC). [Courtesy Rainin Instrument Co., Emeryville, CA.]

stationary and mobile phases (decreasing the term C in the van Deemter equation, 20-7). Also, smaller particles reduce the magnitude of irregular flow paths (the term A in the van Deemter equation).

Stationary Phase

Normal-phase chromatography uses a polar stationary phase and a less polar solvent. *Eluent strength is increased by adding a more polar solvent.* **Reversed-phase chromatography** is the more common scheme in which the stationary phase is nonpolar or weakly polar and the solvent is more polar. *Eluent strength is increased by adding a less polar solvent.* Reversed-phase chromatography eliminates tailing arising from adsorption of polar compounds by polar packings (see Figure 20-11b). Reversed-phase chromatography is also relatively insensitive to polar impurities (such as water) in the eluent.

Microporous particles of silica with diameters of 3–10 μm are the most common solid stationary phase support. These particles are permeable to solvent and have a surface area up to 500 m^2 per gram of silica. Adsorption chromatography occurs directly on the silica surface.

More commonly, liquid-liquid partition chromatography is conducted with a **bonded stationary phase** covalently attached to silanol groups on the silica surface.

Normal-phase chromatography: polar stationary phase and less polar solvent

Reversed-phase chromatography: low-polarity stationary phase and polar solvent

Bonded polar phases		Bonded nonpolar phases	
R = $(CH_2)_3NH_2$	Amino	R = $(CH_2)_{17}CH_3$	Octadecyl
R = $(CH_2)_3C\equiv N$	Cyano	R = $(CH_2)_7CH_3$	Octyl
R = $(CH_2)_2OCH_2CH(OH)CH_2OH$	Diol	R = $(CH_2)_3C_6H_5$	Phenyl

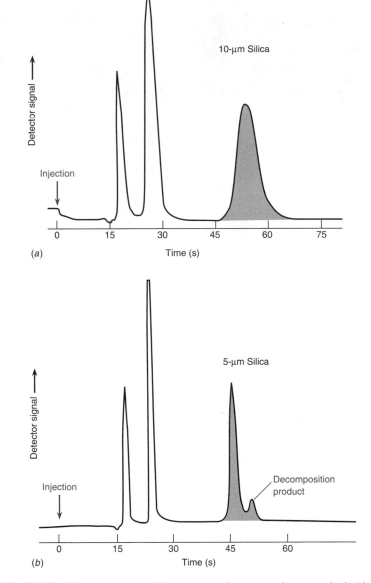

(a)

(b)

FIGURE 21-13 Chromatograms of the same sample run on columns packed with (a) 10-μm and (b) 5-μm particle diameter silica. [From R. E. Majors, *J. Chromatogr. Sci.* **1973**, *11*, 88.]

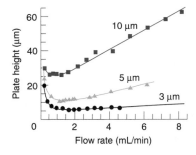

FIGURE 21-14 Plate height as a function of flow rate for stationary phase particle sizes of 10, 5, and 3 μm. Smaller particles give a smaller optimal plate height, thereby producing sharper chromatographic peaks. [Courtesy Perkin-Elmer Corp., Norwalk, CT.]

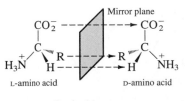

Optical isomers

The octadecyl (C_{18}) stationary phase is by far the most common in HPLC. The Si — O — Si bond that attaches the stationary phase to the silica is stable only over the pH range 2–8. Strongly acidic or basic eluents generally cannot be used with silica.

Optical isomers are mirror image compounds such as D- and L-amino acids. Most compounds with four different groups attached to one tetrahedral carbon atom exist in two mirror image (optical) isomers. Optical isomers can be separated from each other by chromatography on a stationary phase containing just one optical isomer of the bonded phase. A major push for separating optical isomers has been made by the drug industry, because one isomer that is biologically active needs to be separated from the other, which is inactive or toxic. Figure 21-15 shows separation of the two optical isomers of the anti-inflammatory drug Naproxen.

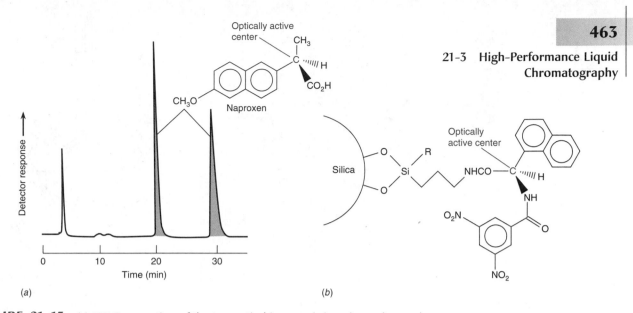

FIGURE 21-15 (*a*) HPLC separation of the two optical isomers (mirror image isomers) of the drug Naproxen eluted with 0.05 M ammonium acetate in methanol. Naproxen is the active ingredient of the anti-inflammatory drug Aleve. (*b*) Structure of the bonded stationary phase. [Courtesy Phenomenex, Torrance, CA.]

The Column

High-performance liquid chromatography columns are expensive and easily degraded by irreversible adsorption of impurities from samples and solvents. Therefore, we place a short, disposable **guard column** containing the same stationary phase as the main column at the entrance to the main column (Figure 21-16). Substances that would bind irreversibly to the main column are instead bound in the guard column, which is periodically discarded.

Because the column is under high pressure, a special technique is required to inject sample. The *injection valve* in Figure 21-17 has interchangeable steel sample loops that hold fixed volumes from 2 μL to 1 000 μL. In the load position, a syringe is used to wash and load the loop with fresh sample at atmospheric pressure. When the valve is rotated 60° counterclockwise, the content of the sample loop is injected into the column at high pressure. You should pass samples through a 0.5- to 2-μm filter prior to injection to avoid contaminating the column with particles, plugging the tubing, and damaging the pump. For liquid chromatography, we use a *blunt nose syringe,* not the pointy needle used in gas chromatography.

Solvents

Elution with a single solvent or a constant solvent mixture is called **isocratic elution.** If one solvent does not discriminate adequately between the components of a mixture or if the solvent does not provide sufficiently rapid elution of all components, then **gradient elution** can be used. In gradient elution, solvent is changed continuously from a weak eluent strength to a strong eluent strength by mixing more and more of a strong solvent into the weak solvent during the chromatography.

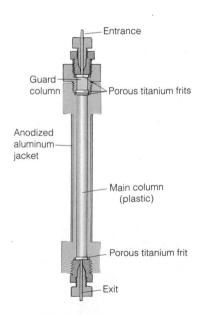

FIGURE 21-16 HPLC column with replaceable guard column to collect irreversibly adsorbed impurities. Titanium frits distribute the liquid evenly over the diameter of the column. [Courtesy Upchurch Scientific, Oak Harbor, WA.]

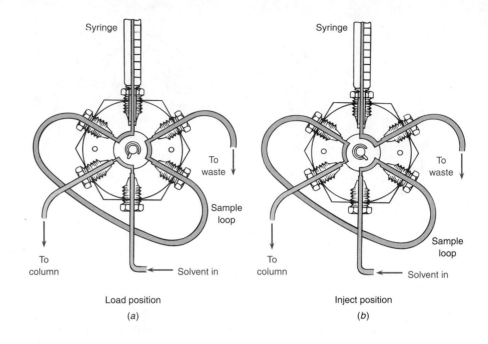

Syringe   Syringe

To waste   To waste

Sample loop   Sample loop

To column   Solvent in   To column   Solvent in

Load position   Inject position
(*a*)   (*b*)

FIGURE 21-17   Injection valve for HPLC. Replaceable sample loop comes in various fixed-volume sizes.

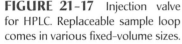

To prepare a mixture of an aqueous buffer with an organic solvent, adjust the pH of the buffer to the desired value *prior* to mixing it with organic solvent. Once you mix in the organic solvent, the meaning of "pH" is not well defined.

Figure 21-18 shows the effect of increasing eluent strength in the *isocratic* elution of eight compounds from a reversed-phase column. In a reversed-phase separation, eluent strength *decreases* as the solvent becomes *more* polar. The first chromatogram (upper left) was obtained with a solvent consisting of 90 vol % acetonitrile and 10 vol % aqueous buffer. Acetonitrile has a high eluent strength, and all compounds are eluted rapidly. In fact, only three peaks are observed because of overlap. It is customary to call the aqueous solvent A and the organic solvent B. The first chromatogram was obtained with 90% B. When the eluent strength is reduced by changing the solvent to 80% B, there is slightly more separation and five peaks are observed. At 60% B, we begin to see a sixth peak. At 40% B, there are eight clear peaks, but compounds 2 and 3 are not fully resolved. At 30% B, all peaks are resolved, but the separation takes too long to be useful. Backing up to 35% B (the bottom trace) separates all peaks in a little over 2 h (which is still too long for many purposes).

From the isocratic elutions in Figure 21-18, the *gradient* in Figure 21-19 was selected to resolve all peaks while reducing the time from 2 h to 38 min. First, 30% B was run for 8 min to separate components 1, 2, and 3. The eluent strength was then increased steadily over 5 min to 45% B and held there for 15 min to elute peaks 4 and 5. Finally, the solvent was changed to 80% B over 2 min and held there to elute the last peaks.

Pure HPLC solvents are expensive, and most organic solvents are expensive to dispose of in an environmentally sound manner. To reduce waste, we can use a device that discards solvent from a column when it is contaminated with solutes but recycles pure solvent emerging between or after the chromatographic peaks.

Detectors

An **ultraviolet detector** is most common, with a flow cell such as that in Figure 21-20. Simple systems employ the intense 254-nm emission of a mercury lamp. More versatile instruments use a deuterium or xenon lamp and monochromator, so you can choose the optimum wavelength for your analytes. In some systems, the

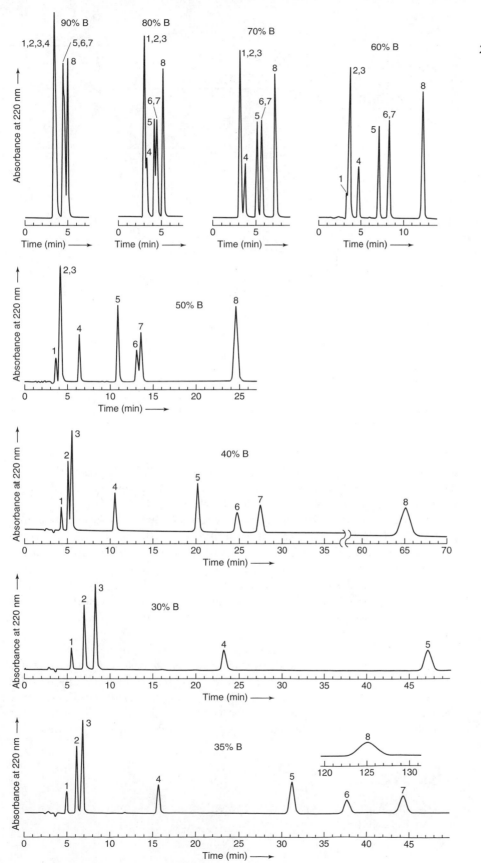

FIGURE 21-18 Isocratic HPLC separation of a mixture of aromatic compounds at 1.0 mL/min on a 0.46 × 25 cm Hypersil ODS column (C_{18} on 5-μm silica) at ambient temperature (~22°C): (1) benzyl alcohol; (2) phenol; (3) 3′,4′-dimethoxyacetophenone; (4) benzoin; (5) ethyl benzoate; (6) toluene; (7) 2,6-dimethoxytoluene; (8) *o*-methoxybiphenyl. Eluent consisted of aqueous buffer (designated A) and acetonitrile (designated B). The notation "90% B" in the first chromatogram means 10 vol % A and 90 vol % B. The buffer contained 25 mM KH_2PO_4 plus 0.1 g/L sodium azide adjusted to pH 3.5 with HCl.

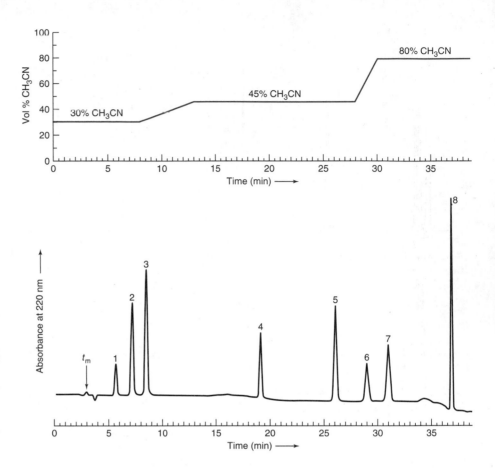

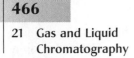

FIGURE 21-19 Gradient elution of the same mixture of aromatic compounds from Figure 21-18 with the same column, flow rate, and solvents. The upper trace is the *segmented gradient* profile, so named because it is divided into several different segments.

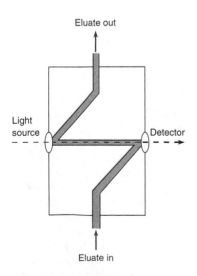

FIGURE 21-20 Light path in a micro flow cell of a spectrophotometric detector. Cells are available with a 0.5-cm pathlength containing only 8 μL of liquid.

ultraviolet-visible absorption spectrum of each solute can be recorded with a photodiode array (Figure 18-10) as it is eluted. Clearly, an ultraviolet detector can only respond to analytes with sufficient ultraviolet absorbance.

A more universal sensor is the **refractive index detector,** but its detection limit is about 1 000 times poorer than that of an ultraviolet detector, and it is not useful for gradient elution because the baseline changes as the solvent changes. In general, a solute has a different refractive index (ability to deflect a ray of light) from that of the solvent. The detector measures the deflection of a light ray by the eluate. An *evaporative light-scattering detector* is another nearly universal detector in which eluate is evaporated to make an aerosol of fine particles of nonvolatile solutes that can be detected by light scattering.

The *electrochemical detector* in Figure 16-6 responds to analytes that can be oxidized or reduced at an electrode over which the eluate passes. Electric current is proportional to solute concentration. *Fluorescence detectors* are especially sensitive but respond only to the few analytes that fluoresce (Figure 18-20). The fluorescence detector works by irradiating eluate at one wavelength and monitoring emission at a longer wavelength. Emission intensity is proportional to solute concentration.

Ask Yourself

21-C. (a) What is the difference between normal-phase and reversed-phase chromatography?

(b) What is the difference between isocratic and gradient elution?

(c) Why does eluent strength increase in normal-phase chromatography when a more polar solvent is added?

(d) Why does eluent strength increase in reversed-phase chromatography when a less polar solvent is added?

(e) What is the purpose of a guard colunn?

21-4   Sample Preparation for Chromatography

Sample preparation is the process of transforming a sample into a form that is suitable for analysis. This process might involve extracting analyte from a complex matrix, *preconcentrating* very dilute analytes to get a concentration high enough to measure, removing or masking interfering species, or chemically transforming (*derivatizing*) the analyte into a more convenient or more easily detected form. *Solid-phase microextraction* and *purge and trap* are sample preparation techniques that are especially important for gas chromatography. *Solid-phase extraction* is a sample preparation technique for liquid chromatography.

Solid-phase microextraction is a simple way to extract compounds for gas chromatography from liquids, air, or even sludge without using any solvent. The key component is a fused-silica fiber coated with a 10- to 100-μm-thick film of nonvolatile liquid stationary phase similar to those used in gas chromatography. Figure 21-21 shows the fiber attached to the base of a syringe with a fixed metal needle. The fiber can be extended from the needle or retracted inside the needle. Figure 21-22 demonstrates the procedure of exposing the fiber to a sample solution (or the gaseous headspace above the liquid) for a fixed length of time while stirring and, perhaps, heating. Only a fraction of the analyte in the sample is extracted into the fiber.

After sampling, the fiber is retracted and the needle is introduced into the inlet of a gas chromatograph. The fiber is extended inside the hot injection liner, where

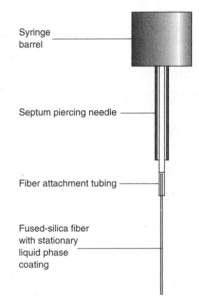

Syringe barrel

Septum piercing needle

Fiber attachment tubing

Fused-silica fiber with stationary liquid phase coating

FIGURE 21-21 Syringe for solid-phase microextraction. The fused-silica fiber is withdrawn inside the steel needle after sample collection and whenever the syringe is used to pierce a septum.

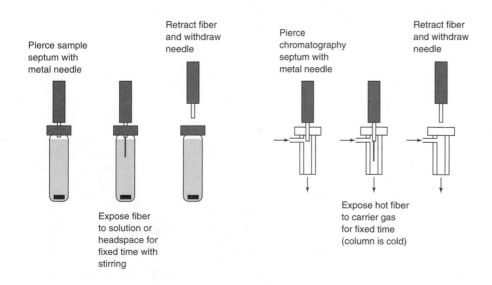

Pierce sample septum with metal needle

Retract fiber and withdraw needle

Expose fiber to solution or headspace for fixed time with stirring

Pierce chromatography septum with metal needle

Retract fiber and withdraw needle

Expose hot fiber to carrier gas for fixed time (column is cold)

FIGURE 21-22 Sampling by solid-phase microextraction and desorption of analyte from the coated fiber into a gas chromatograph. [Adapted from Supelco Chromatography Products catalog, Bellefonte, PA.]

analyte is thermally desorbed from the fiber in the splitless mode for a fixed time. *Cold trapping* (page 456) focuses the desorbed analyte at the head of the column prior to chromatography. Solid-phase microextraction can also be adapted to liquid chromatography by inserting the fiber into an injection port in which the fiber is washed with a strong solvent.

Purge and trap is a method for removing volatile analytes from liquids or solids (such as groundwater or soil), concentrating the analytes, and introducing them into a gas chromatograph. In contrast to solid-phase microextraction, which only removes a portion of analyte from the sample, the goal in purge and trap is to remove 100% of the analyte from the sample. Quantitative removal of polar analytes from polar matrices can be difficult.

Figure 21-23 shows apparatus for extracting volatile flavor components from beverages. Helium purge gas from a stainless steel needle is bubbled through the beverage in the sample vial, which is heated to 50°C to aid evaporation of analytes. Gas exiting the sample vial passes through an adsorption tube containing three layers of adsorbent compounds with increasing adsorbent strength. For example, the moderate adsorbent could be a nonpolar phenylmethylpolysiloxane, the stronger adsorbent could be the polymer Tenax, and the strongest adsorbent could be carbon molecular sieves.

During the purge and trap process, gas flows through the adsorbent tube from end A to end B in Figure 21-23. After purging all the analyte from the sample, gas flow is reversed to go from B to A and the trap is purged at 25°C to remove as much water or other solvent as possible from the adsorbents. Outlet A of the adsorption tube is then directed to the injection port of a gas chromatograph operating in splitless mode and the trap is heated to ~200°C. Desorbed analytes flow into the chromatography column, where they are concentrated by cold trapping. After complete desorption from the trap, the chromatography column is warmed up to initiate the separation.

Solid-phase extraction uses a liquid chromatography solid phase in a short, open column to separate one or more analytes from a liquid mixture. For example, Figure 21-24 shows C_{18}-silica being used to collect a nonpolar organic molecule from aqueous solution. If urine containing traces of steroid drugs is passed through a short, open column of C_{18}-silica, the steroids are retained in the C_{18} coating. Upon washing the column with a small volume of nonpolar or weakly polar solvent such as hexane or dichloromethane, the steroid is released from the column into the solvent, where it can be analyzed by chromatography.

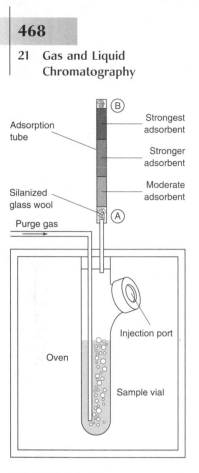

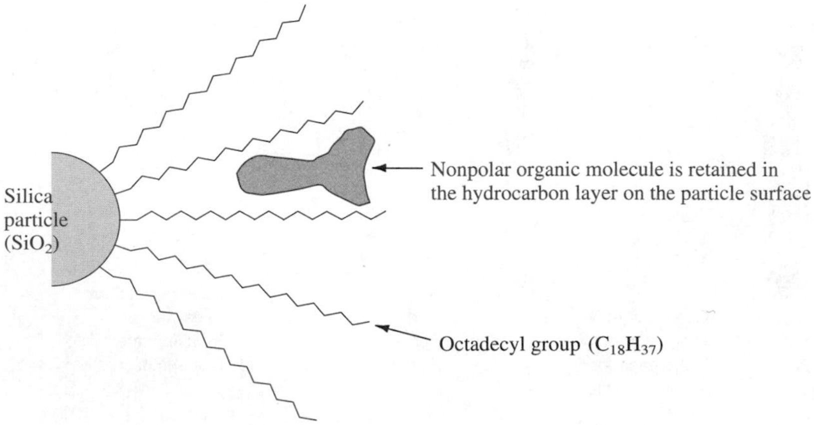

FIGURE 21–23 Purge and trap apparatus for extracting volatile substances from a liquid or solid by flowing gas. You need to establish the time and temperature required to purge 100% of the analyte from the sample in separate control experiments.

FIGURE 21–24 In solid-phase extraction with C_{18}-silica, nonpolar solutes from an aqueous sample are retained in the stationary phase in a short, open column. The retained solutes are later eluted with an organic solvent.

Silica particle (SiO_2)

Nonpolar organic molecule is retained in the hydrocarbon layer on the particle surface

Octadecyl group ($C_{18}H_{37}$)

Sample cleanup refers to the removal of undesirable components of the unknown that might interfere with measurement of the analyte. In the urine-steroid analysis, the polar molecules that are not retained by the C_{18}-silica wash right through the column and are separated from the analyte.

Ask Yourself

21-D. (a) Why is it necessary to use cold trapping on the gas chromatography column if sample is introduced from a solid-phase microextraction syringe or from a purge and trap absorption tube?
    (b) What is the purpose of solid-phase extraction?
    (c) What kind of sample cleanup does solid-phase extraction accomplish in the analysis of steroids in urine?

Important Terms

bonded stationary phase	high-performance liquid chromatography	sample preparation
cold trapping		solid-phase extraction
dialysis	isocratic elution	solid-phase microextraction
electron capture detector	normal-phase chromatography	solvent trapping
eluent strength	on-column injection	split injection
flame ionization detector	purge and trap	splitless injection
gas chromatography	refractive index detector	temperature programming
gradient elution	reversed-phase chromatography	thermal conductivity detector
guard column	sample cleanup	ultraviolet detector

Problems

21-1. (a) Explain the difference between wall-coated, support-coated, and porous-layer open tubular columns for gas chromatography.

(b) What is the advantage of a bonded stationary phase in gas chromatography?

(c) Why do we use a makeup gas for some gas chromatography detectors?

21-2. Explain why plate height increases at **(a)** very low and **(b)** very high flow rates in Figure 21-14. (*Hint:* Refer to Section 20-3.)

21-3. Consider a narrow-bore (0.25 mm diameter) open tubular gas chromatography column coated with a thin film (0.10 μm thick) of stationary phase that provides 5 000 plates

per meter. Consider also a wide-bore (0.53 mm diameter) column with a thick film (5.0 μm thick) of stationary phase that provides 1 500 plates per meter.

(a) Why does the thin film provide more theoretical plates per meter than the thick film?

(b) The density of stationary phase is approximately 1.0 g/mL. What mass of stationary phase is in each column in a length equivalent to one theoretical plate?

(c) How many nanograms of analyte can be injected into each column if the mass of analyte is not to exceed 1.0% of the mass of stationary phase in one theoretical plate?

21-4. Nonpolar aromatic compounds were separated by HPLC on a bonded phase containing octadecyl groups

$[-(CH_2)_{17}CH_3]$ covalently attached to silica particles. The eluent was 65 vol % methanol in water. How would the retention times be affected if 90% methanol were used instead?

21-5. Polar solutes were separated by HPLC, using a bonded phase containing polar diol substituents $[-CH(OH)CH_2OH]$. How would the retention times be affected if the eluent were changed from 40 vol % to 60 vol % acetonitrile in water? Acetonitrile $(CH_3C \equiv N)$ is less polar than water.

21-6. Draw the chemical structures of two nonpolar bonded phases and two polar bonded phases in HPLC. Begin with a silicon atom at the surface of a silica particle.

21-7. (a) Why is high pressure needed in HPLC?

(b) Why does the efficiency (decreased plate height) of liquid chromatography increase as the stationary phase particle size is reduced?

21-8. A 15-cm-long HPLC column packed with 5-μm particles has an optimum plate height of 10.0 μm in Figure 21-14. What will be the half-width of a peak eluting at 10.0 min? If the particle size were 3 μm and the plate height were 5.0 μm, what would be the half-width?

21-9. Octanoic acid and 1-aminooctane were separated by HPLC on a bonded phase containing octadecyl groups $[-(CH_2)_{17}CH_3]$. The eluent was 20 vol % methanol in water adjusted to pH 3.0 with HCl.

$$CH_3(CH_2)_6CO_2H \qquad CH_3(CH_2)_7NH_2$$
Octanoic acid   1-Aminooctane

(a) Draw the predominant form (neutral or ionic) of a carboxylic acid and an amine at pH 3.0.

(b) State which compound is expected to be eluted first and why.

21-10. (a) When you try separating an unknown mixture by reversed-phase chromatography with 50% acetonitrile-50% water, the peaks are eluted between 1 and 3 min, and they are too close together to be well resolved for quantitative analysis. Should you use a higher or lower percentage of acetonitrile in the next run?

(b) Suppose that you try separating an unknown mixture by normal-phase chromatography with the solvent mixture 50% hexane-50% methyl *t*-butyl ether (which is more polar than hexane). The peaks are too close together and are eluted rapidly. Should you use a higher or lower percentage of hexane in the next run?

21-11. A known mixture of compounds C and D gave the following HPLC results:

Compound	Concentration (mg/mL) in mixture	Peak area (cm^2)
C	1.03	10.86
D	1.16	4.37

A solution was prepared by mixing 12.49 mg of D plus 10.00 mL of unknown containing just C, and diluting to 25.00 mL. Peak areas of 5.97 and 6.38 cm^2 were observed for C and D, respectively. Find the concentration of C (mg/mL) in the unknown. (*Hint:* Review Section 20-5 on internal standards.)

21-12. Spherical, microporous silica particles used in chromatography have a density of 2.20 g/mL, a diameter of 10.0 μm, and a measured surface area of 300 m^2/g.

(a) The volume of a spherical particle is $\frac{4}{3}\pi r^3$, where r is the radius. The mass of the sphere is volume $\times$ density ($= mL \times g/mL$). How many particles are in 1.00 g of silica?

(b) The surface area of a sphere is $4\pi r^2$. Calculate the surface area of 1.00 g of solid, spherical silica particles.

(c) By comparing the calculated and measured surface areas, what can you say about the porosity of the particles?

21-13. Why is splitless injection used with purge and trap sample preparation?

How Would You Do It?

21-14. Ideally, the octadecyl $(-C_{18}H_{37})$ groups of a C_{18}-silica HPLC stationary phase stand almost straight out from the silica particles like the hairs of a brush:

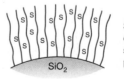

Stationary phase C_{18} chains extended straight in strongly solvating (organic) mobile phase, S

If a column is washed with an aqueous phase containing no organic solvent, the C_{18} groups are insoluble in the mobile phase and collapse on themselves:

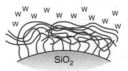

Collapsed C_{18} chains in weakly solvating (aqueous) mobile phase, W

Chromatogram 1 shows a standard mixture separated on a 4.6-mm-diameter × 15-cm-long C_{18} column at 75°C with a mobile phase of 10:90 acetonitrile-50 mM phosphate buffer (pH 2.6) at a flow rate of 2 mL/min. Peak 7 is eluted at 9.42 min with a plate number of 12 800. After recording the chromatogram, the column was flushed with water, stored overnight, and then flushed for 20 min with 170 column volumes of mobile phase (10:90 acetonitrile-50 mM phosphate). Injection of the standard mixture gave chromatogram 2 in which peak 7 has an apparent plate number of 7 600. After washing with 90 more column volumes of mobile phase, chromatogram 3 was obtained in which peak 7 has a plate number of 11 500.

(a) Suggest an explanation for what is observed in the three chromatograms.

(b) Here are guidelines for flushing a C_{18} column. Explain the rationale for each.

    1. Do not use 100% aqueous phase.

    2. To remove strongly retained solutes, flush the column initially with 5–10 column volumes of the most recent mobile phase (such as 60:40 H_2O-acetonitrile) *without* buffer. That is, replace buffer with pure water.

    3. Then flush with 10–20 volumes of strong solvent (such as 10:90 H_2O-acetonitrile).

    4. Store the column in the strong solvent.

    5. Equilibrate with 10–20 volumes of new, desired, buffered mobile phase.

    6. Use a standard mixture to check for correct retention times and plate numbers.

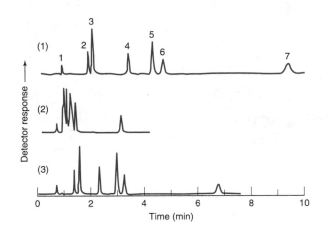

Chromatograms from R. G. Wolcott and J. W. Dolan, *LC/GC* **1999**, *17*, 316.

Further Reading

H. M. McNair and J. M. Miller, *Basic Gas Chromatography* (New York: Wiley, 1998).

L. R. Snyder, J. J. Kirkland, and J. L. Glajch, *Practical HPLC Method Development* (New York: Wiley, 1997).

R. L. Grob, *Modern Practice of Gas Chromatography* (New York: Wiley, 1995).

M. McMaster and C. McMaster, *GC/MS: A Practical User's Guide* (New York: Wiley, 1998).

J. V. Hinshaw and L. S. Ettre, *Introduction to Open Tubular Gas Chromatography* (Cleveland, OH: Advanstar Communications, 1994).

Capillary Electrophoresis in Biology

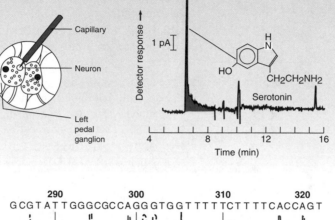

Analyzing the contents of a single nerve cell. [From T. M. Olefirowicz and A. G. Ewing, *Anal. Chem.* **1990**, *62*, 1872.]

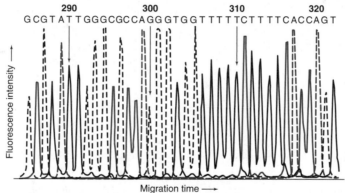

A portion of the sequence of a strand of DNA analyzed by capillary electrophoresis. The letters A, T, C, and G stand for nucleotide bases. [From S. Lui, Y. Shi, W. W. Ja, and R. A. Mathies, *Anal. Chem.* **1999**, *71*, 566. For a review, see I. Kheterpal and R. A. Mathies, *Anal. Chem.* **1999**, *71*, 31A.]

*C*apillary electrophoresis separates and measures ions on the basis of their different migration rates in an electric field. The tiny sample required for an analysis is in the picoliter to nanoliter (10^{-12} to 10^{-9} L) range. For example, a silica capillary with an inner diameter of 5 μm can be drawn to a fine point and inserted into a giant nerve cell of a pond snail. Then just 1% of the cell volume is drawn into the capillary and the contents separated with a 25-kV electric field and detected by an electrochemical detector. The first peak is the neurotransmitter serotonin, whose concentration is 3 μM inside the cell.

Electrophoresis is a key tool in the effort to sequence the entire human genome. DNA (deoxyribonucleic acid) can be analyzed on a glass chip etched with microscopic capillaries that separate strands with lengths from 1 to 500 nucleotide bases in 20 min, with the shortest DNA migrating fastest. Each strand ending in a different nucleotide base is chemically labeled so that it will fluoresce at a specific wavelength. The fluorescence wavelength of each peak reaching the detector tells us which of the four possible nucleotides is at the end of the DNA.

Chapter 22

CHROMATOGRAPHIC METHODS AND CAPILLARY ELECTROPHORESIS

*L*iquid chromatography, discussed in the last chapter, separates solutes by *adsorption* or *partition* mechanisms. In this chapter, we describe separations by *ion-exchange, molecular exclusion,* and *affinity* chromatography, which were illustrated in Figure 20-2. We also consider *capillary electrophoresis,* which separates species on the basis of their different rates of migration in an electric field. Capillary electrophoresis is especially suited for extremely small volume samples. The opening of Chapter 0 showed a single vesicle from a cell being manipulated with a laser ("optical tweezers") to position it at the mouth of a capillary. Electrophoresis showed that individual vesicles store different chemicals. This conclusion could not be reached from studies of the total contents of whole populations of vesicles.

22-1 Ion–Exchange Chromatography

Ion-exchange chromatography is based on the attraction between solute ions and charged sites in the stationary phase (Figure 20-2). **Anion exchangers** have positively charged groups on the stationary phase that attract solute <u>anions</u>. **Cation exchangers** contain negatively charged groups that attract solute <u>cations</u>.

The stationary phase for ion-exchange chromatography is usually a *resin,* such as polystyrene, which consists of amorphous (noncrystalline) particles. Polystyrene

Anion exchangers contain bound *positive* groups.

Cation exchangers contain bound *negative* groups.

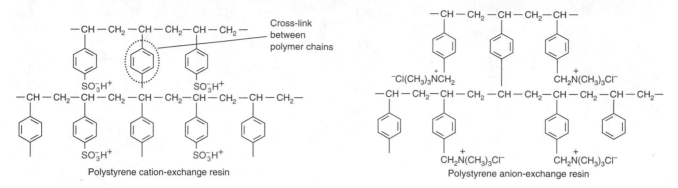

Polystyrene cation-exchange resin

Polystyrene anion-exchange resin

FIGURE 22-1 Structures of polystyrene ion-exchange resins. *Cross-links* are covalent bridges between polymer chains.

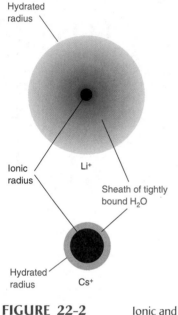

FIGURE 22-2 Ionic and hydrated radii of Li⁺ and Cs⁺. The ionic radius is the size of the bare cation. The hydrated radius includes the effective size of the ion plus its sheath of water molecules, which are tightly bound by electrostatic attraction. Smaller bare ions have larger hydrated radii because they attract water molecules more strongly.

Electrostatic attraction

A large excess of one ion will displace another ion from the resin.

is made into a cation exchanger when negative sulfonate ($-SO_3^-$) or carboxylate ($-CO_2^-$) groups are attached to the benzene rings (Figure 22-1). Polystyrene is an anion exchanger if ammonium groups ($-NR_3^+$) are attached. Cross-links (covalent bonds) between polystyrene chains in Figure 22-1 control the pore sizes into which solutes can diffuse.

Gel particles are softer than resin particles. Cellulose and dextran ion-exchange gels, which are polymers of the sugar glucose, possess larger pore sizes and lower charge densities. Gels are better suited than resins for ion exchange of macromolecules, such as proteins.

Ion-Exchange Selectivity

Consider the competition of Cs^+ and Li^+ for sites on the cation-exchange resin, R^-:

$$R^-Cs^+ + Li^+ \rightleftharpoons R^-Li^+ + Cs^+ \qquad K = \frac{[R^-Li^+][Cs^+]}{[R^-Cs^+][Li^+]} \qquad (22\text{-}1)$$

The equilibrium constant is called the *selectivity coefficient,* because it describes the relative affinities of the resin for Li^+ and Cs^+. Discrimination between different ions tends to increase with the extent of cross-linking, because the resin pore size shrinks as cross-linking increases.

The **hydrated radius** of an ion is the effective size of the ion plus its tightly bound sheath of water molecules, which are attracted by the positive or negative charge of the ion (Figure 22-2). The large species $Li(H_2O)_x^+$ does not have as much access to the resin as the smaller species $Cs(H_2O)_y^+$.

More highly charged ions bind more tightly to ion-exchange resins. For ions of the same charge, the larger the hydrated radius, the less tightly the ion is bound. An approximate order of selectivity for some cations is

$$Pu^{4+} \gg La^{3+} > Y^{3+} > Sc^{3+} > Al^{3+} \gg Ba^{2+} > Pb^{2+} > Sr^{2+} >$$
$$Ca^{2+} > Ni^{2+} > Cd^{2+} > Cu^{2+} > Co^{2+} > Zn^{2+} > Mg^{2+} \gg Tl^+ >$$
$$Ag^+ > Cs^+ > Rb^+ > K^+ > NH_4^+ > Na^+ > H^+ > Li^+$$

Reaction 22-1 can be driven in either direction. Washing a column containing Cs^+ with a substantial excess of Li^+ will replace Cs^+ with Li^+. Washing a column in the Li^+ form with excess Cs^+ will convert it to the Cs^+ form.

An ion exchanger loaded with one ion will bind a small amount of a different ion nearly quantitatively (completely). A resin loaded with Cs^+ binds small amounts of Li^+ quantitatively, even though the selectivity is greater for Cs^+. The same resin binds large quantities of Ni^{2+} because the selectivity for Ni^{2+} is greater than that for Cs^+. Even though Fe^{3+} is bound more tightly than H^+, Fe^{3+} can be quantitatively removed from the resin by washing with excess acid.

To separate one ion from another by ion-exchange chromatography, *gradient elution* with increasing ionic strength (ionic concentration) in the eluent is extremely valuable. In Figure 22-3, a gradient of $[H^+]$ was used to separate lanthanide cations (M^{3+}). The more strongly bound metal ions require a higher concentration of H^+ for elution. An ionic strength gradient is analogous to a solvent gradient in HPLC or a temperature gradient in gas chromatography. Box 22-1 shows some applications of ion-exchange chromatography.

"Quantitative" is chemists' jargon for "complete."

What Is Deionized Water?

Deionized water is prepared by passing water through an anion-exchange resin loaded with OH^- and a cation-exchange resin loaded with H^+. Suppose, for example, that $Cu(NO_3)_2$ is present in the water. The cation-exchange resin binds Cu^{2+} and replaces it with $2H^+$. The anion-exchange resin binds NO_3^- and replaces it with OH^-. The H^+ and OH^- combine, so the eluate is pure water:

Home water softeners use ion exchange to remove Ca^{2+} and Mg^{2+} from "hard" water (Box 12-3).

$$\left. \begin{array}{l} Cu^{2+} \xrightarrow{\ H^+ \ ion \ exchange\ } 2H^+ \\ 2NO_3^- \xrightarrow{\ OH^- \ ion \ exchange\ } 2OH^- \end{array} \right\} 2H^+ + 2OH^- \longrightarrow pure\ H_2O$$

Charge is conserved during ion exchange. One Cu^{2+} displaces $2H^+$ from a cation-exchange column. It takes $3H^+$ to displace one Fe^{3+}. One SO_4^{2-} displaces $2OH^-$ from an anion-exchange column.

Preconcentration

Measuring extremely low levels of analyte is called **trace analysis.** Trace analysis is especially important for environmental problems in which low concentrations of substances, such as mercury in fish, can become concentrated over many years in people who eat large quantities of fish. For trace analysis, analyte concentration may

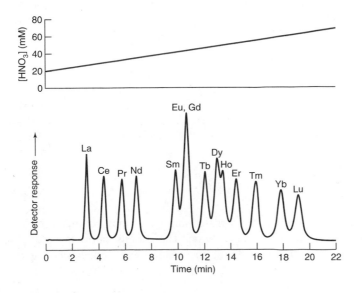

FIGURE 22-3 Elution of lanthanide(III) ions from a cation exchanger, using a gradient of H^+ (20 to 80 mM HNO_3 over 25 min) to drive off more strongly retained cations. The higher the atomic number of the lanthanide, the smaller its ionic radius and the more strongly it binds to chelating groups on the resin. Lanthanides were detected spectrophotometrically by reaction with a color-forming reagent after elution. [From Y. Inoue, H. Kumagai, Y. Shimomura, T. Yokoyama, and T. M. Suzuki, *Anal. Chem.* **1996,** *68,* 1517.]

Box 22-1 *Informed Citizen*

Applications of Ion Exchange

In the rush to create an atomic bomb during World War II, it was necessary to isolate significant quantities of pure lanthanide elements (rare earth elements 57 to 71).[1] The photograph shows some of 12 ion-exchange columns (10 cm diameter × 3.0 m long) in a pilot plant at Iowa State College. Each column took several *weeks* to separate a 50- to 100-g mixture of rare earth chlorides.

    Among its analytical applications, ion chromatography is used today to measure ions in water droplets in clouds. The table shows the mean composition of nonprecipitating clouds for six summer months in 1996 at the summit of Mt. Brocken in the Harz Mountains of Germany.

Ionic composition of clouds

Ion	Concentration (μM)	Ion	Concentration (μM)
Cl^-	101	H^+	131
NO_3^-	360	Na^+	100
SO_4^{2-}	156	NH_4^+	472
		K^+	1.3
		Ca^{2+}	26
		Mg^{2+}	12

Data from K. Acker, D. Möller, W. Wieprecht, D. Kalaß, and R. Auel, *Fresenius J. Anal. Chem.* **1998**, *361*, 59.

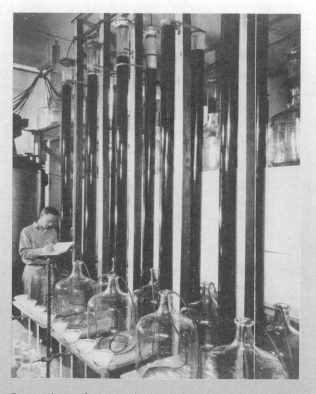

Preparative-scale ion-exchange columns used to separate rare earths for the Manhattan Project during World War II. [Iowa State University Library, Special Collections Department.]

be so low that it cannot be measured without **preconcentration,** a process in which analyte is brought to a higher concentration prior to analysis.

    Metals in natural waters can be preconcentrated with a cation-exchange column:

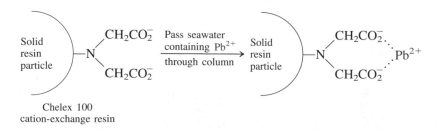

When a large volume of water is passed through a small volume of resin, the cations are concentrated into the small column. The cations can then be displaced into a small volume of solution by eluting the column with concentrated acid:

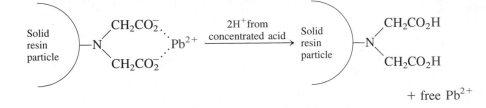

+ free Pb^{2+}

When lead in seawater was preconcentrated by a factor of 50, the detection limit was reduced from 15 ng of lead per liter of seawater down to 0.3 ng/L. How many grams is 1 ng?

Ask Yourself

22-A. (a) What is deionized water? What kind of impurities are not removed by deionization?

   (b) Why is gradient elution used in Figure 22-3?

   (c) Explain how preconcentration of cations with an ion exchanger works. Why must the concentrated acid eluent be very pure?

22-2   Ion Chromatography

Ion chromatography is a high-performance version of ion-exchange chromatography, with a key modification that removes eluent ions before detecting analyte ions. Ion chromatography is the method of choice for anion analysis. It is used in the semiconductor industry to monitor anions and cations at 0.1-ppb levels in deionized water. Figure 22-4 shows an example of anion chromatography in environmental analysis.

In ion chromatography, anions are separated by ion exchange and detected by their electrical conductivity. The conductivity of the electrolyte in the eluent is ordinarily high enough to make it difficult or impossible to detect the conductivity change when analyte ions are eluted. Therefore, the key feature of *suppressed-ion* chromatography is removal of unwanted electrolyte prior to conductivity measurement.

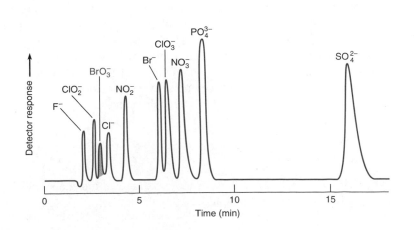

FIGURE 22-4 Converting one human hazard into another. Chlorination of drinking water converts some organic compounds into potential carcinogens, such as $CHCl_3$. To reduce this risk, ozone (O_3) has replaced Cl_2 in some municipal purification systems. Unfortunately, O_3 converts bromide (Br^-) into bromate (BrO_3^-), another carcinogen that must be monitored. The figure shows an anion chromatographic separation of ions found in drinking water. With preconcentration of the water, the detection limit for bromate is ~2 ppb. [From R. J. Joyce, *Am. Environ. Lab.* May 1994, 1.]

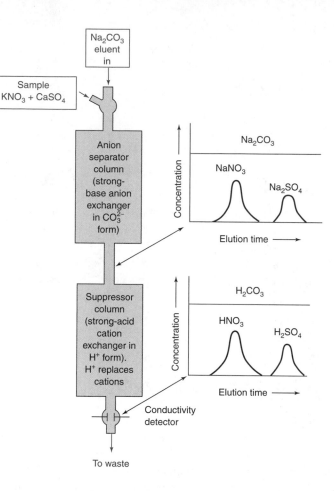

FIGURE 22-5 Suppressed-ion anion chromatography. [Adapted from H. Small, *Anal. Chem.* **1983,** *55,* 235A.]

The separator column separates the analytes, and the suppressor column replaces the ionic eluent with a non-ionic species.

For the sake of illustration, consider Figure 22-5 in which KNO_3 and $CaSO_4$ are injected into the *separator column*—an anion-exchange column in the carbonate form—followed by elution with Na_2CO_3. NO_3^- and SO_4^{2-} equilibrate with the resin and are slowly displaced by the CO_3^{2-} eluent. K^+ and Ca^{2+} cations are not retained and simply wash through. After a period of time, $NaNO_3$ and Na_2SO_4 are eluted from the separator column, as shown in the upper graph of Figure 22-5. These species cannot be easily detected, however, because the solvent contains a high concentration of Na_2CO_3, whose high conductivity obscures that of the analyte species.

To remedy this problem, the solution is next passed through a *suppressor column,* in which cations are replaced by H^+ from a cation-exchange resin loaded with H^+. When $NaNO_3$ and Na_2CO_3 from the separator column pass through the suppressor, Na^+ is replaced by H^+, making a solution of HNO_3 and H_2CO_3 ($\rightleftharpoons CO_2 + H_2O$). In the absence of analyte, only H_2CO_3, which has very low conductivity, emerges from the suppressor. When analyte is present, either HNO_3 or H_2SO_4, each with high conductivity, is produced and detected. Between runs, the suppressor column is regenerated by passing electrolytically generated H^+ through the resin.

Ask Yourself

22-B. What are the purposes of the separator column and the suppressor column in suppressed-ion chromatography?

In **molecular exclusion chromatography** (also called *gel filtration* or *gel permeation chromatography*), molecules are separated according to their size. Small molecules enter the small pores in the stationary phase, but large molecules do not (Figure 20-2). Because small molecules must pass through an effectively larger volume in the column, large molecules are eluted first (Figure 22-6). This technique is widely used in biochemistry and polymer chemistry to purify macromolecules and to measure molecular mass (Figure 22-7).

In molecular exclusion chromatography, the volume of mobile phase (the solvent) in the column outside the stationary phase is called the *void volume, V_0*. Large molecules that are excluded from the stationary phase are eluted in the void volume. Void volume is measured by passing through the column a molecule that is too large to enter the pores. The dye Blue Dextran (molecular mass $= 2 \times 10^6$) is commonly used.

Large molecules pass through the column *faster* than small molecules do.

Molecular Mass Determination

Retention volume is the volume of mobile phase required to elute a particular solute from the column. Each stationary phase has a range over which there is a logarithmic relation between molecular mass and retention volume. We can estimate the molecular mass of an unknown by comparing its retention volume with those of standards. For proteins, it is important to use eluent with an ionic strength high enough (such as 0.05 M NaCl) to eliminate electrostatic adsorption of solute by occasional charged sites on the gel.

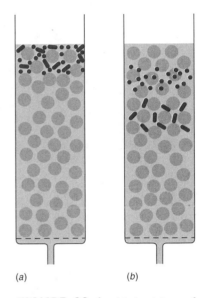

(a)   (b)

FIGURE 22-6 (a) A mixture of large and small molecules is applied to the top of a molecular exclusion chromatography column. (b) Large molecules cannot penetrate the pores of the stationary phase, but small molecules can. Therefore less of the volume is available to large molecules and they move down the column faster.

EXAMPLE Molecular Mass Determination by Gel Filtration

The proteins below were chromatographed on a gel filtration column and retention volumes (V_r) were measured. Estimate the molecular mass (MM) of the unknown.

Compound	V_r (mL)	Molecular mass	log(molecular mass)
Blue Dextran 2000	17.7	2×10^6	6.301
Aldolase	35.6	158 000	5.199
Catalase	32.3	210 000	5.322
Ferritin	28.6	440 000	5.643
Thyroglobulin	25.1	669 000	5.825
Unknown	30.3	?	

SOLUTION Figure 22-8 plots V_r versus log(MM). The least-squares fit to the five calibration standards is shown. Putting the retention volume of unknown protein into the least-squares equation allows us to solve for the molecular mass of the unknown:

$$V_r \text{ (mL)} = -15.75[\log(\text{MM})] + 117.0$$
$$30.3 = -15.75[\log(\text{MM})] + 117.0$$
$$\Rightarrow \quad \log(\text{MM}) = 5.505 \quad \Rightarrow \quad \text{MM} = 10^{5.505} = 320\ 000$$

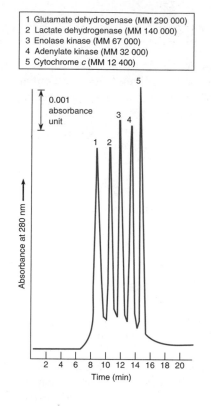

FIGURE 22-7 Separation of proteins by molecular exclusion chromatography, using a TSK 3000SW HPLC column. The highest molecular masses are eluted first.

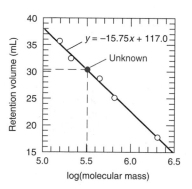

FIGURE 22-8 Calibration curve used to estimate the molecular mass of an unknown protein by molecular exclusion chromatography.

Ask Yourself

22-C. A gel filtration column has a radius (r) of 0.80 cm and a length (l) of 20.0 cm.

(a) Calculate the total volume of the column, which is equal to $\pi r^2 l$.

(b) Blue Dextran was eluted in a volume of 18.2 mL. What volume is occupied by the stationary phase plus the solvent inside the pores of the stationary phase?

(c) Suppose that the pores occupy 60.0% of the stationary phase volume. Over what volume range (x mL for the largest molecules to y mL for the smallest molecules) are all solutes expected to be eluted?

22-4 Affinity Chromatography

Affinity chromatography is used to isolate a single compound from a complex mixture. The technique is based on specific binding of that one compound to the stationary phase (Figure 20-2). When sample is passed through the column, only one solute is bound. After everything else has washed through, the one adhering solute is eluted by changing conditions such as pH or ionic strength to weaken its binding. Affinity chromatography is especially applicable in biochemistry and is based on specific interactions between enzymes and substrates, antibodies and antigens, or receptors and hormones.

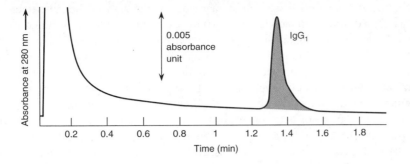

Figure 22-9 shows the isolation of the protein immunoglobulin G (IgG) by affinity chromatography on a column containing covalently bound *protein A*. Protein A binds to one specific region of IgG at pH $\gtrsim$ 7.2. When a crude mixture containing IgG and other proteins was passed through the column at pH 7.6, everything except IgG was eluted within 0.3 min. At 1 min, the eluent pH was lowered to 2.6 and IgG was cleanly eluted at 1.3 min.

FIGURE 22-9 Purification of monoclonal antibody IgG by affinity chromatography on a column (4.6 mm diameter $\times$ 5 cm long) containing protein A covalently attached to a polymer support. Other proteins in the sample are eluted from 0 min to 0.3 min at pH 7.6. When the eluent pH is lowered to 2.6, IgG is freed from protein A and emerges from the column. [From B. J. Compton and L. Kreilgaard, *Anal. Chem.* **1994**, *66*, 1175A.]

22-5 What Is Capillary Electrophoresis?

Electrophoresis is the migration of ions in an electric field. Anions are attracted to the anode and cations are attracted to the cathode. Different ions migrate at different speeds. **Capillary electrophoresis** is an extremely high resolution separation technique conducted with solutions of ions in a narrow capillary tube. As we shall see shortly, a clever modification of the technique allows us to separate neutral analytes as well as ions. Capillary electrophoresis applies with equal ease to the separation of both macromolecules, such as proteins and DNA, and small species, such as Na^+ and benzene. The opening of this chapter showed how capillary electrophoresis could analyze the contents of a single cell.

The typical experiment in Figure 22-10 features a fused-silica (SiO_2) capillary that is 50 cm long and has an inner diameter of 25–75 μm. The capillary is immersed

Cations are attracted to the negative terminal (the cathode).

Anions are attracted to the positive terminal (the anode).

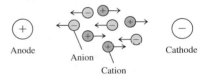

Background electrolyte, also called *run buffer,* is the solution in the electrode reservoirs. It controls pH and ionic composition in the capillary.

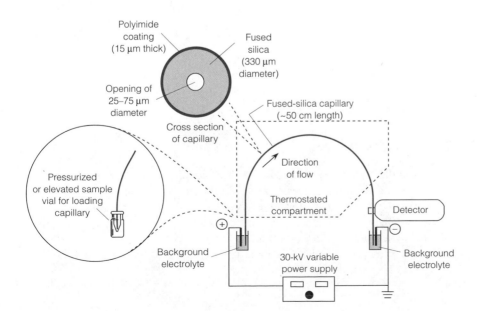

FIGURE 22-10 Capillary electrophoresis. Sample is injected by elevating or applying pressure to the sample vial or applying suction at the outlet of the capillary.

The greater the charge on the ion, the faster it migrates in the electric field. The greater the size of the molecule, the slower it migrates.

Working ranges for chloride analysis:

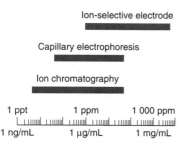

in *background electrolyte* solution at each end. At the start of the experiment, one end is dipped into a sample vial and ~10 nL (nanoliters, 10^{-9} L) of liquid is introduced by applying pressure to the sample vial. After the capillary is placed back into the electrolyte solution, 20–30 kV is applied to the electrodes to cause ions in the capillary to migrate. Different ions migrate at different speeds, so they separate from one another as they travel through the capillary. Ions are detected inside the capillary near the far end with an ultraviolet absorbance monitor (or some other detector). The graph of detector response versus time in Figure 22-11 is called an *electropherogram*. (In chromatography, we call the same graph a *chromatogram*.) Capillary electrophoresis is not as sensitive as ion chromatography, but it is more sensitive than ion-selective electrodes.

Capillary electrophoresis can provide extremely narrow bands. The three mechanisms of band broadening in chromatography are longitudinal diffusion (*B* in the van Deemter equation 20-7), the finite rate of mass transfer between the stationary and mobile phases (*C* in the van Deemter equation), and multiple flow paths around particles (*A* in the van Deemter equation). An open tubular column in chromatography or electrophoresis reduces band broadening (relative to that of a packed column) by eliminating multiple flow paths (the *A* term). Capillary electrophoresis further reduces broadening by eliminating the mass transfer problem (the *C* term) because *there is no stationary phase*. The only source of broadening under ideal conditions is longitudinal diffusion of solute as it migrates through the capillary. Capillary electrophoresis routinely achieves 50 000 to 500 000 theoretical plates, which is an order of magnitude higher than chromatography.

FIGURE 22-11 Measurement of nitrate in aquarium water (Box 5-1) by capillary electrophoresis. At the detection wavelength of 222 nm, many species in the water have too little absorbance to be observed. (*a*) Standard mixture containing 15 μg/mL nitrate (NO_3^-), 5 μg/mL nitrite (NO_2^-), and 10 μg/mL of the internal standard, periodate (IO_4^-). (*b*) 1:100 dilution of aquarium water with distilled water to which internal standard was added. The aquarium has a high concentration of nitrate, but no detectable nitrite in this analysis. [From D. S. Hage, A. Chattopadhyay, C. A. C. Wolfe, J. Grundman, and P. B. Kelter, *J. Chem. Ed.* **1998**, *75*, 1588.] Periodate is a questionable standard because it is a strong oxidizing agent that might react with organic matter in the aquarium water.

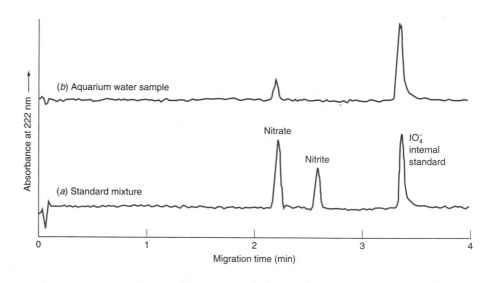

Ask Yourself

22-D. Capillary electrophoresis is noteworthy for analyzing very small volumes of sample and for producing very high resolution separations.

(a) A typical injected sample occupies a 5-mm length of the capillary. What volume of sample is this if the inside diameter of the capillary is 25 μm? 50 μm? (The volume of a cylinder of radius *r* is $\pi r^2 \times$ length.)

(b) Which mechanisms of band broadening that operate in chromatography are absent in capillary electrophoresis?

Capillary electrophoresis involves two simultaneous processes called *electrophoresis* and *electroosmosis*. Electrophoresis is the migration of ions in an electric field. Electroosmosis pumps the entire solution through the capillary from the anode toward the cathode. Superimposed on this one-way flow are the flow of cations, which are attracted to the cathode, and the flow of anions, which are attracted to the anode. In Figure 22-10, cations swim from the injection end at the left toward the detection end at the right. Anions swim toward the left in Figure 22-10. Both cations and anions are swept from left to right by electroosmosis. Cations arrive at the detector before anions. Neutral molecules swept along by electroosmosis arrive at the detector after the cations and before the anions.

Two processes operate in capillary electrophoresis:

- *Electrophoresis:* migration of cations to the cathode and anions to the anode
- *Electroosmosis:* migration of the entire bulk fluid toward the cathode

Electroosmosis

Electroosmosis is a pumping action caused by the applied electric field that propels the fluid inside a fused-silica capillary from the anode toward the cathode. To understand electroosmosis, consider what happens at the inside wall of the capillary. The wall is covered with silanol ($Si-OH$) groups that are negatively charged ($Si-O^-$) above pH 2. Figure 22-12a shows that the capillary wall and the solution immediately adjacent to the wall form an *electric double layer.* The double layer

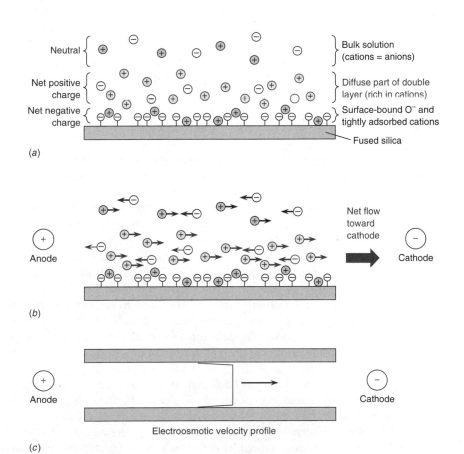

(a)

(b)

(c)

FIGURE 22-12 (a) Electric double layer is created by negatively charged silica surface and excess cations in the diffuse part of the double layer in the solution near the wall. The wall is negative and the diffuse part of the double layer is positive. (b) Predominance of cations in diffuse part of the double layer produces net electroosmotic flow toward the cathode when an external field is applied. (c) Electroosmotic velocity profile is uniform over more than 99.9% of the cross section of the capillary. Capillary tubing is required to maintain constant temperature in the liquid. Temperature variation in larger diameter tubing causes bands to broaden.

Ions in the diffuse part of the double layer adjacent to the capillary wall are the "pump" that drives electro-osmotic flow.

Migration time in electrophoresis is analogous to retention time in chromatography.

is composed of (1) a negative charge fixed to the wall and (2) an equal positive charge in solution adjacent to the wall. The thickness of the positive layer, called the *diffuse part of the double layer,* is approximately 1 nm. When an electric field is applied, cations are attracted to the cathode and anions are attracted to the anode. Excess cations in the diffuse part of the double layer drive the entire solution in the capillary toward the cathode (Figure 22-12b). The greater the applied electric field, the faster the flow.

Color Plate 19 shows a critical distinction between electroosmotic flow induced by an electric field and ordinary hydrodynamic flow induced by a pressure difference. Because it is driven by ions on the walls of the capillary, electroosmotic flow is uniform across the diameter of the liquid, as shown schematically in Figure 22-12c. The only mechanism that broadens the moving band is diffusion. By contrast, hydrodynamic flow has a parabolic velocity profile, with fastest motion at the center of the capillary and little velocity at the wall. The parabolic profile would create broad bands.

Electroosmosis decreases at low pH because the wall loses its negative charge when $Si-O^-$ is converted to $Si-OH$ and the number of cations in the double layer diminishes. Electroosmotic velocity is measured by adding an ultraviolet-absorbing neutral solute, such as methanol, to the sample and measuring the time it takes (called the *migration time*) to reach the detector. In one experiment with 30 kV across a 50-cm capillary, the electroosmotic velocity was 4.8 mm/s at pH 9 and 0.8 mm/s at pH 3.

In Figures 22-10 and 22-12, electroosmosis is from left to right because cations in the double layer are attracted to the cathode. Superimposed on electroosmosis of the bulk fluid, electrophoresis transports cations to the right and anions to the left. At neutral or high pH, electroosmosis is faster than electrophoresis and the net flow of anions is to the right. At low pH, electroosmosis is weak and anions may flow to the left and never reach the detector. To separate anions at low pH, you can reverse the polarity to make the sample side negative and the detector side positive.

Detectors

The most common detector is an *ultraviolet absorbance monitor* set to a wavelength near 200 nm, where many solutes absorb. It is not possible to use such short wavelengths with larger diameter columns, because the solvent absorbs too much of the radiation. A *fluorescence detector* works for fluorescent analytes or fluorescent derivatives. *Electrochemical detection* is sensitive to analytes that can gain or lose electrons at an electrode. Eluate can also be directed into a *mass spectrometer* (Box 20-2), which provides information on the quantity and molecular structure of analyte. *Conductivity detection* with ion-exchange suppression of the background electrolyte (as in ion chromatography, Figure 22-5) gives 1–10 ppb sensitivity for small ions.

In contrast to the direct detection of analyte discussed so far, **indirect detection** relies on measuring a strong signal from the background electrolyte and a weak signal from the analyte as it passes the detector. Figure 22-13 illustrates indirect fluorescence detection, but the same principle applies to any type of detection. A fluorescent ion with the same sign of charge as the analyte is added to background electrolyte to provide a steady background signal. When the analyte ion emerges, the concentration of background ion necessarily decreases, because electroneutrality must be preserved. If the analyte ion is not fluorescent, the fluorescence level decreases when analyte emerges. What we observe is a *negative* signal.

Figure 22-14 shows indirect ultraviolet detection of Cl^- anion in the presence of the ultraviolet-absorbing chromate anion, CrO_4^{2-}. In the absence of analyte, CrO_4^{2-}

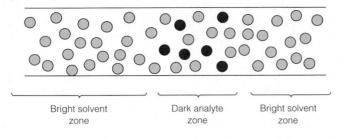

Bright solvent
zone

Dark analyte
zone

Bright solvent
zone

FIGURE 22-13   *Principle of indirect detection:* When analyte emerges from the capillary, the strong background signal decreases.

gives a steady absorbance at 254 nm. When Cl^- reaches the detector, there is less CrO_4^{2-} present and Cl^- does not absorb; therefore the detector signal *decreases*.

Wall Effects in the Separation of Proteins

Capillary electrophoresis is well suited for the analytical separation of proteins in biochemistry. However, some proteins are tightly adsorbed by the negatively charged capillary wall. Figure 22-15 shows that this problem can be largely overcome by attaching hydrophilic ("water-loving") polymers to the silanol groups on the wall.

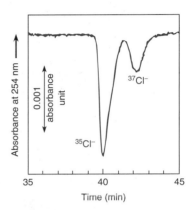

FIGURE 22-14   Separation of natural isotopes of 0.56 mM Cl^- by capillary electrophoresis with indirect spectrophotometric detection at 254 nm. The background electrolyte contains 5 mM CrO_4^{2-} to provide absorbance at 254 nm. There are few ways to separate isotopes so cleanly. This impressive separation was done by an undergraduate student at the University of Calgary. [From C. A. Lucy and T. L. McDonald, *Anal. Chem.* **1995**, *67*, 1074.]

Ask Yourself

22-E. (a) Capillary electrophoresis was conducted with a solution whose pH was 9, at which the electroosmotic velocity is greater than the electrophoretic velocity. Draw a picture of the capillary, showing the placement of the anode, cathode, injector, and detector. Show the direction of electroosmotic flow and the direction of electrophoretic flow of a cation and an anion. Show the direction of net flow for each ion.

(b) If the pH is reduced to 3, the electroosmotic velocity is less than the electrophoretic velocity. In what directions will cations and anions migrate?

(c) Explain why the detector signal is negative in Figure 22-14.

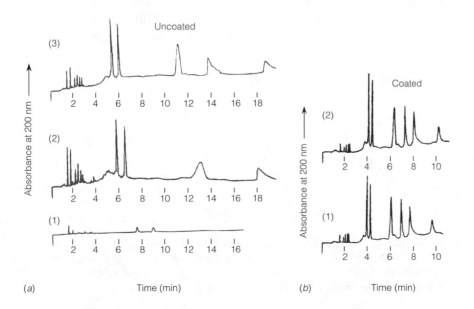

FIGURE 22-15 *Wall effects.* Consecutive injections of proteins in (*a*) uncoated and (*b*) covalently coated capillaries at pH 8.5. The coating is a hydrophilic polymer. Most of the protein from the first injection (1) into the uncoated column stuck to the walls and never reached the detector. Adsorption gives irreproducible migration times and peak areas, as well as asymmetric peak shapes. The coated column gives reproducible results with normal peak shapes. [Courtesy of Bio-Rad Laboratories, Richmond, CA.]

Order of elution in capillary zone electrophoresis:

1. Cations (highest mobility first)
2. All neutrals (unseparated)
3. Anions (highest mobility last)

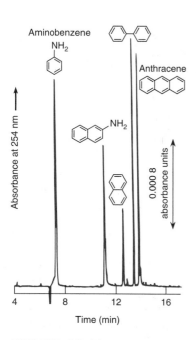

FIGURE 22-16 Separation of neutral molecules by micellar electrokinetic capillary electrophoresis. The average plate count in this experiment is 250 000 plates in 50 cm of capillary length. [From J. T. Smith, W. Nashabeh, and Z. E. Rassi, *Anal. Chem.* **1994,** *66,* 1119.]

Micellar electrokinetic capillary electrophoresis: The more time a solute spends inside the micelle, the longer its migration time.

The type of electrophoresis we have been discussing so far is called **capillary zone electrophoresis,** in which separation is based on different electrophoretic velocities of different ions. Electroosmotic flow of the bulk fluid is toward the cathode (Figure 22-12b). Cations migrate faster than the bulk fluid and anions migrate slower than the bulk fluid. Therefore the order of elution is cations before neutrals before anions. If the electrode polarity is reversed, the order of elution is anions before neutrals before cations. Neither scheme separates neutral molecules from one another.

Micellar Electrokinetic Capillary Chromatography

This mouthful of words describes a form of capillary electrophoresis that separates neutral molecules as well as ions (Figure 22-16). The key modification in **micellar electrokinetic capillary chromatography** is that the capillary solution contains *micelles,* described in Box 22-2 (which you should read now).

To understand how neutral molecules are separated, suppose that the background electrolyte contains negatively charged micelles. In Figure 22-17, electroosmotic flow is to the right. Electrophoretic migration of the negatively charged micelles is to the left, but net motion is to the right because the electroosmotic flow is faster than the electrophoretic flow.

In the absence of micelles, all neutral molecules would reach the detector together at a time we designate t_0. Micelles injected with the sample reach the detector at time t_{mc}, which is longer than t_0 because the micelles are anions that migrate upstream. If a neutral molecule equilibrates between free solution and the inside of the micelles, its migration time is increased, because it migrates at the slower rate of the micelle part of the time. In this case, the neutral molecule reaches the detector at a time between t_0 and t_{mc}.

The more soluble the neutral molecule is in the micelle, the more time it spends inside the micelle and the longer its migration time. The nonpolar interior of a sodium dodecyl sulfate micelle dissolves nonpolar solutes best. Polar solutes are not as soluble in the micelles and have a shorter retention time than nonpolar solutes do. Migration times of cations and anions may also be affected by micelles because ions can dissolve in some micelles. Micellar electrokinetic capillary chromatography is truly a form of chromatography because micelles behave like a pseudostationary

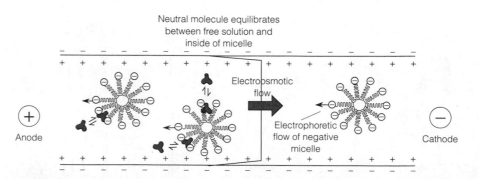

FIGURE 22-17 Negatively charged sodium dodecyl sulfate micelles migrate upstream against the electroosmotic flow. Neutral molecules are in dynamic equilibrium between free solution and the inside of the micelle. The more time spent in the micelle, the more the neutral molecule lags behind the electroosmotic flow.

Box 22-2 *Explanation*

What Is a Micelle?

A **micelle** is an aggregate of molecules with ionic head-groups and long, nonpolar tails. Such molecules are called *surfactants,* an example of which is sodium dodecyl sulfate:

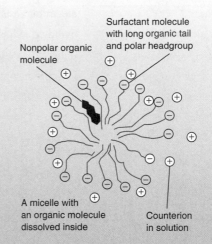

Surfactant molecule with long organic tail and polar headgroup

Nonpolar organic molecule

A micelle with an organic molecule dissolved inside

Counterion in solution

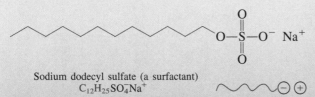

Sodium dodecyl sulfate (a surfactant)
$C_{12}H_{25}SO_4^-Na^+$

The polar headgroups of a micelle face outward, where they are surrounded by polar water molecules. The non-polar tails face inward, where they form a little pocket resembling a nonpolar hydrocarbon solution. *Nonpolar solutes are soluble inside the micelle.*

At low concentrations, surfactant molecules are not associated with one another. When the concentration ex-ceeds the *critical micelle concentration,* spontaneous aggregation into micelles occurs. Isolated surfactant molecules exist in equilibrium with micelles.

phase. Solutes partition between the mobile phase (the aqueous solution) and the pseudostationary micelles.

Capillary Gel Electrophoresis

In **capillary gel electrophoresis,** macromolecules are separated by *sieving* as they migrate through a gel inside a capillary tube. Small molecules travel faster than large molecules through the gel. Large molecules become entangled in the gel and their motion is slowed. (This behavior is the opposite of that in molecular exclusion chromatography, in which large molecules are excluded from the gel and move faster than small ones.) Capillary gel electrophoresis is used to sequence DNA, which was shown at the opening of this chapter. More than 500 different lengths of DNA are separated in less than 20 min by this technique.

Ask Yourself

22-F. **(a)** Explain why neutral solutes are eluted between times t_0 and t_{mc} in micellar electrokinetic capillary chromatography, where t_0 is the elution time of neutral molecules in the absence of micelles and t_{mc} is the elution time of the micelles.

(b) Micellar electrokinetic capillary chromatography in Figure 22-16 was conducted at pH 10 with anionic micelles and the anode on the sample side, as in Figure 22-10.

(i) Is aminobenzene ($C_6H_5NH_2$) a cation, an anion, or a neutral molecule in this experiment?

(ii) Explain how Figure 22-16 allows you to decide which compound, aminobenzene or anthracene, is more soluble in the micelles?

Important Terms

affinity chromatography

anion exchanger

capillary electrophoresis

capillary gel electrophoresis

capillary zone electrophoresis

cation exchanger

deionized water

electroosmosis

electrophoresis

hydrated radius

indirect detection

ion chromatography

ion-exchange chromatography

micellar electrokinetic capillary chromatography

micelle

molecular exclusion chromatography

preconcentration

retention volume

trace analysis

Problems

22-1. **(a)** Hexanoic acid and 1-aminohexane, adjusted to pH 12 with NaOH, were passed through a cation-exchange column loaded with NaOH at pH 12. State the principal species that will be eluted and the order in which they are expected.

$$CH_3CH_2CH_2CH_2CH_2CO_2H$$
Hexanoic acid

$$CH_3CH_2CH_2CH_2CH_2CH_2NH_2$$
1-Aminohexane

(b) Hexanoic acid and 1-aminohexane, adjusted to pH 3 with HCl, were passed through a cation-exchange column loaded with HCl at pH 3. State the principal species that will be eluted and the order in which they are expected.

22-2. The exchange capacity of an ion-exchange resin is defined as the number of moles of charged sites per gram of dry resin. Describe how you would measure the exchange capacity of an anion-exchange resin by using standard NaOH, standard HCl, or any other reagent you wish.

22-3. Commercial vanadyl sulfate ($VOSO_4$, FM 163.00) is contaminated with H_2SO_4 and H_2O. A solution was prepared by dissolving 0.244 7 g of impure $VOSO_4$ in 50.0 mL of water. Spectrophotometric analysis indicated that the concentration of the blue VO^{2+} ion was 0.024 3 M. A 5.00-mL sample was passed through a cation-exchange column loaded with H^+. VO^{2+} is exchanged for $2H^+$ by this process. H_2SO_4 is unchanged by the cation-exchange column.

$$VOSO_4 \xrightarrow[\text{loaded with } H^+]{\text{Cation-exchange column}} H_2SO_4$$

$$H_2SO_4 \xrightarrow[\text{loaded with } H^+]{\text{Cation-exchange column}} H_2SO_4$$

H^+ eluted from the column required 13.03 mL of 0.022 74 M NaOH for titration. Find the weight percents of $VOSO_4$, H_2SO_4, and H_2O in the vanadyl sulfate.

22-4. Consider a protein with a net negative charge tightly adsorbed on an anion-exchange gel at pH 8.

(a) How will a gradient from pH 8 to some lower pH be useful for eluting the protein?

(b) How would a gradient of increasing ion concentration (at constant pH) be useful for eluting the protein?

22-5. Look up the pK_a values for trimethylamine, dimethylamine, methylamine, and ammonia. Predict the order of elution of these compounds from a cation-exchange column eluted with a gradient of increasing pH, beginning at pH 7.

22-6. One mole of 1,2-ethanediol consumes one mole of periodate in the reaction

$$\begin{array}{c} CH_2OH \\ | \\ CH_2OH \end{array} + IO_4^- \longrightarrow 2CH_2O + H_2O + IO_3^-$$

1,2-Ethanediol Periodate Formaldehyde Iodate
(FW 62.068)

To analyze 1,2-ethanediol, oxidation with excess IO_4^- is followed by passage of the reaction solution through an anion-exchange resin that binds both IO_4^- and IO_3^-. IO_3^- is then quantitatively removed from the resin by elution with NH_4Cl. The absorbance of eluate is measured at 232 nm to find the quantity of IO_3^- (molar absorptivity (ε) = 900 M^{-1} cm^{-1}) produced by the reaction. In one experiment, 0.213 9 g of aqueous 1,2-ethanediol was dissolved in 10.00 mL. Then 1.000 mL of the solution was treated with 3 mL of 0.15 M KIO_4 and subjected to ion-exchange separation of IO_3^- from unreacted IO_4^-. The eluate (diluted to 250.0 mL) had an absorbance of $A_{232} = 0.521$ in a 1.000-cm cell, and a blank had

488

$A_{232} = 0.049$. Find the weight percent of 1,2-ethanediol in the original sample.

22-7. Box 22-1 gives the mean composition of clouds at a mountain top in Germany.

(a) What is the pH of the cloud water?

(b) Do the anion charges equal the magnitude of the cation charges? What does your answer suggest about the quality of the analysis?

(c) What is the total mass of dissolved ions in each milliliter of water?

22-8. In *ion-exclusion chromatography,* ions are separated from nonelectrolytes (uncharged molecules) by an ion-exchange column. Nonelectrolytes penetrate the stationary phase, whereas ions with the same sign charge as the stationary phase are repelled by the stationary phase. Because electrolytes have access to less of the column volume, they are eluted before nonelectrolytes. A mixture of trichloroacetic acid (TCA, $pK_a = 0.66$), dichloroacetic acid (DCA, 1.30), and monochloroacetic acid (MCA, 2.86) was separated by passage through a cation-exchange resin eluted with 0.01 M HCl. The order of elution was TCA < DCA < MCA. Explain why the three acids are separated and the order of elution.

22-9. Polystyrene standards of known molecular mass gave the following calibration data in a molecular exclusion column. Prepare a plot of log (molecular mass) versus retention time (t_r) and find the equation of the line by the method of least squares. Find the molecular mass of an unknown with a retention time of 13.00 min.

Molecular mass	Retention time, t_r (min)
8.50×10^6	9.28
3.04×10^6	10.07
1.03×10^6	10.88
3.30×10^5	11.67
1.56×10^5	12.14
6.60×10^4	12.74
2.85×10^4	13.38
9.20×10^3	14.20
3.25×10^3	14.96
5.80×10^2	16.04

22-10. A molecular exclusion column has a diameter of 7.8 mm and a length of 30 cm. The solid portion of the particles occupies 20% of the volume, the pores occupy 40%, and the volume between particles occupies 40%.

(a) At what volume would totally excluded molecules be expected to emerge?

(b) At what volume would the smallest molecules be expected?

(c) A mixture of polyethylene glycols of various molecular masses is eluted between 23 and 27 mL. What does this imply about the retention mechanism for these solutes on the column?

22-11. (a) If a capillary is set up as in Figure 22-10 with the injector end positive and the detector end negative, in what order will cations, anions, and neutral molecules be eluted?

(b) The electropherograms in Figure 22-11 were obtained with the injector end *negative* and a run buffer of pH 4.0. Will there be high or low electroosmotic flow? Are the anions moving with or against the electroosmotic flow? Will cations injected with the sample reach the detector before or after the anions?

22-12. (a) What is electroosmosis?

(b) Why is the electroosmotic flow in a silica capillary five times faster at pH 9 than at pH 3?

(c) When the Si—OH groups on a silica capillary wall are converted to Si—$O(CH_2)_{17}CH_3$ groups, the electroosmotic flow is small and nearly independent of pH. Explain why.

22-13. Why is the detector response *negative* in indirect spectrophotometric detection?

22-14. Explain how neutral molecules can be separated by micellar electrokinetic capillary chromatography. Why is this a form of chromatography?

22-15. A van Deemter plot for capillary electrophoresis is a graph of plate height versus migration velocity, where migration velocity is governed by the net sum of electroosmotic flow and electrophoretic flow.

(a) What is the principal source of band broadening in ideal capillary zone electrophoresis? Sketch what you expect the van Deemter curve to look like.

(b) What are the sources of band broadening in micellar electrokinetic capillary chromatography? Sketch what you expect the van Deemter curve to look like.

22-16. (a) Measure the peak width of $^{35}Cl^-$ in Figure 22-14 and calculate the number of theoretical plates.

(b) The distance from injection to the detector is 40 cm in Figure 22-14. From your answer to **(a),** find the plate height.

(c) Why are the peaks negative in Figure 22-14?

22-17. The water-soluble vitamins niacinamide (a neutral compound), riboflavin (a neutral compound), niacin (an anion), and thiamine (a cation) were separated by micellar electrokinetic capillary chromatography in 15 mM borate buffer (pH 8.0) with 50 mM sodium dodecyl sulfate. The migration times were niacinamide, 8.1 min; riboflavin, 13.0 min; niacin, 14.3 min; and thiamine, 21.9 min. What would the order have been in the absence of sodium dodecyl sulfate? Which compound is most soluble in the micelles?

How Would You Do It?

22-18. 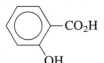   **(a)** To obtain the best separation of two weak acids by electrophoresis, it makes sense to use the pH at which their charge difference is greatest. Explain why.

(b) Prepare a spreadsheet to examine the charges of *ortho*- and *para*-hydroxybenzoic acids as a function of pH. At what pH is the difference greatest?

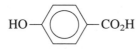

o-Hydroxybenzoic acid
$pK_a = 2.97$

p-Hydroxybenzoic acid
$pK_a = 4.58$

22-19. An aqueous solution of NaCl, $NaNO_3$, and Na_2SO_4 was passed through a C_{18}-silica reversed-phase liquid chromatography column eluted with water.[2] None of the cations or anions is retained by the C_{18} stationary phase, so all three salts were eluted in a single, sharp band with a retention time of 0.9 min. In a second experiment, the column was equilibrated with aqueous 10 mM pentylammonium formate, whose hydrophobic tail is soluble in the C_{18} stationary phase.

Hydrocarbon tail soluble
in C_{18} stationary phase $\overset{+}{N}H_3$ HCO_2^-

When the mixture of NaCl, $NaNO_3$, and Na_2SO_4 was passed through the column and eluted with 10 mM pentylammonium formate, the cations all came out in a single peak with a retention time of 0.9 min. However, the anions were separated, with retention times of 1.9 min (Cl^-), 2.1 min (NO_3^-), and 4.1 min (SO_4^{2-}). Explain why this column behaves as an anion exchanger. Why is sulfate eluted last?

Notes and References

1. L. S. Ettre, *LC/GC* **1999,** *17,* 1104; F. H. Spedding, *Disc. Faraday Soc.* **1949,** *7,* 214.

2. J.-P. Mercier, P. Chaimbault, C. Elfakir, and M. Dreux, *Am. Lab. News Ed.* October 1999, p. 22.

Further Reading

C. L. Cooper, *J. Chem. Ed.* **1998,** *75,* 343, 347 (basic discussion of capillary electrophoresis).

R. Weinberger, *Practical Capillary Electrophoresis,* 2nd ed. (San Diego: Academic Press, 2000).

M. G. Khaledi, Ed., *High Performance Capillary Electrophoresis: Theory, Techniques, and Applications* (New York: Wiley, 1998).

P. Camilleri, Ed., *Capillary Electrophoresis: Theory and Practice* (Boca Raton, FL: CRC Press, 1997).

J. P. Landers, Ed., *Handbook of Capillary Electrophoresis* (Boca Raton, FL: CRC Press, 1997).

Chapter 23

EXPERIMENTS

Experiments in this chapter illustrate major analytical techniques described in the text. Although the procedures are safe when carried out with reasonable care, *all chemical experiments are potentially hazardous.* Any solution that fumes (such as concentrated HCl) and all nonaqueous solvents should be handled in a fume hood. Pipetting should never be done by mouth. Spills on your body should immediately be flooded with water and your instructor should be notified for possible further action. Spills on the benchtop should be cleaned immediately. Many chemicals cannot be flushed down the drain. Your instructor should establish a safe procedure for disposing of each chemical that you use (see Box 2-1).

When equipment permits, it is environmentally friendly and usually economical to reduce the scale of an experiment. For example, 50-mL burets can be replaced by 10-mL burets at some loss in analytical precision, but little educational loss. An even smaller scale is possible with microburets made from 2-mL pipets.[1] Micropipets and small volumetric flasks can replace large glass pipets and large volumetric flasks (again at a sacrifice of precision).

With properly chosen electrodes, potentiometric titrations can be performed on a scale of a few milliliters, instead of tens of milliliters, by using a syringe or a micropipet to deliver titrant. Instructions in this chapter were written for the commonly available, large-size lab equipment, but your modification to a smaller scale is encouraged.

The San Jose State University Laboratory Approach

What is the purpose of the introductory analytical chemistry laboratory? It should help students develop the sense and skill required to obtain precise and accurate results in quantitative measurements (whether or not they are in chemistry). It should help students develop a sense of the reliability and limitations of quantitative measurements. The laboratory should illustrate concepts taught in the classroom and expose students to modern methods used by practicing analytical chemists. The vast majority of analytical measurements made by professional analysts are done with sophisticated instruments. However, the most reliable standards used to calibrate these instruments are usually analyzed by classical "wet" chemical methods.[2]

At San Jose State University, an effort was made to provide beginning students with "a more comprehensive and realistic exposure to modern analytical science" (including instrumental methods and quality assurance practices) while still fostering the development of quantitative laboratory skills.[3] The 15-week laboratory was divided into two segments. The first five weeks were devoted to "analyst certification." Students had to demonstrate proficiency in the basic techniques of gravimetric and titrimetric analysis and in photometric procedures before they were allowed to move on to more advanced experiments. As part of the analyst certification process, students prepared and standardized materials and reagents that would be used as standards in the next 10 weeks of "Studies in Contemporary Analytical Science."

Clearly, the exposure to instrumentation in the 10-week segment was based on what was available at the university and on the instructors' interests. At San Jose, the instrumental experiments included acid-base potentiometric measurements, ion-selective electrodes, atomic absorption, voltammetry, and high-performance liquid chromatography. Samples for analysis were chosen from commercial, environmental, and clinical materials of significance to students. Exposure to instruments was at the "functional" level, rather than at the more detailed level expected for chemistry majors in an instrumental analysis course.

"Good Laboratory Practice" was emphasized throughout the course. This practice requires documentation of procedures and results to meet legal standards. Students were responsible for calibration and certification of instruments to ensure that results were reliable. Control charts for instruments were kept and updated by students as part of quality assurance. Statistical analysis of results included the calculation of confidence intervals, the t test to compare results from different analyses, the F test to compare variance from different analyses, and the Q test for rejection of discrepant data. Detection limits for instrumental methods were estimated, when appropriate. Least-squares regression, spreadsheets, and computer graphing were used freely.

In our experience, virtually every university and almost every instructor teaches a unique set of laboratory experiments based on instructor interest, available equipment, prescribed curricula, and inertia. Experiments in this chapter are mainly intended to contribute to the menu from which instructors choose. (And, of course, many instructors develop new experiments.) Many of the experiments were drawn from the excellent selection published in the *Journal of Chemical Education.*[4]

A sequence of three experiments (23-21 through 23-23) is adapted from the San Jose State University laboratory manual to illustrate the San Jose approach. In the first experiment, students prepare and standardize a Mn^{2+} solution by an EDTA titration as part of the "analyst certification" process. The second experiment, which is also part of analyst certification, is the spectrophotometric analysis of Mn in steel. The procedure involves sample preparation and a standard addition, using the standard Mn^{2+} from the previous experiment. The third experiment measures Mn from the same steel sample by atomic absorption. Confidence intervals are used to compare results from the two methods, and a detection limit is estimated for the atomic absorption method. Experiment 23-14, voltammetry of vitamin C, is also taken from the San Jose lab manual.

23-1 Penny Statistics

U.S. pennies minted after 1982 have a Zn core with a Cu overlayer. Prior to 1982, pennies were made of brass, with a uniform composition (95 wt % Cu / 5 wt % Zn). In 1982, both the heavier brass coins and the lighter zinc coins were made. In this experiment,[5] your class will weigh many coins and pool the data to answer the following questions: (1) Do pennies from different years have the same mass? (2) Do pennies from different mints have the same mass? Read Section 2-3 on the analytical balance and Chapter 4 on statistics prior to doing this experiment.

Gathering Data

Each student should collect and weigh enough pennies to the nearest milligram to provide a total set that contains 300 to 500 brass coins and a similar number of zinc coins. Instructions are given for a spreadsheet, but the same operations can be carried out with a calculator. Compile all the class data in a spreadsheet. Each column should list the masses of pennies from only one calendar year. Use the spreadsheet "sort" function to sort each column so that the lightest mass is at the top of the column and the heaviest is at the bottom. There will be two columns for 1982, in which both types of coins were made. Select a year other than 1982 for which you have many coins and divide the coins into those made in Denver (with a "D" beneath the year) and those minted in Philadelphia (with no mark beneath the year).

Discrepant Data

At the bottom of each column, list the mean (Equation 4-1) and standard deviation (Equation 4-2). Retain at least one extra digit beyond the milligram place to avoid round-off errors in your calculations.

Damaged or corroded coins may have masses different from those of the general population. Discard grossly discrepant masses lying ≥ 4 standard deviations from the mean in any one year. (For example, if one column has an average mass of 3.000 g and a standard deviation of 0.030 g, the 4-standard-deviation limit is $\pm(4 \times 0.030) = \pm 0.120$ g. A mass that is ≤ 2.880 or ≥ 3.120 g should be discarded.) After rejecting discrepant data, recompute the average and standard deviation for each column.

Confidence Intervals and t Test

Select the two years (≥ 1982) in which the zinc coins have the highest and lowest average masses. For each of the two years, compute the 95% confidence interval by using Equa-

tion 4-3. Use the t test (Equation 4-4) to compare the two mean values at the 95% confidence level. Are the two average masses significantly different? Try the same for two years of brass coins (≤ 1982). Try the same for the one year whose coins you segregated into those from Philadelphia and those from Denver. Do the two mints produce coins with the same mass?

Gaussian Distribution of Masses

List the masses of all pennies made in or after 1983 in a single column, sorted from lowest to highest mass. There should be at least 300 masses listed. Divide the data into 0.01-g intervals (e.g., 2.480 to 2.489 g) and prepare a bar graph, like that shown in Figure 23-1. Find the mean ($\bar{x}$), median, and standard deviation (s) for all coins in the graph. For random (Gaussian) data, only 3 out of 1 000 measurements should lie outside of $\bar{x} \pm 3s$. Indicate which bars (if any) lie beyond $\pm 3s$. In Figure 23-1, two bars at the right are outside of $\bar{x} \pm 3s$.

Least-Squares Analysis: Do Pennies Have the Same Mass Each Year?

Prepare a graph like that in Figure 23-2 in which the ordinate (y-axis) is the mass of zinc pennies minted each year since 1982 and the abscissa (x-axis) is the year. For

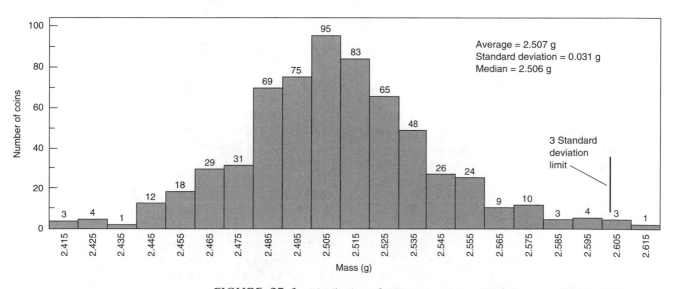

FIGURE 23-1 Distribution of 613 penny masses (made in years 1982 to 1992) measured in Dan's house by Jimmy Kusznir and Doug Harris in December 1992.

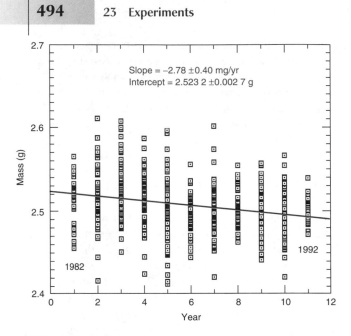

Slope = −2.78 ±0.40 mg/yr
Intercept = 2.523 2 ±0.002 7 g

FIGURE 23-2   Penny mass versus year for 440 coins. Because the slope of the least-squares line is *significantly* less than 0, we conclude that the average mass of older pennies is greater than the average mass of newer pennies.

simplicity, let 1982 be year 1, 1983 be year 2, and so on. If the mass of a penny increases systematically from year to year, then the least-squares line through the data will have a positive slope. If the mass decreases, the slope will be negative. If the mass is constant, the slope will be 0. Even if the mass is really constant, your selection of coins is random and the slope is not exactly 0.

We want to know whether the slope is *significantly* different from 0. Suppose that you have data for 14 years. Enter all of the data into two columns of the least-squares spreadsheet in Figure 4-8. Column B (x_i) is the year (1 to 14) and column C (y_i) is the mass of each penny. Your table will have several hundred entries. (If you are not using a spreadsheet, just tabulate the average mass for each year. Your table will have only 14 entries.) Calculate the slope (m) and intercept (b) of the best straight line through all points and find the uncertainties in slope (s_m) and intercept (s_b).

Use Student's *t* to find the 95% confidence interval for the slope:

$$\text{confidence interval for slope} = m \pm ts_m \quad (23\text{-}1)$$

where *t* is from Table 4-2 for $n - 2$ degrees of freedom. For example, if you have $n = 300$ pennies, $n - 2 = 298$,

and it would be reasonable to use the value of *t* $(= 1.960)$ at the bottom of the table for $n = \infty$. If the least-squares slope is $m \pm s_m = -2.78 \pm 0.40$ mg/year, then the 95% confidence interval is $m \pm ts_m = -2.78 \pm (1.960)(0.40) = -2.78 \pm 0.78$ mg/year.

The 95% confidence interval is $-2.78 \pm 0.78 = -3.56$ to -2.00 mg/year. We are 95% confident that the true slope is in this range and is, therefore, not 0. We conclude that older zinc pennies are heavier than newer zinc pennies.

Reporting Your Results

Gathering Data

1. Attach a table of masses, with one column sorted by mass for each year.

2. Divide 1982 into two columns, one for light (zinc) and one for heavy (brass) pennies.

3. One year should be divided into one column from Denver and one from Philadelphia.

Discrepant Data

1. List the mean $(\bar{x})$ and standard deviation (s) for each column.

2. Discard data that lie outside of $\bar{x} \pm 4s$ and recompute $\bar{x}$ and s.

Confidence Intervals and t Test

1. For the year ≥ 1982 with highest average mass:
  95% confidence interval $(= \bar{x} \pm ts/\sqrt{n}) = $ _____

For the year ≥ 1982 with lowest average mass:
  95% confidence interval = _____
Comparison of means with *t* test:
  $t_{\text{calculated}}$ (Equation 4-4) = _____
  t_{table} (Table 4-2) = _____
  Is the difference significant? _____

2. For the year ≤ 1982 with highest average mass:
  95% confidence interval = _____
For the year ≤ 1982 with lowest average mass:
  95% confidence interval = _____
Comparison of means with *t* test:
  $t_{\text{calculated}}$ (Equation 4-4) = _____
  t_{table} (Table 4-2) = _____
  Is the difference significant? _____

3. For Philadelphia versus Denver coins in one year:
Philadelphia 95% confidence interval =

_____

Denver 95% confidence interval =

_____

Comparison of means with t test:
$t_{calculated}$ (Equation 4-4) = _____
t_{table} (Table 4-2) = _____
Is the difference significant? _____

Gaussian Distribution of Masses

Prepare a graph analogous to Figure 23-1, with labels showing the $\pm 3s$ limits.

Least-Squares Analysis

Prepare a graph analogous to Figure 23-2 and construct a least-squares table or a spreadsheet analogous to Table 4-5 or Figure 4-8.

$m \pm s_m$ = _____

$t_{95\% \text{ confidence}}$ = _____

$t_{99\% \text{ confidence}}$ = _____

95% confidence: $m \pm ts_m$ = _____.
Does interval include 0? _____
99% confidence: $m \pm ts_m$ = _____.
Does interval include 0? _____
Is there a systematic increase or decrease of penny mass with year? _____

23-2 Statistical Evaluation of Acid–Base Indicators

This experiment introduces you to the use of indicators and to the statistical concepts of mean, standard deviation, Q test, and t test.[6] You will compare the accuracy of different indicators in locating the end point in the titration of the base "tris" with hydrochloric acid:

$$(HOCH_2)_3CNH_2 + H^+ \longrightarrow (HOCH_2)_3CNH_3^+ \quad (23\text{-}2)$$
Tris(hydroxymethyl)aminomethane
"tris"

Study Section 2-3 on using a balance and Section 2-4 on the buret before this experiment.

Other nice introductory experiments on statistics deal with the measurement of density using common laboratory equipment[7] and with the measurement of glucose using commercial glucometers for diabetics.[8]

Reagents

~0.1 M HCl: Each student needs ~500 mL of unstandardized solution, all from a single batch that will be analyzed by the whole class.

Tris: Solid, primary standard powder should be available (~4 g/student).

Indicators: Bromothymol blue (BB), methyl red (MR), bromocresol green (BG), methyl orange (MO), and erythrosine (E) should be available in dropper bottles.
   BB: Dissolve 100 mg in 16.0 mL of 0.01 M NaOH and dilute to 250 mL with distilled water.
   MR: Dissolve 20 mg in 60 mL of ethanol and add 40 mL of distilled water.
   BG: Dissolve 100 mg in 14.3 mL of 0.01 M NaOH and dilute to 250 mL with distilled water.
   MO: Dissolve 25 mg in 250 mL of distilled water.
   E: Dissolve 25 mg in 250 mL of distilled water.
Color changes for the titration of tris with HCl are
   BB: blue (pH 7.6) → yellow (pH 6.0) (end point is disappearance of green)
   MR: yellow (pH 6.0) → red (pH 4.8) (end point is disappearance of orange)
   BG: blue (pH 5.4) → yellow (pH 3.8) (end point is green)
   MO: yellow (pH 4.4) → red (pH 3.1) (end point is first appearance of orange)
   E: red (pH 3.6) → orange (pH 2.2) (end point is first appearance of orange)

Procedure

Each student should carry out the following procedure with *one* of the indicators. Different students should be assigned different indicators so that at least four students evaluate each of the indicators.

1. Calculate the molecular mass of tris and the mass required to react with 35 mL of 0.10 M HCl. Weigh this much tris into a 125-mL flask.

2. It is good practice to rinse a buret with a new solution to wash away traces of previous reagents. Wash your 50-mL buret with three 10-mL portions of 0.1 M HCl and discard the washings. Tilt and rotate the buret so that the liquid washes the walls, and drain the liquid through the stopcock. Then fill the buret with 0.1 M HCl to near the 0-mL mark, allow a minute for the liquid to settle, and record the reading to the nearest 0.01 mL.

3. The first titration will be rapid, to allow you to find the approximate end point. Add ~20 mL of HCl from the buret to the flask and swirl to dissolve the tris. Add 2–4 drops of indicator and titrate with ~1-mL aliquots of HCl to find the end point.

4. From the first titration, calculate how much tris is required to cause each succeeding titration to require 35–40 mL of HCl. Weigh this much tris into a clean flask (it need not be dry). Refill your buret to near 0 mL and record the reading. Repeat the titration in step 3, but use 1 drop at a time near the end point. When you are very near the end point, use less than a drop at a time. To do this, carefully suspend a fraction of a drop from the buret tip and touch it to the inside wall of the flask. Tilt the flask so that the bulk solution overtakes the droplet and then swirl the flask to mix the solution. Record the volume of HCl required to reach the end point to the nearest 0.01 mL. Calculate the true mass of tris with the buoyancy equation 2-1 (density of tris = 1.327 g/mL). Calculate the molarity of HCl.

5. Repeat the titration to obtain at least six accurate measurements of the HCl molarity.

6. Use the Q test (Section 4-3) to decide whether any results should be discarded. Report your retained values, their mean (Equation 4-1), their standard deviation (Equation 4-2), and the percent relative standard

deviation (= 100 × standard deviation/mean). Learn to compute standard deviation with the function built into your calculator.

Analyzing Class Data

Pool the data from your class to fill in Table 23-1, which shows two possible results. The quantity s_x is the standard deviation of all results reported by many students. The pooled standard deviation, s_p, is derived from the standard deviations reported by each student. If two students see the end point differently, each result might be very reproducible, but their reported molarities will be different. Together, they will generate a large value of s_x (because their results are so different), but a small value of s_p (because each one was reproducible).

Select the pair of indicators giving average HCl molarities that are furthest apart. Use the t test (Equation 4-4) to decide whether the average molarities are significantly different from each other at the 95% confidence level.[9] When you calculate the pooled standard deviation for Equation 4-4, the values of s_1 and s_2 in Equation 4-5 are the values of s_x (not s_p) in Table 23-1. Select the pair of indicators giving the second most different molarities and use the t test again to see whether or not this second pair of results is significantly different.

	TABLE 23-1	**Pooled Data**				
Indicator	Number of measurements (n)	Number of students (S)	Mean HCl molarity $(M)^a$ $(\bar{x})$	s_x $(M)^b$	Relative standard deviation (%) $100 \times s_x/\bar{x}$	s_p $(M)^c$
BB MR	28	5	0.095 65	0.002 25	2.35	0.001 09
BG MO E	29	4	0.086 41	0.001 13	1.31	0.000 99

a. Computed from all values that were not discarded with the Q test.
b. s_x = standard deviation of all n measurements (degrees of freedom = $n - 1$).
c. s_p = pooled standard deviation for S students (degrees of freedom = $n - S$). Computed with a modification of Equation 4-5:

$$s_p = \sqrt{\frac{s_1^2(n_1 - 1) + s_2^2(n_2 - 1) + s_3^2(n_3 - 1) + \cdots}{n - S}}$$

where there is one term in the numerator for each student using that indicator.

Reporting Your Results

Your Individual Data

Trial	Mass of tris from balance (g)	True mass of tris (Eq. 2-1)	HCl volume (mL)	Calculated HCl molarity (M)
1				
2				
3				
4				
5				
6				

On the basis of the Q test, should any molarity be discarded? If so, which one? _____

Mean value of retained results: _____

Standard deviation: _____

Relative standard deviation (%): _____

Pooled Class Data

1. Attach your copy of Table 23-1, with all entries filled in.

2. Compare the two most different molarities in Table 23-1. (Show your *t* test and state your conclusion.)

3. Compare the two second most different molarities in Table 23-1.

23-3   Calibration of Volumetric Glassware

For most accurate results in future experiments, you should calibrate your own volumetric glassware (burets, pipets, flasks, etc.) to measure the exact volumes delivered or contained. This experiment develops basic skills in using a balance and handling volumetric glassware. Before beginning, you should study Sections 2-3, 2-4, and 2-9.

Calibration of a 50-mL Buret

In this procedure, you will construct a graph (like Figure 3-2) needed to convert the measured volume delivered by your buret to the true volume delivered.

1. Fill the buret with distilled water and force any air bubbles out the tip. See that the buret drains without leaving drops on the walls. If drops are left, clean the buret with soap and water or soak it with peroxydisulfate-sulfuric acid cleaning solution.[10] Adjust the meniscus to be at or slightly below 0.00 mL and touch the buret tip to a beaker to remove the suspended drop of water. Allow the buret to stand for 5 min while you weigh a 125-mL flask fitted with a rubber stopper. (Hold the flask with a tissue or paper towel, not with your hands, to avoid changing its mass with fingerprint residue.) If the level of the liquid in the buret has changed, tighten the stopcock and repeat the procedure.

2. Drain ~10 mL of water (at a rate of < 20 mL/min) into the weighed flask and cap it tightly to prevent evaporation. Allow 30 s for the liquid on the walls to descend before you read the buret. Estimate all readings to the nearest 0.01 mL. Weigh the flask again to determine the mass of water delivered.

3. Now drain the buret from 10 to ~20 mL and measure the mass of water delivered. Repeat the procedure for 30, 40, and 50 mL. Then do the entire procedure (10, 20, 30, 40, 50 mL) a second time.

4. Use Table 2-4 to convert the mass of water to the volume delivered. Repeat any set of duplicate buret corrections that do not agree to within 0.04 mL. Prepare a calibration graph like that in Figure 3-2, showing the correction factor at each 10-mL level.

EXAMPLE   Buret Calibration

When draining the buret at 24°C, you observe the following values:

Final reading	10.01	10.08 mL
Initial reading	0.03	0.04
Difference	9.98	10.04 mL
Mass	9.984	10.056 g
Actual volume delivered	10.02	10.09 mL
Correction	+0.04	+0.05 mL
Average correction		+0.045 mL

To calculate the actual volume delivered when 9.984 g of water is delivered at 24°C, use the correction factor in Table 2-4: volume = (9.984 g)(1.003 8 mL/g) = 10.02 mL. The average correction for both sets of data is +0.045 mL.

To obtain the correction for a volume greater than 10 mL, add successive masses of water collected in the flask. Suppose that the following masses were measured:

Volume interval (mL)	Mass delivered (g)
0.03 to 10.01	9.984
10.01 to 19.90	9.835
19.90 to 30.06	10.071
Sum 30.03 mL	29.890 g

The actual volume of water delivered is $(29.890 \text{ g}) \times (1.003\,8 \text{ mL/g}) = 30.00$ mL. Because the indicated volume is 30.03 mL, the buret correction at 30 mL is -0.03 mL.

What does this mean? Suppose that Figure 3-2 applies to your buret. If you begin a titration at 0.04 mL and end at 29.00 mL, you would deliver 28.96 mL if the buret were perfect. In fact, Figure 3-2 tells you that the buret delivers 0.03 mL less than the indicated amount, so only 28.93 mL were actually delivered. To use the calibration curve, you should begin all titrations near 0.00 mL. Use your calibration curve whenever you use your buret.

Other Calibrations

Pipets are calibrated by weighing water delivered from them. A volumetric flask is calibrated by weighing it empty and then weighing it filled to the mark with distilled water. Perform each procedure at least twice. Compare your results with the tolerances in Tables 2-1, 2-2, and 2-3.

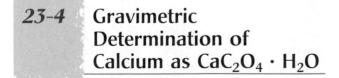

23-4   Gravimetric Determination of Calcium as $CaC_2O_4 \cdot H_2O$

Calcium ion can be analyzed by precipitation with oxalate in basic solution to form $CaC_2O_4 \cdot H_2O$.[11] The precipitate is soluble in acidic solution because oxalate reacts with H^+ to make $HC_2O_4^-$. Large, easily filtered, pure crystals of product are obtained if the precipitation is carried out slowly. This goal can be achieved by dissolving Ca^{2+} and $C_2O_4^{2-}$ in acidic solution and gradually raising the pH by thermal decomposition of urea, as described in Section 6-2. Read Sections 6-1 to 6-3 for this experiment.

Reagents

Ammonium oxalate solution: Make 1 L of solution containing 40 g of $(NH_4)_2C_2O_4$ plus 25 mL of 12 M HCl. Each student will need 80 mL of this solution.

Methyl red indicator: Dissolve 20 mg in 60 mL of ethanol and add 40 mL of distilled water.

0.1 M HCl: (225 mL/student) Dilute 8.3 mL of 37 wt % HCl up to 1 L.

Urea: 45 g/student.

Unknowns: Prepare 1 L of solution containing 15–18 g of $CaCO_3$ plus 38 mL of 12 M HCl. Each student will need 100 mL of this solution. Alternatively, solid unknowns are available from Thorn Smith.[12]

Procedure

1. Dry three medium-porosity, sintered-glass filter crucibles for 1–2 h at 105°C; cool them in a desiccator for 30 min and weigh them. Repeat the procedure with 30-min heating periods until successive weighings agree to within 0.3 mg. Use a paper towel or tongs, not your fingers, to handle the crucibles. Alternatively, use a 900-W kitchen microwave oven to dry your crucibles to constant mass in two heating periods of 4 min each, followed by heating periods of 2 min each (with 15 min allowed for cooldown after each cycle).

2. Use a few small portions of unknown to rinse a 25-mL transfer pipet and discard the washings. Use a rubber bulb or other mechanical device, not your mouth, to provide suction. Transfer exactly 25 mL of unknown to each of three 250- or 400-mL beakers and dilute each with ~75 mL of 0.1 M HCl. Add 5 drops of methyl red indicator to each beaker. This indicator is red below pH 4.8 and yellow above pH 6.0.

3. Add ~25 mL of ammonium oxalate solution to each beaker while stirring with a glass rod. Remove the rod and rinse it into the beaker. Add ~15 g of solid urea to each sample, cover it with a watchglass, and boil gently for ~30 min until the indicator turns yellow.

4. Filter each hot solution through a weighed crucible, using suction (Figure 2-13). Add ~3 mL of ice-cold, distilled water to the beaker and use a rubber policeman to help transfer the remaining solid to the crucible. Repeat this procedure with small portions of ice-cold water until all of the precipitate has been transferred. Finally, use two 10-mL portions of ice-cold water to rinse each beaker and pour the washings over the precipitate.

5. Dry the precipitate, first with aspirator suction for 1 min, then in an oven at 105°C for 1–2 h. Bring each filter crucible to constant mass. The product is somewhat hygroscopic, so only one filter at a time should be removed from the desiccator and weighed rapidly. Alternatively, dry the precipitate in a microwave oven once for 4 min, followed by several 2-min periods, with cooling for 15 min before weighing. The water of crystallization is not lost.

6. Calculate the average molarity ($\bar{x}$) of Ca^{2+} in the unknown solution or the average wt % of Ca in the unknown solid. Report the standard deviation (s) and percent relative standard deviation ($= 100 \times s/\bar{x}$).

23-5   Gravimetric Measurement of Phosphorus in Plant Food

Nutrients in plant foods—nitrogen, phosphorus, and potassium—are indicated on the label by three numbers, such as 15-30-20. This notation means that the fertilizer contains 15 wt % N, 30 wt % P_2O_5, and 20 wt % K_2O. (This is a fictitious notation. Phosphorus is not in the form P_2O_5 and potassium is not in the form K_2O in the fertilizer.)

This experiment[13] measures the phosphorus content of plant food by the reaction

$$\underset{\text{From plant food}}{HPO_4^{2-}} + NH_3 + Mg^{2+} + 6H_2O \longrightarrow$$

$$\underset{\text{FM 245.41}}{MgNH_4PO_4 \cdot 6H_2O(s)} \quad (23\text{-}3)$$

The experiment is not as accurate as others in this book, but it can be carried out with tap water and chemicals from the grocery store. The product is dried at room temperature to avoid loss of water of crystallization above 40°C. In more accurate determinations of phosphate, the solid $MgNH_4PO_4 \cdot 6H_2O$ is *ignited* (heated strongly) and decomposes to $Mg_2P_2O_7$. Read Sections 6-1 to 6-3 for this experiment.

Reagents

Epsom salts: $MgSO_4 \cdot 7H_2O$ (FM 246.52), 5–15 g/student
Ammonia: (200 mL/student) Use aqueous ammonia from the grocery ($\sim$4 wt % NH_3) or prepare 4.0 wt % NH_3 ($= 2.3$ M) by diluting 160 mL of concentrated (28 wt %) NH_3 up to 1.0 L in the hood.
Isopropyl alcohol: Each student needs 100 mL of rubbing alcohol.

Procedure

1. Weigh 10 to 11 g of commercial plant food to the nearest 0.01 g and dissolve it in 140 mL of water in a 250-mL beaker. Stir the mixture well and filter it (Figures 2-14 and 2-15) into a 1-L beaker to remove insoluble residue. A coffee basket filter can be used instead of laboratory filter paper. Rinse the 250-mL beaker twice with 5-mL portions of water and pour the washings through the filter into the 1-L beaker.

2. From the plant food label, estimate how many moles of phosphorus are in the solid that you weighed. (The middle number of the label gives the wt % of P_2O_5 (FM 141.94).) Calculate how many grams of Epsom salts ($MgSO_4 \cdot 7H_2O$, FM 246.52) are required for a 50% molar excess of Mg^{2+} for Reaction 23-3. Dissolve the required amount of Epsom salts in water, using 10 mL of H_2O per gram of $MgSO_4 \cdot 7H_2O$, and pour the resulting solution into the filtered plant food solution.

3. In a fume hood or a well-ventilated area, gradually add 200 mL of ammonia solution (measured with a beaker) to the plant food solution while stirring well with a glass rod. Allow the white precipitate to stand for at least 15 min.

4. Collect the precipitate by filtering through two layers of fluted filter paper or two layers of coffee filters. Then add 50 mL of isopropyl alcohol to the 1-L beaker and swirl to dislodge particles of solid. Pour the mixture over the precipitate in the filter and allow the liquid to drain. Repeat the process once more with another 50 mL of alcohol.

5. Spread the filter paper on several layers of paper towel to dry overnight, protected from dust. The next day, use a spatula to break up lumps. Then allow an additional day of drying, so that the solid is a free-flowing powder.

6. Weigh an empty 250-mL beaker. Scrape the solid from the filter into the beaker and weigh the beaker again to find the mass of solid precipitate. To be certain that the product is dry, allow the beaker to stand for at least another day protected from dust and weigh it again. If successive weighings agree to within ±0.1 g, the product is dry enough.

7. From the mass of dry product, calculate the moles of product in Reaction 23-3 and the moles of phosphorus analyzed. From the moles of phosphorus, compute the moles of P_2O_5 and grams of P_2O_5 in the solid plant food. Report the wt % P_2O_5 in the plant food. Pool the class results to find the mean and standard deviation for the wt % P_2O_5 in each type of plant food analyzed. Compare the class results to the package labels.

23-6 Preparing Standard Acid and Base

Hydrochloric acid and sodium hydroxide are the most common strong acids and bases used in the laboratory. Both reagents need to be standardized to learn their exact concentration. Section 9-5 provides background information for the procedures described below.

Reagents

50 wt % NaOH: (3 mL/student) Dissolve 50 g of reagent-grade NaOH in 50 mL of distilled water and allow the suspension to settle overnight. Na_2CO_3 is insoluble in the solution and precipitates. Store the solution in a tightly sealed polyethylene bottle and handle it gently to avoid stirring the precipitate when liquid is withdrawn.

Phenolphthalein indicator: Dissolve 50 mg in 50 mL of ethanol and add 50 mL of distilled water.

Bromocresol green indicator: Dissolve 100 mg in 14.3 mL of 0.01 M NaOH and dilute to 250 mL with distilled water.

Concentrated (37 wt %) HCl: 10 mL/student.

Primary standards: Potassium hydrogen phthalate (~2.5 g/student) and sodium carbonate (~1.0 g/student).

0.05 M NaCl: 50 mL/student.

Standardizing NaOH

1. Dry primary standard-grade potassium hydrogen phthalate for 1 h at 105°C and store it in a capped bottle in a desiccator.

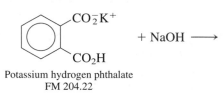

Potassium hydrogen phthalate
FM 204.22

2. Boil 1 L of distilled water for 5 min to expel CO_2. Pour the water into a polyethylene bottle, which should be tightly capped whenever possible. Calculate the volume of 50 wt % NaOH needed to prepare 1 L of 0.1 M NaOH. (The density of 50 wt % NaOH is 1.50 g per milliliter of solution.) Use a graduated cylinder to transfer this much NaOH to the bottle of water. (**CAUTION:** 50 wt % NaOH eats people. Flood any spills on your skin with water.) Mix well and cool the solution to room temperature (preferably overnight).

3. Weigh four samples of solid potassium hydrogen phthalate and dissolve each in ~25 mL of distilled water in a 125-mL flask. Each sample should contain enough solid to react with ~25 mL of 0.1 M NaOH. Add 3 drops of phenolphthalein to each flask and titrate one rapidly to find the end point. The buret should have a loosely fitted cap to minimize entry of CO_2 from the air.

4. Calculate the volume of NaOH required for each of the other three samples and titrate them carefully. During each titration, periodically tilt and rotate the flask to wash all liquid from the walls into the bulk solution. Near the end, deliver less than 1 drop of titrant at a time. To do this, carefully suspend a fraction of a drop from the buret tip, touch it to the inside wall of the flask, wash it into the bulk solution by careful tilting, and swirl the solution. The end point is the first appearance of faint pink color that persists for 15 s. (The color will slowly fade as CO_2 from the air dissolves in the solution.)

5. Calculate the average molarity ($\bar{x}$), the standard deviation (s), and the percent relative standard deviation

$(= 100 \times s/\bar{x})$. If you were careful, the relative standard deviation should be $< 0.2\%$.

Standardizing HCl

1. Use the table inside the cover of this book to calculate the volume of ~37 wt % HCl that should be added to 1 L of distilled water to produce 0.1 M HCl and prepare this solution.

2. Dry primary standard-grade sodium carbonate for 1 h at 105°C and cool it in a desiccator.

3. Weigh four samples, each containing enough Na_2CO_3 to react with ~25 mL of 0.1 M HCl and place each in a 125-mL flask. When you are ready to titrate each one, dissolve it in ~25 mL of distilled water. Add 3 drops of bromocresol green indicator and titrate one rapidly to a green color to find the approximate end point.

$$2HCl + Na_2CO_3 \longrightarrow CO_2 + 2NaCl + H_2O$$
$$\text{FM } 105.99$$

4. Carefully titrate each sample until it turns from blue to green. Then boil the solution to expel CO_2. The color should return to blue. Carefully add HCl from the buret until the solution turns green again and report the volume of acid at this point.

5. Perform one blank titration of 50 mL of 0.05 M NaCl containing 3 drops of indicator. Subtract the volume of HCl needed for the blank from that required to titrate Na_2CO_3.

6. Calculate the mean HCl molarity, standard deviation, and percent relative standard deviation.

23-7 Analysis of a Mixture of Carbonate and Bicarbonate

This procedure involves two titrations. First, total alkalinity $(= [HCO_3^-] + 2[CO_3^{2-}])$ is measured by titrating the mixture with standard HCl to a bromocresol green end point:

$$HCO_3^- + H^+ \longrightarrow H_2CO_3$$
$$CO_3^{2-} + 2H^+ \longrightarrow H_2CO_3$$

A separate aliquot of unknown is treated with excess standard NaOH to convert HCO_3^- to CO_3^{2-}:

$$HCO_3^- + OH^- \longrightarrow CO_3^{2-} + H_2O$$

Then all the carbonate is precipitated with $BaCl_2$:

$$Ba^{2+} + CO_3^{2-} \longrightarrow BaCO_3(s)$$

The excess NaOH is immediately titrated with standard HCl to determine how much HCO_3^- was present. A blank titration is required for the Ba^{2+} addition, because Ba^{2+} precipitates some of the NaOH:

$$Ba^{2+} + 2OH^- \rightleftharpoons Ba(OH)_2(s)$$

From the total alkalinity and bicarbonate concentration, you can calculate the original carbonate concentration.

Reagents

Standard NaOH and standard HCl: From Experiment 23-6.

CO₂-free water: Boil 500 mL of distilled water to expel carbon dioxide and pour the water into a 500-mL plastic bottle. Screw the cap on tightly and allow the water to cool to room temperature. Keep tightly capped when not in use.

10 wt % aqueous BaCl₂: 35 mL/student.

Bromocresol green and phenolphthalein indicators: See Experiment 23-6.

Unknowns: Solid unknowns (2.5 g/student) can be prepared from reagent-grade sodium or potassium carbonate and bicarbonate. Unknowns should be stored in a desiccator but should not be heated. Even mild heating at 50°–100°C converts $NaHCO_3$ to Na_2CO_3.

Procedure

1. Accurately weigh 2.0–2.5 g of unknown into a 250-mL volumetric flask by weighing the sample in a capped weighing bottle, delivering some to a funnel in the volumetric flask, and reweighing the bottle. Continue this process until the desired mass of reagent has been transferred to the funnel. Rinse the funnel repeatedly with small portions of CO_2-free water to dissolve the sample. Remove the funnel, dilute to the mark, and mix well.

2. *Total alkalinity:* Pipet a 25.00-mL aliquot of unknown solution into a 250-mL flask and titrate with standard 0.1 M HCl, using bromocresol green indicator as described in the previous experiment for standardizing HCl. Repeat this procedure with two more 25.00-mL aliquots.

3. *Bicarbonate content:* Pipet 25.00 mL of unknown and 50.00 mL of standard 0.1 M NaOH into a 250-mL flask. Swirl and add 10 mL of 10 wt % $BaCl_2$, using a graduated cylinder. Swirl again to precipitate $BaCO_3$, add 2 drops of phenolphthalein indicator, and immediately titrate with standard 0.1 M HCl. Repeat this procedure with two more 25.00-mL samples of unknown.

4. *Blank titration:* Titrate with standard 0.1 M HCl a mixture of 25 mL of water, 10 mL of 10 wt % $BaCl_2$, and 50.00 mL of standard 0.1 M NaOH. Repeat this procedure twice more. The difference in moles of HCl needed in steps 4 and 3 equals the moles of bicarbonate in the mixture.

5. From the results of step 2, calculate the total alkalinity and its standard deviation. From the results of steps 3 and 4, calculate the bicarbonate concentration and its standard deviation. Using the standard deviations as estimates of uncertainty, calculate the concentration (and uncertainty) of carbonate in the sample. Express the composition of the solid unknown in a form such as 63.4 ($\pm$0.5) wt % K_2CO_3 and 36.6 ($\pm$0.2) wt % $NaHCO_3$.

23-8 Variance of Sampling a Heterogeneous Material—An Acid–Base Titration[14]

Variance is the square of the standard deviation of a measurement. When we analyze a *heterogeneous* material—one whose composition varies from place to place—the overall variance (s_o^2) is the sum of the variance of the sampling operation (s_s^2) and the variance of the analytical procedure (s_a^2):

Additivity of variance: $\qquad s_o^2 = s_s^2 + s_a^2 \qquad$ (23-4)

If either the sampling or analytical standard deviation is much larger than the other, then the overall variance is dominated by the larger variance. For example, if s_s is three

times larger than s_a, then 90% of the overall variance comes from s_s (because $s_o^2 = s_s^2 + s_a^2 = 3^2 + 1^2 = 10$).

In this experiment, you will measure the content of strong base (mainly NaOH) in the household drain cleaner, Drano. This product consists of solid pellets of NaOH, some aluminum shavings, some insoluble material, and, undoubtedly, some Na_2CO_3 from reaction of NaOH with atmospheric CO_2. If you weigh out replicate portions of Drano, the quantity of strong base varies substantially from portion to portion. In this experiment, you will evaluate the analytical variance and the sampling variance. A possible class project is to find out how the sampling variance depends on the sample size.

A more advanced experiment on the variance from sampling, sample preparation, and analysis of sodium in foods is available.[15]

Reagents

Bromocresol green indicator: Dissolve 100 mg in 14.3 mL of 0.01 M NaOH and dilute to 250 mL with distilled water.

Phenolphthalein indicator: Dissolve 50 mg in 50 mL of ethanol and dilute to 100 mL with distilled water.

Standard hydrochloric acid: ~0.1 M HCl. Prepare by diluting ~8.2 mL of 37 wt % HCl to 1.0 L with water. Store in a tightly capped plastic bottle. Standardize as described in steps 1–4 of Experiment 23-2 with primary standard "tris" (tris(hydroxymethyl)-aminomethane), using the green end point of bromocresol green indicator. Perform the titration carefully three times.

Solid Drano: (~4 g/student) We recommend having several tightly capped cans of the commercial product in the lab. Students can weigh solid directly out of the can.

Measuring the Analytical Variance

1. Weigh out 1.9–2.1 g of solid Drano in a 50-mL beaker on a top-loading balance accurate to 0.001 g. Add 20 mL of distilled water to the beaker and swirl to dissolve the solid. Any aluminum shavings present will not dissolve. Decant the liquid into a 100-mL volumetric flask, leaving the solid shavings in the beaker. Add ~10 mL of water to the beaker, swirl to rinse the inside of the beaker well, and decant the liquid into the volumetric flask. Repeat this process with several 10-mL portions of water to complete a quantitative transfer of the Drano (minus the metal shavings) into the volumetric flask. Finally, dilute to the mark, cap the flask, and mix well by inverting 20 times. There will be small particles of undissolved solid suspended in the liquid.

2. Dip a clean, dry 10-mL transfer pipet into the volumetric flask and take up ~5 mL of dissolved Drano. *Be extremely careful not to suck liquid into the rubber bulb on the pipet.* Tip the pipet to wash it out with liquid and discard the liquid down the drain.

3. Use the pipet from step 2 to transfer 10.00 mL of Drano solution from the volumetric flask into a 125-mL flask equipped with a stirring bar and add 2 drops of phenolphthalein indicator. Titrate with standard 0.1 M HCl from a 50-mL buret to the colorless end point. The flask should be magnetically stirred during the titration. A sheet of white paper placed beneath the flask will help you see the color change better. The phenolphthalein color may gradually fade to become nearly imperceptible during the titration prior to the end point. If so, add one more drop of indicator and continue the titration. If you add indicator and it turns pink, you have not passed the end point. This first titration can be done rapidly (~5 mL/min) to locate the approximate end point.

4. Repeat the previous step three times to obtain three accurate measurements of the end point. If the equivalence point is <15 mL HCl, titrate 20.00 mL of Drano solution instead of 10.00 mL. In each case, add 90–95% of the required HCl rapidly, and then titrate slowly to find the end point.

5. From each equivalence volume of HCl, compute the wt % of NaOH in the solid Drano, assuming that the only base present is NaOH. Calculate the standard deviation in wt % NaOH. As an example, your result might be $35.3_3 \pm 0.4_2$ wt % NaOH. The *relative* standard deviation is $0.4_2/35.3_3 = 0.011_9$. This number is the relative analytical standard deviation, s_a, in the weight percent of NaOH in the Drano sample that you used. The relative analytical variance is $s_a^2 = 0.011_9{}^2 = 1.4 \times 10^{-4}$. The term "relative" means that each standard deviation is divided by the mean to which it applies.

Measuring the Sampling Variance

1. Weigh 0.28–0.32 g (measured accurately to 0.001 g) of solid Drano into a 125-mL flask containing a magnetic stirring bar. Note the level of standard HCl in your buret and add ~5 mL rapidly. Stir to dissolve the Drano. (Metal shavings will remain undissolved.) Add 2 drops of phenolphthalein indicator and titrate to the colorless end point. Because you do not know where the end point will be, try titrating at a constant rate of ~2 mL/min (~1.5 s/drop). You will need to pay careful attention to the color to try to find the end point to within ~2–3 drops. If you need to add more indicator because it fades, stop the titration while you add another drop of indicator. From the required volume of HCl, compute wt % NaOH in the Drano. It is likely that your answer will differ substantially from what you observed in the first part of this experiment.

2. Repeat step 1 twice more with fresh 0.27–0.33 g portions of solid Drano.

3. From the three titrations of solid Drano, express the mean and relative standard deviation for wt % NaOH in Drano. This standard deviation is the relative overall standard deviation, s_o, because it includes contributions from the sampling operation and from the analytical titration. The square of s_o is the overall variance.

4. Compute the relative sampling variance, s_s^2 in Equation 23-4. Express your answer with an appropriate number of significant figures.

A Possible Class Project

Different groups of students can measure s_s^2 for different masses of solid Drano. The instructions above apply to a quantity of 0.30 g of solid (± 0.02 g). Different groups can titrate the following quantities of solid: 0.10, 0.30, 1.0, and 3.0 g. Use 0.033, 0.10, 0.33, and 1.0 M standard HCl, respectively, to titrate the different masses of solid, so that a 50-mL buret can always be used. From the class data, prepare a graph of relative sampling standard deviation versus mass of solid analyzed. Discuss the shape of this curve.

23-9 Kjeldahl Nitrogen Analysis

The Kjeldahl nitrogen analysis described in Section 9-6 is widely used to measure the nitrogen content of pure organic compounds and complex substances such as milk, cereal, and flour. Digestion in boiling H_2SO_4 with a catalyst converts organic nitrogen into NH_4^+. The solution is then made basic, and the liberated NH_3 is distilled into a known amount of HCl (Reactions 9-5 to 9-7). Unreacted HCl is titrated with NaOH to determine how much HCl was consumed by NH_3. Because the solution to be titrated contains both HCl and NH_4^+, we choose an indicator that permits titration of HCl without beginning to titrate NH_4^+. Problem 9-26 shows that bromocresol green (transition range 3.8–5.4) fulfills this purpose.

Reagents

Standard NaOH and standard HCl: From Experiment 23-6.
Bromocresol green and phenolphthalein indicators: See Experiment 23-6.
Potassium sulfate: 10 g/student.
Selenium-coated boiling chips: Hengar selenium-coated granules are convenient. Alternative catalysts are 0.1 g of Se, 0.2 g of $CuSeO_3$, or a crystal of $CuSO_4$.
Concentrated (98 wt %) H_2SO_4: 25 mL/student.
Unknowns: Unknowns are pure acetanilide, *N*-2-hydroxyethylpiperazine-*N'*-2-ethanesulfonic acid (HEPES buffer, Table 8-2), tris(hydroxymethyl)-aminomethane (tris buffer, Table 8-2), or the *p*-toluenesulfonic acid salts of ammonia, glycine, or nicotinic acid. Each student needs enough unknown to produce 2–3 mmol of NH_3.

Digestion

1. Dry your unknown at 105°C for 45 min and accurately weigh an amount that will produce 2–3 mmol of NH_3. Place the sample in a *dry* 500-mL Kjeldahl flask (Figure 9-7), so that as little as possible sticks to the walls. Add 10 g of K_2SO_4 (to raise the boiling temperature) and three selenium-coated boiling chips. Pour in 25 mL of 98 wt % H_2SO_4, washing down any solid from the walls. (CAUTION: Concentrated H_2SO_4 eats people. If you get any on your skin, flood it immediately with water, followed by soap and water.)

2. In a fume hood, clamp the flask at a 30° angle away from you. Heat gently with a burner until foaming ceases and the solution becomes homogeneous. Continue boiling gently for an additional 30 min.

3. Cool the flask for 30 min *in the air,* and then in an ice bath for 15 min. Slowly, and with constant stirring,

add 50 mL of ice-cold, distilled water. Dissolve any solids that crystallize. Transfer the liquid to the 500-mL distillation flask in Figure 23-3. Wash the Kjeldahl flask five times with 10-mL portions of distilled water, and pour the washings into the distillation flask.

Distillation

1. Set up the apparatus in Figure 23-3 and tighten the connections well. Pipet 50.00 mL of standard 0.1 M HCl into the receiving beaker and clamp the funnel in place below the liquid level.

2. Add 5–10 drops of phenolphthalein indicator to the three-neck flask shown in Figure 23-3 and secure the stoppers. Pour 60 mL of 50 wt % NaOH into the adding funnel and drip this into the distillation flask over a period of 1 min until the indicator turns pink. (CAUTION: 50 wt % NaOH eats people. Flood any spills on your skin with water.) Do not let the last 1 mL through the stopcock, so that gas cannot escape from the flask. Close the stopcock and heat the flask gently until two-thirds of the liquid has distilled.

3. Remove the funnel from the receiving beaker *before* removing the burner from the flask (to avoid sucking distillate back into the condenser). Rinse the funnel well with distilled water and catch the rinses in the beaker. Add 6 drops of bromocresol green indicator solution to the beaker and carefully titrate to the blue end point with standard 0.1 M NaOH. You are looking for the first appearance of light blue color. (Practice titrations with HCl and NaOH beforehand to familiarize yourself with the end point.)

4. Calculate the wt % of nitrogen in the unknown.

23-10 Using a pH Electrode for an Acid–Base Titration

This experiment uses a pH electrode to follow the course of an acid-base titration. You will observe how pH changes slowly during most of the reaction and rapidly near the equivalence point. You will compute the first and second derivatives of the titration curve to locate the end point. From the mass of unknown acid or base and the moles of titrant, you can calculate the molecular mass of the unknown. Section 9-4 provides background for this experiment.

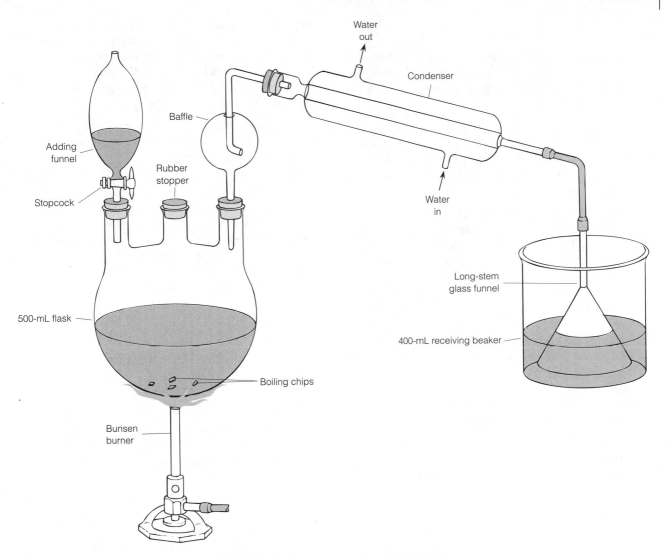

FIGURE 23-3   Apparatus for Kjeldahl distillation.

Reagents

Standard NaOH and standard HCl: From Experiment 23-6.

Bromocresol green and phenolphthalein indicators: See Experiment 23-6.

pH calibration buffers: pH 7 and pH 4. Use commercial standards.

Unknowns: Unknowns should be stored in a desiccator by your instructor.

Suggested acid unknowns: potassium hydrogen phthalate (FM 204.22), 2-(*N*-morpholino)ethanesulfonic acid (MES, Table 8-2, FM 195.24), imidazole hydrochloride (Table 8-2, FM 104.54, hygroscopic), potassium hydrogen iodate (Table 9-4, FM 389.91).

Suggested base unknowns: tris (Table 9-4, FM 121.14), imidazole (FM 68.08), disodium hydrogen phosphate (Na_2HPO_4, FM 141.96), sodium glycinate (may be found in chemical catalogs as glycine, sodium salt hydrate, $H_2NCH_2CO_2Na \cdot xH_2O$, FM 97.05 + x(18.015)). For sodium glycinate, one objective of the titration is to find the number of waters of hydration from the molecular mass.

Procedure

1. Your instructor will recommend a mass of unknown (5–8 mmol) for you to weigh accurately and dissolve in distilled water in a 250-mL volumetric flask. Dilute to the mark and mix well.

2. Following instructions for your particular pH meter, calibrate a meter and glass electrode, using buffers with pH values near 7 and 4. Rinse the electrodes well with distilled water and blot them dry with a tissue before immersing in any new solution.

3. The first titration is rough, so you will know the approximate end point in the next titration. For the rough titration, pipet 25.0 mL of unknown into a 125-mL flask. If you are titrating an unknown acid, add 3 drops of phenolphthalein indicator and titrate with standard 0.1 M NaOH to the pink end point, using a 50-mL buret. If you are titrating an unknown base, add 3 drops of bromocresol green indicator and titrate with standard 0.1 M HCl to the green end point. Add 0.5 mL of titrant at a time so that you can estimate the equivalence volume to within 0.5 mL. Near the end point, the indicator temporarily changes color as titrant is added. If you recognize this, you can slow down the rate of addition and estimate the end point to within a few drops.

4. Now comes the careful titration. Pipet 100.0 mL of unknown solution into a 250-mL beaker containing a magnetic stirring bar. Position the electrode(s) in the liquid so that the stirring bar will not strike the electrode. If a combination electrode is used, the small hole near the bottom on the side must be immersed in the solution. This hole is the salt bridge to the reference electrode. Allow the electrode to equilibrate for 1 min with stirring and record the pH.

5. Add 1 drop of indicator and begin the titration. The equivalence volume will be four times greater than it was in step 3. Add ~1.5-mL aliquots of titrant and record the exact volume, the pH, and the color 30 s after each addition. When you are within 2 mL of the equivalence point, add titrant in 2-drop increments. When you are within 1 mL, add titrant in 1-drop increments. Continue with 1-drop increments until you are 0.5 mL past the equivalence point. The equivalence point has the most rapid change in pH. Add five more 1.5-mL aliquots of titrant and record the pH after each.

Data Analysis

1. Construct a graph of pH versus titrant volume. Mark on your graph where the indicator color change(s) was observed.

2. Following the example in Table 9-3 and Figure 9-4, compute the first derivative (the slope, $\Delta pH/\Delta V$), for each data point within ± 1 mL of the equivalence volume. From your graph, estimate the equivalence volume as shown in Figure 23-4.

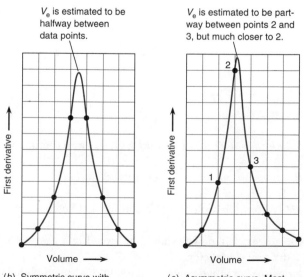

V_e is estimated to be halfway between data points.

V_e is estimated to be partway between points 2 and 3, but much closer to 2.

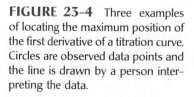

FIGURE 23-4 Three examples of locating the maximum position of the first derivative of a titration curve. Circles are observed data points and the line is drawn by a person interpreting the data.

(a) Symmetric curve with an obvious maximum.

(b) Symmetric curve with no maximum.

(c) Asymmetric curve. Most real data look like this.

3. Following the example in Table 9-3, compute the second derivative (the slope of the slope, $\Delta(\text{slope})/\Delta V$). Prepare a graph like Figure 9-5 and locate the equivalence volume.

4. Go back to your graph from step 1 and mark where the indicator color changes were observed. Compare the indicator end point to the end point estimated from the first and second derivatives.

5. From the equivalence volume and the mass of unknown, calculate the molecular mass of the unknown.

23-11 EDTA Titration of Ca^{2+} and Mg^{2+} in Natural Waters

The most common multivalent metal ions in streams, lakes, and oceans are Ca^{2+} and Mg^{2+}. In this experiment, you will find the total concentration of metal ions that can react with EDTA (Chapter 12), and you will assume that this equals the concentration of Ca^{2+} and Mg^{2+}. In a second experiment, you will analyze Ca^{2+} separately after precipitating $Mg(OH)_2$ with strong base. This experiment (and the last part of Experiment 23-15) could be the basis for a class project to study the composition of natural waters collected in different seasons and different locations.

Reagents

EDTA: $Na_2H_2EDTA \cdot 2H_2O$ (FM 372.24), 0.6 g/student.
Buffer (pH 10): Add 142 mL of 28 wt % aqueous NH_3 to 17.5 g of NH_4Cl and dilute to 250 mL with distilled water.
Eriochrome black T indicator: Dissolve 0.2 g of the solid indicator in 15 mL of triethanolamine plus 5 mL of absolute ethanol.
Hydroxynaphthol blue indicator: 0.5 g/student.
Unknowns: Collect water from streams or lakes or from the ocean. To minimize bacterial growth, plastic jugs should be filled to the top and tightly sealed. Refrigeration is recommended.

Procedure

1. Dry $Na_2H_2EDTA \cdot 2H_2O$ at 80°C for 1 h and cool in a desiccator. Accurately weigh out ~0.6 g and dissolve it with heating in 400 mL of distilled water in a 500-mL volumetric flask. Cool to room temperature, dilute to the mark, and mix well.

2. Pipet a sample of unknown into a 250-mL flask. A 1.000-mL sample of seawater or a 50.00-mL sample of tap water is usually reasonable. If you use 1.000 mL of seawater, add 50 mL of distilled water. To each sample, add 3 mL of pH 10 buffer and 6 drops of Eriochrome black T indicator. Titrate with EDTA from a 50-mL buret and note when the color changes from wine red to blue. You may need to practice finding the end point several times by adding a little tap water and titrating with more EDTA. Save a solution at the end point to use as a color comparison for other titrations.

3. Repeat the titration with three samples to accurately measure the total Ca^{2+} and Mg^{2+} concentration. Perform a blank titration with 50 mL of distilled water and subtract the blank from each result.

4. For the determination of Ca^{2+}, pipet four samples of unknown into clean flasks (adding 50 mL of distilled water if you use 1.000 mL of seawater). Add 30 drops of 50 wt % NaOH to each solution and swirl for 2 min to precipitate $Mg(OH)_2$ (which may not be visible). Add ~0.1 g of solid hydroxynaphthol blue to each flask. (This indicator is used because it remains blue at higher pH than Eriochrome black T does.) Titrate one sample rapidly to find the end point; practice finding it several times, if necessary.

5. Titrate the other three samples carefully. After reaching the blue end point, allow each sample to sit for 5 min with occasional swirling so that any $Ca(OH)_2$ precipitate can redissolve. Then titrate back to the blue end point. (Repeat this procedure if the blue color turns to red upon standing.) Perform a blank titration with 50 mL of distilled water.

6. Calculate the total concentration of Ca^{2+} and Mg^{2+}, as well as the individual concentrations of each ion. Calculate the relative standard deviation of replicate titrations.

23-12 Synthesis and Analysis of Ammonium Decavanadate

The species in a solution of V^{5+} are a delicate function of both pH and concentration.

$$VO_4^{3-} \rightleftharpoons HVO_4^{2-} \rightleftharpoons V_2O_7^{4-} \rightleftharpoons H_2V_2O_7^{2-}$$
In strong base

$$VO_2^+ \rightleftharpoons H_2V_{10}O_{28}^{4-} \rightleftharpoons \begin{bmatrix} V_4O_{12}^{4-} \\ V_3O_9^{3-} \end{bmatrix}$$
In strong acid

$\downarrow$ NH_4^+ alcohol

$$(NH_4)_6V_{10}O_{28} \cdot 6H_2O$$
FM 1 173.7

The decavanadate ion ($V_{10}O_{28}^{6-}$), whose ammonium salt you will isolate, consists of 10 VO_6 octahedra sharing edges with one another (Figure 23-5).

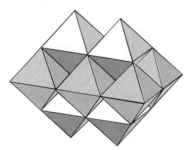

FIGURE 23-5 Structure of $V_{10}O_{28}^{6-}$ anion, consisting of 10 VO_6 octahedra sharing edges with one another. Each octahedron has a vanadium atom at the center and an oxygen atom at each vertex.

After preparing this salt, you will determine the vanadium content by a redox titration and NH_4^+ by the Kjeldahl method.[16] For the redox titration, V^{5+} is first reduced to V^{4+} with sulfurous acid and then titrated with standard permanganate (Color Plate 20).

$$\underset{V^{5+}}{V_{10}O_{28}^{6-}} + 5\underset{\text{Sulfurous acid}}{H_2SO_3} + 26H^+ \longrightarrow$$

$$\underset{V^{4+}}{10VO^{2+}} + 5H_2SO_4 + 13H_2O$$

$$\underset{\text{Blue}}{5VO^{2+}} + \underset{\text{Purple}}{MnO_4^-} + H_2O \longrightarrow \underset{\text{Yellow}}{5VO_2^+} + \underset{\text{Colorless}}{Mn^{2+}} + 2H^+$$

Reagents

Ammonium metavanadate: 3g NH_4VO_3/student.
50 vol % aqueous acetic acid: 4 mL/student.

95 wt % aqueous ethanol: 180 mL/student.
Potassium permanganate: 1.6 g $KMnO_4$/student.
Sodium oxalate: 1 g $Na_2C_2O_4$/student.
0.9 M H_2SO_4: 1 L/student.
1.5 M H_2SO_4: 80 mL/student.
Sodium hydrogen sulfite: 2 g $NaHSO_3$/student.
For Kjeldahl analysis: Standard HCl, standard NaOH, and bromocresol green indicator, as in Experiment 23-6.

Synthesis of $(NH_4)_6V_{10}O_{28} \cdot 6H_2O$

1. Heat 3.0 g of ammonium metavanadate (NH_4VO_3) in 100 mL of distilled water with constant stirring (but not boiling) until most or all of the solid has dissolved. Filter the solution and add 4 mL of 50 vol % aqueous acetic acid with stirring.

2. Add 150 mL of 95 wt % ethanol with stirring and then cool the solution in a refrigerator or ice bath.

3. After maintaining a temperature of 0°–10°C for 15 min, filter the orange product with suction and wash with two 15-mL portions of ice-cold 95 wt % ethanol.

4. Dry the product in the air (protected from dust) for 2 days. Typical yield is 2.0–2.5 g.

Preparation and Standardization of $KMnO_4$

1. Prepare a 0.02 M permanganate solution by dissolving 1.6 g of $KMnO_4$ in 500 mL of distilled water. Boil gently for 1 h, cover, and allow the solution to cool overnight. Filter through a clean, fine sintered-glass funnel, discarding the first 20 mL of filtrate. Store the solution in a clean glass amber bottle. Do not let the solution touch the cap.

2. Dry sodium oxalate ($Na_2C_2O_4$) at 105°C for 1 h, cool in a desiccator, and weigh three ~0.25-g samples into 500-mL flasks or 400-mL beakers. To each, add 250 mL of 0.9 M H_2SO_4 that has been recently boiled and cooled to room temperature. Stir with a thermometer to dissolve the sample and add 90–95% of the theoretical amount of $KMnO_4$ solution needed for the titration. (This amount can be calculated from the mass of $KMnO_4$ used to prepare the permanganate solution. The chemistry is Reaction 5-1.)

3. Leave the solution at room temperature until it is colorless. Then heat it to 55°–60°C and complete the titration by adding $KMnO_4$ until the first pale pink color persists. Proceed slowly near the end, allowing 30 s for each drop to lose its color.

4. As a blank, titrate 250 mL of 0.9 M H_2SO_4 to the same pale pink color. Subtract the blank volume from each titration volume. Compute the average molarity of $KMnO_4$.

Vanadium Analysis

1. Accurately weigh two 0.3-g samples of ammonium decavanadate into 250-mL flasks and dissolve each in 40 mL of 1.5 M H_2SO_4 (with warming, if necessary).

2. In a fume hood, add 50 mL of distilled water and 1 g of $NaHSO_3$ to each sample and dissolve with swirling. After 5 min, boil gently for 15 min to remove excess H_2SO_3 by the reaction $H_2SO_3(aq) \rightarrow SO_2(g)\uparrow + H_2O(l)$.

3. Titrate the warm solution with standard 0.02 M $KMnO_4$ from a 50-mL buret. The end point is when the yellow color of VO_2^+ takes on a dark shade (from excess MnO_4^-) that persists for 15 s.

4. Calculate the average wt % of vanadium in the ammonium decavanadate and compare your result to the theoretical value.

Ammonium Ion Analysis

Ammonium ion is analyzed by a modification of the Kjeldahl procedure. Transfer 0.6 g of accurately weighed ammonium decavanadate to the distillation flask in Figure 23-3 and add 200 mL of distilled water. Then proceed as described in the section labeled "Distillation" in Experiment 23-9. Calculate the wt % of nitrogen in the ammonium decavanadate and compare your result to the theoretical value.

23-13   Iodimetric Titration of Vitamin C

Read Section 15-3 before doing this experiment. Ascorbic acid (vitamin C) is a mild reducing agent that reacts rapidly with triiodide (Box 15-2). In this experiment,[17] you will generate a known excess of I_3^- by the reaction of iodate with iodide (Reaction 15-20), allow the reaction with ascorbic acid to proceed, and then back-titrate the excess I_3^- with thiosulfate (Reaction 15-21 and Color Plate 11).

Reagents

Starch indicator: Make a paste of 5 g of soluble starch and 5 mg of Hg_2I_2 in 50 mL of distilled water. Pour the paste into 500 mL of boiling distilled water and boil until it is clear.
Sodium thiosulfate: 9 g $Na_2S_2O_3 \cdot 5H_2O$/student.
Sodium carbonate: 50 mg Na_2CO_3/student.
Potassium iodate: 1 g KIO_3/student.
Potassium iodide: 12 g KI/student.
0.5 M H_2SO_4: 30 mL/student.
Vitamin C: Dietary supplement containing 100 mg of vitamin C per tablet is suitable. Each student needs six tablets.
0.3 M H_2SO_4: 180 mL/student.

Preparation and Standardization of Thiosulfate Solution

1. Prepare 0.07 M $Na_2S_2O_3$ by dissolving 8.7 g of $Na_2S_2O_3 \cdot 5H_2O$ in 500 mL of freshly boiled, distilled water containing 0.05 g of Na_2CO_3. Store in a tightly capped amber bottle. Sodium thiosulfate is not a primary standard, so the molarity of the solution is not exactly known.

2. Prepare ~0.01 M KIO_3 by accurately weighing ~1 g of solid reagent and dissolving it in a 500-mL volumetric flask. From the mass of KIO_3 (FM 214.00), compute the molarity of the solution.

3. To standardize the thiosulfate solution, pipet 50.00 mL of KIO_3 solution into a 250-mL flask. Add 2 g of solid KI and 10 mL of 0.5 M H_2SO_4. Immediately titrate with thiosulfate until the solution is pale yellow. Then add 2 mL of starch indicator and complete the titration. Repeat the titration twice more with 50.00-mL volumes of KIO_3. From the stoichiometries of Reactions 15-20 and 15-21, compute the average molarity of thiosulfate and the relative standard deviation.

Analysis of Vitamin C

1. Dissolve two tablets in 60 mL of 0.3 M H_2SO_4, using a glass rod to help break the solid. (Some solid binding material will not dissolve.)

2. Add 2 g of solid KI and 50.00 mL of standard KIO_3. Titrate with standard thiosulfate as above. Add 2 mL of starch just before the end point.

3. Repeat the analysis two more times. Find the mean and relative standard deviation for the number of milligrams of vitamin C per tablet.

23-14 Measuring Vitamin C in Fruit Juice by Voltammetry with Standard Addition

This experiment is described in Sections 16-3 and 16-4. The three-electrode cell in Figure 16-8 employs a wax-impregnated graphite working electrode, an Ag|AgCl reference electrode,[18] and a graphite auxiliary electrode. The electrochemistry at the working electrode is the oxidation of ascorbic acid (vitamin C) in Reaction 16-9.

Reagents

Nitric acid: ~3 M HNO_3.

Standard ascorbic acid: This standard should be freshly prepared each lab period. Follow these directions if standard additions are to be made in 1-mL aliquots: Accurately weigh ~0.5 g of ascorbic acid (FM 176.12) into a 100-mL volumetric flask. Add ~50 mL of water and swirl to dissolve. Add ~1 mL of ~3 M HNO_3, swirl, and dilute to 100 mL with water to obtain ~0.03 ascorbic acid standard. The purpose of the nitric acid is to slow the reaction of ascorbic acid with oxygen. If standard additions are to be made in ~100 μL aliquots, prepare the ~0.3 M standard in a 10-mL volumetric flask from 0.5 g ascorbic acid (accurately weighed) and ~100 μL of ~3 M HNO_3.

Unknowns: ~100 mL of fruit juice for each student or group of students working together. Possible unknowns include powdered drink mixes, canned drinks, different brands of orange juice, or vegetable juices.

Preparation of Working Electrode

Most instructors will want to prepare the inexpensive auxiliary and working electrodes in advance and provide them to students. The auxiliary electrode is made of very pure graphite, which is a form of carbon.[19] It is used as received. The working electrode, called a *wax-impregnated graphite electrode,* is made of the same pure graphite sharpened to a point in a pencil sharpener. Soak the sharpened rod for 3 h in molten paraffin at 100°C to fill up pores with the electrically inert wax. Then use forceps to withdraw the rod slowly from the wax bath and allow an electrically insulating layer of paraffin to form on the outside of the rod. Slice the sharp tip with a razor blade to expose a ~$\frac{1}{2}$-mm-diameter section of graphite and gently grind the tip with fine sand paper to leave a small, flat, exposed graphite surface, as shown in the inset of Figure 16-8. Prior to beginning the experiment, clean the graphite tip by rubbing it on filter paper with the electrode perpendicular to the paper. (If the electrode becomes fouled, cut off the tip with a razor to expose fresh surface.)

Procedure

1. Use small quantities of unknown fruit juice to rinse the inside of a pipet and discard the washings. Then pipet 50 mL of unknown juice into a clean, dry 100-mL beaker containing a stirring bar.

2. Clean the tip of the working electrode by rubbing it perpendicular to a piece of filter paper on a flat surface.

3. Set up the apparatus in Figure 16-8 on top of a magnetic stirrer. Use clamps with insulated claws to hold the electrodes in place. (If the claws are bare metal, use tubing or tape to insulate them.) The potentiostat varies the voltage between the working and reference electrodes and measures the current between the working and auxiliary electrodes. The voltammogram (a graph of current versus potential) is recorded on a computer or strip chart.

4. Following instructions for your potentiostat, apply the following sequence of voltages:

a. Condition the working electrode at −1.5 V (versus Ag|AgCl) for 2 min with stirring. This voltage reduces and removes organic material from the tip of the electrode.

b. Change the potential to −0.4 V and continue stirring for 30 s. Dislodge any bubbles of gas from the tip of the electrode by gentle tapping. Discontinue stirring and let the solution become calm for an additional 30 s.

c. Scan the potential from -0.4 V to $+1.2$ V at a rate of $+33$ mV/s while measuring current to record a voltammogram such as the lowest trace in Figure 16-9.

5. *Standard addition:* Using a transfer pipet or a microliter pipet, add 1.00 mL of $\sim$0.03 M standard ascorbic acid (or 100 μL of $\sim$0.3 M standard) into the beaker and turn on the stirrer. Repeat the sequence in step 4 to condition the electrode and record a new voltammogram.

6. Add another increment of standard and record another voltammogram by repeating the sequence in step 4. The objective is to increase the current from that of the unknown by a factor of 1.5 to 3 by using at least four standard additions. The total volume of added standard should not exceed $\sim$10 mL. Depending on your unknown, it may be necessary to adjust the concentration of standard or volume of each increment to achieve this goal.

Data Analysis

1. Following the example in Figure 16-9, extrapolate the baseline from the region where there is no reaction (-0.4 to 0 V) into the region of the current plateau ($\sim$0.8 V). Measure the peak current on each curve with respect to the baseline. The very shallow peak shifts to higher potential as more acid is added.

2. Prepare a graph of Equation 16-12, such as Figure 16-11. The spreadsheet in Figure 16-10 is handy for this purpose.

3. Use the method of least squares to find the equation of the straight line and compute the *x*-intercept (where $y = 0$), which is the concentration of ascorbic acid in the original unknown.

4. Estimate the uncertainty in the *x*-intercept with the following equation:

standard deviation of *x*-intercept =

$$\frac{s_y}{m}\sqrt{\frac{1}{n} + \frac{\bar{y}^2}{m^2\Sigma(x_i - \bar{x})^2}}$$

where s_y is the standard deviation of y (Equation 4-12), m is the slope of the least-squares line (Equation 4-9), n is the number of data points (including the unknown plus the standard additions), $\bar{y}$ is the average value of y for all

data points, and $\bar{x}$ is the average value of x for all data points. The sum inside the square root extends over all data points. If you measured the unknown plus five standard additions, $n = 6$ and the sum includes six terms.

23-15 Potentiometric Halide Titration with Ag$^+$

Mixtures of halides can be titrated with Ag$^+$ by using the apparatus in Figure 5-4 to monitor the course of the reaction as described in Sections 5-5 and 14-1. In this experiment, you will measure the Cl$^-$ and I$^-$ content of a carefully prepared solid unknown. The second procedure can be used to measure Cl$^-$ in natural waters for a class project.

Reagents

Unknowns: Each student receives a vial containing 0.22–0.44 g of KCl plus 0.50–1.00 g of KI (weighed accurately). The object is to determine the quantity of each salt in the mixture.

Buffer (pH 2): (6 mL/student) Titrate 1 M H$_2$SO$_4$ with 1 M NaOH to a pH near 2.0.

Silver nitrate: 1.2 g AgNO$_3$/student.

Procedure

1. Pour your unknown into a 50- or 100-mL beaker. Dissolve the solid in $\sim$20 mL of distilled water and pour the solution into a 100-mL volumetric flask. Rinse the vial and beaker five times with small portions of H$_2$O and transfer the washings to the flask. Dilute to the mark and mix well.

2. Dry 1.2 g of AgNO$_3$ (FM 169.87) at 105°C for 1 h and cool in a desiccator for 30 min with minimal exposure to light. Accurately weigh 1.2 g and dissolve it in a 100-mL volumetric flask.

3. Set up the apparatus shown in Figure 5-4. The 3-cm length of silver wire is soldered to copper wire. The copper wire has a jack that goes to the reference socket of a pH meter. The reference electrode is a glass pH electrode connected to its usual socket on the meter. If a combination pH electrode is employed, the reference jack of the combination electrode is not used. The silver electrode should be taped to the inside of the 100-mL beaker so that the Ag/Cu junction remains dry for the entire titration. The stirring bar should not hit either electrode.

4. Pipet 25.00 mL of unknown into the beaker, add 3 mL of pH 2 buffer, and begin magnetic stirring. Record the initial level of $AgNO_3$ in a 50-mL buret and add ~1 mL of titrant to the beaker. Turn the pH meter to the millivolt scale and record the volume and voltage. It is convenient (but not essential) to set the initial reading to +800 mV by adjusting the meter.

5. Titrate the solution with ~1-mL aliquots until 50 mL of titrant have been added or until you can see two abrupt voltage changes. You need not allow more than 15–30 s for each point. Record the volume and voltage at each point. Make a graph of millivolts versus milliliters to find the approximate positions (± 1 mL) of the two end points.

6. Turn the pH meter to standby, remove the beaker, rinse the electrodes well with distilled water, and blot them dry with a tissue.[20] Clean the beaker and set up the titration apparatus again. (The beaker need not be dry.)

7. Perform an accurate titration using 1-drop aliquots near the end points (and 1-mL aliquots elsewhere). Do not allow more than 30 s per point for equilibration.

8. Prepare a graph of millivolts versus milliliters and locate the end points, as illustrated in Figure 5-5. The I^- end point is taken as the intersection of the two dashed lines in the inset of Figure 5-5. The Cl^- end point is the steepest part of the curve at the second break. You can use the first derivative (see Figure 23-4) to locate the Cl^- end point. Calculate the wt % of KI and wt % of KCl in your solid unknown.

Analyzing Cl⁻ in Streams, Lakes, or Salt Water

A variation of the preceding procedure could make a class project in environmental analysis. For example, you could study changes in streams as a function of the season or recent rainfall. You could study stratification (layering) of water in lakes. The measurement responds to all ions that precipitate with Ag^+, of which Cl^- is, by far, the dominant ion in natural waters.

You can use the electrodes in Figure 5-4, or you can construct the rugged combination electrode shown in Figure 23-6.[21] The indicator electrode in Figure 23-6 is a bare silver wire in contact with analyte solution. The inner (reference) chamber of the combination electrode in Figure 23-6 contains a copper wire dipped into $CuSO_4$ solution. The inner electrode maintains a constant potential because the concentration of Cu^{2+} in the solution is constant. The solution makes electrical contact at the septum, where a piece of thread leaves enough space for solution to slowly drain from the electrode into the external sample solution.

1. Collect water from streams or lakes or from the ocean. To minimize bacterial growth, plastic jugs should be filled to the top and tightly sealed. Refrigeration is recommended.

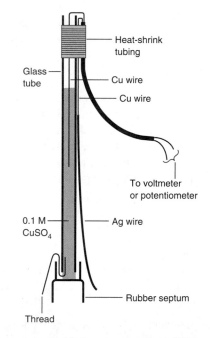

FIGURE 23-6 Combination electrode constructed from 8-mm–outer-diameter glass tubing. Strip the insulation off the ends of a two-wire copper cable. One bare Cu wire extends into the glass tube. A silver wire soldered to the other Cu wire is left outside the glass tube. The assembly is held together at the top with heat-shrink tubing. When ready for use, a cotton thread is inserted into the bottom of the tube and a 14/20 serum cap moistened with 0.1 M $CuSO_4$ is fitted over the end so that part of the thread is inside and part is outside. After adding 0.1 M $CuSO_4$ to the inside of the tube, the electrode is ready for use.

2. Prepare 4 mM AgNO₃ solution with an accurately known concentration, as in step 2 above. One student can prepare enough reagent for five people by using 0.68 g of AgNO₃ in a 1-L volumetric flask.

3. Measure with a graduated cylinder 100 mL of a natural water sample and pour it into a 250-mL beaker. Position the electrode (Figure 23-6) in the beaker so that a magnetic stirring bar will not hit the electrode.

4. Carry out a rough titration by adding 1.5-mL increments of titrant to the unknown and reading the voltage after 30 s to the nearest millivolt. Prepare a graph of voltage versus volume of titrant to locate the end point, which will not be as abrupt as those in Figure 5-5. If necessary, adjust the volume of unknown in future steps so that the end point comes at 20–40 mL. If you need less unknown, make up the difference with distilled water.

5. Carry out a more careful titration with fresh unknown. Add three-quarters of the titrant volume required to reach the equivalence point all at once. Then add ~0.4-mL increments (8 drops from a 50-mL buret) of titrant until you are 5 mL past the equivalence point. Allow 30 s (or longer, if necessary) for the voltage to stabilize after each addition. The end point is the steepest part of the curve, which can be estimated by the method shown in Figure 23-4.

6. Calculate the molarity and parts per million (μg/mL) of Cl^- in the unknown. Use data from several students with the same sample to find the mean and standard deviation of ppm Cl^-.

23-16 Spectrophotometric Determination of Nitrite in Aquarium Water

Background discussion for this experiment[22,23] is found in Box 5-1 and Section 17-4.

Reagents

Sodium nitrite: 1 g NaNO₂ (FM 68.995)/student.
Sodium oxalate: 4 g Na₂C₂O₄ (FM 134.00)/student.
Potassium permanganate: 1 g KMnO₄ (FM 158.03)/student.

0.9 M H₂SO₄: 1 L/student.
Concentrated (96 wt %) H₂SO₄: 20 mL/student.
Color-forming reagent: (Reaction 17-7; 15 mL/student) Mix 1.0 g of sulfanilamide, 0.10 g of *N*-(1-naphthyl)ethylenediamine dihydrochloride, and 10 mL of 85 wt % H₃PO₄ and dilute with distilled water to 100 mL. Store in a dark bottle in the refrigerator.

Procedure

1. Prepare and standardize 0.01 M KMnO₄ solution as described in Experiment 23-12. Reduce the quantities of KMnO₄ and Na₂C₂O₄ by a factor of 2.

2. Prepare 0.018 M NaNO₂ by dissolving 0.62 g of NaNO₂ in 500 mL of distilled water.

3. Follow steps 1–4 in Section 5-3 (page 101) to standardize the NaNO₂.

4. Withdraw ~30 mL of aquarium water and filter it to remove suspended solids. Analyze duplicate 10.00-mL aliquots of freshly drawn aquarium water by the procedure in Section 17-4 (page 377). Use three or four standard points for the calibration curve such that the unknown lies within the range of the standards (Figure 17-9).

23-17 Measuring Ammonia in a Fishtank with an Ion-Selective Electrode

Reagents

Standard ammonia solution: Dissolve ~0.382 g of NH₄Cl (FM 53.491) in a 1-L volumetric flask to obtain a solution containing ~100 μg of nitrogen per milliliter. From the mass of NH₄Cl that you weighed, calculate the nitrogen concentration in μg/mL.

Procedure[22, 23]

1. Using appropriate dilutions of your standard ammonia solution, prepare standards containing 3, 1, 0.3, and 0.1 μg N/mL in 100-mL volumetric flasks.

2. Pour the 100-mL standard solution containing 0.1 μg N/mL into a clean, dry 150-mL beaker. Immerse an ammonia ion-selective electrode and a reference electrode in the solution and begin magnetic stirring. Do not stir the solution so rapidly that air bubbles are drawn into the liquid. Add 1.0 mL of 10 M NaOH (to convert NH_4^+ into NH_3) and record the voltage to the nearest millivolt when the reading stabilizes. Do not allow more than 5 min before reading the voltage, because the temperature of the solution will increase from the stirring and the reading will change.

3. Repeat the same procedure for the other three standards. Prepare a graph of mV versus log[N], where [N] is the concentration of nitrogen in μg/mL. You should observe a reasonable straight line.

4. Measure 100 mL of freshly collected aquarium water in a graduated cylinder. Pour it into a clean, dry 150-mL beaker and repeat the process in step 2.

5. From the least-squares slope and intercept of the calibration curve from step 3, and the potential measured in step 4, compute the concentration of ammonia nitrogen in the aquarium water.

6. Repeat steps 4 and 5 once more with a fresh sample to obtain a duplicate measurement.

23-18 Spectrophotometric Determination of Iron in Dietary Tablets

In this experiment, iron from dietary supplement tablets is dissolved in acid, reduced to Fe^{2+} with hydroquinone, and complexed with o-phenanthroline to form an intensely colored complex (Color Plate 12).[24]

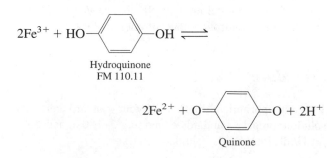

$$2Fe^{3+} + HO-\text{⟨⟩}-OH \rightleftharpoons$$

Hydroquinone
FM 110.11

$$2Fe^{2+} + O=\text{⟨⟩}=O + 2H^+$$

Quinone

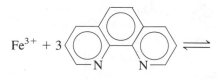

o-Phenanthroline
FM 180.21

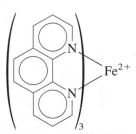

Red complex (λ_{max} = 512 nm)

Reagents

Hydroquinone: (20 mL/student) Freshly prepared solution containing 10 g/L in distilled water. Store in an amber bottle.

Trisodium citrate: (20 mL/student) 25 g/L of Na_3citrate $\cdot$ $2H_2O$ (FM 294.10) in distilled water.

o-Phenanthroline: (25 mL/student) Dissolve 2.5 g in 100 mL of ethanol and add 900 mL of distilled water. Store in an amber bottle.

6 M HCl: (25 mL/student) Dilute 124 mL of 37 wt % HCl up to 250 mL with distilled water.

Standard Fe (40 μg Fe/mL): (35 mL/student) Dissolve 0.281 g of reagent-grade $Fe(NH_4)_2(SO_4)_2 \cdot 6H_2O$ (FM 392.14) in distilled water in a 1-L volumetric flask containing 1 mL of 98 wt % H_2SO_4.

Procedure

1. Place one tablet in a 125-mL flask or 100-mL beaker and boil gently (*in a fume hood*) with 25 mL of 6 M HCl for 15 min. Filter the solution into a 100-mL volumetric flask. Wash the beaker and filter several times with small portions of distilled water to complete a quantitative transfer. Allow the solution to cool, dilute to the mark, and mix well. Dilute 5.00 mL of this solution to 100.0 mL in a fresh volumetric flask. If the label indicates that the tablet contains <15 mg of Fe, use 10.00 mL instead of 5.00 mL.

2. Pipet 10.00 mL of standard Fe solution into a beaker and measure the pH (with pH paper or a glass electrode). Add sodium citrate solution 1 drop at a time until a pH of ~3.5 is reached. Count the drops needed. (It should require ~30 drops.)

3. Pipet a fresh 10.00-mL aliquot of Fe standard into a 100-mL volumetric flask and add the same number of drops of citrate required in step 2. Add 2.00 mL of hydroquinone solution and 3.00 mL of o-phenanthroline solution, dilute to the mark with distilled water, and mix well.

4. Prepare three more solutions from 5.00, 2.00, and 1.00 mL of Fe standard and prepare a blank containing no Fe. Use sodium citrate solution in proportion to the volume of Fe solution. (If 10 mL of Fe requires 30 drops of citrate solution, then 5 mL of Fe requires 15 drops of citrate solution.)

5. Determine how many drops of citrate solution are needed to bring 10.00 mL of the iron tablet solution to pH 3.5. This adjustment will require about 3.5 or 7 mL of citrate, depending on whether 5 or 10 mL of unknown was diluted in the second part of step 1.

6. Transfer 10.00 mL of the solution containing the dissolved tablet to a 100-mL volumetric flask. Add the required amount of citrate solution, 2.00 mL of hydroquinone solution, and 3.00 mL of o phenanthroline solution. Dilute to the mark and mix well.

7. Allow the solutions to stand for at least 15 min. Then measure the absorbance of each at 512 nm in a 1-cm cell. (The color is stable, so all solutions should be prepared and all the absorbances measured at once.) Use distilled water in the reference cuvet and subtract the absorbance of the blank from the absorbance of the Fe standards.

8. Make a graph of absorbance versus micrograms of Fe in the standards. Find the slope and intercept (and standard deviations) by the method of least squares (Section 4-4). Calculate the molarity of Fe(o-phenanthroline)$_3^{2+}$ in each solution and find the average molar absorptivity (ε in Beer's law) from the four absorbances. (All the iron has been converted to the phenanthroline complex.) If you use a Spectronic 20, assume that the pathlength is 1.1 cm for the sake of this calculation.

9. From the calibration curve, find the number of milligrams of Fe in the tablet. Use Equation 4-16 to find the uncertainty in the number of milligrams of Fe.

23-19 Microscale Spectrophotometric Measurement of Iron in Foods by Standard Addition

This microscale experiment[25] uses the same chemistry as that of Experiment 23-18 to measure iron in foods such as broccoli, peas, cauliflower, spinach, beans, and nuts. A possible project is to compare processed (canned or frozen) vegetables to fresh vegetables. Instructions are provided for small volumes, but the experiment can be scaled up to fit available equipment. Section 16-4 describes the method of standard addition. You do not need to read the rest of Chapter 16 to understand Section 16-4.

Reagents

2.0 M HCl: (15 mL/student) Dilute 165 mL of concentrated (37 wt %) HCl up to 1 L.
Hydroquinone: 4 mL/student (Experiment 23-18).
Trisodium citrate dihydrate: 4 g/student.
o-Phenanthroline: 4 mL/student (Experiment 23-18).
Standard Fe (40 μg Fe/mL): 4 mL/student (Experiment 23-18).
6 M HCl: Dilute 500 mL of 37 wt % HCl up to 1 L with distilled water. Store in a bottle and reuse many times for soaking crucibles.

Procedure

1. Fill a clean porcelain crucible with 6 M HCl in the hood and allow it to stand for 1 h to remove traces of iron from previous use. Rinse well with distilled water and dry. After weighing the empty crucible, add 5–6 g of finely chopped food sample and weigh again to obtain the mass of food. (Some foods, like frozen peas, should not be chopped because they will lose their normal liquid content.)

2. This step could require 3 h, during which you can be doing other lab work. Carefully heat the crucible with a Bunsen burner in a hood (Figure 23-7). Use a low flame to *dry* the sample, being careful to avoid spattering.

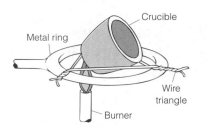

FIGURE 23-7   Positioning a crucible above a burner.

Increase the flame temperature to *char* the sample. Keep the crucible lid and tongs nearby. If the sample bursts into flames, use tongs to place the lid on the crucible to smother the flame. After charring, use the hottest possible flame (bottom of crucible should be red hot) to *ignite* the black solid, converting it to white ash. Continue ignition until all traces of black disappear.

3. After cooling the crucible to room temperature, add 10.00 mL of 2.0 M HCl by pipet and swirl gently for 5 min to dissolve the ash. Filter the mixture through a small filter and collect the filtrate in a vial or small flask. You need to recover >8 mL for the analysis.

4. Weigh 0.71 g of trisodium citrate dihydrate into each of four 10-mL volumetric flasks. Using a 2-mL volumetric pipet or a 1-mL micropipet, add 2.00 mL of ash solution to each flask. Add 4 mL of distilled water and swirl to dissolve the citrate. The solution will have a pH near 3.6. Using a micropipet, add 0.20 mL of hydroquinone solution and 0.30 mL of phenanthroline solution to each flask.

5. Label the volumetric flasks 0 through 3. Add no Fe standard to Flask 0. Using a micropipet, add 0.250 mL of Fe standard to Flask 1. Add 0.500 mL of Fe standard to Flask 2 and 0.750 mL to Flask 3. The four flasks now contain 0, 1, 2, and 3 μg Fe/mL, in addition to Fe from the food. Dilute each to the mark with distilled water, mix well, and allow 15 min to develop full color.

6. Prepare a blank by mixing 0.71 g of trisodium citrate dihydrate, 2.00 mL of 2.0 M HCl, 0.20 mL of hydroquinone solution, and 0.30 mL of phenanthroline solution and diluting to 10 mL. The blank does not require a volumetric flask.

7. Measure the absorbance of each solution at 512 nm in a 1-cm cell with distilled water in the reference cell. Before each measurement, remove all liquid from the cuvet with a Pasteur pipet. Then use ~1 mL of your next solution (delivered with a clean, dry Pasteur pipet)

to wash the cuvet. Remove and discard the washing. Repeat the washing once more with fresh solution and discard the washing. Finally, add your new solution to the cuvet for measuring absorbance.

8. Subtract the absorbance of the blank from each reading and make a graph like that in Figure 16-11 to find the Fe content of the unknown solution. When all solutions have the same final volume (as they do in this experiment), the functions to plot in the standard addition graph are the absorbance on the y-axis and the final concentration of added standard Fe on the x-axis. From the x-intercept of your graph, find wt % Fe in the food.

23-20   Spectrophotometric Analysis of a Mixture: Caffeine and Benzoic Acid in a Soft Drink

In this experiment, we use ultraviolet absorbance (Figure 23-8) to measure two major species in soft drinks.[26] Caffeine is added as a stimulant and sodium benzoate is a preservative.

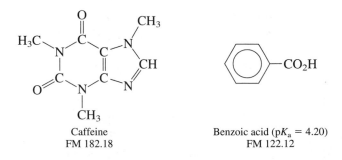

Caffeine
FM 182.18

Benzoic acid (pK_a = 4.20)
FM 122.12

All solutions will contain 0.010 M HCl, so the sodium benzoate is protonated to make benzoic acid. Caffeine has no appreciable basicity, so it is neutral at pH 2.

We restrict ourselves to non-diet soft drinks because the sugar substitute aspartame in diet soda has some ultraviolet absorbance that slightly interferes in the present experiment. We also avoid darkly colored drinks because the colorants have ultraviolet absorbance. Mountain Dew, Mello Yello, and, probably, other lightly colored drinks are suitable for this experiment. There is undoubtedly some ultraviolet absorbance from colorants in these beverages that contributes systematic error to this experiment.

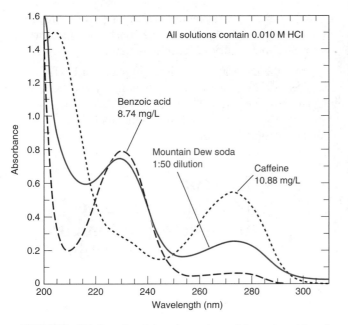

FIGURE 23-8 Ultraviolet absorption of benzoic acid, caffeine, and a 1:50 dilution of Mountain Dew soft drink. All solutions contain 0.010 M HCl.

The procedure we describe includes the construction of calibration curves. The experiment could be shortened by recording just one spectrum of caffeine (20 mg/L) and one of benzoic acid (10 mg/L) and assuming that Beer's law is obeyed. The experiment could be expanded to use high-performance liquid chromatography (HPLC) and/or capillary electrophoresis to obtain independent measurements of caffeine and benzoic acid (and aspartame in diet drinks).[26]

Reagents

Stock solutions: An accurately known solution containing ~100 mg benzoic acid/L in water and another containing ~200 mg caffeine/L should be available.
0.10 M HCl: Dilute 8.2 mL of 37 wt % HCl to 1 L.

Procedure

1. *Calibration standards:* Prepare benzoic acid solutions containing 2, 4, 6, 8, and 10 mg/L in 0.010 M HCl. To prepare a 2 mg/L solution, mix 2.00 mL of benzoic acid standard plus 10.0 mL of 0.10 M HCl in a 100-mL

volumetric flask and dilute to the mark with water. Use 4, 6, 8, and 10 mL of benzoic acid to prepare the other standards. In a similar manner, prepare caffeine standards containing 4, 8, 12, 16, and 20 mg/L in 0.010 M HCl.

2. *Soft drink:* Warm ~20 mL of soft drink in a beaker on a hot plate to expel CO_2 and filter the warm liquid through filter paper to remove any particles. After cooling to room temperature, pipet 4.00 mL into a 100-mL volumetric flask. Add 10.0 mL of 0.10 M HCl and dilute to the mark. Prepare a second sample containing 2.00 mL of soft drink instead of 4.00 mL.

3. *Verifying Beer's law:* Record an ultraviolet baseline from 350 to 210 nm with water in the sample and reference cuvets (1.000 cm pathlength). Record the ultraviolet spectrum of each of the 10 standards with water in the reference cuvet. Note the wavelength of peak absorbance for benzoic acid (λ') and the wavelength for the peak absorbance of caffeine (λ''). Measure the absorbance of each standard at both wavelengths and subtract the baseline absorbance (if your instrument does not do this automatically). Prepare a calibration graph of absorbance versus concentration (M) for each compound at each of the two wavelengths. Each graph should go through 0. The least-squares slope of the graph is the molar absorptivity at that wavelength.

4. *Unknowns:* Measure the ultraviolet absorption spectrum of the 2:100 and 4:100 dilutions of the soft drink. With the absorbance at the wavelengths λ' and λ'', use Equation 18-5 to find the concentrations of benzoic acid and caffeine in the original soft drink. Report results from both dilute solutions.

5. *Synthetic unknown:* If your instructor chooses, measure the spectrum of a synthetic, unknown mixture of benzoic acid and caffeine prepared by the instructor. Use Equation 18-5 to find the concentration of each component in the synthetic unknown.

23-21 Mn^{2+} Standardization by EDTA Titration

Experiments 23-21 to 23-23 illustrate a sequence in which students (1) prepare and standardize a Mn^{2+} solution by EDTA titration and then (2) use this standard in the analysis of Mn in steel by two different instrumental techniques.

Reagents

$MnSO_4 \cdot H_2O$: (1 g/student) This material is not a primary standard.

EDTA: $Na_2H_2EDTA \cdot 2H_2O$; 1 g/student.

0.5 M NH_3/NH_4^+ buffer (pH 9.3): Mix 6.69 g of NH_4Cl (FM 53.491) plus 7.60 g of 28% aqueous NH_3 (FM 17.031) with enough water to give a total volume of 250 mL.

Hydroxylamine hydrochloride: ($NH_3OH^+Cl^-$, FM 69.49) 1 g/student. (CAUTION: Do not breathe dust from $NH_3OH^+Cl^-$; avoid contact with skin and eyes.)

Pyrocatechol violet indicator: Dissolve 0.1 g in 100 mL H_2O.

Procedure

1. *Standard 0.005 M EDTA:* Dry $Na_2H_2EDTA \cdot 2H_2O$ (FM 372.25) at 80°C for 1 h and cool in a desiccator. Accurately weigh out ~0.93 g and dissolve it with heating in 400 mL of distilled water in a 500-mL volumetric flask. Cool to room temperature, dilute to the mark, and mix well.

2. *Mn^{2+} stock solution:* Prepare a solution containing ~1.0 mg Mn/mL (~0.018 M) by dissolving ~0.77 g $MnSO_4 \cdot H_2O$ (FM 169.01) in a clean plastic screw-cap bottle with 250 mL water delivered from a graduated cylinder. Masses and volumes need not be accurate because you will standardize this solution.

3. Rinse a clean 50-mL pipet several times with small volumes of Mn^{2+} stock solution and discard the washings into a chemical waste container. Then pipet 50.00 mL of Mn^{2+} stock solution into a 250-mL volumetric flask. Add ~0.8 g (not accurately weighed) of solid hydroxylamine hydrochloride to the flask, and swirl to dissolve the solid. Add ~400 mL of water and swirl to mix the contents. Dilute to the mark with water, place the cap firmly in place, and invert 20 times to mix the solution. This solution contains ~0.003 6 M Mn^{2+}. The reducing agent, hydroxylamine, maintains manganese in the +2 state.

4. Rinse a 50-mL pipet several times with small volumes of the diluted Mn^{2+} solution from step 3. Pipet 50 mL of the diluted Mn^{2+} solution into a 250-mL Erlenmeyer flask, add 5 mL of pH 9.3 buffer (by graduated cylinder), and add 3–5 drops of pyrocatechol violet indicator. Titrate with standard EDTA from a 50-mL buret and note the end point when the color changes from blue to violet.

5. Repeat step 4 twice more to obtain a total of three replicate titrations. The Erlenmeyer flask must be clean, but it need not be dry for each new titration.

6. From the molarity and volume of standard EDTA required for titration, calculate the molarity and standard deviation of the original ~0.018 M $MnSO_4$ stock solution. Express your answer with an appropriate number of significant digits.

23-22 Measuring Manganese in Steel by Spectrophotometry with Standard Addition

Experiments 23-21 to 23-23 illustrate a sequence in which students (1) prepare and standardize a Mn^{2+} solution and then (2) use this standard in the analysis of Mn in steel by two different instrumental techniques. In this experiment, steel is dissolved in acid and its Mn is oxidized to the violet-colored permanganate ion, whose absorbance is measured with a spectrophotometer:

$$2Mn^{2+} + 5IO_4^- + 3H_2O \longrightarrow 2MnO_4^- + 5IO_3^- + 6H^+$$

Periodate (colorless) (colorless) → Permanganate (violet) Iodate (colorless)

$\lambda_{max} \approx 525$ nm

Steel is an alloy of iron that typically contains ~0.5 wt % Mn plus numerous other elements. When steel is dissolved in hot nitric acid, the iron is converted to Fe(III). Spectrophotometric interference in the measurement of MnO_4^- by Fe(III) is minimized by adding H_3PO_4 to form a nearly colorless complex with Fe(III). Interference by most other colored impurities is eliminated by subtracting the absorbance of a reagent blank from that of the unknown. Appreciable Cr in the steel will interfere with the present procedure. Carbon from the steel is eliminated by oxidation with peroxydisulfate ($S_2O_8^{2-}$):

$$C(s) + 2S_2O_8^{2-} + 2H_2O \longrightarrow CO_2 + 4SO_4^{2-} + 4H^+$$

Reagents

3 M nitric acid: (150 mL/student) Dilute 190 mL of 70 wt % HNO_3 to 1 L with water.

0.05 M nitric acid: (300 mL/student) Dilute 3.2 mL of 70 wt % HNO_3 to 1 L with water.

Ammonium hydrogen sulfite: (0.5 mL/student) 45 wt % NH_4HSO_3 in water.

Potassium periodate (KIO_4): 1.5 g/student.

Unknowns: Steel, ~2 g/student. Analyzed samples are available from Thorn Smith.[12]

Procedure

1. Steel can be used as received or, if it appears to be coated with oil or grease, it should be rinsed with acetone and dried at 110°C for 5 min, and cooled in a desiccator.

2. Weigh duplicate samples of steel to the nearest 0.1 mg into 250-mL beakers. The mass of steel should be chosen to contain ~2–4 mg of Mn. For example, if the steel contains 0.5 wt % Mn, a 0.6-g sample will contain 3 mg of Mn. Your instructor should give you guidance on how much steel to use.

3. Dissolve each steel sample separately in 50 mL of 3 M HNO_3 by gently boiling in the hood, while covered with a watchglass. If undissolved particles remain, stop boiling after 1 h. Replace the HNO_3 as it evaporates.

4. *Standard* Mn^{2+} *(~0.1 mg Mn/mL):* While the steel is dissolving, pipet 10.00 mL of standard Mn^{2+} (~1 mg Mn/mL) from Experiment 23-21 into a 100-mL volumetric flask, dilute to the mark with water, and mix well. You will use this solution in Experiments 23-22 and 23-23. Keep it stoppered, and wrap the stopper with Parafilm or tape to minimize evaporation.

5. Cool the beakers from step 3 for 5 min. Then carefully add ~1.0 g of $(NH_4)_2S_2O_8$ or $K_2S_2O_8$ and boil for 15 min to oxidize carbon to CO_2.

6. If traces of pink color (MnO_4^-) or brown precipitate $(MnO_2(s))$ are observed, add 6 drops of 45 wt % NH_4HSO_3 and boil for 5 min to reduce all manganese to Mn(II):

$$2MnO_4^- + 5HSO_3^- + H^+ \longrightarrow$$
$$2Mn^{2+} + 5SO_4^{2-} + 3H_2O$$
$$MnO_2 + HSO_3^- + H^+ \longrightarrow Mn^{2+} + SO_4^{2-} + H_2O$$

(The purpose of removing colored species at this time is that the solution from step 6 is eventually going to serve as a colorimetric reagent blank.)

7. After cooling the solutions to near room temperature, filter each solution quantitatively through

#41 filter paper into a 250-mL volumetric flask. (If gelatinous precipitate is present, use #42 filter paper.) To complete a "quantitative" transfer, wash the beaker many times with small volumes of hot 0.05 M HNO_3 and pass the washings through the filter to wash liquid from the precipitate into the volumetric flask. Finally, allow the volumetric flasks to cool to room temperature, dilute to the mark with water, and mix well.

8. Transfer ~100 mL of solution from each 250-mL volumetric flask to clean, dry Erlenmeyer flasks and stopper the flasks tightly. Label these solutions A and B and save them for atomic absorption analysis in Experiment 23-23. To help prevent evaporation, it is a good idea to seal around the stoppers with a few layers of Parafilm or tape.

9. Carry out the following spectrophotometric analysis with one of the unknown steel solutions prepared in step 7:

a. Pipet 25.00 mL of liquid from the 250-mL volumetric flask in step 7 into each of three clean, dry 100-mL beakers designated "blank," "unknown," and "standard addition." Add 5 mL of 85 wt % H_3PO_4 (from a graduated cylinder) into each beaker. Then add standard Mn^{2+} (0.1 mg/mL from step 4, delivered by pipet) and solid KIO_4 as follows:

Beaker	Volume of Mn^{2+} (mL)	Mass of KIO_4 (g)
Blank	0	0
Unknown	0	0.4
Standard addition	5.00	0.4

b. Boil the unknown and standard addition beakers gently for 5 min to oxidize Mn^{2+} to MnO_4^-. Continue boiling, if necessary, until the KIO_4 dissolves.

c. Quantitatively transfer the contents of each of the three beakers into 50-mL volumetric flasks. Wash each beaker many times with small portions of water and transfer the water to the corresponding volumetric flask. Dilute each flask to the mark with water and mix well.

d. Fill one 1.000-cm-pathlength cuvet with unknown solution and another cuvet with blank solution. It is always a good idea to rinse the cuvet a few times with small quantities of the solution to be measured and discard the rinses.

e. Measure the absorbance of the unknown at 525 nm with blank solution in the reference cuvet. For best results, measure the absorbance at several wavelengths to locate the maximum absorbance. Use this wavelength for subsequent measurements.

f. Measure the absorbance of the standard addition with the blank solution in the reference cuvet. The absorbance of the standard addition will be ~0.45 absorbance units greater than the absorbance of the unknown (based on adding ~0.50 mg of standard Mn^{2+} to the unknown).

10. Repeat step 9 with the other unknown steel solution from step 7.

Data Analysis

1. From the known concentration of the Mn standard in step 4, calculate the concentration of added Mn in the 50-mL volumetric flask containing the standard addition.

2. All of the Mn^{2+} is converted to MnO_4^- in step 9. From the difference between the absorbance of the standard addition and the unknown, calculate the molar absorptivity of MnO_4^-. Compute the average molar absorptivity from steps 9 and 10.

3. From the absorbance of each unknown and the average molar absorptivity of MnO_4^-, calculate the concentration of MnO_4^- in each 50-mL unknown solution.

4. Calculate the weight percent of Mn in each unknown steel sample and the percent relative range of your results:

% relative range =
$$\frac{100 \times [\text{wt \% in steel 1} - \text{wt \% in steel 2}]}{\text{mean wt \%}}$$

23-23 Measuring Manganese in Steel by Atomic Absorption Using a Calibration Curve

This experiment complements the results of the spectrophotometric analysis in Experiment 23-22. In principle, the spectrophotometric analysis and the atomic absorption analysis should give the same value for the weight percent of Mn in the unknown steel. You will use Mn^{2+} that you standardized in Experiment 23-21 as the standard for the atomic absorption analysis.

Reagents

0.05 M nitric acid: (600 mL/student) Dilute 3.2 mL of 70 wt % HNO_3 to 1 L with water.
Unknown steel: Solutions A and B from step 8 of Experiment 23-22.
Standard manganese: ~0.1 mg Mn/mL from step 4 of Experiment 23-22. This concentration corresponds to ~100 μg/mL = ~100 ppm.

Calibration Curve

1. Prepare standard solutions containing ~1, 2, 3, 4, and 5 ppm Mn (= μg Mn/mL). Use your standard solution containing ~100 ppm Mn from step 4 of Experiment 23-22. Pipet 1.00 mL of the standard into a 100-mL volumetric flask and dilute to 100 mL with 0.05 M HNO_3 to prepare a 1-ppm standard. Similarly, pipet 2.00, 3.00, 4.00, and 5.00 mL into the other flasks and dilute each to 100 mL with 0.05 M HNO_3. Calculate the concentration of Mn in μg/mL in each standard. (The purpose of the HNO_3 is to provide H^+ ions to compete with Mn^{2+} ions for binding sites on the glass surface. Without excess acid, some fraction of metal ions from a dilute solution can be lost to the glass surface. To avoid adding impurity metal ions, we use a dilute solution of the purest available acid.)

2. Measure the atomic absorption signal from each of the five standards in step 1. Use a Mn hollow-cathode lamp and a wavelength of 279.48 nm. Measure each standard three times.

3. Measure the atomic absorption signal from a blank (0.05 M HNO_3). We will use this signal later to estimate the detection limit from Mn. For this purpose, measure a blank seven separate times and compute the mean and standard deviation of the seven measurements.

Measuring the Unknown

1. Immediately after measuring the points on the calibration curve, measure the atomic absorption signal from unknown steel solutions A and B from step 8 of Experiment 23-22. Measure the absorption of each solution three times. (If the signals from A and B do not lie in the calibration range, dilute them as necessary so that they do lie in the calibration range. Dilutions must be done accurately with volumetric pipets and volumetric flasks.)

Data Analysis

1. Make a calibration graph showing the blank plus 5 standards (7 blank readings and $3 \times 5 = 15$ standard readings, for a total of $n = 22$ points). Compute the least-squares slope and intercept from Equations 4-9 and 4-10 and show the least-squares line on the graph. Estimate the standard deviations by using Equations 4-12 through 4-14. Express the equation of the calibration curve in the form $y (\pm s_y) = [m (\pm s_m)]x + [b (\pm s_b)]$, where y is the atomic absorbance signal and x is the concentration of Mn in ppm.

2. Use the mean value of the three readings for each unknown to calculate the concentration of Mn solutions A and B.

3. Calculate the uncertainty in Mn concentration in each unknown from Equation 4-16. Because you have measured each unknown three times, the first term in the radical in Equation 4-16 should be 1/3 instead of 1/1. In Equation 4-16, x is the mean atomic absorption signal for the unknown and there are 22 values of x_i for the points on the standard curve.

4. From the Mn concentrations (and uncertainties) in solutions A and B, calculate the wt % Mn (and its uncertainty) in the two replicate steel samples.

5. The uncertainty in wt % Mn is the standard deviation. Find the 95% confidence interval for wt % Mn in each of the two steel samples that you analyzed. For example, suppose that you find the wt % of Mn in steel to be 0.43_3, with a standard deviation of 0.01_1. (The subscripted digits are not significant but are retained to avoid round-off errors.) The standard deviation was derived from three replicate measurements of one solution of dissolved steel. The equation for confidence interval is $\mu = \bar{x} \pm ts/\sqrt{n}$, where μ is the true mean, $\bar{x}$ is the measured mean, s is the standard deviation, n is the number of measurements (3 in this case), and t is Student's t for 95% confidence and $n - 1 = 2$ degrees of freedom. In Table 4-2, we find $t = 4.303$. Therefore,

the 95% confidence interval is $0.43_3 \pm ts/\sqrt{n} = 0.43_3 \pm (4.303)(0.01_1)/\sqrt{3} = 0.43_3 \pm 0.02_7$.

6. Use the t test (Equation 4-4) to compare the two atomic absorption results to each other. Are they significantly different at the 95% confidence level?

7. Use the mean wt % Mn for the two samples and the pooled standard deviation (Equation 4-5) to estimate a 95% confidence interval around the mean value. Does the mean spectrophotometric value for wt % Mn from Experiment 23-22 lie within the 95% confidence interval for the atomic absorption results? (We cannot use the t test to compare Experiments 23-22 and 23-23 because we do not have enough samples in Experiment 23-22 to find a standard deviation. Otherwise, we would use the t test.)

8. *Detection limit:* The detection limit of an analytical method is the minimum concentration of analyte that can be "reliably" distinguished from 0. Different statistical criteria for the word "reliably" lead to different definitions of detection limit. If you have measured points on a calibration curve, one common definition of detection limit is

$$\text{detection limit (ppm)} = \frac{\bar{y}_B + 3s_B}{b}$$

where $\bar{y}_B$ is the mean atomic absorbance reading for the blank, s_B is the standard deviation for the blank, and b is the least-squares slope of the calibration curve (absorbance/ppm). In this experiment, you measured a blank solution seven times. Use the mean and standard deviation from these seven readings to calculate the detection limit. (If you subtracted the mean value of the blank from each absorbance reading when you constructed the standard curve, then $\bar{y}_B = 0$.)

23-24 Properties of an Ion-Exchange Resin

This experiment[27] explores properties of a cation-exchange resin, which is an organic polymer containing sulfonic acid groups ($-SO_3H$). When a cation, such as Cu^{2+}, flows into the resin, the cation is tightly bound by sulfonate groups, releasing one H^+ for each positive charge bound to the resin (Figure 23-9). Cu^{2+} can be displaced from the

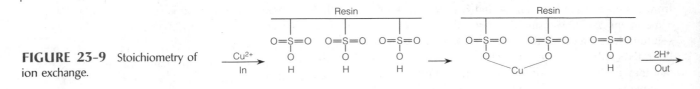

FIGURE 23–9 Stoichiometry of ion exchange.

resin by a large excess of H^+ or by an excess of any other cation for which the resin has some affinity.

First, known quantities of NaCl, $Fe(NO_3)_3$, and NaOH will be passed through the resin in the H^+ form. The H^+ released by each cation will be measured by titration with NaOH.

In the second part of the experiment, you will analyze a sample of impure vanadyl sulfate ($VOSO_4 \cdot 2H_2O$). As supplied commercially, this salt contains $VOSO_4$, H_2SO_4, and H_2O. A solution will be prepared from a known mass of reagent. The VO^{2+} content can be assayed spectrophotometrically, and the total cation (VO^{2+} and H^+) content can be assayed by ion exchange. Together, these measurements enable you to establish the quantities of $VOSO_4$, H_2SO_4, and H_2O in the sample.

Reagents

Bio-Rad Dowex 50W-X2 (100/200 mesh) cation-exchange resin: 1.1 g/student.
0.3 M NaCl: (5–10 mL/student) Accurately weigh ~17.5 g of NaCl (FM 58.44) and dissolve it in a 1-L volumetric flask.
0.1 M Fe(NO₃)₃ · 6H₂O: (5–10 mL/student) Accurately weigh ~35 g of $Fe(NO_3)_3 \cdot 6H_2O$ (FM 349.95) and dissolve it in a 1-L volumetric flask.
VOSO₄: The common grade (usually designated "purified") is used for this experiment. Students can make their own solutions and measure the absorbance at 750 nm, or a bottle of stock solution (25 mL per student) can be supplied. The stock should contain 8 g/L (accurately weighed) and be labeled with the absorbance.
0.02 M NaOH: Each student should prepare an accurate 1/5 dilution of standard 0.1 M NaOH. For example, 50.0 mL of 0.1 M NaOH can be pipetted into a 250-mL volumetric flask and diluted to volume with distilled water.
0.1 M HCl: 50 mL/student.
Phenolphthalein indicator: See Experiment 23-6.

Procedure

1. Prepare a chromatography column from glass tubing that has a 0.7-cm diameter and is 15 cm long by fitting the tubing at the bottom with a cork that has a small hole to serve as the outlet. Place a small ball of glass wool above the cork to retain the resin. Use a glass rod to plug the outlet temporarily. (Alternatively, an inexpensive column such as 0.7 × 15 cm Econo-Column from Bio-Rad Laboratories[28] works well in this experiment.) Fill the column with distilled water, close it off, and test for leaks. Then drain the water until 2 cm remains and close the column again.

2. Make a slurry of 1.1 g of Bio-Rad Dowex 50W-X2 (100/200 mesh) cation-exchange resin in 5 mL of distilled water and pour it into the column. If the resin cannot be poured into the column all at once, allow some to settle, remove the supernatant liquid with a pipet, and pour in the rest of the resin. If the column is stored between laboratory periods, it should be upright and capped, and should contain distilled water above the level of the resin.[29]

3. The general procedure for analysis of a sample is as follows:

a. Generate the H^+-saturated resin by passing ~10 mL of 1 M HCl through the column. Apply the liquid sample to the glass wall so as not to disturb the resin.

b. Wash the column with ~15 mL of distilled water. Use the first few milliliters to wash the glass walls and allow the water to soak into the resin before continuing the washing.[30]

c. Place a clean 125-mL flask under the outlet and pipet the sample onto the column.

d. After reagent has soaked in, wash it through with 10 mL of H_2O, collecting all eluate.

e. Add 3 drops of phenolphthalein to the flask and titrate with standard 0.02 M NaOH.

4. Analyze 2.000-mL aliquots of 0.3 M NaCl and 0.1 M $Fe(NO_3)_3$, following the procedure in step 3. Calculate the theoretical volume of NaOH needed for each titration. If you do not come within 2% of this volume, repeat the analysis.

5. Pass 10.0 mL of your 0.02 M NaOH through the column as directed in step 3 and titrate the eluate. Explain what you observe.

6. Analyze 10.00 mL of $VOSO_4$ solution as described in step 3.

7. Step 6 gives the total cation content ($= VO^{2+} + 2H_2SO_4$) of 10.00 mL of $VOSO_4$ solution. From the absorbance at 750 nm and the molar absorptivity of VO^{2+} ($\varepsilon = 18.0 \ M^{-1} \ cm^{-1}$), calculate the concentration of VO^{2+} in the solution. By difference, calculate the concentration of H_2SO_4 in $VOSO_4$ reagent. From the difference between the mass of $VOSO_4$ reagent that was weighed out and the content of $VOSO_4$ and H_2SO_4, calculate the mass of H_2O in the $VOSO_4$ reagent. Express the composition of the vanadyl sulfate in the form $(VOSO_4)_{1.00}(H_2SO_4)_x(H_2O)_y$.

23-25 Using an Internal Standard in High-Performance Liquid Chromatography

You will receive an unknown solution containing three compounds from the set toluene, biphenyl, naphthalene, 2-naphthol, and 9-fluorenone. The objective is to identify the components of your unknown by high-performance liquid chromatography (HPLC) and to measure the concentration of one component by using an internal standard (Section 20-5).

Reagents

Standards: Toluene (10 mg/mL), biphenyl (2 mg/mL), naphthalene (2 mg/mL), 2-naphthol (2 mg/mL), and 9-fluorenone (2 mg/mL) in methanol in tightly capped bottles. Exact concentrations should be labeled.

Unknowns: (50 mL/student) Mixtures of any three of the standard compounds at concentrations of 0.1–1.5 mg/mL in methanol in a tightly capped bottle.

HPLC solvent: Methanol:water (65:35 vol/vol).

Procedure

0. CAUTION: All organic solvents should be handled in a fume hood. Vessels should be sealed whenever possible to minimize evaporation of methanol. If solvent evaporates from the solutions you use for quantitative analysis, the analysis becomes inaccurate.

1. *Establish a separation:* Inject 10 μL of the unknown onto a C_{18}-silica column and elute it at 1.0 mL/min with 65 vol % methanol in water. Monitor the eluate with an ultraviolet detector set to 254 nm or a shorter wavelength. If the peaks are too intense, inject a smaller volume or dilute the unknown as necessary to bring them on scale. If you make a dilution, you must know the exact dilution factor. If peaks overlap, try reducing the percentage of methanol in the mobile phase (to, say, 60%) or use a lower flow rate. If there is good separation, you could try increasing the flow rate or the percentage of methanol in the mobile phase to reduce the run time. Do not exceed the allowed pressure limit for your chromatograph.

2. *Identification of unknown by co-chromatography:* Prepare five samples for co-chromatography by mixing 1 drop of each pure standard with 20 drops of unknown in a test tube or vial. By observing which peaks grow or whether new peaks appear, you should be able to identify each peak in your unknown. It may be necessary to use ratios other than 1:20 to produce reasonable effects in the chromatogram. Identify the three components of your unknown.

3. *Chromatographic parameters:* For each component of the unknown, measure the retention time, the number of plates, and the plate height from the chromatogram in step 1. Also report the resolution between the two most closely spaced peaks.

4. *Quantitative analysis with an internal standard:* Chose one component (designated X) of the unknown to measure. X should not overlap other components. Select one of the two compounds that is not part of your unknown to be the internal standard (designated S). S must be separated from other peaks in the chromatogram.

a. By trial and error, prepare a mixture of unknown plus S such that the peak areas of X and S are within a factor of 2 of each other. For this purpose, use syringes or pipets to make accurately known mixtures of S and your unknown in tightly capped vials. When you mix standards and unknown in this experiment, we assume that volumes are additive. That is, if you mix 0.100 mL of S with 1.000 mL of unknown, assume that the volume is 1.100 mL. (Additivity is not true for concentrated solutions.)

b. By trial and error, prepare a known mixture of pure X and pure S that has approximately the same ratio of peak areas as the mixture in step a. From the known concentrations and measured peak areas, calculate the response factor in Equation 20-9. Prepare two more solutions in which the concentration ratio [X]/[S] is approximately half as great and twice as great as in the first known mixture and measure the response factor for each solution. The response factors from all three solutions should be the same (within some small experimental uncertainty) if the analysis is within the linear response range.

c. Using the mean response factor from step b and the peak areas of X and S from step a, calculate the concentration of compound X in your unknown.

23-26 Analysis of Sulfur in Coal by Ion Chromatography

When coal is burned, sulfur in the coal is converted to $SO_2(g)$, which is further oxidized to H_2SO_4 in the atmosphere. Rainfall laden with H_2SO_4 is harmful to plant life. Measuring the sulfur content of coal is therefore important to efforts to limit man-made sources of acid rain. This experiment[31] measures sulfur in coal by first heating the coal in the presence of air in a flux (Section 2-10) containing Na_2CO_3 and MgO, which converts the sulfur to Na_2SO_4. The product is dissolved in water, and sulfate ion is measured by ion chromatography. Although you will be given a sample of coal to analyze, consider how you would obtain a representative sample from an entire trainload of coal being delivered to a utility company.

Reagents

Coal: (1 g /student) Coal can be obtained from electric power companies and some heavy industries. Coal is also available as a Standard Reference Material.[32]
Flux: (4 g/student) 67 wt % MgO/33 wt % Na_2CO_3.

6 M HCl: 25 mL/student.
Phenolphthalein indicator: See Experiment 23-6.
Ammonium sulfate: Solid reagent for preparing standards.

Procedure

1. Grind the coal to a fine powder with a mortar and pestle. Mix 1 g of coal (accurately weighed) with ~3 g of flux in a porcelain crucible. Mix thoroughly with a spatula and tap the crucible to pack the powder. Gently cover the mixture with ~1 g of additional flux. Cover the crucible and place it in a muffle furnace. Then turn on the furnace and set the temperature to 800°C and leave the sample in the furnace at 800°C overnight. The reaction is over when all black particles have disappeared. A burner can be used in place of the furnace.

2. After cooling to room temperature, place the crucible into a 150-mL beaker and add 100 mL of distilled water. Heat the beaker on a hot plate to just below boiling for 20 min to dissolve as much solid as possible. Pour the liquid through filter paper in a conical funnel directly into a 250-mL volumetric flask. Wash the crucible and beaker three times with 25-mL portions of distilled water and pour the washings through the filter. Add 5 drops of phenolphthalein indicator to the flask and neutralize with 6 M HCl until the pink color disappears. Dilute to 250 mL and mix well.

3. Pipet 25.00 mL of sulfate solution from the 250-mL volumetric flask in step 2 into a 100-mL volumetric flask and dilute to volume to prepare a fourfold dilution for ion chromatography. Inject a sample of this solution into an ion chromatograph[33] to be sure that the concentration is in a reasonable range for analysis. More or less dilution may be necessary. Prepare the correct dilution for your equipment.

4. Assuming that the coal contains 3 wt % sulfur, calculate the concentration of SO_4^{2-} in the solution in step 3. Using ammonium sulfate and appropriate volumetric glassware, prepare five standards containing 0.1, 0.5, 1.0, 1.5, and 2.0 times the calculated concentration of SO_4^{2-} in the unknown.

5. Analyze all solutions by ion chromatography. Prepare a calibration curve from the standards, plotting peak area versus SO_4^{2-} concentration. Use the least-squares fit to find the concentration of SO_4^{2-} in the unknown. Calculate the wt % of S in the coal.

23-27 Measuring Carbon Monoxide in Automobile Exhaust by Gas Chromatography

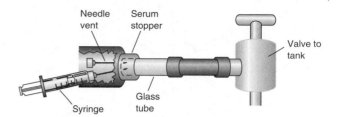

FIGURE 23-10 Collecting a sample of standard gas mixture from a lecture bottle. Remove the vent needle when withdrawing gas into the syringe. Do not open the tank valve so much that the connections pop open. (**CAUTION:** *Handle CO only in a hood.*)

Carbon monoxide is a colorless, odorless, poisonous gas emitted from automobile engines because of incomplete combustion of fuel to CO_2. A well-tuned car with a catalytic converter might emit 0.01 vol % CO, whereas an old "clunker" could emit as much as 15 vol % CO! In this experiment, you will collect samples of auto exhaust and measure the CO content by gas chromatography.[34] A possible class project is to compare different types of cars and different states of maintenance of vehicles. CO emission is greatest within the first few minutes after starting a cold engine. After warm-up, a well-tuned vehicle may emit too little CO to detect with an inexpensive gas chromatograph. CO can be measured as a function of time after start-up.

Chromatography is performed at 50°–60°C with a 2-m-long packed column containing 5A molecular sieves with He carrier gas and thermal conductivity detection. The column should be flushed periodically by disconnecting it from the detector and flowing He through for 30 h. Flushing after 50–100 injections desorbs H_2O and CO_2 from the sieves.

Reagents

CO gas standard: Lecture bottle containing 1 vol % CO in N_2.[35]

Procedure

1. Attach with heavy tape a heat-resistant hose to the exhaust pipe of a car. (**CAUTION:** *Avoid breathing the exhaust.*) Use the free end of the hose to collect exhaust in a heavy-walled, 0.5-L plastic zipper-type bag from the grocery. Flush the bag well with exhaust before sealing it tightly. Allow the contents to come to room temperature for analysis.

2. Inject a 1-mL sample of air into the gas chromatograph by using a gas-tight syringe and adjust the temperature and/or flow rate so that N_2 is eluted within 2 min. You should see peaks for O_2 and N_2.

3. Inject 1.00 mL of standard 1 vol % CO in N_2. One way to obtain gas from a lecture bottle is to attach a

hose to the tank (**IN A HOOD**) with a serum stopper on a glass tube at the end of the hose (Figure 23-10). Place a needle in the serum stopper and slowly bleed gas from the tank to flush the hose. Insert a gas-tight syringe into the serum stopper, remove the vent needle, and slowly withdraw gas into the syringe. Then close the tank to prevent pressure buildup in the tubing. When you inject the standard into the chromatograph, you should see a peak for CO with about three times the retention time of N_2. Adjust the detector attenuation so the CO peak is near full scale. Reinject the standard twice and measure the peak area each time.

4. Inject two 1.00-mL samples of auto exhaust and measure the peak area each time. Compute the vol % of CO in the unknown from its average peak area:

$$\frac{\text{vol \% CO in the unknown}}{\text{vol \% CO in the standard}} =$$

$$\frac{\text{peak area of unknown/detector attenuation}}{\text{peak area of standard/detector attenuation}}$$

23-28 Analysis of Analgesic Tablets by High–Performance Liquid Chromatography

Nonprescription headache medications such as Excedrin or Vanquish contain mixtures of acetaminophen and aspirin for relief and caffeine as a stimulant. This experiment describes conditions for separating and measuring the components by high-performance liquid chromatography (HPLC).[36] Instructions are given for measuring caffeine, but any and all of the components could be measured.

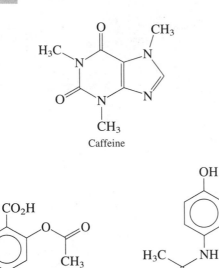

Caffeine

Aspirin
(acetylsalicylic acid)

Acetaminophen

Reagents

HPLC solvent: Organic solvents should be handled in a fume hood. All solvents in this experiment should be HPLC-grade. Mix 110 mL of acetonitrile, 4.0 mL of triethylamine, and 4.0 mL of acetic acid in a 2-L volumetric flask and dilute to the mark with HPLC-grade water. Filter through a 0.45-μm filter and store in a tightly capped amber bottle.

Caffeine stock solution (100 μg/mL): Dissolve 1.000 g of caffeine in 50 mL of HPLC solvent in a 100-mL volumetric flask with gentle heating (in the hood). Cool to room temperature and dilute to the mark with HPLC solvent. Dilute 10.00 mL to 100 mL with HPLC solvent in a volumetric flask to obtain 1 000 μg/mL. Dilute once again to obtain 100 μg/mL.

Acetaminophen and aspirin samples: Prepare two solutions, each containing one of the analytes at a concentration of ~50 μg/mL in HPLC solvent. Filter through 0.22 μm nylon syringe filters and store in capped amber bottles.

Procedure

1. *Caffeine quantitative analysis standards:* Dilute the 100 μg/mL stock solution down to 50, 10, and 5 μg/mL with HPLC solvent. Filter ~3 mL of each solution

through a 0.22-μm syringe filter into a capped vial. Filter ~3 mL of the 100 μg/mL solution into a fourth vial.

2. *Sample preparation:* Grind the analgesic tablet into a fine powder with a clean mortar and pestle. Dissolve ~0.5 g (weighed accurately) in 50 mL of HPLC solvent with gentle heating. Cool to room temperature and dilute to volume with HPLC solvent. Dilute 10.00 mL of this solution to 100 mL with HPLC solvent in a volumetric flask. Filter ~3 mL of the dilute solution through a 0.22-μm syringe filter into a capped vial.

3. *Chromatography conditions:* Use a 2.1-mm-diameter $\times$ 10-cm-long C_{18}-silica column with 5-μm particle size and ultraviolet detection at 254 nm. With a flow rate of 1.5 mL/min, each run is complete in 4 min.

4. *Calibration curve:* Inject 10 μL of each of the caffeine standards (5, 10, 50, and 100 μg/mL) into the HPLC and measure the peak area. Repeat this process twice more and use the average areas from the three runs to construct a calibration curve of area versus concentration. Compute the least-squares slope and intercept for the line through points.

5. *Qualitative analysis:* Record a chromatogram of 10 μL of the analgesic tablet solution. Then mix 2 drops of the tablet solution with 2 drops of 50 μg/mL caffeine solution in a test tube or vial. Inject 10 μL of the mixture into the chromatograph and observe which peak grows. Repeat the process again by adding 50 μg/mL acetaminophen and 50 μg/mL aspirin and identify which peaks in the analgesic are acetaminophen and aspirin.

6. *Quantitative analysis:* Inject 10 μL of the analgesic tablet solution and measure the area of the caffeine peak. Repeat this process twice more and take the average from three injections. Using your calibration graph, determine the concentration of caffeine in the solution and the weight percent of caffeine in the original tablet.

23-29 Anion Content of Drinking Water by Capillary Electrophoresis

Chloride, sulfate, and nitrate are the major anions in fresh water. Fluoride is a minor species added to some drinking water at a level near 1.6 ppm to help prevent tooth decay. In this experiment, you will measure the three major anions by capillary electrophoresis.[37] Possible class projects are to compare water from different sources (homes, lakes, rivers, ocean) and various bottled drinking waters.

You need to read and understand Sections 22-5 and 22-6 prior to doing this experiment. Your equipment should be similar to that in Figure 22-10. Convenient capillary dimensions are a diameter of 75 μm and a length of 40 cm from the inlet to the detector (total length = 50 cm). Because the anions have little ultraviolet absorbance at wavelengths above 200 nm, we add chromate anion (CrO_4^{2-}) to the buffer and use indirect ultraviolet detection at 254 nm. The principle of indirect detection is explained in Figure 22-13.

One other significant condition for a successful separation in this experiment is to reduce the electroosmotic flow rate to permit a better separation of the anions based on their different electrophoretic mobilities. At pH 8, electroosmotic flow is so fast that the anions are swept from the injector to the detector too quickly to be separated well from one another. To reduce the electroosmotic flow, we could lower the pH to protonate some of the $-O^-$ groups on the wall. Alternatively, what we do in this experiment is to add the cationic surfactant tetradecyl(trimethyl)ammonium ion, $CH_3(CH_2)_{13}\overset{+}{N}(CH_3)_3$, which is attracted to the $-O^-$ groups on the wall and partially neutralizes the negative charge of the wall. This cationic surfactant is abbreviated OFM$^+$, for "osmotic flow modifier."

Reagents

Run buffer: 4.6 mM CrO_4^{2-} + 2.5 mM OFM$^+$ at pH 8. Dissolve 1.08 g $Na_2CrO_4 \cdot 4H_2O$ (FM 234.02) plus 25.0 mL of 100 mM tetradecyl(trimethyl)ammonium hydroxide[38] in 800 mL of HPLC-grade H_2O. Place a pH electrode in the solution and add solid boric acid (H_3BO_3) (with magnetic stirring) to reduce the pH to 8.0. Dilute to 1.00 L with HPLC-grade H_2O, mix well, filter through a 0.45-μm filter, and store in the refrigerator in a tightly capped plastic bottle. Degas prior to use.

Quantitative standards: Prepare one stock solution containing 1 000 ppm Cl$^-$, 1 000 ppm NO$_3^-$, and 1 000 ppm SO$_4^{2-}$ by dissolving the following salts in HPLC-grade H_2O: 2.103 g KCl (FM 74.551), 1.631 g KNO$_3$ (FM 101.103), and 1.814 g K$_2$SO$_4$ (FM 174.260). (Concentration refers to the mass of the anion. For example, 1 000 ppm sulfate means 1 000 μg of SO$_4^{2-}$ per mL of solution, not 1 000 μg of K$_2$SO$_4$.) Dilute the stock solution with HPLC-grade H_2O to make standards with concentrations of 2, 5, 10, 20, 50, and 100 ppm of the anions. Store the

solutions in tightly capped plastic bottles.

Standards for qualitative analysis: Prepare four separate 1.00-L solutions, each containing just one anion at a concentration of ~50 ppm. To do this, dissolve ~0.105 g KCl, ~0.082 g KNO$_3$, ~0.091 g K$_2$SO$_4$, or ~0.153 g KF (FM 58.097) in 1.00 L.

Procedure

0. CAUTION: *Electrophoresis uses a dangerously high voltage. Be sure to follow all safety procedures for the instrument.*

1. When preparing a capillary for its first use, wash through 1 M NaOH for 15 min, followed by 0.1 M NaOH for 15 min, followed by run buffer for 15 min. In this experiment, wash the column with run buffer for 1 min between sample injections.

2. *Identify the peaks:* Inject a 50-ppm mixture of Cl$^-$, NO$_3^-$, and SO$_4^{2-}$ by applying a pressure of 0.3 bar for 5 s. Then insert the sample end of the capillary back in run buffer and perform a separation for 5 min at 10 kV with the capillary thermostatted near 25°C. The voltage should be positive at the injector and negative at the detector. The detector should be set at 254 nm. After the run, wash the column with run buffer for 1 min. Mix the 50-ppm anion mixture with an equal volume of 50-ppm Cl$^-$ and run an electropherogram of the mixture. The Cl$^-$ peak should be twice the size it was in the first run. Repeat the procedure with additions of NO$_3^-$, SO$_4^{2-}$, and F$^-$. This process tells you which peak belongs to each anion and where to look for F$^-$ in drinking water.

3. *Calibration curves:* Inject each of the standard mixtures from lowest concentration to highest concentration (2, 5, 10, 20, 50, and 100 ppm) and measure the area of each peak in each run. Repeat the sequence twice more and use the average peak area at each concentration to construct a calibration curve for each anion. Find the least-squares straight line to fit the graph of area versus concentration for each anion.

4. *Unknowns:* Make three replicate injections of each unknown water sample and measure the areas of the peaks. Use the average area of each peak and the calibration curves to find the concentrations of the anions in the water. If you analyze any saltwater samples, they should be diluted by a factor of 100 with HPLC-grade water to bring the anion concentrations down to the range of fresh waters.

23-30 Amino Acid Analysis by Capillary Electrophoresis

In this experiment,[39] you will hydrolyze a protein with HCl to break the protein into its component amino acids (see Table 10-1 and page 227). After adding an internal standard, the amino acids and the internal standard will be derivatized (chemically converted) to a form that absorbs light strongly at 420 nm.

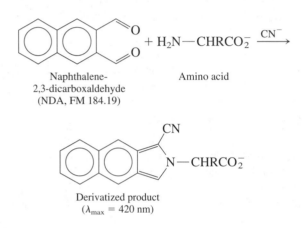

Naphthalene-
2,3-dicarboxaldehyde
(NDA, FM 184.19)

Amino acid

Derivatized product
(λ_{max} = 420 nm)

The mixture will then be separated by capillary electrophoresis, and the quantity of each component will be measured relative to that of the internal standard. The goal is to find the relative number of each amino acid in the protein.

Reagents

Protein: (6 mg/student) Use a pure protein such as lysozyme or cytochrome *c*. The amino acid content should be available in the literature for you to compare with your results.[40]

6 M HCl: Dilute 124 mL of concentrated (37 wt %) HCl up to 250 mL with distilled water in the hood.

0.05 M NaOH: Dissolve 0.50 g of NaOH (FM 40.00) in 250 mL of distilled water.

1.5 M NH₃: Dilute 26 mL of 28 wt % NH_3 up to 250 mL with distilled water in the hood.

10 mM KCN: Dissolve 16 mg of KCN (FM 65.12) in 25 mL of distilled water.

Borate buffer (pH 9.0): To prepare 20 mM buffer, dissolve 0.76 g of sodium tetraborate

($Na_2B_4O_7 \cdot 10H_2O$, FM 381.37) in 70 mL of distilled water. Using a pH electrode, adjust the pH to 9.0 with 0.3 M HCl (a 1:20 dilution of 6 M HCl with distilled water) and dilute to 100 mL with distilled water.

Run buffer (20 mM borate–50 mM sodium dodecyl sulfate, pH 9.0): Prepare this as you prepared borate buffer, but add 1.44 g of $CH_3(CH_2)_{11}OSO_3Na$ (FM 288.38) prior to adjusting the pH with HCl.

Amino acids: Prepare 100 mL of standard solution containing all 15 of the amino acids in Figure 23-11, each at a concentration near 2.5 mM in a solvent of 0.05 M NaOH. Table 10-1 gives molecular masses of amino acids.

α-Aminoadipic acid internal standard: Prepare a 5 mM solution by dissolving 40 mg of $HO_2C(CH_2)_3CH(NH_2)CO_2H$ (FM 161.16) in 50 mL of distilled water.

Naphthalene-2,3-dicarboxaldehyde (NDA): Prepare a 10 mM solution by dissolving 18 mg of NDA in 10 mL of acetonitrile.

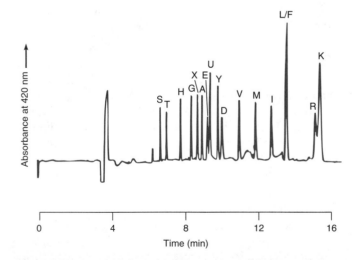

FIGURE 23-11 Electropherogram of standard mixture of NDA-derivatized amino acids plus internal standard, all at equal concentrations. Abbreviations for amino acids are given in Table 10-1. The internal standard, designated X, is α-aminoadipic acid. U is an unidentified peak. Acid hydrolysis converts Q into E and converts N into D, so Q and N are not observed. The analysis of lysine (K) is not reliable because its NDA derivative is unstable. Cysteine (C) and tryptophan (W) are degraded during acid hydrolysis and are not observed. Proline (P) is not a primary amine, so it does not react with NDA to form a detectable product. [From P. L. Weber and D. R. Buck, *J. Chem. Ed.* **1994**, *71*, 609.]

Hydrolysis of the Protein

1. Use a glass tube (17 × 55 mm) sealed at one end. Fit the open end with a rubber septum and evacuate it through a needle. Dissolve 6 mg of protein in 0.5 mL of 6 M HCl in a small vial. Purge the solution with N_2 for 1 min to remove O_2 and immediately transfer the liquid by syringe into the evacuated tube. Add 0.5 mL of fresh HCl to the vial, purge, and transfer again into the tube. Heat the part of the glass tube containing the liquid at 100°–110°C in an oil bath for 18–24 h, preferably behind a shield.

2. After cooling, remove the septum and transfer the liquid to a 25-mL round-bottom flask. Evaporate the solution to dryness with gentle heat and suction from a water aspirator. Use a trap like that in Figure 2-13 whenever you use an aspirator. Rinse the hydrolysis tube with 1 mL of distilled water, add it to the flask, and evaporate to dryness again. Dissolve the residue in 1.0 mL of 0.05 M NaOH and filter it through a 0.45-μm pore size syringe filter. The total concentration of all amino acids in this solution is ~50 mM.

Derivatization

3. Using a micropipet, place 345 μL of 20 mM borate buffer into a small screw-cap vial. Add 10 μL of the standard amino acid solution. Then add 10 μL of 5 mM α-aminoadipic acid (the internal standard), 90 μL of 10 mM KCN, and 75 μL of 10 mM naphthalene-2,3-dicarboxaldehyde. The fluorescent yellow-green color of the amino acid-NDA product should appear within minutes. After 25 min, add 25 μL of 1.5 M NH_3 to react with excess NDA. Wait 15 min before electrophoresis.

4. Place 345 μL of 20 mM borate buffer (pH 9.0) into a small screw-cap vial. Add 10 μL of the hydrolyzed protein solution in 0.05 M NaOH from step 2. Then add 10 μL of 5 mM α-aminoadipic acid (the internal standard), 60 μL of 10 mM KCN, and 50 μL of 10 mM naphthalene-2,3-dicarboxaldehyde. After 25 min, add 25 μL of 1.5 M NH_3 to react with excess NDA. After 15 min, begin electrophoresis. (Precise timing reduces variations between the standard and the unknown.)

Electrophoresis and Analysis of Results

5. (**CAUTION:** *Electrophoresis uses a dangerously high voltage of 20–24 kV. Be sure to follow all safety procedures.*) Conduct the electrophoresis in a 50-μm-inner-diameter uncoated silica capillary with spectrophotometric detection at 420 nm. Precondition the column by injecting 30 μL of 0.1 M NaOH and flushing 15 min later with 30 μL each of distilled water and then run buffer. The column should be reconditioned in the same manner after every 3–4 runs.

6. Inject 3 nL of the derivatized standard amino acid mixture from step 3. The electropherogram should be similar to Figure 23-11. Measure the area of each peak (or the height, if you do not have a computer for measurement of area). Repeat the injection and measure the areas again.

7. Find the quotient

$$\frac{\text{area of amino acid peak}}{\text{area of internal standard peak}}$$

for each amino acid in the standard mixture. (Use peak heights if area is not available.) Prepare a table showing the relative areas for each peak in each injection of standard and find the average quotient for each amino acid from both runs.

8. Inject 3 nL of derivatized, hydrolyzed protein from step 4 and measure the same quotient as measured in step 7. Repeat the injection and find the average quotient from both injections.

9. Find the concentration of each amino acid in the protein hydrolysate by using Equation 23-5:

$$\frac{\text{concentration ratio (X/S) in unknown}}{\text{concentration ratio in standard mixture}} =$$

$$\frac{\text{area ratio (X/S) in unknown}}{\text{area ratio in standard mixture}} \quad (23\text{-}5)$$

in which X refers to an amino acid and S is the internal standard. The numerator on the right side of Equation 23-5 was found in step 8, and the denominator on the right side was found in step 7. (Use the average values from the two sets of runs.) The denominator on the left side is known from the masses of amino acids and internal standard weighed into the standard solution. For each amino acid, you can now solve Equation 23-5 to find the numerator on the left side, which is the quotient.

$$\frac{\text{concentration of amino acid in protein hydrolysate}}{\text{concentration of internal standard in protein hydrolysate}}$$

But you know the concentration of internal standard in the protein hydrolysate from the volume and concentration of internal standard used in step 4. Therefore, you can calculate the concentration of each amino acid in the hydrolysate.

10. Find the mole ratio of amino acids in the protein. If there were no experimental error, you could divide all concentrations by the lowest one. Because

the lowest concentration has a large relative error, pick an amino acid with two or three times the concentration of the least concentrated amino acid. Define this concentration to be exactly 2 or 3. Then compute the molarities of other amino acids relative to the chosen amino acid. Your result is a formula such as $S_{10.6}T_{6.6}H_{0.82}G_{12.5}A_{11.2}E_{5.3}Y_{\equiv 3}D_{19.8}V_{6.1}M_{2.2}I_{5.7}(L + F)_{11.5}$.

Notes and References

1. M. M. Singh, C. McGowan, Z. Szafran, and R. M. Pike, *J. Chem. Ed.* **1998,** *75,* 371; ibid. **2000,** *77,* 625.

2. C. M. Beck, *Anal. Chem.* **1994,** *66,* 224A.

3. S. P. Perone, J. Pesek, C. Stone, and P. Englert, *J. Chem. Ed.* **1998,** *75,* 1444.

4. For a computerized list of experiments published in the *Journal of Chemical Education,* see *J. Chem. Ed.* **1994,** *71,* 450.

5. T. H. Richardson, *J. Chem. Ed.* **1991,** *68,* 310. In a related experiment, students measure the mass of copper in the penny as a function of the year of minting: see R. J. Stolzberg, *J. Chem. Ed.* **1998,** *75,* 1453.

6. D. T. Harvey, *J. Chem. Ed.* **1991,** *68,* 329.

7. R. S. Herrick, L. P. Nestor, and D. A. Benedetto, *J. Chem. Ed.* **1999,** *76,* 1411.

8. P. L. Edmiston and T. R. Williams, *J. Chem. Ed.* **2000,** *77,* 377.

9. A condition for using the *t* test with Equations 4-4 and 4-5 is that the standard deviations for both sets of measurements should not be significantly different from each other. Comparison of standard deviations in Table 23-1 requires the *F* test, which is beyond the scope of this book.

10. Cleaning solution is prepared by your instructor by dissolving 36 g of ammonium peroxydisulfate, $(NH_4)_2S_2O_8$, in a loosely stoppered 2.2-L ("one-gallon") bottle of 98 wt % sulfuric acid. Add more peroxydisulfate every few weeks, as necessary, to maintain the oxidizing capacity of the solution. CAUTION: Cleaning solution eats dirt, grease, clothing, and people. Wear gloves and pour it very carefully. Place used reagent back into the original bottle. It can be reused many times.

11. C. H. Hendrickson and P. R. Robinson, *J. Chem. Ed.* **1979,** *56,* 341.

12. Thorn Smith, Inc., 7755 Narrow Gauge Road, Beulah, MI 49617. Phone: 231-882-4672; e-mail: tslabs@benzie.com. The following analyzed unknowns are available from Thorn Smith

(as of 2000): Al-Mg alloy (for Al, Mg), Al-Zn alloy (for Al, Zn), Sb ore (for Sb), brasses (for Sn, Cu, Pb, Zn), calcium carbonate (for Ca), cast iron (for P, Mn, S, Si, C), cement (for Si, Fe, Al, Ca, Mg, S, ignition loss), Cr ore (for Cr), Cu ore (for Cu), copper oxide (for Cu), ferrous ammonium sulfate (for Fe), Fe ore (for Fe), iron oxide (for Fe), limestone (for Ca, Mg, Si, ignition loss), magnesium sulfate (for Mg), Mn ore (for Mn), Monel metal (for Si, Cu, Ni), nickel silver (for Cu, Ni, Zn), nickel oxide (for Ni), phosphate rock (for P), potassium hydrogen phthalate (for H^+), silver alloys (for Ag, Cu, Zn, Ni), soda ash (for Na_2CO_3), soluble antimony (for Sb), soluble chloride, soluble oxalate, soluble phosphate, soluble sulfate, steels (for C, Mn, P, S, Si, Ni, Cr, Mo), Zn ore (for Zn). Primary standards are also available: potassium hydrogen phthalate (to standardize NaOH), arsenious acid (to standardize I_2), $CaCO_3$ (to standardize EDTA), $Fe(NH_4)_2(SO_4)_2$ (to standardize $Cr_2O_7^{2-}$ or MnO_4^-), $K_2Cr_2O_7$ (to standardize $S_2O_3^{2-}$), $AgNO_3$, Na_2CO_3 (to standardize acid), NaCl (to standardize $AgNO_3$), $Na_2C_2O_4$ (to standardize $KMnO_4$).

13. S. Solomon, A. Lee, and D. Bates, *J. Chem. Ed.* **1993,** *70,* 410.

14. Experiment suggested by Steven D. Brown, University of Delaware. For a related experiment in the titration of liquid household cleaners, see V. T. Lieu and G. E. Kalbus, *J. Chem. Ed.* **1988,** *65,* 184.

15. F. A. Settle and M. Pleva, *Anal. Chem.* **1999,** *71,* 538A.

16. G. G. Long, R. L. Stanfield, and C. F. Hentz, Jr., *J. Chem. Ed.* **1979,** *56,* 195.

17. D. N. Bailey, *J. Chem. Ed.* **1974,** *51,* 488.

18. To prepare your own Ag|AgCl reference electrode, a very simple recipe is given by G. A. East and M. A. del Valle, *J. Chem. Ed.* **2000,** *77,* 97.

19. Graphite rods (6.15 mm diameter × 102 mm long) can be purchased from Alpha/Aesar, catalog number 40766, 99.999 5% carbon, ~$1 per rod for a quantity of 50 rods. Phone: 800-343-0660; http://www.alfa.com.

20. Silver halide adhering to the glass electrode can be removed by soaking the electrode in concentrated sodium thiosulfate solution after the experiment is finished. Thorough cleaning is not necessary between steps 6 and 7. Silver halides can be saved and converted back to pure $AgNO_3$ by the procedure of E. Thall, *J. Chem. Ed.* **1981,** *58,* 561.

21. G. Lisensky and K. Reynolds, *J. Chem. Ed.* **1991,** *68,* 334; R. Ramette, *Chemical Equilibrium* (Reading, MA: Addison-Wesley, 1981), p. 649.

22. K. D. Hughes, *Anal. Chem.* **1993,** *65,* 883A.

23. *Standard Methods for the Examination of Water and Wastewater,* 18th ed. (Washington, DC: American Public Health Association, 1992).

24. R. C. Atkins, *J. Chem. Ed.* **1975,** *52,* 550.

25. Idea based on P. E. Adams, *J. Chem. Ed.* **1995,** *72,* 649.

26. V. L. McDevitt, A. Rodriquez, and K. R. Williams, *J. Chem. Ed.* **1998,** *75,* 625.

27. Part of this experiment is from M. W. Olson and J. M. Crawford, *J. Chem. Ed.* **1975,** *52,* 546.

28. Bio-Rad Laboratories, 2000 Alfred Nobel Drive, Hercules, CA 94547. Phone: 510-741-1000.

29. When the experiment is finished, the resin can be collected, washed with 1 M HCl and water, and reused.

30. Unlike most other chromatography resins, the one used in this experiment retains water when allowed to run "dry." Ordinarily, you must not let liquid fall below the top of the solid phase in a chromatography column.

31. E. Koubek and A. E. Stewart, *J. Chem. Ed.* **1992,** *69,* A146.

32. Standards containing 0.5–5 wt % S are available from NIST Standard Reference Materials Program, Room 204, Building 202, Gaithersburg, MD 20899-0001. Phone: 301-975-6776; e-mail: SRMINFO@enh.nist.gov.

33. For example, chromatography can be done with 25–50 μL of a sample on a 4-mm-diameter $\times$ 250-mm-long Dionex IonPac AS5 analytical column and an AG5 guard column using 2.2 mM Na_2CO_3/2.8 mM $NaHCO_3$ eluent at 2.0 mL/min with ion suppression. SO_4^{2-} is eluted near 6 min and is detected by its conductivity on a full-scale setting of 30 microsiemens. Many combinations of column and eluent are suitable for this analysis.

34. D. Jaffe and S. Herndon, *J. Chem. Ed.* **1995,** *72,* 364.

35. Available, for example, from Scott Specialty Gases, Route 611, Plumsteadville, PA 18949. Phone: 215-766-8861.

36. G. K. Ferguson, *J. Chem. Ed.* **1998,** *75,* 467.

37. S. Demay, A. Martin-Girardequ, and M.-F. Gonnord, *J. Chem. Ed.* **1999,** *76,* 812.

38. 100 mM tetradecyl(trimethyl)ammonium hydroxide (Catalog number WAT049387) is available from Waters Corp., 34 Maple Street, Milford, MA 01757. Phone: 508-478-2000; www.waters.com. The surfactant is sold under the trade name "Osmotic Flow Modifier," abbreviated OFM^+OH^-.

39. P. L. Weber and D. R. Buck, *J. Chem. Ed.* **1994,** *71,* 609.

40. For hen egg white lysozyme, the amino acid content is $S_{10}T_7H_1G_{12}A_{12}(E_2Q_3)Y_3(D_8N_{13})V_6M_2I_6L_8F_3R_{11}K_6C_8W_6P_2$ (R. E. Canfield and A. K. Liu, *J. Biol. Chem.* **1965,** *240,* 2000; D. C. Philips, *Scientific American,* May 1966). For horse cytochrome c, the composition is $S_0T_{10}H_3G_{12}A_6(E_9Q_3)Y_4(D_3N_5)V_3M_2I_6L_6F_4R_2K_{19}C_2W_1P_4$ (E. Margoliash and A. Schejter, *Adv. Protein Chem.* **1966,** *21,* 114). Both proteins are available from Sigma Chemical Co., P.O. Box 14508, St. Louis, MO 63178. Phone: 314-771-5750.

GLOSSARY

abscissa The horizontal (x) axis of a graph.

absolute uncertainty The uncertainty associated with a measurement.

absorbance, A Defined as $A = \log(P_0/P)$, where P_0 is the radiant power of light striking a sample on one side and P is the radiant power emerging from the other side.

absorption Occurs when a substance is taken up *inside* another. See also **adsorption.**

absorption spectrum A graph of absorbance or transmittance of light as a function of wavelength, frequency, or wavenumber.

accuracy How close the measured value is to the "true" value.

acid A substance that increases the concentration of H^+ when added to water.

acid dissociation constant, K_a The equilibrium constant for the reaction of an acid, HA, with H_2O: HA + H_2O $\rightleftharpoons$ A^- + H_3O^+

acid wash Procedure in which glassware is soaked in 3–6 M HCl for >1 h to remove traces of cations adsorbed on the surface of the glass and to replace them with H^+. The soaking in acid is followed by rinsing well with distilled water and soaking in distilled water to remove the acid.

acidic solution One in which the concentration of H^+ (actually the activity of H^+) is greater than the concentration (activity) of OH^-.

activity, $\mathcal{A}$ The value that replaces concentration in a thermodynamically correct equilibrium expression. The activity of X is given by $\mathcal{A}_X = [X]\gamma_X$, where γ_X is the activity coefficient and [X] is the concentration.

activity coefficient, γ The number by which the concentration must be multiplied to give activity.

adsorption Occurs when a substance becomes attached to the *surface* of another substance. See also **absorption.**

adsorption chromatography A technique in which the solute equilibrates between the mobile phase and adsorption sites on the stationary phase.

adsorption indicator Used for precipitation titrations, it becomes attached to a precipitate and changes color when the surface charge of the precipitate changes sign at the equivalence point.

aerosol A suspension of very fine liquid or solid particles in air or gas. Examples include fog and smoke.

affinity chromatography A technique in which a particular solute is retained on a column by a specific interaction with a molecule bound to the stationary phase.

aliquot Portion.

amalgam A solution of anything in mercury.

amine A compound with the general formula RNH_2, R_2NH, or R_3N, where R is any group of atoms.

amino acid One of 20 building blocks of proteins, having the general structure

$$\overset{\displaystyle R}{\underset{\displaystyle {}^+H_3NCHCO_2^-}{|}}$$

where R is a different substituent for each acid.

ammonium ion *The* ammonium ion is NH_4^+. *An* ammonium ion is any ion of the type RNH_3^+, $R_2NH_2^+$, R_3NH^+, or R_4N^+, where R is an organic substituent.

ampere, A One ampere is the current that will produce a force of exactly 2×10^{-7} N/m when that current flows through two "infinitely" long, parallel conductors of negligible cross section, with a spacing of 1 m, in a vacuum. An electric current of one ampere corresponds to a flow of one coulomb of charge per second through a circuit.

533

amperometry The measurement of electric current for analytical purposes.

amphiprotic molecule One that can act as both a proton donor and a proton acceptor. The intermediate species of polyprotic acids are amphiprotic.

analyte The substance being analyzed.

anion A negatively charged ion.

anion exchanger An ion exchanger with positively charged groups covalently attached to the support. It reversibly binds anions.

anode The electrode at which oxidation occurs. In electrophoresis, it is the positively charged electrode.

antibody A protein manufactured by an organism for sequestering and marking foreign molecules for destruction.

antigen A molecule that is foreign to an organism and stimulates the production of antibodies.

antilogarithm The antilogarithm of a is b if $10^a = b$.

aqueous solution One in which water is the solvent.

ashless filter paper Specially treated paper that leaves a negligible residue after ignition. It is used for gravimetric analysis.

atomic absorption spectroscopy A technique in which the absorption of light by gaseous atoms in a flame or furnace is used to measure the concentration of atoms.

atomic emission spectroscopy A technique in which the emission of light by thermally excited atoms in a flame or furnace is used to measure the concentration of atoms.

atomic mass The number of grams of an element containing Avogadro's number of atoms.

atomization The process in which a compound is decomposed into its atoms at high temperature.

autoprotolysis The reaction of a solvent in which two molecules of the same species transfer a proton from one to the other; e.g., $CH_3OH + CH_3OH \rightleftharpoons CH_3OH_2^+ + CH_3O^-$.

auxiliary complexing agent A species, such as ammonia, that is added to a solution to stabilize another species and keep that other species in solution. It binds loosely enough to be displaced by a titrant.

auxiliary electrode The current-carrying partner of the working electrode in electrolysis.

average The sum of measured values divided by the number of values.

back titration One in which an excess of standard reagent is added to react with analyte. Then the excess reagent is titrated with a second standard reagent or with a standard solution of analyte.

background correction In atomic spectroscopy, a means of distinguishing signal due to analyte from signal due to absorption, emission, or scattering by the flame, furnace, plasma, or sample matrix.

base A substance that decreases the concentration of H^+ when added to water.

base hydrolysis constant or **base "dissociation" constant,** K_b The equilibrium constant for the reaction of a base, B, with H_2O: $B + H_2O \rightleftharpoons BH^+ + OH^-$.

basic solution One in which the concentration (actually, the activity) of OH^- is greater than the concentration (activity) of H^+.

Beer's law Relates the absorbance (A) of a sample to the concentration (c) of the absorbing species, the pathlength (b), and the molar absorptivity (ε) of the absorbing species: $A = \varepsilon bc$.

biosensor An analytical device that combines electronic or optical components with a biological component such as an enzyme or antibody for highly specific sensing of one substance.

blank A solution containing all the reagents for an analysis, but no deliberately added analyte.

blank titration One in which a solution containing all the reagents except analyte is titrated. The volume of titrant used in the blank titration should be subtracted from the volume used to titrate unknown.

blocking Occurs when metal binds tightly to a metal ion indicator and is not readily released to the titrant (such as EDTA). A blocked indicator is unsuitable for a titration because no color change is observed at the end point.

Boltzmann distribution The relative population of two states at thermal equilibrium:

$$\frac{N_2}{N_1} = \left(\frac{g_2}{g_1}\right) e^{-(E_2 - E_1)/kT}$$

where N_i is the population of the state, g_i is the degeneracy of the state, E_i is the energy of the state, k is Boltzmann's constant, and T is temperature in kelvins. Degeneracy refers to the number of states with the same energy.

bomb Sealed vessel for conducting high-temperature, high-pressure reactions.

bonded stationary phase In chromatography, a stationary liquid phase covalently attached to the solid support.

buffer A mixture of an acid and its conjugate base. A buffered solution is one that resists changes in pH when acid or base is added.

buoyancy Occurs when an object is weighed in air and the observed mass is less than the true mass because the object has displaced an equal volume of air from the balance pan.

buret A calibrated glass tube with a stopcock at the bottom. Used to deliver known volumes of liquid.

calibration curve A graph showing the value of some property versus concentration of analyte. When the corresponding property of an unknown is measured, its concentration can be determined from the graph.

capacitor current See **charging current.**

capillary electrophoresis Separation of a mixture into its components by using a strong electric field imposed between the two ends of a narrow capillary tube filled with electrolyte solution.

capillary gel electrophoresis Same as capillary electrophoresis, but the tube is filled with a gel through which solutes can migrate.

capillary zone electrophoresis A form of capillary electrophoresis in which ionic solutes are separated because of differences in their electrophoretic mobility.

carboxylate anion The conjugate base (RCO_2^-) of a carboxylic acid.

carboxylic acid A molecule with the general structure RCO_2H, where R is any group of atoms.

cathode The electrode at which reduction occurs. In electrophoresis, it is the negatively charged electrode.

cation A positively charged ion.

cation exchanger An ion exchanger with negatively charged groups covalently attached to the support. It reversibly binds cations.

characteristic The part of a logarithm to the left of the decimal point.

charge balance A statement that the sum of all positive charges in solution equals the magnitude of the sum of all negative charges in solution.

charging current Electric current arising from flow of electrons into or out of an electrode and flow of ions toward or away from the electrode due to electrostatic attraction when the electrode potential is changed. Also called *capacitor current*.

chelating ligand A ligand that binds to a metal through more than one atom.

chemical interference In atomic spectroscopy, any chemical reaction that decreases the efficiency of atomization.

chemiluminescence Emission of light by an excited-state product of a chemical reaction.

chromatogram A graph showing the concentration of solutes emerging from a chromatography column as a function of elution time or volume.

chromatography A technique in which molecules in the mobile phase are separated because of their different affinities for a stationary phase. The greater the affinity for the stationary phase, the longer the molecule is retained.

Clark electrode Amperometric oxygen sensor using a Pt cathode and Ag|AgCl reference electrode.

co-chromatography Simultaneous chromatography of a known compound with an unknown. If a known and an unknown have the same retention time on several different columns, they are probably identical.

cold trapping Injection technique for splitless gas chromatography in which solute is trapped far below its boiling point in a narrow band at the beginning of the chromatography column.

colloid Particles with dimensions in the range 1–100 nm.

combination electrode Consists of a glass pH electrode with a concentric reference electrode built on the same body.

combustion analysis A technique in which a sample is heated in an atmosphere of O_2 to oxidize it to CO_2 and H_2O, which are collected and measured. Modifications permit the simultaneous analysis of N, S, and halogens.

common ion effect Occurs when a salt is dissolved in a solution already containing one of the ions of the salt. The salt is less soluble than it would be in a solution without that ion. An application of Le Châtelier's principle.

complexometric titration One in which the reaction between analyte and titrant involves complex formation.

composite sample A representative sample prepared from a heterogeneous material. If the material consists of distinct regions, the composite is made by taking portions from each region, with relative amounts proportional to the size of each region.

compound electrode An ion-selective electrode consisting of a conventional electrode surrounded by a barrier that is selectively permeable to the analyte of interest. Alternatively, the barrier region might convert external analyte into a different species, to which the inner electrode is sensitive.

concentration An expression of the quantity per unit volume or unit mass of a substance. Common measures of concentration are molarity (mol/L) and molality (mol/kg of solvent).

conditional formation constant Equilibrium constant for the reaction of a metal with a ligand to form a complex at a specified pH. Also called *effective formation constant.*

confidence interval The range of values within which there is a specified probability that the true value will occur.

conjugate acid-base pair An acid and a base that differ through the gain or loss of a single proton.

coprecipitation Occurs when a substance whose solubility is not exceeded precipitates along with one whose solubility is exceeded.

coulomb, C The amount of charge per second that flows past any point in a circuit when the current is one ampere.

coulometry A technique in which analyte participating in an electrochemical reaction is measured by counting the number of electrons required for complete reaction.

cumulative formation constant See **overall formation constant.**

current, I The amount of charge flowing through a circuit per unit time.

cuvet A cell used to hold samples for spectrophotometric measurements.

Debye-Hückel equation Gives the activity coefficient (γ) as a function of ionic strength (μ). The extended Debye-Hückel equation, applicable to ionic strengths up to about 0.1 M, is $\log \gamma = [-0.51z^2\sqrt{\mu}]/[1 + (\alpha\sqrt{\mu}/305)]$, where z is the ionic charge and α is the effective hydrated radius in picometers.

decant To pour liquid off of a solid and leave the solid behind.

deionized water Water that has been passed through a cation exchanger in the H^+ form and an anion exchanger in the OH^- form to remove ions from the solution.

density The mass per unit volume of a substance.

derivatization Chemical alteration of an analyte so that it can be detected conveniently or separated conveniently.

desiccant A drying agent.

desiccator A sealed chamber in which samples can be dried in the presence of a desiccant and/or by vacuum pumping.

detection limit That concentration of an element that gives a signal equal to twice the peak-to-peak noise level of the baseline.

determinate error See **systematic error.**

dialysate Liquid retained by the membrane in dialysis. For a microdialysis probe, dialysate is the liquid exiting the outlet of the probe.

dialysis A technique in which solutions are placed on either side of a semipermeable membrane that allows small molecules, but not large molecules, to cross. The small molecules in the two solutions diffuse across and equilibrate with each other. The large molecules are retained on their original side.

diffraction Bending of light when it is reflected from or passes through a row of closely spaced slits. Interference from adjacent slits spreads the light out into different wavelengths at different diffraction angles.

diffusion current The limiting current determined by the rate at which analyte diffuses to the electrode in polarography.

digestion The process in which a precipitate is left (usually warm) in the presence of mother liquor to promote recrystallization and particle growth. Purer, more easily filterable crystals result. Also, any chemical treatment in which a substance is decomposed into a form suitable for analysis.

direct titration One in which the analyte is treated with titrant, and the volume of titrant required for complete reaction is measured.

displacement titration An EDTA titration procedure in which analyte is treated with excess $MgEDTA^{2-}$ to displace Mg^{2+}: $M^{n+} + MgEDTA^{2-} \rightleftharpoons MEDTA^{n-4} + Mg^{2+}$. The liberated Mg^{2+} is then titrated with EDTA. This procedure is useful if there is no suitable indicator for direct titration of M^{n+}.

disproportionation An oxidation reaction in which an element in one oxidation state produces the same element in higher and lower oxidation states (e.g., $Cu^+ \rightleftharpoons Cu^{2+} + Cu(s)$).

$E^{\circ\prime}$ The effective standard reduction potential at pH 7 (or at some other specified conditions).

electric field The potential difference (volts) between two points divided by the distance (meters) separating the points.

electric potential The potential difference (volts) between two points is the energy (joules) needed to transport one coulomb of positive charge from the negative point to the positive point.

electroactive species Any species that can be oxidized or reduced at an electrode.

electrochemical detector Chromatography detector that measures current when an electroactive solute emerges from the column and passes over a working electrode held at a fixed potential with respect to a reference electrode. Also called *amperometric detector.*

electrode A device at which or through which electrons flow into or out of chemical species involved in a redox reaction.

electrogravimetric analysis Technique in which the mass of an electrolytic deposit is used to quantify the analyte.

electrokinetic injection In capillary electrophoresis, the use of an electric field to inject sample into the capillary. Because different species have different mobilities, the injected sample does not have the same composition as the original sample.

electrolysis Process in which passage of an electric current causes a chemical reaction to occur.

electromagnetic spectrum The spectrum of all electromagnetic radiation (visible light, radio waves, X-rays, etc.).

electron capture detector Gas chromatography detector that is particularly sensitive to compounds with halogen atoms and nitro groups. Makeup gas (N_2 or 5% CH_4 in Ar) added to eluate is ionized by high-energy electrons from ^{63}Ni to liberate extra electrons that create a small, steady current to a positively charged collector. High electron affinity compounds eluted from the chromatography column are detected because they capture electrons and reduce the current at the collector.

electronic balance A balance that uses an electromagnet to balance the load on the pan. The mass of the load is proportional to the current needed to balance it.

electronic transition One in which an electron is transferred from one energy level to another.

electroosmosis Bulk flow of fluid in a capillary tube caused by an electric field. Ions in the diffuse part of the double layer at the wall of the capillary serve as the "pump."

electrophoresis Migration of ions in solution in an electric field. Cations move toward the cathode and anions move toward the anode.

eluate or **effluent** What comes out of a chromatography column.

eluent The solvent applied to the beginning of a chromatography column.

eluent strength Also called *solvent strength.* A measure of the adsorption energy of a solvent on the stationary phase in chromatography. The greater the eluent strength, the more rapidly will the solvent displace solutes from the column.

elution The process of passing a liquid or a gas through a chromatography column.

end point The point in a titration at which there is a sudden change in a physical property, such as indicator color, pH, conductivity, or absorbance. Used as a measure of the equivalence point.

equilibrium constant, K For the reaction $aA + bB \rightleftharpoons cC + dD$, $K = [C]^c[D]^d/[A]^a[B]^b$, where $[X]$ is the concentration of species X. The equilibrium constant is more correctly written in terms of activities, instead of concentrations.

equivalence point The point in a titration at which the quantity of titrant is exactly sufficient for stoichiometric reaction with the analyte.

excited state Any state of an atom or a molecule having more than its minimum possible energy.

extended Debye-Hückel equation See **Debye-Hückel equation.**

extraction The process in which a solute is allowed to equilibrate between two phases, usually for the purpose of separating solutes from one another.

extrapolation The estimation of a value that lies beyond the range of measured data.

Fajans titration A precipitation titration in which the end point is signaled by adsorption of a colored indicator on the precipitate.

faradaic current Current that flows at an electrode and is due to redox reactions.

Faraday constant $9.648\ 534\ 15 \times 10^4$ C/mol of charge.

filtrate Liquid that passes through a filter.

flame ionization detector A gas chromatography detector in which solute is burned in a H_2-air flame to produce CHO^+ ions. The current carried through the flame by these ions is proportional to the concentration of susceptible species in the eluate.

fluorescence Process in which a molecule emits a photon 10^{-8} to 10^{-4} s after absorbing a photon. The molecule goes from an excited singlet state (S_1) to the ground electronic state (S_0).

flux The medium that is melted in sample preparation to decompose and dissolve the sample.

formal concentration or **analytical concentration** The molarity of a substance were it not to change its chemical form upon being dissolved. This measure represents the total number of moles of substance dissolved in a liter of solution, regardless of any reactions that do take place when the solute is dissolved.

formal potential The potential of a half-reaction (relative to a standard hydrogen electrode) when the formal concentrations of reactants and products are unity. Any other conditions (such as pH, ionic strength, and concentrations of ligands) must also be specified.

formation constant or **stability constant** The equilibrium constant for the reaction of a metal with a ligand to form a metal-ligand complex.

formula mass, FM The mass containing one mole of the indicated chemical formula of a substance. For example, the formula mass of $CuSO_4 \cdot 5H_2O$ is the sum of the masses of copper, sulfate, and five water molecules.

frequency The number of oscillations of a wave per second.

fusion Process in which an otherwise insoluble substance is dissolved in a molten salt such as Na_2CO_3, Na_2O_2, or KOH. Once the substance has dissolved, the melt is cooled, dissolved in aqueous solution, and analyzed.

galvanic cell One that produces electricity by means of a spontaneous chemical reaction.

gas chromatography A form of chromatography in which the mobile phase is a gas.

gathering A process in which a trace constituent of a solution is intentionally coprecipitated with a major constituent.

Gaussian distribution Theoretical bell-shaped distribution of measurements when all error is random. The center of the curve is the mean (μ) and the width is characterized by the standard deviation (σ). A normalized Gaussian distribution, also called the *normal error curve*, has an area of unity and is given by

$$y = \left(\frac{1}{\sigma\sqrt{2\pi}}\right)e^{-(x-\mu)^2/2\sigma^2}$$

glass electrode One that has a thin glass membrane across which a pH-dependent voltage develops. The voltage (and hence pH) is measured by a pair of reference electrodes on either side of the membrane.

gradient elution Chromatography in which the composition of the mobile phase is progressively changed to increase the eluent strength of the solvent.

graphite furnace A hollow graphite rod that can be heated electrically to about 2 500 K and is used to decompose and atomize a sample for atomic spectroscopy.

grating Either a reflective or a transmitting surface with closely spaced lines to disperse light into its component wavelengths.

gravimetric analysis Any analytical method that relies on measuring the mass of a substance (such as a precipitate) to complete the analysis.

ground state The state of an atom or a molecule with the minimum possible energy.

guard column In HPLC, a short column that is packed with the same material as the main column and is placed between the injector and the main column. The guard column removes impurities that might irreversibly bind to and degrade the main column. Also called *precolumn.*

half-wave potential The potential at the midpoint of the rise in current in a polarographic wave.

Henderson-Hasselbalch equation A logarithmic, rearranged form of the acid dissociation equilibrium equation: $pH = pK_a + \log([A^-]/[HA])$.

hertz, Hz The unit of frequency, s^{-1}.

heterogeneous sample A sample that is not uniform throughout.

high-performance liquid chromatography (HPLC) A chromatographic technique using very small stationary phase particles and high pressure to force solvent through the column.

hollow-cathode lamp One that emits sharp atomic lines characteristic of the element from which the cathode is made.

homogeneous precipitation A technique in which a precipitating agent is generated slowly by a reaction in homogeneous solution, effecting a slow crystallization instead of a rapid precipitation of product.

homogeneous sample A sample that has the same composition everywhere.

hydrated radius The effective radius of an ion or a molecule plus its associated water molecules in solution.

hydrodynamic injection In capillary electrophoresis, the use of a pressure difference between the two ends of the capillary to inject sample into the capillary.

hydrolysis Reaction with water. The reaction $B + H_2O \rightleftharpoons BH^+ + OH^-$ is called hydrolysis of a base.

hydronium ion, H_3O^+ What we really mean when we write $H^+(aq)$.

hydrophilic substance One that is soluble in water or attracts water to its surface.

hydrophobic substance One that is insoluble in water or repels water from its surface.

hygroscopic substance One that readily picks up water from the atmosphere.

ignition The heating to high temperature of a gravimetric precipitate to convert it to a known, constant composition that can be weighed.

immunoassay An analytical measurement based on the use of antibodies.

indeterminate error See **random error.**

indicator A compound with a physical property (usually color) that changes abruptly near the equivalence point of a chemical reaction.

indicator electrode One whose potential depends on the concentration (actually the activity) of one or more species in contact with the electrode.

indicator error The difference between the indicator end point of a titration and the true equivalence point.

indirect detection Chromatographic detection based on the *absence* of signal when analyte passes through the detector. For example, in ion chromatography a light-absorbing ionic species can be added to the eluent. Analyte that does not absorb light replaces an equivalent amount of light-absorbing eluent when the analyte emerges from the column, thereby decreasing the observed absorbance.

indirect titration One that is used when the analyte cannot be directly titrated. For example, analyte A may be precipitated with excess reagent R to make AR. The product is filtered and the excess R washed away. Then AR is dissolved in a new solution, and R is titrated.

inductively coupled plasma A high-temperature plasma that derives its energy from an oscillating radio frequency field. It is used to atomize a sample for atomic emission spectroscopy.

intercept For a straight line whose equation is $y = mx + b$, b is the intercept. It is the value of y when $x = 0$.

interference The effect when the presence of one substance changes the signal in the analysis of another substance.

internal standard A known quantity of a compound added to a solution containing an unknown quantity of analyte. The concentration of analyte is then measured relative to that of the internal standard.

interpolation The estimation of the value of a quantity that lies between two known values

ion chromatography A high-performance version of ion-exchange chromatography, with a key modification that removes eluent ions before detecting analyte ions.

ion-exchange chromatography A technique in which solute ions are retained by oppositely charged sites in the stationary phase.

ionic atmosphere The region of solution around an ion or a charged particle. It contains an excess of oppositely charged ions.

ionic strength, μ Given by $\mu = \frac{1}{2}\Sigma\, c_i z_i^2$, where c_i is the concentration of the ith ion in solution and z_i is the charge on that ion. The sum extends over all ions in solution, including the ions whose activity coefficients are being calculated.

ion-selective electrode One whose potential is selectively dependent on the concentration of one particular ion in solution.

ionization interference In atomic spectroscopy, a lowering of signal intensity as a result of ionization of analyte atoms.

isocratic elution Chromatography using a single solvent for the mobile phase.

isoelectric focusing A technique in which a sample containing polyprotic molecules is subjected to a strong electric field in a medium with a pH gradient. Each species migrates until it reaches the region of its isoelectric pH. In that region, the molecule has no net charge, ceases to migrate, and remains focused in a narrow band.

isosbestic point A wavelength at which the absorption spectra of two species cross each other. The appearance of an isosbestic point during a chemical reaction is evidence that there are only two species present at a constant total concentration.

joule, J SI unit of energy. One joule is expended when a force of 1 N acts over a distance of 1 m. This energy is equivalent to that required to raise 102 g (about $\frac{1}{4}$ pound) by 1 m.

junction potential An electric potential difference that exists at the junction between two different electrolyte solutions or substances. It arises in solutions as a result of unequal rates of diffusion of different ions.

Karl Fischer titration A sensitive technique to measure water based on its reaction with an amine, I_2, SO_2, and an alcohol.

Kjeldahl nitrogen analysis Procedure for the analysis of nitrogen in organic compounds. The compound is digested with boiling H_2SO_4 to convert nitrogen to NH_4^+, which is treated with base and distilled as NH_3 into a standard acid solution. The number of moles of acid consumed equals the number of moles of NH_3 liberated from the compound.

Le Châtelier's principle States that if a system at equilibrium is disturbed, the direction in which it proceeds back to equilibrium is such that the disturbance is partially offset.

least squares See **method of least squares.**

Lewis acid One that can form a chemical bond by sharing a pair of electrons donated by another species.

Lewis base One that can form a chemical bond by sharing a pair of its electrons with another species.

ligand An atom or a group attached to a central atom in a molecule. The term is often used to mean any group attached to anything else of interest.

linear voltage ramp A constantly increasing voltage applied to an electrode in voltammetry.

liquid-based ion-selective electrode One that has a hydrophobic membrane separating an inner reference electrode from the analyte solution. The membrane contains a liquid ion exchanger that transports analyte across the membrane to produce an electric potential difference between the two sides of the membrane.

liquid chromatography Chromatography with a liquid mobile phase.

liter, L A volume that is exactly 1 000 cm^3.

logarithm The base 10 logarithm of n is a if $10^a = n$ (which means log $n = a$).

luminescence Any emission of light by a molecule.

mantissa The part of a logarithm to the right of the decimal point.

masking The process of adding a chemical substance (a masking agent) to a sample to prevent one or more components from interfering in a chemical analysis.

masking agent A reagent that selectively reacts with one (or more) component(s) of a solution to prevent the component(s) from interfering in a chemical analysis.

mass balance A statement that the sum of the moles of any element in all of its forms in a solution must equal the moles of that element delivered to the solution.

mass spectrometer A device that ionizes (and often fragments) gaseous molecules and then separates them by mass and measures the quantity of each ion.

matrix The medium containing analyte. For many analyses, it is important that standards be prepared in the same matrix as the unknown.

matrix modifier Substance added to sample for atomic spectroscopy to retard analyte evaporation until the matrix is fully charred.

mean The average of a set of all results.

mechanical balance A balance having a beam that pivots on a fulcrum. Standard masses are used to measure the mass of an unknown.

mediator A substance added to an electrochemical cell to carry electrons between the electrode and the species of interest.

meniscus The curved surface of a liquid.

metal ion indicator A compound whose color changes when it binds to a metal ion.

method of least squares Process of fitting a mathematical function to a set of measured points by minimizing the sum of the squares of the distances from the points to the curve.

micellar electrokinetic capillary chromatography A form of capillary electrophoresis in which a micelle-forming surfactant is present. Migration times of solutes depend on the fraction of time spent in the micelles.

micelle An aggregate of molecules with ionic headgroups and long, nonpolar tails. The inside of the micelle resembles hydrocarbon solvent, whereas the outside interacts strongly with aqueous solution.

microdialysis probe A needle with a semipermeable (dialysis) membrane near the tip. When inserted into a living organism, small molecules from the organism enter the probe. Fluid from the probe is analyzed by chromatography, electrochemistry, etc.

microporous particles Chromatographic stationary phase consisting of porous particles 3–10 μm in diameter, with high efficiency and high capacity for solute.

mobile phase In chromatography, the phase that travels through the column.

Mohr titration Argentometric titration in which the end point is signaled by the formation of red $Ag_2CrO_4(s)$.

molality The number of moles of solute per kilogram of solvent.

molar absorptivity, ε, or **extinction coefficient** The constant of proportionality in Beer's law: $A = \varepsilon bc$, where A is absorbance, b is pathlength, and c is the molarity of the absorbing species.

molarity, M The number of moles of solute per liter of solution.

molecular exclusion chromatography or **gel filtration** or **gel permeation chromatography** or **molecular sieve chromatography** A technique in which the stationary phase has a porous structure into which small molecules can enter but large molecules cannot. Molecules are separated by size, with larger molecules moving faster than smaller ones.

molecular mass The number of grams of a substance that contains Avogadro's number of molecules.

molecular orbital Describes the distribution of an electron in a molecule.

molecular sieve A solid particle, with pores the size of small molecules, used to separate small molecules in gas chromatography and used as a desiccant.

monochromatic light Light of a single color; or electromagnetic radiation, in general, of a single wavelength.

monochromator A device (usually a prism, grating, or filter) for selecting a single wavelength of light.

monodentate ligand One that binds to a metal ion through only one atom.

mortar and pestle A mortar is a hard ceramic or steel vessel in which a solid sample is ground with a hard pestle.

mother liquor The solution from which a substance has crystallized.

multidentate ligand One that binds to a metal ion through more than one atom.

nebulization The process of breaking a liquid into a mist of fine droplets.

Nernst equation For the half-reaction $aA + ne^- \rightleftharpoons bB$, the Nernst equation giving the half-cell potential, E, is

$$E = E° - \left(\frac{0.059\ 16}{n}\right) \log\left(\frac{[B]^b}{[A]^a}\right) \text{(at 25°C)}$$

where $E°$ is the standard reduction potential that applies when $[A] = [B] = 1$ M (actually when the activities of A and B are unity).

neutralization The process in which a stoichiometric equivalent of acid (or base) is added to a base (or acid).

nonpolar compound A compound with little or no charge separation (little or no dipole moment), such as a compound containing only hydrogen and carbon.

normal-phase chromatography A chromatographic separation utilizing a polar stationary phase and a less polar mobile phase.

nucleation The process whereby molecules in solution come together randomly to form small aggregates.

on-column injection Used in gas chromatography to place a thermally unstable sample directly onto the column without excessive heating in the injection port. Solute is condensed at the beginning of the column by cold trapping. Then the temperature is raised to intiate chromatography.

open tubular column In chromatography, a capillary column whose walls are coated with stationary phase.

ordinate The vertical (y) axis of a graph.

overall formation constant Equilibrium constant for the addition of several ligands (X) to a metal (M), such as $M + 3X \rightleftharpoons MX_3$. Also called *cumulative formation constant.*

oxidant See **oxidizing agent.**

oxidation A loss of electrons or an increase of the oxidation state.

oxidizing agent or **oxidant** A substance that takes electrons in a chemical reaction.

packed column A chromatography column filled with stationary phase particles.

parallax error The apparent displacement of an object when the observer changes position. Occurs when the scale of an instrument is viewed from a position that is not perpendicular to the scale; consequently, the apparent reading is not the true reading.

partition chromatography A technique in which separation is achieved by equilibration of solute between two phases.

parts per billion, ppb An expression of concentration that refers to nanograms (10^{-9} g) of solute per gram of solution.

parts per million, ppm An expression of concentration that refers to micrograms (10^{-6} g) of solute per gram of solution.

peptization Occurs when washing some ionic precipitates with distilled water washes away the ions that neutralize the charges of individual particles and help to hold the particles together. The washed particles then disintegrate and pass through the filter with the wash liquid.

pH Defined as pH $= -\log \mathcal{A}_{H^+}$, where $\mathcal{A}_{H^+}$ is the activity of H^+. For all applications in this book, we use the approximation pH $= -\log [H^+]$.

phosphorescence Relatively long-lived emission of light occurring 10^{-4} to 10^2 s after absorbing a photon. The molecule goes from an excited triplet state (T_1) to the ground electronic state (S_0).

photochemistry Chemistry initiated by absorption of a photon.

photodiode array An array of semiconductor diodes used to detect light. The array is normally used to detect light that has been spread out into its component wavelengths by a polychromator. Only one narrow band of wavelengths falls on each element of the array.

photomultiplier tube A sensitive detector in which a cathode emits electrons when bombarded by light. The electrons then strike a series of dynodes (plates held at progressively more positive potential) and more electrons are released at each successive dynode. More than 10^6 electrons might reach the anode for each photon striking the cathode.

photon A "particle" of light with energy $h\nu$, where h is Planck's constant and ν is the frequency of the light.

pipet A glass tube calibrated to deliver a fixed or variable volume of liquid.

pK The negative logarithm of an equilibrium constant: $pK = -\log K$.

plate height The length of a chromatography column divided by the number of theoretical plates in the column.

polar compound A compound with a significant amount of charge separated on different atoms. Polar compounds tend to attract other polar compounds fairly strongly by electrostatic forces and, sometimes, hydrogen bonding.

polarogram The graph of current versus potential of the working electrode in a polarography experiment.

polarographic wave The S-shaped increase in current during a redox reaction in polarography.

polarography Voltammetry conducted at a dropping mercury electrode.

polychromator A device that spreads light out into its component wavelengths and directs each small band of wavelengths to a different region.

polyprotic acids and bases Compounds that can donate or accept more than one proton.

potentiometry An analytical method in which an electric potential difference (a voltage) of a cell is measured.

potentiostat A device that maintains a constant voltage or a desired voltage profile between two electrodes.

power The amount of energy per unit time (J/s) being expended.

precipitant A reagent that causes a precipitate to form.

precision The reproducibility of a measured result.

preconcentration The process of concentrating trace components of a mixture prior to their analysis.

primary standard A reagent that is pure enough and stable enough to be used directly after weighing. The entire mass is considered to be pure reagent.

purge and trap Method for removing volatile analytes from liquids or solids by bubbling (*purging*) a gas through the sample. The gas is then passed through an absorbent trap that retains analytes. To inject the analyte into a gas chromatograph, the trap is heated and purged in the reverse direction into the chromatograph. Analyte is concentrated at the beginning of the chromatography column by cold trapping.

pyrolysis Thermal decomposition of a substance.

Q **test** A statistical test used to decide whether a discrepant datum should be rejected or retained from a set of measurements.

qualitative analysis The process of determining the identity of the constituents of a substance.

quantitative analysis The process of measuring how much of a constituent is present in a substance.

quantitative transfer To transfer the entire contents from one vessel to another. This is usually accomplished by rinsing the first vessel several times with fresh liquid and pouring each rinse into the receiving vessel.

random error or **indeterminate error** A type of error that can be either positive or negative and cannot be eliminated. It arises from the ultimate limitations on a physical measurement.

random sample Bulk sample constructed by taking portions of the entire lot at random.

reagent Refers to any substance used in a chemical reaction.

redox indicator A compound used to find the end point of a redox titration because it changes color when it goes from its oxidized to its reduced state.

redox reaction A chemical reaction involving transfer of electrons from one species to another.

redox titration A titration based on an oxidation-reduction reaction between analyte and titrant.

reducing agent or **reductant** A substance that donates electrons in a chemical reaction.

reductant See **reducing agent.**

reduction A gain of electrons or a lowering of the oxidation state.

reference electrode One that maintains a constant potential against which the potential of another half-cell can be measured.

refractive index detector Liquid chromatography detector that measures the change in the refractive index of eluate as solutes emerge from a column.

relative uncertainty The uncertainty of a quantity divided by the value of the quantity. It is usually expressed as a percentage of the measured quantity.

residual current The current measured in a voltammetry experiment in the absence of analyte.

resistance A measure of the retarding force opposing the flow of electric current.

resolution How close two bands in a spectrum or a chromatogram can be to each other and still be seen as two peaks. In chromatography, it is defined as the difference in retention times of adjacent peaks divided by their width at the base.

response factor An empirical factor measuring the relative response of a detector to an analyte and a standard.

retention time The time, measured from injection, needed for a solute to be eluted from a chromatography column.

retention volume The volume of mobile phase required to elute a particular solute from a chromatography column.

reversed-phase chromatography A technique in which the stationary phase is less polar than the mobile phase.

rotational transition Occurs when a molecule changes its rotational energy by rotating faster or slower.

rubber policeman A glass rod with a flattened piece of rubber on the tip. The rubber is used to scrape solid particles from glass surfaces in gravimetric analysis.

salt An ionic solid.

salt bridge A conducting ionic medium in contact with two electrolyte solutions. It allows ions to flow without immediately contaminating one electrolyte solution with the other.

sample cleanup Removal of portions of the sample that do not contain analyte and may interfere with analysis.

sample preparation The process of converting a heterogeneous bulk sample into a homogeneous laboratory sample prior to analysis. Sample preparation also refers to concentrating a very dilute sample or eliminating species that interfere with the analysis.

sampled current polarography A polarography experiment that uses a staircase waveform with a current measurement for a short time at the end of each voltage step.

sampling The process of obtaining a small bulk sample whose composition is representative of the lot to be analyzed.

saturated calomel electrode (S.C.E.) A calomel electrode saturated with KCl. The electrode half-reaction is $Hg_2Cl_2(s) + 2e^- \rightleftharpoons 2Hg(l) + 2Cl^-$.

saturated solution One containing the maximum amount of a compound that can dissolve at equilibrium.

segregated material A material containing distinct regions of different composition.

selectivity coefficient With respect to an ion-selective electrode, a measure of the relative response of the electrode to two different ions. In ion-exchange chromatography, this is the equilibrium constant for displacement of one ion by another from the resin.

semipermeable membrane A barrier across which some molecules, but not others, can cross.

septum A disk, usually made of silicone rubber, covering the injection port of a gas chromatograph. The sample is injected by syringe through the septum.

SI units The units of an international system of measurement based on the meter, kilogram, second, ampere, kelvin, candela, mole, radian, and steradian.

significant figure The number of significant figures in a quantity is the minimum number of figures needed to express the quantity in scientific notation. In experimental data, the first uncertain figure is the last significant figure.

silver-silver chloride electrode A common reference electrode containing a silver wire coated with AgCl paste and dipped in a solution saturated with AgCl and (usually) KCl. The half-reaction is $AgCl(s) + e^- \rightleftharpoons Ag(s) + Cl^-$.

singlet state Electronic state in which all electron spins are paired.

slope For a straight line whose equation is $y = mx + b$, the value of m is the slope. It is the ratio $\Delta y/\Delta x$ for any segment of the line.

slurry A suspension of a solid in a liquid.

solid-phase extraction A procedure that uses a chromatographic solid phase in a short column to separate one or more analytes from a mixture. Solutes adsorbed on the column can be eluted with a small volume of solvent.

solid-phase microextraction Extraction of analyte from liquid or gas into a coated fiber from a syringe needle. After extraction of some of the analyte for a fixed period of time, the fiber is withdrawn into the needle and the needle is injected into a chromatograph. The fiber is extended inside the injection port and adsorbed solutes are desorbed by heating (for gas chromatography) or by a strong solvent (for liquid chromatography).

solid-state ion-selective electrode An ion-selective electrode with a membrane made of an inorganic crystal. Ion-exchange equilibrium between the solution and the crystal surface creates a voltage across the crystal that depends on the concentration of analyte on each side.

solubility product, K_{sp} The equilibrium constant for the dissolution of a solid salt to give its ions in solution. For the reaction $M_mN_n(s) \rightleftharpoons mM^{n+} + nN^{m-}$, $K_{sp} = [M^{n+}]^m[N^{m-}]^n$.

solute A minor component of a solution.

solvent The major constituent of a solution.

solvent trapping Splitless gas chromatography injection technique in which solvent is condensed below its boiling point at the start of the column. Solutes dissolve in this narrow band of solvent. Chromatography is initiated by raising the column temperature.

species Chemists refer to any molecule or ion as a "species." The term is both singular and plural.

spectral interference In atomic spectroscopy, any physical process that affects the intensity of radiation at the analytical wavelength. Created by substances that absorb, scatter, or emit radiation of the analytical wavelength.

spectrophotometer A device that measures absorption of electromagnetic radiation. It includes a source, a wavelength selector (monochromator), and a detector.

spectrophotometric titration One in which absorption of electromagnetic radiation is used to monitor the progress of the chemical reaction.

spectrophotometry Any method using electromagnetic radiation to measure chemical concentrations.

split injection Used in open tubular gas chromatography to inject a small fraction ($\sim$1%) of the sample while the rest is purged out of the injection port to waste.

splitless injection Used in open tubular gas chromatography for trace analysis and quantitative analysis. Nearly the entire vaporized sample from the injection port is directed onto the column and focused in a narrow band by solvent trapping or cold trapping. The column is then warmed to initiate chromatography.

square wave voltammetry A form of voltammetry with a waveform consisting of a square wave superimposed on a staircase wave. The technique is faster and more sensitive than other forms of voltammetry.

stability constant See **formation constant.**

staircase voltage ramp Voltage applied to an electrode in a series of discrete steps.

standard addition Analytical procedure in which known quantities of analyte are added to the unknown; the accompanying increase in the signal allows the operator to determine how much analyte was in the original unknown.

standard curve A graph showing the response of an analytical technique to known quantities of analyte.

standard deviation Measures how closely data are clustered about the mean value. For a finite set of data, the standard deviation, s, is computed from the formula

$$s = \sqrt{\frac{\Sigma(x_i - \bar{x})^2}{n-1}}$$

where n is the number of results, x_i is an individual result, and $\bar{x}$ is the mean result.

standard hydrogen electrode (S.H.E.) or **normal hydrogen electrode (N.H.E.)** One in which $H_2(g)$ bubbles over a catalytic Pt surface in contact with aqueous H^+. The activities of H_2 and H^+ are both unity in the hypothetical standard electrode. The concentrations are approximately 1 bar for H_2 and 1 M for H^+. The cell reaction is $H^+ + e^- \rightleftharpoons \frac{1}{2}H_2(g)$.

standard reduction potential, $E°$ The voltage that would be measured when the hypothetical cell containing the desired half-reaction (with all species at unit activity—approximately 1 M, 1 bar, or pure solid or liquid) is connected to a standard hydrogen electrode.

standard solution A solution whose composition is known by virtue of the way it was made from a reagent of known purity or by virtue of its reaction with a known quantity of a standard reagent.

standardization The process whereby the concentration of a reagent is determined by reaction with a known quantity of a second reagent.

stationary phase In chromatography, the phase that does not move.

stepwise formation constant Equilibrium constant for each successive addition of a ligand (X) to a metal (M), such as $M + X \rightleftharpoons MX$ or $MX_2 + X \rightleftharpoons MX_3$.

stripping analysis A very sensitive polarographic technique in which analyte is concentrated from a dilute solution by reduction into a single drop of Hg. It is then measured polarographically during anodic redissolution.

strong acids and bases Those that are completely dissociated (to H^+ or OH^-) in water.

strong electrolyte One that is mostly dissociated into ions when dissolved.

Student's t A statistical tool used to express confidence intervals and to compare results from different experiments.

supernatant liquid The liquid above a solid that has settled to the bottom of a vessel.

supersaturated solution One that contains more dissolved solute than would be present at equilibrium.

supporting electrolyte An unreactive salt added in high concentration to most solutions for voltammetry experiments. The supporting electrolyte carries most of the charging current (but none of the faradaic current).

suppressed-ion chromatography Separation of ions by using an ion-exchange column and subsequently removing the ionic eluent.

surfactant A molecule, such as soap, with an ionic or polar headgroup and a long, nonpolar tail. Surfactants aggregate in aqueous solution to form micelles.

systematic error or **determinate error** A type of error due to procedural or instrumental factors that cause a measurement to be reproducibly too large or too small. If you repeat the measurement in the same way, you will observe the same error in the same direction (probably with some random error superimposed on the systematic error). Systematic error can, in principle, be discovered and corrected.

t **test** Used to decide whether the results of two experiments are within experimental uncertainty of each other.

tare The mass of an empty vessel used for weighing. When a balance is "tared," the empty mass is set to 0.

temperature programming Technique in gas chromatography in which the temperature of the column is raised during the separation of a mixture.

theoretical plate An imaginary segment of a chromatography column in which one equilibration of solute occurs between the stationary and mobile phases. The number of theoretical plates on a column with Gaussian bandshapes is $N = 5.55\ t_r^2/w_{1/2}^2 = t_r^2/\sigma^2$, where t_r is the retention time of a band, $w_{1/2}$ is the width at half-height, and σ is the standard deviation of the band.

thermal conductivity detector A device that detects bands eluted from a gas chromatography column by measuring changes in the thermal conductivity of the gas stream.

thermogravimetric analysis A technique in which the mass of a substance is measured as the substance is heated. Changes in mass reflect decomposition of the substance, often to well-defined products.

titrant The substance added to the analyte in a titration.

titration A procedure in which one substance (titrant) is carefully added to another (analyte) until complete reaction has occurred. The quantity of titrant required for complete reaction tells how much analyte is present.

titration error The difference between the observed end point and the true equivalence point in a titration.

trace analysis The measurement of extremely low levels of analyte.

transmittance, *T* Defined as $T = P/P_0$, where P_0 is the radiant power of light striking the sample on one side and P is the radiant power of light emerging from the other side of the sample.

triplet state Electronic state in which two electrons have parallel spins.

ultraviolet detector Liquid chromatography detector that measures ultraviolet absorbance of solutes emerging from the column.

van Deemter equation Equation relating the plate height in chromatography to the flow rate. The smaller the plate height, the narrower the chromatographic peaks.

vibrational transition Occurs when a molecule changes its vibrational energy by oscillating with greater or lesser amplitude.

volatile Easily vaporized.

Volhard titration Titration of Ag^+ with SCN^-, in which the formation of the red complex $Fe(SCN)^{2+}$ marks the end point.

volt, V Unit of electric potential. If the potential difference between two points is one volt, it requires one joule of energy to move one coulomb of charge between the two points.

voltammetry Analytical method in which the relationship between current and voltage is observed during an electrochemical reaction.

voltammogram The graph of current versus working electrode potential in a voltammetry experiment.

volume percent Expression of composition defined as (volume of solute/volume of solution) $\times$ 100.

volumetric analysis A technique in which the volume of material needed to react with analyte is measured.

volumetric flask One having a tall, thin neck with a calibration mark. When the liquid level is at the calibration mark, the flask contains its specified volume of liquid.

watt, W The SI unit of power, equal to an energy flow of one joule per second. When an electric current of one ampere flows through a potential difference of one volt, the power is one watt.

wavelength, λ The distance between consecutive crests of a wave.

wavenumber, $\tilde{\nu}$ The reciprocal of the wavelength, λ.

weak acids and bases Those whose dissociation constants are not large.

weak electrolyte One that partially dissociates into ions when it dissolves.

weight percent Expression of composition defined as (mass of solute/mass of solution) $\times$ 100.

working electrode The one at which the reaction of interest occurs in an electrolysis experiment.

*y***-intercept** See **intercept.**

zwitterion A molecule with a positive charge localized in one position and a negative charge localized at another position.

SOLUBILITY PRODUCTS[†]

Formula	K_{sp}	Formula	K_{sp}
Azides: L = N$_3^-$		Hg$_2$L$_2$	1.2×10^{-18}
CuL	4.9×10^{-9}	TlL	1.8×10^{-4}
AgL	2.8×10^{-9}	PbL$_2$	1.7×10^{-5}
Hg$_2$L$_2$	7.1×10^{-10}	**Chromates: L = CrO$_4^{2-}$**	
TlL	2.2×10^{-4}	BaL	2.1×10^{-10}
PdL$_2(\alpha)$	2.7×10^{-9}	CuL	3.6×10^{-6}
Bromates: L = BrO$_3^-$		Ag$_2$L	1.2×10^{-12}
BaL · H$_2$O	7.8×10^{-6}	Hg$_2$L	2.0×10^{-9}
AgL	5.5×10^{-5}	Tl$_2$L	9.8×10^{-13}
TlL	1.7×10^{-4}	**Cobalticyanides: L = Co(CN)$_6^{3-}$**	
PbL$_2$	7.9×10^{-6}	Ag$_3$L	3.9×10^{-26}
Bromides: L = Br$^-$		(Hg$_2$)$_3$L$_2$	1.9×10^{-37}
CuL	5×10^{-9}	**Cyanides: L = CN$^-$**	
AgL	5.0×10^{-13}	AgL	2.2×10^{-16}
Hg$_2$L$_2$	5.6×10^{-23}	Hg$_2$L$_2$	5×10^{-40}
TlL	3.6×10^{-6}	ZnL$_2$	3×10^{-16}
HgL$_2$	1.3×10^{-19}	**Ferrocyanides: L = Fe(CN)$_6^{4-}$**	
PbL$_2$	2.1×10^{-6}	Ag$_4$L	8.5×10^{-45}
Carbonates: L = CO$_3^{2-}$		Zn$_2$L	2.1×10^{-16}
MgL	3.5×10^{-8}	Cd$_2$L	4.2×10^{-18}
CaL (calcite)	4.5×10^{-9}	Pb$_2$L	9.5×10^{-19}
CaL (aragonite)	6.0×10^{-9}	**Fluorides: L = F$^-$**	
SrL	9.3×10^{-10}	LiL	1.7×10^{-3}
BaL	5.0×10^{-9}	MgL$_2$	6.6×10^{-9}
Y$_2$L$_3$	2.5×10^{-31}	CaL$_2$	3.9×10^{-11}
La$_2$L$_3$	4.0×10^{-34}	SrL$_2$	2.9×10^{-9}
MnL	5.0×10^{-10}	BaL$_2$	1.7×10^{-6}
FeL	2.1×10^{-11}	ThL$_4$	5×10^{-29}
CoL	1.0×10^{-10}	PbL$_2$	3.6×10^{-8}
NiL	1.3×10^{-7}	**Hydroxides: L = OH$^-$**	
CuL	2.3×10^{-10}	MgL$_2$ (amorphous)	6×10^{-10}
Ag$_2$L	8.1×10^{-12}	MgL$_2$ (brucite crystal)	7.1×10^{-12}
Hg$_2$L	8.9×10^{-17}	CaL$_2$	6.5×10^{-6}
ZnL	1.0×10^{-10}	BaL$_2$ · 8H$_2$O	3×10^{-4}
CdL	1.8×10^{-14}	YL$_3$	6×10^{-24}
PbL	7.4×10^{-14}	LaL$_3$	2×10^{-21}
Chlorides: L = Cl$^-$		CeL$_3$	6×10^{-22}
CuL	1.9×10^{-7}	UO$_2$ ($\rightleftharpoons$ U^{4+} + 4OH$^-$)	6×10^{-57}
AgL	1.8×10^{-10}	UO$_2$L$_2$ ($\rightleftharpoons$ UO$_2^{2+}$ + 2OH$^-$)	4×10^{-23}
		MnL$_2$	1.6×10^{-13}
		FeL$_2$	7.9×10^{-16}
		CoL$_2$	1.3×10^{-15}

[†]Solubility products generally apply at 25°C and zero ionic strength. The designations α, β, or γ after some formulas refer to particular crystalline forms.

Formula	K_{sp}	Formula	K_{sp}
NiL_2	6×10^{-16}	$UO_2L \, (\rightleftharpoons UO_2^{2+} + C_2O_4^{2-})$	2.2×10^{-9}
CuL_2	4.8×10^{-20}	Phosphates: $L = PO_4^{3-}$	
VL_3	4.0×10^{-35}	$MgHL \cdot 3H_2O \, (\rightleftharpoons Mg^{2+} + HL^{2-})$	1.7×10^{-6}
CrL_3	1.6×10^{-30}	$CaHL \cdot 2H_2O \, (\rightleftharpoons Ca^{2+} + HL^{2-})$	2.6×10^{-7}
FeL_3	1.6×10^{-39}	$SrHL \, (\rightleftharpoons Sr^{2+} + HL^{2-})$	1.2×10^{-7}
CoL_3	3×10^{-45}	$BaHL \, (\rightleftharpoons Ba^{2+} + HL^{2-})$	4.0×10^{-8}
$VOL_2 \, (\rightleftharpoons VO^{2+} + 2OH^-)$	3×10^{-24}	LaL	3.7×10^{-23}
PdL_2	3×10^{-29}	$Fe_3L_2 \cdot 8H_2O$	1×10^{-36}
ZnL_2 (amorphous)	3.0×10^{-16}	$FeL \cdot 2H_2O$	4×10^{-27}
$CdL_2 \, (\beta)$	4.5×10^{-15}	$(VO)_3L_2 \, (\rightleftharpoons 3VO^{2+} + 2L^{3-})$	8×10^{-26}
HgO (red) $(\rightleftharpoons Hg^{2+} + 2OH^-)$	3.6×10^{-26}	Ag_3L	2.8×10^{-18}
$Cu_2O \, (\rightleftharpoons 2Cu^+ + 2OH^-)$	4×10^{-30}	$Hg_2HL \, (\rightleftharpoons Hg_2^{2+} + HL^{2-})$	4.0×10^{-13}
$Ag_2O \, (\rightleftharpoons 2Ag^+ + 2OH^-)$	3.8×10^{-16}	$Zn_3L_2 \cdot 4H_2O$	5×10^{-36}
AuL_3	3×10^{-6}	Pb_3L_2	3.0×10^{-44}
$AlL_3 \, (\alpha)$	3×10^{-34}	GaL	1×10^{-21}
GaL_3 (amorphous)	10^{-37}	InL	2.3×10^{-22}
InL_3	1.3×10^{-37}	Sulfates: $L = SO_4^{2-}$	
$SnO \, (\rightleftharpoons Sn^{2+} + 2OH^-)$	6×10^{-27}	CaL	2.4×10^{-5}
PbO (yellow) $(\rightleftharpoons Pb^{2+} + 2OH^-)$	8×10^{-16}	SrL	3.2×10^{-7}
PbO (red) $(\rightleftharpoons Pb^{2+} + 2OH^-)$	5×10^{-16}	BaL	1.1×10^{-10}
Iodates: $L = IO_3^-$		RaL	4.3×10^{-11}
CaL_2	7.1×10^{-7}	Ag_2L	1.5×10^{-5}
SrL_2	3.3×10^{-7}	Hg_2L	7.4×10^{-7}
BaL_2	1.5×10^{-9}	PbL	6.3×10^{-7}
YL_3	7.1×10^{-11}	Sulfides: $L = S^{2-}$	
LaL_3	1.0×10^{-11}	MnL (pink)	3×10^{-11}
CeL_3	1.4×10^{-11}	MnL (green)	3×10^{-14}
ThL_4	2.4×10^{-15}	FeL	8×10^{-19}
$UO_2L_2 \, (\rightleftharpoons UO_2^{2+} + 2IO_3^-)$	9.8×10^{-8}	$CoL \, (\alpha)$	5×10^{-22}
CrL_3	5×10^{-6}	$CoL \, (\beta)$	3×10^{-26}
AgL	3.1×10^{-8}	$NiL \, (\alpha)$	4×10^{-20}
Hg_2L_2	1.3×10^{-18}	$NiL \, (\beta)$	1.3×10^{-25}
TlL	3.1×10^{-6}	$NiL \, (\gamma)$	3×10^{-27}
ZnL_2	3.9×10^{-6}	CuL	8×10^{-37}
CdL_2	2.3×10^{-8}	Cu_2L	3×10^{-49}
PbL_2	2.5×10^{-13}	Ag_2L	8×10^{-51}
Iodides: $L = I^-$		Tl_2L	6×10^{-22}
CuL	1×10^{-12}	$ZnL \, (\alpha)$	2×10^{-25}
AgL	8.3×10^{-17}	$ZnL \, (\beta)$	3×10^{-23}
$CH_3HgL \, (\rightleftharpoons CH_3Hg^+ + I^-)$	3.5×10^{-12}	CdL	1×10^{-27}
$CH_3CH_2HgL \, (\rightleftharpoons CH_3CH_2Hg^+ + I^-)$	7.8×10^{-5}	HgL (black)	2×10^{-53}
TlL	5.9×10^{-8}	HgL (red)	5×10^{-54}
Hg_2L_2	1.1×10^{-28}	SnL	1.3×10^{-26}
SnL_2	8.3×10^{-6}	PbL	3×10^{-28}
PbL_2	7.9×10^{-9}	In_2L_3	4×10^{-70}
Oxalates: $L = C_2O_4^{2-}$		Thiocyanates: $L = SCN^-$	
CaL	1.3×10^{-8}	CuL	4.0×10^{-14}
SrL	4×10^{-7}	AgL	1.1×10^{-12}
BaL	1×10^{-6}	Hg_2L_2	3.0×10^{-20}
La_2L_3	1×10^{-25}	TlL	1.6×10^{-4}
ThL_2	4.2×10^{-22}	HgL_2	2.8×10^{-20}

ACID DISSOCIATION CONSTANTS

Name	Structure[†]	pK_a[‡]	K_a
Acetic acid (ethanoic acid)	CH_3CO_2H	4.757	1.75×10^{-5}
Alanine	$\overset{\overset{NH_3^+}{\mid}}{\underset{\underset{CO_2H}{\mid}}{CHCH_3}}$	2.348 (CO_2H) 9.867 (NH_3)	4.49×10^{-3} 1.36×10^{-10}
Aminobenzene (aniline)	$\text{⟨benzene⟩}-NH_3^+$	4.601	2.51×10^{-5}
2-Aminobenzoic acid (anthranilic acid)	$\text{⟨benzene⟩}\overset{NH_3^+}{\underset{CO_2H}{}}$	2.08 (CO_2H) 4.96 (NH_3)	8.3×10^{-3} 1.10×10^{-5}
2-Aminoethanol (ethanolamine)	$HOCH_2CH_2NH_3^+$	9.498	3.18×10^{-10}
2-Aminophenol	$\text{⟨benzene⟩}\overset{OH}{\underset{NH_3^+}{}}$	4.78 (NH_3) (20°) 9.97 (OH) (20°)	1.66×10^{-5} 1.05×10^{-10}
Ammonia	NH_4^+	9.244	5.70×10^{-10}
Arginine	$\overset{\overset{NH_3^+}{\mid}}{\underset{\underset{CO_2H}{\mid}}{CHCH_2CH_2CH_2NHC}}\overset{NH_2^+}{\underset{NH_2}{}}$	1.823 (CO_2H) 8.991 (NH_3) (12.48) (NH_2)	1.50×10^{-2} 1.02×10^{-9} 3.3×10^{-13}
Arsenic acid (hydrogen arsenate)	$\overset{\overset{O}{\parallel}}{HO-As-OH}\underset{OH}{}$	2.24 6.96 11.50	5.8×10^{-3} 1.10×10^{-7} 3.2×10^{-12}
Arsenious acid (hydrogen arsenite)	$As(OH)_3$	9.29	5.1×10^{-10}
Asparagine	$\overset{\overset{NH_3^+}{\mid}}{\underset{\underset{CO_2H}{\mid}}{CHCH_2\overset{\overset{O}{\parallel}}{C}NH_2}}$	2.14 (CO_2H) 8.72 (NH_3)	7.2×10^{-3} 1.9×10^{-9}

[†]Each acid is written in its protonated form. The acidic protons are indicated in **bold** type.
[‡]pK_a values refer to 25°C unless otherwise indicated. Values in parentheses are considered to be less reliable.

Name	Structure[†]	pK_a[‡]	K_a
Aspartic acid	(structure: NH_3^+ / β / $CHCH_2CO_2H$ / $\alpha \rightarrow CO_2H$)	1.990 (α-CO_2H) 3.900 (β-CO_2H) 10.002 (NH_3)	1.02×10^{-2} 1.26×10^{-4} 9.95×10^{-11}
Benzene-1,2,3-tricarboxylic acid (hemimellitic acid)	(benzene ring with three CO_2H)	2.88 4.75 7.13	1.32×10^{-3} 1.78×10^{-5} 7.4×10^{-8}
Benzoic acid	(benzene ring with CO_2H)	4.202	6.28×10^{-5}
2,2'-Bipyridine	(bipyridine structure, N N / H^+)	4.35	4.5×10^{-5}
Boric acid (hydrogen borate)	$B(OH)_3$	9.236 (12.74) (20°) (13.80) (20°)	5.81×10^{-10} 1.82×10^{-13} 1.58×10^{-14}
Bromoacetic acid	$BrCH_2CO_2H$	2.902	1.25×10^{-3}
Butane-2,3-dione dioxime (dimethylglyoxime)	(structure: HON NOH / CH_3 CH_3)	10.66 12.0	2.2×10^{-11} 1×10^{-12}
Butanoic acid	$CH_3CH_2CH_2CO_2H$	4.819	1.52×10^{-5}
cis-Butenedioic acid (maleic acid)	(structure with two CO_2H, cis)	1.910 6.332	1.23×10^{-2} 4.66×10^{-7}
trans-Butenedioic acid (fumaric acid)	(structure: CO_2H / HO_2C, trans)	3.053 4.494	8.85×10^{-4} 3.21×10^{-5}
Butylamine	$CH_3CH_2CH_2CH_2NH_3^+$	10.640	2.29×10^{-11}
Carbonic acid (hydrogen carbonate)	(structure: $HO-C(=O)-OH$)	6.352 10.329	4.45×10^{-7} 4.69×10^{-11}
Chloroacetic acid	$ClCH_2CO_2H$	2.865	1.36×10^{-3}
Chlorous acid (hydrogen chlorite)	$HOCl=O$	1.95	1.12×10^{-2}
Chromic acid (hydrogen chromate)	(structure: $HO-Cr(=O)(=O)-OH$)	−0.2 (20°) 6.51	1.6 3.1×10^{-7}
Citric acid (2-hydroxypropane-1,2,3- tricarboxylic acid)	(structure: CO_2H / $HO_2CCH_2CCH_2CO_2H$ / OH)	3.128 4.761 6.396	7.44×10^{-4} 1.73×10^{-5} 4.02×10^{-7}
Cyanoacetic acid	$NCCH_2CO_2H$	2.472	3.37×10^{-3}

(continued)

Name	Structure[†]	$pK_a^{‡}$	K_a
Cyclohexylamine	cyclohexyl–NH_3^+	10.64	2.3×10^{-11}
Cysteine	NH_3^+ | $CHCH_2SH$ | CO_2H	(1.71) (CO_2H) 8.36 (SH) 10.77 (NH_3)	1.95×10^{-2} 4.4×10^{-9} 1.70×10^{-11}
Dichloroacetic acid	Cl_2CHCO_2H	1.30	5.0×10^{-2}
Diethylamine	$(CH_3CH_2)_2NH_2^+$	10.933	1.17×10^{-11}
1,2-Dihydroxybenzene (catechol)		9.40 12.8	4.0×10^{-10} 1.6×10^{-13}
1,3-Dihydroxybenzene (resorcinol)		9.30 11.06	5.0×10^{-10} 8.7×10^{-12}
D-2,3-Dihydroxybutanedioic acid (D-tartaric acid)	$HO_2CCHCHCO_2H$ with OH above and OH below	3.036 4.366	9.20×10^{-4} 4.31×10^{-5}
Dimethylamine	$(CH_3)_2NH_2^+$	10.774	1.68×10^{-11}
2,4-Dinitrophenol		4.11	7.8×10^{-5}
Ethane-1,2-dithiol	$HSCH_2CH_2SH$	8.85 (30°) 10.43 (30°)	1.4×10^{-9} 3.7×10^{-11}
Ethylamine	$CH_3CH_2NH_3^+$	10.636	2.31×10^{-11}
Ethylenediamine (1,2-diaminoethane)	$H_3^+NCH_2CH_2^+NH_3$	6.848 9.928	1.42×10^{-7} 1.18×10^{-10}
Ethylenedinitrilotetraacetic acid (EDTA)	$(HO_2CCH_2)_2^+NHCH_2CH_2^+NH(CH_2CO_2H)_2$	0.0 (CO_2H) 1.5 (CO_2H) 2.0 (CO_2H) 2.66 (CO_2H) 6.16 (NH) 10.24 (NH)	1.0 0.032 0.010 0.002 2 6.9×10^{-7} 5.8×10^{-11}
Formic acid (methanoic acid)	HCO_2H	3.745	1.80×10^{-4}
Glutamic acid	NH_3^+ | $CHCH_2CH_2CO_2H$ (γ) | $α \rightarrow CO_2H$	2.23 (α-CO_2H) 4.42 (γ-CO_2H) 9.95 (NH_3)	5.9×10^{-3} 3.8×10^{-5} 1.12×10^{-10}
Glutamine	NH_3^+ O | ‖ $CHCH_2CH_2CNH_2$ | CO_2H	2.17 (CO_2H) 9.01 (NH_3)	6.8×10^{-3} 9.8×10^{-10}

Name	Structure[†]	pK_a[‡]	K_a
Glycine (aminoacetic acid)	NH$_3^+$ | CH$_2$ | CO$_2$H	2.350 (CO$_2$H) 9.778 (NH$_3$)	4.47×10^{-3} 1.67×10^{-10}
Guanidine	$^+$NH$_2$ || H$_2$N—C—NH$_2$	13.54 (27°)	2.9×10^{-14}
1,6-Hexanedioic acid (adipic acid)	HO$_2$CCH$_2$CH$_2$CH$_2$CH$_2$CO$_2$H	4.42 5.42	3.8×10^{-5} 3.8×10^{-6}
Histidine	NH$_3^+$ | CHCH$_2$ | CO$_2$H (imidazole ring, NH)	1.7 (CO$_2$H) 6.02 (NH) 9.08 (NH$_3$)	2×10^{-2} 9.5×10^{-7} 8.3×10^{-10}
Hydrazoic acid (hydrogen azide)	HN$=$$\overset{+}{N}$$=$$\overset{-}{N}$	4.65	2.2×10^{-5}
Hydrogen cyanate	HOC$\equiv$N	3.48	3.3×10^{-4}
Hydrogen cyanide	HC$\equiv$N	9.21	6.2×10^{-10}
Hydrogen fluoride	HF	3.17	6.8×10^{-4}
Hydrogen peroxide	HOOH	11.65	2.2×10^{-12}
Hydrogen sulfide	H$_2$S	7.02 ~18	9.5×10^{-8} ~10^{-18}
Hydrogen thiocyanate	HSC$\equiv$N	0.9	0.13
Hydroxyacetic acid (glycolic acid)	HOCH$_2$CO$_2$H	3.831	1.48×10^{-4}
Hydroxybenzene (phenol)	C$_6$H$_5$—OH	9.98	1.05×10^{-10}
2-Hydroxybenzoic acid (salicylic acid)	C$_6$H$_4$(CO$_2$H)(OH)	2.97 (CO$_2$H) 13.74 (OH)	1.07×10^{-3} 1.82×10^{-14}
Hydroxylamine	HO$\overset{+}{N}$H$_3$	5.96	1.10×10^{-6}
8-Hydroxyquinoline (oxine)	quinoline, NH$^+$, HO	4.91 (NH) 9.81 (OH)	1.23×10^{-5} 1.55×10^{-10}
Hypochlorous acid (hydrogen hypochlorite)	HOCl	7.53	3.0×10^{-8}
Hypophosphorous acid (hydrogen hypophosphite)	O || H$_2$POH	1.23	5.9×10^{-2}
Imidazole (1,3-diazole)	ring —NH$^+$ | N | H	6.993	1.02×10^{-7}
Iminodiacetic acid	H$_2$$\overset{+}{N}$(CH$_2CO_2$H)$_2$	1.82 (CO$_2$H) 2.84 (CO$_2$H) 9.79 (NH$_2$)	1.51×10^{-2} 1.45×10^{-3} 1.62×10^{-10}

(continued)

Name	Structure†	p$K_a^\ddagger$	K_a
Iodic acid (hydrogen iodate)	$\underset{\displaystyle HO-\overset{\textstyle O}{\overset{\|}{I}}=O}{}$	0.77	0.17
Iodoacetic acid	ICH_2CO_2H	3.175	6.68×10^{-4}
Isoleucine	$\overset{\displaystyle NH_3^+}{\underset{\displaystyle CO_2H}{CHCH(CH_3)CH_2CH_3}}$	2.319 (CO$_2$H) 9.754 (NH$_3$)	4.80×10^{-3} 1.76×10^{-10}
Leucine	$\overset{\displaystyle NH_3^+}{\underset{\displaystyle CO_2H}{CHCH_2CH(CH_3)_2}}$	2.329 (CO$_2$H) 9.747 (NH$_3$)	4.69×10^{-3} 1.79×10^{-10}
Lysine	$\overset{\alpha \rightarrow \displaystyle NH_3^+}{\underset{\displaystyle CO_2H}{CHCH_2CH_2CH_2CH_2NH_3^+}} \;\; \epsilon$	2.04 (CO$_2$H) 9.08 (α-NH$_3$) 10.69 (ε-NH$_3$)	9.1×10^{-3} 8.3×10^{-10} 2.0×10^{-11}
Malonic acid (propanedioic acid)	$HO_2CCH_2CO_2H$	2.847 5.696	1.42×10^{-3} 2.01×10^{-6}
Mercaptoacetic acid (thioglycolic acid)	$HSCH_2CO_2H$	(3.60) (CO$_2$H) 10.55 (SH)	2.5×10^{-4} 2.82×10^{-11}
2-Mercaptoethanol	$HSCH_2CH_2OH$	9.72	1.91×10^{-10}
Methionine	$\overset{\displaystyle NH_3^+}{\underset{\displaystyle CO_2H}{CHCH_2CH_2SCH_3}}$	2.20 (CO$_2$H) 9.05 (NH$_3$)	6.3×10^{-3} 8.9×10^{-10}
Methylamine	$CH_3\overset{+}{N}H_3$	10.64	2.3×10^{-11}
4-Methylaniline (p-toluidine)	$CH_3-\!\!\!\left\langle \bigcirc \right\rangle\!\!\!-\overset{+}{N}H_3$	5.084	8.24×10^{-6}
2-Methylphenol (o-cresol)		10.09	8.1×10^{-11}
4-Methylphenol (p-cresol)	$CH_3-\!\!\!\left\langle \bigcirc \right\rangle\!\!\!-OH$	10.26	5.5×10^{-11}
Morpholine (perhydro-1,4-oxazine)	$O\left\langle \bigcirc \right\rangle\overset{+}{N}H_2$	8.492	3.22×10^{-9}
1-Naphthoic acid		3.70	2.0×10^{-4}
2-Naphthoic acid		4.16	6.9×10^{-5}

Name	Structure[†]	p$K_a^{\ddagger}$	K_a
1-Naphthol	OH (structure)	9.34	4.6×10^{-10}
2-Naphthol	OH (structure)	9.51	3.1×10^{-10}
Nitrilotriacetic acid	$H\overset{+}{N}(CH_2CO_2H)_3$	1.1 (CO_2H) (20°) 1.650 (CO_2H) (20°) 2.940 (CO_2H) (20°) 10.334 (NH) (20°)	8×10^{-2} 2.24×10^{-2} 1.15×10^{-3} 4.63×10^{-11}
4-Nitrobenzoic acid	$O_2N-\!\!\!\bigcirc\!\!\!-CO_2H$	3.442	3.61×10^{-4}
Nitroethane	$CH_3CH_2NO_2$	8.57	2.7×10^{-9}
4-Nitrophenol	$O_2N-\!\!\!\bigcirc\!\!\!-OH$	7.15	7.1×10^{-8}
N-Nitrosophenylhydroxylamine (cupferron)	(structure with NO, N, OH)	4.16	6.9×10^{-5}
Nitrous acid	$HON\!=\!O$	3.15	7.1×10^{-4}
Oxalic acid (ethanedioic acid)	HO_2CCO_2H	1.252 4.266	5.60×10^{-2} 5.42×10^{-5}
Oxoacetic acid (glyoxylic acid)	$\overset{O}{\overset{\|}{HCCO_2H}}$	3.46	3.5×10^{-4}
Oxobutanedioic acid (oxaloacetic acid)	$\overset{O}{\overset{\|}{HO_2CCH_2CCO_2H}}$	2.56 4.37	2.8×10^{-3} 4.3×10^{-5}
2-Oxopentanedioic (α-ketoglutaric acid)	$\overset{O}{\overset{\|}{HO_2CCH_2CH_2CCO_2H}}$	1.85 4.44	1.41×10^{-2} 3.6×10^{-5}
2-Oxopropanoic acid (pyruvic acid)	$\overset{O}{\overset{\|}{CH_3CCO_2H}}$	2.55	2.8×10^{-3}
1,5-Pentanedioic acid (glutaric acid)	$HO_2CCH_2CH_2CH_2CO_2H$	4.34 5.43	4.6×10^{-5} 3.7×10^{-6}
1,10-Phenanthroline	(structure with NH+, N)	4.86	1.38×10^{-5}

(continued)

Name	Structure[†]	$pK_a^{‡}$	K_a
Phenylacetic acid	C_6H_5—CH_2CO_2H	4.310	4.90×10^{-5}
Phenylalanine	$\overset{NH_3^+}{\underset{CO_2H}{CHCH_2}}$—$C_6H_5$	2.20 (CO_2H) 9.31 (NH_3)	6.3×10^{-3} 4.9×10^{-10}
Phosphoric acid (hydrogen phosphate)	HO—P(=O)(OH)—OH	2.148 7.199 12.15	7.11×10^{-3} 6.32×10^{-8} 7.1×10^{-13}
Phosphorous acid (hydrogen phosphite)	HP(=O)(OH)—OH	1.5 6.79	3×10^{-2} 1.62×10^{-7}
Phthalic acid (benzene-1,2-dicarboxylic acid)		2.950 5.408	1.12×10^{-3} 3.90×10^{-6}
Piperazine (perhydro-1,4-diazine)	H_2^+N—NH_2^+	5.333 9.731	4.65×10^{-6} 1.86×10^{-10}
Piperidine	NH_2^+	11.123	7.53×10^{-12}
Proline		1.952 (CO_2H) 10.640 (NH_2)	1.12×10^{-2} 2.29×10^{-11}
Propanoic acid	$CH_3CH_2CO_2H$	4.874	1.34×10^{-5}
Propenoic acid (acrylic acid)	$H_2C=CHCO_2H$	4.258	5.52×10^{-5}
Propylamine	$CH_3CH_2CH_2NH_3^+$	10.566	2.72×10^{-11}
Pyridine (azine)	NH^+	5.229	5.90×10^{-6}
Pyridine-2-carboxylic acid (picolinic acid)		1.01 (CO_2H) 5.39 (NH)	9.8×10^{-2} 4.1×10^{-6}
Pyridine-3-carboxylic acid (nicotinic acid)		2.05 (CO_2H) 4.81 (NH)	8.9×10^{-3} 1.55×10^{-5}
Pyridoxal-5-phosphate		1.4 (POH) 3.44 (OH) 6.01 (POH) 8.45 (NH)	0.04 3.6×10^{-4} 9.8×10^{-7} 3.5×10^{-9}

Name	Structure[†]	$pK_a^{\ddagger}$	K_a
Pyrophosphoric acid (hydrogen diphosphate)	$\begin{matrix} O & O \\ \| & \| \\ (HO)_2POP(OH)_2 \end{matrix}$	0.8 2.2 6.70 9.40	0.16 6×10^{-3} 2.0×10^{-7} 4.0×10^{-10}
Serine	$\begin{matrix} \overset{+}{N}H_3 \\ \| \\ CHCH_2OH \\ \| \\ CO_2H \end{matrix}$	2.187 (CO_2H) 9.209 (NH_3)	6.50×10^{-3} 6.18×10^{-10}
Succinic acid (butanedioic acid)	$HO_2CCH_2CH_2CO_2H$	4.207 5.636	6.21×10^{-5} 2.31×10^{-6}
Sulfuric acid (hydrogen sulfate)	$\begin{matrix} O \\ \| \\ HO-S-OH \\ \| \\ O \end{matrix}$	1.99 (pK_2)	1.02×10^{-2}
Sulfurous acid (hydrogen sulfite)	$\begin{matrix} O \\ \| \\ HOSOH \end{matrix}$	1.91 7.18	1.23×10^{-2} 6.6×10^{-8}
Thiosulfuric acid (hydrogen thiosulfate)	$\begin{matrix} O \\ \| \\ HOSSH \\ \| \\ O \end{matrix}$	0.6 1.6	0.3 0.03
Threonine	$\begin{matrix} \overset{+}{N}H_3 \\ \| \\ CHCHOHCH_3 \\ \| \\ CO_2H \end{matrix}$	2.088 (CO_2H) 9.100 (NH_3)	8.17×10^{-3} 7.94×10^{-10}
Trichloroacetic acid	Cl_3CCO_2H	0.66	0.22
Triethanolamine	$(HOCH_2CH_2)_3NH^+$	7.762	1.73×10^{-8}
Triethylamine	$(CH_3CH_2)_3NH^+$	10.715	1.93×10^{-11}
1,2,3-Trihydroxybenzene (pyrogallol)	(benzene ring with OH, OH, OH)	8.94 11.08 (14)	1.15×10^{-9} 8.3×10^{-12} 10^{-14}
Trimethylamine	$(CH_3)_3NH^+$	9.800	1.58×10^{-10}
Tris(hydroxymethyl)aminomethane (tris or tham)	$(HOCH_2)_3CNH_3^+$	8.075	8.41×10^{-9}
Tryptophan	$\begin{matrix} \overset{+}{N}H_3 \\ \| \\ CHCH_2 \\ \| \\ CO_2H \end{matrix}$ (indole ring)	2.35 (CO_2H) 9.33 (NH_3)	4.5×10^{-3} 4.7×10^{-10}
Tyrosine	$\begin{matrix} \overset{+}{N}H_3 \\ \| \\ CHCH_2 \\ \| \\ CO_2H \end{matrix}$ (benzene ring)—OH	2.17 (CO_2H) 9.19 (NH_3) 10.47 (OH)	6.8×10^{-3} 6.5×10^{-10} 3.4×10^{-11}
Valine	$\begin{matrix} \overset{+}{N}H_3 \\ \| \\ CHCH(CH_3)_2 \\ \| \\ CO_2H \end{matrix}$	2.286 (CO_2H) 9.718 (NH_3)	5.18×10^{-3} 1.91×10^{-10}

STANDARD REDUCTION POTENTIALS

Reaction[†]	$E°$ (volts)
Aluminum	
$Al^{3+} + 3e^- \rightleftharpoons Al(s)$	-1.677
$Al(OH)_4^- + 3e^- \rightleftharpoons Al(s) + 4OH^-$	-2.328
Arsenic	
$H_3AsO_4 + 2H^+ + 2e^- \rightleftharpoons H_3AsO_3 + H_2O$	0.575
$H_3AsO_3 + 3H^+ + 3e^- \rightleftharpoons As(s) + 3H_2O$	$0.247\ 5$
$As(s) + 3H^+ + 3e^- \rightleftharpoons AsH_3(g)$	-0.238
Barium	
$Ba^{2+} + 2e^- \rightleftharpoons Ba(s)$	-2.906
Beryllium	
$Be^{2+} + 2e^- \rightleftharpoons Be(s)$	-1.968
Boron	
$2B(s) + 6H^+ + 6e^- \rightleftharpoons B_2H_6(g)$	-0.150
$B_4O_7^{2-} + 14H^+ + 12e^- \rightleftharpoons 4B(s) + 7H_2O$	-0.792
$B(OH)_3 + 3H^+ + 3e^- \rightleftharpoons B(s) + 3H_2O$	-0.889
Bromine	
$BrO_4^- + 2H^+ + 2e^- \rightleftharpoons BrO_3^- + H_2O$	1.745
$HOBr + H^+ + e^- \rightleftharpoons \frac{1}{2}Br_2(l) + H_2O$	1.584
$BrO_3^- + 6H^+ + 5e^- \rightleftharpoons \frac{1}{2}Br_2(l) + 3H_2O$	1.513
$Br_2(aq) + 2e^- \rightleftharpoons 2Br^-$	1.098
$Br_2(l) + 2e^- \rightleftharpoons 2Br^-$	1.078
$Br_3^- + 2e^- \rightleftharpoons 3Br^-$	1.062
$BrO^- + H_2O + 2e^- \rightleftharpoons Br^- + 2OH^-$	0.766
$BrO_3^- + 3H_2O + 6e^- \rightleftharpoons Br^- + 6OH^-$	0.613
Cadmium	
$Cd^{2+} + 2e^- \rightleftharpoons Cd(s)$	-0.402
$Cd(NH_3)_4^{2+} + 2e^- \rightleftharpoons Cd(s) + 4NH_3$	-0.613
Calcium	
$Ca(s) + 2H^+ + 2e^- \rightleftharpoons CaH_2(s)$	0.776
$Ca^{2+} + 2e^- \rightleftharpoons Ca(s)$	-2.868
$Ca(acetate)^+ + 2e^- \rightleftharpoons Ca(s) + acetate^-$	-2.891
$CaSO_4(s) + 2e^- \rightleftharpoons Ca(s) + SO_4^{2-}$	-2.936
Carbon	
$C_2H_2(g) + 2H^+ + 2e^- \rightleftharpoons C_2H_4(g)$	0.731
$O=\bigcirc=O + 2H^+ + 2e^- \rightleftharpoons HO-\bigcirc-OH$	0.700

Reaction[†]	$E°$ (volts)
$CH_3OH + 2H^+ + 2e^- \rightleftharpoons CH_4(g) + H_2O$	0.583
dehydroascorbic acid $+ 2H^+ + 2e^-$ $\rightleftharpoons$ ascorbic acid $+ H_2O$	0.390
$(CN)_2(g) + 2H^+ + 2e^- \rightleftharpoons 2HCN(aq)$	0.373
$H_2CO + 2H^+ + 2e^- \rightleftharpoons CH_3OH$	0.237
$C(s) + 4H^+ + 4e^- \rightleftharpoons CH_4(g)$	$0.131\ 5$
$HCO_2H + 2H^+ + 2e^- \rightleftharpoons H_2CO + H_2O$	-0.029
$CO_2(g) + 2H^+ + 2e^- \rightleftharpoons CO(g) + H_2O$	$-0.103\ 8$
$CO_2(g) + 2H^+ + 2e^- \rightleftharpoons HCO_2H$	-0.114
$2CO_2(g) + 2H^+ + 2e^- \rightleftharpoons H_2C_2O_4$	-0.432
Cerium	
$Ce^{4+} + e^- \rightleftharpoons Ce^{3+}$	$\begin{cases} 1.72 \\ 1.70 \quad 1\ F\ HClO_4 \\ 1.44 \quad 1\ F\ H_2SO_4 \\ 1.61 \quad 1\ F\ HNO_3 \\ 1.47 \quad 1\ F\ HCl \end{cases}$
$Ce^{3+} + 3e^- \rightleftharpoons Ce(s)$	-2.336
Cesium	
$Cs^+ + e^- \rightleftharpoons Cs(s)$	-3.026
Chlorine	
$HClO_2 + 2H^+ + 2e^- \rightleftharpoons HOCl + H_2O$	1.674
$HClO + H^+ + e^- \rightleftharpoons \frac{1}{2}Cl_2(g) + H_2O$	1.630
$ClO_3^- + 6H^+ + 5e^- \rightleftharpoons \frac{1}{2}Cl_2(g) + 3H_2O$	1.458
$Cl_2(aq) + 2e^- \rightleftharpoons 2Cl^-$	1.396
$Cl_2(g) + 2e^- \rightleftharpoons 2Cl^-$	$1.360\ 4$
$ClO_4^- + 2H^+ + 2e^- \rightleftharpoons ClO_3^- + H_2O$	1.226
$ClO_3^- + 3H^+ + 2e^- \rightleftharpoons HClO_2 + H_2O$	1.157
$ClO_3^- + 2H^+ + e^- \rightleftharpoons ClO_2 + H_2O$	1.130
$ClO_2 + e^- \rightleftharpoons ClO_2^-$	1.068
Chromium	
$Cr_2O_7^{2-} + 14H^+ + 6e^- \rightleftharpoons 2Cr^{3+} + 7H_2O$	1.36
$CrO_4^{2-} + 4H_2O + 3e^-$ $\rightleftharpoons Cr(OH)_3\ (s, hydrated) + 5OH^-$	-0.12
$Cr^{3+} + e^- \rightleftharpoons Cr^{2+}$	-0.42
$Cr^{3+} + 3e^- \rightleftharpoons Cr(s)$	-0.74
$Cr^{2+} + 2e^- \rightleftharpoons Cr(s)$	-0.89
Cobalt	
$Co^{3+} + e^- \rightleftharpoons Co^{2+}$	$\begin{cases} 1.92 \\ 1.817 \quad 8\ F\ H_2SO_4 \\ 1.850 \quad 4\ F\ HNO_3 \end{cases}$

[†]All species are aqueous unless otherwise indicated.

Reaction[†]	$E°$ (volts)
$Co(NH_3)_6^{3+} + e^- \rightleftharpoons Co(NH_3)_6^{2+}$	0.1
$CoOH^+ + H^+ + 2e^- \rightleftharpoons Co(s) + H_2O$	0.003
$Co^{2+} + 2e^- \rightleftharpoons Co(s)$	-0.282
$Co(OH)_2(s) + 2e^- \rightleftharpoons Co(s) + 2OH^-$	-0.746
Copper	
$Cu^+ + e^- \rightleftharpoons Cu(s)$	0.518
$Cu^{2+} + 2e^- \rightleftharpoons Cu(s)$	0.339
$Cu^{2+} + e^- \rightleftharpoons Cu^+$	0.161
$CuCl(s) + e^- \rightleftharpoons Cu(s) + Cl^-$	0.137
$Cu(IO_3)_2(s) + 2e^- \rightleftharpoons Cu(s) + 2IO_3^-$	-0.079
$Cu(ethylenediamine)_2^+ + e^-$ $\rightleftharpoons Cu(s) + 2\ ethylenediamine$	-0.119
$CuI(s) + e^- \rightleftharpoons Cu(s) + I^-$	-0.185
$Cu(EDTA)^{2-} + 2e^- \rightleftharpoons Cu(s) + EDTA^{4-}$	-0.216
$Cu(OH)_2(s) + 2e^- \rightleftharpoons Cu(s) + 2OH^-$	-0.222
$Cu(CN)_2^- + e^- \rightleftharpoons Cu(s) + 2CN^-$	-0.429
$CuCN(s) + e^- \rightleftharpoons Cu(s) + CN^-$	-0.639
Fluorine	
$F_2(g) + 2e^- \rightleftharpoons 2F^-$	2.890
$F_2O(g) + 2H^+ + 4e^- \rightleftharpoons 2F^- + H_2O$	2.168
Gallium	
$Ga^{3+} + 3e^- \rightleftharpoons Ga(s)$	-0.549
$GaOOH(s) + 3H_2O + 3e^- \rightleftharpoons Ga(s) + 3OH^-$	-1.320
Germanium	
$Ge^{2+} + 2e^- \rightleftharpoons Ge(s)$	0.1
$H_4GeO_4 + 4H^+ + 4e^- \rightleftharpoons Ge(s) + 4H_2O$	-0.039
Gold	
$Au^+ + e^- \rightleftharpoons Au(s)$	1.69
$Au^{3+} + 2e^- \rightleftharpoons Au^+$	1.41
$AuCl_2^- + e^- \rightleftharpoons Au(s) + 2Cl^-$	1.154
$AuCl_4^- + 2e^- \rightleftharpoons AuCl_2^- + 2Cl^-$	0.926
Hydrogen	
$2H^+ + 2e^- \rightleftharpoons H_2(g)$	0.000 0
$H_2O + e^- \rightleftharpoons \frac{1}{2}H_2(g) + OH^-$	$-0.828\ 0$
Indium	
$In^{3+} + 3e^- \rightleftharpoons In(s)$	-0.338
$In^{3+} + 2e^- \rightleftharpoons In^+$	-0.444
$In(OH)_3(s) + 3e^- \rightleftharpoons In(s) + 3OH^-$	-0.99
Iodine	
$IO_4^- + 2H^+ + 2e^- \rightleftharpoons IO_3^- + H_2O$	1.589
$H_5IO_6 + 2H^+ + 2e^- \rightleftharpoons HIO_3 + 3H_2O$	1.567
$HOI + H^+ + e^- \rightleftharpoons \frac{1}{2}I_2(s) + H_2O$	1.430
$ICl_3(s) + 3e^- \rightleftharpoons \frac{1}{2}I_2(s) + 3Cl^-$	1.28
$ICl(s) + e^- \rightleftharpoons \frac{1}{2}I_2(s) + Cl^-$	1.22
$IO_3^- + 6H^+ + 5e^- \rightleftharpoons \frac{1}{2}I_2(s) + 3H_2O$	1.210
$IO_3^- + 5H^+ + 4e^- \rightleftharpoons HOI + 2H_2O$	1.154
$I_2(aq) + 2e^- \rightleftharpoons 2I^-$	0.620
$I_2(s) + 2e^- \rightleftharpoons 2I^-$	0.535
$I_3^- + 2e^- \rightleftharpoons 3I^-$	0.535
$IO_3^- + 3H_2O + 6e^- \rightleftharpoons I^- + 6OH^-$	0.269
Iron	
$Fe(phenanthroline)_3^{3+} + e^-$ $\rightleftharpoons Fe(phenanthroline)_3^{2+}$	1.147
$Fe(bipyridyl)_3^{3+} + e^-$ $\rightleftharpoons Fe(bipyridyl)_3^{2+}$	1.120
$FeOH^{2+} + H^+ + e^- \rightleftharpoons Fe^{2+} + H_2O$	0.900
$FeO_4^{2-} + 3H_2O + 3e^- \rightleftharpoons FeOOH(s) + 5OH^-$	0.80

Reaction[†]	$E°$ (volts)
$Fe^{3+} + e^- \rightleftharpoons Fe^{2+}$	$\begin{cases} 0.771 \\ 0.732\ \ 1\ F\ HCl \\ 0.767\ \ 1\ F\ HClO_4 \\ 0.746\ \ 1\ F\ HNO_3 \end{cases}$
$FeOOH(s) + 3H^+ + e^- \rightleftharpoons Fe^{2+} + 2H_2O$	0.74
$ferricinium^+ + e^- \rightleftharpoons ferrocene$	0.400
$Fe(CN)_6^{3-} + e^- \rightleftharpoons Fe(CN)_6^{4-}$	0.356
$FeOH^+ + H^+ + 2e^- \rightleftharpoons Fe(s) + H_2O$	-0.16
$Fe^{2+} + 2e^- \rightleftharpoons Fe(s)$	-0.44
$FeCO_3(s) + 2e^- \rightleftharpoons Fe(s) + CO_3^{2-}$	-0.756
Lanthanum	
$La^{3+} + 3e^- \rightleftharpoons La(s)$	-2.379
Lead	
$Pb^{4+} + 2e^- \rightleftharpoons Pb^{2+}$	1.69 1 F HNO_3
$PbO_2(s) + 4H^+ + SO_4^{2-} + 2e^-$ $\rightleftharpoons PbSO_4(s) + 2H_2O$	1.685
$PbO_2(s) + 4H^+ + 2e^- \rightleftharpoons Pb^{2+} + 2H_2O$	1.458
$3PbO_2(s) + 2H_2O + 4e^- \rightleftharpoons Pb_3O_4(s)$ $+ 4OH^-$	0.269
$Pb_3O_4(s) + H_2O + 2e^-$ $\rightleftharpoons 3PbO(s, red) + 2OH^-$	0.224
$Pb_3O_4(s) + H_2O + 2e^-$ $\rightleftharpoons 3PbO(s, yellow) + 2OH^-$	0.207
$Pb^{2+} + 2e^- \rightleftharpoons Pb(s)$	-0.126
$PbF_2(s) + 2e^- \rightleftharpoons Pb(s) + 2F^-$	-0.350
$PbSO_4(s) + 2e^- \rightleftharpoons Pb(s) + SO_4^{2-}$	-0.355
Lithium	
$Li^+ + e^- \rightleftharpoons Li(s)$	-3.040
Magnesium	
$Mg(OH)^+ + H^+ + 2e^- \rightleftharpoons Mg(s) + H_2O$	-2.022
$Mg^{2+} + 2e^- \rightleftharpoons Mg(s)$	-2.360
$Mg(C_2O_4)(s) + 2e^- \rightleftharpoons Mg(s) + C_2O_4^{2-}$	-2.493
$Mg(OH)_2(s) + 2e^- \rightleftharpoons Mg(s) + 2OH^-$	-2.690
Manganese	
$MnO_4^- + 4H^+ + 3e^- \rightleftharpoons MnO_2(s) + 2H_2O$	1.692
$Mn^{3+} + e^- \rightleftharpoons Mn^{2+}$	1.56
$MnO_4^- + 8H^+ + 5e^- \rightleftharpoons Mn^{2+} + 4H_2O$	1.507
$Mn_2O_3(s) + 6H^+ + 2e^- \rightleftharpoons 2Mn^{2+} + 3H_2O$	1.485
$MnO_2(s) + 4H^+ + 2e^- \rightleftharpoons Mn^{2+} + 2H_2O$	1.230
$Mn(EDTA)^- + e^- \rightleftharpoons Mn(EDTA)^{2-}$	0.825
$MnO_4^- + e^- \rightleftharpoons MnO_4^{2-}$	0.56
$3Mn_2O_3(s) + H_2O + 2e^-$ $\rightleftharpoons 2Mn_3O_4(s) + 2OH^-$	0.002
$Mn_3O_4(s) + 4H_2O + 2e^-$ $\rightleftharpoons 3Mn(OH)_2(s) + 2OH^-$	-0.352
$Mn^{2+} + 2e^- \rightleftharpoons Mn(s)$	-1.182
$Mn(OH)_2(s) + 2e^- \rightleftharpoons Mn(s) + 2OH^-$	-1.565
Mercury	
$2Hg^{2+} + 2e^- \rightleftharpoons Hg_2^{2+}$	0.908
$Hg^{2+} + 2e^- \rightleftharpoons Hg(l)$	0.852
$Hg_2^{2+} + 2e^- \rightleftharpoons 2Hg(l)$	0.796
$Hg_2SO_4(s) + 2e^- \rightleftharpoons 2Hg(l) + SO_4^{2-}$	0.614
$Hg_2Cl_2(s) + 2e^- \rightleftharpoons 2Hg(l) + 2Cl^-$	$\begin{cases} 0.268 \\ 0.241 \end{cases}$ (saturated calomel electrode)
$Hg(OH)_3^- + 2e^- \rightleftharpoons Hg(l) + 3OH^-$	0.231
$Hg(OH)_2 + 2e^- \rightleftharpoons Hg(l) + 2OH^-$	0.206
$Hg_2Br_2(s) + 2e^- \rightleftharpoons 2Hg(l) + 2Br^-$	0.140

(continued)

Reaction[†]	$E°$ (volts)
$HgO(s, \text{yellow}) + H_2O + 2e^-$ $\rightleftharpoons Hg(l) + 2OH^-$	0.098 3
$HgO(s, \text{red}) + H_2O + 2e^- \rightleftharpoons Hg(l) + 2OH^-$	0.097 7
Molybdenum	
$MoO_4^{2-} + 2H_2O + 2e^- \rightleftharpoons MoO_2(s) + 4OH^-$	−0.818
$MoO_4^{2-} + 4H_2O + 6e^- \rightleftharpoons Mo(s) + 8OH^-$	−0.926
$MoO_2(s) + 2H_2O + 4e^- \rightleftharpoons Mo(s) + 4OH^-$	−0.980
Nickel	
$NiOOH(s) + 3H^+ + e^- \rightleftharpoons Ni^{2+} + 2H_2O$	2.05
$Ni^{2+} + 2e^- \rightleftharpoons Ni(s)$	−0.236
$Ni(CN)_4^{2-} + e^- \rightleftharpoons Ni(CN)_3^{2-} + CN^-$	−0.401
$Ni(OH)_2(s) + 2e^- \rightleftharpoons Ni(s) + 2OH^-$	−0.714
Nitrogen	
$HN_3 + 3H^+ + 2e^- \rightleftharpoons N_2(g) + NH_4^+$	2.079
$N_2O(g) + 2H^+ + 2e^- \rightleftharpoons N_2(g) + H_2O$	1.769
$2NO(g) + 2H^+ + 2e^- \rightleftharpoons N_2O(g) + H_2O$	1.587
$NO^+ + e^- \rightleftharpoons NO(g)$	1.46
$2NH_3OH^+ + H^+ + 2e^- \rightleftharpoons N_2H_5^+ + 2H_2O$	1.40
$NH_3OH^+ + 2H^+ + 2e^- \rightleftharpoons NH_4^+ + H_2O$	1.33
$N_2H_5^+ + 3H^+ + 2e^- \rightleftharpoons 2NH_4^+$	1.250
$HNO_2 + H^+ + e^- \rightleftharpoons NO(g) + H_2O$	0.984
$NO_3^- + 4H^+ + 3e^- \rightleftharpoons NO(g) + 2H_2O$	0.955
$NO_3^- + 3H^+ + 2e^- \rightleftharpoons HNO_2 + H_2O$	0.940
$NO_3^- + 2H^+ + e^- \rightleftharpoons \frac{1}{2}N_2O_4(g) + H_2O$	0.798
$N_2(g) + 8H^+ + 6e^- \rightleftharpoons 2NH_4^+$	0.274
$N_2(g) + 5H^+ + 4e^- \rightleftharpoons N_2H_5^+$	−0.214
$N_2(g) + 2H_2O + 4H^+ + 2e^-$ $\rightleftharpoons 2NH_3OH^+$	−1.83
$\frac{3}{2}N_2(g) + H^+ + e^- \rightleftharpoons HN_3$	−3.334
Oxygen	
$OH + H^+ + e^- \rightleftharpoons H_2O$	2.56
$O(g) + 2H^+ + 2e^- \rightleftharpoons H_2O$	2.430 1
$O_3(g) + 2H^+ + 2e^- \rightleftharpoons O_2(g) + H_2O$	2.075
$H_2O_2 + 2H^+ + 2e^- \rightleftharpoons 2H_2O$	1.763
$HO_2 + H^+ + e^- \rightleftharpoons H_2O_2$	1.44
$\frac{1}{2}O_2(g) + 2H^+ + 2e^- \rightleftharpoons H_2O$	1.229 1
$O_2(g) + 2H^+ + 2e^- \rightleftharpoons H_2O_2$	0.695
$O_2(g) + H^+ + e^- \rightleftharpoons HO_2$	−0.05
Palladium	
$Pd^{2+} + 2e^- \rightleftharpoons Pd(s)$	0.915
$PdO(s) + 2H^+ + 2e^- \rightleftharpoons Pd(s) + H_2O$	0.79
$PdCl_6^{4-} + 2e^- \rightleftharpoons Pd(s) + 6Cl^-$	0.615
$PdO_2(s) + H_2O + 2e^- \rightleftharpoons PdO(s) + 2OH^-$	0.64
Phosphorus	
$\frac{1}{4}P_4(s, \text{white}) + 3H^+ + 3e^- \rightleftharpoons PH_3(g)$	−0.046
$\frac{1}{4}P_4(s, \text{red}) + 3H^+ + 3e^- \rightleftharpoons PH_3(g)$	−0.088
$H_3PO_4 + 2H^+ + 2e^- \rightleftharpoons H_3PO_3 + H_2O$	−0.30
$H_3PO_4 + 5H^+ + 5e^- \rightleftharpoons \frac{1}{4}P_4(s, \text{white}) + 4H_2O$	−0.402
$H_3PO_3 + 2H^+ + 2e^- \rightleftharpoons H_3PO_2 + H_2O$	−0.48
$H_3PO_2 + H^+ + e^- \rightleftharpoons \frac{1}{4}P_4(s) + 2H_2O$	−0.51
Platinum	
$Pt^{2+} + 2e^- \rightleftharpoons Pt(s)$	1.18
$PtO_2(s) + 4H^+ + 4e^- \rightleftharpoons Pt(s) + 2H_2O$	0.92
$PtCl_4^{2-} + 2e^- \rightleftharpoons Pt(s) + 4Cl^-$	0.755
$PtCl_6^{2-} + 2e^- \rightleftharpoons PtCl_4^{2-} + 2Cl^-$	0.68
Potassium	
$K^+ + e^- \rightleftharpoons K(s)$	−2.936

Reaction[†]	$E°$ (volts)
Rubidium	
$Rb^+ + e^- \rightleftharpoons Rb(s)$	−2.943
Scandium	
$Sc^{3+} + 3e^- \rightleftharpoons Sc(s)$	−2.09
Selenium	
$SeO_4^{2-} + 4H^+ + 2e^- \rightleftharpoons H_2SeO_3 + H_2O$	1.150
$H_2SeO_3 + 4H^+ + 4e^- \rightleftharpoons Se(s) + 3H_2O$	0.739
$Se(s) + 2H^+ + 2e^- \rightleftharpoons H_2Se(g)$	−0.082
$Se(s) + 2e^- \rightleftharpoons Se^{2-}$	−0.67
Silicon	
$Si(s) + 4H^+ + 4e^- \rightleftharpoons SiH_4(g)$	−0.147
$SiO_2(s, \text{quartz}) + 4H^+ + 4e^-$ $\rightleftharpoons Si(s) + 2H_2O$	−0.990
$SiF_6^{2-} + 4e^- \rightleftharpoons Si(s) + 6F^-$	−1.24
Silver	
$Ag^{2+} + e^- \rightleftharpoons Ag^+$	1.989
$Ag^{3+} + 2e^- \rightleftharpoons Ag^+$	1.9
$AgO(s) + H^+ + e^- \rightleftharpoons \frac{1}{2}Ag_2O(s) + \frac{1}{2}H_2O$	1.40
$Ag^+ + e^- \rightleftharpoons Ag(s)$	0.799 3
$Ag_2C_2O_4(s) + 2e^- \rightleftharpoons 2Ag(s) + C_2O_4^{2-}$	0.465
$AgN_3(s) + e^- \rightleftharpoons Ag(s) + N_3^-$	0.293
$AgCl(s) + e^- \rightleftharpoons Ag(s) + Cl^-$	$\begin{cases} 0.222 \quad \text{(saturated} \\ 0.197 \quad \text{KCl)} \end{cases}$
$AgBr(s) + e^- \rightleftharpoons Ag(s) + Br^-$	0.071
$Ag(S_2O_3)_2^{3-} + e^- \rightleftharpoons Ag(s) + 2S_2O_3^{2-}$	0.017
$AgI(s) + e^- \rightleftharpoons Ag(s) + I^-$	−0.152
$Ag_2S(s) + H^+ + 2e^- \rightleftharpoons 2Ag(s) + SH^-$	−0.272
Sodium	
$Na^+ + \frac{1}{2}H_2(g) + e^- \rightleftharpoons NaH(s)$	−2.367
$Na^+ + e^- \rightleftharpoons Na(s)$	−2.714 3
Strontium	
$Sr^{2+} + 2e^- \rightleftharpoons Sr(s)$	−2.889
Sulfur	
$S_2O_8^{2-} + 2e^- \rightleftharpoons 2SO_4^{2-}$	2.01
$S_2O_6^{2-} + 4H^+ + 2e^- \rightleftharpoons 2H_2SO_3$	0.57
$4SO_2 + 4H^+ + 6e^- \rightleftharpoons S_4O_6^{2-} + 2H_2O$	0.539
$SO_2 + 4H^+ + 4e^- \rightleftharpoons S(s) + 2H_2O$	0.450
$2H_2SO_3 + 2H^+ + 4e^- \rightleftharpoons S_2O_3^{2-} + 3H_2O$	0.40
$S(s) + 2H^+ + 2e^- \rightleftharpoons H_2S(g)$	0.174
$S(s) + 2H^+ + 2e^- \rightleftharpoons H_2S(aq)$	0.144
$S_4O_6^{2-} + 2H^+ + 2e^- \rightleftharpoons 2HS_2O_3^-$	0.10
$5S(s) + 2e^- \rightleftharpoons S_5^{2-}$	−0.340
$2S(s) + 2e^- \rightleftharpoons S_2^{2-}$	−0.50
$2SO_3^{2-} + 3H_2O + 4e^- \rightleftharpoons S_2O_3^{2-} + 6OH^-$	−0.566
$SO_3^{2-} + 3H_2O + 4e^- \rightleftharpoons S(s) + 6OH^-$	−0.659
$SO_4^{2-} + 4H_2O + 6e^- \rightleftharpoons S(s) + 8OH^-$	−0.751
$SO_4^{2-} + H_2O + 2e^- \rightleftharpoons SO_3^{2-} + 2OH^-$	−0.936
$2SO_3^{2-} + 2H_2O + 2e^- \rightleftharpoons S_2O_4^{2-} + 4OH^-$	−1.130
$2SO_4^{2-} + 2H_2O + 2e^- \rightleftharpoons S_2O_6^{2-} + 4OH^-$	−1.71
Thallium	
$Tl^{3+} + 2e^- \rightleftharpoons Tl^+$	$\begin{cases} 1.280 \\ 0.77 \quad 1\text{ F HCl} \\ 1.22 \quad 1\text{ F H}_2\text{SO}_4 \\ 1.23 \quad 1\text{ F HNO}_3 \\ 1.26 \quad 1\text{ F HClO}_4 \end{cases}$
$Tl^+ + e^- \rightleftharpoons Tl(s)$	−0.336

Reaction[†]	$E°$ (volts)
Tin	
$Sn(OH)_3^+ + 3H^+ + 2e^- \rightleftharpoons Sn^{2+} + 6H_2O$	0.142
$Sn^{4+} + 2e^- \rightleftharpoons Sn^{2+}$	0.139 1 F HCl
$SnO_2(s) + 4H^+ \rightleftharpoons Sn^{2+} + 2H_2O$	−0.094
$Sn^{2+} + 2e^- \rightleftharpoons Sn(s)$	−0.141
$SnF_6^{2-} + 4e^- \rightleftharpoons Sn(s) + 6F^-$	−0.25
$Sn(OH)_6^{2-} + 2e^- \rightleftharpoons Sn(OH)_3^- + 3OH^-$	−0.93
$Sn(s) + 4H_2O + 4e^- \rightleftharpoons SnH_4(g) + 4OH^-$	−1.316
$SnO_2(s) + H_2O + 2e^- \rightleftharpoons SnO(s) + 2OH^-$	−0.961
Titanium	
$TiO^{2+} + 2H^+ + e^- \rightleftharpoons Ti^{3+} + H_2O$	0.1
$Ti^{3+} + e^- \rightleftharpoons Ti^{2+}$	−0.9
$TiO_2(s) + 4H^+ + 4e^- \rightleftharpoons Ti(s) + 2H_2O$	−1.076
$TiF_6^{2-} + 4e^- \rightleftharpoons Ti(s) + 6F^-$	−1.191
$Ti^{2+} + 2e^- \rightleftharpoons Ti(s)$	−1.60
Tungsten	
$W(CN)_8^{3-} + e^- \rightleftharpoons W(CN)_8^{4-}$	0.457
$W^{6+} + e^- \rightleftharpoons W^{5+}$	0.26 12 F HCl
$WO_3(s) + 6H^+ + 6e^- \rightleftharpoons W(s) + 3H_2O$	−0.091
$W^{5+} + e^- \rightleftharpoons W^{4+}$	−0.3 12 F HCl
$WO_2(s) + 2H_2O + 4e^- \rightleftharpoons W(s) + 4OH^-$	−0.982
$WO_4^{2-} + 4H_2O + 6e^- \rightleftharpoons W(s) + 8OH^-$	−1.060
Uranium	
$UO_2^+ + 4H^+ + e^- \rightleftharpoons U^{4+} + 2H_2O$	0.39
$UO_2^{2+} + 4H^+ + 2e^- \rightleftharpoons U^{4+} + 2H_2O$	0.273

Reaction[†]	$E°$ (volts)
$UO_2^{2+} + e^- \rightleftharpoons UO_2^+$	0.16
$U^{4+} + e^- \rightleftharpoons U^{3+}$	−0.577
$U^{3+} + 3e^- \rightleftharpoons U(s)$	−1.642
Vanadium	
$VO_2^+ + 2H^+ + e^- \rightleftharpoons VO^{2+} + H_2O$	1.001
$VO^{2+} + 2H^+ + e^- \rightleftharpoons V^{3+} + H_2O$	0.337
$V^{3+} + e^- \rightleftharpoons V^{2+}$	−0.255
$V^{2+} + 2e^- \rightleftharpoons V(s)$	−1.125
Xenon	
$H_4XeO_6 + 2H^+ + 2e^- \rightleftharpoons XeO_3 + 3H_2O$	2.38
$XeF_2 + 2H^+ + 2e^- \rightleftharpoons Xe(g) + 2HF$	2.2
$XeO_3 + 6H^+ + 6e^- \rightleftharpoons Xe(g) + 3H_2O$	2.1
Yttrium	
$Y^{3+} + 3e^- \rightleftharpoons Y(s)$	−2.38
Zinc	
$ZnOH^+ + H^+ + 2e^- \rightleftharpoons Zn(s) + 2H_2O$	−0.497
$Zn^{2+} + 2e^- \rightleftharpoons Zn(s)$	−0.762
$Zn(NH_3)_4^{2+} + 2e^- \rightleftharpoons Zn(s) + 4NH_3$	−1.04
$ZnCO_3(s) + 2e^- \rightleftharpoons Zn(s) + CO_3^{2-}$	−1.06
$Zn(OH)_3^- + 2e^- \rightleftharpoons Zn(s) + 3OH^-$	−1.183
$Zn(OH)_4^{2-} + 2e^- \rightleftharpoons Zn(s) + 4OH^-$	−1.199
$Zn(OH)_2(s) + 2e^- \rightleftharpoons Zn(s) + 2OH^-$	−1.249
$ZnO(s) + H_2O + 2e^- \rightleftharpoons Zn(s) + 2OH^-$	−1.260
$ZnS(s) + 2e^- \rightleftharpoons Zn(s) + S^{2-}$	−1.405

Appendix *D*

OXIDATION NUMBERS AND BALANCING REDOX EQUATIONS

*T*he *oxidation number,* or *oxidation state,* is a bookkeeping device used to keep track of the number of electrons formally associated with a particular element. The oxidation number is meant to tell how many electrons have been lost or gained by a neutral atom when it forms a compound. Because oxidation numbers have no real physical meaning, they are somewhat arbitrary, and not all chemists will assign the same oxidation number to a given element in an unusual compound. However, there are some ground rules that provide a useful start.

1. The oxidation number of an element by itself—e.g., $Cu(s)$ or $Cl_2(g)$—is 0.

2. The oxidation number of H is almost always $+1$, except in metal hydrides—e.g., NaH, in which H is -1.

3. The oxidation number of oxygen is almost always -2. The only common exceptions are peroxides, in which two oxygen atoms are connected and each has an oxidation number of -1. Two examples are hydrogen peroxide ($H-O-O-H$) and its anion ($H-O-O^-$). The oxidation number of oxygen in gaseous O_2 is, of course, 0.

4. The alkali metals (Li, Na, K, Rb, Cs, Fr) almost always have an oxidation number of $+1$. The alkaline earth metals (Be, Mg, Ca, Sr, Ba, Ra) are almost always in the $+2$ oxidation state.

5. The halogens (F, Cl, Br, I) are usually in the -1 oxidation state. Exceptions occur when two different halogens are bound to each other or when a halogen is bound to more than one atom. When different halogens are bound to each other, we assign the oxidation number -1 to the more electronegative halogen.

The sum of the oxidation numbers of each atom in a molecule must equal the charge of the molecule. In H_2O, for example, we have

$$
\begin{array}{lr}
2 \text{ hydrogen} = 2(+1) = & +2 \\
\text{oxygen} & = -2 \\
\hline
\text{net charge} & 0
\end{array}
$$

In SO_4^{2-}, sulfur must have an oxidation number of $+6$ so that the sum of the oxidation numbers will be -2:

$$
\begin{array}{lr}
\text{oxygen} = 4(-2) = & -8 \\
\text{sulfur} & = +6 \\
\hline
\text{net charge} & -2
\end{array}
$$

In benzene (C_6H_6), the oxidation number of each carbon must be -1 if hydrogen is assigned the number $+1$. In cyclohexane (C_6H_{12}), the oxidation number of each carbon must be -2, for the same reason. The carbons in benzene are in a higher oxidation state than those in cyclohexane.

The oxidation number of iodine in ICl_2^- is $+1$. This is unusual, because halogens are usually -1. However, because chlorine is more electronegative than iodine, we assign Cl as -1, thereby forcing I to be $+1$.

The oxidation number of As in As_2S_3 is $+3$, and the value for S is -2. This is arbitrary, but reasonable. Because S is more electronegative than As, we make S negative and As positive; and because S is in the same family as oxygen, which is usually -2, we assign S as -2, thus leaving As as $+3$.

The oxidation number of S in $S_4O_6^{2-}$ (tetrathionate) is $+2.5$. The *fractional oxidation state* comes about because six O atoms contribute -12. Because the charge is -2, the four S atoms must contribute $+10$. The average oxidation number of S must be $+\frac{10}{4} = 2.5$.

The oxidation number of Fe in $K_3Fe(CN)_6$ is $+3$. To make this assignment, we first recognize cyanide (CN^-) as a common ion that carries a charge of -1. Six cyanide ions give -6, and

three potassium ions (K^+) give $+3$. Therefore, Fe should have an oxidation number of $+3$ for the whole formula to be neutral. In this approach, it is not necessary to assign individual oxidation numbers to carbon and nitrogen, as long as we recognize that the charge of CN is -1.

Problems

Answers are given at the end of this appendix.

1. Write the oxidation state of the boldface atom in each of the following species.

(a) **Ag**Br
(b) **S**$_2$O$_3^{2-}$
(c) **Se**F$_6$
(d) H**S**$_2$O$_3^-$
(e) HO$_2$
(f) **N**O

(g) **Cr**$^{3+}$
(h) **Mn**O$_2$
(i) **Pb**(OH)$_3^-$
(j) **Fe**(OH)$_3$
(k) **Cl**O$^-$
(l) K$_4$**Fe**(CN)$_6$

(m) **Cl**O$_2$
(n) **Cl**O$_2^-$
(o) **Mn**(CN)$_6^{4-}$
(p) **N**$_2$
(q) **N**H$_4^+$
(r) **N**$_2$H$_5^+$

(s) H**As**O$_3^{2-}$
(t) (CH$_3$)$_4$**Li**$_4$
(u) **P**$_4$O$_{10}$
(v) **C**$_2$H$_6$O
(w) **V**O(SO$_4$)
(x) **Fe**$_3$O$_4$

2. Identify the oxidizing agent and the reducing agent on the left side of each of the following reactions.

(a) $Cr_2O_7^{2-} + 3Sn^{2+} + 14H^+ \rightarrow 2Cr^{3+} + 3Sn^{4+} + 7H_2O$

(b) $4I^- + O_2 + 4H^+ \rightarrow 2I_2 + 2H_2O$

(c) $5CH_3\overset{\overset{\displaystyle O}{\|}}{C}H + 2MnO_4^- + 6H^+ \rightarrow$

$5CH_3\overset{\overset{\displaystyle O}{\|}}{C}OH + 2Mn^{2+} + 3H_2O$

Balancing Redox Reactions

To balance a reaction involving oxidation and reduction, we must first identify which element is oxidized and which is reduced. We then break the net reaction into two imaginary *half-reactions,* one of which involves only oxidation and the other only reduction. Although free electrons never appear in a balanced net reaction, they do appear in balanced half-reactions. If we are dealing with aqueous solutions, we proceed to balance each half-reaction, using H_2O and either H^+ or OH^-, as necessary. *A reaction is balanced when the number of atoms of each element is the same on both sides, and the net charge is the same on both sides.*

Acidic Solutions

Here are the steps we will follow:

1. Assign oxidation numbers to the elements that are oxidized or reduced.

2. Break the reaction into two half-reactions, one involving oxidation and the other reduction.

3. For each half-reaction, balance the number of atoms that are oxidized or reduced.

4. Balance the electrons to account for the change in oxidation number by adding electrons to one side of each half-reaction.

5. Balance oxygen atoms by adding H_2O to one side of each half-reaction.

6. Balance the H atoms by adding H^+ to one side of each half-reaction.

7. Multiply each half-reaction by the number of electrons in the other half-reaction so that the number of electrons on each side of the total reaction will cancel. Then add the two half-reactions and simplify to the smallest integral coefficients.

EXAMPLE   Balancing a Redox Equation

Balance the following equation, using H^+, but not OH^-:

$$Fe^{2+} + MnO_4^- \rightleftharpoons Fe^{3+} + Mn^{2+}$$
$$+2 \qquad +7 \qquad\quad +3 \qquad +2$$
$$\text{Permanganate}$$

SOLUTION

1. *Assign oxidation numbers.* These are assigned for Fe and Mn in each species in the above reaction.

2. *Break the reaction into two half-reactions.*

$$\text{oxidation half-reaction:}\quad Fe^{2+} \rightleftharpoons Fe^{3+}$$
$$+2 \qquad\quad +3$$

$$\text{reduction half-reaction:}\quad MnO_4^- \rightleftharpoons Mn^{2+}$$
$$+7 \qquad\quad +2$$

3. *Balance the atoms that are oxidized or reduced.* Because there is only one Fe or Mn in each species on each side of the equation, the atoms of Fe and Mn are already balanced.

4. *Balance electrons.* Electrons are added to account for the change in each oxidation state.

$$Fe^{2+} \rightleftharpoons Fe^{3+} + e^-$$
$$MnO_4^- + 5e^- \rightleftharpoons Mn^{2+}$$

In the second case, we need $5e^-$ on the left side to take Mn from $+7$ to $+2$.

5. *Balance oxygen atoms.* There are no oxygen atoms in the Fe half-reactions. There are four oxygen atoms on the left side of the Mn reaction, so we add four molecules of H_2O to the right side:

$$MnO_4^- + 5e^- \rightleftharpoons Mn^{2+} + 4H_2O$$

6. *Balance hydrogen atoms.* The Fe equation is already balanced. The Mn equation needs $8H^+$ on the left.

$$MnO_4^- + 5e^- + 8H^+ \rightleftharpoons Mn^{2+} + 4H_2O$$

At this point, each half-reaction must be completely balanced (the same number of atoms and charge on each side), *or you have made a mistake.*

7. *Multiply and add the reactions.* We multiply the Fe equation by 5 and the Mn equation by 1 and add:

$$5Fe^{2+} \rightleftharpoons 5Fe^{3+} + 5\cancel{e^-}$$
$$\underline{MnO_4^- + 5\cancel{e^-} + 8H^+ \rightleftharpoons Mn^{2+} + 4H_2O}$$
$$5Fe^{2+} + MnO_4^- + 8H^+ \rightleftharpoons 5Fe^{3+} + Mn^{2+} + 4H_2O$$

The total charge on each side is $+17$, and we find the same number of atoms of each element on each side. The equation is balanced.

Basic Solutions

The method many people prefer for basic solutions is to balance the equation first with H^+. The answer can then be converted to one in which OH^- is used instead. This is done by adding to each side of the equation a number of hydroxide ions equal to the number of H^+ ions appearing in the equation. For example, to balance Equation D-1 with OH^- instead of H^+, proceed as follows:

$$2I_2 + IO_3^- + 10Cl^- + 6H^+ \rightleftharpoons 5ICl_2^- + 3H_2O \qquad \text{(D-1)}$$
$$\underline{+ \, 6OH^- \qquad\qquad\qquad\qquad + \, 6OH^-}$$
$$2I_2 + IO_3^- + 10Cl^- + \underbrace{6H^+ + 6OH^-}_{\substack{6H_2O \\ \Downarrow \\ 3H_2O}} \rightleftharpoons 5ICl_2^- + 3H_2O + 6OH^-$$

Realizing that $6H^+ + 6OH^- = 6H_2O$, and canceling $3H_2O$ on each side, gives the final result:

$$2I_2 + IO_3^- + 10Cl^- + 3H_2O \rightleftharpoons 5ICl_2^- + 6OH^-$$

Problems

3. Balance the following reactions, using H^+, but not OH^-.

(a) $Fe^{3+} + Hg_2^{2+} \rightleftharpoons Fe^{2+} + Hg^{2+}$

(b) $Ag + NO_3^- \rightleftharpoons Ag^+ + NO$

(c) $VO^{2+} + Sn^{2+} \rightleftharpoons V^{3+} + Sn^{4+}$

(d) $SeO_4^{2-} + Hg + Cl^- \rightleftharpoons SeO_3^{2-} + Hg_2Cl_2$

(e) $CuS + NO_3^- \rightleftharpoons Cu^{2+} + SO_4^{2-} + NO$

(f) $S_2O_3^{2-} + I_2 \rightleftharpoons I^- + S_4O_6^{2-}$

(g) $Cr_2O_7^{2-} + CH_3\overset{\overset{\displaystyle O}{\|}}{C}H \rightleftharpoons CH_3\overset{\overset{\displaystyle O}{\|}}{C}OH + Cr^{3+}$

(h) $MnO_4^{2-} \rightleftharpoons MnO_2 + MnO_4^-$

(i) $ClO_3^- \rightleftharpoons Cl_2 + O_2$

4. Balance the following equations by using OH^-, but not H^+.

(a) $PbO_2 + Cl^- \rightleftharpoons ClO^- + Pb(OH)_3^-$

(b) $HNO_2 + SbO^+ \rightleftharpoons NO + Sb_2O_5$

(c) $Ag_2S + CN^- + O_2 \rightleftharpoons S + Ag(CN)_2^- + OH^-$

(d) $HO_2^- + Cr(OH)_3^- \rightleftharpoons CrO_4^{2-} + OH^-$

(e) $ClO_2 + OH^- \rightleftharpoons ClO_2^- + ClO_3^-$

(f) $WO_3^- + O_2 \rightleftharpoons HW_6O_{21}^{5-} + OH^-$

Answers

1. (a) $+1$ **(i)** $+2$ **(q)** -3

(b) $+2$ **(j)** $+3$ **(r)** -2

(c) $+6$ **(k)** $+1$ **(s)** $+3$

(d) $+2$ **(l)** $+2$ **(t)** -4

(e) $-1/2$ **(m)** $+4$ **(u)** $+5$

(f) $+2$ **(n)** $+3$ **(v)** -2

(g) $+3$ **(o)** $+2$ **(w)** $+4$

(h) $+4$ **(p)** 0 **(x)** $+8/3$

2.

	Oxidizing agent	Reducing agent
(a)	$Cr_2O_7^{2-}$	Sn^{2+}
(b)	O_2	I^-
(c)	MnO_4^-	CH_3CHO

3. (a) $2Fe^{3+} + Hg_2^{2+} \rightleftharpoons 2Fe^{2+} + 2Hg^{2+}$

(b) $3Ag + NO_3^- + 4H^+ \rightleftharpoons 3Ag^+ + NO + 2H_2O$

(c) $4H^+ + 2VO^{2+} + Sn^{2+} \rightleftharpoons 2V^{3+} + Sn^{4+} + 2H_2O$

(d) $2Hg + 2Cl^- + SeO_4^{2-} + 2H^+ \rightleftharpoons$
$$Hg_2Cl_2 + SeO_3^{2-} + H_2O$$

(e) $3CuS + 8NO_3^- + 8H^+ \rightleftharpoons$
$$3Cu^{2+} + 3SO_4^{2-} + 8NO + 4H_2O$$

(f) $2S_2O_3^{2-} + I_2 \rightleftharpoons S_4O_6^{2-} + 2I^-$

(g) $Cr_2O_7^{2-} + 3CH_3CHO + 8H^+ \rightleftharpoons$
$$2Cr^{3+} + 3CH_3CO_2H + 4H_2O$$

(h) $4H^+ + 3MnO_4^{2-} \rightleftharpoons MnO_2 + 2MnO_4^- + 2H_2O$

(i) $2H^+ + 2ClO_3^- \rightleftharpoons Cl_2 + \frac{5}{2}O_2 + H_2O$

4. (a) $H_2O + OH^- + PbO_2 + Cl^- \rightleftharpoons Pb(OH)_3^- + ClO^-$

(b) $4HNO_2 + 2SbO^+ + 2OH^- \rightleftharpoons 4NO + Sb_2O_5 + 3H_2O$

(c) $Ag_2S + 4CN^- + \frac{1}{2}O_2 + H_2O \rightleftharpoons$
$$S + 2Ag(CN)_2^- + 2OH^-$$

(d) $2HO_2^- + Cr(OH)_3^- \rightleftharpoons CrO_4^{2-} + OH^- + 2H_2O$

(e) $2ClO_2 + 2OH^- \rightleftharpoons ClO_2^- + ClO_3^- + H_2O$

(f) $12WO_3^- + 3O_2 + 2H_2O \rightleftharpoons 2HW_6O_{21}^{5-} + 2OH^-$

SOLUTIONS TO "ASK YOURSELF" QUESTIONS

Chapter 0

0-A. **(a)** A heterogeneous material has different compositions in different regions. A homogeneous material is the same everywhere.

(b) The composition of a random heterogeneous material varies randomly from place to place. A segregated heterogeneous material has relatively large regions of distinctly different compositions.

(c) A random sample is selected by taking material at random from the lot. A composite sample is selected deliberately by taking predetermined portions of the lot from selected regions. Random sampling is appropriate for a random heterogeneous lot. Composite sampling is appropriate when the lot is segregated into regions of different composition.

Chapter 1

1-A. See Table 1-3.

1-B. **(a)** Office worker: $2.2 \times 10^6 \dfrac{\text{cal}}{\text{day}} \times 4.184 \dfrac{\text{J}}{\text{cal}}$

$$= 9.2 \times 10^6 \text{ J/day}$$

Mountain climber: $3.4 \times 10^6 \dfrac{\text{cal}}{\text{day}} \times 4.184 \dfrac{\text{J}}{\text{cal}}$

$$= 14.2 \times 10^6 \text{ J/day}$$

(b) $60 \dfrac{\text{s}}{\text{min}} \times 60 \dfrac{\text{min}}{\text{h}} \times 24 \dfrac{\text{h}}{\text{day}} = 8.64 \times 10^4 \text{ s/day}$

(c) $\dfrac{9.2 \times 10^6 \text{ J/day}}{8.64 \times 10^4 \text{ s/day}} = 1.1 \times 10^2 \text{ W};$

$$14.2 \times 10^6 \text{ J/day} = 1.6 \times 10^2 \text{ W}$$

(d) The office worker consumes more power than the light bulb.

1-C. **(a)** $\left(1.67 \dfrac{\text{g solution}}{\text{mL}}\right)\left(1\,000 \dfrac{\text{mL}}{\text{L}}\right) = 1\,670 \dfrac{\text{g solution}}{\text{L}}$

(b) $\left(0.705 \dfrac{\text{g HClO}_4}{\text{g solution}}\right)\left(1\,670 \dfrac{\text{g solution}}{\text{L}}\right)$

$$= 1.18 \times 10^3 \dfrac{\text{g HClO}_4}{\text{L}}$$

(c) $\left(1.18 \times 10^3 \dfrac{\text{g}}{\text{L}}\right)\Big/\left(100.458 \dfrac{\text{g}}{\text{mol}}\right) = 11.7 \dfrac{\text{mol}}{\text{L}}$

1-D. **(a)** $\left(1.50 \dfrac{\text{g solution}}{\text{mL solution}}\right)\left(1\,000 \dfrac{\text{mL solution}}{\text{L solution}}\right)$

$$= 1.50 \times 10^3 \dfrac{\text{g solution}}{\text{L solution}}$$

(b) $\left(0.480 \dfrac{\text{g HBr}}{\text{g solution}}\right)\left(1.50 \times 10^3 \dfrac{\text{g solution}}{\text{L solution}}\right)$

$$= 7.20 \times 10^2 \dfrac{\text{g HBr}}{\text{L solution}}$$

(c) $\left(7.20 \times 10^2 \dfrac{\text{g HBr}}{\text{L solution}}\right)\Big/\left(\dfrac{80.912 \text{ g HBr}}{\text{mol}}\right)$

$$= 8.90 \text{ M}$$

(d) $M_{con} \cdot V_{con} = M_{dil} \cdot V_{dil}$

$(8.90 \text{ M})(x \text{ mL}) = (0.160 \text{ M})(250 \text{ mL}) \quad \Rightarrow$

$$x = 4.49 \text{ mL}$$

1-E. **(a)** and **(b)**

$$
\begin{array}{lll}
\text{HOBr} + \text{OCl}^- \rightleftharpoons \text{HOCl} + \text{OBr}^- & K_1 = 1/15 \\
\underline{\text{HOCl} \rightleftharpoons \text{H}^+ + \text{OCl}^-} & K_2 = 3.0 \times 10^{-8} \\
\text{HOBr} \rightleftharpoons \text{H}^+ + \text{OBr}^- & K = K_1 K_2 = \\
& \quad 2 \times 10^{-9}
\end{array}
$$

(c) Consumption of a product drives the reaction forward (to the right).

Chapter 2

2-A. The lab notebook must (1) state what was done; (2) state what was observed; and (3) be understandable to a stranger.

2-B. **(a)** $m = \dfrac{(24.913 \text{ g})\left(1 - \dfrac{0.001\,2 \text{ g/mL}}{8.0 \text{ g/mL}}\right)}{\left(1 - \dfrac{0.001\,2 \text{ g/mL}}{1.00 \text{ g/mL}}\right)} = 24.939 \text{ g}$

(b) $(24.939 \text{ g})/(0.998\,00 \text{ g/mL}) = 24.989 \text{ mL}$

2-C. Dissolve $(0.250\,0 \text{ L})(0.150\,0 \text{ mol/L}) = 0.037\,50 \text{ mol}$ of K_2SO_4 in less than 250 mL of water in a 250-mL volumetric flask. Add more water and mix. Dilute to the 250.0-mL mark and invert the flask 20 times for complete mixing.

2-D. Transfer pipet. The last liquid should be blown out of the serological pipet because it is calibrated down to the tip. With the measuring pipet, simply drain liquid from the 0-mL mark to the 1-mL mark.

2-E. $(10.000\ 0\ \text{g})(1.002\ 0\ \text{mL/g}) = 10.020\ \text{mL}$

2-F. $S^{2-} + 2H^+ \rightarrow H_2S$. $H_2S(g)$ is lost when the solution is boiled to dryness.

Chapter 3

3-A. **(a)** 5   **(b)** 4   **(c)** 3

3-B. **(a)** 3.71   **(b)** 10.7   **(c)** 4.0×10^1 **(d)** 2.85×10^{-6}
(e) 12.625 1 **(f)** 6.0×10^{-4} **(g)** 242

3-C. **(a)** Carmen **(b)** Cynthia   **(c)** Chastity   **(d)** Cheryl

3-D. **(a)** percent relative uncertainty in mass =
$$(0.002/4.635) \times 100 = 0.04_3\%$$
percent relative uncertainty in volume =
$$(0.05/1.13) \times 100 = 4.4\%$$
(b) density $= \dfrac{4.635 \pm 0.002\ \text{g}}{1.13 \pm 0.05\ \text{mL}} = \dfrac{4.635\ (\pm 0.04_3\%)\ \text{g}}{1.13\ (\pm 4._4\%)\ \text{mL}}$
$$= 4.10 \pm ?\ \text{g/mL}$$
uncertainty $= \sqrt{(0.04_3)^2 + (4._4)^2} = 4._4\%$
$4._4\%$ of $4.10 = 0.18$.
The answer can be written 4.1 ± 0.2 g/mL

3-E. $-196°C = 77.15\ \text{K} = -320.8°F$

Chapter 4

4-A. $\bar{x} = \dfrac{821 + 783 + 834 + 855}{4} = 823._2$

$$s = \sqrt{\dfrac{\begin{array}{c}(821 - 823._2)^2 + (783 - 823._2)^2 \\ + (834 - 823._2)^2 + (855 - 823._2)^2\end{array}}{4 - 1}}$$
$$= 30._3$$
relative standard deviation =
$$(30._3/823._2) \times 100 = 3.6_8\%$$
median $= (821 + 834)/2 = 827._5$
range $= 855 - 783 = 72$

4-B. **(a)** $\bar{x} = \dfrac{117 + 119 + 111 + 115 + 120}{5} = 116._4$

$$s = \sqrt{\dfrac{\begin{array}{c}(117 - 116._4)^2 + (119 - 116._4)^2 \\ + \ldots + (120 - 116._4)^2\end{array}}{5 - 1}}$$
$$= 3._{58}$$

(b) $s_{\text{pooled}} = \sqrt{\dfrac{2.8^2(4 - 1) + 3.58^2(5 - 1)}{4 + 5 - 2}} = 3.27$

$t_{\text{calculated}} = \dfrac{|111.0 - 116.4|}{3.27} \sqrt{\dfrac{4 \cdot 5}{4 + 5}} = 2.46$

$t_{\text{table}} = 2.365$ for 95% confidence and $4 + 5 - 2$
$= 7$ degrees of freedom
$t_{\text{calculated}} > t_{\text{table}}$, so the difference is significant at the 95% confidence interval

4-C. $Q = (216 - 204)/(216 - 192) = 0.50 < 0.64$. Retain 216.

4-D.

x_i	y_i	$x_i y_i$	x_i^2	$d_i\ (= y_i - mx_i - b)$	d_i^2
1	3	3	1	-0.167	0.027 89
3	2	6	9	0.333	0.110 89
5	0	0	25	-0.167	0.027 89
$\Sigma x_i = 9$	$\Sigma y_i = 5$	$\Sigma x_i y_i = 9$	$\Sigma(x_i^2) = 35$		$\Sigma(d_i^2) = 0.166\ 67$

$D = n\,\Sigma(x_i^2) - (\Sigma x_i)^2 = 3 \cdot 35 - 9^2 = 24$
$m = [n\,\Sigma x_i y_i - \Sigma x_i \Sigma y_i]/D$
$$= [3 \cdot 9 - 9 \cdot 5]/24 = -0.750$$
$b = [\Sigma(x_i^2)\Sigma y_i - \Sigma(x_i y_i)\Sigma x_i]/D$
$$= [35 \cdot 5 - 9 \cdot 9]/24 = 3.917$$
$s_y = \sqrt{\dfrac{\Sigma(d_i^2)}{3 - 2}} = \sqrt{\dfrac{0.166\ 67}{1}} = 0.408\ 2$
$s_m = s_y\sqrt{\dfrac{n}{D}} = 0.408\ 2\sqrt{\dfrac{3}{24}} = 0.144$
$s_b = s_y\sqrt{\dfrac{\Sigma(x_i^2)}{D}} = 0.408\ 2\sqrt{\dfrac{35}{24}} = 0.493$
$y\,(\pm 0.4_{08}) = -0.75_0\,(\pm 0.14_4)x + 3.9_{17}\,(\pm 0.4_{93})$

4-E. $x = \dfrac{y - b}{m} = \dfrac{1.00 - 3.9_{17}}{-0.750} = 3.89$

uncertainty in $x = \dfrac{s_y}{|m|}\sqrt{1 + \dfrac{x^2 n}{D} + \dfrac{\Sigma(x_i^2)}{D} - \dfrac{2x\Sigma x_i}{D}}$

$$= \dfrac{0.408}{0.75}\sqrt{1 + \dfrac{(3.89)^2(3)}{24} + \dfrac{35}{24} - \dfrac{2(3.89)(9)}{24}} = 0.65$$

Chapter 5

5-A. **(a)** The concentrations of reagents used in an analysis are determined either by weighing out pure primary standards or by reaction with such standards. If the standards are not pure, none of the concentrations will be correct.

(b) In a blank titration, the quantity of titrant required to reach the end point in the absence of analyte is measured. This quantity is subtracted from the quantity of titrant needed in the presence of analyte.

(c) In a direct titration, titrant reacts directly with analyte. In a back titration, a known excess of reagent that reacts with analyte is used. The excess is then titrated.

(d) The uncertainty in the equivalence point is constant at ±0.04 mL. The relative uncertainty in delivering 20 mL is $0.04/20 = 0.2\%$. The relative uncertainty in delivering 40 mL is only half as great: $0.04/40 = 0.1\%$.

5-B. (a) $0.197\,0$ g of ascorbic acid $= 1.118\,5 \times 10^{-3}$ mol, which requires $1.118\,5 \times 10^{-3}$ mol I_3^-. $[I_3^-] = 1.118\,5 \times 10^{-3}$ mol$/0.029\,41$ L $= 0.038\,03$ M

(b) $(0.031\,63$ L $I_3^-)(0.038\,03$ M$)$
$$= 1.203 \times 10^{-3} \text{ mol } I_3^-$$
$$= 1.203 \times 10^{-3} \text{ mol ascorbic acid}$$

(c) $(1.203 \times 10^{-3}$ mol ascorbic acid$)(176.126$ g/mol$)$
$$= 0.211\,9 \text{ g ascorbic acid}$$
$$\frac{0.211\,9 \text{ g ascorbic acid}}{0.424\,2 \text{ g tablet}} \times 100 = 49.94 \text{ wt \%}$$

5-C. (a) total volume $KMnO_4 = 50.00 + 9.68 - 0.07$
$$= 59.61 \text{ mL}$$
total mol $KMnO_4 = (0.009\,22$ M$)(0.059\,61$ L$)$
$$= 5.496 \times 10^{-4} \text{ mol}$$

(b) volume $Na_2C_2O_4 = 30.00$ mL
mol $Na_2C_2O_4 = (0.025\,00$ M$)(0.030\,00$ L$)$
$$= 7.500 \times 10^{-4} \text{ mol}$$

(c) mol $KMnO_4$ reacting with $Na_2C_2O_4$
$$= (7.500 \times 10^{-4} \text{ mol } Na_2C_2O_4)\left(\frac{2 \text{ mol } KMnO_4}{5 \text{ mol } Na_2C_2O_4}\right)$$
$$= 3.000 \times 10^{-4} \text{ mol}$$

(d) $KMnO_4$ reacting with $NaNO_2$
$$= (\text{total mol } KMnO_4 - KMnO_4 \text{ reacting}$$
$$\text{with } Na_2C_2O_4)$$
$$= 5.496 \times 10^{-4} - 3.00 \times 10^{-4}$$
$$= 2.496 \times 10^{-4} \text{ mol}$$

(e) mol $NaNO_2$ reacting with $KMnO_4$
$$= (2.496 \times 10^{-4} \text{ mol } KMnO_4)\left(\frac{5 \text{ mol } NaNO_2}{2 \text{ mol } KMnO_4}\right)$$
$$= 6.240 \times 10^{-4} \text{ mol}$$

(f) $[NaNO_2] = \dfrac{6.24 \times 10^{-4} \text{ mol}}{0.050\,00 \text{ L}} = 0.012\,48$ M

5-D. (a) $PbBr_2(s) \underset{x \quad 2x}{\overset{K_{sp}}{\rightleftharpoons}} Pb^{2+} + 2Br^-$
$$x(2x)^2 = 2.1 \times 10^{-6} \Rightarrow$$
$$x = [Pb^{2+}] = 8.0_7 \times 10^{-3} \text{ M}$$

(b) $[Pb^{2+}](0.100)^2 = 2.1 \times 10^{-6} \Rightarrow$
$$[Pb^{2+}] = 2.1 \times 10^{-4} \text{ M}$$

5-E. (a)
$$Ag^+ + Br^- \rightleftharpoons AgBr(s) \quad K = 1/K_{sp}(AgBr) = 2.0 \times 10^{12}$$
$$Ag^+ + Cl^- \rightleftharpoons AgCl(s) \quad K = 1/K_{sp}(AgCl) = 5.6 \times 10^9$$
AgBr precipitates first.

(b) mol Br^- = mol Ag^+ at first end point
$$= (0.015\,55 \text{ L})(0.033\,33 \text{ M})$$
$$= 5.183 \times 10^{-4} \text{ mol}$$
$$[Br^-] = (5.183 \times 10^{-4} \text{ mol})/(0.025\,00 \text{ L})$$
$$= 0.020\,73 \text{ M}$$

(c) mol Cl^- = mol Ag^+ at second end point $-$ mol Ag^+ at first end point
$$= (0.042\,23 - 0.015\,55 \text{ L})(0.033\,33 \text{ M})$$
$$= 8.892 \times 10^{-4} \text{ mol}$$
$$[Cl^-] = (8.892 \times 10^{-4} \text{ mol})/(0.025\,00 \text{ L})$$
$$= 0.035\,57 \text{ M}$$

5-F. (a) AgCl is more soluble than AgSCN and will slowly dissolve in the presence of excess SCN^-. This consumes the SCN^- and causes the red end point color to fade. Nitrobenzene coats the AgCl and inhibits reaction with SCN^-.

(b) Consider the titration of C^+ (in a flask) by A^- (from a buret). Before the equivalence point, there is excess C^+ in solution. Selective adsorption of C^+ on the CA crystal surface gives the crystal a positive charge. After the equivalence point, there is excess A^- in solution. Selective adsorption of A^- on the CA crystal surface gives it a negative charge.

(c) Beyond the equivalence point, there is excess $Fe(CN)_6^{4-}$ in solution. Adsorption of this anion on the precipitate will make the particles *negative*.

Chapter 6

6-A. (a) $\dfrac{0.214\,6 \text{ g AgBr}}{187.772 \text{ g AgBr/mol}} = 1.142\,9 \times 10^{-3}$ mol AgBr

(b) $[NaBr] = \dfrac{1.142\,9 \times 10^{-3} \text{ mol}}{50.00 \times 10^{-3} \text{ L}} = 0.022\,86$ M

6-B. (a) Absorbed impurities are taken into a substance. Adsorbed impurities are found on the surface of a substance.

(b) Included impurities occupy lattice sites of the host crystal. Occluded impurities are trapped in a pocket inside the host.

(c) An ideal gravimetric precipitate should be insoluble, easily filtered, possess a known, constant composition, and be stable to heat.

(d) High supersaturation often leads to formation of colloidal product with a large amount of impurities.

(e) Supersaturation can be decreased by increasing temperature (for most solutions), mixing well during addition of precipitant, and using dilute reagents. Homogeneous precipitation also reduces supersaturation.

(f) Washing with electrolyte preserves the electric double layer and prevents peptization.

(g) The volatile HNO_3 evaporates during drying. $NaNO_3$ is nonvolatile and will lead to a high mass for the precipitate.

(h) During the first precipitation, the concentration of impurities in the solution is high, giving a relatively high concentration of impurities in the precipitate. In the reprecipitation, the level of solution impurities is reduced, thus giving a purer precipitate.

(i) In thermogravimetric analysis, the mass of a sample is measured as the sample is heated. The mass lost during decomposition provides information about the composition of the sample.

6-C. **(a)** $\dfrac{0.104 \text{ g } CeO_2}{172.115 \text{ g } CeO_2/\text{mol}} = 6.043 \times 10^{-4} \text{ mol } CeO_2$

$6.043 \times 10^{-4} \text{ mol } CeO_2 \times \dfrac{1 \text{ mol } Ce}{1 \text{ mol } CeO_2}$
$$= 6.043 \times 10^{-4} \text{ mol } Ce$$
$(6.043 \times 10^{-4} \text{ mol } Ce)(140.116 \text{ g } Ce/\text{mol } Ce)$
$$= 0.084\,66 \text{ g } Ce$$

(b) weight % Ce $= \dfrac{0.084\,66 \text{ g } Ce}{4.37 \text{ g unknown}} \times 100$
$$= 1.94 \text{ wt } \%$$

6-D. **(a)** In *combustion,* a substance is heated in the presence of excess O_2 to convert carbon to CO_2 and hydrogen to H_2O. In *pyrolysis,* the substance is decomposed by heating in the absence of added O_2. All oxygen in the sample is converted to CO by passage through a suitable catalyst.

(b) WO_3 catalyzes the complete combustion of C to CO_2 in the presence of excess O_2. Cu reduces SO_3 to SO_2 and removes excess O_2 from the gas stream.

(c) The tin capsule melts and is oxidized to SnO_2 to liberate heat and crack the sample. Tin uses the available oxygen immediately, ensures that sample oxidation occurs in the gas phase, and acts as an oxidation catalyst.

(d) By dropping the sample in before very much O_2 is present, pyrolysis of the sample to give gaseous products occurs prior to oxidation. This practice minimizes the formation of nitrogen oxides.

(e) $C_8H_7NO_2SBrCl + 9\frac{1}{4}O_2 \rightarrow$
$$8CO_2 + \tfrac{5}{2}H_2O + \tfrac{1}{2}N_2 + SO_2 + HBr + HCl$$

Chapter 7

7-A. neutralize ... conjugate

7-B. **(a)** $[Mg^{2+}][OH^-]^2 = K_{sp} = 7.1 \times 10^{-12}$
$0.050 \text{ M} \quad ? \qquad = (0.050)[OH^-]^2$
$$\Rightarrow \quad [OH^-] = 1.1_9 \times 10^{-5} \text{ M}$$

(b) $[H^+] = K_w/[OH^-] = 8.3_9 \times 10^{-10} \text{ M} \Rightarrow$
$$\text{pH} = -\log[H^+] = 9.08$$

7-C. The stronger acid is **A,** which has the larger value of K_a.
$$Cl_2CHCO_2H \overset{K_a}{\rightleftharpoons} Cl_2CHCO_2^- + H^+$$
$$ClCH_2CO_2H \overset{K_a}{\rightleftharpoons} ClCH_2CO_2^- + H^+$$
The stronger base is **C,** which has the larger value of K_b.
$$H_2NNH_2 + H_2O \overset{K_b}{\rightleftharpoons} H_2NNH_3^+ + OH^-$$
$$H_2NCONH_2 + H_2O \overset{K_b}{\rightleftharpoons} H_2NCONH_3^+ + OH^-$$

7-D. **(a)** **(i)** $\text{pH} = -\log(1.0 \times 10^{-3}) = 3.00$
(ii) $\text{pH} = -\log(1.0 \times 10^{-14}/1.0 \times 10^{-2}) = 12.00$

(b) **(i)** $\text{pH} = -\log(3.2 \times 10^{-5}) = 4.49$
(ii) $\text{pH} = -\log(1.0 \times 10^{-14}/0.007\,7) = 11.89$

(c) $[H^+] = 10^{-4.44} = 3.6 \times 10^{-5} \text{ M}$

(d) $[H^+] = 1.0 \times 10^{-14}/0.007\,7 = 1.3 \times 10^{-12} \text{ M}$
It is derived from $H_2O \rightleftharpoons H^+ + OH^-$

(e) The concentration of the strong base is too low to perturb the pH of pure water. The pH is very close to 7.00.

7-E. **(a)** $HCO_2H \rightleftharpoons HCO_2^- + H^+$

(b) HCO_2^-

(c) $K_a = \dfrac{[HCO_2^-][H^+]}{[HCO_2H]} = 1.80 \times 10^{-4}$

(d) $K_b = \dfrac{[HCO_2H][OH^-]}{[HCO_2^-]}$

(e) $K_b = \dfrac{K_w}{K_a} = 5.6 \times 10^{-11}$

7-F. **(a)** Let $x = [H^+] = [A^-]$ and $0.100 - x = [HA]$.
$$\dfrac{x^2}{0.100 - x} = 1.00 \times 10^{-5} \Rightarrow$$
$$x = 9.95 \times 10^{-4} \text{ M} \Rightarrow \text{pH} = -\log x = 3.00$$
$$\dfrac{[A^-]}{[A^-] + [HA]} = \dfrac{9.95 \times 10^{-4}}{0.100} = 9.95 \times 10^{-3}$$

(b) $\begin{array}{ccccc} HA & \rightleftharpoons & H^+ & + & A^- \\ 0.045\,0 - 10^{-2.78} & & 10^{-2.78} & & 10^{-2.78} \end{array}$
$$K_a = \dfrac{(10^{-2.78})^2}{0.045\,0 - 10^{-2.78}} = 6.4 \times 10^{-5} \Rightarrow pK_a = 4.19$$

(c) $\begin{array}{ccccc} HA & \rightleftharpoons & H^+ & + & A^- \\ F - x & & x & & x \end{array}$
$$\dfrac{[A^-]}{[HA] + [A^-]} = \dfrac{x}{F - x + x} = 0.006\,0$$
With $F = 0.045\,0$ M,
$$x = 2.7 \times 10^{-4} \text{ M} \Rightarrow K_a = \dfrac{x^2}{F - x}$$
$$= 1.6 \times 10^{-6} \Rightarrow pK_a = 5.79$$

7-G. **(a)** Let $x = [OH^-] = [BH^+]$ and $0.100 - x = [B]$.
$$\dfrac{x^2}{0.100 - x} = 1.00 \times 10^{-5}$$
$$\Rightarrow x = 9.95 \times 10^{-4} \text{ M}$$
$$\Rightarrow [H^+] = \dfrac{K_w}{x} = 1.005 \times 10^{-11} \Rightarrow \text{pH} = 11.00$$

$$\frac{[BH^+]}{[B] + [BH^+]} = \frac{9.95 \times 10^{-4}}{0.100} = 9.95 \times 10^{-3}$$

(b)

$$\begin{array}{ccccc} B & + & H_2O & \rightleftharpoons & BH^+ & + & OH^- \\ 0.10 - (K_w/10^{-9.28}) & & & & K_w/10^{-9.28} & & K_w/10^{-9.28} \end{array}$$

$$K_b = \frac{(K_w/10^{-9.28})^2}{0.10 - (K_w/10^{-9.28})} = 3.6 \times 10^{-9}$$

(c)

$$\begin{array}{ccccccc} B & + & H_2O & \overset{K_b}{\rightleftharpoons} & BH^+ & + & OH^- \\ 0.10 - x & & & & x & & x \end{array}$$

$$\frac{[BH^+]}{[BH^+] + [B]} = 0.020$$

$$\frac{x}{0.10 - x + x} = 0.020 \quad \Rightarrow \quad x = 0.002\,0$$

$$K_b = \frac{x^2}{0.10 - x} = \frac{(0.002\,0)^2}{0.10 - 0.002\,0} = 4.1 \times 10^{-5}$$

Chapter 8

8-A. **(a)** $pH = pK_a + \log\left(\dfrac{[A^-]}{[HA]}\right)$

$$= 5.00 + \log\left(\frac{0.050}{0.100}\right) = 4.70$$

(b) $pH = 3.745 + \log\left(\dfrac{[HCO_2^-]}{[HCO_2H]}\right)$

pH:	3.000	3.745	4.000
$[HCO_2^-]/[HCO_2H]$:	0.180	1.00	1.80

8-B. **(a)** $pH = pK_a + \log\left(\dfrac{[tris]}{[trisH^+]}\right)$

$$= 8.075 + \log\left(\frac{(10.0\ g)/(121.136\ g/mol)}{(10.0\ g)/(157.597\ g/mol)}\right) = 8.19$$

(b)

	tris	+	H^+	$\rightarrow$	$trisH^+$	+	H_2O
Initial mmol:	82.55		5.25		63.45		
Final mmol:	77.30		—		68.70		

$$pH = 8.075 + \log\left(\frac{77.30}{68.70}\right) = 8.13$$

(c)

	$trisH^+$	+	OH^-	$\rightarrow$	tris	+	H_2O
Initial mmol:	63.45		5.25		82.55		
Final mmol:	58.20		—		87.80		

$$pH = 8.075 + \log\left(\frac{87.80}{58.20}\right) = 8.25$$

8-C. **(a)** mol tris $= 10.0\ g/(121.136\ g/mol) = 0.082\,55$ mol
$= 82.55$ mmol

	tris	+	H^+	$\rightarrow$	$trisH^+$	+	H_2O
Initial mmol:	82.5_5		x		—		
Final mmol:	$82.5_5 - x$		—		x		

$$pH = pK_a + \log\left(\frac{[tris]}{[trisH^+]}\right)$$

$$7.60 = 8.075 + \log\left(\frac{82.5_5 - x}{x}\right)$$

$$-0.475 = \log\left(\frac{82.5_5 - x}{x}\right) \quad \Rightarrow$$

$$10^{-0.475} = 10^{\log[(82.5_5 - x)/x]}$$

$$0.335_0 = \frac{82.5_5 - x}{x} \quad \Rightarrow \quad x = 61.8_4\ \text{mmol}$$

volume of HCl required
$$= (0.061\,8_4\ mol)/(1.20\ M) = 51.5\ mL$$

(b) I would weigh out 0.020 0 mol of acetic acid ($= 1.201$ g) and place it in a beaker with $\sim$75 mL of water. While monitoring the pH with a pH electrode, I would add 3 M NaOH ($\sim$4 mL is required) until the pH is exactly 5.00. I would then pour the solution into a 100-mL volumetric flask and wash the beaker several times with a few milliliters of distilled water. Each washing would be added to the volumetric flask, to ensure quantitative transfer from the beaker to the flask. After swirling the volumetric flask to mix the solution, I would carefully add water up to the 100-mL mark, insert the cap, and invert 20 times to ensure complete mixing.

8-D. **(a)**

Compound	pK_a	
hydroxybenzene	9.98	
propanoic acid	4.87	
cyanoacetic acid	2.47	$\leftarrow$ Most suitable because
sulfuric acid	1.99	pK_a is closest to pH

(b) Buffer capacity is based on the ability of buffer to react with added acid or base, without making a large change in the ratio of concentrations $[A^-]/[HA]$. The greater the concentration of each component, the less relative change is brought about by reaction with a small increment of added acid or base.

(c)

Compound	pK_a (for conjugate acid)	
ammonia	9.24	$\leftarrow$ Most suitable because
aniline	4.60	pK_a is closest to pH
hydrazine	8.48	
pyridine	5.23	

8-E. **(a)** The quotient $[HIn]/[In^-]$ changes from 10:1 when $pH = pK_{HIn} - 1$ to 1:10 when $pH = pK_{HIn} + 1$. This change is generally sufficient to cause a complete color change.

(b) red, orange, yellow

Chapter 9

9-A. The titration reaction is $H^+ + OH^- \rightarrow H_2O$ and
$V_e = 5.00$ mL
because
$$\underbrace{(V_e(mL))(0.100\ M)}_{\text{mmol HCl at } V_e} = \underbrace{(50.00\ mL)(0.010\,0\ M)}_{\text{Initial mmol NaOH}} \Rightarrow V_e = 5.00\ mL$$

Representative calculations:
$V_a = 1.00$ mL:
$OH^- = \underbrace{(50.00 \text{ mL})(0.010\,0 \text{ M})}_{\text{Initial mmol NaOH}} -$

$\underbrace{(1.00 \text{ mL})(0.100 \text{ M})}_{\text{Added mmol HCl}} = 0.400$ mmol

$[OH^-] = (0.400 \text{ mmol})/(51.00 \text{ mL}) = 0.007\,84$ M
$[H^+] = K_w/(0.007\,84 \text{ M}) = 1.28 \times 10^{-12}$ M
$pH = -\log(1.28 \times 10^{-12}) = 11.89$
$V_a = V_e = 5.00$ mL:
$H_2O \rightleftharpoons \underset{x}{H^+} + \underset{x}{OH^-}$

$K_w = x^2 \Rightarrow x = 1.00 \times 10^{-7}$ M $\Rightarrow$ pH = 7.00
$V_a = 5.01$ mL:
excess $H^+ = (0.01 \text{ mL})(0.100 \text{ M}) = 0.001$ mmol
$[H^+] = (0.001 \text{ mmol})/(55.01 \text{ mL}) = 1._8 \times 10^{-5}$ M
$pH = -\log(1._8 \times 10^{-5}) = 4.74$

V_a (mL)	pH	V_a (mL)	pH	V_a (mL)	pH
0.00	12.00	4.50	10.96	5.10	3.74
1.00	11.89	4.90	10.26	5.50	3.05
2.00	11.76	4.99	9.26	6.00	2.75
3.00	11.58	5.00	7.00	8.00	2.29
4.00	11.27	5.01	4.74	10.00	2.08

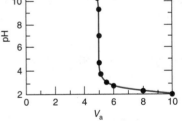

9-B. The titration reaction is
$$HCO_2H + OH^- \rightarrow HCO_2^- + H_2O$$
$\underbrace{(V_e(\text{mL}))(0.050\,0 \text{ M})}_{\text{mmol KOH at } V_e} = \underbrace{(50.0 \text{ mL})(0.050\,0 \text{ M})}_{\text{Initial mmol HCO}_2\text{H}} \Rightarrow$
$$V_e = 50.0 \text{ mL}$$

Representative calculations:
$V_b = 0$ mL: $\underset{0.050\,0 - x}{HA} \rightleftharpoons \underset{x}{H^+} + \underset{x}{A^-}$
$K_a = 1.80 \times 10^{-4}$ p$K_a = 3.745$

$\dfrac{x^2}{0.050\,0 - x} = K_a \Rightarrow x = 2.91 \times 10^{-3} \Rightarrow$ pH = 2.54
$V_b = 48.0$ mL:

	HA	+	OH^-	$\rightarrow$	A^-	+	H_2O
Initial mmol:	2.50		2.40		—		
Final mmol:	0.10		—		2.40		

$pH = pK_a + \log\left(\dfrac{[A^-]}{[HA]}\right) = 3.745 + \log\left(\dfrac{2.40}{0.10}\right) = 5.13$
$V_b = 50.0$ mL: formal conc. of A^-
$$= \frac{(50.0 \text{ mL})(0.050\,0 \text{ M})}{100.0 \text{ mL}} = 0.025\,0 \text{ M}$$

$\underset{0.025\,0 - x}{A^-} + H_2O \rightleftharpoons \underset{x}{HA} + \underset{x}{OH^-}$ $K_b = K_w/K_a = 5.56 \times 10^{-11}$

$\dfrac{x^2}{0.025\,0 - x} = K_b \Rightarrow x = 1.18 \times 10^{-6} \Rightarrow$

$$pH = -\log\left(\frac{K_w}{x}\right) = 8.07$$

$V_b = 60.0$ mL: excess $[OH^-] = \dfrac{(10.0 \text{ mL})(0.050\,0 \text{ M})}{110.0 \text{ mL}}$
$$= 4.55 \times 10^{-3} \text{ M}$$
$$\Rightarrow \text{ pH} = 11.66$$

V_b (mL)	pH	V_b (mL)	pH	V_b (mL)	pH
0.0	2.54	45.0	4.70	50.5	10.40
10.0	3.14	48.0	5.13	51.0	10.69
20.0	3.57	49.0	5.44	52.0	10.99
25.0	3.74	49.5	5.74	55.0	11.38
30.0	3.92	50.0	8.07	60.0	11.66
40.0	4.35				

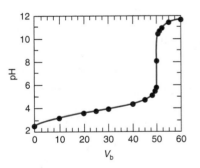

At $V_b = \frac{1}{2}V_e$, the pH should equal pK_a, which it does.

9-C. (a) Titration of a weak base, B, produces the conjugate acid, BH^+, which is necessarily acidic.

(b) The titration reaction is $B + H^+ \rightarrow BH^+$.
Find V_e:
$(V_e)(0.200 \text{ M}) = (100.0 \text{ mL})(0.100 \text{ M}) \Rightarrow V_e = 50.0$ mL
Representative calculations:
$V_a = 0.0$ mL: $\underset{0.100-x}{B} + H_2O \rightleftharpoons \underset{x}{BH^+} + \underset{x}{OH^-}$
$$K_b = 2.6 \times 10^{-6}$$

$\dfrac{x^2}{0.100 - x} = K_b = 2.6 \times 10^{-6} \Rightarrow x = 5.09 \times 10^{-4}$

$pH = -\log\left(\dfrac{K_w}{x}\right) = 10.71$

$V_a = 20.0$ mL:

	B	+	H$^+$	$\rightarrow$	BH$^+$
Initial mmol:	10.00		4.00		—
Final mmol:	6.00		—		4.00

$$pH = pK_a \text{ (for BH}^+) + \log\left(\frac{[B]}{[BH^+]}\right)$$

$$\underset{\uparrow}{K_a = K_w/K_b} \qquad = 8.41 + \log\left(\frac{6.00}{4.00}\right) = 8.59$$

$V_a = V_e = 50$ mL: All B has been converted to the conjugate acid, BH$^+$. The formal concentration of BH$^+$ is

$$F' = \frac{(100.0 \text{ mL})(0.100 \text{ M})}{150.0 \text{ mL}} = 0.066\,7 \text{ M}$$

The pH is determined by the reaction

$$\underset{0.066\,7 - x \quad x \quad x}{BH^+ \rightleftharpoons B + H^+}$$

$$\frac{x^2}{0.066\,7 - x} = K_a = \frac{K_w}{K_b} \quad \Rightarrow \quad x = 1.60 \times 10^{-5} \Rightarrow$$
$$pH = 4.80$$

$V_a = 51.0$ mL: There is excess [H$^+$].

$$\text{excess } [H^+] = \frac{(1.0 \text{ mL})(0.200 \text{ M})}{(151.0 \text{ mL})}$$
$$= 1.32 \times 10^{-3} \quad \Rightarrow \quad pH = 2.88$$

V_a (mL)	pH	V_a	pH	V_a	pH
0.0	10.71	30.0	8.23	50.0	4.80
10.0	9.01	40.0	7.81	50.1	3.88
20.0	8.59	49.0	6.72	51.0	2.88
25.0	8.41	49.9	5.71	60.0	1.90

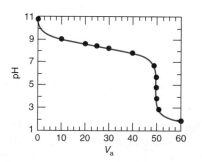

9-D. (a) Figure 9-1: bromothymol blue: blue → yellow
Figure 9-2: thymol blue: yellow → blue
Figure 9-10 ($pK_a = 8$): thymolphthalein:
colorless → blue

(b) The derivatives are shown in the spreadsheet. In the first derivative graph, the maximum is near 119 mL. In Figure 9-5, the second derivative graph gives an end point of 118.9 μL.

9-E. (a) Tris(hydroxymethyl)aminomethane ($H_2NC(CH_2OH)_3$), sodium carbonate (Na_2CO_3), or borax ($Na_2B_4O_7 \cdot 10H_2O$)

	A	B	C	D	E	F
1			First Derivative		Second Derivative	
2	μL NaOH	pH	μL	Derivative	μL	Derivative
3	107	6.921				
4	110	7.117	108.5	6.533E-02		
5	113	7.359	111.5	8.067E-02	110	5.11E-03
6	114	7.457	113.5	9.800E-02	112.5	8.67E-03
7	115	7.569	114.5	1.120E-01	114	1.40E-02
8	116	7.705	115.5	1.360E-01	115	2.40E-02
9	117	7.878	116.5	1.730E-01	116	3.70E-02
10	118	8.09	117.5	2.120E-01	117	3.90E-02
11	119	8.343	118.5	2.530E-01	118	4.10E-02
12	120	8.591	119.5	2.480E-01	119	-5.00E-03
13	121	8.794	120.5	2.030E-01	120	-4.50E-02
14	122	8.952	121.5	1.580E-01	121	-4.50E-02
15						
16	C4 = (A4+A3)/2				E5 = (C5+C4)/2	
17	D4 = (B4-B3)/(A4-A3)				F5 = (D5-D4)/(C5-C4)	

can be used to standardize HCl. Potassium hydrogen phthalate ($HO_2C - C_6H_4 - CO_2^-K^+$) or potassium hydrogen iodate ($KH(IO_3)_2$) can be used to standardize NaOH.

(b) 30 mL of 0.05 M OH$^-$ = 1.5 mmol OH$^-$ = 1.5 mmol potassium hydrogen phthalate. $(1.5 \times 10^{-3} \text{ mol}) \times (204.233 \text{ g/mol}) = 0.30$ g of potassium hydrogen phthalate.

9-F. (a) 5.00 mL of 0.033 6 M HCl = 0.168$_0$ mmol
6.34 mL of 0.010 0 M NaOH = 0.063 4 mmol
HCl consumed by NH$_3$ = 0.168$_0$ - 0.063 4
$$= 0.104_6 \text{ mmol}$$
mol NH$_3$ = mol HCl = 0.104$_6$ mmol

(b) mol nitrogen = mol NH$_3$ = 0.104$_6$ mmol
$(0.104_6 \times 10^{-3} \text{ mol N})(14.006\ 7 \text{ g/mol})$
$$= 1.46_5 \text{ mg of nitrogen}$$

(c) $(256 \ \mu L)\left(\dfrac{1 \text{ mL}}{1\ 000 \ \mu L}\right)(37.9 \text{ mg protein/mL})$
$$= 9.70_2 \text{ mg protein}$$

(d) wt % N = $\dfrac{1.46_5 \text{ mg N}}{9.70_2 \text{ mg protein}} \times 100 = 15.1$ wt %

9-G. (a) Your spreadsheet should reproduce the results in Figure 9-10.

(b)

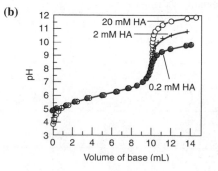

Chapter 10

10-A. **(a)** **(i)** $H_3\overset{+}{N}CH_2CH_2\overset{+}{N}H_3 \xrightarrow{K_{a1} = 1.42 \times 10^{-7}}$

$$H_2NCH_2CH_2\overset{+}{N}H_3 + H^+$$

$H_2NCH_2CH_2\overset{+}{N}H_3 \xrightarrow{K_{a2} = 1.18 \times 10^{-10}}$

$$H_2NCH_2CH_2NH_2 + H^+$$

(ii)

$^-O_2CCH_2CO_2^- + H_2O \xrightarrow{K_{b1} = K_w/K_{a2} = 4.98 \times 10^{-9}}$

$$HO_2CCH_2CO_2^- + OH^-$$

$HO_2CCH_2CO_2^- + H_2O \xrightarrow{K_{b2} = K_w/K_{a1} = 7.04 \times 10^{-12}}$

$$HO_2CCH_2CO_2H + OH^-$$

(b) **(iii)**

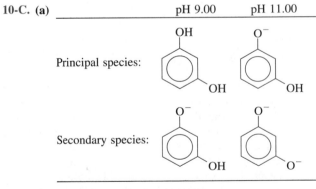

Aspartic acid

Arginine

10-B. **(a)** $H_2SO_3 \rightleftharpoons HSO_3^- + H^+$

$ 0.050 - x \quad x \quad x$

$\dfrac{x^2}{0.050 - x} = K_1 = 1.23 \times 10^{-2} \implies$

$$x = 1.94 \times 10^{-2}$$

$[HSO_3^-] = [H^+] = 1.94 \times 10^{-2}\,M \implies pH = 1.71$
$[H_2SO_3] = 0.050 - x = 0.031\,M$

$[SO_3^{2-}] = \dfrac{K_2[HSO_3^-]}{[H^+]} = K_2 = 6.6 \times 10^{-8}\,M$

(b) $pH \approx \frac{1}{2}(pK_1 + pK_2) = \frac{1}{2}(1.91 + 7.18) = 4.54$
$[H^+] = 10^{-pH} = 2.9 \times 10^{-5}\,M;$
$[HSO_3^-] \approx 0.050\,M$

$[H_2SO_3] = \dfrac{[H^+][HSO_3^-]}{K_1} = \dfrac{(2.9 \times 10^{-5})(0.050)}{1.23 \times 10^{-2}}$
$$= 1.2 \times 10^{-4}\,M$$

$[SO_3^{2-}] = \dfrac{K_2[HSO_3^-]}{[H^+]} = \dfrac{(6.6 \times 10^{-8})(0.050)}{2.9 \times 10^{-5}}$
$$= 1.1 \times 10^{-4}\,M$$

(c) $\underset{0.050 - x}{SO_3^{2-}} + H_2O \rightleftharpoons \underset{x}{HSO_3^-} + \underset{x}{OH^-}$

$\dfrac{x^2}{0.050 - x} = K_{b1} = \dfrac{K_w}{K_{a2}} = 1.52 \times 10^{-7} \implies$
$$x - 8.7 \times 10^{-5}$$

$[HSO_3^-] = 8.7 \times 10^{-5}\,M; [H^+] = \dfrac{K_w}{x}$
$$= 1.15 \times 10^{-10}\,M \implies pH = 9.94$$

$[SO_3^{2-}] = 0.050 - x = 0.050\,M$

$[H_2SO_3] = \dfrac{[H^+][HSO_3^-]}{K_1} = 8.1 \times 10^{-13}\,M$

10-C. **(a)**

	pH 9.00	pH 11.00
Principal species:		
Secondary species:		

(b) HC^-: $pH \approx \frac{1}{2}(pK_2 + pK_3)$
$$= \frac{1}{2}(8.36 + 10.77) = 9.56$$

10-D. **(a)** $(50.0\,mL)(0.050\,0\,M) = V_{e1}(0.100\,M) \implies$
$$V_{e1} = 25.0\,mL$$

$V_{e2} = 2V_{e1} = 50.0\,mL$

(b) 0 mL: $\underset{0.050\,0 - x}{H_2A} \rightleftharpoons \underset{x}{H^+} + \underset{x}{HA^-}$

$\dfrac{x^2}{0.050\,0 - x} = K_1 = 1.42 \times 10^{-3}$

$\implies x = 7.75 \times 10^{-3} \implies pH = 2.11$

12.5 mL: $pH = pK_1 = 2.85$

25.0 mL: pH $\approx \frac{1}{2}(pK_1 + pK_2)$
$$= \frac{1}{2}(2.847 + 5.696) = 4.27$$

37.5 mL: pH $= pK_2 = 5.70$

50.0 mL: H_2A has been converted to A^{2-} at a formal concentration of

$$F = \frac{(50.0\ \text{mL})(0.050\ 0\ \text{M})}{100.0\ \text{mL}} = 0.025\ 0\ \text{M}$$

$$\begin{array}{ccccc} A^{2-} & + & H_2O & \rightleftharpoons & HA^- + OH^- \\ 0.025\ 0 - x & & & & x \quad\quad x \end{array}$$

$$\frac{x^2}{0.025\ 0 - x} = K_{b1} = \frac{K_w}{K_{a2}} \Rightarrow x = 1.12 \times 10^{-5}$$

$$\Rightarrow \quad pH = -\log\left(\frac{K_w}{x}\right) = 9.05$$

55.0 mL: There is an excess of 5.0 mL of OH^-:

$$[OH^-] = \frac{(5.0\ \text{mL})(0.100\ \text{M})}{105.0\ \text{mL}}$$
$$= 0.004\ 76\ \text{M} \Rightarrow pH = 11.68$$

(c)

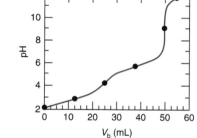

Chapter 11

11-A. $PbI_2(s) \overset{K_{sp}}{\rightleftharpoons} \underset{x}{Pb^{2+}} + \underset{2x}{2I^-}$

$x(2x)^2 = K_{sp} = 7.9 \times 10^{-9} \Rightarrow 2x = [I^-] = 2.5\ \text{mM}$
The observed concentration of dissolved iodine in pure water is approximately 3.8 mM, or 50% higher than predicted. One reason for this is that there are more species in solution than just Pb^{2+} and I^-, such as PbI^+. A second reason is that as PbI_2 dissolves, it increases its own solubility by adding ions to the solution to create ionic atmospheres around the dissolved ions and decreasing their attraction for each other. When KNO_3 is added, the number of ions in the ionic atmosphere increases further, thereby decreasing the attraction of Pb^{2+} and I^- for each other, and increasing the solubility of PbI_2.

11-B. When $\mu = 0$, the activity coefficients are 1 and we have
$$H_2O \rightleftharpoons \underset{x}{H^+} + \underset{x}{OH^-} \Rightarrow$$

$$K_w = \mathcal{A}_{H^+}\mathcal{A}_{OH^-} = [H^+]\gamma_{H^+}[OH^-]\gamma_{OH^-} =$$
$$1.0 \times 10^{-14} = (x)(1)(x)(1) \Rightarrow x = 1.0 \times 10^{-7}\ \text{M}$$
$$pH = -\log[H^+]\gamma_{H^+} = -\log[1.0 \times 10^{-7}](1) = 7.00$$
If the ionic strength is 0.1 M, the activity coefficients are

$\gamma_{H^+} = 0.83$ and $\gamma_{OH^-} = 0.76$. The concentrations of H^+ and OH^- are no longer 10^{-7} M:
$$K_w = \mathcal{A}_{H^+}\mathcal{A}_{OH^-} = [H^+]\gamma_{H^+}[OH^-]\gamma_{OH^-}$$
$$1.0 \times 10^{-14} = (x)(0.83)(x)(0.76) \Rightarrow x = 1.2_6 \times 10^{-7}\ \text{M}$$
The concentration of H^+ increases by 26% when 0.1 M NaCl is added to the water. However, the pH is not changed very much:
$$pH = -\log[H^+]\gamma_{H^+} = -\log[1.2_6 \times 10^{-7}](0.83) = 6.98$$

11-C. **(a)** $C_2O_4^{2-} \overset{H^+}{\rightarrow} HC_2O_4^- \overset{H^+}{\rightarrow} H_2C_2O_4$
The species are Na^+, $C_2O_4^{2-}$, $HC_2O_4^-$, $H_2C_2O_4$, Cl^-, H^+, OH^-, and H_2O.

(b) Charge balance: $[Na^+] + [H^+]$
$$= 2[C_2O_4^{2-}] + [HC_2O_4^-] + [Cl^-] + [OH^-]$$

(c) One mass balance is that the total concentration of
Na must be $2 \times \dfrac{5.00\ \text{mmol}}{0.100\ \text{L}} \Rightarrow [Na^+] = 0.100\ \text{M}.$

A second mass balance is that the total moles of oxalate are $\dfrac{5.00\ \text{mmol}}{0.100\ \text{L}} \Rightarrow 0.050\ 0\ \text{M} = [C_2O_4^{2-}] +$
$$[HC_2O_4^-] + [H_2C_2O_4].$$
A third mass balance is that $[Cl^-] = 0.025\ 0\ \text{M}$.

(d) We are adding 2.50 mmol H^+ to 5.00 mmol of the base, oxalate. The predominant species are $C_2O_4^{2-} + HC_2O_4^-$ with a negligible amount of $H_2C_2O_4$. The pH is going to be near pK_2 for oxalic acid, which is 4.27. Therefore $[H^+] \approx 10^{-4.27}$ M and $[OH^-] = K_w/[H^+] \approx 10^{-9.73}$ M in this solution. The charge and mass balances can be simplified by neglecting $[H_2C_2O_4]$, $[H^+]$, and $[OH^-]$ in comparison to the concentrations of major species:
Charge balance:
$$[Na^+] \approx 2[C_2O_4^{2-}] + [HC_2O_4^-] + [Cl^-]$$
Mass balance: $0.050\ 0\ \text{M} \approx [C_2O_4^{2-}] + [HC_2O_4^-]$

11-D. **(a)** *Pertinent reactions:* Two given in the problem plus $H_2O \rightleftharpoons H^+ + OH^-$.
Charge balance: invalid because pH is fixed
Mass balance: $[Ag^+] = [CN^-] + [HCN]$ (A)
Equilibrium constants:
$$K_{sp} = [Ag^+][CN^-] = 2.2 \times 10^{-16} \quad\quad (B)$$
$$K_b = \frac{[HCN][OH^-]}{[CN^-]} = 1.6 \times 10^{-5} \quad\quad (C)$$
$$K_w = [H^+][OH^-] = 1.0 \times 10^{-14} \quad\quad (D)$$
Count equations and unknowns: There are four equations (A–D) and four unknowns: $[Ag^+]$, $[CN^-]$, $[HCN]$, and $[OH^-]$. $[H^+]$ is known. *Solve:* Because $[H^+] = 10^{-9.00}$ M, $[OH^-] = 10^{-5.00}$ M. Putting this value of $[OH^-]$ into Equation C gives
$$[HCN] = \frac{K_b}{[OH^-]}[CN^-] = 1.6[CN^-]$$
Substituting into Equation A gives
$$[Ag^+] = [CN^-] + [HCN] = [CN^-] + 1.6[CN^-]$$
$$= 2.6[CN^-]$$

Substituting into Equation B gives

$$K_{sp} = 2.2 \times 10^{-16} = [Ag^+][CN^-]$$
$$= (2.6[CN^-])[CN^-] \Rightarrow [CN^-] = 9.2_0 \times 10^{-9} \text{ M}$$
$$[Ag^+] = K_{sp}/[CN^-] = 2.2 \times 10^{-16}/9.2_0 \times 10^{-9}$$
$$= 2.3_9 \times 10^{-8} \text{ M}$$
$$[HCN] = 1.6[CN^-] = 1.6(9.2_0 \times 10^{-9})$$
$$= 1.4_7 \times 10^{-8} \text{ M}$$

(b) *Mass balance:* Because all species are derived from $AgCN(s)$, the moles of silver must equal the moles of cyanide:

$$[Ag^+] + [AgCN(s)] + [Ag(CN)_2^-] + [AgOH(aq)]$$
$$= [CN^-] + [HCN] + [AgCN(s)] + 2[Ag(CN)_2^-]$$

which simplifies to $[Ag^+] + [AgOH(aq)]$

$$= [CN^-] + [HCN] + [Ag(CN)_2^-]$$

11-E. At pH 2.00: $\alpha_{HA} = \dfrac{10^{-2.00}}{10^{-2.00} + 10^{-3.00}} = 0.90_9$

$$\alpha_{A^-} = \dfrac{10^{-3.00}}{10^{-2.00} + 10^{-3.00}} = 0.090_9$$

$$\dfrac{[HA]}{[A^-]} = \dfrac{0.90_9}{0.090_9} = 10._0$$

The results for all three pH values are

	α_{HA}	α_{A^-}	$[HA]/[A^-]$
pH = 2.00	0.90_9	0.090_9	$10._0$
pH = 3.00	0.50_0	0.50_0	1.0_0
pH = 4.00	0.090_9	0.90_9	0.10_0

Of course, you already knew these results from the Henderson-Hasselbalch equation.

Chapter 12

12-A. A monodentate ligand binds to a metal ion through one ligand atom. A multidentate ligand binds through more than one ligand atom. A chelating ligand is a multidentate ligand.

12-B. **(a)** $M^{n+} + Y^{4-} \rightleftharpoons MY^{n-4}$

$$K_f = [MY^{n-4}]/([M^{n+}][Y^{4-}])$$

(b) At low pH, H^+ competes with M^{n+} in binding to the ligand atoms of EDTA.

(c) The auxiliary complexing agent prevents the metal ion from precipitating with hydroxide at high pH. EDTA displaces the auxiliary complexing agent from the metal ion.

(d)

H_6Y^{2+}	H_5Y^+	H_4Y	H_3Y^-	H_2Y^{2-}	HY^{3-}	Y^{4-}

$$\uparrow \qquad \uparrow \qquad \uparrow \qquad \uparrow \qquad \uparrow \qquad \uparrow$$
$$0.0 \quad 1.5 \quad 2.0 \quad 2.66 \quad 6.16 \quad 10.24$$
pH: $\quad$ 0.75 1.75 2.33 4.41 8.20

12-C. **(a)** Only a small amount of indicator is employed. Most of the Mg^{2+} is not bound to indicator. The free Mg^{2+} reacts with EDTA before MgIn reacts. Therefore the concentration of MgIn is constant until all of the Mg^{2+} has been consumed. Only when MgIn begins to react does the color change.

(b) **(i)** H_3In^{3-} **(ii)** yellow   **(iii)** red
      (iv) violet $\rightarrow$ red

12-D. **(a)** $(25.0 \text{ mL})(0.050\ 0 \text{ M}) = 1.25 \text{ mmol EDTA}$

(b) $(5.00 \text{ mL})(0.050\ 0 \text{ M}) = 0.25 \text{ mmol Zn}^{2+}$

(c) mmol Ni^{2+} = mmol EDTA − mmol Zn^{2+}
$$= 1.25 - 0.25 = 1.00 \text{ mmol Ni}^{2+}$$
$$[Ni^{2+}] = (1.00 \text{ mmol})/(50.0 \text{ mL}) = 0.020\ 0 \text{ M}$$

12-E. **(a)** At pH 5.00:
$$K_f' = (3.7 \times 10^{-7})(10^{10.69}) = 1.8 \times 10^4$$
$$Ca^{2+} + EDTA \rightleftharpoons CaY^{2-}$$

	Ca^{2+}	EDTA	CaY^{2-}
Initial concentration (M):	0	0	0.010
Final concentration (M):	x	x	$0.010 - x$

$$\dfrac{[CaY^{2-}]}{[Ca^{2+}][EDTA]} = \dfrac{0.010 - x}{x^2} = K_f' = 1.8 \times 10^4$$
$$\Rightarrow x = [Ca^{2+}] = 7.2 \times 10^{-4} \text{ M}$$

(b) Fraction of bound calcium $- \dfrac{[CaY^{2-}]}{[CaY^{2-}] + [Ca^{2+}]}$
$$= \dfrac{[0.010 - 0.000\ 72]}{0.010} = 0.93$$

93% of Ca is bound to EDTA at the equivalence point at pH 5.00.

(c) At pH 9.00: $\quad K_f' = (0.054)(10^{10.69}) = 2.6 \times 10^9$

$$\dfrac{[CaY^{2-}]}{[Ca^{2+}][EDTA]} = \dfrac{0.010 - x}{x^2}$$
$$= 2.6 \times 10^9 \Rightarrow x = [Ca^{2+}] = 2.0 \times 10^{-6} \text{ M}$$

Fraction of bound calcium $= \dfrac{[CaY^{2-}]}{[CaY^{2-}] + [Ca^{2+}]}$
$$= \dfrac{[0.010 - 2.0 \times 10^{-6}]}{0.010} = 0.999\ 8$$

99.98% of Ca is bound to EDTA at the equivalence point at pH 9.00.

12-F. At pH 10.0, $K_f' = (0.36)(10^{10.69}) = 1.7_6 \times 10^{10}$
At $V_{EDTA} = 5.00$ mL, calculations are identical to those of Mg^{2+}:
initial mmol $Ca^{2+} = (0.050\ 0 \text{ M Ca}^{2+})(50.0 \text{ mL})$
$$= 2.50 \text{ mmol}$$
mmol remaining $= (0.900)(2.50 \text{ mmol}) = 2.25 \text{ mmol}$

$$[Ca^{2+}] = \dfrac{2.25 \text{ mmol}}{55.0 \text{ mL}} = 0.040\ 9 \text{ M} \Rightarrow$$
$$pCa^{2+} = -\log[Ca^{2+}] = 1.39$$

$V_{EDTA} = 50.00$ mL is the equivalence point:

$$[Ca^{2+}] = \dfrac{2.50 \text{ mmol}}{100.0 \text{ mL}} = 0.025\ 0 \text{ M}$$

	Ca^{2+}	+ EDTA	$\rightleftharpoons$	CaY^{2-}
Initial concentration (M):	—	—		0.025 0
Final concentration (M):	x	x		$0.025\ 0 - x$

$$\frac{[CaY^{2-}]}{[Ca^{2+}][EDTA]} = K_f' = 1.7_6 \times 10^{10}$$

$$\frac{0.025\ 0 - x}{x^2} = 1.7_6 \times 10^{10} \Rightarrow$$

$$x = 1.2 \times 10^{-6}\ M \Rightarrow pCa^{2+} = -\log x = 5.92$$

At $V_{EDTA} = 51.00\ mL$, there is 1.00 mL of excess EDTA:

$$[EDTA] = \frac{0.050\ 0\ mmol}{101.0\ mL} = 0.000\ 49_5\ M$$

$$[CaY^{2-}] = \frac{2.50\ mol}{101.0\ mL} = 0.024\ 8\ M$$

$$\frac{[CaY^{2-}]}{[Ca^{2+}][EDTA]} = K_f' = 1.7_6 \times 10^{10}$$

$$\frac{[0.024\ 8]}{[Ca^{2+}](0.000\ 49_5)} = 1.7_6 \times 10^{10}$$

$$\Rightarrow [Ca^{2+}] = 2.8 \times 10^{-9}\ M \Rightarrow pCa^{2+} = 8.55$$

Chapter 13

13-A. **(a)** $I_2 + 2e^- \rightleftharpoons 2I^-$
Oxidant

(b) $2S_2O_3^{2-} \rightleftharpoons S_4O_6^{2-} + 2e^-$
Reductant

(c) $1.00\ g\ S_2O_3^{2-}/(112.13\ g/mol)$
$\qquad = 8.92\ mmol\ S_2O_3^{2-} = 8.92\ mmol\ e^-$
$(8.92 \times 10^{-3}\ mol)(9.649 \times 10^4\ C/mol) = 861\ C$

(d) current (A) = coulombs/s = 861 C/60 s = 14.3 A

(e) work = $E \cdot q$ = (0.200 V)(861 C) = 172 J

13-B. **(a)** $Pt(s) \mid Br_2(l) \mid HBr(aq,\ 0.10\ M) \parallel$
$\qquad\qquad\qquad Al(NO_3)_3\ (aq,\ 0.010\ M) \mid Al(s)$

(b) right: $\quad Ag(S_2O_3)_2^{3-} + e^- \rightleftharpoons Ag(s) + 2S_2O_3^{2-}$
left: $\qquad\qquad Fe(CN)_6^{4-} \rightleftharpoons Fe(CN)_6^{3-} + e^-$

net: $Fe(CN)_6^{4-} + Ag(S_2O_3)_2^{3-} \rightleftharpoons$
$\qquad\qquad\qquad Fe(CN)_6^{3-} + Ag(s) + 2S_2O_3^{2-}$

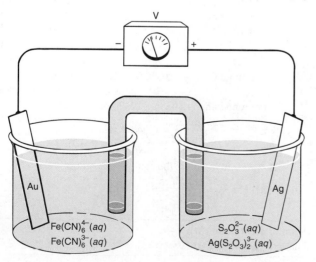

13-C. $Pt(s) \mid H_2(g,\ 1\ bar) \mid H^+(aq,\ 1\ M)$
$\qquad \parallel Fe^{2+}(aq,\ 1\ M),\ Fe^{3+}(aq,\ 1\ M) \mid Pt(s)$
$E°$ for the reaction $Fe^{3+} + e^- \rightleftharpoons Fe^{2+}$ is 0.771 V.
Therefore electrons flow from left to right through the meter.

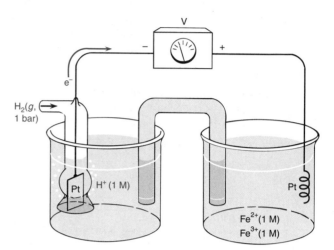

13-D. **(a)** $E = E° - \left(\dfrac{0.059\ 16}{3}\right)\log\left(\dfrac{P_{AsH_3}}{[H^+]^3}\right)$

$$E = -0.238 - \left(\frac{0.059\ 16}{3}\right)\log\left(\frac{0.010\ 0}{(10^{-3.00})^3}\right)$$
$$= -0.376\ V$$

(b) $Zn(s) \mid Zn^{2+}(0.1\ M) \parallel Cu^{2+}(0.1\ M) \mid Cu(s)$
right half-cell: $\quad Cu^{2+} + 2e^- \rightleftharpoons Cu(s)$
$\qquad\qquad\qquad\qquad\qquad E°_+ = 0.339\ V$
left hand-cell: $\quad Zn^{2+} + 2e^- \rightleftharpoons Zn(s)$
$\qquad\qquad\qquad\qquad\qquad E°_- = -0.762\ V$

$$E = \left\{0.339 - \left(\frac{0.059\ 16}{2}\right)\log\left(\frac{1}{0.1}\right)\right\} -$$
$$\left\{-0.762 - \left(\frac{0.059\ 16}{2}\right)\log\left(\frac{1}{0.1}\right)\right\} = 1.101\ V$$

Because the voltage is positive, electrons are transferred from Zn to Cu. The net reaction is $Cu^{2+} + Zn(s) \rightleftharpoons Cu(s) + Zn^{2+}$.

13-E. **(a)** right half-cell: $Cu^{2+} + 2e^- \rightleftharpoons Cu(s)\ E°_+ = 0.339\ V$
left half-cell: $Zn^{2+} + 2e^- \rightleftharpoons Zn(s)\ \ E°_- = -0.762\ V$
$E° = E°_+ - E°_- = 1.101\ V$
$K = 10^{nE°/0.059\ 16} = 10^{2(1.101)/0.059\ 16} = 1.7 \times 10^{37}$

(b) $AgBr(s) + e^- \rightleftharpoons Ag(s) + Br^-\quad E°_+ = 0.071\ V$
$\underline{\quad Ag^+ + e^- \rightleftharpoons Ag(s)\qquad\qquad E°_- = 0.799\ V}$
$AgBr(s) \rightleftharpoons Ag^+ + Br^-\qquad E° = 0.071 - 0.799$
$\qquad\qquad\qquad\qquad\qquad\qquad = -0.728\ V$
$K_{sp} = 10^{1E°/0.059\ 16} = 5 \times 10^{-13}$

13-F. (a) $E = E_+ - E_- = \left\{ \underbrace{E_+^\circ}_{0.771\ V} - (0.059\ 16) \log\left(\frac{[Fe^{2+}]}{[Fe^{3+}]}\right) \right\} \underbrace{- 0.197}_{\substack{\text{Reference} \\ \text{electrode} \\ \text{voltage}}}$

$0.703 = \left\{ 0.771 - 0.059\ 16 \log\left(\frac{[Fe^{2+}]}{[Fe^{3+}]}\right) \right\} - 0.197$

$0.059\ 16 \log\left(\frac{[Fe^{2+}]}{[Fe^{3+}]}\right) = -0.129 \Rightarrow \log\left(\frac{[Fe^{2+}]}{[Fe^{3+}]}\right)$
$= -2.18$

$\Rightarrow \frac{[Fe^{2+}]}{[Fe^{3+}]} = 10^{-2.18} = 6.6 \times 10^{-3}$

(b) (i) 0.326 V   **(ii)** 0.463 V

Chapter 14

14-A. (a) At 0.10 mL:

initial Cl^- = 2.00 mmol; added Ag^+ = 0.020 mmol

$[Cl^-] = \dfrac{(2.00 - 0.020)\ mmol}{(40.0 + 0.10)\ mL} = 0.049\ 4\ M$

$[Ag^+] = K_{sp}/[Cl^-] = (1.8 \times 10^{-10})/(0.049\ 4)$
$= 3.6 \times 10^{-9}\ M$

In a similar manner, we find

2.50 mL: $[Cl^-] = 0.035\ 3\ M$ $[Ag^+] = 5.1 \times 10^{-9}\ M$
5.00 mL: $[Cl^-] = 0.022\ 2\ M$ $[Ag^+] = 8.1 \times 10^{-9}\ M$
7.50 mL: $[Cl^-] = 0.010\ 5\ M$ $[Ag^+] = 1.7 \times 10^{-8}\ M$
9.90 mL: $[Cl^-] = 0.000\ 401\ M$
$[Ag^+] = 4.5 \times 10^{-7}\ M$

(b) At V_e: $[Ag^+][Cl^-] = x^2 = K_{sp} \Rightarrow$
$[Cl^-] = [Ag^+] = \sqrt{K_{sp}} = 1.3 \times 10^{-5}\ M$

(c) At 10.10 mL: $[Ag^+] = \dfrac{(0.10\ mL)(0.200\ M)}{50.1\ mL}$
$= 4.0 \times 10^{-4}\ M$

At 12.00 mL: $[Ag^+] = \dfrac{(2.00\ mL)(0.200\ M)}{52.0\ mL}$
$= 7.7 \times 10^{-3}\ M$

(d) At 0.10 mL:
$E = 0.558 + (0.059\ 16) \log[Ag^+]$
$E = 0.558 + (0.059\ 16) \log(3.6 \times 10^{-9}) = 0.059\ V$
In a similar manner, we find the results below:

mL AgNO₃	E (V)
0.10	0.059
2.50	0.067
5.00	0.079
7.50	0.099
9.90	0.182
10.00	0.270
10.10	0.357
12.00	0.433

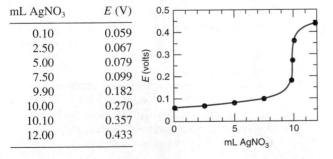

14-B. Cl^- diffuses into the $NaNO_3$, and NO_3^- diffuses into the NaCl. The mobility of Cl^- is greater than that of NO_3^-, so the NaCl region is depleted of Cl^- faster than the $NaNO_3$ region is depleted of NO_3^-. The $NaNO_3$ side becomes negative, and the NaCl side becomes positive.

14-C. As Ca^{2+} diffuses from right to left, the solution on the left side becomes more positive and the right side becomes more negative. The positive charge at the left repels Ca^{2+} and opposes further diffusion of Ca^{2+} across the membrane.

14-D. (a) Uncertainty in pH of standard buffers, junction potential, alkaline or acid errors at extreme pH values, equilibration time for electrode.

(b) $(4.63)(0.059\ 16\ V) = 0.274\ V$

(c) If the strongly alkaline solution has a high concentration of Na^+ (as in NaOH), the Na^+ cation competes with H^+ for cation exchange sites on the glass surface. The glass responds as if some H^+ were present, and the apparent pH is lower than the actual pH.

14-E. (a)

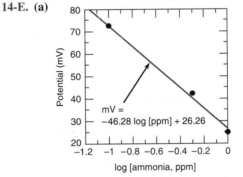

(b) Plugging the experimental values of potential into the equation of the calibration curve above gives
$106 = -46.28 \log[ppm] + 26.26 \Rightarrow$
$[ppm] = 0.019\ ppm$
$115 = -46.28 \log[ppm] + 26.26 \Rightarrow$
$[ppm] = 0.012\ ppm$
(Note that the unknown values lie beyond the calibration points, which is not the best practice. It would be a good idea to obtain calibration points at lower concentration to include the full range of unknown points.)

(c) $56 = -46.28 \log[ppm] + 26.26 \Rightarrow$
$[ppm] = 0.23\ ppm$

Chapter 15

15-A. Titration reaction:
$Sn^{2+} + 2Ce^{4+} \rightarrow Sn^{4+} + 2Ce^{3+}$ $V_e = 10.0\ mL$
Representative calculations:
At 0.100 mL: The ratio $[Sn^{2+}]/[Sn^{4+}]$ is 9.90/0.100

$$E_+ = 0.139 - \frac{0.059\ 16}{2} \log\left(\frac{[Sn^{2+}]}{[Sn^{4+}]}\right)$$

$$= 0.139 - \frac{0.059\ 16}{2} \log\left(\frac{9.90}{0.100}\right) = 0.080\ V$$

$$E = E_+ - E_- = 0.080 - 0.241 = -0.161\ V$$

At 10.00 mL:

$$2E_+ = 2(0.139) - 0.059\ 16 \log\left(\frac{[Sn^{2+}]}{[Sn^{4+}]}\right)$$

$$E_+ = 1.47 - 0.059\ 16 \log\left(\frac{[Ce^{3+}]}{[Ce^{4+}]}\right)$$

$$3E_+ = 1.748 - 0.059\ 16 \log\left(\frac{[Sn^{2+}][Ce^{3+}]}{[Sn^{4+}][Ce^{4+}]}\right) \qquad (A)$$

At the equivalence point, $[Ce^{3+}] = 2[Sn^{4+}]$ and $[Ce^{4+}] = 2[Sn^{2+}]$.
Putting these equalities into Equation A makes the log term 0.
Therefore $3E_+ = 1.748$ and $E_+ = 0.583\ V$.
$$E = E_+ - E_- = 0.583 - 0.241 = 0.342\ V$$
At 10.10 mL: The ratio $[Ce^{3+}]/[Ce^{4+}]$ is 10.00/0.10

$$E_+ = 1.47 - 0.059\ 16 \log\left(\frac{[Ce^{3+}]}{[Ce^{4+}]}\right)$$

$$E_+ = 1.47 - 0.059\ 16 \log\left(\frac{10.00}{0.10}\right) = 1.35_2\ V$$

$$E = E_+ - E_- = 1.35_2 - 0.241 = 1.11\ V$$

mL	E (V)	mL	E (V)	mL	E (V)
0.100	−0.161	9.50	−0.064	10.10	1.11
1.00	−0.130	10.00	0.342	12.00	1.19
5.00	−0.102				

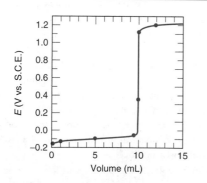

15-B. $Fe(CN)_6^{3-} + e^- \rightleftharpoons Fe(CN)_6^{4-}$ $E° = 0.356\ V$
$Tl^{3+} + 2e^- \rightleftharpoons Tl^+$ $E° = 0.77\ V$ (in 1 M HCl)
The end point will be between 0.356 and 0.77 V. Methylene blue with $E° = 0.53\ V$ is closest to the midpoint of the steep part of the titration curve. The color change would be from blue to colorless.

15-C. **(a)** 50.00 mL contains exactly 1/10 of the $KIO_3 = 0.102\ 2\ g = 0.477\ 5_7$ mmol KIO_3. Each mol of iodate

makes 3 mol of triiodide, so $I_3^- = 3(0.477\ 5_7) = 1.432_7$ mmol. Each mol of I_3^- is equivalent to one mol of I_2.

(b) Two moles of thiosulfate react with one mole of I_2. Therefore there must have been $2(1.432_7) = 2.865_4$ mmol of thiosulfate in 37.66 mL; so the concentration is $(2.865_4$ mmol)/(37.66 mL) = 0.076\ 08_7$ M.

(c) 50.00 mL of KIO_3 makes 1.432$_7$ mmol I_2. The unreacted I_2 requires 14.22 mL of sodium thiosulfate $= (14.22\ mL)(0.076\ 08_7\ M) = 1.082_0$ mmol, which reacts with $\frac{1}{2}(1.082_0$ mmol) = 0.541$_0$ mmol I_2. The ascorbic acid must have consumed the difference $= 1.432_7 - 0.541_0 = 0.891_7$ mmol I_2. Each mole of ascorbic acid consumes one mole of I_2, so mol ascorbic acid = 0.891$_7$ mmol, which has a mass of $(0.891_7 \times 10^{-3}$ mol)(176.13 g/mol) = 0.157$_1$ g. The weight percent of ascorbic acid in the unknown is $100 \times (0.157_1$ g)/(1.223 g) = 12.8 wt%.

Chapter 16

16-A. **(a)** Cathode reaction at (−) electrode:
$H_2O + e^- \rightarrow \frac{1}{2}H_2 + OH^-$
Anode reaction at (+) electrode:
$H_2O \rightarrow \frac{1}{2}O_2 + 2H^+ + 2e^-$

(b) coulombs = (89.2 mA)(666 s)
$= (89.2 \times 10^{-3}\ C/s)(666\ s) = 5.94_1\ C$
mol e^- = mol OH^- = (5.94$_1$ C)/(96 485 C/mol)
$= 0.615_7$ mmol
HA in unknown = (0.615$_7$ mmol)/(5.00 mL)
$= 0.123\ M$

(c) We could use a pH electrode or an indicator in the flask of acid to find the end point.

16-B. **(a)** The glucose monitor has a test strip with two carbon indicator electrodes and a silver-silver chloride reference electrode. Indicator electrode 1 is coated with glucose oxidase and a mediator. When a drop of blood is placed on the test strip, glucose from the blood is oxidized near indicator electrode 1 by mediator to gluconolactone and the mediator is reduced. With a potential of +0.2 V (with respect to the Ag | AgCl electrode) on the indicator electrode, reduced mediator is reoxidized at the indicator electrode. The electric current between indicator electrode 1 and the reference electrode is proportional to the rate of oxidation of the mediator, which is proportional to the concentration of glucose plus any interfering species in the blood. Indicator electrode 2 has mediator, but no glucose oxidase. Current measured between indicator electrode 2 and the reference electrode is proportional to the concentration of interfering species in the blood. The difference between the two currents is proportional to the concentration of glucose in the blood.

(b) In the absence of a mediator, the rate of oxidation of glucose depends on the concentration of O_2 in the blood. If $[O_2]$ is low, the current will be low and the monitor will give an incorrect, low reading for the glucose concentration. A mediator such as 1,1′-dimethylferrocene can replace O_2 in the glucose oxidation and be subsequently reduced at the indicator electrode. The concentration of mediator is constant and high enough so that variations in electrode current are due mainly to variations in glucose concentration. Also, by lowering the required electrode potential for oxidation of the mediator, there is less possible interference by other species in the blood.

16-C.

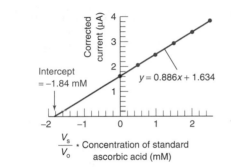

	A	B	C	D	E
1	Vitamin C standard addition experiment				
2	Add 25.0 mM ascorbic acid to 50.0 mL of orange juice				
3					
4		Vs =			
5	Vo (mL) =	mL ascorbic	x-axis function	I(s+x) =	y-axis function
6	50	acid added	Si*Vs/Vo	signal (µA)	I(s+x)*V/Vo
7	[S]i (mM) =	0.000	0.000	1.66	1.660
8	25	1.000	0.500	2.03	2.071
9		2.000	1.000	2.39	2.486
10		3.000	1.500	2.79	2.957
11		4.000	2.000	3.16	3.413
12		5.000	2.500	3.51	3.861
13					
14	C7 = $A$8*B7/$A$6		E7 = D7*($A$6+B7)/$A$6		

The equation of the least-squares line through the data points is $y = 0.886x + 1.634$. To find the x-intercept, set $y = 0$.

$$0 = 0.886x + 1.634 \quad \Rightarrow \quad x = -1.84 \text{ mM}$$

The concentration of ascorbic acid in the orange juice is 1.84 mM.

16-D. (a) Faradaic current arises from redox reactions at the electrode. It is what we are trying to measure. Charging current comes from the flow of ions toward or away from the electrode as electrons flow into or out of the electrode when a potential step is applied. Charging current is unrelated to any electrochemical reactions.

(b) Charging current decays more rapidly than faradaic current after a potential step. By waiting 1 s after a step before measuring current, the charging current has largely decayed and there is still significant faradaic current.

(c) Square wave polarography is much faster than other forms of polarography. It gives increased signal because current is measured from both reduction and oxidation of analyte in each cycle. It gives a derivative peak shape that makes it easier to resolve closely spaced signals.

(d) In anodic stripping voltammetry, analyte is reduced and concentrated at the working electrode at a controlled potential for a constant time. The potential is then ramped in a positive direction to reoxidize the analyte, during which current is measured. The height of the oxidation wave is proportional to the original concentration of analyte. Stripping is the most sensitive polarographic technique because analyte is concentrated from a dilute solution. The longer the period of concentration, the more sensitive is the analysis.

Chapter 17

17-A. (a) $\nu = c/\lambda = (2.998 \times 10^8 \text{ m/s})/(100 \times 10^{-9} \text{ m})$
$= 2.998 \times 10^{15} \text{ s}^{-1} = 2.998 \times 10^{15} \text{ Hz}$
$\tilde{\nu} = 1/\lambda = 1/(100 \times 10^{-9} \text{ m}) = 10^7 \text{ m}^{-1}$

$$(10^7 \text{ m}^{-1})\left(\frac{1 \text{ m}}{100 \text{ cm}}\right) = 10^5 \text{ cm}^{-1}$$

$E = h\nu = (6.626\ 2 \times 10^{-34} \text{ J s}) \times$
$\qquad (2.998 \times 10^{15} \text{ s}^{-1}) = 1.986 \times 10^{-18} \text{ J}$
$(1.986 \times 10^{-18} \text{ J/photon}) \times$
$\qquad (6.022 \times 10^{23} \text{ photons/mol}) = 1\ 196 \text{ kJ/mol}$

(b), (c), (d) These are done in an analogous manner. The final results for all four parts are presented in the following table:

λ	ν (Hz)	$\tilde{\nu}$ (cm^{-1})	E (kJ/mol)	Spectral region	Molecular process
100 nm	2.998×10^{15}	10^5	1.196×10^3	ultraviolet	electronic excitation
500 nm	5.996×10^{14}	2×10^4	239.3	visible	electronic excitation
10 µm	2.998×10^{13}	$1\ 000$	11.96	infrared	molecular vibration
1 cm	2.998×10^{10}	1	0.011 96	microwave	molecular rotation

17-B. (a) $A = \varepsilon b c = (1.05 \times 10^3 \text{ M}^{-1} \text{ cm}^{-1})(1.00 \text{ cm}) \times$
$$(2.33 \times 10^{-4} \text{ M}) = 0.245$$

(b) $T = 10^{-A} = 0.569 = 56.9\%$

(c) Doubling b will double $A \Rightarrow$
$$A = 0.489 \Rightarrow T = 10^{-A} = 32.4\%$$

(d) Doubling c also doubles $A \Rightarrow$
$$A = 0.489 \Rightarrow T = 10^{-A} = 32.4\%$$

(e) $A = \varepsilon b c = (2.10 \times 10^3 \text{ M}^{-1} \text{ cm}^{-1})(1.00 \text{ cm}) \times$
$$(2.33 \times 10^{-4} \text{ M}) = 0.489$$

17-C. (a) Don't touch the cuvet with your fingers. Wash the cuvet as soon as you are finished with it and drain out the rinse water. Use matched cuvets. Place the cuvet in the instrument with the same orientation each time. Cover the cuvet to prevent evaporation and to keep dust out.

(b) If absorbance is too high, too little light reaches the detector for accurate measurement. If absorbance is too low, there is too little difference between sample and reference for accurate measurement.

17-D. (a) $\varepsilon = \dfrac{A}{cb} = \dfrac{0.267 - 0.019}{(3.15 \times 10^{-6} \text{ M})(1.000 \text{ cm})}$
$$= 7.87 \times 10^4 \text{ M}^{-1} \text{ cm}^{-1}$$

(b) $c = \dfrac{A}{\varepsilon b} = \dfrac{0.175 - 0.019}{(7.87 \times 10^4 \text{ M}^{-1} \text{ cm}^{-1})(1.000 \text{ cm})}$
$$= 1.98 \times 10^{-6} \text{ M}$$

Chapter 18

18-A. The source contains a visible lamp and an ultraviolet lamp, only one of which is used at a time. Grating 1 apparently selects some limited range of wavelengths to leave the source. In the monochromator, grating 2 is the principal element that selects the wavelength to reach the exit slit. The width of the exit slit determines how wide a range of wavelengths leaves the monochromator. The rotating chopper is a mirror that alternately directs the monochromatic light through the sample or the reference cuvet in the sample compartment. The rotating chopper after the sample compartment directs light from each path to the detector, which is a photomultiplier tube.

18-B. We use Equations 18-5 with $b = 0.100$ cm:
$$D = b(\varepsilon'_X \varepsilon''_Y - \varepsilon'_Y \varepsilon''_X)$$
$$= (0.100)[(720)(274) - (212)(479)]$$
$$= 9.57_3 \times 10^3 \text{ (the units are M}^{-2} \text{ cm}^{-1})$$

$$[X] = \frac{1}{D}(A'\varepsilon''_Y - A''\varepsilon'_Y)$$
$$= \frac{(0.233)(274) - (0.200)(212)}{9.57_3 \times 10^7} = 2.24 \text{ mM}$$

$$[Y] = \frac{1}{D}(A''\varepsilon'_X - A'\varepsilon''_X)$$
$$= \frac{(0.200)(720) - (0.233)(479)}{9.57_3 \times 10^7} = 3.38 \text{ mM}$$

18-C. (a) $163 \times 10^{-6} \text{ L of } 1.43 \times 10^{-3} \text{ M Fe(III)}$
$$= 2.33 \times 10^{-7} \text{ mol Fe(III)}$$

(b) 1.17×10^{-7} mol transferrin in 2.00×10^{-3} L $\Rightarrow$ 5.83×10^{-5} M transferrin

(c) Prior to the equivalence point, all added Fe(III) binds to the protein to form a red complex whose absorbance is shown in the figure. After the equivalence point, no more protein binding sites are available. The slight increase in absorbance arises from the color of the iron reagent in the titrant.

18-D. (a) In electronic transitions, energy is absorbed or liberated when the distribution of electrons in a molecule is changed. In vibrational transitions, the amplitude of vibrations increases or decreases. A rotational transition causes the molecule to rotate faster or slower.

(b) A molecule in an excited state can collide with other molecules and transfer kinetic energy without emitting a photon. The molecule that absorbs the energy in the collision ends up moving faster or vibrating with greater amplitude or rotating faster.

(c) Fluorescence is a transition from an excited singlet electronic state to the ground singlet state. Phosphorescence is a transition from an excited triplet state to the ground singlet state. Fluorescence is at higher energy than phosphorescence and fluorescence occurs faster than phosphorescence.

(d) Fluorescence involves a set of transitions opposite those for absorption. Instead of going from the ground vibrational state of S_0 to various states of S_1 in absorption, fluorescence takes a molecule from the ground vibrational state of S_1 to various states of S_0. Absorption has a series of peaks from λ_0 to higher energy. Fluorescence has a series of peaks from λ_0 to lower energy.

(e) Photochemistry is the breaking of chemical bonds initiated by absorption of a photon. Chemiluminescence is the emission of light as a result of a chemical reaction.

18-E. The calibration graph is shown below. The least-squares equation of the straight line is $y = (6.142 \times 10^{-3})x + 1.804$, where y is detector signal and x is cells/mL. For the unknown signal $y = 15.9$, we solve the equation $15.9 = (6.142 \times 10^{-3})x + 1.804$ to find $x = 2\,295$ cells/mL.

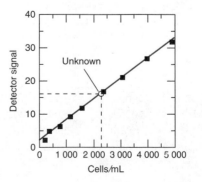

Chapter 19

19-A. In atomic absorption spectroscopy, light of a specific frequency is passed through a flame containing free atoms. The absorbance of light is measured and is proportional to the concentration of atoms. In atomic emission spectroscopy, no lamp is used. The intensity of light emitted by excited atoms in the flame is measured and is proportional to the concentration of atoms.

19-B. Atoms in a furnace are confined in a small volume for a relatively long time. The detection limit for an element in the unknown is small because the concentration of atoms in the gas phase is relatively high. A large sample volume is not required because gaseous atoms do not rapidly escape from the furnace. In a flame or plasma, the volume of gas phase is relatively large, so the concentration of atoms is relatively low. A great deal of sample is required because the atoms are rapidly moving through and escaping from the flame or plasma.

19-C. (a) First find the frequency of light:
$$\nu = c/\lambda = (2.998 \times 10^8 \text{ m/s})/(400 \times 10^{-9} \text{ m})$$
$$= 7.495 \times 10^{14} \text{ s}^{-1}$$
$$\text{energy} = h\nu = (6.626 \times 10^{-34} \text{ J} \cdot \text{s}) \times$$
$$(7.495 \times 10^{14} \text{ s}^{-1}) = 4.966 \times 10^{-19} \text{ J}$$

(b) $\dfrac{N^*}{N_0} = \left(\dfrac{g^*}{g_0}\right)e^{-\Delta E/kT} =$

$$\left(\frac{1}{1}\right)e^{-(4.966 \times 10^{-19} \text{ J})/[(1.381 \times 10^{-23} \text{ J/K})(2\,500 \text{ K})]} = 5.7 \times 10^{-7}$$

19-D. *Absorption* is measured in the presence and absence of a strong magnetic field. The Zeeman effect splits the analyte absorption peak when a magnetic field is applied and very little light is absorbed at the hollow-cathode lamp wavelength. The difference between absorption with and without the magnetic field is due to analyte. *Emission* is measured at the peak wavelength and at slightly lower and higher wavelengths. The peak emission relative to the baseline drawn between the two off-wavelength measurements is the emission due to analyte.

19-E. (a) *Spectral interference* arises from overlap of analyte absorption or emission lines with absorptions or emissions from other elements or molecules in the sample or the flame.

(b) *Chemical interference* is caused by any substance that decreases the extent of atomization of analyte.

(c) *Ionization interference* is a reduction of the concentration of free atoms by ionization of the atoms.

Chapter 20

20-A. 1-C, 2-D, 3-A, 4-E, 5-B

20-B. Your answer to part **(a)** will be different from mine, which was measured on a larger figure than the one in your textbook. However, your answers to parts **(b)**–**(e)** should be the same as mine.

(a) t_r: octane, 200 units; nonane, 260 units

w: octane, 40 units; nonane, 52 units

(b) octane: $N = \dfrac{5.55(200)^2}{(40)^2} = 139$

nonane: $N = \dfrac{5.55(260)^2}{(52)^2} = 139$

(c) octane: $H = L/N = (1.00 \text{ m}/139) = 7.2 \text{ mm}$

nonane: $H = (1.00 \text{ m}/139) = 7.2 \text{ mm}$

(d) resolution $= \dfrac{\Delta t_r}{w_{av}} = \dfrac{260 - 200}{\frac{1}{2}(40 + 52)} = 1.30$

(e) $\dfrac{\text{large load}}{\text{small load}} = \left(\dfrac{\text{large column radius}}{\text{small column radius}}\right)^2$

$\dfrac{27.0 \text{ mg}}{3.0 \text{ mg}} = \left(\dfrac{\text{large column radius}}{2.0 \text{ mm}}\right)^2 \Rightarrow$

large column radius = 6.0 mm

large column diameter = 12.0 mm

Flow rate should be proportional to the cross-sectional area of the column, which is proportional to the square of the radius.

$\dfrac{\text{large column flow rate}}{\text{small column flow rate}} = \left(\dfrac{\text{large column radius}}{\text{small column radius}}\right)^2$

$$= \left(\frac{6.0 \text{ mm}}{2.0 \text{ mm}}\right)^2 = 9.0$$

If the small column flow rate is 7.0 mL/min, the large column flow rate should be $(9.0)(7.0 \text{ mL/min}) = 63 \text{ mL/min}$.

20-C. (a) Molecules in the gas phase move faster than molecules in liquid. Therefore, longitudinal diffusion in the gas phase is much faster than longitudinal diffusion in the liquid phase, so band broadening by this mechanism is more rapid in gas chromatography than in liquid chromatography.

(b) (i) Optimum flow rate gives minimum plate height (23 cm/s for He). **(ii)** When flow is too fast, there is not adequate time for solute to equilibrate between the phases as the mobile phase streams past the stationary phase. This also broadens the band. **(iii)** When flow is too slow, plate height increases (i.e., band broadening gets worse) because solute spends a long time on the column and is broadened by longitudinal diffusion.

(c) There is no broadening by multiple flow paths in an open tubular column.

(d) The longer the column, the better the resolution. We can use a longer open tubular column than a packed column because the particles in the packed column resist flow and require high pressures for high flow rate.

20-D. The peak at 136 mass units is $C_4H_9{}^{79}Br^+$ and the peak at 138 is $C_4H_9{}^{81}Br^+$. Natural bromine consists of 50.5% ^{79}Br and 49.5% ^{81}Br, so the intensities of the peaks are almost equal. Peaks at 107 and 109 mass units are $C_2H_4{}^{79}Br^+$ and

$C_2H_4{}^{81}Br^+$. The peak at 57 mass units is $C_4H_9^+$, which has no Br and therefore has no isotopic partner at 59 mass units. The peak at 41 mass units is $C_3H_5^+$.

20-E. First evaluate the response factor from the known mixture:

$$\frac{A_X}{[X]} = F\left(\frac{A_S}{[S]}\right)$$

$$\frac{0.644}{52.4 \text{ nM}} = F\left(\frac{1.000}{38.9 \text{ nM}}\right) \Rightarrow F = 0.478_1$$

For the unknown mixture, we can write

$$\frac{A_X}{[X]} = F\left(\frac{A_S}{[S]}\right)$$

$$\frac{1.093}{[X]} = 0.478_1\left(\frac{1.000}{742 \text{ nM}}\right) \Rightarrow$$

$$[X] = 1.70 \times 10^3 \text{ nM} = 1.70 \ \mu M$$

Chapter 21

21-A. (a) Low-boiling solutes are separated well at low temperature, and the retention of high-boiling solutes is reduced to a reasonable time at high temperature.

(b) An open tubular column gives higher resolution than a packed column. The narrower the column, the higher the resolution that can be attained. A packed column can handle much more sample than an open tubular column, which is vital for preparative separations in which we are trying to isolate some quantity of the separated components.

(c) Diffusion of solute in H_2 and He is more rapid than in N_2. Therefore equilibration of solute between mobile phase and stationary phase is faster. The column can be run faster without excessive broadening from the finite rate of mass transfer between the mobile and stationary phases.

(d) Split injection is the ordinary mode for open tubular columns. Splitless injection is useful for trace and quantitative analysis. On-column injection is useful for thermally sensitive solutes that might decompose during a high-temperature injection.

(e) (i) all analytes; **(ii)** carbon atoms bearing hydrogen atoms; **(iii)** molecules with halogens, conjugated $C{=}O$, CN, NO_2; **(iv)** P and S; **(v)** P and N; **(vi)** S, **(vii)** all analytes.

21-B. Solvent is competing with solute for adsorption sites. The strength of the solvent-adsorbant interaction is independent of solute.

21-C. (a) In normal-phase chromatography, the stationary phase is more polar than the mobile phase. In reversed-phase chromatography, the stationary phase is less polar than the mobile phase.

(b) In isocratic elution, the composition of eluent is constant. In gradient elution, the composition of the eluent is changed—usually continuously—from low eluent strength to high eluent strength.

(c) In normal-phase chromatography, polar solvent must compete with analyte for polar sites on the stationary phase. The more polar the solvent, the better it binds to the stationary phase and the greater is its ability to displace analyte from the stationary phase.

(d) In reversed-phase chromatography, nonpolar analyte adheres to the nonpolar stationary phase. A polar solvent does not compete with analyte for nonpolar sites on the stationary phase. Making the solvent less polar gives it greater ability to displace analyte from the stationary phase.

(e) A guard column is a small, disposable column containing the same stationary phase as the main column. Sample and solvent pass through the guard column first. Any irreversibly bound impurities stick to the guard column, which is eventually thrown away. This guard column prevents junk from being irreversibly bound to the expensive main column and eventually ruining it.

21-D. (a) Analytes are released into the chromatography column over a long period of time (possibly many minutes) from the heated fiber or the heated absorption tube. If the analytes were not cold trapped on the column prior to beginning chromatography, analytes would be eluted in extremely broad bands instead of sharp peaks.

(b) Solid-phase extraction uses a short column containing any of the stationary phases used in HPLC. The column carries out gross separations of one type of analyte from other types of analyte (e.g., separating nonpolar from polar analytes).

(c) In the urine steroid analysis, the solid-phase extraction column retains relatively nonpolar solutes from urine and lets everything else, such as polar molecules and salts, pass right through. When the solid-phase extraction column is washed with a nonpolar solvent, the steroids are eluted and can be concentrated into a small volume for application to the HPLC column. The polar molecules and salts from the urine were removed in the sample cleanup and not applied to the HPLC column.

Chapter 22

22-A. (a) Deionized water has been passed through ion-exchange columns to convert cations to H^+ and anions to OH^-, making H_2O. Nonionic impurities (such as neutral organic compounds) are not removed by this process.

(b) Gradient elution with increasing concentration of H^+ is required to displace more and more strongly bound cations from the column.

(c) Cations from a large volume of water are collected on a small ion-exchange column and then eluted in a small volume of concentrated acid. If the acid were not very pure, impurities in the acid could be greater than the concentration of trace species collected from the large volume of water.

22-B. The separator column separates ions by ion exchange, whereas the suppressor exchanges the counterion to reduce the conductivity of eluent.

22-C. (a) total volume $= \pi(0.80 \text{ cm})^2(20.0 \text{ cm}) = 40.2 \text{ mL}$

(b) If the void volume ($=$ volume of mobile phase excluded from gel) is 18.2 mL, the stationary phase plus its included solvent must occupy $40.2 - 18.2 = 22.0 \text{ mL}$.

(c) If pores occupy 60.0% of the stationary phase volume, the pore volume is $(0.600)(22.0 \text{ mL}) = 13.2 \text{ mL}$. Large molecules excluded from the pores are eluted in the void volume of $x = 18.2 \text{ mL}$. The smallest molecules (which can enter all of the pores) are eluted in a volume of $y = 18.2 + 13.2 = 31.4 \text{ mL}$. (In fact, there is usually some adsorption of solutes on the stationary phase, so retention volumes in real columns can be greater than 31.4 mL.)

22-D. (a) 25 μm diameter: volume $= \pi r^2 \times$ length $=$
$\pi(12.5 \times 10^{-6} \text{ m})^2 (5 \times 10^{-3} \text{ m}) = 2.5 \times 10^{-12} \text{ m}^3$

$(2.5 \times 10^{-12} \text{ m}^3)\left(\dfrac{1 \text{ L}}{10^{-3} \text{ m}^3}\right)$
$= 2.5 \times 10^{-9} \text{ L} = 2.5 \text{ nL}$

50 μm diameter:
volume $= \pi(25 \times 10^{-6} \text{ m})^2(5 \times 10^{-3} \text{ m}) = 9.8 \text{ nL}$

(b) Capillary electrophoresis eliminates peak broadening from (1) the finite rate of mass transfer between the mobile and stationary phases and (2) multiple flow paths around stationary phase particles.

22-E. (a)

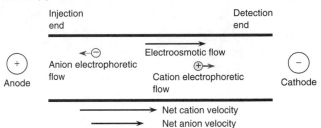

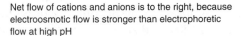

Net flow of cations and anions is to the right, because electroosmotic flow is stronger than electrophoretic flow at high pH

(b) At pH 3, the net flow of cations is to the *right* and the net flow of anions is to the *left*.

(c) When there is no analyte present, the constant concentration of chromate in the background buffer gives a steady ultraviolet absorbance at 254 nm. When Cl^- emerges, it displaces some of the chromate anion (to maintain electroneutrality). Because Cl^- does not absorb at 254 nm, the absorbance *decreases*.

22-F. (a) In the absence of micelles, all neutral molecules would move along with the electroosmotic speed of the bulk solvent and arrive at the detector at time t_0. Negative micelles migrate upstream and arrive at time $t_{mc} > t_0$. Neutral molecules partition between the bulk solvent and the micelles, so they arrive between t_0 and t_{mc}. The more time a neutral molecule spends in the micelles, the closer is its migration time to t_{mc}.

(b) (i) pK_a for $C_6H_5NH_3^+$ is 4.60. At pH 10, the predominant form is neutral.

(ii) Because anthracene arrives at the detector last, it must spend more time inside the micelles and therefore it is more soluble in the micelles.

ANSWERS TO PROBLEMS

Complete solutions to all problems can be found in the *Solutions Manual for Exploring Chemical Analysis*. Numerical answers and other short answers are given here.

Chapter 1

1-2. (a) milliwatt $= 10^{-3}$ watt
(b) picometer $= 10^{-12}$ meter
(c) kiloohm $= 10^3$ ohm
(d) microcoulomb $= 10^{-6}$ coulomb
(e) terajoule $= 10^{12}$ joule
(f) nanosecond $= 10^{-9}$ second
(g) femtogram $= 10^{-15}$ gram
(h) decipascal $= 10^{-1}$ pascal

1-3. (a) 100 fJ or 0.1 pJ   (b) 43.172 8 nC
(c) 299.79 THz   (d) 0.1 nm or 100 pm
(e) 21 TW   (f) 0.483 amol or 483 zmol

1-4. (a) 7.457×10^4 W   (b) 7.457×10^4 J/s
(c) 1.782×10^4 cal/s   (d) 6.416×10^7 cal/h

1-5. (a) 0.025 4 m, 39.37 inches
(b) 0.214 mile/s, 770 mile/h
(c) 1.04×10^3 m, 1.04 km, 0.643 mile

1-7. 1.10 M

1-8. (a) 0.054 8 ppm   (b) 54.8 ppb

1-9. 4.4×10^{-3} M, 6.7×10^{-3} M

1-10. (a) 70.5 g   (b) 29.5 g   (c) 0.702 mol

1-11. 6.18 g

1-12. (a) 6.6×10^3 L   (b) 5.8×10^5 g

1-13. 8.0 g

1-14. (a) 55.6 mL   (b) 1.80 g/mL

1-15. 5.48 g

1-16. 10^{-3} g/L, 10^3 μg/L, 1 μg/mL, 1 mg/L

1-17. 7×10^{-10} M

1-18. (a) 764 g ethanol   (b) 16.6 M

1-19. (a) 0.228 g Ni   (b) 1.06 g/mL

1-20. 1.235 M

1-21. Dilute 8.26 mL of 12.1 M HCl to 100.0 mL.

1-22. (a) 3.40 M   (b) 14.7 mL

1-23. Shredded Wheat: 3.6 Cal/g, 102 Cal/oz; Doughnut: 3.9, 111; Hamburger: 2.8, 79; Apple: 0.48, 14

1-25. (a) $K = 1/[Ag^+]^3[PO_4^{3-}]$ (b) $K = P_{CO_2}^6/P_{O_2}^{15/2}$

1-26. (a) $P_A = 0.028$ bar, $P_E = 48._0$ bar   (b) 1.2×10^{10}

1-27. unchanged

1-28. 4.5×10^3

1-29. (a) 3.6×10^{-7} (b) 3.6×10^{-7} M   (c) 3.0×10^4

1-31. Assuming mean concentration of 2.3 mg nitrate nitrogen/L, flow $\approx$ 5 000 tons/yr.

Chapter 2

2-6. 5.403 1 g
2-7. 14.85 g
2-8. 0.296 1 g
2-9. 9.980 mL
2-10. 5.022 mL
2-11. 15.631 mL

Chapter 3

3-1. (a) 1.237   (b) 1.238   (c) 0.135
(d) 2.1   (e) 2.00

3-2. (a) 0.217   (b) 0.216   (c) 0.217

3-3. (a) 4   (b) 4   (c) 4

3-4. (a) 12.3   (b) 75.5   (c) 5.520×10^3
(d) 3.04   (e) 3.04×10^{-10} (f) 11.9
(g) 4.600   (h) 4.9×10^{-7}

3-5. (a) 12.01   (b) 10.9   (c) 14   (d) 14.3
(e) -17.66 (f) 5.97×10^{-3} (g) 2.79×10^{-5}

3-6. (a) 208.232   (b) 560.594

3-8. (b) 25.031, systematic; ±0.009, random
(c) 1.98 and 2.03, systematic; ±0.01 and ±0.02, random
(d) random
(e) random
(f) mass is systematically low because empty funnel was not dried

3-9. (a) 3.124 (±0.005) or 3.123_6 ($\pm0.005_2$)
(b) 3.124 ($\pm0.2\%$) or 3.123_6 ($\pm0.1_7\%$)

3-10. (a) 2.1 ±0.2 (or 2.1 $\pm11\%$)
(b) 0.151 ±0.009 (or 0.151 $\pm6\%$)

(c) $0.22_3 \pm 0.02_4 \, (\pm 11\%)$

(d) $0.097_1 \pm 0.002_2 \, (\pm 2._3\%)$

3-11. (a) $21.0_9 \, (\pm 0.1_6)$ or $21.1 \, (\pm 0.2)$; relative uncertainty $= 0.8\%$

(b) $27.4_3 \, (\pm 0.8_6)$; relative uncertainty $= 3._1\%$

(c) $(14._9 \pm 1._3) \times 10^4$ or $(15 \pm 1) \times 10^4$; relative uncertainty $= 9\%$

3-12. (a) $10.18 \, (\pm 0.07) \, (\pm 0.7\%)$

(b) $174 \, (\pm 3) \, (\pm 2\%)$

(c) $0.147 \, (\pm 0.003) \, (\pm 2\%)$

(d) $7.86 \, (\pm 0.01) \, (\pm 0.1\%)$

(e) $2 \, 185.8 \, (\pm 0.8) \, (\pm 0.04\%)$

3-13. (a) $6.0 \pm 0.2 \, (\pm 3._7\%)$

(b) $1.30_8 \pm 0.09_2 \, (\pm 7.0\%)$

(c) $1.30_8 \, (\pm 0.09_2) \times 10^{-11} \, (\pm 7._0\%)$

(d) $2.7_2 \pm 0.7_8 \, (\pm 29\%)$

3-14. 78.112 ± 0.005

3-15. 95.978 ± 0.009

3-16. (a) $58.442 \, 5 \pm 0.000 \, 9$ g/mol

(b) $0.450 \, 7 \, (\pm 0.000 \, 5)$ M

3-17. (a) $0.020 \, 76_6 \pm 0.000 \, 02_8$ M   (b) yes

3-18. (a) 16.66 mL   (b) $0.169 \, (\pm 0.002)$ M

3-19. 100 J $= 23.90$ cal $= 0.094 \, 78$ Btu $= 6.241 \times 10^{20}$ eV

3-20. Method 2 is more accurate. The relative uncertainty in mass in method 1 is far greater than any other uncertainty in either procedure. Method 1 gives $0.002 \, 72_6 \pm 0.000 \, 01_8$ M $AgNO_3$. Method 2 gives $0.002 \, 726 \pm 0.000 \, 006_5$ M $AgNO_3$.

Chapter 4

4-2. $0.683, 0.955, 0.997$

4-3. (a) $1.527 \, 67$ (b) $0.001 \, 26, 0.082 \, 5\%$

(c) $1.527 \, 93 \pm 0.000 \, 10$

4-5. $108.6_4, 7.1_4, 108.6_4 \pm 6.8_1$

4-6. yes

4-7. $2.299 \, 47 \pm 0.001 \, 15, 2.299 \, 47 \pm 0.001 \, 71$

4-8. no

4-9. yes, no

4-10. yes

4-11. discard 0.195

4-12. retain 0.169

4-13. $y = -2x + 15$

4-14. $-1.299 \, (\pm 0.001) \times 10^4, 3 \, (\pm 3) \times 10^2$

4-15. (a) $2.0_0 \pm 0.3_8$ (b) $2.0_0 \pm 0.2_6$

4-16. (a) $y \, (\pm 0.005_7) = 0.021 \, 7_7 \, (\pm 0.000 \, 1_9)x + 0.004_6 \, (\pm 0.004_4)$

(c) protein $= 23.0_7 \pm 0.2_9 \, \mu g$

4-17. average $= 2.5$, standard deviation $= 1.290 \, 99$

4-19. It does not appear to be $CuCO_3$ or $CuCO_3 \cdot xH_2O$. The 99% confidence intervals for the class and instructor data exceed 51.43 wt % Cu expected in $CuCO_3$. If the material were a hydrate, the Cu content would be even less.

Chapter 5

5-2. 43.2 mL, 270.0 mL

5-3. 4.300×10^{-2} M

5-4. 0.149 M

5-5. 32.0 mL

5-6. (a) $0.045 \, 00$ M   (b) 36.42 mg/mL

5-7. 947 mg

5-8. 1.72 mg

5-9. (a) $0.020 \, 34$ M   (b) $0.125 \, 7$ g

(c) $0.019 \, 83$ M

5-10. (a) $0.105 \, 3$ mol/kg solution   (b) 0.305 M

5-11. 9.066 mM

5-12. 3.555 mM

5-13. 30.5 wt %

5-14. $0.020 \, 6 \, (\pm 0.000 \, 7)$ M

5-15. (a) $7._1 \times 10^{-5}$ M   (b) $1._0 \times 10^{-3}$ g/100 mL

5-16. (a) $6.6_9 \times 10^{-5}$ M   (b) 14.4 ppm

5-17. $1 \, 400$ ppb, 76 ppb, 0.98 ppb

5-18. (a) $[Hg_2^{2+}] = 6.8_8 \times 10^{-7}$ M $[IO_3^-] = 1.3_8 \times 10^{-6}$ M

(b) $[Hg_2^{2+}] = 1.3 \times 10^{-14}$ M

5-19. I^- before Br^- before Cl^- before CrO_4^{2-}

5-20. (a) 0.500 mmol tris, 0.545 mmol pyridine

(b) liberates heat

Chapter 6

6-1. 2.03

6-2. $0.085 \, 38$ g

6-3. 50.79 wt % Ni

6-4. 7.22 mL

6-5. 8.665 wt %

6-6. 0.339 g

6-7. (a) 5.5_1 mg/100 mL   (b) 5.834 mg, yes

6-8. Ba, 47.35 wt %; K, 8.279 wt %; Cl, 31.95 wt %

6-9. 11.69 mg CO_2, 2.051 mg H_2O

6-10. (a) 51.36 wt % C, 3.639 wt % H   (b) C:H $\approx 6:5$

6-11. $C_4H_9NO_2$

6-12. 104.1 ppm

6-13. (a) $0.027 \, 36$ M   (b) systematic

6-14. 75.40 wt %

6-15. $C_8H_{9.06\pm0.17}N_{0.997\pm0.010}$

6-16. $s_{pooled} = 0.036_{36}$ and $t = 5.1_3 \implies$ difference is significant above 99% confidence level

Chapter 7

7-1. (a) HCN/CN^-, HCO_2H/HCO_2^-

(b) H_2O/OH^-, HPO_4^{2-}/PO_4^{3-}

(c) H_2O/OH^-, HSO_3^-/SO_3^{2-}

7-2. $[H^+] > [OH^-]$, $[OH^-] > [H^+]$

7-3. (a) 4 (b) 9 (c) 3.24 (d) 9.76

7-4. (a) 7.46 (b) 2.9×10^{-7} M

7-5. 2.5×10^{-5} M

7-6. 7.8

7-7. see Table 7-1

7-8. carboxylic acids and ammonium salts, carboxylate anions and amines

7-9. (a) 0.010 M, 2.00   (b) $2.8_6 \times 10^{-13}$ M, 12.54
(c) 0.030 M, 1.52   (d) 3.0 M, -0.48
(e) 1.0×10^{-12} M, 12.00

7-10. (b) trichloroacetic acid

7-11. (b) sodium 2-mercaptoethanol

7-12. $2H_2SO_4 \rightleftharpoons H_3SO_4^+ + HSO_4^-$

7-13.

7-14. $OCl^- + H_2O \rightleftharpoons HOCl + OH^-, 3.3 \times 10^{-7}$

7-15. 9.78

7-16. 2.2×10^{-12} M

7-17. 3.02, 9.51×10^{-2}

7-18. 5.40, 2.65×10^{-5}

7-19. 3.15, 7.05×10^{-4} M, 0.084 3 M, 0.085 0 M

7-20. 3.70

7-21. 7.30

7-22. 5.51, 3.1×10^{-6} M, 0.060 M

7-23. (a) 3.03, 0.094   (b) 7.00, 0.999

7-24. 5.50

7-25. $K_a = 2.3 \times 10^{-11}$, $K_b = 4.3 \times 10^{-4}$

7-27. 11.28, 0.058 M, $1.9_2 \times 10^{-3}$ M

7-28. pH = 8.88, 0.007 56%;
pH = 8.38, 0.023 9%; pH = 7.00, 0.568%

7-29. 10.95

7-30. 9.97, 0.003 6

7-31. $3.3_6 \times 10^{-6}$

7-32. 2.2×10^{-7}

7-34. 2.93, 0.118

7-35. 9.95×10^{-4} M

Chapter 8

8-2. 4.13

8-3. (b) 1/1000   (c) $pK_a - 4$

8-4. (a) 1.5×10^{-7} (b) 0.15

8-5. (a) 14   (b) 1.4×10^{-7}

8-6. (a) 2.3×10^{-7} (b) 1.00   (c) 23

8-8. 3.59

8-9. (a 8.37   (b) 0.423 g
(c) 8.33   (d) 8.41

8-10. (b) 7.18   (c) 7.00   (d) 6.86 mL

8-11. (a) 2.56   (b) 2.86

8-12. 3.27 mL

8-13. 13.7 mL

8-14. 4.68 mL

8-15. ii

8-16. (b) NaOH

8-17. (a) 90.9 mL HCl

8-18. (a) $p = 0.940\ 6$ mol, $q = 0.059\ 35$ mol
(b) $\Delta(\text{pH}) = -0.077$

8-19. (a) red   (b) orange   (c) yellow   (d) red

Chapter 9

9-5. $V_e = 10.0$ mL;
pH = 13.00, 12.95, 12.68, 11.96, 10.96, 7.00, 3.04, 1.75

9-6. $V_e = 12.5$ mL;
pH = 1.30, 1.35, 1.60, 2.15, 3.57, 7.00, 10.43, 11.11

9-7. $V_e = 5.00$ mL;
pH = 2.66, 3.40, 4.00, 4.60, 5.69, 8.33, 10.96, 11.95

9-8. $V_e = 5.00$ mL; pH = 6.32, 10.64, 11.23, 11.85

9-9. 2.80, 3.65, 4.60, 5.56, 8.65, 11.96

9-10. 8.18

9-11. 3.72

9-12. 0.091 8 M

9-13. (a) 0.025 92 M   (b) 0.020 31 M
(c) 9.69   (d) 10.07

9-14. $V_e = 10.0$ mL;
pH = 11.00, 9.95, 9.00, 8.05, 7.00, 5.02, 3.04, 1.75

9-15. $V_e = 47.79$ mL; pH = 8.74, 5.35, 4.87, 4.40, 3.22, 2.58

9-16. (a) 2.2×10^9
(b) 10.92, 9.57, 9.35, 8.15, 5.53, 2.74

9-17. (a) 9.44   (b) 2.55   (c) 5.15

9-18. no

9-19. yellow, green, blue

9-21. (a) colorless → pink
(b) systematically requires too much NaOH

9-22. cresol red (orange → red) or phenolphthalein (colorless → red)

9-23. 10.727 mL

9-24. 0.063 56 M

9-25. (a) 0.087 99   (b) 25.74 mg   (c) 2.860 wt %

9-26. (a) 5.62   (b) methyl red

9-29. tribasic, 0.015 3 M

Chapter 10

10-2. $K_{a2} = 1.02 \times 10^{-2}$, $K_{b2} = 1.79 \times 10^{-13}$

10-3. $7.09 \times 10^{-3}, 6.33 \times 10^{-8}, 7.1 \times 10^{-13}$

10-5. piperazine: $K_{b1} = 5.38 \times 10^{-5}$, $K_{b2} = 2.15 \times 10^{-9}$
phthalate: $K_{b1} = 2.56 \times 10^{-9}$, $K_{b2} = 8.93 \times 10^{-12}$

10-7. $2.49 \times 10^{-8}, 5.78 \times 10^{-10}, 1.34 \times 10^{-11}$

10-8. $1.62 \times 10^{-5}, 1.54 \times 10^{-12}$

10-9. (a) 1.95, 0.089 M, 1.12×10^{-2} M, 2.01×10^{-6} M
(b) 4.27, 3.8×10^{-3} M, 0.100 M, 3.8×10^{-3} M
(c) 9.35, 7.04×10^{-12} M, 2.23×10^{-5} M, 0.100 M

10-10. (a) 11.00, 0.099 0 M, 9.95×10^{-4} M, 1.00×10^{-9} M
(b) 7.00, 1.0×10^{-3} M, 0.100 M, 1.0×10^{-3} M
(c) 3.00, 1.00×10^{-9} M, 9.95×10^{-4} M, 0.099 0 M

10-11. 11.60, $[B] = 0.296$ M, $[BH^+] = 3.99 \times 10^{-3}$ M,
$[BH_2^{2+}] = 2.15 \times 10^{-9}$ M

10-12. 7.53, $[BH_2^{2+}] = 9.48 \times 10^{-4}$ M, $[BH^+] \approx 0.150$ M,
$[B] = 9.49 \times 10^{-4}$ M

10-13. 5.59

10-14. **(a)** HA   **(b)** A$^-$ **(c)** 1.0, 0.10

10-15. **(a)** 4.00   **(b)** 8.00   **(c)** H$_2$A
(d) HA$^-$ **(e)** A^{2-}

10-16. **(a)** 9.00   **(b)** 9.00
(c) BH$^+$ **(d)** 1.0×10^3

10-18. **(a)** 5.59   **(b)** 10.74

10-20. 5.41, $[H_3Arg^{2+}] = 1.3 \times 10^{-5}$ M,
$[H_2Arg^+] = 0.050$ M, $[HArg] = 1.3 \times 10^{-5}$ M,
$[Arg^-] = 1.1 \times 10^{-12}$ M

10-21. -2

10-22. $V_{e1} = 10.0$ mL, $V_{e2} = 20.0$ mL;
pH = 2.51, 4.00, 6.00, 8.00, 10.46, 12.21

10-23. $V_{e1} = 10.0$ mL, $V_{e2} = 20.0$ mL;
pH = 11.49, 10.00, 8.00, 6.00, 3.54, 1.79

10-24. **(a)** phenolphthalein (colorless → red)
(b) p-nitrophenol (colorless → yellow)
(c) bromothymol blue (yellow → green)
or bromocresol purple (yellow → yellow + purple)
(d) thymolphthalein (colorless → blue)

10-25. $V_{e1} = 40.0$ mL, $V_{e2} = 80.0$ mL;
pH = 11.36, 9.73, 7.53, 5.33, 3.41, 1.85

10-26. $V_{e1} = 25.0$ mL, $V_{e2} = 50.0$ mL;
pH = 7.55, 6.02, 3.86, 1.97

10-27. **(a)** 2.56, 3.46, 4.37, 8.42, 11.45   **(b)** second
(c) thymolphthalein; first trace of blue

10-28. mean molecular mass = 327.0
mean formula is
$HOCH_2CH_2 - [OCH_2CH_2]_5 - OCH_2CH_2OH$

Chapter 11

11-4. **(a)** true   **(b)** true   **(c)** true

11-9. **(a)** 0.2 mM   **(b)** 0.6 mM   **(c)** 2.4 mM

11-10. **(a)** 0.660   **(b)** 0.54   **(c)** 0.18   **(d)** 0.83

11-11. 0.004 6

11-12. 0.88_7

11-13. **(a)** 0.42_2 **(b)** 0.43_2

11-14. **(a)** 1.3×10^{-6} M   **(b)** 2.9×10^{-11} M

11-15. $\gamma_{H^+} = 0.86$, pH = 2.07

11-16. $[H^+] = 1.2 \times 10^{-7}$ M; pH = 6.99

11-17. 11.94, 12.00

11-18. $[OH^-] = 1.1 \times 10^{-4}$ M; pH = 9.92

11-19. 6.6×10^{-7} M

11-20. 2.5×10^{-3} M

11-21. **(a)** 9.2 mM   **(b)** Remainder is $CaSO_4(aq)$

11-22. $[H^+] + 2[Ca^{2+}] + [Ca(HCO_3)^+] + [Ca(OH)^+] +$
$[K^+] = [OH^-] + [HCO_3^-] + 2[CO_3^{2-}] + [ClO_4^-]$

11-23. $[H^+] = [OH^-] + [HSO_4^-] + 2[SO_4^{2-}]$

11-24. $[H^+] = [OH^-] + [H_2AsO_4^-] + 2[HAsO_4^{2-}] +$
$3[AsO_4^{3-}]$

11-25. **(a)** $2[Mg^{2+}] + [H^+] = [Br^-] + [OH^-]$
(b) $2[Mg^{2+}] + [H^+] + [MgBr^+] = [Br^-] + [OH^-]$

11-26. $[CH_3CO_2^-] + [CH_3CO_2H] = 0.1$ M

11-27. **(a)** 0.20 M $= [Mg^{2+}]$ **(b)** 0.40 M $= [Br^-]$
(c) 0.20 M $= [Mg^{2+}] + [MgBr^+]$
(d) 0.40 M $= [Br^-] + [MgBr^+]$

11-28. **(a)** $[F^-] + [HF] = 2[Ca^{2+}]$
(b) $[F^-] + [HF] + 2[HF_2^-] = 2[Ca^{2+}]$

11-29. $2[Ca^{2+}] = 3\{[PO_4^{3-}] + [HPO_4^{2-}] + [H_2PO_4^-] + [H_3PO_4]\}$

11-30. $[Y^{2-}] = [X_2Y_2^{2+}] + 2[X_2Y^{4+}]$

11-31. $[OH^-] = 1.5 \times 10^{-7}$ M; $[H^+] = 6.8 \times 10^{-8}$ M;
$[Mg^{2+}] = 4.0 \times 10^{-8}$ M

11-32. 5.8×10^{-4} M

11-33. **(a)** $[Ag^+] = 2.4 \times 10^{-8}$ M; $[CN^-] = 9.2 \times 10^{-9}$ M;
$[HCN] = 1.5 \times 10^{-8}$ M
(b) 2.9×10^{-8} M, 1.3×10^{-8} M, 1.6×10^{-8} M

11-34. **(a)** 5.0×10^{-9} mol   **(b)** 4.0×10^{-6} M
(c) 1.1×10^{-8} M

11-35. **(a)** $5.57 \times 10^{-2}, 0.94_4$ **(b)** $0.37_1, 0.62_9$
(c) $0.85_5, 0.14_5$

11-36. **(a)** $0.98_0, 0.019_6$ **(b)** $0.83_4, 0.16_6$
(c) $0.61_3, 0.38_7$

11-37. **(a)** $0.090_9, 0.90_9$ **(b)** $0.50_0, 0.50_0$
(c) $0.66_6, 0.33_4$

Chapter 12

12-1. **(a)** 10.0 mL   **(b)** 10.0 mL

12-8. 0.010 3 M

12-9. $[Ni^{2+}] = 0.012 4$ M, $[Zn^{2+}] = 0.007 18$ M

12-10. 0.024 30 M

12-11. $1.256 (\pm 0.003)$ mM

12-12. 0.014 68 M

12-13. 0.092 6 M

12-14. 5.150 mg Mg, 20.89 mg Zn, 69.64 mg Mn

12-15. 32.7 wt % (theoretical = 32.90 wt %)

12-16. **(a)** 3.4×10^{-10} **(b)** 0.64

12-17. **(a)** $Co^{2+} + 4NH_3 \rightleftharpoons Co(NH_3)_4^{2+}$
(b) $Co(NH_3)_3^{2+} + NH_3 \rightleftharpoons Co(NH_3)_4^{2+}$,
log $K_4 = 0.64$

12-18. **(a)** 3.3×10^7 **(b)** 3.9×10^{-5} M

12-19. **(a)** 100.0 mL   **(b)** 0.016 7 M   **(c)** 0.054
(d) 5.4×10^{10} **(e)** 6.8×10^{-7} M
(f) 1.9×10^{-10} M

12-20. **(a)** 1.70   **(b)** 2.18   **(c)** 2.81
(d) 3.87   **(e)** 4.87   **(f)** 5.70
(g) 6.53   **(h)** 8.23   **(i)** 8.53

12-21. **(a)** (∞) **(b)** 10.42   **(c)** 9.64
(d) 8.56   **(e)** 7.55   **(f)** 6.21
(g) 4.88   **(h)** 3.20   **(i)** 2.93

12-23. 5.00 g

12-24. Reaction with NH_3 and OH^- raises the titration curve by 0.48 log units before V_e and leaves it unchanged after V_e.

Chapter 13

13-1. **(b)** 6.242×10^{18} e^-/C   **(c)** 9.649×10^4 C/mol

13-2. **(a)** oxidant: TeO_3^{2-}; reductant: $S_2O_4^{2-}$
(b) 3.02×10^3 C   **(c)** 0.840 A

13-3. 0.015 9 J

13-4. **(a)** $71._5$ A   **(b)** 6.8×10^6 J

13-7. **(a)** 0.572 V   **(b)** 0.568 V

13-8. **(a)** $Pt(s)|Cr^{2+}(aq), Cr^{3+}(aq)\|Tl^+(aq)|Tl(s)$
(b) 0.08_4 V   **(c)** $Tl^+ + Cr^{2+} \rightleftharpoons Tl(s) + Cr^{3+}$
(d) Pt

13-9. **(b)** 0.430 V; reduction occurs at right-hand electrode

13-10. **(a)** electrons flow right to left;
$\frac{3}{2}Br_2(l) + Al(s) \rightleftharpoons 3Br^- + Al^{3+}$
(b) Br_2 **(c)** 1.31 kJ   **(d)** 2.69×10^{-8} g/s

13-11. **(a)** -0.357 V; 9.2×10^{-7} **(b)** 0.722 V; 7×10^{48}

13-12. 7×10^{-12}

13-13. **(a)** 1.9×10^{-6} **(b)** -0.386 V; oxidized

13-14. 34 g/L

13-15. **(b)** 0.675 V; 10^{114} **(c)** 0.178 V   **(d)** 8.5

13-16. **(b)** 0.044 V

13-17. **(a)** 0.086 V   **(b)** -0.021 V   **(c)** 0.021 V

13-18. 0.684 V

13-20. **(a)** $E_{cell} = 0.059\ 16 \log(c_r/c_1)$. If $c_r = c_1$, $E_{cell} = 0$.

13-21. lead-acid battery; 83.42 A · h/kg; hydrogen-oxygen fuel cell: 2 975 A · h/kg

Chapter 14

14-1. **(a)** $Cu^{2+} + 2e^- \rightleftharpoons Cu(s)$ **(c)** 0.112 V

14-2. **(a)** $Br_2(aq) + 2e^- \rightleftharpoons 2Br^-$, $E_+ = 1.057$ V
(b) 0.816 V

14-3. 0.1 mL: $[Ag^+] = 1.1 \times 10^{-11}$ M, $E = -0.090$ V
10.0 mL: $[Ag^+] = 2.2 \times 10^{-11}$ M; $E = -0.073$ V
25.0 mL: $[Ag^+] = 1.0_5 \times 10^{-6}$ M; $E = 0.204$ V
30.0 mL: $[Ag^+] = 0.012\ 5$ M; $E = 0.445$ V

14-4. (0.1 mL, $E = 0.481$ V), (10.0 mL, 0.445 V),
(20.0 mL, 0.194 V), (30.0 mL, -0.039 V)

14-5. left side

14-6. 10.67

14-7. **(a)** $+0.10$ **(b)** 13%

14-10. small

14-11. $+0.029\ 6$ V

14-12. 0.211 mg/L

14-13. **(a)** -0.407 V   **(b)** $1.5_5 \times 10^{-2}$ M

14-14. -0.332 V

14-15. **(a)** 1.60×10^{-4} M   **(b)** 3.59 wt %

14-16. K^+

14-17. **(a)** $E = \text{constant} + \beta \left(\dfrac{0.059\ 16}{3} \right) \log([La^{3+}]_{outside})$
(b) 19.7 mV   **(c)** $+25.1$ mV   **(d)** $+100.7$ mV

14-18. **(a)** $E = 51.10\ (\pm 0.24) + 28.14\ (\pm 0.08_5) \log[Ca^{2+}]$
$(s_y = 0.2_7)$
(b), (c) $2.43\ (\pm 0.06) \times 10^{-3}$ M

14-19. $1.22_1 \pm 0.02_9 = 1.19_2$ to 1.25_0

14-20. **(c)** ~ 4.5 **(d)** $\sim 3\ \mu M$

14-21. 1.2×10^7

Chapter 15

15-2. **(d)** 0.490, 0.526, 0.626, 0.99, 1.36, 1.42, 1.46 V

15-3. **(d)** 1.58, 1.50, 1.40, 0.733, 0.065, 0.005, -0.036 V

15-4. **(d)** -0.120, -0.102, -0.052, 0.21, 0.48, 0.53 V

15-5. 0.371, 0.439, 0.507, 1.128, 1.252, 1.266 V

15-6. 0.409, 0.468, 0.620, 0.858, 0.918 V

15-7. **(c)** 0.570, 0.307, 0.184 V

15-8. solid curve: diphenylbenzidine sulfonic acid
(colorless → violet) or
tris(2,2′-bipyridine)iron (red → pale blue)
dashed curve: diphenylamine sulfonic acid
(colorless → red violet) or diphenylbenzidine sulfonic
acid (colorless → violet)

15-11. **(a)** $0.029\ 14$ M   **(b)** no

15-12. **(a)** colorless → pale red   **(b)** 35.50 mg
(c) It does matter.

15-13. **(a)** 26   **(b)** 4.02 mg

15-14. mol $NH_3 = 2$ (initial mol $H_2SO_4 - \frac{1}{2} \times$ mol thiosulfate)

15-15. **(a)** 700   **(b)** 1.0   **(c)** 0.34 g I_2/L

15-16. **(a)** 8 nmol   **(b)** maybe just barely

15-17. **(a)** 0.125   **(b)** $6.87_5 \pm 0.03_8$

Chapter 16

16-3. V_2

16-4. 0.342 M

16-5. 6.04

16-6. 12.567 5 g

16-7. 54.77 wt %

16-8. 151 μg/mL

16-9. **(a)** anode   **(b)** 52.0 g/mol   **(c)** 0.039 6 M

16-10. **(a)** $5._2 \times 10^{-9}$ mol e^- **(b)** $0.000\ 2_6$ mL

16-11. **(a)** 1.946 mmol H_2 **(b)** 0.041 09 M
(c) 4.750 h

16-12. trichloroacetic acid = 26.3%;
dichloroacetic acid = 49.5%

16-13. organohalide = 16.7 μM; 592 μg Cl/L

16-14. $96\ 486.6_7 \pm 0.2_8$ C/mol

16-15. **(b)** 0.25 mM

16-17. 0.01% ascorbic acid consumed in 10 min; yes

16-18. 0.658 mM

16-19. 1.21 mM

16-20. 2.37 (± 0.02) mM

16-21. 0.096 mM

16-22. 0.000 35 wt %

16-23. (a) $Cu^{2+} + 2e^- \rightarrow Cu(s)$
(b) $Cu(s) \rightarrow Cu^{2+} + 2e^-$
(c) 313 ppb

16-24. (a) 31.2 μg nitrite/g bacon
(b) 67.6 μg nitrate/g bacon

Chapter 17

17-1. (a) double　　(b) halve　　(c) double

17-2. (a) 3.06×10^{-19} J/photon, 184 kJ/mol
(b) 4.97×10^{-19} J/photon, 299 kJ/mol

17-3. (a) orange　　(b) violet-blue or violet
(c) blue-green

17-5. absorbance or molar absorptivity versus wavelength

17-7. violet-blue

17-8. (a) 1.20×10^{15} Hz, 4.00×10^4 cm^{-1},
7.95×10^{-19} J/photon, 479 kJ/mol
(b) 1.20×10^{14} Hz, 4 000 cm^{-1},
7.95×10^{-20} J/photon, 47.9 kJ/mol

17-9. 0.004 4, 0.046, 0.30, 1.00, 2.00, 3.00, 4.00

17-10. 3.56×10^4 M^{-1} cm^{-1}

17-11. (a) 2.76×10^{-5} M　　(b) 2.24 g/L

17-12. (a) 7.80×10^{-3} M　　(b) 7.80×10^{-4} M
(c) 1.63×10^3 M^{-1} cm^{-1}

17-13. (a) 0.52　(b) 30%

17-14. (a) 0.347　(b) 20.2%

17-15. 2.30×10^3 M^{-1} cm^{-1}

17-16. (b) 1.74×10^5 M^{-1} cm^{-1}

17-17. (a) 1.50×10^3 M^{-1} cm^{-1}　　(b) 2.31×10^{-4} M
(c) 4.69×10^{-4} M　　(d) 5.87×10^{-3} M

17-18. (a) 6.97×10^{-5} M　　(b) 6.97×10^{-4} M
(c) 1.02 mg

17-19. 2.19×10^{-4} M

17-20. (a) 1.735 ppm　　(b) 1.239×10^{-4} M

17-21. ± 0.017 ppm

17-23. (a) $4.49_3 \times 10^3$ M^{-1} cm^{-1}　　(b) $1.00_6 \times 10^{-4}$ M
(c) $5.03_0 \times 10^{-4}$ M　　(d) 16.1 wt %

17-24. (a) 1.57×10^{-5} M　　(b) 0.180　(c) 6.60 mg

17-25. (a) $2.42_5 \times 10^4$ M^{-1} cm^{-1}　　(b) 1.26 wt %

Chapter 18

18-2. deuterium

18-4. (a) 2.38×10^3　　(b) 143

18-9. (a) $[X] = 4.42 \times 10^{-5}$ M
(b) $[Y] = 5.96 \times 10^{-5}$ M

18-11. $[MnO_4^-] = 8.35 \times 10^{-5}$ M; $[Cr_2O_7^{2-}] = 1.78 \times 10^{-4}$ M

18-12. (a) [transferrin] = 8.99 mg/mL, [Fe] = 12.4 μg/mL
(b) fraction of Fe in transferrin = 73.7%

18-13. (a) A = 2 080[HIn] + 14 200[In$^-$]　　(b) 6.79

18-14. 3.95

18-15. [A] = 0.009 11 M; [B] = 0.004 68 M

18-16. ~2 and 0

18-17. 1.73×10^{-5} M

Chapter 19

19-6. (a) 4.699×10^{-19} J　　(b) 3.67×10^{-6}
(c) +8.4%　　(d) 0.010 3

19-7. (a) 6.07×10^{-19} J　　(b) 3.3×10^{-8}
(c) +12%　　(d) 0.002 0

19-8. 0.005 7 μg/mL

19-9. 17.4 μg/mL

19-10. (a) 3.00 mM　　(b) 3.60 mM

19-11. 0.120 M

19-12. 1.04 ppm

19-13. (a) 0, 10.0, 20.0, 30.0, 40.0 μg/mL
(b) 204 μg/mL

19-14. 1.64 μg/mL

19-15. (a) 7.49 μg/mL　　(b) 25.6 μg/mL

19-16. 8.33×10^{-5} M

Chapter 20

20-2. 0.1 mm

20-8. (a) $4.3_5 \times 10^4$　　(b) 3.6 μm

20-9. (a) $w_{1/2}$ = 0.8 mm for ethyl acetate and
2.6 mm for toluene
(b) $N = 1.1 \times 10^3$ for ethyl acetate and 1.1×10^3
for toluene

20-10. 6.8 cm diameter $\times$ 25 cm long, 17 mL/min

20-11. (a) $w_{1/2}$ = 0.172 min, $N = 3.58 \times 10^4$
(b) 0.838 mm
(c) w (measured) = 0.311 min;
$w/w_{1/2}$ (measured) = 1.81

20-12. (a) heptane: 7.41×10^4 plates, 0.404 mm plate height
$C_6H_4F_2$: 8.55×10^4 plates, 0.351 mm plate height
(b) resolution = 1.01

20-13. (a) 1.11 cm diameter $\times$ 32.6 cm long
(b) 0.069 1 mL　　(c) 0.256 mL/min

20-14. 1.47 mL/min

20-15. (a) 0.168_4　　(b) 0.847 mM　　(c) 6.16 mM
(d) 12.3 mM

20-16. 161 μg/mL

20-17. 0.47 mmol

20-18. 31 = CH_2OH^+; 41 = $C_3H_5^+$; 43 = $C_3H_7^+$; 56 = $C_4H_8^+$

20-19. (a) $u_{optimum}$ = 31.6 mL/min
(b) $u_{optimum}$ = 44.7 mL/min, $H_{optimum}$ increases
(c) $u_{optimum}$ increases, $H_{optimum}$ decreases

20-20. (a) 3.25 mm　　(b) 615 plates　　(c) 0.760 min

Chapter 21

21-3. **(b)** narrow bore: 16 ng; wide bore: 5.6 μg

(c) narrow bore: 0.16 ng; wide bore: 56 ng

21-4. retention time decreases

21-5. retention time increases

21-8. 0.19_2 min, 0.13_6 min

21-9. **(b)** amine will be eluted first

21-10. **(a)** lower   **(b)** higher

21-11. 0.418 mg/mL

21-12. **(a)** 8.68×10^8 **(b)** 0.273 m^2

Chapter 22

22-1. **(a)** mixture contains RCO_2^-, RNH_2, Na^+, and OH^-; all pass directly through the cation-exchange column

(b) mixture contains RCO_2H, RNH_3^+, H^+, and Cl^-; RNH_3^+ is retained and the others are eluted

22-3. $VOSO_4 = 80.9$ wt %, $H_2SO_4 = 10._7$ wt %, $H_2O = 8._4$ wt %

22-5. $NH_3 < (CH_3)_3N < CH_3NH_2 < (CH_3)_2NH$

22-6. 38.0 wt %

22-7. **(a)** 3.88   **(b)** $-773 \ \mu M$ and $+780 \ \mu M$

(c) 0.053 2 mg/mL

22-9. 4.9×10^4

22-10. **(a)** 5.7 mL   **(b)** 11.5 mL

(c) adsorption occurs

22-11. **(a)** cations < neutrals < anions

(b) low, against, never!

22-15. **(a)** longitudinal diffusion: $H \approx B/u$

(b) longitudinal diffusion and mass transfer: $H \approx B/u + Cu$

22-16. **(a)** $w_{1/2} = 0.75$ min, 1.6×10^4 plates

(b) plate height = 25 μm

22-17. thiamine < (niacinamide + riboflavin) < niacin; thiamine is most soluble

22-18. **(b)** 3.78

INDEX

Abbreviations

b = box
d = demonstration
e = experiment

i = illustration
m = marginal note
p = problem

r = reference
t = table

587